Fundamentals of Multisite Radar Systems

Fundamentals of Multisite Radar Systems

Multistatic Radars and Multiradar Systems

Victor S. Chernyak
Scientific Research Institute of Radio Device Engineering (NIIRP), Moscow, Russia

GORDON AND BREACH SCIENCE PUBLISHERS
Australia • Canada • China • France • Germany • India • Japan • Luxembourg
Malaysia • The Netherlands • Russia • Singapore • Switzerland

Originally published in Russian in 1993 as Многопозиционная радиолокация by The Radio and Communication (Радио и Связь) Publishing House, Moscow.

Printed in Singapore.

Amsteldijk 166
1st Floor
1079 LH Amsterdam
The Netherlands

British Library Cataloguing in Publication Data

A Catalogue record for this book is available from the British Library.

ISBN: 90-5699-165-5

CONTENTS

PREFACE

The progress in radar engineering during the last few decades was urged on by rapidly growing requirements of fundamental radar characteristics. Among those requirements are the increase in range and coverage (with simultaneous decrease in target radar 'visibility'), the significant increase in measurement accuracy, handling capacity in multitarget environments, proofing against different kinds of interference and so on. Requirements have arisen from so called 'signal information' used for target discrimination and recognition. The necessity to estimate coordinates and tracks of radiation sources (e.g., jammers) has given rise to the development of the passive location.

In spite of impressive achievements in radar element technology (antennas, transmitters, receivers, processors) modern high requirements cannot in many cases be met by traditional radars. There is a need for advancement in radar system design and construction principles.

One promising way is to move from the individual radar with a single transmitting station and a single receiving station (usually colocated) to the multisite radar system (MSRS) which includes several spatially separated transmitting and receiving stations (or monostatic radars) coupled together for cooperative target observation.

The fundamental idea behind MSRSs is to make more effective use of the information contained in the spatial characteristics of the electromagnetic field. It is well known that an electromagnetic field scattered by illuminated targets propagates through the whole space (except for some shielded regions). A monostatic radar extracts information from a single small field region corresponding to a receiving antenna aperture. In a MSRS information is extracted from several spatially separated regions of scattered (or radiated by signal sources) fields. This allows improved information gathering, interference proofing and some other important characteristics.

The development of MSRSs is in agreement with the general trend in modern engineering: to integrate individual technical means into systems where fundamental characteristics are enriched due to the cooperative performance and interaction between system elements.

In recent years MSRSs have been under active study. Many technical papers, reports to conferences and symposia have been published where the theory of MSRSs is developed and some results of design and experiments are presented.

Analyzing the future of radar, many authors consider multistatic radars and multiradar (netted) systems to be one of the principal areas of modern radar development for the next few years [68]. Such opinions are based not only on increasing requirements for the radar information but also on the significant progress in adjacent engineering fields which gains MSRS feasibility. The most important for MSRSs are achievements in multichannel antennas with electronic beam steering, high speed digital processors and computers, transmission lines with high capacity, and precise synchronization systems.

Several books have been published where important space–time processing problems were considered which concern MSRSs to some extent [18,23,24,42,52,72, 191,192].

Some valuable information about MSRSs is presented in [1,46] and especially in [173,174,208,244,257]. However, there is no summarizing book specially devoted to fundamentals of multisite radar systems (including multistatic radars and multiradar, netted radar systems), both active and passive (MSRSs).

This book is an attempt to fill the gap and to give a systematic account of the theoretical foundations of MSRSs with a unified approach. The book has been written on the basis of the author's experience in this field for more than 20 years in the Soviet Union and Russia. The author sought to make the book intelligible for a wide range of specialists and useful for their practical activities. Therefore much attention is paid to the physical sense of the results obtained. Principal algorithms and performance analysis are presented in the form which allows for direct use in practice.

The book is organized so that a reader firstly gets a general idea of MSRSs, their performance principles, advantages and drawbacks, and most important characteristics. Part 1 serves that purpose, especially Chapter 1, where the material is presented at an 'elementary' level without involving complex equations.

Such preliminary acquaintance with the general facilities and features of MSRSs allows the reader (especially, a reader-engineer) to determine his attitude to MSRSs and to approach more deliberately the study of the MSRS signal detection and parameter estimation statistical theory derived in Part 2 and Part 3.

As with a monostatic radar, the ultimate goal of a MSRS is usually to measure target coordinates and tracks or trajectories (for moving targets). Therefore, the signal detection and parameter estimation should be considered as an integrated statistical problem (see, for example, [126,127]). For a monostatic radar, a multitarget environment gives rise to the problem of data association, i.e. association between targets and measurements obtained in different time intervals. Besides it is necessary in a MSRS to associate data obtained by spatially separated stations, i.e. to determine from which target, if any, a particular station measurement originates (interstation data association). Thus for a MSRS in a multitarget environment it would be considered an integrated statistical problem 'detection and measurement –data association – target location and tracking'.

However the monostatic radar experience shows that separate signal detection and parameter estimation optimization does not lead to noticeable losses. It is well known that optimal and close to optimal signal detectors and parameter estimators have common parts and are implemented by similar equipment and algorithms. Therefore, much simpler separate consideration of detection and estimation problems in the majority of radar theory text-books and manuals (for example, [47,48,72]) is warranted from both methodological and practical points of view. This is valid for the data association problem too. In accordance with this, the target detection theory for MSRSs is considered in Part 2 while the target position estimation and tracking theory is considered in Part 3. Principles of the data association in MSRSs are also discussed in Part 3.

Throughout the book the 'classical' statistical approach to signal detection, parameter estimation and filtration is used with Gaussian models for noises and interferences. From the author's experience that approach is adequate for the majority of practical radar engineering problems.

The limited space available in the book does not permit us to consider certain important and actual MSRSs issues including radioimaging (see, e.g., [62, 63, 114–116]), measurement of geometrical characteristics and motion parameters for spatially distributed targets, optimization of MSRS performance control under limited resource conditions and various others. Specific matters of MSRS implementation and examples of MSRS design and performance are described in [67].

ACKNOWLEDGEMENTS

The author would like to thank Dr L.P. Zaslavsky and Dr L.V. Osipov. Our team-work for many years has promoted the appearance of this book by no small degree. I also thank my collaborators: S.M. Yelistratov, A.V. Baikov, Ye.V. Tshadilov and G.E. Lobanova for some calculations and simulations and Ye.K. Kulikov for his help in getting together the manuscript. I am indebted to Professor I.B. Fedorov and Dr G.P. Slukin (Moscow State Technical University) for their criticism upon reviewing the Russian edition of the book.

This English edition is a revised and expanded translation of the Russian edition. I am grateful to the reviewers of this edition – Dr Alfonso Farina (Alenia, Italy), Professor Hugh Griffiths (University College London) and Dr Eberhard Hanle (FGAN FFM, Germany) – whose valuable recommendations have been conducive to improving the book. I am especially thankful to Dr Alfonso Farina for his continuous support and encouragement during the publication of this book.

LIST OF ABBREVIATIONS

AA	antenna array
ABM	antiballistic missile
ADP	antenna directivity pattern
AGSR	advanced ground surveillance radar
AMJCA	adaptive mainlobe jamming cancellation algorithm
AOA	angle of arrival
AS	antenna system
ASFIR	active swept frequency interferometer radar
ATC	air traffic control
ATCAS	air traffic control automated system
CDF	cumulative distribution function
CDS	cloud of dipole scatterers
CFAR	constant false alarm rate
CMR	collection of masking reflectors
CNR	clutter-to-noise ratio
CW	continuous wave
DDF	differential Doppler frequency
DF	Doppler frequency
DS	dipole scatterer
DTL	data transmission line
ECM	electronic countermeasures *or* error covariance matrix
FC	fusion centre
FIM	Fisher information matrix
FIR	flight information region
FMCW	frequency modulated continuous wave
FS	flare spot
INR	interference-to-noise ratio
ISF	interference suppression factor
JORN	Jindalee over-the-horizon operational radar network
JPDA	joint probabilistic data association
KREMS	Kiernan reentry measurement site
LF	likelihood function *or* likelihood functional
LFM	linear frequency modulation
LOP	line of position
LOS	line of sight
MGF	moment generating function
MMS	multistatic measurement system
MSRS	multisite radar system
MTI	moving target indication
OTH	over-the-horizon
PAA	phased antenna array
PDA	probabilistic data association
PDF	probability density function
PLSS	precision location/strike system
PRI	pulse repetition interval
PSM	polarization scattering matrix
RAM	radar absorbent material
RDP	resultant directivity pattern
r.f.	radio frequency
r.m.s.	root mean square
RSRP	resultant signal reception pattern
SCNR	signal-to-clutter-plus-noise ratio
SCR	signal-to-clutter ratio
SDP	spatial discrimination pattern
SINR	signal-to-interference-plus-noise ratio
SNR	signal-to-noise ratio
SOP	surface of position
SRC	space resolution cell
SSR	secondary surveillance radar
TDOA	time difference of arrival
TOA	time of arrival
TPUR	target position uncertainty region
UHF	ultra high frequency

PART 1

General Characteristics, Radar Cross Section of Targets, Coverage of MSRSs

1. GENERAL CHARACTERISTICS

1.1. DEFINITION AND CLASSIFICATION

Multisite Radar Systems (MSRSs) have been developing for many years. However, there is no generally accepted definition of MSRSs. In some works [42,102] MSRSs are considered in a "narrow sense": a MSRS is defined as a single radar with, as a rule, one transmitting and several spatially separated receiving stations. Such MSRSs are usually called "Multistatic Radars" [86,92,102]. Many papers consider arbitrary systems of spatially separated radars where all received information concerning observed objects (targets) is fused and jointly processed. Such systems are often called "Multiradar (or Netted Radar) Systems" (e.g., [77,85,103]).

We introduce a "wide" definition of MSRSs which covers both Multistatic Radars and Multiradar (Netted Radar) Systems. We define a MSRS as *a radar system including several spatially separated transmitting, receiving and (or) transmitting–receiving facilities where information of each target from all sensors are fused and jointly processed*. Thus a MSRS has two principal distinctions: several spatially separated stations and fusion (joint processing) of received target information. It is the combination of these two peculiar features that gives rise to the main benefits of MSRSs.

It follows from the above definition that so called "bistatic radars" (e.g. [171,178, 190,192]) with spatially separated single transmitting and single receiving stations do not belong to MSRSs. The joint information processing is possible if a radar system includes at least two receiving stations and one transmitting station (which may coincide with one of receiving stations) or at least two transmitting stations and one receiving station (which in its turn may coincide with one of transmitting stations). However, bistatic radars may be considered as elements ("cells") of certain types of MSRSs that utilize their properties and characteristics. MSRSs intended for radiation sources (e.g. jammers) observation contain at least two receiving stations.

Many different types of MSRSs are known. They differ in purpose, number and type of included stations, main characteristics. It is clear from discussing the definition of the MSRS that there is no generally accepted classification of MSRSs. Known suggestions [1,23,24] do not take into account a number of important properties which have a significant influence on performance characteristics of MSRSs.

It is hardly reasonable to construct a unified "tree" for the MSRS classification. Apparently, it will be better to extract several essential attributes and classify MSRSs in accordance with them [65,67] (see Fig. 1.1).

1. Depending on *the type of targets of interest*, MSRSs may be divided into three classes: active, passive and active–passive MSRSs. Nonradiating targets are "served" by *active* MSRSs including at least one transmitting station. Target information is extracted from reflected signals (echoes). Radiating targets (radiation sources, e.g. jammers) are "served" by *passive* MSRSs which comprise receiving stations only. Target information is extracted from received radiation. *Active–passive* MSRSs "serve" both nonradiating and radiating targets (in active or passive mode

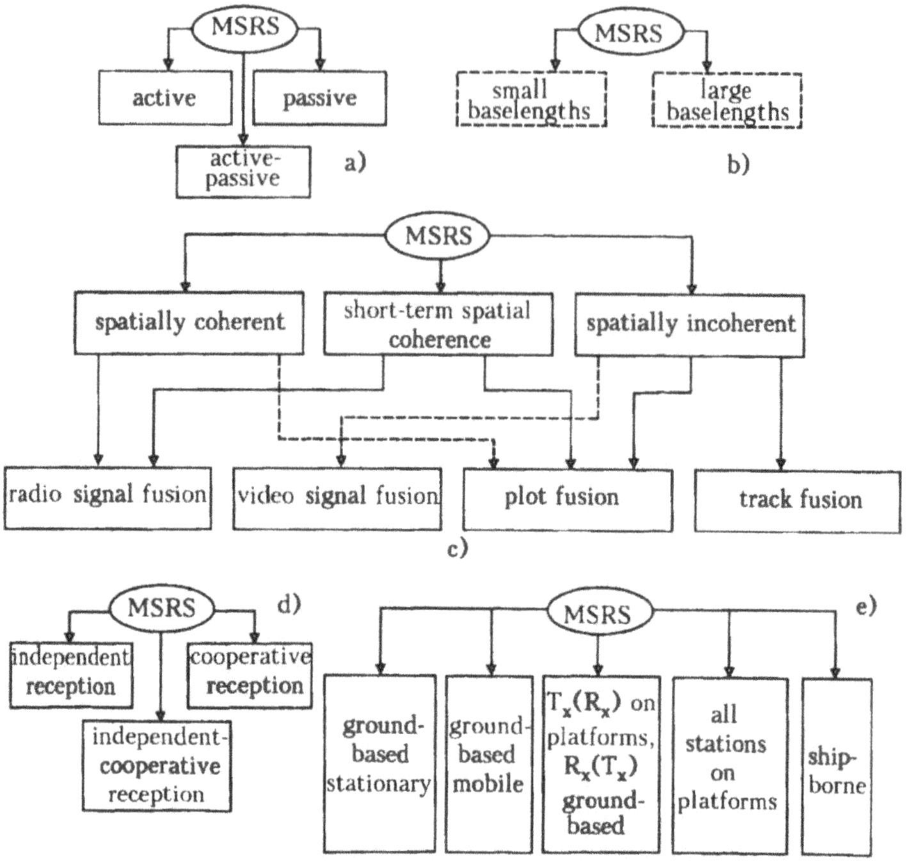

Figure 1.1 Classification of MSRSs. Classification characters: (a) type of targets to be served – radiating or nonradiating targets; (b) baselengths between stations; (c) degree of spatial coherence and information fusion level; (d) degree of signal reception autonomy; (e) station location and mobility

respectively). For example, if a jammer is concealing its own echo and echoes from other nonradiating targets, such a MSRS can estimate the jammer's coordinates and track parameters. (This is the "principle of complementation" of active and passive modes). An active–passive MSRS must include at least one transmitting station.

Two of the most important features which determine capabilities of MSRS, are *the degree of spatial coherence and the information integration* (*fusion*) *level*. These two features are coupled. The *spatial coherence* of a MSRS means its ability to maintain a strong dependence between signal r.f. phases in separated stations and, consequently, to utilize useful information contained in those phase relations. One should distinguish between the spatial coherence of a MSRS and the spatial coherence of signals at the inputs of the MSRS' receiving stations (see Section 4.1). The latter depends on baselengths between stations, signal wavelength, target size (see Section 4.2) and fluctuations of the propagation medium characteristics. Spatial coherence of a MSRS represents, in effect, phase stability of its equipment. Spatial coherence of a MSRS and that of signals at the inputs of receiving stations may be considered as independent characteristics.

2. Depending on *degree of spatial coherence*, all MSRSs may be placed into three classes. Phase shifts of signals and interferences in equipment of *spatially coherent* MSRSs are known and maintained during the time intervals much greater than the duration of signals used (usually during several hours and even days). If a MSRS is designed for operating within certain time intervals, the equipment phase stability is maintained during those intervals. Equipment phase shifts may be measured periodically (for example, in gaps between operational time intervals) with the help of some reference signals (e.g. radiostars [104] or point reflectors [115]). Results of these measurements may be used for phase alignment or simply taken into account in signal processing. Thus spatially coherent MSRSs require not only precise time synchronization and frequency control but phase control (phase synchronization). Spatially coherent MSRSs may be referred to as sparsely populated phase antenna arrays. In principle, such MSRSs can utilize information contained in the electromagnetic field spatial structure most fully. However, many stations (several dozens or more [95,104,114]) are required for achieving an acceptable form of Spatial Discrimination Pattern – SDP (the three-dimensional analogue of the Antenna Directivity Pattern – ADP). For this reason and because of difficulties of the interstation phasing implementation, spatially coherent MSRS are the most complicated and expensive. It is clear, that such MSRSs are useless to built and deploy if expected echoes at the inputs of separated stations are incoherent. Therefore, the baselengths between stations of a spatially coherent active MSRS are usually small enough to ensure maintaining the scattered signal spatial coherence. The smaller are baselengths, the simpler MSRSs turn out to be in practice, but shortening the baselines leads to losses in the MSRS' informativeness. This reduces advantages of the spatial coherence of MSRSs.

In MSRSs with *short-term spatial coherence*, the equipment phase stability is maintained within short time intervals but no less than the used signal duration. Usually these time intervals do not exceed some fractions of a second. Slow uncontrolled phase shifts are admitted. It may be assumed that at the beginning of each time interval of signal reception and processing, interstation phase shifts are random and mutually independent. Hence they do not contain useful information. Joint signal processing in MSRSs with short-term spatial coherence can use all information contained in signal complex envelopes and, of course, in plots and tracks from different stations. For instance, this permits using interstation phase shift changes during the signal duration for tangential velocity estimation by the differential Doppler method. However, it is impossible to take bearing of a target by the phase direction finding method as in an interferometer. Giving up the information which may contain r.f. initial phase relations of signals at the inputs of separated stations, makes the interstation phase control in such a MSRS needless. Only time synchronization and frequency control may be necessary. A MSRS with short-term spatial coherence may comprise a few stations only. The SDP is determined by complex envelope relations (in particular, by signal DTOAs – Differences of Time Of Arrival) and hence is unambiguous in contrast with that of a spatially coherent MSRS where SDP is formed by phase relations. All those features of MSRSs with short-term spatial coherence decrease complexity and cost significantly as compared with spatially coherent MSRSs. True, the resolution and accuracy characteristics depend no longer on carrier frequency but on frequency bandwidth. However, these losses may be compensated in many cases thanks to the possibility of lengthening the baselines.

In *spatially incoherent* MSRSs both interstation phase information and its changes in time are not used. All that information is eliminated, e.g. in envelope detectors, before signal or data fusion. Such MSRSs can utilize information containing only in signal real envelope relations, plots and tracks from separated stations. In connection with this only time synchronization of separated stations is usually necessary[1].

Spatially noncoherent MSRSs are much simpler than MSRSs with short-term spatial coherence and, of course, than spatially coherent MSRSs. However, the phase information elimination leads to certain power and especially information losses. In particular, joint coherent processing of mainlobe jamming (received by separated stations – see Chapter 8) for mainlobe jamming cancellation is impossible. It also makes impossible direct measuring of differential Doppler frequency shifts of signals in spatially separated stations for the target and jammer tangential velocity estimation. It should be noted that spatial incoherence of a MSRS does not rule out the possibility of temporal coherence of each station before information fusion. For example, if a spatial incoherent MSRS is comprised of several radars, it may be possible to measure Doppler frequency shifts and consequently target radial velocities in each radar.

3. According *to information integration* (*fusion*) *level*, MSRSs may be divided into four classes. In each class both analogous and digital Data Transmission Lines (DTLs) may be used. When *the radio signal integration level* is used, all signals, noises and interferences from spatially separated stations are subjected to joint processing. These mixtures are transmitted via DTLs for fusion either immediately from the inputs of the stations or after preliminary linear filtration in each station. As a rule, wideband DTLs (with large handling capacity) are required. If *the video signal integration level* is used, all signals, noises and interferences are subjected to joint processing too, but after phase elimination in each station. The video signal fusion does not reduce required DTL capacity significantly, as compared with the radio signal fusion, but leads to certain power and especially information losses. These losses are the result of giving up even the short-term spatial coherence. This is why the video signal fusion is seldom used.

The required DTL handling capacity is reduced drastically when *the plot*[2] *integration level* is used. "Primary" information processing is accomplished in each station completely including thresholding and parameter estimation of detected signals. Only information considered "useful" after "primary" processing is transmitted via DTLs for fusion. At the same time, partial processing and control decentralization in MSRSs are achieved in this case. In particular, only preliminary decisions regarding the presence or the absence of a target may be made in each station, while a final decision is made in a Fusion Centre (FC) as a result of combining the preliminary decisions coming from all stations. This is so called decentralized (or distributed) detection (see Chapter 6 and Section 10.5).

When *the track* (*trajectory*) *integration level* is used, not only "primary", but "secondary" information processing is accomplished in each station which ends with target track formation. Track parameter estimates coming from spatially separated stations are fused. "False" tracks are further eliminated and "true" tracks' parameters are estimated more accurately at the fusion process. Common tracks are built by combining information from local tracks. DTL handling capacity requirements are of the same order as with the plot fusion. Partial decentralization of information processing and control is achieved too.

---

[1] Interstation frequency control is necessary in MSRSs with so-called cooperative signal reception (see below) for coherent signal processing in each receiver (i.e. signal processing using Doppler frequency shifts of received signals).

[2] A plot (a contact, a report) is the result of an individual target detection and parameter measurement, or a measurement of target coordinates and their derivatives, when an "instantaneous" target state is estimated without taking into account preceding measurement results.

In general, the higher information integration level is used (i.e. the less information is lost in each station before fusion), the better are power and information performance characteristics of a MSRS, but the more complicated is the system and the higher DTL handling capacity is required. As mentioned above, spatial coherence degree and information integration level of a MSRS are connected with each other. These connections are shown in Fig. 1.1. In spatially coherent MSRSs, as a rule, the radio signal fusion is used. In some cases the plot fusion may be employed (with maintained phase information). The same information integration levels may be used in MSRSs with short-term spatial coherence. In incoherent MSRS video signal fusion, plot fusion or track fusion is possible. Usually plot fusion or track fusion is employed. Actual MSRSs may be of a hybrid type where information can be fused at several different levels. For example, if there are no external interferences, plot fusion may be used in a MSRS. In the presence of jamming radio signal fusion may be employed in the same MSRS. This allows to determine jammer positions and tracks by hyperbolic method of passive location with the help of interstation jamming correlation processing.

4. One more important attribute of MSRSs may be called *degree of autonomy of signal reception.* If a MSRS is comprised of several monostatic or bistatic radars, each radar may be designed for receiving scattered signals from targets illuminated by the transmitter of the same radar only (dedicated transmitter). This is a MSRS with *independent (autonomous) signal reception.* The independent signal reception is usually used in spatially incoherent MSRSs with information fusion at the plot or track level. Such MSRSs are often called Netted Radars or Radar Networks. Different radars may operate in different frequency ranges.

Substantially better power and information characteristics are demonstrated by MSRSs with *the cooperative signal reception* where each radar or receiving facility can receive and process echoes from targets illuminated by any radar or transmitting facility of the MSRS. The particular case is a MSRS with one transmitting (or transmitting–receiving) station and several receiving stations. The spatial coherence of MSRSs with cooperative signal reception may be arbitrary with appropriate information integration level.

A MSRS with *the independent-cooperative signal reception* contains both receiving stations exploiting transmissions of their own radars and receiving stations which can also exploit transmissions of other radars of the MSRS.

5. According to *station locations and mobility* in the measurement process five classes may be considered. *Ground-based MSRSs with stationary stations* comprise not only actual stationary MSRSs, but MSRSs where involved units may be transferred from one location to another. It is important that station locations are only not to be changed in the operation (measurement) process. Sometimes such MSRSs are called "systems with stationary baselines". It should be noted (see Fig. 1.1), that *MSRSs with transmitting facilities* (T_x) *on air or space platforms and ground-based receiving facilities* (R_x) are put down to the same class as *MSRS with* R_x *on air* or *space platforms and ground-based* T_x. *Shipborne MSRS* may be placed on one or several ships.

It is clear, that ground-based, airborne, spaceborne or shipborne MSRSs differ not only in construction, but in technical characteristics, capabilities and limitations. Apparently, ground-based MSRS with stationary baselines were proposed, studied and built earlier than others. However, in the last years much attention was paid to different MSRSs with moving baselines. It is associated with noticeable achievements in precise radionavigational techniques and systems, data transmission and accurate synchronization of remotely located facilities.

In the classification described we do not divide MSRS into different classes according to *baselenghts* (distances between stations), though in certain works [1,24] the baselengths are considered to be one of the main distinction features of a MSRS. Of course, the baseline lengths strongly influence the MSRS' characteristics. However, in contrast to the distinctions discussed above, the baselength is a continuous value. It is difficult to set up sharp boundaries so that the transition through them will lead to abrupt changes in features and characteristics of MSRS. It should be noted, that baselengths between transmitting and receiving stations (which are used in [1] as one of the most important attribute for the MSRS classification), and baselengths between receiving stations have a different impact on MSRS' technical characteristics. Some important characteristics, such as measurement accuracy, depend not upon the baselength itself, but upon the so-called effective baselength. *The effective baselength is the length of the baseline's projection on the plane orthogonal to the bisector of the angle between directions from a target to stations of interest.* Besides that, influence of baselengths on MSRS' features depends both on included units and on the degree of signal reception autonomy. For example, if a MSRS contains one transmitting and several receiving stations, one part of technical characteristics (detection characteristics, coverage area) depends mainly upon the baselengths between transmitting and receiving stations, whereas the other part (measurement accuracy, resolving power) is determined by effective baselengths between receiving stations. The other MSRS classification according to baseline distances is based upon the interstation correlation of signal fluctuations [24]. This is inconvenient in many cases too. As it is known (see Section 4.2), that correlation depends not only on the baselengths (or the effective baselengths) of a MSRS, but on target dimensions, signal wavelength and target distance from the MSRS. Signals from a target at a large distance can be spatially coherent and amplitude fluctuations may be completely correlated at the inputs of separated stations. But as the target nears the MSRS, signals may become incoherent and amplitude fluctuations completely uncorrelated. Therefore fluctuation correlation degree cannot be considered as a characteristic of a MSRS itself. Besides, this criterion cannot be applied to MSRSs comprised of several radars with independent signal reception, if a target is illuminated by different radars at different frequencies, or (and) if a moving target is illuminated at different instants of time. In such cases, signal fluctuations at the inputs of spatially separated stations may be uncorrelated regardless of the baselengths. This criterion does not cover an important case of passive MSRSs or passive mode of active–passive MSRSs either.

Thus, because of difficulties in establishing quantitative criteria and well-defined boundaries between classes, it is not reasonable to use the discussed feature as one of the main classification attributes. Nevertheless, sometimes it is convenient in practical work to divide MSRSs coarsely into two classes: with "short" baseline distances and with "long" baseline distances. The first class includes MSRSs where effective baselengths are essentially less than expected target ranges in the main part of coverage area. On the contrary, if effective baselengths are of the same order or greater than expected target ranges (in the main part of coverage area), such MSRSs may be considered as MSRSs with long baseline distances. It concerns both nonradiating and radiating targets. The "hardware" and "software" of MSRSs with short baseline distances are simpler and, as a rule, less expensive. Geometrical relationships between signal parameters and target coordinates as well as most processing algorithms are simpler. However, MSRSs with long baseline distances can, for example, make higher accuracy of target coordinate and track parameter estimation available.

For the reasons discussed above, we have not included baseline distances in the list of main classification attributes. Therefore, breaking MSRSs down into two classes according to this feature is shown in Fig. 1.1 by dashed lines.

The described classification covers only some main distinctive features of MSRSs and as any classification is rather conventional. At the same time, it allows to have a systematic picture of many different MSRSs and to consider and study features and characteristics of the whole classes of MSRSs.

***In this section**, we have introduced a wide definition of Multisite Radar Systems (MSRSs) including both Multistatic Radars and Multiradar (Netted Radar) Systems. We have suggested a classification scheme for MSRSs taking into account their principal attributes which determine most important features and performance characteristics. This classification permits to consider whole classes of MSRSs instead of each MSRS individually. It will be used throughout the book.*

1.2. MAIN ADVANTAGES OF MSRSs

Owing to information fusion from spatially separated stations, a MSRS presents a series of significant advantages over both a monostatic radar and a collection of radars not integrated in a system. We note here the main advantages only so that a reader could have a general idea about them. Some of these advantages are discussed in more detail in the following chapters in the course of analysis of MSRS' characteristics. It is obvious, that actual significance of either advantage depends upon the designated purpose of a MSRS and requirements to it. On the other hand, the possibilities of taking certain advantages are not the same for different types of MSRSs (see Section 1.1).

Capability to Form Coverage Area of Required Configuration for Expected Environments

The additional factors determining coverage area of a MSRS, as compared with a monostatic radar, are system geometry and fusion algorithms (see Chapter 3). In particular, changing these factors permits the extension of coverage area in required directions. If a MSRS consists of mobile units, the flexible reconfiguration of the coverage area is easy to achieve.

Power Advantages

Evidently, addition of any number of transmitting or (and) receiving stations to a monostatic radar upgrades the total power or (and) sensitivity of the system. However, MSRSs have some extra power advantages. First of all, the cooperative signal reception where each receiving station (or radar) can exploit the transmission energy of all transmitting stations (radars), enjoys significant power benefit. If baseline distances are sufficiently long, scattered signal fluctuations are statistically independent at different receiving stations. The same result may be obtained if a target is illuminated by sufficiently separated transmitting stations. In this case information fusion may lead to an additional power gain due to fluctuation smoothing, especially if high detection probabilities are required (see Sections 5.4, 5.5, 6.2). This gain may also be obtained in MSRSs with independent signal reception including systems consisting of radars with different carrier frequencies. Simultaneous target observation from different directions make it possible to defeat Stealth technologies [89,140,171,192]. When stations are so separated that the angle between directions from a target to a transmitting and a receiving stations nears 180°, the Radar Cross Section (RCS), and hence the scattered signal intensity at the input of the receiving station, may drastically increase. It is important that this increase cannot be reduced by Stealth technologies including body-shaping and Radar Absorbent Material (RAM) coating (see Section 2.2). Besides, there are some additional reasons for power advantages. For example, when transmitting and receiving stations are spatially separated, then signal power losses decrease, since a duplexer and other receiver protection assemblies are needless [81].

High Accuracy of the Position Estimation of a Target

Target position determination by a usual monostatic radar is much less accurate in cross-range than in down-range direction, especially for distant targets. A MSRS allows to estimate all the three target coordinates through range measurements from several spatially separated monostatic radars or through range-sum measurements relative to several spatially separated transmitting and receiving stations. Figure 1.2(a) shows sections of two error centroids obtained as a result of target position measurement by each of the two radars. Each centroid in 3D space represents an ellipsoid which is usually flattened like a "pancake". The intersection of those ellipsoids may represent a resultant error centroid after information fusion (joint processing) from two radars. It can be seen, that there is a noticeable gain in the target position estimation accuracy mainly due to the range measurements. It may be considered that range measurements in a MSRS give an increase of the angle coordinate estimate accuracy as compared with a monostatic radar. For rough calculations of angle accuracy provided by range measurements in each spatially

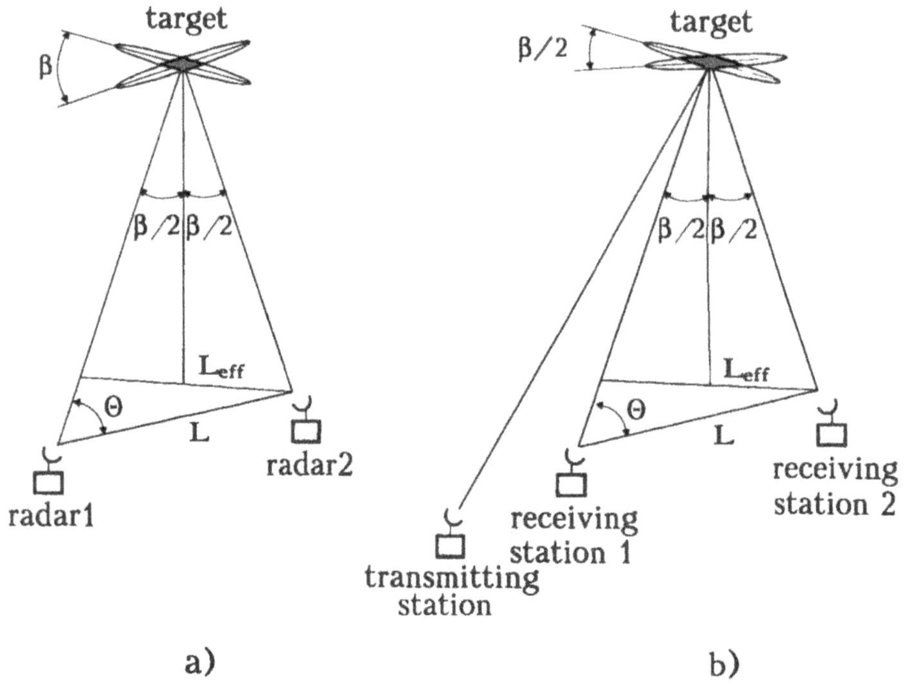

Figure 1.2 Increase of angular coordinate measurement accuracy: (a) a MSRS consists of two radars with autonomous signal reception; (b) a MSRS consists of a single transmitting and two receiving stations

separated pair of radars, the following approximate expression is convenient to use

$$\sigma_\theta = \sigma_R \sqrt{2}/L \sin\theta \approx \sigma_R \sqrt{2}/L_{eff}. \quad (1.1)$$

Here σ_θ denotes the r.m.s. error of the angle estimation in the bistatic plane passed through the target and both radars; σ_R denotes the r.m.s. error of range measurements (assuming these errors to be statistically independent in different radars with equal r.m.s. value); L denotes the baseline distance (baselength); L_{eff} denotes the effective baselength.

When a MSRS consists of one transmitting–receiving station (a monostatic radar) and one receiving station or one transmitting station and two receiving stations, then instead of (1.1) we have (see Fig. 1.2(b))

$$\sigma_\theta = \sigma_{R\Sigma} \sqrt{2}/L_{eff} = 2\sqrt{2}\,\sigma_R / L_{eff}. \quad (1.2)$$

Here, $\sigma_{R\Sigma}$ denotes the r.m.s. error of range-sum (transmitting station–target–receiving station) measurements; $\sigma_{R\Sigma} = c\sigma_\tau$ where c is the light velocity, σ_τ is the r.m.s. error of the Time Of Arrival (TOA) measurements ($\sigma_R = c\sigma_\tau/2$). It is seen that, with other conditions being equal, replacing a MSRS consisting of two radars with independent signal reception by a MSRS consisting of one transmitting and two receiving stations (one of which may coincide with transmitting station) is equivalent to cutting effective baselength in half. This means that only half of the effective baselength $L_{eff}/2$ is actually "working". Expressions (1.1) and (1.2) have been obtained under the condition of large target range/effective baselength ratio ($R/L_{eff} \gg 1$)[3].

It follows from equations (1.1) and (1.2) that with range measurements of high accuracy (i.e. if wideband signals are used) and if effective baselengths are sufficiently large, the r.m.s. error may be much less than that of a usual bearing measurement by a monostatic radar.

Example 1.1. Let $\sigma_R = 5\,\text{m}$, $L_{eff} = 30\,\text{km}$. Then (1.1) yields $\sigma_\theta = 0.8'$, and (1.2) yields $\sigma_\theta = 1.6'$.

In certain cases this feature of MSRSs permits large and expensive radar antennas to be replaced by small, weakly directional antennas without accuracy losses of target position location [100,107].

However, equations (1.1), (1.2) and Fig. 1.2 show that if L_{eff} is small (error ellipsoids from two stations are nearly parallel) or (and) if range measurement errors are large, these measurements cannot improve angular accuracy significantly. In this case bearing measurements by spatially separated stations play the main role in angle accuracy improvement. For instance, such a situation occurs when a ground based MSRS measures a small elevation angle of a target, since effective baselength is proportional to the sine of the elevation angle.

In the general case, when total number of "primary coordinates" (ranges, range-sums, bearings) of each target measured by all spatially separated stations exceeds the necessary minimum determining the target position (in a space of considered dimensions), then redundant measurements are used for position accuracy refinement.

The track updating rate may often be higher in a MSRS than in a monostatic radar. This leads to higher tracking accuracy too [75,83].

---

[3] As mentioned above (see Section 1.1), MSRSs with short baselines are must simpler than those with long baselines (including the geometric relations), but enjoy main advantageous features of MSRSs. We present in this section some simple equations for rough estimation of MSRS' characteristics. Though obtained under the $R/L_{eff} \gg 1$ condition, they often may be used even when $R > (2\ 3)L_{eff}$.

Possibility of Estimation Target Velocity and Acceleration Vectors by the Doppler Method

Doppler frequency shift measurements at several spatially separated stations allow one to estimate the velocity vector of a target. This may be of great importance for accurate target tracking, especially along the manoeuvre portions of target paths (e.g. [85,257]), for ballistic target observation when trajectory parameters of a target are to be estimated with high accuracy and in a minimal time interval etc. In a simple system comprising two radars with the baselength L between them and with independent (autonomous) signal reception the measured Doppler frequencies (DFs) are $F_{D1}=2\mathbf{v}\mathbf{r}_1/\lambda$ and $F_{D2}=2\mathbf{v}\mathbf{r}_2/\lambda$. In these formulas: $\mathbf{v}$ is the target velocity vector; $\mathbf{r}_1, \mathbf{r}_2$ are the unit vectors directed from the target to Radar 1 and Radar 2, respectively. If $\mathbf{v}$ lies in the plane passing through the target and both radars or if it is a projection of the velocity vector onto this plane, simple equations may be easily obtained from Fig. 1.3 for the r.m.s. errors of the radial and tangential (in the same plane) velocity estimation. These equations are convenient for preliminary calculations

$$\begin{aligned}\sigma_{vR}&=[\lambda/2\sqrt{2}\cos(\beta/2)]\sigma_F \approx (\lambda/2\sqrt{2})\sigma_F;\\ \sigma_{v\tau}&=[\lambda/2\sqrt{2}\sin(\beta/2)]\sigma_F \approx (\lambda/\sqrt{2})(R/L_{eff})\sigma_F.\end{aligned} \tag{1.3}$$

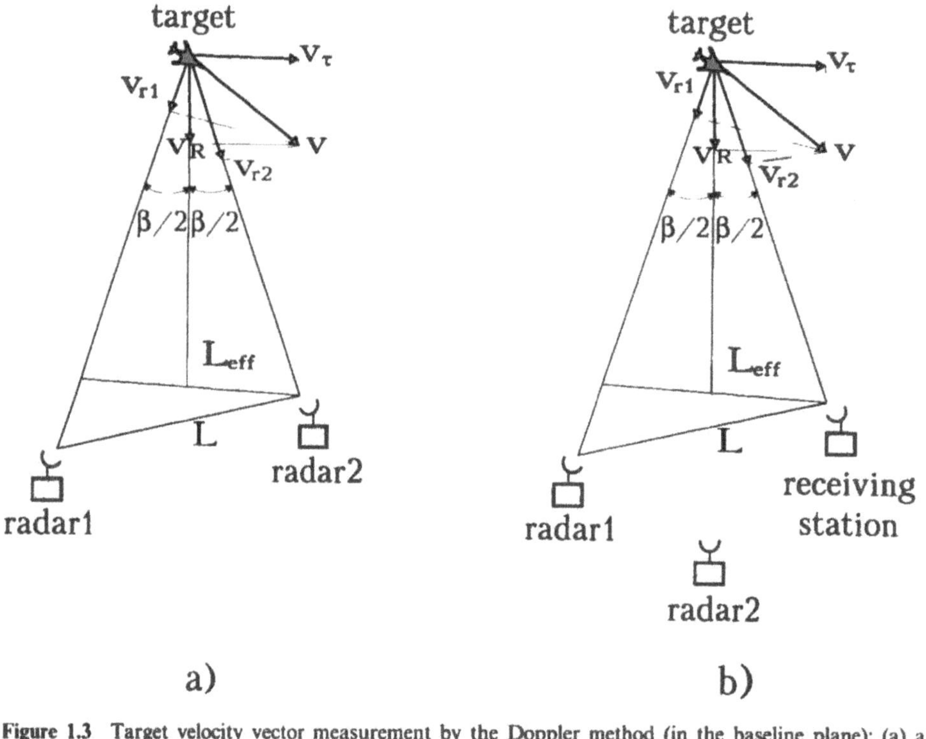

Figure 1.3 Target velocity vector measurement by the Doppler method (in the baseline plane): (a) a MSRS consists of two radars with autonomous signal reception; (b) a MSRS consists of a single transmitting and two receiving stations ($\mathbf{v}$ = velocity vector, $\mathbf{v}_{r1}, \mathbf{v}_{r2}$ = radial velocities with respect to radar 1 and radar 2; $\mathbf{v}_R, \mathbf{v}_\tau$ = radial and tangential velocities with respect to the MSRS)

In equation (1.3) σ_F is the r.m.s. error of DF measurements from each radar (assumed equal); λ is the wavelength; R is the target range; β and L_{eff} are the same as in (1.1). The approximate equalities in the right part of (1.3) are valid under the "short baseline distances" condition: $R/L_{eff} \gg 1$, when $\cos(\beta/2) \approx 1$, $\sin(\beta/2) \approx (\beta/2) \approx L_{eff}/2R$. It is seen that for the fixed DF measurement accuracy the r.m.s. error of the tangential velocity estimate is greater by the factor $\mathrm{ctg}(\beta/2) \approx 2R/L_{eff}$ than that of the radial velocity. If a receiving station is substituted for one radar, for example Radar 2 (see Fig. 1.3(b)), $F_{D2} = \mathbf{v}(\mathbf{r}_1 + \mathbf{r}_2)/\lambda = 2v_r \cos(\beta/2)/\lambda$. Then

$$\begin{aligned} \sigma_{vR} &= [\lambda/2\cos(\beta/2)]\sigma_F \approx (\lambda/2)\sigma_F; \\ \sigma_{vt} &= [\lambda/\sqrt{2}\sin(\beta/2)]\sigma_F \approx (\lambda\sqrt{2})(R/L_{eff})\sigma_F. \end{aligned} \tag{1.4}$$

Comparing (1.4) with (1.3) we note the increase of σ_{vR} by the factor $\sqrt{2}$. Besides that, only one half of the effective baselength "works" for the tangential velocity estimation, as in the case of the angle estimation [see (1.1) and (1.2)]. The same result may be obtained if Radar 2 could be placed in the midpoint of the effective baseline (in the bisector of the bistatic angle) instead of the receiving station.

The more rigorous consideration of Doppler velocity measurement problems are presented in Chapters 11, 13 and 14. By estimating the speed of Doppler shift variations or by differentiating the velocity vector components one can obtain the target acceleration vector. The use of Doppler velocity and acceleration estimates for target tracking, gains track accuracy and general quality of the tracking process, especially in the time intervals where the target velocity changes abruptly (the manoeuvre of an aircraft or the deceleration of a warhead of a ballistic missile in the atmosphere). Under certain conditions MSRSs can track targets using Doppler shifts only and range derivatives of higher order [23,46,85].

Capability to Measure Three Coordinates and Velocity Vector of Radiation Sources

It is well known, that both monostatic and bistatic radars can determine only Directions Of signal Arrival (DOAs) in a passive mode, i.e. bearings on radiation sources. Unlike those radars, MSRSs can obtain three coordinates and their derivatives. This may be achieved by triangulation, or hyperbolic methods, or their combination [18,67]. The triangulation method determines the position of a radiation source in 3D space from the intersection of DOAs from spatially separated receiving stations. The hyperbolic method determines the position of a source from the intersection of hyperboloids of revolution which have their foci at receiving stations (see Section 11.1). A fixed Time Difference Of signal Arrival (TDOA) at a pair of stations (corresponding to the fixed source range difference relative to those stations) determines a hyperboloid of revolution on which surface the source must lie. (A hyperboloid of revolution is a body obtained by revolving of a hyperbola about the axis passing through its foci.) The TDOA is estimated by the signal delay in one station which is necessary to maximize the mutual correlation of signals received by the two stations. It is important to note, that when range R of a radiation source is several times greater than effective baselength L_{eff} between stations, then angular errors of both methods are independent of the source range, so that linear cross-range errors are proportional to range, while range errors are proportional to squared range. A simple relationship can be written for the approximate comparison of source position fix accuracy attainable by triangulation and hyperbolic method. Under the $R/L_{eff} \gg 1$ condition a range difference measurement with the r.m.s. error $\sigma_{\Delta R}$ (for the hyperbolic method) is approximately equivalent to a bearing measurement (for the triangulation) with the r.m.s. error

$$\sigma_{\theta\,eq} \approx \sigma_{\Delta R}/L_{eff} \tag{1.5}$$

where L_{eff} is, as earlier, the effective baselength between two receiving stations.

Example 1.2. A pair of stations where $L_{eff} = 30\,\text{km}$ and $\sigma_{\Delta R} = 10\,\text{m}$ is approximately equivalent to a direction finder placed in the midpoint of the baseline if its bearing accuracy (r.m.s. value) in the plane passing through the source and both stations is $\sigma_{\theta\,eq} \approx 3.3 \times 10^{-4}\,\text{rad} \approx 1.2'$.

The Doppler shift measurements of the mutual correlation function of signals received by a pair of spatially separated stations from a moving radiation source makes it available to estimate the source radial velocity difference relative to these stations. A MSRS containing four or more stations can obtain all the three components of the source velocity vector by Doppler frequency shift measurements. Using triangulation one can estimate source velocity by differentiating the position estimates only. The MSRS's capability to determine three space coordinates and velocity vector of a radiation source is an important feature for tracking such sources. It concerns jammers too, when they prevent tracking targets being concealed by those jammers (including the case of "self-screening" where a jammer is placed on a target of interest). The passive mode of MSRSs may also be used to reconnoitre radar locations of hostile air defence systems (Electronic Support Measures – ESM) [98].

Increase of Resolution Capability

Full characterization of the resolution capability of a radar or a MSRS is possible in terms of detection probability and measurement accuracy in the presence of neighbouring targets and other interference sources (see Chapters 8–12). A simpler "deterministic" approach based on the Rayleigh criterion of resolution is widely used for engineering calculations. A measure of the radar resolution capability in an arbitrary parameter (range, angle coordinates, velocity) is adopted to be the extent (the "width") of the response to echo from a point target. It is assumed that two point targets can be resolved, i.e. detected and measured separately, if the distance between them in any parameter exceeds the extent of the response to echo from each target. Intensities of those echoes are assumed to be approximately equal. The extent at a certain level (for example, at the level of $-3\,\text{dB}$ relative to the response maximum) is usually called a resolution cell in corresponding parameter. Using the Rayleigh criterion allows us to show the resolution capability advantage of a MSRS in a form easy to grasp. Let us consider first active MSRSs (or an active mode of active–passive MSRSs). Figure. 1.4 shows a MSRS comprising two monostatic radars. It is seen that two targets are not resolved by Radar 1. Both targets fall into one resolution cell in range and angle. If, as is normal the practice, range resolution of a radar is much better than cross-range resolution, the angle difference between targets may be quite enough to resolve targets in range by Radar 2. This effect may be treated as a capability of MSRSs to resolve targets in angle within the main beams of receiving antennas. The equivalent angular resolution capability of a system consisting of two radars may be obtained from the range resolution capability of each radar $\delta R = c/2\Delta f_s$ where c is the light velocity, Δf_s is the signal bandwidth. It is easy to show that when the target range R is by several times greater than the

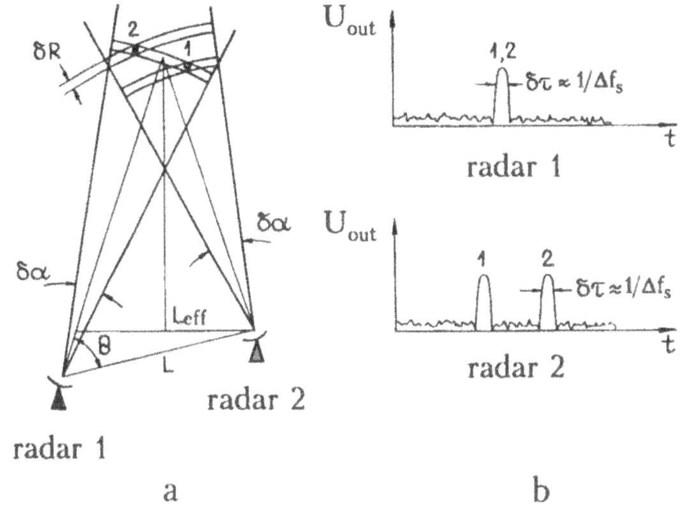

Figure 1.4 Resolution of targets that are not resolved by one radar: (a) arrangement of radars and targets 1 and 2; (b) receiver output signals of radar 1 and radar 2, respectively ($\delta\alpha$ = angle resolution cell = antenna beamwidth, δR = range resolution cell, $\delta\tau$ = TOA resolution cell, Δf_s = signal bandwidth)

effective baselength L_{eff} between radars, then

$$\delta\theta \approx \delta R/L_{eff} = c/2L_{eff}\Delta f_s. \tag{1.6}$$

When Radar 2 is replaced by a receiving station so that two targets are resolved by this station in range-sum (the transmitting station–a target–the receiving station), then instead of (1.6) we have

$$\delta\theta \approx c/L_{eff}\Delta f_s. \tag{1.7}$$

The quantity $\delta\theta$ in (1.6) and (1.7) may be considered as the beamwidth of a "Resultant Directivity Pattern (RDP)" of a pair of radars or a pair of receiving stations, respectively, in the plane passing through the target and both radars or both receiving stations. When the product $L_{eff}\Delta f_s$ is large enough, then the beamwidth of a RDP is much less than the beamwidth of a usual antenna.

Example 1.3. Let again $L_{eff} = 30$ km and $\Delta f_s = 10$ MHz. Then (1.6) yields $\delta\theta = 0.5 \times 10^{-3}$ rad $\approx 1.7'$ and (1.7) yields $\delta\theta \approx 10^{-3}$ rad $\approx 3.4'$.

However, when the angle between a baseline and a target direction is small, the effective baselength decreases which leads to broadening of RDP beamwidth and consequently to angular resolution deterioration. Such a situation usually takes place when a ground based MSRS has to resolve in angle of elevation several targets appearing near the horizon.

In passive triangulation MSRSs, the resolution cell is determined from the intersection of antenna beamwidths. Unlike monostatic radars, two spatially separated receiving stations have resolution capability in range which may be expressed by the following approximate formula (for $R/L_{eff} \gg 1$)

$$\delta R \approx (R^2/L_{eff})\delta\alpha \tag{1.8}$$

where $\delta\alpha$ is the antenna beamwidth of receiving stations (see Fig. 1.5). It follows from (1.8) that this resolution capability is usually poor. The cross range resolution is of the order $R\,\delta\alpha$.

Example 1.4. Let again $L_{eff}=30\,\text{km}$ and $\delta\alpha=10^{-2}\,\text{rad}\approx 34'$. From (1.8) we have $\delta R\approx 30\,\text{km}$ for $R=300\,\text{km}$ and $\delta R\approx 16.7\,\text{km}$ for $R=200\,\text{km}$. The cross range resolution under the same conditions yields 3 km and 2 km respectively.

The range resolution can achieve the value $R\,\delta\alpha$ when the angle between directions from a source to both stations nears 90°. However, in this case range should be estimated not from the midpoint of the baseline, as when $R/L_{eff}\gg 1$, but from each station. In the beamwidth intersection cross range resolution relative to one station means the range resolution relative to the other.

In passive MSRSs where signals from spatially separated stations are undergone correlation processing (hyperbolic method) the resolution capability is determined by the extent of the envelope's mainlobe of the signal mutual correlation function. This mainlobe width can be expressed in TDOA (delay) $\delta\tau\approx 1/\Delta f_s$ as well as in range difference $\delta\Delta R\approx c/\Delta f_s$. For the RDP of two receiving stations (1.7) is valid if $R/L_{eff}\gg 1$. When the product $L_{eff}\Delta f_s$ is sufficiently large (see the above example 1.3

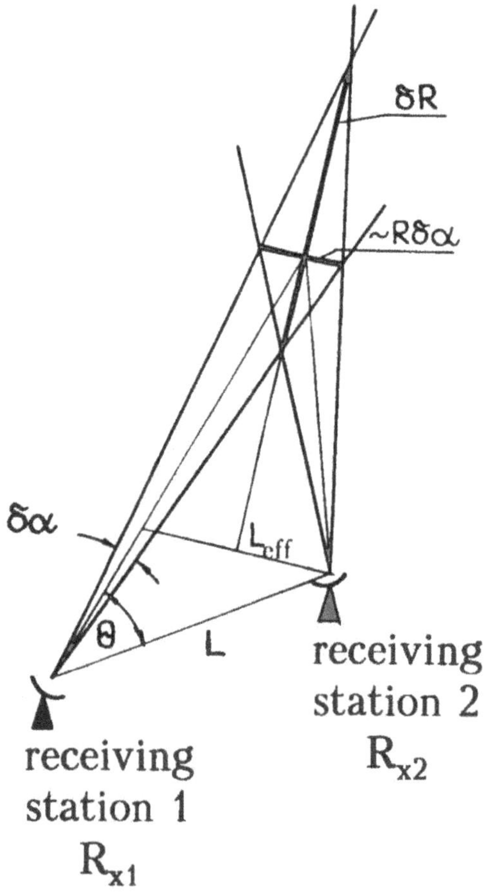

Figure 1.5 Angular and range resolution capability of a passive triangulation MSRS

often called "the graceful degradation" [75]. Obviously, this concerns not only destroying a station as a result of a hostile physical attack, but a failure of equipment as well. Therefore, graceful degradation means higher reliability of MSRSs. The possibility of reconfiguration of MSRSs, if some stations fail, aids in achieving graceful degradation.

As was mentioned above, spatial separation of transmitting and receiving stations makes it difficult to reveal positions of "silent" receiving stations, especially when they are mobile or can be re-located in a short time interval. Because of that, receiving stations of such MSRSs are essentially less vulnerable to a direct physical attack, in particular when Anti-Radiation Missiles (ARMs) are used that locate and home in on the source of radiation to destroy the radar. Unlike "silent" receiving stations, not providing a source of radiation for homing ARMs, transmitting stations broadcast their presence to hostile forces. For transmitting stations protection several techniques were suggested including removal far enough from a dangerous zone, for example from a border or a battle line [122], use of Low-Probability-of-Intercept (LPI) waveforms, decoy transmitters, several netted transmitters with irregular switching between them ("winking" transmitter mode) [67,122].

Survivability of MSRSs can be further enhanced if signal and data processing decentralization is used (see Chapters 6, 7 and [67]).

Technical and Operational Advantages

MSRSs have certain technical and operational advantages, especially when transmitting stations are spatially separated from receiving ones. As was mentioned above in such stations a duplexer and other receiver protection assemblies are needless. There is no dead zone near to receiving stations owing to transmission pulse duration and recovery time of protection circuits. Required dynamic range of receivers may be significantly reduced, since dynamic range of the reflected signals, particularly clutter, collected by remote receiving stations is much lower as compared with monostatic radars [102,171,192].

Some advantages of MSRSs are considered, e.g., in [139,257].

In Section 1.2 main advantages of MSRSs have been outlined:

- *capability to form coverage area of required configuration for expected environments;*
- *power advantages;*
- *high accuracy of the position estimation of a target;*
- *possibility of estimation target velocity and acceleration vectors by the Doppler method;*
- *capability to measure three coordinates and velocity vector of radiation sources;*
- *increase of resolution capability;*
- *improvement of jamming and clutter resistance;*
- *increase of target-handling capacity;*
- *increase of "signal information" body;*
- *increase of survivability and reliability (the "graceful degradation" feature).*

We have presented not only brief descriptions, but simple formulas (for approximate estimation of some important characteristics) and numerical examples. The main goal of this section is to give a general idea of principal advantages of MSRS without complex mathematics.

1.3. MAIN DRAWBACKS OF MSRSs

Main drawbacks of MSRSs are, as a rule, additional difficulties which one have to overcome when creating a MSRS. They may be considered as the price that has to be paid for all advantages briefly discussed in Section 1.2.

Centralized Control of Spatially Separated Stations

Depending on the type of a MSRS this control can be reduced to target distribution among several groups of radars or can solve more complex tasks: coordinated scanning of space, choice of operational frequencies for different transmitters and receivers, waveforms, processing algorithms etc. If a MSRS contains mobile stations, the stations' position control may be necessary [99]. For increasing survivability of MSRSs, partial decentralization of system control is important.

Necessity of Data Transmission Lines

Each MSRS must contain Data Transmission Lines (DTLs) for signal or data transmission from spatially separated stations to information Fusion Centre (FC). These DTLs are used for command and control information transmission too. DTLs with required characteristics are not a difficult engineering problem now. But they increase complexity and cost of a MSRS. It should be taken into account that DTLs need protection against interferences and, in certain cases, against direct physical attack and destruction [102,122].

When plot or track integration levels is used in a MSRS, then DTLs may be of low handling capacity, e.g. as telephone channels [75,77,96]. For radio signal integration level wideband DTLs (of high handling capacity) are necessary. Several multiplexing techniques can be used to reduce handling capacity requirements, for example, data transmission during time gates (in strobes) [82,102]. For command and control information transmission DTLs with low handling capacity are usually quite adequate.

Additional Requirements to Synchronization, Phasing of Spatially Separated Stations, Transmission of Reference Frequencies and Signals

For joint information processing (information fusion) in MSRSs some kind of synchronization between different stations and FC is necessary. Specific requirements depend on the type of a MSRS. Highly accurate target coordinate measurements by the elliptic or hyperbolic methods (see Chapter 11) require precise synchronization. Synchronization errors should be no more than a small fraction of the inverse value of the signal bandwidth. Though this is not a trivial problem, it has been solved in operational systems. For example, in the Multistatic Measurement System (MMS) deployed at the Kwajalein Range of the USA, the synchronization errors are of the order of 0.5 nanoseconds [107]. Recently a method based on meteor trails has been successfully used for precise synchronization of remote sites [151].

When cooperative signal reception is used in a MSRS (see Section 1.1), then frequency and signal waveform emitted by any transmitting station (or any radar) must be known at receiving stations (or other radars). It can be achieved both by signal transmission via DTLs from transmitting to receiving stations (or between all radars) and by transmission of special commands providing alignment of the receiving stations (or radars) to correct operational frequency and signal waveform. For coherent signal processing in each receiver (MTI systems, Doppler measurements) a common reference frequency is necessary at each station to couple the transmitters' and receiver heterodynes' frequencies. A common reference frequency at all receiving stations is necessary also for correlation processing of received signals in passive or active–passive MSRSs if, as usually the practice, frequency transformation is used at each station before signal transmission to FC for mutual correlation processing. The common frequency may be transmitted to each station from a master oscillator (e.g. from an FC) or may be obtained by using a high stable frequency standard at each station. Frequency standards should be periodically matched to each other or to a reference standard. In spatially coherent MSRSs additional phasing of spaced stations (phase synchronization) is required. It means maintaining of phase relationships between all stations.

Increased Requirements to Signal and Data Processors and Computer Systems

This is in fact the reverse of one of the principal advantages of MSRSs: a significant increase of the total information body from a target as compared with monostatic radars. For processing all available information in real time high-speed processors and computers are usually necessary. However, there are certain peculiar procedures increasing computational burden. First, it is coordinate conversion of radar data from a local coordinate system connected with each station into a common coordinate system of a MSRS. Second, it is interstation data association between measurements (plots, tracks) obtained by different stations, on the one hand, and targets, on the other (interstation data identification). Besides that, most geometrical relationships and tracking algorithms are more complicated than for monostatic radars (see Chapters 11–15). At the same time it should be noted that the current state of signal processing techniques and computer engineering is quite sufficient to meet all actual requirements.

Necessity of Accurate Station Positioning and Mutual Alignment

For fusion of coordinate information coming from spatially separated stations and resulting target track construction, the stations' positions are to be known and alignment of stations is necessary. Errors of station position determination and of orientation of local coordinate system axes influence directly the accuracy of output information of a MSRS. Positions of stationary stations in ground-based MSRSs can be determined by geodetic methods with sufficient accuracy. More difficult is position determination for mobile stations. However, modern navigational methods (e.g. using the Global Positioning Systems – GPSs "NAVSTAR" or "GLONASS") can be quite adequate both for stationary and mobile stations. Misalignment of spatially separated stations in target coordinate measurements (i.e. in azimuth, elevation, range, etc.) as well as station positioning inaccuracy lead to systematic and (or) slowly varying errors of target location determination by a MSRS and make the interstation measurement association more difficult [75,76,80,182,252]. Several methods and algorithms have been proposed for alignment of spaced stations (or radars) in MSRSs [75,173,250–252,254]. Remaining misalignment errors may be considered as unknown parameters and estimated together with target coordinates [75,110,118,153,253,255].

Need for Direct Lines of Sight between Stations and Targets

Except for Over-The-Horizon (OTH) radar systems coverage of a MSRS is limited by the need for direct lines of sight between stations and targets. If a target is not simultaneously visible to several transmitting and receiving stations (or to several radars) of a MSRS, no information originated from the same target comes from spatially separated stations (or radars) and information fusion becomes senseless. This may be an important constraint for ground-based MSRS when low-level targets are to be detected and tracked. Hilly and mountainous terrain may additionally obstructs low-level detection [190].

MSRSs containing several spatially separated stations (or radars), data transmission lines (DTLs), information fusion centres (FCs) *are more complex and expensive than monostatic radars.* Specifically, for simultaneous target observation by several stations (or radars), antennas with electronic scanning, preferably multibeam, are often necessary [82,102,159,155]. However, comparison in complexity and cost is correct if capabilities and performance characteristics are equal. Certain characteristics of MSRSs are not achievable in monostatic radars while realization of some other characteristics requires a drastic increase of complexity and cost (for example, the employment of phased antenna arrays of enormous size [114,115]). In [95,100] it was noted that a MSRS containing simple stations of the same type are less expensive than a monostatic radar with similar technical characteristics. Of course, deployment of MSRSs is reasonable if usual monostatic radars cannot meet imposed requirements, i.e. when high informativeness, interference resistance and survivability are the principal requirements [67].

In many cases significant benefits can be obtained at the low cost when a MSRS is created by integrating operational monostatic radars [77,80,97] or by adding remote receiving stations to these radars [107].

In this section*, main drawbacks of MSRSs have been discussed. As a rule, they represent additional difficulties which one have to overcome when creating a MSRS and may be considered as the price that has to be paid for all advantages briefly discussed in Section 1.2. These main drawbacks are:*

- *centralized control of spatially separated stations;*
- *necessity of data transmission lines;*
- *additional requirements to synchronization, phasing of spatially separated stations, transmission of reference frequencies and signals;*
- *increased requirements to signal and data processors and computer systems;*
- *necessity of accurate station positioning and mutual alignment;*
- *need for direct lines of sight between stations and targets (except for over-the-horizon radar systems);*
- *greater complexity and cost than those of monostatic radars.*

It has been emphasized that deployment of MSRSs is reasonable when usual monostatic radars cannot meet imposed requirements.

1.4. BRIEF HISTORICAL OUTLINE

As it is well known, the earliest radars were bistatic Continuous Wave (CW) radars. They detected an object as it crossed the transmitting station–receiving station baseline by measuring beat frequency of its Doppler-shifted reflection and direct signal propagating from the transmitting station to the receiving station. After the invention of the duplexer at the US Naval Research Laboratory in 1936, which provided a means of using a common antenna for both transmitting and receiving, monostatic radars became practical and interest in bistatic radars became dormant. The interest in bistatic radars was revived only in 1950s [162,192].

In the Soviet Union the successful field tests of the first CW bistatic radar were accomplished in the summer of 1934 and first five radars were in production in the autumn of that year. The first operational radar was RUS-1 (Radio Ulavlivatel Samoletov – Radio Catcher of Aircrafts), a bistatic CW radar. Transmitting and receiving stations were separated by 35 km, the wavelength was 4 m. This radar was accepted by the Red Army in 1939. Forty-five of these systems were employed in the Far East and Caucasus at the time of the Great Patriotic War against Nazism. The first pulsed radar Redut (in production RUS-2), tested in 1937, was bistatic as well, though with a shorter baseline (up to 1000 m). It provided not only aircraft detection (as CW radars) but range and angle coordinate measurement. The duplexer was invented in the Soviet Union independently in 1940. This invention permitted to build the first true monostatic pulse radar Pegmatit (RUS-2s) which played an important role in the Great Patriotic War [158,162,172,186].

Though a MSRS with spatial separation between transmitting and receiving stations or/and with cooperative signal reception includes bistatic radars as its parts (elements or cells), the bistatic radar itself does not belong to MSRSs (see Section 1.1). For this reason we shall not further discuss the history of the bistatic radar. The reader can find an interesting and rather full review of this topic in [162,171,192]. A brief outline of the history of High Frequency (HF) Over-The-Horizon (OTH) bistatic radars in USSR see in [168].

There is only scanty information in open literature concerning operational MSRSs for military applications.

In USSR the earliest MSRS Vega was constructed and manufactured in 1936. It was a CW MSRS for aircraft detection based on the same principles as the bistatic radar RUS-1. The MSRS Vega contained one transmitting station with radiation power 5–10 kW and five receiving stations separated from the transmitting one by several scores of kilometers in different directions. Radio or phone lines connected all receiving stations with the transmitting station, where the command post was situated. The wavelength was 3.5–4 m. This MSRS, however, did not go into quantity production [172].

In the earlier 1950s first MSRSs for precise ballistic missile trajectory measurements at the test ranges were developed and deployed. The MSRSs (trajectory measurement complexes) designed for beacon tracking included several pulse monostatic ground-based radars separated by tens and hundreds kilometers and located along the ground projection of the expected ballistic missile trajectory. Each tested missile was equipped with a transponder which responded to interrogations from the radars. Received signals from the transponder at each radar site were used for missile range measurements relative to that site. These measurements transmitted via narrowband (telephone) links to the information fusion centre where they were jointly processed for trajectory parameter estimation. The principles described above formed the basis for the command and measurement complex used for artificial satellite tracking. Such a complex including a MSRS with several spatially separated pulse monostatic radars made it possible to track the first Soviet "sputnik" in 1957. After several modernizations it is in use up to now. Apart from range measuring radars phase measurement facilities were and are utilized in MSRS for satellite orbit parameter estimation. In particular, in the early 1980s passive correlation interferometers "Raduga" ("Rainbow"), later called "Ritm" ("Rhythm"), were deployed. They do not need in dedicated transmitter on a satellite but utilize any radiation from the satellite of interest within 1–6 GHz frequency band [201].

One of the first operational MSRS of the postwar generation which used target echo ("skin tracking") was developed as a part of the first Soviet experimental Anti-Ballistic Missile (ABM) system which is known under the name "System A".

This ABM system was deployed at the Sari-Shagan Test Range (Kazakhstan), near the lake Balkhash, in 1958–1960. In March 4, 1961, the "System A" for the first time in the world had realized interception and defeat of the warhead of a ballistic missile (type R-12). The warhead's velocity was more than 3 km/s and the interceptor V-1000 had a high-explosive fragmentation (nonnuclear) charge. Therefore precise tracking of the warhead and interceptor was a very important but difficult problem. This problem had been solved by a MSRS containing three monostatic pulse radars, located at the vertices of a correct triangle of sides about 150 km. Each radar comprised two channels with individual parabolic antennas: the target channel and the interceptor channel. The range measurement accuracy of each radar was better than 5 m (r.m.s. value), and the trilateration method was used for warhead and interceptor coordinate measurement and tracking. According to the classification described in Section 1.1, it was an active ground-based spatially incoherent MSRS with stationary stations' positions, independent signal reception and information fusion at the plot level. The MSRS could detect and begin tracking warheads at a range up to 700 km. For initial target detection and target pointing a separate radar was used which achieved detection ranges (for warheads) up to 1200 km. All components of this MSRS were connected with each other by radio relay links. Fully automated control of the MSRS (and of the total System A) was realized by the computer M-40. Its speed was 40 thousands operations per second [160].

The first Soviet experimental ground-based passive system used correlation processing for jammers detection and location was deployed near Kharkov, Ukraine, in 1965. The system contained two receiving stations separated by 10 km. Noise-like signals at the wavelength about 10 cm received from a jammer by the remote station, were transmitted to the "main" station via radio link. At the "main" station these signals were undergone mutual correlation processing together with the signals received by the "main" station from the same jammer. The receivers' bandwidths were about 8 MHz while the output bandwidth of the correlator was about 1 kHz, so that "integration (accumulation) factor" was near 8000. Successful field tests demonstrated possibility of reliable detection and accurate location of jammers in conformity with the results predicted by the previous theoretical research [161].

During the subsequent years the work in the field of MSRSs in USSR (later in Russia) was continued.

As an example, the passive air defence complex Baza (Base) developed in 1970s may be mentioned. That complex was intended for detection and location of jammers (by all the three space coordinate measurements). It contained a central station, several remote stations and connecting radio links [186].

At the same period an active–passive complex was built for the Ballistic Missile Attack Warning System. The complex integrated the operational radar Dnepr with a new separated receiving station Daugava. Information fusion permitted increase of coordinate measurement accuracy, resolving measurement ambiguities and improving some other characteristics [187].

In 1980s new active–passive air defence MSRSs were developed within the framework of the research and experimental work Gamma. At present a new MSRS is in the development stage [186].

In the United States MSRSs have found important applications for precision measurements of missile trajectories at the Eastern Test Range, which extends from the Florida mainland to the Indian Ocean. These MSRS include the Azuza, the Mistram and the Udop. All systems employ a cooperative beacon transponder on the observed target and a ground-based transmitting station and several receiving stations at separate, precisely located sites. The C-band CW interferometric MSRS Azuza, in operation from 1950s, contains one transmitter and nine receivers located along two crossed baselines with the total lengths of about 500 m. Intermediate receivers spaced at 5 and 50 m are used for phase ambiguities resolution. The Azuza system measures range by phase measurement of sideband frequencies modulating the carrier, coherent range by Doppler count, two direction cosines, and two cosine rates. Errors of less than 3 m in range and 20 ppm in direction cosine are obtainable [46,162,192].

The MSRS Mistram (Missile Trajectory Measurement) is also a CW interferometric system with receiving stations situated along two mutually perpendicular baselines spaced at 3 and 30 km. This MSRS can measure range, four range differences, range rate and four range difference rates of a target. The range error is less than 0.8 m.

The MSRS Udop (UHF Doppler) utilizes the Doppler measurements. A target is illuminated at 450 MHz. Five receiving stations are located along the baselines with the lengths from 25 to 75 miles and receive signals from the target's transponder at 900 MHz. These five stations yield slant-range rate. To compute range or position, an initial position is required from some other tracking system. The random error is 0·06 m but with the systematic error of 2.7 m plus the initial error. Udop is of relatively low cost compared with other high-accuracy systems [46,162,192].

One of the first MSRS in the USA is the Navspasur (Navy Space Surveillance System). It is a CW MSRS, in operation since 1960, that detects orbiting objects as they pass through the electronic "fence" over the continental United States. The system includes three groups of stations interspersed along a great circle at 33.5°N latitude from Fort Stewart, Georgia, to San Diego, California about 3500 km apart. Each group contains one central transmitting station and two receiving stations, one of each side of the transmitting station and separated from the transmitting station by 400–500 km. Operational frequency lies in the range of 214–219 MHz. The large linear arrays of the transmitting stations generate stationary, vertical, coplanar fan beams, oriented along the great circle. The length of the linear transmitting arrays of the east and west groups are 1600 m, while the transmitting array of the central group located at Kickapoo Lake, Texas, is about of 3300 m long. Four receiving stations contain eight linear phase-measuring arrays each of 120 m in length and one additional alerting array. The main purpose of these stations is to detect objects at low and medium orbits. The alerting array detects the satellite before it comes within the range of phase-measuring arrays to measure the satellite's Doppler shift and to select appropriate narrow band receiver. Two receiving stations intended for detection of objects at high orbits contain seven parallel phase-measuring linear arrays of 740 m long and two additional alerting arrays of 1480 m long. All receiving stations are multiple array interferometers measuring the Angle Of Arrival (AOA) of signals reflected from orbiting objects. The receiving arrays are located non-uniformly so that to resolve ambiguities which arise in the interferometric measurements. Data from the six receiving stations are forwarded via land lines to the Space Surveillance Operations Centre in Dahlgren, Virginia, where orbital computations are made using measured AOAs and precisely known stations' positions. The final data are sent to the North American Aerospace Defence Command (Norad). In 1986 the Navspasur was being upgraded to double its range from 7500 to 15 000 nm [46,162,184,192]. According to the classification presented in Section 1.1, the Navspasur system is an active, stationary, ground-based, incoherent MSRS (though coherent interferometer subsystems are employed at each receiving site) with plot integration level and cooperative signal reception.

A second multistatic fan beam CW radar fence for satellite tracking, called Doploc (Doppler Phase Lock), was deployed in an interim configuration in the late 1950s. It was intended for both beacon and skin tracking. Receiving sites were located at White Sands Missile Range, New Mexico, and at Forrest City, Arkansas, about

1600 km apart. The transmitting station using a 50-kW transmitter operating on 108 MHz, was located at Fort Sill, Oklahoma, roughly midway between receiving sites. Each receiving site generated three fan beams, one directly overhead and one to either side of vertical. The Doploc receivers used a narrow band (1–50 Hz) phased-lock loop that track the Doppler frequency shift from the satellite. The time-varying Doppler shift was recorded as the satellite passed through each beam. From the single-pass data, the satellite ephemeris was estimated using iterative calculations, while imposing satellite elliptic motion as a constraint. The system was exercised against many satellites in the late 1950s but operational configuration of Doploc was not deployed [192].

In 1964 at the Rome Air Development Center the experimental Active Swept Frequency Interferometer Radar (ASFIR) was deployed. The system included one "main" transmitting–receiving station and one remote receiving station. The baselength was 13.6 km and the wavelength was 10 cm. The ASFIR could detect and locate space targets with the RCS of 0.1 m^2 at the range up to 2000 km. It could work in two modes: in the pulse mode with Linear Frequency Modulated (LFM) pulses and in the pulse mode without internal modulation. The pulse duration was 2 ms in either mode. The frequency difference of echoes received by two stations from a target is proportional to the TDOA in the first mode and to the rate of the TDOA variation in the second. When the target range to the effective baselength ratio is large enough, then those values, in their turn, determine the angle coordinate and the angle velocity of the target in the plane passing through the target and the stations [130].

In 1977 the Lincoln Laboratory of the Massachusetts Institute of Technology began working on a research programme concerning MSRSs. The Netted Radar Program was a joint DARPA[4] /Army effort. The principal goals of the programme were to improve battlefield surveillance, target acquisition, and battle management capabilities for the Army through automated and adaptive netting of spatially separated battlefield radars into a MSRS. At the first stage in 1978–1979 the experimental system contained two ground-based radars and a fusion centre. Narrowband data links (2400 bit/s) were used for data transmission [131]. At the second stage in 1980–1981 the experimental MSRS containing five radars was tested. Three ground-based ground-surveillance radars included two TPS-5X 16 GHz mechanically-scanned radars and one multifunction 5 GHz Advanced Ground Surveillance Radar (AGSR) with a cylindrical electronically-scanned antenna array. The two TRS-5X radars were US Army AN/PPS-5 radars modified with the digital signal and data processing capabilities required for netting. Apart from those radars a ground-based Army Firefinder radar and an airborne 16 GHz electronically-scanned surveillance radar were used. Output information from all surveillance radars at the track level was sent over narrowband data links to the fusion centre (Target Integration Centre – TIC). The resulting picture was transferred via similar data links to several remote displays including a mobile displays located at various places within continental USA. The information was displayed in real time with the updating interval about 5 s. For the transfer of output "raw" information from the airborne radar to the ground-based TIC a data link with handling capacity 12 Mbit/s was utilized. In those experiments the interaction between all facilities was verified and developed. The experiments successfully demonstrated feasibility of automatical integration of diverse tactical radars using technology which was available at that time. Radars were remotely controlled over existing narrowband radio links. High effectiveness of MSRSs was shown. A real-time comprehensive picture of the

[4] DARPA is the Defense Advanced Research Project Agency.

battle-field, made available simultaneously to many tactical organizations and individuals, resulted in a host of new capabilities and opportunities [96].

The Multistatic Measurement System (MMS) was deployed in 1978–1980 at the Kwajalein Missile Test Range (Marshall Islands) as a part of the Kiernan Re-entry Measurement Site (KREMS). MMS was developed by the Lincoln Laboratory of the Massachusetts Institute of Technology to collect bistatic signature data and perform high-accuracy tracking of re-entry targets. The high-power Tradex L-band and Altair UHF radars (with the central frequencies 1320 MHz and 415 MHz, respectively) at Roi-Namur Island are utilized to illuminate targets and to receive target echoes. Two remote unmanned receiving stations are installed on Gellinam Island, about 40 km away, and on Illegini Island about 35 km away. The first remote station can receive echoes from targets illuminated by both radars while the other receiving station (on Illegini Island) can receive echoes in the L-band only. The monostatic radars acquire and track the target and point high-gain (36 dB at L-band and 24 dB at UHF) antennas of the remote receiving stations. Parabolic antennas of 6.1 m in diameter are used. Signal bandwidth is up to 20 MHz. The amplitudes and phases of target echoes received by remote stations are transmitted in digital form via radio links to the master radar site on Roi-Namur Island, where these data are combined with the data from radars for joint processing (information fusion). The system measures dual-frequency target signature data, three-dimensional target position and Doppler velocity with accuracies better than 4 m and 0.1 m/s, respectively (r.m.s. values) throughout re-entry. To achieve such high accuracies the target echo TDOAs at the inputs of the each pair of stations are measured with errors of the order of 0.5 ns owing to the precise synchronization system. For MMS alignment correlation processing of noise-like signals from radio stars is used. The effective bandwidth at the correlator inputs is 60 MHz and integration time at the correlator output is 100–1000 s. The maximum range of the re-entry targets is 500–700 km [107,162].

In the late 1970s under the sponsorship of the US Air Force (USAF) the Precision Location/Strike System (PLSS) was developed. The purpose of this system is to reconnoitre hostile air defence means, to locate and identify hostile radars and to point striking weapons with high accuracy. The system works on the principle of hyperbolic location. At three spatially separated points TDOAs of pulse signals from a hostile radar are measured by three aircrafts flying at a height of 15–20 km. Aircraft positions are known with high accuracy achieved by the radio navigational system. Obtained information is transmitted to the ground-based data fusion and control centre. The prototype of the PLSS is the ALSS developed in 1972 [98,179,181].

The Jindalee Over-the-horizon Operational Radar Network (JORN) is a new MSRS which is under construction now in Australia. This MSRS is based on the operational OTH experimental radar Jindalee at the Alice Springs. The JORN comprises two remote OTH skywave radars and a centralized control centre known as the JORN Coordination Centre (JCC). The radars are located near Longreach in central Queensland and near Laverton in Western Australia while the JCC is situated near Adelaide in South Australia. Frequency Modulated Continuous Wave (FMCW) signals are used. At each radar site, the transmitter is isolated from the receiver to prevent interference between the two and protect the sensitivity of the receiver. This isolation is accomplished by separating the transmitting station and the receiving station of each radar by about 100 km. Thus each radar of the JORN is a bistatic one. Each radar can measure range, azimuth, velocity and Signal-to-Noise Ratio (SNR) of observed targets. Signal processing (primary information processing) is carried out in each receiving station while target tracking (secondary information processing) is realized at the JCC. For increase of the total survivability target tracking is possible by each radar individually. Surveillance information can be passed from the Alice Spring radar to JCC too for integration with information from the two JORN radars. The JORN also incorporates an extensive network of beacons and sounders, as part of the frequency management system (which is an important peculiarity of OTH radars), at widely separated sites around the northern coast line, islands and national offshore territories. Accordingly to the classification presented in Section 1.1, the JORN is a ground-based active spatially incoherent MSRS with fixed station positions, long baselines and information integration at the plot (or track) level. Probably, cooperative signal reception will be used [135–137,166,167].

During the 1970s–1980s much attention was paid in many countries to the improvement of Air Traffic Control (ATC) systems by integrating operational radars into MSRSs. This was caused by the increased volume of air traffic and more stringent system safety, capacity and productivity requirements. These requirements often cannot be met by conventional radar systems. On the other hand, the existence of several spatially separated operational radars with overlapping coverages usually used in each Flight Information Region (FIR) permits their integration in a MSRS at relatively low cost.

In 1972–1975 the Applied Physics Laboratory of the Johns Hopkins University (USA) conducted two empirical investigations into the advantages and the problems of MSRSs created by integration of radar and beacon airspace surveillance systems. At the first stage an experiment was conducted using ATC radars and beacons located in the Washington–Baltimore area, which is geographically characterized by fairly flat terrain. The MSRS included two sites separated by 25 nm. At each site a radar and a beacon were used. A second and more comprehensive experiment was conducted at the second stage in the Los Angeles basin which is characterized by mountainous terrain. Information was fused from 12 radar and beacon sensors located at seven sites with the baselengths from 15 to 38 nm. The results of those two experiments have confirmed the advantages of MSRS. It was established that a MSRS can significantly improve airspace area surveillance coverage [133,134].

From the late 1970s to the early 1980s a MSRS was developed for the Air Traffic Control Automated System (ATCAS) of the Rome Flight Information Region (Italy). This MSRS integrates five operational radar systems sited in the west side of the centre and south of Italy. The relative distance (baselength) between adjacent sites is approximately 250 km. The higher coverage overlapping is in the Rome terminal area, which is more crowded. Each site is mainly formed of a primary radar, a secondary radar, a combiner, a formatting device and a modem for the transmission of the plots via duplicated telephone lines (2400 bits/s) to a common centre. In this centre plot information coming from all sites is fused and a single track for each target is estimated [75,197].

In the late of 1970s MSRSs with information fusion at the track level were employed by the Eurocontrol Agency for ATC systems in Europe: the Maastricht Automatic Data Processing and Display System – MADAP and the Karlsruhe Automatic Data Processing and Display System – KARLDAP. A similar MSRS was at that time developed by Selenia S.p.A. (Italy) for the en-route ATC system of Mazatlan (Mexico) [75,77]. Creating of such MSRSs with "multi-radar tracking" was a result of the systems' development. In the previous stage the mono-radar tracking process was made independently for the plots coming from each radar, and one of these local, a mono-radar track was selected as the common track. This selection was based on a hierarchy of radars defined by a mosaic over the area controlled by the air traffic control centre. The role of correct information fusion in a MSRS for the MADAP system became more important when the number of radars inputting information was increased to six which ensured almost triple coverage over the entire MADAP airspace. In connection with this theoretical and experimental investigations were conducted into advantages and disadvantages of information integration at the plot level and at the track level [77,132].

Now MSRSs are suggested for perspective automated ATC systems using Secondary Surveillance Radars (SSRs) mode S [148].

In the last years several MSRSs were developed for surface movement control on airport manoeuvring areas. Those MSRSs consist of several small pulsed radars (called miniradars for their small dimensions and weight) working in millimeter or short centimeter wave regions. The small antennas may be installed at well suited locations with respect to the avoidance of shadowing and multipath effects. The MSRSs have low cost and provide coverage of the whole aerodrome surface with detection, location, tracking, classification and orientation (with image processing) of aircrafts and other vehicles. These functions are used for guidance and control as well as for detection and management of runway incursions and of erroneous dangerous movements. MSRSs have some important advantages as compared to monostatic radars which usually suffer from shadowing, multipath propagation and degraded resolution with increased measurement distance [146,198,199,200].

In this section*, we have presented a brief historical outline of MSRSs. Much attention has been paid to systems designed and deployed in Russia (USSR).*

2. RADAR CROSS SECTION (RCS) OF TARGETS

2.1. BISTATIC RCS OF TARGETS

The most important radar characteristic of a target is its Radar Cross Section (RCS). For monostatic radars RCS is a quantitative characteristic of the target ability to scatter energy in the direction opposite to the incident wave direction. For MSRSs a more general characteristic is necessary which takes into account possible difference between directions from a target to the transmitting and receiving stations. Such a situation takes place when transmitting and receiving stations are spatially separated or when in a MSRS consisting of monostatic radars cooperative signal reception is used. The required characteristic is called *the bistatic RCS* of a target.

From the general accepted definition, the RCS of a target is equal to the surface area of a symbolic object which scatters total incident energy isotropically and creates at a distant receiving point the same power flux density as the target (see e.g., [20]). In terms of the electric field strength (which linearly relates to the instantaneous value or amplitude of a signal), the RCS of a target can be expressed as

$$\sigma = \lim_{R\to\infty} 4\pi R^2(|E_r|^2/|E_{in}|^2) \cong 4\pi R^2(|E_r|^2/|E_{in}|^2) \qquad (2.1)$$

where E_{in} is the electric field strength of the incident plane wave at a target; E_r is the electric field strength of the receiving antenna's preferred polarization at the distant receiving point, R is the target distance from the receiving station. When R is much greater than the target dimensions, then the symbol "approximate equal" in the right equation of (2.1) may be replaced by the symbol "equal". Since $|E_r|^2$ is inversely proportional to R^2, the value R^2 is cancelled out in the process of calculations. If signals of both orthogonal polarizations are received, it is to be taken into account in $|E_r|^2$. Equation (2.1) remains valid if the electric components of field E_r and E_{in} are replaced by the magnetic components H_r and H_{in}. It is assumed in (2.1) that the target size is much less than the resolution cell of a radar.

The definition given above holds both for the monostatic and the bistatic target RCS. However, RCS of most targets depends both on the illumination and reception directions, so that the bistatic RCS, σ_b, is not equal to the monostatic one, σ_m, in the general case. The angle β_b between the directions from a target to the transmitting and receiving stations is called *a bistatic angle*. This is one of the main geometrical characteristics of bistatic radars including the case where they are cells of MSRSs.

A strict solution of the problem of the electromagnetic field diffraction at a target and a precise derivation of the monostatic and the more bistatic RCS are possible only for a few types of simple shaped targets. Therefore, approximate methods and simplified models are used for RCS practical calculations. These calculations are usually verified at the special indoor or outdoor test facilities. For real complex targets (e.g. aircrafts, missiles) direct measurements of full-scale targets or of scaled target models is the most important method to obtain the RCS in practice [46,183].

As a result of the monostatic radar development for many years a large body of information concerning the monostatic RCS, σ_m, of different types of targets is accumulated. There is much less information available about the bistatic RCS, σ_b. Hence it is important to reveal the existing relation between σ_m and σ_b. It may help to obtain the unknown σ_b of a target if σ_m and, for instance, the bistatic angle β_b is known. Let us consider this problem for some simple models of targets.

A Convex Continuous Perfectly Conducting Target Whose Dimensions are Much Greater Than the Wavelength

For such targets the approximate technique based on the physical-optics method is applicable [20,32,109].

Let the origin of the coordinate system be placed inside a target. An arbitrary point on the target surface is determined by the radius-vector $\boldsymbol{\rho}$. The directions from the target to the transmitting and receiving stations are denoted by the unit vectors $\mathbf{r}_0$ and $\mathbf{r}$, whereas the corresponding distances are denoted by R_0 and R, respectively. Then $R_0\mathbf{r}_0 = \mathbf{R}_0$, $R\mathbf{r} = \mathbf{R}$. The electric field strength of the incident plane wave at the point $\boldsymbol{\rho}$ is

$$\mathbf{E}_{in}(\boldsymbol{\rho}) = \mathbf{E}_0 \exp[-j(2\pi/\lambda)(R_0 - \boldsymbol{\rho}\mathbf{r}_0)] \tag{2.2}$$

where $\mathbf{E}_0 = E_0\mathbf{e}$ is the field strength vector (equal for all $\boldsymbol{\rho}$) determined by the transmitting station characteristics; E_0 is the complex amplitude; $\mathbf{e}$ is the unit vector in the $\mathbf{E}_0$ direction; λ is the wavelength.

According to the physical optics approximation, the scattered field at a distant receiving point $\mathbf{R}$, in the far-field zone, may be presented in the form [32]

$$\mathbf{E}_r(\mathbf{R}) = j\frac{\exp[-j(2\pi/\lambda)R]}{\lambda R} \int_{S_0} [\mathbf{n}(\mathbf{r}\mathbf{E}_{in}) - \mathbf{E}_{in}(\mathbf{r}\mathbf{n})] \exp[j(2\pi/\lambda)\boldsymbol{\rho}\mathbf{r}]\, dS \tag{2.3}$$

where $\mathbf{n} = \mathbf{n}(\boldsymbol{\rho})$ is the unit external normal to the target surface at the point $\boldsymbol{\rho}$; the integral is taken over the part of the target's surface S_0 which is illuminated by the transmitting station and seen by the receiving station. Substituting $\mathbf{E}_{in}$ from (2.2) into (2.3), we obtain

$$\mathbf{E}_r(\mathbf{R}) = j\frac{\exp[-j(2\pi/\lambda)(R_0 + R)]E_0}{\lambda R} \int_{S_0} \boldsymbol{\zeta}(\boldsymbol{\rho}, \mathbf{r}_0, \mathbf{r}) \exp[j(2\pi/\lambda)\boldsymbol{\rho}(\mathbf{r}_0 + \mathbf{r})]\, dS. \tag{2.4}$$

Here the vector function is introduced

$$\boldsymbol{\zeta}(\boldsymbol{\rho}, \mathbf{r}_0, \mathbf{r}) = \mathbf{n}(\mathbf{r}\mathbf{e}) - \mathbf{e}(\mathbf{r}\mathbf{n}) \tag{2.5}$$

which describes the "scattering features" of each point $\boldsymbol{\rho}$ (since $\mathbf{n} = \mathbf{n}(\boldsymbol{\rho})$). This function may be called scattering function of the target. It is seen that the scattering function depends on the illumination direction $\mathbf{r}_0$ (which is perpendicular to $\mathbf{e}$) and on the receiving direction $\mathbf{r}$. In the general case $\boldsymbol{\zeta}/|\boldsymbol{\zeta}| \neq \mathbf{e}$ i.e. the polarizations of the incident and scattered waves are not equal.

Substitution of (2.2) and (2.4) into (2.1) yields the bistatic RCS of a target

$$\sigma_b = (4\pi/\lambda^2) \left| \int_{S_0} \boldsymbol{\zeta}(\boldsymbol{\rho}, \mathbf{r}_0, \mathbf{r}) \exp[j(2\pi/\lambda)\boldsymbol{\rho}(\mathbf{r}_0 + \mathbf{r})]\, dS \right|^2.$$

Notice that $\mathbf{r}_0+\mathbf{r}=2\cos(\beta_b/2)\mathbf{r}_b$ where β_b is the bistatic angle (between $\mathbf{r}_0$ and $\mathbf{r}$); $\mathbf{r}_b$ is the unit vector along the bisector of the angle β_b. Then

$$\sigma_b=(4\pi/\lambda^2)\left|\int_{S_0}\zeta(\boldsymbol{\rho},\mathbf{r}_0,\mathbf{r})\exp[j(4\pi/\lambda)\cos(\beta_b/2)\boldsymbol{\rho}\mathbf{r}_b]\,dS\right|^2. \tag{2.6}$$

In a monostatic radar the transmitting and receiving stations are placed at one site, e.g. at the point **R**. Under this condition $\mathbf{r}=\mathbf{r}_0$, so that $\mathbf{re}=\mathbf{r}_0\mathbf{e}=0$. From (2.5) we have

$$\zeta(\boldsymbol{\rho},\mathbf{r}_0,\mathbf{r})=\zeta(\boldsymbol{\rho},\mathbf{r})=-\mathbf{e}(\mathbf{rn})=-\mathbf{e}\cos\angle(\mathbf{r},\mathbf{n}) \tag{2.7}$$

and for the monostatic RCS we obtain

$$\sigma_m=(4\pi/\lambda^2)\left|\int_{S_0}\zeta(\boldsymbol{\rho},\mathbf{r})\exp[j(4\pi/\lambda)\boldsymbol{\rho}\mathbf{r}]\,dS\right|^2. \tag{2.8}$$

The expressions (2.6) and (2.8) differ in the form of the function ζ and in the indexes of exponents. It is clear that when β_b is small, then equation (2.7) approximately holds for targets considered in the bistatic case as well. It is exactly true when $\mathbf{E}_{in}$ and consequently $\mathbf{e}$ are perpendicular not only to $\mathbf{r}_0$ but also to $\mathbf{r}$, i.e. when $\mathbf{re}=0$ in (2.5). If the condition (2.7) is satisfied, the polarization of the scattered wave at the point **R** is the same as that of the incident wave (except, possibly, for the sign).

Thus, the scattering functions of a target for a monostatic and a bistatic radars may be considered as practically equal in many situations. In these situations the substitution of $\zeta(\boldsymbol{\rho},\mathbf{r})=-\mathbf{e}(\mathbf{rn})$ for $\zeta(\boldsymbol{\rho},\mathbf{r}_0,\mathbf{r})$ in (2.6) is possible.

It follows from (2.5)–(2.8) that when $l/\lambda\gg 1$ (where l is the characteristic dimension of a target), then the integrand is the product of a slowly varying scattering function and a rapidly oscillating exponent. As a result, all the contributions of different parts of the target surface to the integral are mutually compensated except those from the immediate vicinity of a stationary-phase point where the oscillation rate has its minimum [109,216]. Such a target is perceived by a radar as one scattering centre (a "flare spot"). In (2.8) it is a point where the plane of constant signal phase $\boldsymbol{\rho}\mathbf{r}=$const. is tangent to the surface of the target [or, which is the same, where an external normal to the target surface coincides with the direction to the radar: $\mathbf{n}(\boldsymbol{\rho})=\mathbf{r}$]. In (2.6) it is a point where the plane $\boldsymbol{\rho}\mathbf{r}_b=$const. is tangent to the surface of the target (or where an external normal to the target surface coincides with the bisector of the angle β_b: $\mathbf{n}(\boldsymbol{\rho})=\mathbf{r}_b$). At the stationary-phase point of the integral in (2.8) $\zeta(\boldsymbol{\rho}_0,\mathbf{r})=-\mathbf{e}(\mathbf{rn})=-\mathbf{e}(\mathbf{rr})=-\mathbf{e}$. At the stationary-phase point of the integral in (2.6), under the assumption of equality of the scattering functions, $\zeta(\boldsymbol{\rho}_0,\mathbf{r})=-\mathbf{e}(\mathbf{rn})=-\mathbf{e}(\mathbf{rr}_b)=-\mathbf{e}\cos(\beta_b/2)$. Substituting these expressions into (2.6) and (2.8) yields

$$\sigma_b\approx\frac{4\pi\cos^2(\beta_b/2)}{\lambda^2}\left|\int_{S_0}\exp[j(4\pi/\lambda)\cos(\beta_b/2)\boldsymbol{\rho}\mathbf{r}_b]\,dS\right|^2, \tag{2.9}$$

$$\sigma_m\approx(4\pi/\lambda^2)\left|\int_{S_0}\exp[j(4\pi/\lambda)\boldsymbol{\rho}\mathbf{r}]\,dS\right|^2. \tag{2.10}$$

For specific targets the RCS σ_b can be calculated using (2.9) and the stationary phase method [109,216]. However, comparing (2.9) with (2.10) one can see that when

S_0 is equal, then *the bistatic target RCS, σ_b, at the wavelength λ can be closely approximated by the monostatic RCS, σ_m, viewed on the bisector of the bistatic angle β_b at the wavelength $\lambda/\cos(\beta_b/2)$*. This is the so-called "monostatic–bistatic equivalence principle" [21,46]. Since $\cos(\beta_b/2) \cong 1$ for small β_b, increasing of λ for the "equivalent" monostatic RCS is insignificant, especially as the RCS of targets considered does not often depend on the wavelength λ.

A Target is a Set of Discrete Scattering Centres ("Flare Spots")

The scattering field from a target of a complex shape can, in many cases, be represented as the sum of fields from several discrete scattering centres ("flare spots"). It is usually valid if characteristic dimensions of a target are much larger than the wavelength. This representation is one of the simplest and widely practised target models. The scattering centres are usually either specular centres on singly or doubly curved surfaces or are actual geometric discontinuities of the target (e.g. boundaries, points of inflection, and so on). In the first case their positions on the target surface depend on the illumination and reception directions whereas in the second case they are rigidly connected with those discontinuities. We assume here that own directivity patterns of discrete scatterers are sufficiently wide.

The monostatic RCS of a such target can be written as follows:

$$\sigma_m = \left| \sum_{k=1}^{N} \sqrt{\sigma_k} \exp\{j[(4\pi/\lambda)\boldsymbol{\rho}_k \mathbf{r} + \varphi_k]\} \right|^2 \tag{2.11}$$

where σ_k is the RCS of the kth scattering centre for the polarization of the receiving part of a radar; φ_k is the initial phase shift introduced by the kth scattering centre into reflected field; $\boldsymbol{\rho}_k$ is the radius-vector of the kth scattering centre; $\mathbf{r}$ is the unit vector directed to the radar. The bistatic RCS of the target can be expressed in the form

$$\sigma_b = \left| \sum_{k=1}^{N} \sqrt{\sigma_k} \exp\{j[(4\pi/\lambda)\cos(\beta_b/2)\boldsymbol{\rho}_k \mathbf{r}_b + \varphi_k]\} \right|^2 . \tag{2.12}$$

In the general case N, σ_k, $\boldsymbol{\rho}_k$ and φ_k in (2.12) may be not the same as in (2.11). However, at the constant target aspect to the $\mathbf{r}_b$ (i.e. to the bisector of the angle β_b) the values of N, σ_k, φ_k and projections $\boldsymbol{\rho}_k \mathbf{r}_b$ often do not change when β_b increases from zero up to a certain value $\beta_{b_{max}}$. It follows from the comparison of (2.11) with (2.12) that in these cases the bistatic RCS of a target, σ_b, at the wavelength λ is equal to its monostatic RCS, σ_m, for a radar located at the bisector of the angle β_b and working at the wavelength $\lambda/\cos(\beta_b/2)$.

Thus, for a target which can be represented as a set of point scatterers, and under the conditions considered, "the monostatic–bistatic equivalence principle" is also valid [21].

This statement is important for bistatic RCS evaluation of real targets whose characteristic dimensions are much larger than the MSRS' wavelength. Unfortunately, it is impossible to determine a single boundary of application of the principle for different targets, i.e. the angle $\beta_{b_{max}}$. This angle is limited by the angular width of individual scatterer patterns, which, in their turn, depend on the type and size of scatterers involved. The angular width of the scatterer patterns is governed to a great extent by the size (in wavelengths) of the associated cophase area. If a scatterer results from direct illumination only, the size of the associated cophase area (in wavelengths) is usually small, and the corresponding angular pattern is broad. This is a so-called "simple scatterer". When a scatterer is a result of multiple reflections from adjacent parts of the target surface, as in a corner reflector, then the size of the associated cophase area can be large, so that the corresponding angular pattern is narrow. This is a so-called "reflex scatterer". For small β_b the conditions of the "statement of the monostatic–bistatic equivalence" are usually satisfied, for example, if $\beta_b < 10°$. For greater β_b a careful examination of both the target geometry and monostatic RCS σ_m is required to judge whether the limits posed by the restricted angular reradiation from individual scattering centres are being exceeded. Monostatic RCS envelope is useful in this examination, because narrow reradiation lobes generally are associated with large RCS, and RCS envelope values in excess of a few square wavelengths suggest immediately that narrow lobes from reflex scatterers may be present [21]. The monostatic–bistatic equivalence principle cannot be applied to depolarizing targets [46].

It was shown by experiments that the monostatic–bistatic RCS equivalence principle is useful for bistatic RCS evaluation of such complex targets as aircrafts and missile if β_b is as large as 60–90° and in some cases even up to 130–150 . However, it concerns not the "fine structure" of bistatic RCS as a function of the target aspect but the averaged values (e.g. medians) as well as maxima and minima over certain intervals of aspect angles [30]. For MSRSs with short baselines (compared with target range, see Section 1.1) when β_b is small, the bistatic RCS of a target may be considered as practically equal to its monostatic RCS.

Experimental data concerning bistatic RCS of real targets are rather scanty in open literature. As was noted in [111], bistatic RCS of a target can be both lower and higher than monostatic. For example, the measured bistatic RCS of an aircraft turned out to be lower on average by 2–5 dB [122] and in other cases even by 6–8 dB [87,89] than monostatic RCS. It can be explained by decreasing the contribution of specular reflections from scattering centres with narrow angular patterns (engine manifolds, wing joints, etc.). At the same time the bistatic target reradiation pattern is more smoothed, the lobes are broader. This result is in accordance with the "equivalent" wavelength increase by the factor $1/\cos(\beta_b/2)$. In Fig. 2.1 reradiation monostatic and bistatic patterns of a generic airborne platform shape measured at the frequency 1250 MHz are shown as a function of the aspect angle. Transmitter and receiver polarization is horizontal so that the E-field is parallel to the wings. Platform characteristic dimensions are much larger than the wavelength. Bistatic angle β_b is 160° [185].

It is important to note that bistatic RCS of certain targets can be much greater than monostatic. It concerns primarily targets with special means for monostatic RCS reduction. The target can be so shaped that at certain aspect angles incident radar signals are reflected away in a different direction. Such shaping is usually designed to give a "cone-in-silence" about the direction of flight of an aircraft or a missile (the *Stealth* technology) (e.g., [140]).

Polarization Scattering Matrices

In the general case scattered field polarization from a real target is not the same as that of the incident field. For this reason the Polarization Scattering Matrix (PSM) is widely used as the most full characteristic of target scattering features at a fixed frequency [17,20]. The PSM shows the relationship between the orthogonal components of incident field at a target and the same scattered far-field components at a receiving point. It is assumed that the propagation medium is ideal. The PSM is defined in terms of a polarization basis.

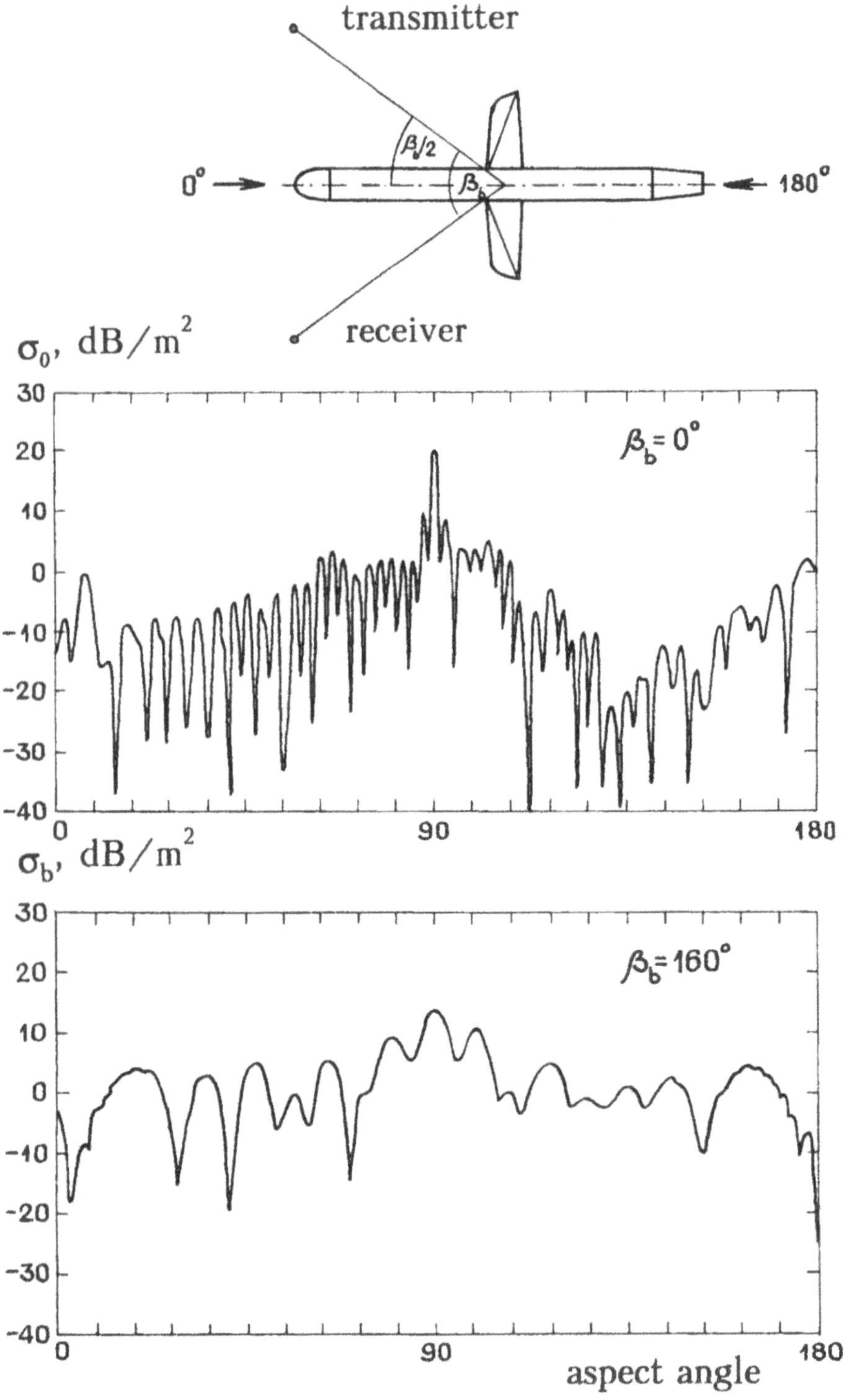

Figure 2.1 Monostatic and bistatic reradiation patterns of a generic airborne platform shape measured at the frequency 1250 MHz ([185], © 1988 IEEE)

Let us consider a target placed at a point O and illuminated by a plane wave from a distant transmitting station located at a point A (see Fig. 2.2). The receiving station is located at a distant point B so that the receiving wave may also be considered as a plane one. Let the point A be the origin of the right-handed Cartesian coordinate system X_1, Y_1, Z_1, and the point B be the origin of the similar coordinate system X_2, Y_2, Z_2.

Without loss of generality, directions of the axes can be chosen as follows: Z_1 coincides with the incident wave direction, Z_2 is opposite to the scattering wave direction, X_1 and X_2 are parallel to each other and perpendicular to the bistatic plane AOB.

We assume the incident wave to be linearly polarized so that the angle between the vector of electric field $\mathbf{E}_{in}^{lin}$ and X_1 is α_1. This vector can be represented by two rectangular components (orthogonal projections on the X_1 and Y_1), namely by the vertical component $\mathbf{E}_{in}^{V}$ and the horizontal component $\mathbf{E}_{in}^{H}$ (relative to the bistatic plane AOB). The polarization scattering matrix (PSM) shows the relationship between $E_{in}^{V}=|\mathbf{E}_{in}^{V}|$ and $E_{in}^{H}=|\mathbf{E}_{in}^{H}|$ on the one hand, with the similar components of the scattered field E_s^V and E_s^H at the receiving point (the projections of $\mathbf{E}_s^{lin}$ on the X_2 and Y_2), on the other.

$$\begin{pmatrix} E_s^V \\ E_s^H \end{pmatrix} = \begin{pmatrix} a_{VV} & a_{VH} \\ a_{HV} & a_{HH} \end{pmatrix} \begin{pmatrix} E_{in}^V \\ E_{in}^H \end{pmatrix}, \tag{2.13}$$

$$\mathbf{E}_s^{lin} = \mathbf{S}_{lin} \mathbf{E}_{in}^{lin}. \tag{2.14}$$

In (2.13) and (2.14) the common factor depending on the distance from the target to the receiving station is omitted.

The same relationship can be written in any other polarization basis. For instance, if polarization is circular, in (2.13) E_{in}^R, E_{in}^L, E_s^R, E_s^L should be substituted for E_{in}^V, E_{in}^H,

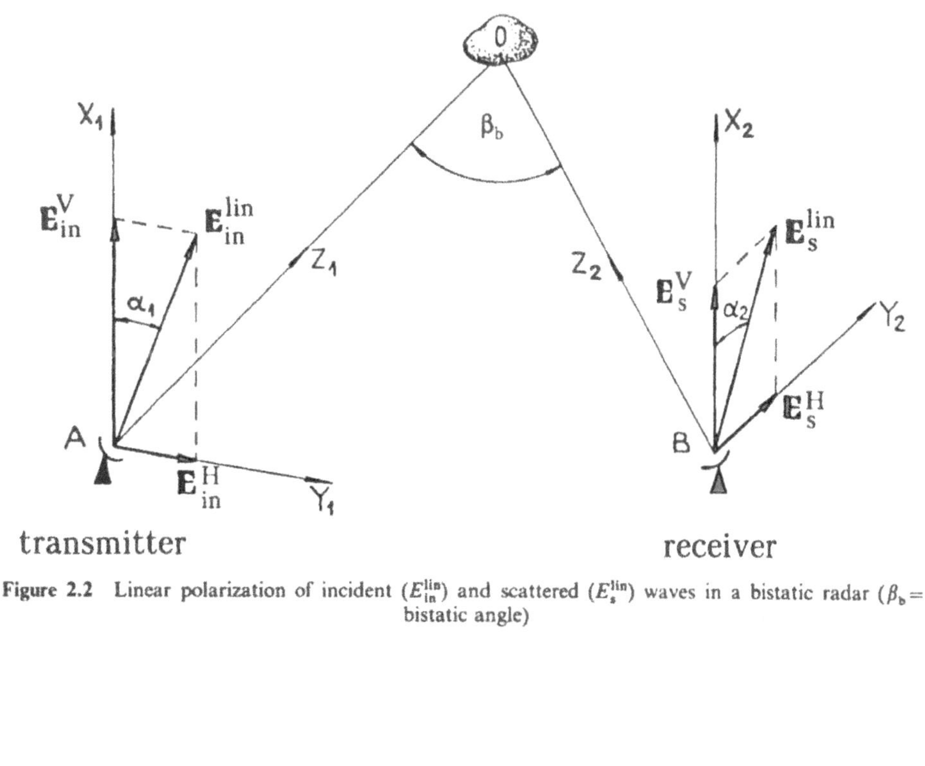

Figure 2.2 Linear polarization of incident (E_{in}^{lin}) and scattered (E_s^{lin}) waves in a bistatic radar (β_b= bistatic angle)

E_s^V, E_s^H, respectively, and elements of the PSM a_{RR}, a_{RL}, a_{LR} and a_{LL} should be substituted for a_{VV}, a_{VH}, a_{HV} and a_{HH}, respectively. Here the subscripts and superscripts "R" and "L" denote the right and the left rotations of the polarization plane relative to Z_1 for the incident wave and to Z_2 for the scattering wave. It is reasonable to remind that for circular and elliptic polarizations the direction of rotation is determined as it is seen from an illuminated object or a receiver (towards the approaching wave).

The elements of the PSM in (2.13) and (2.14) are complex values. Their squared moduli are equal to the target RCS for the corresponding combination of the incident and scattered field polarization.

The expressions (2.13) and (2.14) hold for any radar. However, the elements of a PSM for monostatic and bistatic radars are different in general. First of all, a monostatic PSM at the fixed frequency depends only on the direction of illumination that is opposite to the direction of reception, i.e. on the target aspect relative to the radar, whereas a bistatic PSM depends on both the illumination and reception directions that are not opposite to each other. At the same time the electromagnetic reciprocity principle is valid[1]: the interchange of the locations of transmitting and receiving stations leads only to transposition of the PSM, i.e. $\mathbf{S}(\mathbf{r}_0, \mathbf{r}) = \mathbf{S}^t(\mathbf{r}, \mathbf{r}_0)$. For a monostatic radar, where $\mathbf{r}_0 = \mathbf{r}$, an important property of the PSM symmetry follows from this principle. Since $\mathbf{S}(\mathbf{r}_0, \mathbf{r}_0) = \mathbf{S}^t(\mathbf{r}_0, \mathbf{r}_0)$, we have $a_{VH} = a_{HV}$, $a_{LR} = a_{RL}$ (in any polarization basis). *For a bistatic radar* $\mathbf{r}_0 \neq \mathbf{r}$ so that *the symmetry of a PSM is violated* and in the general case $\mathbf{S}(\mathbf{r}_0, \mathbf{r}) \neq \mathbf{S}^t(\mathbf{r}_0, \mathbf{r})$; $a_{VH} \neq a_{HV}$, $a_{LR} \neq a_{RL}$. Therefore, if a monostatic PSM for the fixed target aspect and at the wavelength λ is determined by three independent complex values (or by corresponding six real values) only, for the bistatic PSM determination four independent complex values (or corresponding eight real values) are necessary in the general case. This leads to a significant complication of PSM calculations and especially of PSM measurements.

Thus the bistatic PSM are more complex than monostatic ones but contain more information from a target.

For monostatic radars it is convenient to use the polarization eigenbasis of a target where the target PSM takes a diagonal form (for a given aspect and at a given wavelength) [17,20]. For a given PSM $\mathbf{S}$ in an arbitrary polarization basis, there exists a unitary matrix $\mathbf{U}$ transforming $\mathbf{S}$ into a diagonal form [25]

$$\mathbf{U}^*\mathbf{S}\mathbf{U} = \Lambda = \mathrm{diag}(\mu_1, \mu_2)$$

where μ_1, μ_2 are the eigenvalues of the PSM $\mathbf{S}$. When a target is illuminated by a wave which is polarized parallel to one of the orthogonal components of the polarization eigenbasis, then a scattered wave do not contain the other (cross-polarized) orthogonal component. For a monostatic radar the two vectors of the polarization eigenbasis of a target are mutually orthogonal [17]. For a bistatic radar the PSM diagonalization leads in general to a nonorthogonal polarization eigenbasis [16]. This reduces its value for the scattering feature analysis of a target.

The PSM of targets with a certain electrodynamic symmetry have some additional symmetry properties [93,94]. If a target exhibits symmetry about a plane (for instance, XZ) and a transmitting station line of sight lies in this plane, the following equations hold for symmetric receiving points and circular and linear

[1] We do not consider propagation effects related to the Earth magnetic field (Faraday rotation).

polarization

$$a_{RR}(x,-y,z)=a_{LL}(x,y,z), \qquad a_{RL}(x,-y,z)=a_{LR}(x,y,z),$$

$$a_{VV}(x,-y,z)=a_{VV}(x,y,z), \qquad a_{VH}(x,-y,z)=-a_{VH}(x,y,z),$$

$$a_{HH}(x,-y,z)=a_{HH}(x,y,z), \qquad a_{HV}(x,-y,z)=a_{HV}(x,y,z).$$

In particular, it is seen from these equations that when $y=0$, i.e. when the bistatic plane AOB coincides with the target symmetry plane, then

$$a_{RR}(x,0,z)=a_{LL}(x,0,z), \qquad a_{RL}(x,0,z)=a_{LR}(x,0,z).$$

The last equations mean that under the conditions considered the bistatic PSM is symmetric as the monostatic PSM. Besides that, diagonal elements are equal. If a target can be represented as a body of revolution that is illuminated along the axis of symmetry, elements of a bistatic PSM have some extra properties of symmetry [93,94].

However, when even a symmetric target is observed by a radar, its aspect angles relative both to the transmitting and receiving stations are usually not known and may change at random. The situation when the transmitting station turns out to be in the target symmetry plane (and the more on the target symmetry axis if the target is a body of revolution) is scarcely probable. For these reasons bistatic PSM properties of symmetry can be used primarily at test ranges for target PSM measurements [93,94].

In this section*, we have noticed that performance of MSRSs containing spatially separated transmitting and receiving stations or monostatic radars with cooperative signal reception is determined by the target bistatic RCS which accounts for possible different illumination and echo reception directions relative to the target. From the general definition of the RCS we have derived the simple approximate equations* (2.6) *and* (2.9) *permitting to calculate bistatic RCS for smooth, convex, perfectly conducting targets whose dimensions are much greater than the wavelength. We have considered the widely used* (*at microwaves*) *target representation as a collection of scattering centres ("flare spots – FSs") and have obtained a simple equation* (2.12) *for bistatic RCS of such targets. We have noted the usefulness of the so-called "monostatic–bistatic equivalence principle"* (*or "theorem"*) *which allows to utilize a lot of known data concerning monostatic RCS of many targets for bistatic RCS estimation.*

We have discussed some essential features of polarization scattering matrices in the bistatic case, in particular, the absence of symmetry and the absence of orthogonality of eigenbasis vectors. We have emphasized that despite some useful calculation techniques available, the most important way for obtaining bistatic RCS of actual complex targets in practice is its measurement at special test ranges. Some measured bistatic RCSs have been presented.

2.2. TARGET BISTATIC RCS AT FORWARD SCATTERING

According to the "monostatic–bistatic equivalence principle" and to experimental data the bistatic RCS σ_b of a target usually does not differ drastically from the monostatic RCS σ_m of the same target at moderate bistatic angles β_b (taking into account averaging over intervals of aspects). However, when bistatic angle β_b nears 180° the situation sharply changes.

It is known from the electromagnetic field theory that when an absolutely black body (which absorbs all the incident energy) is placed on a path of wave propagation and the dimensions of this body are large compared with the wavelength (though limited), then a scattered field exists behind the body (a "shadow" field). This field is a result of primary field disturbances ("shielding" of a part of the incident wave front). It is important to note that within the physical optics approximation the shadow scattered field from an absolutely black body does not depend on its surface shape and is completely determined by the incident field and the area of the body's shadow, or its silhouette. The shadow field polarization is the same as that of the incident wave [20].

Around a real illuminated target there is not only the shadow field but the "usual" scattered field being generated by electric surface currents induced by the incident wave. Let us denote electrical components of this scattered field (generated by the surface currents) and shadow field by $\mathbf{E}_c$ and $\mathbf{E}_{sh}$, respectively. Then the total scattered field is $\mathbf{E}_s = \mathbf{E}_c + \mathbf{E}_{sh}$ and the sum of the incident and scattered fields is $\mathbf{E}_\Sigma = \mathbf{E}_{in} + \mathbf{E}_s$. If target dimensions are large in comparison with the wavelength, then the shadow and the "usual" scattered fields are spatially separated. The shadow field concentrates within a small solid angle near $\beta_b = 180°$ (on the opposite side of the target from the transmitting station), so that it may be called "forward scattered field". In this region the scattered field generated by target surface currents is much less than the shadow field. Therefore, analyzing the field near $\beta_b = 180°$, one may neglect the influence of target surface currents, i.e. consider a target as an absolutely black body creating a shadow field only. In a similar manner we neglected the shadow field while discussing the scattered field at small and moderate values of β_b in Section 2.1.

It follows from the above that within the physical optics approximation *the shadow field of a target does not depend on the target surface shape and is completely determined by the incident field and the target's shadow, or silhouette*, like for an absolutely black body. From this it is clear that target surface material including the RAM coating, drastically reducing the scattered field generated by surface currents, and hence the monostatic RCS, cannot influence the shadow field.

To calculate the target RCS at forward scattering the Babinet principle may be used. According to that principle a plane absorbing screen of limited dimensions may be replaced by a complementary infinite plane screen with an aperture shaped exactly like the original screen (the complementary screen has openings where the original screen is closed and vice versa). The incident field diffracted at the aperture gives rise to the field coinciding with the shadow field of the original absorbing screen (except for the sign) [5,20]. If the incident wave is a plane one, then the shadow field of a target at a distant receiving point, $\mathbf{E}_{rt}$, (required for the RCS estimation) turns out to be radiation field of a plane in-phase (cophase) aperture placed perpendicular to the incident wave propagation direction and determined by the targets shadow (silhouette) on this plane.

Figure 2.3 shows how to reduce analysis of a target shadow field at first to the analysis of the shadow field of the plane opaque screen whose shape is identical to the shadow, or silhouette, of the target and then to the classical problem of plane wave diffraction at the aperture (whose shape is the same as the shape of the screen) in an infinite opaque plane [90].

If we denote A_t the "radiating" aperture shaped exactly like the silhouette of the target, then taking into account that A_t is uniformly and in-phase illuminated, the target RCS, $\sigma_b(\mathbf{r})$, at a distant receiving point and at bistatic angles near $\beta_b = 180°$

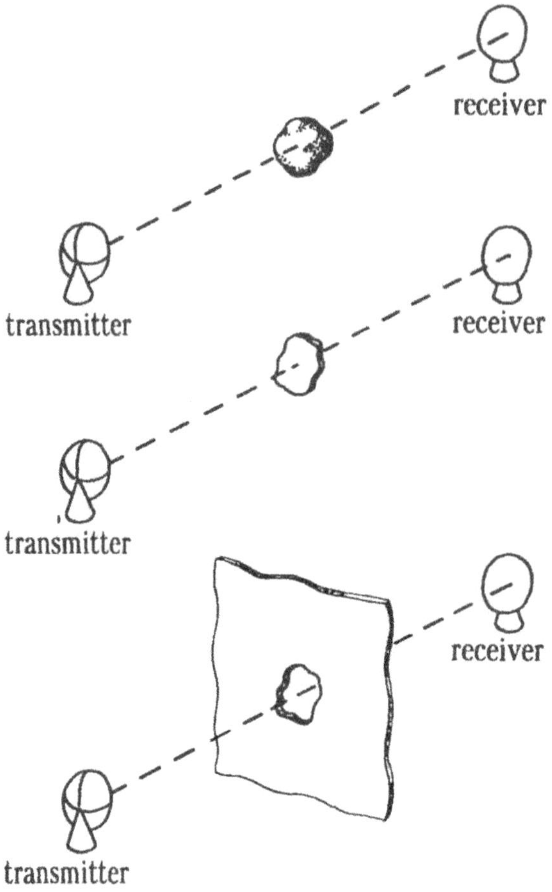

Figure 2.3 To the target forward scatter RCS evaluation based on the Babinet principle

can be expressed as follows:

$$\sigma_b(\mathbf{r}) = 4\pi R^2(|E_{rt}|^2/|E_{in}|^2) = (4\pi/\lambda^2)\left|\int_{A_t} \exp[j(2\pi/\lambda)\boldsymbol{\rho}\mathbf{r}]\mathrm{d}S\right|^2 \tag{2.15}$$

where $\boldsymbol{\rho}$ is the radius vector of an arbitrary point of the aperture A_t; $\mathbf{r}$ is the unit vector in the receiving station direction. The origin of the coordinate system may be chosen at the "centre" of the aperture A_t. It is seen from (2.15) that at $\beta_b = 180°$, when $\mathbf{r}$ is perpendicular to $\boldsymbol{\rho}$, the RCS σ_b has its maximum

$$\sigma_b(180°) = 4\pi(S_t/\lambda)^2 \tag{2.16}$$

where S_t is the area of the aperture A_t. It is convenient to represent (2.16) in the form: $\sigma_b(180°) = G_t S_t$ where $G_t = 4\pi S_t/\lambda^2$ is the peak antenna gain of the uniformly illuminated aperture A_t having the area S_t. Thus the RCS $\sigma_b(180°)$ is greater than the geometrical area S_t of the aperture A_t by a factor of G_t, i.e. by the antenna gain of the shadow area. If the dimensions of a target are much larger than the wavelength, then $S_t \gg \lambda^2$, $G_t \gg 1$ and $\sigma_b(180°) \gg S_t$. Under this condition $\sigma_b(180°)$ is much greater than the monostatic RCS, σ_m.

Example 2.1. The monostatic (and bistatic at $\beta_b < 140$–$150°$) RCS of a perfectly conducting sphere of radius $r_s = 20\lambda$ is $\sigma_m = \sigma_b = S_t = \pi r_s^2 = 400\pi\lambda^2$. According to (2.16) the RCS at forward scattering $\sigma_b(180°) = 4\pi(\pi r_s^2/\lambda)^2 = 64\pi^3 10^4 \lambda^2$. Hence $\sigma_b(180°)$ is greater than σ_m (and σ_b at $\beta_b < 140$–$150°$) by a factor $1600\pi^2$ or 42 dB.

In some cases such a drastic enhancement of target RCS at forward scattering permits lowering the required radiation energy of a radar. It is especially important, that the "shadow" RCS $\sigma_b(180°)$ of a target cannot be reduced by using RAM coating or body-shaping which reduce effectively the monostatic RCS of aircrafts and other targets.

However, large values of RCS determined by (2.26) maintain only in the narrow solid angle about the baseline between the transmitting and receiving stations. The bistatic angle β_b may decrease from 180° no more than by half of the beamwidth of the radiation pattern of aperture A_t. As it is known, this beamwidth at the -3 dB level is equal to $\Delta\theta \approx \lambda/l_t$, where l_t is a linear dimension (in the direction considered) of the aperture A_t. The 3 dB reduction of the RCS of the sphere from Example 2.1 (its radius is equal to 20λ) takes place when β_b differs from 180° by $\pm 0.75°$ only. It means that a target, and transmitting and receiving stations must lie nearly exactly on a single straight line. Such a configuration is difficult to realize in practice, especially in a MSRS with several spatially separated stations.

If $\sigma_b(180°)$ is much greater than the monostatic RCS, σ_m, the significant gain in RCS can remain in the sidelobe region of the aperture A_t radiation pattern. Equation (2.15) permits estimating σ_b when $\beta_b \neq 180°$. It is convenient to convert (2.15) into the Cartesian coordinate system

$$\sigma_b(\cos\theta_x, \cos\theta_y) = (4\pi/\lambda^2)\left|\iint_{A_t} \exp[j(2\pi/\lambda)(x\cos\theta_x + y\cos\theta_y)]\,dx\,dy\right|^2 \tag{2.17}$$

where x, y are the projections of $\mathbf{p}$ on the X and Y axes in the plane of the aperture A_t; $\cos\theta_x$, $\cos\theta_y$ are the direction cosines of the unit vector $\mathbf{r}$ relative to the same axes (i.e. θ_x and θ_y are the angles between $\mathbf{r}$ and the X axis and Y axis, respectively). It is assumed that the incident wave propagates in the positive direction of the Z axis. The double integral in (2.17) presents a double Fourier transformation of the function which is equal to one inside the aperture A_t and to zero outside this aperture [39].

The RCS calculations may be significantly simplified if the receiving station lies in the plane XZ or YZ. Then in (2.17) $\cos\theta_y = 0$ or $\cos\theta_x = 0$, respectively. For instance, if $\cos\theta_y = 0$, integration over y gives $y_1(x) - y_2(x)$, where $y_1(x)$ and $y_2(x)$ are the "upper" and "lower" boundaries of the aperture A_t along the Y axis. Note, that $\theta_x = \beta_b - \pi/2$, so that $\cos\theta_x = \sin\beta_b$ and $\sigma_b(\cos\theta_x, 0)$ may be denoted as $\sigma_b(\beta_b)$. Equation (2.17) is reduced to the one-dimensional Fourier transformation

$$\sigma_b(\beta_b) = (4\pi/\lambda^2)\left|\int_{x_{min}}^{x_{max}} \exp[j(2\pi/\lambda)x\sin\beta_b][y_1(x) - y_2(x)]\,dx\right|^2 \tag{2.18}$$

where x_{min}, x_{max} are the "left" and the "right" boundaries of the aperture A_t along the X axis (see Fig. 2.4).

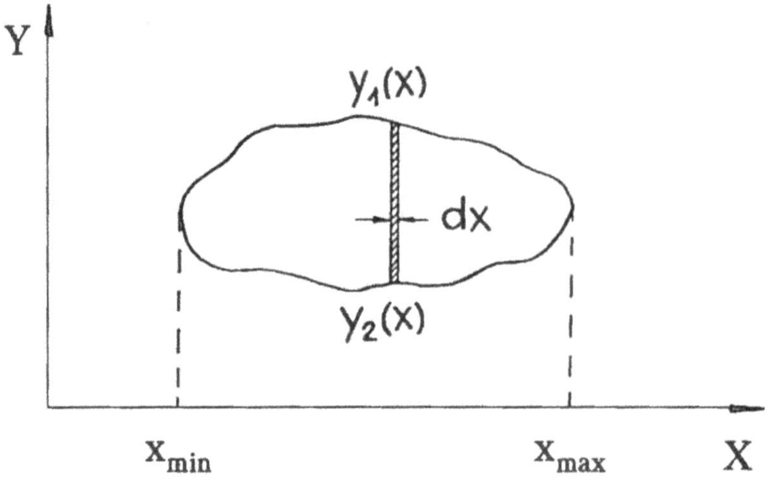

Figure 2.4 To the target forward scatter RCS computation using equation (2.18)

A simple approximate technique using equation (2.18) is described in [90] for the "shadow" RCS evaluation of complex shaped targets. It is based on the piecewise linear approximation of the aperture A_t boundaries i.e. curves $y_1(x)$ and $y_2(x)$. The comparison with the results obtained via exact techniques for a sphere and a cylinder has shown good agreement. When $\beta_b > 90°$ the described technique leads to errors of the order of 1–6 dB. These errors are the less the nearer is β_b to 180°. We present here a slightly generalized technique from [90]. The extent of the aperture in the X direction from x_{min} to x_{max} is to be divided into N_1 segments so that in each segment the curve $y_1(x)$ can be sufficiently well approximated by a straight line. Then the following sum is calculated:

$$I_1 = \sum_{i=1}^{N_1} \int_{x_i}^{x_{i+1}} \left[y_1(x_i) + \frac{y_1(x_{i+1}) - y_1(x_i)}{x_{i+1} - x_i} (x - x_i) \right] \exp[j(2\pi/\lambda)x \sin \beta_b]\, dx. \quad (2.19)$$

Once I_1 is calculated, a new division of the aperture's extent in the X direction from x_{min} to x_{max} into N_2 segments is to be accomplished so that in each segment the curve $y_2(x)$ can be sufficiently well approximated by a straight line. Then the sum I_2 similar to (2.18) but for $y_2(x)$ instead of $y_1(x)$ and N_2 instead of N_1 is calculated. The bistatic RCS is

$$\sigma_b(\beta_b) = (4\pi/\lambda^2)|I_1 - I_2|^2.$$

Example 2.2. Figure 2.5 shows the example of the missile shape "shadow" RCS calculated with the help of the technique described above [90]. Owing to the symmetry about the X axis, $y_2(x) = -y_1(x)$ and $N_1 = N_2$. It can be seen that the RCS has a $(\sin x/x)^2$ variation for the near-in sidelobes, which results from the abruptness of the fins and missile ends. Variable amplitude ripples in the far sidelobe region are due to interference between contributions from the fins, ends, and other parts of the missile. The sidelobe level in the far sidelobe region falls off faster with decreasing bistatic angle than $(\sin x/x)^2$. It follows from Fig. 2.5 that though the beamwidth

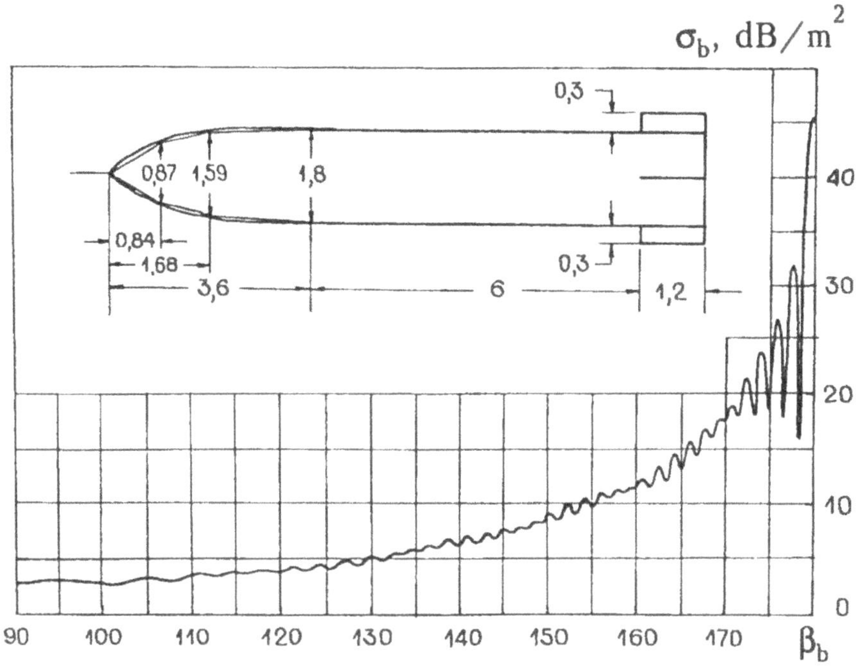

Figure 2.5 Forward scatter RCS of a missile calculated with the help of the piecewise linear contour approximation ([90] © 1985 IEEE). All dimensions are in meters. Wavelength $\lambda = 30\,\text{cm}$

does not exceed 3° (at the $-3\,\text{dB}$ level), there is a sector of about 20° where $\sigma_b(\beta_b \geqslant 170°) \geqslant 18\,\text{dB}$ relative to a square meter which may be significantly greater than the monostatic RCS.

The most interesting in the sidelobe region is the dependence upon β_b of the RCS averaged over several adjacent lobes since it reflects a general character of target RCS rolling off as the difference between β_b and 180 increases. It was shown in [55] that the "shadow" RCS $\sigma_b(\beta_b)$ of convex simply shaped conducting bodies in the sidelobe region at the fixed aspect angle may be estimated as follows:

$$\sigma_b(\beta_b) \approx \lambda l_t / \pi^2 |\pi - \beta_b|^3 \tag{2.20}$$

where l_t is the length of the shadow contour of the target (the contour of the target silhouette A_t cast on the plane perpendicular to the incident wave direction). If we apply the estimate (2.20) to the "shadow" RCS of the missile shape plotted in Fig. 2.5, one can see good agreement down to $\beta_b \approx 130$–$140°$ (assuming averaging over adjacent lobes). However, when $\beta_b < 130$–$140°$ the "shadow" RCS falls down by more than 40 dB with respect to its maximum value, so that this region is not of great interest.

The aspect angle of a target is usually unknown and may vary at random within certain limits. Then l_t in (2.20) should be replaced by $\bar{l}_t$ which is l_t averaged over possible target aspects. The following expression is suitable for the coarse estimation of "shadow" RCS [55]:

$$\bar{\sigma}_b(\beta_b) \approx \begin{cases} 4\pi(\bar{S}_t/\lambda)^2, & \text{if } |\pi - \beta_b| < \beta_b^*, \\ \lambda \bar{l}_t / \pi^2 |\pi - \beta_b|^3, & \text{if } |\pi - \beta_b| > \beta_b^*. \end{cases} \tag{2.21}$$

where

$$\beta_b^* = (\lambda/\pi)\sqrt[3]{\bar{l}_t/4\bar{S}_t^2}. \quad (2.22)$$

In (2.21) and (2.22) $\bar{S}_t$ is the area of the aperture A_t averaged over possible target aspects relative to the incident wave direction.

Unfortunately, one can rather seldom utilize in MSRSs the phenomenon of drastic target RCS enhancement at forward scattering. To obtain such an enhancement a target must be near the baselines between compact groups of transmitting and receiving stations (in the particular case between one transmitting and several receiving stations or vice versa). Though such situations are possible, they are rather exclusions. If the condition of forward scattering is satisfied for an arbitrary pair of stations but the bistatic angles relative to other stations are not near 180°, then the "shadow" RCSs for different stations, and hence the signal intensities at the inputs of those stations, turns out to be essentially different. Under such circumstances, as will be shown in Chapters 5 and 6, the information fusion in MSRSs becomes of little efficiency. However, for a given station geometry and moving targets, if the condition $\beta_b \approx 180°$ is satisfied for only one receiving station, the RCS enhancement relative to that station may be used for target detection by that station and for pointing other stations at the target. It may be used for saving power resources of MSRSs.

The forward scattering regime has important drawbacks. The coordinate and velocity measurement accuracy and resolution are significantly reduced, since the signal propagation time from a transmitting station to a target and then to a receiving station at $\beta_b \approx 180°$ weakly depends on the target position (at $\beta_b = 180$ does not depend at all). Together with the target RCS enhancement, the RCS of ground interfering reflectors (clutter) may substantially increase too. The forward scattered signal from a target arrives at the receiving station nearly at the same time as all the ground clutter echoes from along the baseline as well as the direct signal from the transmitting station [67,171]. Nevertheless, there are interesting suggestions to use forward scattering in MSRSs, for example, by employing satellites of U.S. GPS NAVSTAR and Russian GPS GLONASS [170].

***In this section**, we have considered one of the interesting features of a bistatic radar (which may be treated as a cell of MSRSs): the dramatic target RCS increase at "forward scattering". When bistatic angle, β_b, is close to 180°, bistatic RCS of a target is determined by the area of the silhouette of this target and the wavelength [see (2.16)]. This RCS cannot be reduced by RAM coating and other special techniques ("Stealth" technologies). Unfortunately, this large RCS holds only in a narrow sector near $\beta_b = 180°$. Nevertheless, if RCS at $\beta_b = 180°$ is much greater than the monostatic RCS of a target, the significant gain in RCS can remain outside that narrow sector. Simple approximate formulas (2.17)–(2.19) and (2.20)–(2.22) have been presented for RCS calculations in this case.*

2.3. BISTATIC RCS OF CHAFF CLOUDS

Chaff is one of the most widely used countermeasures designed to reduce the effectiveness of radar. It is so-called passive interference, or clutter, and consists of a large number of scatterers dispensed in the atmosphere (or space) to form a "cloud". Radar illumination energy scattered by a chaff cloud and received by the radar, is to be large enough to mask the presence of targets the chaff is to protect. Each scatterer is a passive dipole made usually of a thin highly conducting (metal or metallized) strip wires [15,105] and may be called a Dipole Scatterer (DS). The most important parameter of a DS is its RCS. The more RCS of one DS, the lower amount of DSs necessary to mask the presence of targets effectively. Therefore, wires typically are cut to a length so that they become resonant dipoles (having a maximum RCS) at the radar's frequency. Half-wavelength DSs ($l \approx \lambda/2$), i.e. resonant DSs of minimal length (and hence of minimal mass), are most widely practised. In some cases full-wave ($l \approx \lambda$) or three-halves wavelength ($l \approx 3\lambda/2$) DSs are used (for example, to make a chaff cloud effective against radars with different frequencies).

The sum of complex signals from a Cloud of DSs (CDS) and from a number of DSs falling within a radar resolution cell may usually be treated as a Gaussian random variable with zero mean[2]. Then the total RCS of these DSs and of a whole CDS is a random variable with an exponential probability distribution completely determined by its mean value. Positions and orientations of different DSs may, as a rule, be considered to be random, mutually independent and equiprobable. Distances between DSs are usually much longer than the wavelength λ. Under these conditions, scattered fields from different DSs at the antenna input of each receiving station are mutually incoherent. Then to estimate the averaged RCS of a CDS (or its part falling within an arbitrary radar resolution cell) it is sufficient to multiply the number N of workable DSs in the volume considered by the averaged RCS of one DS (dipole).

It is important to reveal how effective chaff can be against MSRSs containing spatially separated transmitting and receiving stations or radars with cooperative signal reception where chaff effectiveness primarily depends on the bistatic RCS of chaff. May be, the bistatic RCS of a chaff cloud is significantly smaller than the monostatic RCS?

To answer this question let us consider at first the bistatic RCS of one DS at fixed orientation and then proceed to averaging over equiprobable orientation of a DS, i.e. take into account uniform orientation distribution over the sphere.

Using different methods of the diffraction theory, several authors have obtained formulas permitting very accurate calculations of the field scattered by a passive cylindrical dipole of finite length and nonzero thickness [20,50] (see also references in [105]). However, for the RCS evaluation of CDSs, simplified and not so accurate formulas are used in practice. Such an approach is warranted by the fact that the number of DSs in a CDS (and hence in a resolution cell) is known, as a rule, only roughly, and some (exactly unknown) fraction of them may be disabled because of adhesion [15].

Thus the chaff considered below consists of a large number of nearly identical conducting wires whose thickness is very small compared to their length and to the wavelength; the interaction and coherence between wires can be ignored as well as their shielding.

The monostatic RCS of a DS has its maximum when the electric field vector **E** of the linearly polarized plane wave incident onto the DS is parallel to the axis of the dipole, and the same polarization is preferred by a receiving antenna at a distant point. In this case for a half-wavelength perfectly conducting DS the following approximate relationship is widely used [15,47]:

$$\sigma_m = \sigma_{max} \approx 0.86\lambda^2. \tag{2.23}$$

This relationship may serve as an initial expression for averaged RCS calculations under different conditions.

[2] Of course, if the quantity of DSs is not too small.

Averaged Bistatic RCS of a Half-Wavelength DS for Linear Polarization

Let us locate transmitting and receiving stations and a DS at points A, B, and O, respectively (see Fig. 2.6). Without loss of generality, we choose a right-handed Cartesian coordinate system X_1, Y_1, Z_1 with the origin at the point A as shown in Fig. 2.6: the Z_1 axis coincides with the incident wave propagation direction whereas the Y_1 axis lies in the bistatic plane AOB. A similar coordinate system X_2, Y_2, Z_2 with the origin at the point B is chosen also without loss of generality so that the Z_2 axis is opposite to the propagation direction of the scattered wave from the DS to the receiving station, and the Y_2 axis lies in the plane AOB. Under these conditions the X_1 and X_2 axes are parallel. Let the DS be illuminated by a linearly polarized plane incident wave from the point A. We determine the electric field vector $\mathbf{E}_{in}$ direction by the unit vector $\mathbf{e}_{in}$ lying in the X_1, Y_1 plane at the angle α_1 with the X_1 axis. We determine orientation of the dipole by the unit vector $\mathbf{p}$ in the dipole's axis direction. These unit vectors can be expressed through their projections on the X_1, Y_1, Z_1 axes (directional cosines) as follows:

$$\mathbf{e}_{in} = (\cos\alpha_1, \sin\alpha_1, 0); \qquad \mathbf{p} = (\sin\theta\cos\varphi, \sin\theta\sin\varphi, \cos\theta). \tag{2.24}$$

The power radiation pattern of a half-wavelength dipole can be written in the form [37,47,78]

$$F_1(\psi_1) = \frac{\cos^2[(\pi/2)\sqrt{1-(\mathbf{p}\mathbf{e}_{in})^2}]}{(\mathbf{p}\mathbf{e}_{in})^2} = \frac{\cos^2[(\pi/2)\sin\psi_1]}{\cos^2\psi_1} \tag{2.25}$$

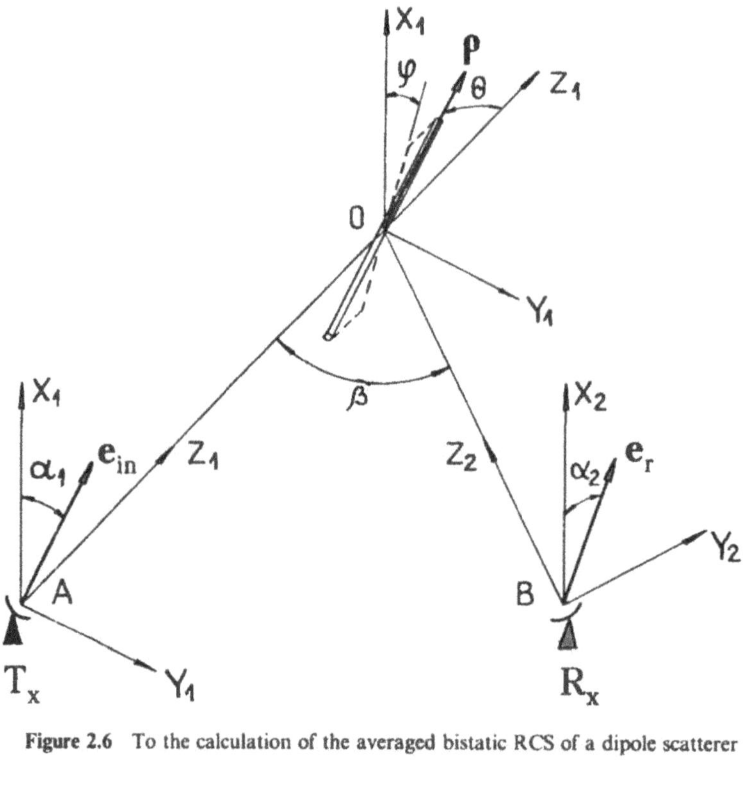

Figure 2.6 To the calculation of the averaged bistatic RCS of a dipole scatterer

where ψ_1 is the angle between the dipole's axis and the vector $\mathbf{E}_{in}$. Expressing the angle between two vectors through their directional cosines [25] we obtain

$$F_1(\alpha_1, \varphi, \theta) = \frac{\cos^2[(\pi/2)\sqrt{1-\cos^2(\varphi-\alpha_1)\sin^2\theta}]}{\cos^2(\varphi-\alpha_1)\sin^2\theta}. \tag{2.26}$$

This function determines a fraction of the total incident power (energy) which is received and reradiated by a lossless DS. Let the receiving antenna be also linearly polarized but, in the general case, the electric field vector $\mathbf{E}_r$ of the preferred antenna polarization forms another angle α_2 with respect to the X_2 axis. The unit vector $\mathbf{e}_r = \mathbf{E}_r/|\mathbf{E}_r|$ can be expressed through its directional cosines in the X_2, Y_2, Z_2 coordinate system like the unit vector $\mathbf{e}_{in}$ in the X_1, Y_1, Z_1 coordinate system

$$\mathbf{e}_r = (\cos\alpha_2, \sin\alpha_2, 0). \tag{2.27}$$

A fraction of the total reradiated power (energy) corresponding to the receiving antenna polarization (i.e. accepted by the receiving antenna) is determined by the power directivity pattern similar to (2.25):

$$F_2(\psi_2) = \frac{\cos^2[(\pi/2)\sqrt{1-(\mathbf{p}\mathbf{e}_r)^2}]}{(\mathbf{p}\mathbf{e}_r)^2} = \frac{\cos^2[(\pi/2)\sin\psi_2]}{\cos^2\psi_2}. \tag{2.28}$$

Here ψ_2 is the angle between the dipole's axis and the vector $\mathbf{E}_r$. To express the power directivity pattern (2.28) as a function of the DS orientation angles φ and θ (as in (2.26)) $\mathbf{e}_r$ is to be converted into the X_1, Y_1, Z_1 system. Note that the X_2, Y_2, Z_2 system is obtained through a rotation of the X_1, Y_1, Z_1 system in the right direction about the X_1 (or X_2) axis through the angle β_b (see Fig. 2.6). Hence $\mathbf{e}_r$ in the X_1, Y_1, Z_1 system is bound up with (2.27) by a rotation transformation [25]:

$$\mathbf{e}_r = \begin{bmatrix} 1 & 0 & 0 \\ 0 & \cos\beta_b & -\sin\beta_b \\ 0 & \sin\beta_b & \cos\beta_b \end{bmatrix} \begin{bmatrix} \cos\alpha_2 \\ \sin\alpha_2 \\ 0 \end{bmatrix} = \begin{bmatrix} \cos\alpha_2 \\ \cos\beta_b \sin\alpha_2 \\ \sin\beta_b \sin\alpha_2 \end{bmatrix}. \tag{2.29}$$

The right part of (2.29) gives the projections of the unit vector $\mathbf{e}_r$ on the X_1, Y_1 and Z_1 axes. Substituting (2.24) and (2.29) in (2.28), we have

$$F_2(\beta_b, \alpha_2, \varphi, \theta) = \frac{\cos^2\{(\pi/2)\sqrt{1-[\sin\theta(\cos\alpha_2\cos\varphi+\cos\beta_b\sin\alpha_2\sin\varphi)+\cos\theta\sin\beta_b\sin\alpha_2]^2}\}}{[\sin\theta(\cos\alpha_2\cos\varphi+\cos\beta_b\sin\alpha_2\sin\varphi)+\cos\theta\sin\beta_b\sin\alpha_2]^2}. \tag{2.30}$$

The averaged bistatic RCS of a dipole (DS) can be obtained from (2.23), (2.26) and (2.30) by averaging over the sphere:

$$\overline{\sigma_b}(\beta_b, \alpha_1, \alpha_2) = \frac{\sigma_{max}}{4\pi}\int_0^{2\pi}\int_0^{\pi} F_1(\alpha_1, \varphi, \theta) F_2(\beta_b, \alpha_2, \varphi, \theta)\sin\theta \, d\theta \, d\varphi. \tag{2.31}$$

For a monostatic radar from (2.30) we have at $\beta_b = 0$

$$F_2(\alpha_2, \varphi, \theta) = \frac{\cos^2[(\pi/2)\sqrt{1-\cos^2(\varphi-\alpha_2)\sin^2\theta}\,]}{\cos^2(\varphi-\alpha_2)\sin^2\theta}. \tag{2.32}$$

The substitution of (2.23), (2.26) and (2.32) in (2.31) gives the averaged monostatic RCS of a DS, $\overline{\sigma_m}(\alpha_1, \alpha_2)$. In this case the integrand is a periodic function of φ with the period 2π. Since an integral of a periodic function over an interval equal to the period of the function does not depend on shifts of this interval, we can write $\overline{\sigma_m}(\alpha_1, \alpha_2) = \overline{\sigma_m}(\Delta\alpha)$. It means that $\overline{\sigma_m}$ depends not on the individual directions of vectors $\mathbf{E}_{in}$ and $\mathbf{E}_r$ but on the difference between their directions (i.e. on the angle $\Delta\alpha = \alpha_1 - \alpha_2$, see Fig. 2.6).

For practical evaluation of the averaged RCS of half-wavelength dipole and of a cloud of such dipoles the power directivity patterns (2.25), (2.28), and hence (2.26), (2.30), are usually replaced by the simplified approximate relationships [47,48,78]:

$$\begin{aligned} &F_1(\psi_1) \approx (\mathbf{pe}_{in})^2 = \cos^2\psi_1, \qquad F_1(\alpha_1, \varphi, \theta) \approx \cos^2(\varphi-\alpha_1)\sin^2\theta, \\ &F_2(\psi_2) \approx (\mathbf{pe}_r)^2 = \cos^2\psi_2, \\ &F_2(\beta_b, \alpha_2, \varphi, \theta) \approx [\sin\theta(\cos\alpha_2\cos\varphi + \cos\beta_b\sin\alpha_2\sin\varphi) + \cos\theta\sin\beta_b\sin\alpha_2]^2 \end{aligned} \tag{2.33}$$

and for a monostatic radar [compare with (2.32)]

$$F_2(\alpha_2, \varphi, \theta) \approx \cos^2(\varphi-\alpha_2)\sin^2\theta. \tag{2.34}$$

Equations (2.33) and (2.34) are precise for a short dipole ($l \ll \lambda$) when the current amplitude is constant along the dipole's length. This is the so-called elementary electric dipole, or Herz's dipole [37]. Current amplitude distribution along the length of a half-wave dipole is close to a sinusoid with its maximum in the centre of the dipole. Just for such a distribution the power radiation patterns (2.25), (2.28) are valid [37]. Nevertheless, in the region near the maximum of radiation patterns, the differences between (2.25), (2.28), (2.32), on the one hand, and (2.33), (2.34), on the other, are practically negligible. Only near the radiation pattern minimum ($\psi \to 90$) equations (2.33) and (2.34) give errors up to $+2$ dB.

After substituting $F_1(\alpha_1, \varphi, \theta)$ and $F_2(\beta_b, \alpha_2, \varphi, \theta)$ from (2.33) in (2.31), the integral can be easily evaluated. Taking into account (2.23), we obtain the averaged bistatic RCS of a half-wavelength DS for linear polarization in the form [78]

$$\overline{\sigma_b}(\beta_b, \alpha_1, \alpha_2) \approx (0.86\lambda^2/15)[1 + 2(\cos\alpha_1\cos\alpha_2 + \cos\beta_b\sin\alpha_1\sin\alpha_2)^2]. \tag{2.35}$$

For a monostatic radar we have from (2.35) at $\beta_b = 0$

$$\overline{\sigma_b}(\Delta\alpha) \approx (0.86\lambda^2/15)(2 + \cos 2\Delta\alpha) = (0.86\lambda^2/15)[2 + \cos 2(\angle\mathbf{E}_{in}, \mathbf{E}_r)]. \tag{2.36}$$

When the transmitting and receiving antennas are polarized in parallel, i.e. $\mathbf{E}_{in} \parallel \mathbf{E}_r$ ($\Delta\alpha = 0$), we obtain from (2.36) a widely used equation [2,15,47]

$$\overline{\sigma_m}(0) = \overline{\sigma_m}_{\max} \approx 0.17\lambda^2. \tag{2.37}$$

When receiving antenna polarization is orthogonal to that of the transmitting antenna, i.e. $\mathbf{E}_{in} \perp \mathbf{E}_r$, $\Delta\alpha = 90°$,

$$\overline{\sigma_m}(90\ ) = \overline{\sigma_m}_{\min} \approx 0.057\lambda^2. \tag{2.38}$$

Thus when the transmitting and receiving antennas' polarizations are mutually orthogonal, the averaged monostatic RCS is one-third (or by -4.8 dB) as large as for parallel polarizations.

The accuracy of the approximate equations (2.35) and (2.36) may be evaluated with the help of plots in Fig. 2.7. It is seen that (2.36) gives slightly overstated values of $\bar{\sigma}_m$ (from 0.2 dB at $\Delta\alpha=0$ to 1.5 dB at $\Delta\alpha=90°$).

As it was mentioned above, the initial equation (2.23) is an approximate one. For this reason one can find in literature slightly different equations for $\bar{\sigma}_m$ as compared with (2.37) and (2.38). For example, according to [105] $\bar{\sigma}_m(0)=0.153\lambda^2$ and $\bar{\sigma}_m(0°)=0.050\lambda^2$. It is clear, however, that these differences do not essentially affect estimates of the averaged RCS of a CDS with unknown exact number of workable DSs, so that simple equations (2.36)–(2.38) may be successfully used.

Let us now return to equation (2.35) for $\bar{\sigma}_b$. Using (2.24) and (2.29), we can rewrite (2.35) as follows:

$$\bar{\sigma}_b(\beta_b, \alpha_1, \alpha_2) \approx (0.86\lambda^2/15)[1+2(\mathbf{e}_{in}\mathbf{e}_r)^2] = (0.86\lambda^2/15)[1+2\cos^2(\angle \mathbf{E}_{in}, \mathbf{E}_r)]$$
$$= (0.86\lambda^2/15)[2+\cos 2(\angle \mathbf{E}_{in}, \mathbf{E}_r)]. \quad (2.39)$$

Thus as with the monostatic RCS, $\bar{\sigma}_m$ [see (2.36)], the bistatic RCS, $\bar{\sigma}_b$, at a fixed wavelength λ is completely specified by the angle between vectors $\mathbf{E}_{in}$ and $\mathbf{E}_r$. Comparing the last expressions of (2.39) and (2.36), one can see their identity.

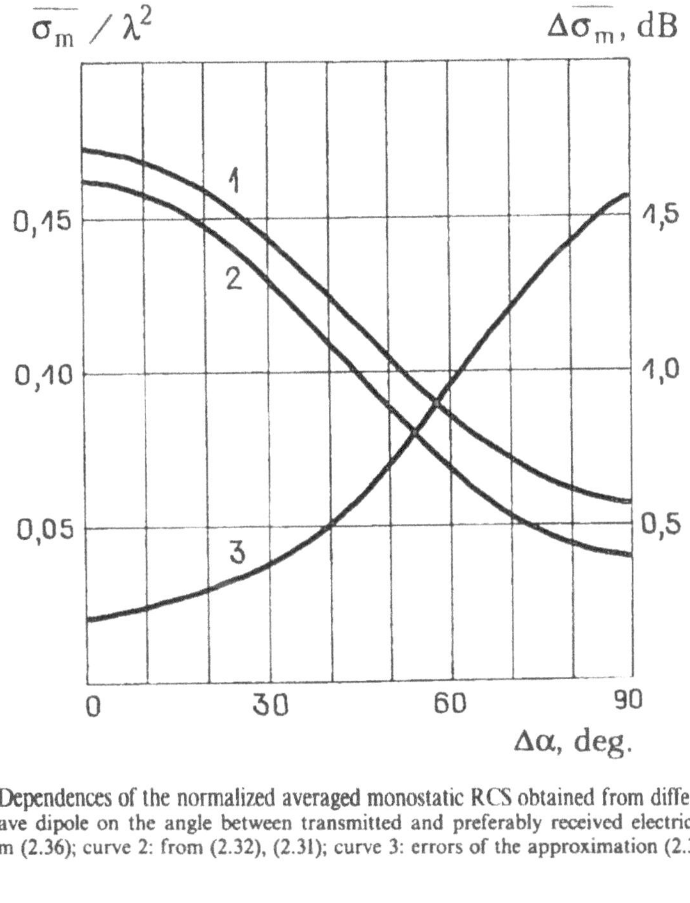

Figure 2.7 Dependences of the normalized averaged monostatic RCS obtained from different equations for a half-wave dipole on the angle between transmitted and preferably received electric field vectors. Curve 1: from (2.36); curve 2: from (2.32), (2.31); curve 3: errors of the approximation (2.36) in decibels

It means that for any possible directions of $\mathbf{E}_{in}$ and $\mathbf{E}_r$ the bistatic RCS of a half-wavelength DS, $\overline{\sigma_b}$, can change within the same limits as the monostatic RCS of the DS, $\overline{\sigma_m}$, i.e. from (2.37) when $\mathbf{E}_{in} \parallel \mathbf{E}_r$ to (2.38) when $\mathbf{E}_{in} \perp \mathbf{E}_r$, or by 4.8 dB. However, in a bistatic radar the angle between $\mathbf{E}_{in}$ and $\mathbf{E}_r$ depends on bistatic angle β_b and individually on the $\mathbf{E}_{in}$ and $\mathbf{E}_r$ directions (i.e. on the angles α_1 and α_2). At the same time one can see that α_1 and α_2 are included in (2.35) in a symmetrical manner so that their changes influence the bistatic RCS $\overline{\sigma_b}$ equally.

Of great practical interest is the dependence of RCS $\overline{\sigma_b}$ on the bistatic angle β_b. It follows from (2.35) that $\overline{\sigma_b}$ is independent of β_b if at least one of the polarization planes (of the transmitting and receiving antennas) is orthogonal to the bistatic plane AOB (see Fig. 2.6). In this case $\alpha_1 = 0$ or $\alpha_2 = 0$. Such a polarization can be called "vertical" (with respect to the plane AOB). Apparently, if $\alpha_1 = 0$ (vertical transmit polarization) at any α_2, the angle $\angle \mathbf{E}_{in}, \mathbf{E}_r = \alpha_2$ and remains constant with variations of β_b caused by the system X_2, Y_2, Z_2 rotation about the X_1 axes, since $\mathbf{E}_r$ moves along the generating straight line of a cone with the cone angle $2\alpha_2$. The same is true if $\alpha_2 = 0$ ("vertical" receive polarization) at any α_1 when the system X_1, Y_1, Z_1 rotates about the X_1 axes (with substituting α_1 for α_2). In these cases the bistatic RCS $\overline{\sigma_b}$ is the same as the monostatic RCS (2.36) where $\Delta\alpha$ should be replaced by α_2 if $\alpha_1 = 0$ or by α_1 if $\alpha_2 = 0$. The values of bistatic RCS $\overline{\sigma_b}$ reach their maximum when both transmit and receive polarizations are parallel and "vertical": $\mathbf{E}_{in} \perp AOB$, $\mathbf{E}_r \perp AOB$, $\alpha_1 = \alpha_2 = 0$. In this case equation (2.37) is valid for bistatic RCS $\overline{\sigma_b}$. The minimum values of $\overline{\sigma_b}$ occur when transmit and receive polarizations are mutually orthogonal: $\mathbf{E}_{in} \perp AOB$, $\mathbf{E}_r \parallel AOB$, $\alpha_1 = 0$, $\alpha_2 = 90°$ or, vice versa, $\mathbf{E}_{in} \parallel AOB$, $\mathbf{E}_r \perp AOB$, $\alpha_1 = 90°$, $\alpha_2 = 0$. For this case $\overline{\sigma_b}$ is determined by equation (2.38).

The dependence of the bistatic RCS $\overline{\sigma_b}$ on β_b is particularly pronounced when both the transmitting and receiving antennas are "horizontally" polarized, i.e. when the common polarization plane coincides with the AOB plane. Then $\mathbf{E}_{in} \parallel AOB$, $\mathbf{E}_r \parallel AOB$, $\alpha_1 = \alpha_2 = 90°$, $\angle \mathbf{E}_{in}, \mathbf{E}_r = \beta_b$ and from (2.35), (2.39) we have

$$\overline{\sigma_b}(\beta_b, 90°, 90°) \approx (0.86\lambda^2/15)(2 + \cos 2\beta_b). \tag{2.40}$$

Here the dependence on β_b is the same as on $\Delta\alpha$ in the case of the "vertical" polarization for a bistatic radar or of the arbitrary linear polarization for a monostatic radar. The maximum and minimum values are achieved when $\beta_b = 0$ and $\beta_b = 90°$, respectively. They are equal to right sides of (2.37) and (2.38), respectively.

The equations obtained for the averaged bistatic RCS $\overline{\sigma_b}$ of a half-wavelength DS permit estimating easily the bistatic RCS of a CDS. If a cloud consists of N operational half-wavelength dipoles, then

$$\overline{\sigma_{\Sigma b}} = N\overline{\sigma_b}. \tag{2.41}$$

Thus for linear polarized fields the bistatic RCS of a CDS weakly depends on the bistatic angle β_b. Even for the most "favorable" combination of transmitting and receiving polarizations the RCS lowers from its maximum only by 4.8 dB at $\beta_b \to 90$. When β_b differs from 90°, the RCS increases, e.g., at $\beta_b = 30$, it is by only 0.8 dB less than at $\beta_b = 0$. Falling off from the most "favorable" polarization combination reduces the RCS's dependence on β_b (down to independence)[3].

---

[3] As mentioned above, one can meet sligthly different values of RCS variations in literature as a result of different accuracy of approximate methods and initial formulas used for RCS estimation. In particular, the difference between the maximum and minimum values of $\overline{\sigma_b}$ obtained in [105] is up to 6 dB. At the same time the difference of 13 dB in [1] is not correct. An error has crept in averaging over the sphere.

Averaged Bistatic RCS for Circular Polarization

A circularly polarized wave can be represented as a sum of two linearly cross-polarized waves with a phase shift of 90° [17,20]. Let the power density flux of the right circular polarization field incident on a DS be Φ_{in}^{R} whereas the receiving-preferred polarization is the left circular one. Then Φ_{in}^{R} and the averaged (over the uniform random orientation of a DS) received power density flux, $\overline{\Phi_r^L}$, can be expressed through the corresponding linearly polarized components: "vertical" Φ_{in}^{V}, $\overline{\Phi_r^V}$ and "horizontal" Φ_{in}^{H}, $\overline{\Phi_r^H}$ (with respect to the *AOB* plane in Fig. 2.6)

$$\Phi_{in}^{V}=0.5\Phi_{in}^{R};\qquad \Phi_{in}^{H}=0.5\Phi_{in}^{R};\qquad \overline{\Phi_r^L}=0.5(\overline{\Phi_r^V}+\overline{\Phi_r^H}). \tag{2.42}$$

Evidently,

$$\begin{aligned}\overline{\Phi_r^V}&=(1/4\pi R^2)[\Phi_{in}^{V}\overline{\sigma_b}(\beta_b,\alpha_1=0,\alpha_2=0)+\Phi_{in}^{H}\overline{\sigma_b}(\beta_b,\alpha_1=90°,\alpha_2=0)],\\ \overline{\Phi_r^H}&=(1/4\pi R^2)[\Phi_{in}^{V}\overline{\sigma_b}(\beta_b,\alpha_1=0,\alpha_2=90°)+\Phi_{in}^{H}\overline{\sigma_b}(\beta_b,\alpha_1=90°,\alpha_2=90°)].\end{aligned} \tag{2.43}$$

Let us denote in (2.43):

$$\begin{aligned}&\overline{\sigma_b}(\beta_b,\alpha_1=0,\alpha_2=0)=\overline{\sigma_b^{VV}}(\beta_b); &&\overline{\sigma_b}(\beta_b,\alpha_1=90°,\alpha_2=0)=\overline{\sigma_b^{HV}}(\beta_b);\\ &\overline{\sigma_b}(\beta_b,\alpha_1=0,\alpha_2=90°)=\overline{\sigma_b^{VH}}(\beta_b); &&\overline{\sigma_b}(\beta_b,\alpha_1=90°,\alpha_2=90°)=\overline{\sigma_b^{HH}}(\beta_b);\end{aligned} \tag{2.44}$$

(for example, $\overline{\sigma_b^{HV}}(\beta_b)$ means the averaged bistatic RCS of a DS illuminated by a horizontally polarized wave when a received field is vertically polarized). Let us now substitute (2.44) in (2.43), express Φ_{in}^{V}, Φ_{in}^{H} from (2.42) through the circular polarization component Φ_{in}^{R}, and substitute them in the equation (2.42) for $\overline{\Phi_r^L}$. Let $\overline{\sigma_b^{RL}}(\beta_b)$ denotes the averaged bistatic RCS of a DS illuminated by a circularly polarized field with the right rotation when a received field is circularly polarized with the left rotation. Then from the definition of RCS (2.1), taking into account that $|\overline{E_r}|^2/|E_{in}|^2=\overline{\Phi_r}/\Phi_{in}$, we obtain

$$\overline{\sigma_b^{RL}}(\beta_b)=4\pi R^2\overline{\Phi_r^L}/\Phi_{in}^{R}=0.25[\overline{\sigma_b^{VV}}(\beta_b)+\overline{\sigma_b^{HH}}(\beta_b)+2\overline{\sigma_b^{HV}}(\beta_b)]. \tag{2.45}$$

This is a general relationship. For a half-wave DS from (2.35) and (2.44) we have

$$\overline{\sigma_b^{VV}}\approx 0.17\lambda^2;\quad \overline{\sigma_b^{HV}}(\beta_b)\approx 0.057\lambda^2;\quad \overline{\sigma_b^{HH}}(\beta_b)=(0.86\lambda^2/15)(2+\cos 2\beta_b). \tag{2.46}$$

Substituting $\overline{\sigma_b^{VV}}(\beta_b)$, $\overline{\sigma_b^{HV}}(\beta_b)$ and $\overline{\sigma_b^{HH}}(\beta_b)$ into (2.45) we obtain

$$\overline{\sigma_b^{RL}}(\beta_b)\approx 0.014\lambda^2(7+\cos 2\beta_b). \tag{2.47}$$

In the particular case when $\beta_b=0$ (i.e. for a monostatic radar)

$$\overline{\sigma_b^{RL}}(\beta_b)\approx 0.11\lambda^2. \tag{2.48}$$

It is seen from (2.42)–(2.48), that for circular polarized fields the averaged bistatic (and monostatic) RCS of a DS is the same when both transmitting and receiving antennas have circular polarizations, regardless of combinations of rotation senses, i.e.

$$\overline{\sigma_b^{RL}}(\beta_b)=\overline{\sigma_b^{LR}}(\beta_b)=\overline{\sigma_b^{RR}}(\beta_b)=\overline{\sigma_b^{LL}}(\beta_b).$$

Comparing (2.47) and (2.48) with (2.35), (2.37), (2.38) and (2.40) we can notice that the averaged monostatic RCS of a half-wavelength DS for a circular polarized field is approximately by 1.9 dB less than for parallel linear polarizations. The averaged bistatic RCS for a circular polarized field of both right and left rotation shows the same dependence on β_b. However, this dependence is weaker than for the horizontal transmitter and receiver polarizations (relative to a bistatic plane). As β_b increases from 0 to 90°, the RCS lowers by 1.25 dB only. At the same time the averaged RCS turns out to be approximately by 1.7 dB larger than the minimal averaged RCS for linear polarization.

Using (2.41) and (2.47), we can obtain the averaged bistatic RCS of a cloud consisting of N operational half-wavelength DSs, if transmitting and receiving antennas are circularly polarized regardless of combinations of rotation senses

$$\overline{\sigma_{\Sigma b}^{\text{circ}}} = 0.014\lambda^2(7+\cos 2\beta_b)N. \tag{2.49}$$

Plots of $\overline{\sigma_b^{VV}}(\beta_b)$, $\overline{\sigma_b^{HV}}(\beta_b)$, $\overline{\sigma_b^{HH}}(\beta_b)$ [see (2.44)] and $\overline{\sigma_b^{\Delta}}(\beta_b)$ versus the bistatic angle β_b are given in Fig. 2.8 for half-wavelength ($l \approx \lambda/2$), full-wave ($l \approx \lambda$) and three-halves wavelength ($l \approx 3\lambda/2$) dipoles [105]. $\overline{\sigma_b^{\Delta}}(\beta_b)$ is half the difference in averaged RCS (per dipole) seen by the receiving station in preferred and orthogonal polarizations when the transmit polarization and preferred receiving antenna polarization are linear and tilted 45° with respect to the bistatic plane AOB (see Fig. 2.6), i.e.

$$\overline{\sigma_b^{\Delta}}(\beta_b) = 0.5[\overline{\sigma_b}(\beta_b, \alpha_1 = 45°, \alpha_2 = 45°) - \overline{\sigma_b}(\beta_b, \alpha_1 = 45°, \alpha_2 = -45°)].$$

For half-wavelength dipoles one can notice a little difference in Fig. 2.8 from the values obtained above. As it was mentioned, this difference is not essential. It is a result of utilizing different initial approximate formulas and calculation methods.

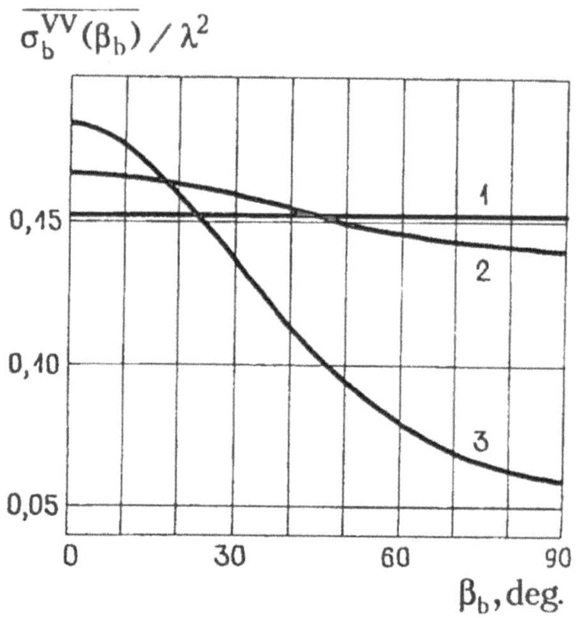

Figure 2.8(a)

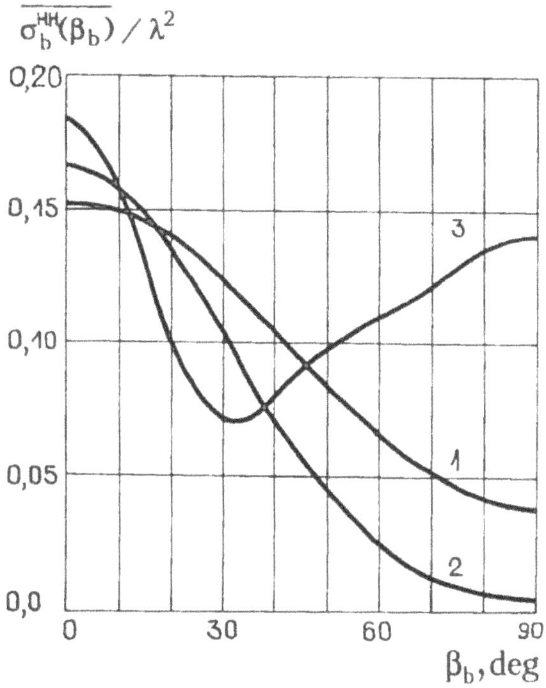

Figure 2.8(b)

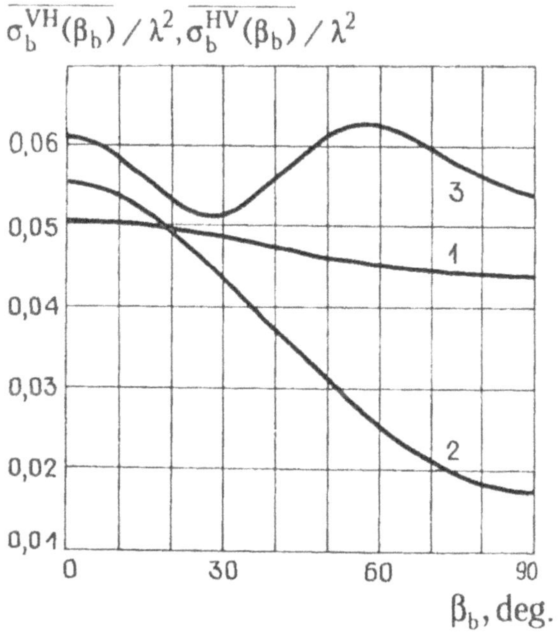

Figure 2.8(c)

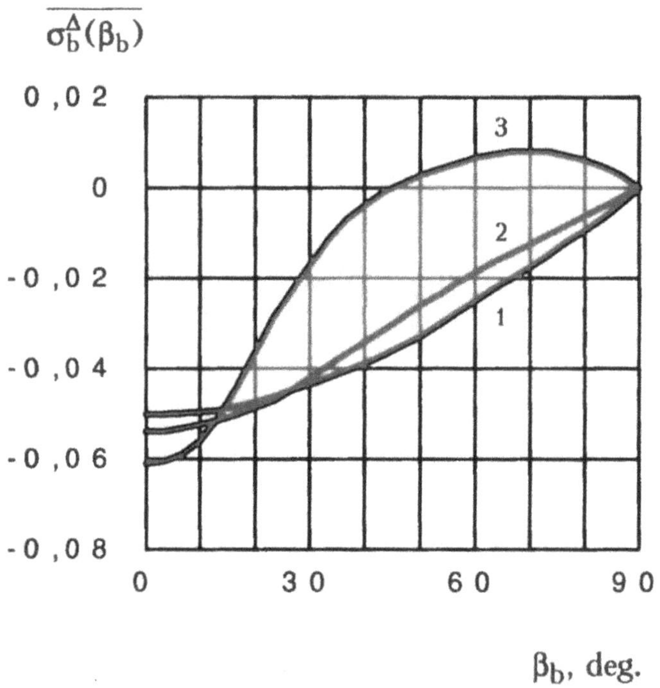

Figure 2.8(d)

Figure 2.8 Dependences of the normalized averaged RCS of a dipole with the length l on the bistatic angle β_b for linear polarization: (a), (b) transmission and reception at the same polarization, vertical and horizontal, respectively; (c) transmission and reception at the orthogonal polarizations; (d) the difference explained in text. Curve 1: $l \approx \lambda/2$; curve 2: $l \approx \lambda$; curve 3: $l \approx 3\lambda/2$. ([105], © 1984 IEEE)

Using the plots from Fig. 2.8 one can calculate the spherical averaged bistatic and monostatic RCS of a DS with $l \approx \lambda$ and $l \approx 3\lambda/2$ (and hence of a cloud consisting of such DSs) for linear, circular and general elliptic polarization.

Taking into account the geometry of Fig. 2.6, the general expression for averaged bistatic RCS of a chaff cloud consisting of N operational DSs of the same type (for receiving antenna preferred polarization) can be written in the form [105]

$$\overline{\sigma_{\Sigma b}} = N[(1+|Q_{in}|^2)(1+|Q_r|^2)]^{-1}\{\overline{\sigma_b^{VV}} + \overline{\sigma_b^{VH}}(|Q_{in}|^2+|Q_r|^2) + \overline{\sigma_b^{HH}}|Q_{in}|^2|Q_r|^2 - 4\overline{\sigma_b^{\Delta}}\,\mathrm{Re}(Q_{in})\,\mathrm{Re}(Q_r)\}. \tag{2.50}$$

In (2.50) Q_{in}, Q_r denote polarization factors for incident and received fields, respectively. For the general case of elliptic polarization $Q = B\exp(j\alpha)/A$ where B and A are complex amplitudes of electric field projections on the Y and X axes, respectively, α is a phase shift between these components [105].

For linear polarization (see Fig. 2.6)

$$Q_{in} = \sin\alpha_1/\cos\alpha_1; \qquad Q_r = \sin\alpha_2/\cos\alpha_2. \tag{2.51}$$

In the particular case of a half-wavelength DS, substituting (2.51) in (2.50), taking into account (2.46) and the approximate equation $\overline{\sigma_b^{\Delta}} \approx -(0.86\lambda^2/15)\cos\beta_b$, one can obtain equation (2.35) again.

For the left and the right circular polarizations, Q_{in} is equal to $-j$ and j, respectively, whereas Q_r has opposite signs (see Fig. 2.6). It means that $|Q_{in}|=|Q_r|=1$, $\mathrm{Re}(Q_{in})=\mathrm{Re}(Q_r)=0$ and from (2.50) we come to (2.45).

Up to now dipoles were assumed to be uniformly randomly oriented, i.e. their directions are uniformly distributed over the sphere. However, when chaff clouds are used against ground based bistatic radars, dipoles often tend to fall with a predominantly horizontal orientation in space [189]. In the general case, if $W(\varphi,\theta)$ is the weight function representing distribution of the wires' orientation in space, averaging over the sphere according to the formula

$$(1/4\pi)\int_0^{2\pi}\int_0^{\pi}[*]\sin\theta\,d\theta\,d\varphi$$

(where the weight function is assumed to be equal to 1) should be replaced by averaging according to the formula

$$\frac{\int_0^{2\pi}\int_0^{\pi}[*]W(\varphi,\theta)\sin\theta\,d\theta\,d\varphi}{\int_0^{2\pi}\int_0^{\pi}W(\varphi,\theta)\sin\theta\,d\theta\,d\varphi} \tag{2.52}$$

where $[*]$ denotes the averaged function. However, this function must be in the same coordinate system as the weight function $W(\varphi,\theta)$. Let $W(\varphi,\theta)$ describe the dipole orientation distribution in horizontal plane (φ) and relative to the vertical direction (θ) in a local coordinate system. The power directivity patterns (2.26), (2.30), (2.32), (2.33), (2.34), though depending on φ and θ, have been derived for the transmitter preferred coordinate system shown in Fig. 2.6 where the Y_1 and Z_1 axes lie in the bistatic plane AOB. If this plane is nearly horizontal, then for using $F_1(\alpha_1,\varphi,\theta)$ and $F_2(\beta_b,\alpha_2,\varphi,\theta)$ in (2.52) it is sufficient to redetermine the orientation of the dipole so that θ be measured from the X_1 axis and φ be measured from the Y_1 axis in the horizontal plane. Then $\mathbf{p}=(\cos\theta,\sin\theta\cos\varphi,\sin\theta\sin\varphi)$. $F_1(\alpha_1,\varphi,\theta)$ and $F_2(\beta_b,\alpha_2,\varphi,\theta)$ are to be transformed before substituting into (2.52) according to this new expression of $\mathbf{p}$. In general when the AOB plane is not horizontal, equations (2.26) and (2.30) should be converted into coordinate system of $W(\varphi,\theta)$. It may be done by rotations of this coordinate system about coordinate axes through corresponding angles.

In some cases a Gaussian weight function in the vertical direction may be used

$$W(\varphi,\theta)=\exp\{-[[(\pi/2)-\theta]^2/\Delta\theta^2]\}$$

where $\Delta\theta$ may be taken, for instance, to be 5°. With this weight function the DSs are oriented equally likely in the azimuth direction and decreasingly from the horizontal to vertical direction according to the Gaussian distribution. As it was shown in [188], when both transmitting and receiving antennas are horizontally polarized, the bistatic RCS is increased compared with the case of uniformly random orientation. This apparently results from the dipoles being predominantly horizontally oriented, which makes for more contributions to the total RCS. In the case of both transmitting and receiving antennas being vertically polarized, the bistatic RCS for orientations that are Gaussian in elevation is much smaller, by about two orders of magnitude, than that for the uniform orientation.

Averaged RCS of the Fraction of a CDS Falling within a Radar Resolution Cell

In most cases a CDS has large dimensions compared with the radar space resolution cell. In order to evaluate the ability of chaff to mask the presence of targets it is necessary to know the averaged RCS of the fraction of a chaff cloud covered by the radar resolution cell.

The averaged RCS of the fraction of a CDS falling within one resolution cell may be easily obtained [see (2.41)] if both the averaged RCS of one DS and the number of DSs within the resolution cell, N_0, is known. However, N_0 is, as a rule, unknown. Instead of N_0, averaged number of dipoles $\overline{N_0}$ can be evaluated assuming, for example, the uniform dipole distribution within a cloud with the density ρ_v. The density value can be usually estimated using known cloud creating technique, number and type of DSs in one dipole magazine, number of magazines, velocities of dipole flying apart from the magazine, adhesion factor and so on [15]. Then the averaged RCS, $\overline{\sigma_c}$, and the averaged number of dipoles in one radar resolution cell, $\overline{N_0}$, can be expressed as

$$\overline{\sigma_c}=\overline{N_0}\bar{\sigma}; \quad \overline{N_0}=\rho_v \Delta V \tag{2.53}$$

where $\bar{\sigma}$ is the averaged RCS of one DS, ΔV is the volume of one radar resolution cell (the "pulse volume"). In (2.53) transmitting and receiving antennas are assumed to have narrow mainbeams in all two angle coordinates, so that the CDS considered covers several radar resolution cells not only in range but in both angle coordinates.

The resolution cell of a monostatic radar may be approximated by a right elliptic cylinder whose height is determined by the resolution capability in range, $h_0=\delta R=c/2\Delta f_s$. The base of this cylinder is an ellipse whose size depends on beamwidths of the radar in the transmitting and receiving modes. In the general case these beamwidths may be different. Let us denote by $\Delta\varphi_T$, $\Delta\theta_T$ and $\Delta\varphi_R$, $\Delta\theta_R$, the beamwidths at the -3 dB level in azimuth and in elevation, respectively, where the subscripts "T" and "R" relate to transmitting and receiving modes, as earlier. It is convenient to replace a radar with different transmitting and receiving beamwidths by an equivalent radar with equal beamwidths. The transmitting antenna directivity pattern (ADP) is to be multiplied with the receiving ADP for a round-trip signal propagation. Assuming for simplicity Gaussian approximation for the mainbeams in all angle coordinates, we can write

$$\begin{aligned}&\exp\{-[2.772(\varphi-\varphi_0)^2/\Delta\varphi^2]-[2.772(\theta-\theta_0)^2/\Delta\theta^2]\}\\&\quad=\exp\{-[1.386(\varphi-\varphi_0)^2/\Delta\varphi_T^2]-[1.386(\theta-\theta_0)^2/\Delta\theta_T^2]\}\\&\qquad\times\exp\{-[1.386(\varphi-\varphi_0)^2/\Delta\varphi_R^2]-[1.386(\theta-\theta_0)^2/\Delta\theta_R^2]\}\end{aligned} \tag{2.54}$$

where $\Delta\varphi$ and $\Delta\theta$ without subscripts are the beamwidths of the equivalent radar at -3 dB level, φ_0 and θ_0 show the antenna pointing direction. From (2.54) we obtain the relationship between beamwidths of the radar of interest and the equivalent radar

$$\Delta\varphi=\sqrt{2}\Delta\varphi_T\Delta\varphi_R/\sqrt{\Delta\varphi_T^2+\Delta\varphi_R^2}; \qquad \Delta\theta=\sqrt{2}\Delta\theta_T\Delta\theta_R/\sqrt{\Delta\theta_T^2+\Delta\theta_R^2}. \tag{2.55}$$

At the range R the ellipse semiaxes (-3 dB two-way "linear beamwidths") for the equivalent radar are

$$a=R\Delta\varphi/2\sqrt{2}; \qquad b=R\Delta\theta/2\sqrt{2}. \tag{2.56}$$

Substituting the right sides of (2.55) for $\Delta\varphi$ and $\Delta\theta$ in (2.56) and taking into account the height of the elliptic cylinder $h_0=\delta R=c/2\Delta f_s$, we obtain the volume of a monostatic radar space resolution cell (at -3 dB level)

$$\Delta V_0 = \pi abh_0 = (\pi c/8\Delta f_s)R^2\Delta\varphi_T\Delta\varphi_R\Delta\theta_T\Delta\theta_R/\sqrt{(\Delta\varphi_T^2+\Delta\varphi_R^2)(\Delta\theta_T^2+\Delta\theta_R^2)}. \quad (2.57)$$

In the particular case where transmitting and receiving antenna beamwidths are equal ($\Delta\varphi_T = \Delta\varphi_R = \Delta\varphi$, $\Delta\theta_T = \Delta\theta_R = \Delta\theta$)

$$\Delta V_0 = (\pi c/16\Delta f_s)R^2\Delta\varphi\Delta\theta. \quad (2.58)$$

Substituting (2.57) or (2.58) for ΔV in (2.53) we obtain $\overline{N_0}$ and hence $\overline{\sigma_c}$.

When the antenna mainbeam of a monostatic radar is narrower than the chaff cloud angle width only in one angle coordinate (usually in azimuth) and broader in the other, then before substituting ΔV_0 for ΔV in (2.53) the latter beamwidth (e.g. $\Delta\theta$) is to be replaced by the angle width of the chaff cloud in this direction.

Let us now consider the volume of the bistatic radar resolution cell. A section of a chaff cloud and mainbeams of transmitting and receiving antennas by the bistatic plane AOB are shown in Fig. 2.9(a), (b).

One resolution cell in range sum $R_T + R_R$ corresponds to the "thickness" h_b of a layer between two prolate spheroids (a body obtained by revolving an ellipse about its major axis) with their foci at the points A and B for which [see Fig. 2.9(b)]

$$(R'_T + R'_R) - (R_T + R_R) = c/\Delta f_s. \quad (2.59)$$

It is reasonable to measure h_b along the bisector of the bistatic angle β_b [$\angle ACB$ in Fig. 2.9(b)] which is perpendicular to the ellipse at the point C.

It follows from elementary geometry that if $(h_b/4)\sin^2(\beta_b/2)(1/R_T + 1/R_R) \ll \cos(\beta_b/2)$, as is usually the practice, the left side of (2.59) is approximately equal to $2h_b\cos(\beta_b/2)$ so that[4] (see Fig. 2.9(b)) [159,190,192]

$$h_b = c/2\Delta f_s \cos(\beta_b/2). \quad (2.60)$$

We assume the mainbeams of transmitting and receiving antennas to be narrow in all two angle coordinates compared with the angle width of a chaff cloud.

As with monostatic radar, a bistatic radar with different transmitting and receiving beamwidths may be replaced by a monostatic radar with equal beamwidths. This equivalent monostatic radar should be placed at the bisector of the bistatic angle. Repeating arguments that have led to (2.54) we have the following relationship

$$\begin{aligned}
&\exp\{-[2.772R^2(\varphi-\varphi_0)^2/R^2\Delta\varphi^2] - [2.772R^2(\theta-\theta_0)^2/R^2\Delta\theta^2]\}\\
&\quad = \exp\{-[1.386R^2(\varphi-\varphi_0)^2/R_T^2\Delta\varphi_T^2\sec^2(\beta_b/2)] - [1.386R^2(\theta-\theta_0)^2/R_T^2\Delta\theta_T^2]\}\\
&\quad\quad \times \exp\{-[1.386R^2(\varphi-\varphi_0)^2/R_R^2\Delta\varphi_R^2\sec^2(\beta_b/2)] - [1.386R^2(\theta-\theta_0)^2/R_R^2\Delta\theta_R^2]\}.
\end{aligned} \quad (2.61)$$

As in (2.54), the left side of equation (2.61) is the product of the transmitting and receiving ADPs of an equivalent monostatic radar with equal transmitting and receiving beamwidths. The right side of (2.61) represents the product of the

[4] The exact equation for $(R'_T + R'_R) - (R_T + R_R)$ that can be used for all values of β and h_b has been presented in [192], but if the condition specified above is satisfied, the difference between that exact equation and the simple approximate equation (2.60) may be neglected.

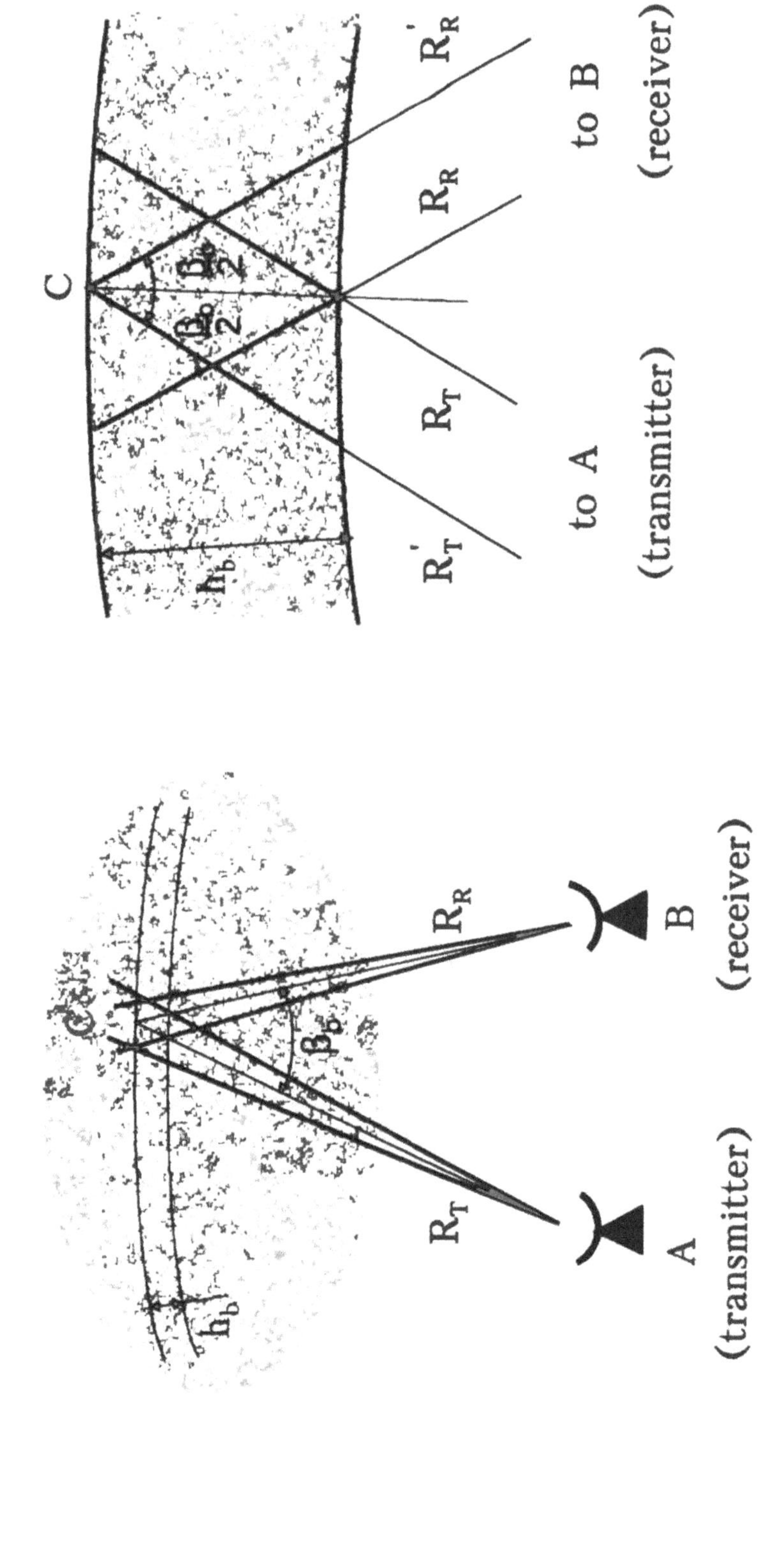

Figure 2.9 To the evaluation of the volume of a bistatic radar resolution cell: (a) a fraction of the CDS in the intersection of transmitting and receiving antenna mainbeams and range resolution cell; (b) the scaled-up part of Fig. 2.9(a)

transmitting and receiving ADPs of a bistatic radar. It is taken into account in (2.61), firstly, that because of different ranges R_T and R_R "linear beamwidths" (cross-range mainbeam dimensions) at the point of interest are to be considered, and secondly, that the two ellipse axes lying in the bistatic plane (in φ direction) are greater than the "linear beamwidths" $R_T\Delta\varphi_T$ and $R_R\Delta\varphi_R$ by a factor $\sec(\beta_b/2)=1/\cos(\beta_b/2)$. The latter is not true for the other ellipse axes (in θ direction).

From (2.61) we obtain

$$\begin{aligned} R\Delta\varphi &= \sqrt{2}R_T\Delta\varphi_T R_R\Delta\varphi_R/\cos(\beta_b/2)\sqrt{R_T^2\Delta\varphi_T^2+R_R^2\Delta\varphi_R^2}, \\ R\delta\theta &= \sqrt{2}R_T\Delta\theta_T R_R\Delta\theta_R/\sqrt{R_T^2\Delta\theta_T^2+R_R^2\Delta\theta_R^2}. \end{aligned} \tag{2.62}$$

Using again expressions (2.56) for the ellipse semiaxes of an equivalent monostatic radar, (2.60) for the height of elliptic cylinder and substituting right sides of (2.62) for $R\Delta\varphi$ and $R\Delta\theta$ in (2.56), we obtain the volume of a bistatic radar resolution cell at -3 dB level

$$\Delta V_b = \pi ab h_b = \frac{\pi c}{8\,\Delta f_s \cos^2(\beta_b/2)}\,\frac{R_T^2R_R^2\Delta\varphi_T\Delta\varphi_R\Delta\theta_T\Delta\theta_R}{\sqrt{(R_T^2\Delta\varphi_T^2+R_R^2\Delta\varphi_R^2)(R_T^2\Delta\theta_T^2+R_R^2\Delta\theta_R^2)}}. \tag{2.63}$$

When cross-range resolution of one station, e.g. the receiving one, is much better than that of the other station (i.e. $R_R\Delta\varphi_R \ll R_T\Delta\varphi_T$, $R_R\Delta\theta_R \ll R_T\Delta\theta_T$) then, as could be expected, ΔV_b is determined by the receiving station

$$\Delta V_b \approx \frac{\pi c}{8\,\Delta f_s \cos^2(\beta_b/2)} R_R^2\Delta\varphi_R\Delta\theta_R. \tag{2.64}$$

In another case, when the cross-range resolution capability of transmitting and receiving mainbeams at the chaff cloud are equal, i.e. $R_T\Delta\varphi_T = R_R\Delta\varphi_R$ and $R_T\Delta\theta_T = R_R\Delta\theta_R$, then from (2.63) we have

$$\Delta V_b = \frac{\pi c R_T\Delta\varphi_T R_T\Delta\theta_T}{16\,\Delta f_s\cos^2(\beta_b/2)} = \frac{\pi c R_R\Delta\varphi_R R_R\Delta\theta_R}{16\,\Delta f_s\cos^2(\beta_b/2)}. \tag{2.65}$$

Comparing (2.65) with (2.58) shows that with $R_T\Delta\varphi_T R_T\Delta\theta_T$ being equal to $R^2\Delta\varphi\Delta\theta$ and with equal bandwidths Δf_s, *the volume of a bistatic radar resolution cell* ΔV_b *is greater than that of a monostatic radar by the factor* $[\cos^2(\beta_b/2)]^{-1}$.

Substituting ΔV_b from (2.63), (2.64) or (2.65) in (2.53), we obtain the average RCS of a fraction of a chaff cloud (CDS) falling within the space resolution cell of a bistatic radar at -3 dB level. The influence of sidelobes of ADP and signals is not considered here.

Evidently, if the "linear beamwidth" of the transmitting and/or receiving stations in one angle coordinate, e.g. in elevation, exceeds the angle width of a chaff cloud ($R_T\Delta\theta_T > R_T\Delta\theta_{CDS}$ and/or $R_R\Delta\theta_R > R_R\Delta\theta_{CDS}$), then $\Delta\theta_T$ and/or $\Delta\theta_R$ in (2.63), (2.65) should be replaced by $\Delta\theta_{CDS}$ before substituting in (2.53).

The cross-range resolution cell of a monostatic radar is often defined as the linear beamwidth at -3 dB level for one-way signal propagation. In this case semiaxes a and b in (2.56) should be multiplied by the factor $\sqrt{2}$. According to this the coefficients 8 in (2.57), (2.63), (2.64) and 16 in (2.58), (2.65) should be replaced by the coefficients 4 and 8 respectively.

The principal conclusion from the above consideration is as follows: *bistatic configuration of a radar does not lead to significant reduction of spurious signal intensity from chaff at the input of a receiver.* Though the averaged RCS of one DS is reduced under certain conditions at the increase of bistatic angle β_b, the volume of a space resolution cell increases simultaneously, so that the averaged RCS of chaff falling within the resolution cell either remains practically constant or even increases.

RCS of Surface Clutter

Apart from chaff that represents clutter of a volumetric type, an important role in radar performance is played by surface clutter, i.e. unwanted echoes from ground (including buildings, trees and so on) and sea surface. As with monostatic RCS, the bistatic RCS of surface clutter is defined as a fraction of the incident energy scattered from a clutter cell area in the direction of the receiver. Instead of (2.53) we have

$$\sigma_{bs} = \rho_{bs}\Delta S_{bc} \tag{2.66}$$

where σ_{bs} is the bistatic RCS of surface clutter falling within a radar resolution cell, ρ_{bs} is the clutter bistatic scattering coefficient, i.e. the clutter bistatic RCS per unit area of the illuminated surface, ΔS_{bc} is the area of one bistatic radar resolution cell. The latter quantity can easily be obtained from the expressions for the volume ΔV_b of a radar resolution cell derived above, by excluding from those expressions the dependence on the elevation angle θ. Specifically, taking into account that $\Delta S_{bc} = 2ah_b$ and substituting a from (2.56), $R\Delta\varphi$ from (2.62), and h_b from (2.60) we obtain the area of one resolution cell at -3 dB level

$$\Delta S_{bc} = 2ah_b = \frac{cR_T R_R \Delta\varphi_T \Delta\varphi_R}{2\Delta f_s \cos^2(\beta_b/2)\sqrt{R_T^2\Delta\varphi_T^2 + R_R^2\Delta\varphi_R^2}}. \tag{2.67}$$

When the cross-range dimension of one mainbeam is substantially less than that of the other, e.g., $R_R\Delta\varphi_R \ll R_T\Delta\varphi_T$, we have from (2.67)

$$\Delta S_{bc} = (c/2\Delta f_s) R_R \Delta\varphi_R / \cos^2(\beta_b/2). \tag{2.68}$$

In another particular case where cross-range dimensions of both mainbeams at the illuminated area are equal, i.e. $R_T\Delta\varphi_T = R_R\Delta\varphi_R = R\Delta\varphi$

$$\Delta S_{bc} = (c/2\sqrt{2}\Delta f_s) R\Delta\varphi / \cos^2(\beta_b/2). \tag{2.69}$$

If, as was mentioned above, the cross-resolution cell of a monostatic radar is considered as the one-way linear beamwidth at -3 dB level, the right side of the equation for a in (2.56) and hence the right sides of (2.67), (2.68), and (2.69) should be multiplied by the factor $\sqrt{2}$.

Values of the surface clutter bistatic scattering coefficient ρ_{bs} depend on the surface features and are obtained by special measurements. The summarized measured and estimated data for ρ_{bs} are presented in [192].

***In this section**, we have examined in detail an important problem of the bistatic RCS of chaff clouds. We have obtained simple approximate expressions* (2.35), (2.39), (2.40) *for averaged bistatic RCS of one half-wave dipole scatterer (DS) at linear polarization and revealed the dependence of its RCS on bistatic angle. The corresponding* *expressions* (2.45)–(2.47) *have been obtained for circular polarization. We have also presented Peeble's results from* [105] *for half-wavelength, full-wave and three-halves wavelength dipoles at different polarizations. Apart from chaff clouds with uniformly randomly oriented dipoles we have considered the case where dipoles have a predominantly specific (for instance, horizontal) orientation in space. Detailed analysis of the volume of one space resolution cell of monostatic and bistatic radars has been carried out and simple formulas* (2.57), (2.58) *and* (2.63)–(2.65) *have been given. All obtained results permit practical calculations of averaged RCS of both chaff clouds and fractions of those clouds falling within a radar resolution cell. The principal conclusion from the consideration of this section is that bistatic configuration of a radar does not lead to significant reduction of spurious signal intensity from chaff at the input of the radar's receiver. We have also briefly considered RCS of surface clutter.*

3. MAXIMUM RANGE AND COVERAGE

3.1. MAXIMUM RANGE AND COVERAGE OF BISTATIC RADARS

If a MSRS consists of spatially separated transmitting and receiving stations or of monostatic radars with the cooperative signal reception (see Section 1.1), the maximum range and coverage estimation for such a MSRS is based on the Signal-to-Noise Ratio (SNR) and coverage calculations for the bistatic radar as a "cell" of the MSRS. The coverage of a bistatic radar we define[1] as *the region of space where a target with given RCS should be positioned so as the SNR at the receiver input be no less than a certain predetermined value* (e.g. which is necessary for the signal detection with required probability at the allowable false alarm probability). The target is assumed to be simultaneously within Line Of Sight (LOS) to both transmitting and receiving stations [190–192]. Evidently, a monostatic radar may be considered as a particular case of a bistatic radar.

Coverage of Bistatic Radars with Fixed Mainbeam Antenna Gain in Surveillance Region

Let arbitrary transmitting and receiving stations be located at the points determined by the radius-vectors $\mathbf{L}_0$ and $\mathbf{L}_i$, respectively. The position of a target with the bistatic RCS σ_b is determined by the radius-vector $\mathbf{R}$. The power SNR, i.e. the ratio of the signal energy to the noise Power Spectral Density (PSD), $q^2_{\text{out}\,i}=E_i/N_i$, at the receiver can be obtained from the range equation. For example, for a pulse radar if a target is in the maximum antenna gain directions relative to both stations [1,46,192]

$$q^2_{\text{out}\,i}=\frac{P\tau G_0 G_i \sigma_b \lambda^2 \eta_i}{(4\pi)^3 k T_{\text{eff}\,i} |\mathbf{R}-\mathbf{L}_0|^2 |\mathbf{R}-\mathbf{L}_i|^2} \tag{3.1}$$

where P and τ are the pulse transmitting power and pulse duration; G_0 and G_i are the maximum antenna gain (in mainbeam peak direction) of the transmitting and receiving stations, respectively; λ is the wavelength; η_i is the total power loss factor ($\eta_i<1$) including losses in transmitting and receiving systems and propagation losses, but excluding antenna losses which are taken into account in G_0 and G_i; k is the Boltzmann's constant ($k=1.38\times 10^{-23}$ W/Hz K); $T_{\text{eff}\,i}$ is the receiver effective noise temperature (K). Equation (3.1) differs from the classical range equation for the monostatic radar [45–48] by the factors $|\mathbf{R}-\mathbf{L}_0|^2|\mathbf{R}-\mathbf{L}_i|^2$ instead of $|\mathbf{R}|^4$.

The inequality determining the coverage can be written from (3.1)

$$|\mathbf{R}-\mathbf{L}_0|^2\,|\mathbf{R}-\mathbf{L}_i|^2=R_T^2 R_R^2 \leqslant A_i^4 \tag{3.2}$$

[1] The coverage of a MSRS may be defined in another way, for example, in terms of accuracy characteristics [47,84,149].

where R_T, R_R are the target ranges relative to the transmitting and receiving stations respectively,

$$A_i^4 = \frac{P\tau G_0 G_i \sigma_b \lambda^2 \eta_i}{(4\pi)^3 kT_{\text{eff}\,i} q_{\text{out}}^2} = \frac{P\tau G_0 \sigma_b S_{\text{eff}} \eta_i}{(4\pi)^2 kT_{\text{eff}\,i} q_{\text{out}}^2}. \tag{3.3}$$

In (3.3) S_{eff} is the effective aperture of the receiving antenna; q_{out}^2 is the minimum SNR value required for target detection with the given detection probability at the allowable false alarm probability.

Example 3.1. Let us consider a bistatic radar with the pulse transmitting power $P = 150\,\text{kW}$, pulse duration $\tau = 10\,\mu\text{s}$, transmitting antenna gain $G_0 = 15 \times 10^3$, receiving antenna effective aperture $S_{\text{eff}} = 12\,\text{m}^2$, thermal noise PSD $kT_{\text{eff}} = 4 \times 10^{-21}\,\text{W/Hz}$, and total loss factor $\eta = 0.2$. Let the bistatic RCS of typical targets that are to be detected be $\sigma_b = 1\,\text{m}^2$. The required SNR, $q_{\text{out}}^2 = 50$ or 17 dB. For a signal with unknown phase and Rayleigh amplitude fluctuations this value of q_{out}^2 provides the detection probability $P_d \approx 0.75$ at the false alarm probability $P_{fa} = 10^{-6}$. From (3.3) and (3.2) we have $|\mathbf{R} - \mathbf{L}_0|^2 |\mathbf{R} - \mathbf{L}_i|^2 = R_T^2 R_R^2 \leqslant 1.71 \times 10^{21}\,\text{m}^4 = 1.71 \times 10^9\,\text{km}^4$ or $R_T R_R \leqslant 41\,350\,\text{km}^2$. If the transmitting and receiving stations of this radar were colocated, the coverage of an equivalent monostatic radar would be determined by the inequality: $R = \sqrt{R_T R_R} \leqslant 203.3\,\text{km}$.

In the spherical coordinate system we denote

$$\mathbf{R} = (R, \beta, \varepsilon); \qquad \mathbf{L}_0 = (L_0, \beta_{a0}, \varepsilon_{a0}); \qquad \mathbf{L}_i = (L_i, \beta_{ai}, \varepsilon_{ai}). \tag{3.4}$$

From the law of cosines and analytical geometry [25]

$$\begin{aligned}
|\mathbf{R} - \mathbf{L}_0|^2 &= R^2 + L_0^2 - 2RL_0 \cos(\angle \mathbf{R}, \mathbf{L}_0); \\
|\mathbf{R} - \mathbf{L}_i|^2 &= R^2 + L_i^2 - 2RL_i \cos(\angle \mathbf{R}, \mathbf{L}_i); \\
\cos(\angle \mathbf{R}, \mathbf{L}_0) &= \cos\varepsilon \cos\varepsilon_{a0} \cos(\beta - \beta_{a0}) + \sin\varepsilon \sin\varepsilon_{a0}; \\
\cos(\angle \mathbf{R}, \mathbf{L}_i) &= \cos\varepsilon \cos\varepsilon_{ai} \cos(\beta - \beta_{ai}) + \sin\varepsilon \sin\varepsilon_{ai}.
\end{aligned} \tag{3.5}$$

Let us substitute (3.5) in (3.2), divide both sides of (3.2) by $L_0^2\, L_i^2$ and introduce $B_i^4 = A_i^4 / L_0^2 L_i^2$. Then for all points (R, β, ε) of the radar coverage the following inequality holds:

$$\begin{aligned}
&\{1 + (R/L_0)^2 - 2(R/L_0)[\cos\varepsilon \cos\varepsilon_{a0} \cos(\beta - \beta_{a0}) + \sin\varepsilon \sin\varepsilon_{a0}]\} \\
&\quad \times \{1 + (R/L_i)^2 - 2(R/L_i)[\cos\varepsilon \cos\varepsilon_{ai} \cos(\beta - \beta_{ai}) + \sin\varepsilon \sin\varepsilon_{ai}]\} \leqslant B_i^4.
\end{aligned} \tag{3.6}$$

For MSRSs with a single transmitting station it is convenient to place the origin of the coordinate system at this station. Then $\mathbf{L}_0 = 0$. Denoting $r_i = R/L_i$, we obtain for a bistatic radar, as the ith cell of a MSRS, instead of (3.6)

$$r_i^2 \{1 + r_i^2 - 2r_i [\cos\varepsilon \cos\varepsilon_{ai} \cos(\beta - \beta_{ai}) + \sin\varepsilon \sin\varepsilon_{ai}]\} \leqslant \tilde{B}_i^4 = A_i^4 / L_i^4. \tag{3.7}$$

Replacing the inequality by the equality in (3.7) yields the equation for the boundary of the radar coverage.

The coverage of a bistatic radar is often considered in the bistatic plane passing through the target of interest and the baseline between stations. Introducing the polar coordinate system (R, α) in this bistatic plane with the origin at the midpoint of the baseline, so that $L = L_i/2$ in (3.7), we can obtain

$$(1+r^2)^2 - 4r^2\cos^2\alpha \leqslant B^4 \tag{3.8}$$

where $r = R/L$, $B = 2\tilde{B}_i$ and the baselength is equal to $2L$. The equality in (3.8) at any fixed B gives the equation for the boundary of the radar coverage area in the chosen bistatic plane. This boundary is Cassini's oval. It is the locus of constant product of distances from two given points (foci) [25] where transmitting and receiving stations are located. In other words, the ovals of Cassini are contours of constant SNR on any bistatic plane assuming the right side of (3.8), B, to be constant at any point of each oval. The latter is not the case in practice since $B = 2\tilde{B}_i$ in accordance with (3.7) and (3.3) includes the target bistatic RCS, σ_b, which depends on the illumination and echo reception directions (see Chapter 2). Besides that, the total power loss factor η_i, including propagation losses, usually depends on the signal position relative to both stations. Therefore, this assumption should be considered as a simplifying one which is useful in understanding basic features of bistatic radars.

The form of Cassini's ovals for the different values of B in (3.8) has been sufficiently investigated and discussed [1,25,46,82,171,190,192]. It can be seen from Fig. 3.1 that the larger B, the more Cassini's ovals resemble the concentric circles which are the coverage area boundaries for a monostatic radar located at the coordinate system origin. The region inside the ovals of Cassini surrounding both the transmitting and receiving stations is sometimes called the *cosite region* [192]. As B

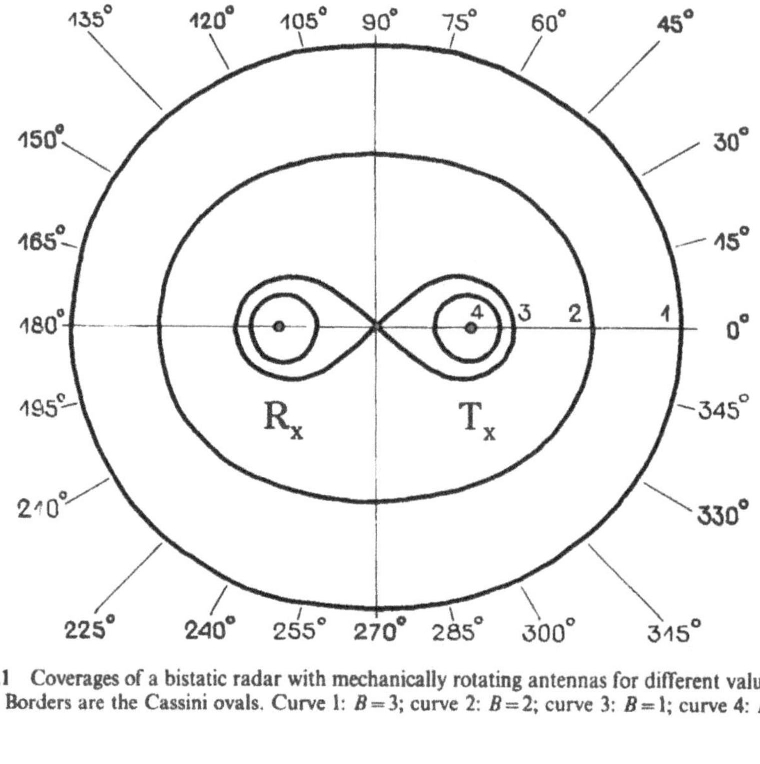

Figure 3.1 Coverages of a bistatic radar with mechanically rotating antennas for different values of B in (3.8). Borders are the Cassini ovals. Curve 1: $B=3$; curve 2: $B=2$; curve 3: $B=1$; curve 4: $B=0.8$

decreases, the ovals become increasingly flattened and then (at $B<1$) split into pairs of curves which encircle transmitting and receiving stations separately. It follows from (3.2), (3.3) and (3.7) that $B^2=(R_T R_R)_{max}/L$ is proportional to the product of maximum ranges relative to the transmitting and receiving stations for a target (with a given bistatic RCS) that can be detected by the radar considered at the required level of confidence.

Example 3.2. Returning to the Example 3.1 we have $B^4=1.71\times10^{21}/L^4$, or $B=203.3/L$ km. If the baselength $2L$ is, for instance, equal to 50 km, $B>8$ and the corresponding oval of Cassini surrounds both the transmitting and receiving stations. It is very close to the circle with the centre at the origin and the radius which is equal to four baselengths, i.e. 200 km. In fact, $r\approx8.06\,L\approx201.5$ km at $\alpha=0$, $\alpha=\pi$ and $r\approx7.94\,L\approx198.5$ km at $\alpha=\pi/2$, $\alpha=3\pi/2$. Increasing the baselength $2L$ and/or decreasing transmitting energy and/or receiver sensitivity reduce B. In particular, increasing the baselength up to 406.6 km with the same transmitting and receiving stations reduces B down to 1. The corresponding oval of Cassini degenerates into a Bernoulli's lemniscate (see Fig. 3.1). The similar result may be obtained with the former baselength $2L=50$ km if, for instance, under other conditions being constant, the pulse transmitting power is reduced to approximately 18.5 kW.

Bistatic radars as elements ("cells") of MSRSs, usually are designed for operating in cosite regions, where $B>1$, since otherwise overlapping coverages of different radars are difficult to achieve.

Unfortunately, if a MSRS contains more than one bistatic radar, bistatic planes of different radars do not, as a rule, coincide in space. Hence for MSRS coverage calculations the coverage of each such radar should be expressed in a common coordinate system so that more complicated general equations (3.6) or (3.7) are to be used.

The expressions derived above relate to a single target illumination and echo reception that rather characterize the target tracking mode. However, those expressions are valid for the space surveillance and search mode as well. For a monostatic radar in this case a surveillance sector (a solid angle) Ω_{sur} and allowable time T_{sur} for one scan are specified. With a bistatic radar similar parameters should be specified usually for a transmitting station. The space surveillance by a receiving station is to be organized in such a manner that in each recurrent period of the transmitter, T_r, the receiving station could receive echoes from all targets illuminated by the transmitting station. Taking into account that during each period, T_r, the transmitting station illuminates a solid angle $\Omega_0=4\pi\eta_{a0}/G_0$ where η_{a0} is the antenna efficiency, the surveillance of the angle Ω_{sur} will require time $T_{sur}=T_r\Omega_{sur}/\Omega_0=\Omega_{sur}G_0T_r/4\pi\eta_{a0}$. Substituting from this expression $G_0=4\pi T_{sur}\eta_{a0}/\Omega_{sur}T_r$ into (3.3) and replacing $P\tau/T_r$ by P_{av} where P_{av} is the average transmitting power, we obtain ($\tilde{\eta}_i=\eta_i\eta_{a0}$)

$$A^4_{sur\,i}=\frac{P_{av}T_{sur}\sigma_b G_i\lambda^2\tilde{\eta}_i}{(4\pi)^2\Omega_{sur}kT_{eff\,i}q^2_{out}}=\frac{P_{av}T_{sur}\sigma_b S_{eff}\tilde{\eta}_i}{4\pi\Omega_{sur}kT_{eff\,i}q^2_{out}}. \tag{3.9}$$

Thus the inequalities (3.2) and (3.6)–(3.8) determine the coverage of a bistatic radar in the surveillance mode too, if we replace the right sides of those equations by $A^4_{sur\,i}$ from (3.9) and by $B^4_{sur\,i}=A^4_{sur\,i}/L_0^2L_i^2$, $\tilde{B}^4_{sur\,i}=A^4_{sur\,i}/L_i^4$, $B^4_{sur}=16\,\tilde{B}^4_{sur\,i}$.

Example 3.3. Let the radar considered in Example 3.1 have the Pulse Repetition Frequency (PRF) $F_r=400$ Hz. Then the averaged transmitting power will be $P_{av}=P\tau F_r=150\times10^3\times10^{-5}\times400=600$ W. If the total loss factor including the transmitting antenna efficiency $\tilde{\eta}_i = 0.15$ and the surveillance sector $\Omega_{sur} = \pi$, the same value of $R_T R_R \leqslant 41\,350\,\text{km}^2$ can be obtained for the allowable surveillance time $T_{sur} = 12.5$ s.

The obtained results are valid under the condition that mainbeam antenna gain of each station, G_0 or G_i, does not change in any mainbeam pointing direction within the scanning sector. Such a condition is satisfied for usual mechanically scanning antennas since in these cases any mainbeam pointing direction is normal to the antenna aperture plane.

Coverage of Bistatic Radars with Mainbeam Antenna Gains Changing within Scanning Sector (Bistatic Radars with Linear or Planar Phased Antenna Arrays)

In the last decades linear and planar electronically scanning Phased Antenna Arrays (PAAs) are widely used in the radar engineering. One of the important features of these PAAs is that the antenna gain in the mainbeam peak direction is changed significantly within the sector of electronic scanning. In the general case

$$G_0 = G_{0\,max} u_0(\beta_0, \varepsilon_0); \qquad G_i = G_{i\,max} u_i(\beta_i, \varepsilon_i) \tag{3.10}$$

where $G_{0\,max}$ and $G_{i\,max}$ are the maximum values of G_0 and G_i which usually occur in the normal direction to the PAA plane; $u_0(\beta_0, \varepsilon_0)$ and $u_i(\beta_i, \varepsilon_i)$ are the functions which describe antenna gain reduction with a mainbeam peak direction being deflected from normal to the PAA plane $[u_0(\beta_0, \varepsilon) \leqslant 1,\ u_i(\beta_i, \varepsilon_i) \leqslant 1,\ u_0(0,0) = u_i(0,0) = 1]$. Clearly, the relationships (3.10) change the configuration of the bistatic radar coverage.

Let $\mathbf{n}_0$ and $\mathbf{n}_i$ be the unit normal vectors to the planes of the transmitting and receiving PAAs, respectively. We assume the functions u_0 and u_i to be symmetrical about $\mathbf{n}_0$ and $\mathbf{n}_i$, specifically

$$u_0(\beta_0, \varepsilon_0) = u_0[\angle(\mathbf{R} - \mathbf{L}_0, \mathbf{n}_0)]; \qquad u_i(\beta_i, \varepsilon_i) = u_i[\angle(\mathbf{R} - \mathbf{L}_i, \mathbf{n}_i)]. \tag{3.11}$$

The substitution of (3.11) in (3.10) and then in (3.1) yields the desired inequality determining the coverage

$$\frac{|\mathbf{R} - \mathbf{L}_0|^2 |\mathbf{R} - \mathbf{L}_i|^2}{u_0[\angle(\mathbf{R} - \mathbf{L}_0, \mathbf{n}_0)] u_i[\angle(\mathbf{R} - \mathbf{L}_i, \mathbf{n}_i)]} \leqslant \tilde{A}_i^4 \tag{3.12}$$

where

$$\tilde{A}_i^4 = \frac{P\tau G_{0\,max} G_{i\,max} \sigma_b \lambda^2 \eta_i}{(4\pi)^3 k T_{eff\,i} q_{out}^2} = \frac{P\tau G_{0\,max} S_{eff\,i\,max} \sigma_b \eta_i}{(4\pi)^2 k T_{eff\,i} q_{out}^2}. \tag{3.13}$$

For a planar PAA with closely spaced elements the gain reduction is usually approximated by the cosine[2] of the angle of the mainbeam pointing direction with

[2] Instead of cosine the squared cosine or cosine to power 3/2 approximations are often used. It may easy be taken into account in (3.14)–(3.19). For instance, if the squared cosine approximation is used, the numerators of the left sides in (3.15), (3.17)–(3.19) are to be raised to the power 4/3 whereas the denominators should be squared.

the normal to the PAA plane [46], i.e.

$$u_0[\angle(\mathbf{R}-\mathbf{L}_0,\mathbf{n}_0)]=\cos[\angle(\mathbf{R}-\mathbf{L}_0,\mathbf{n}_0)]=[(\mathbf{R}-\mathbf{L}_0)\mathbf{n}_0]/|\mathbf{R}-\mathbf{L}_0|;$$
$$u_i[\angle(\mathbf{R}-\mathbf{L}_i,\mathbf{n}_i)]=\cos[\angle(\mathbf{R}-\mathbf{L}_i,\mathbf{n}_i)]=[(\mathbf{R}-\mathbf{L}_i)\mathbf{n}_i]/|\mathbf{R}-\mathbf{L}_i| \tag{3.14}$$

where in the numerators are the scalar products. Substituting (3.14) in (3.12) yields the inequality determining the coverage

$$\frac{|\mathbf{R}-\mathbf{L}_0|^2|\mathbf{R}-\mathbf{L}_i|^2}{[(\mathbf{R}-\mathbf{L}_0)\mathbf{n}_0][(\mathbf{R}-\mathbf{L}_i)\mathbf{n}_i]}\leqslant\tilde{A}_i^4. \tag{3.15}$$

Let us express the obtained relationships in the spherical coordinate system. We denote

$$\mathbf{n}_0=(1,\beta_{v0},\varepsilon_{v0});\qquad \mathbf{n}_i=(1,\beta_{vi},\varepsilon_{vi}). \tag{3.16}$$

Taking into account that

$$[(\mathbf{R}-\mathbf{L}_0)\mathbf{n}_0]=\mathbf{R}\mathbf{n}_0-\mathbf{L}_0\mathbf{n}_0=R\cos[\angle(\mathbf{R},\mathbf{n}_0)]-L_0\cos[\angle(\mathbf{L}_0,\mathbf{n}_0)],$$
$$[(\mathbf{R}-\mathbf{L}_i)\mathbf{n}_i]=\mathbf{R}\mathbf{n}_i-\mathbf{L}_i\mathbf{n}_i=R\cos[\angle(\mathbf{R},\mathbf{n}_i)]-L_i\cos[\angle(\mathbf{L}_i,\mathbf{n}_i)],$$

and using the relationships similar to (3.5), we obtain the general expression determining the coverage of a bistatic radar with planar PAAs in the spherical coordinate system

$$\begin{aligned}&\{R^2+L_0^2-2RL_0[\cos\varepsilon\cos\varepsilon_{a0}\cos(\beta-\beta_{a0})+\sin\varepsilon\sin\varepsilon_{a0}]\}^{3/2}\\&\times\{[R^2+L_i^2-2RL_i[\cos\varepsilon\cos\varepsilon_{ai}\cos(\beta-\beta_{ai})+\sin\varepsilon\sin\varepsilon_{ai}]\}^{3/2}\\&\times\{R[\cos\varepsilon\cos\varepsilon_{v0}\cos(\beta-\beta_{v0})+\sin\varepsilon\sin\varepsilon_{v0}]\\&-L_0[\cos\varepsilon_{a0}\cos\varepsilon_{v0}\cos(\beta_{a0}-\beta_{v0})+\sin\varepsilon_{a0}\sin\varepsilon_{v0}]\}^{-1}\\&\times\{R[\cos\varepsilon\cos\varepsilon_{vi}\cos(\beta-\beta_{vi})+\sin\varepsilon\sin\varepsilon_{vi}]\\&-L_i[\cos\varepsilon_{ai}\cos\varepsilon_{vi}\cos(\beta_{ai}-\beta_{vi})+\sin\varepsilon_{ai}\sin\varepsilon_{vi}]\}^{-1}\leqslant\tilde{A}_i^4.\end{aligned} \tag{3.17}$$

For each value of $\tilde{A}_i^4$, [see (3.13)] a set of points (R,β,ε) satisfying (3.17) form the coverage of a bistatic radar. Those points for which the left side of (3.17) is exactly equal to $\tilde{A}_i^4$, form the boundary surface of the coverage.

The inequality (3.17) may be simplified if one of spatially separated stations, for example the transmitting one, is placed at the origin of the coordinate system, i.e. if $L_0=0$, and if both stations lie in the horizontal plane, i.e. $\varepsilon_{a0}=\varepsilon_{ai}=0$. In the latter case $(\varepsilon_{a0}=\varepsilon_{ai}=0)$ it is convenient to place the origin of the coordinate system at the midpoint of the baseline [so that $L_0=L_i=L$ and the baselength is equal to $2L$ as in (3.8)], and to measure the azimuth from the direction to the transmitting station so that $\beta_{a0}=0$, $\beta_{ai}=\pi$. Then denoting $r=R/L$, $b_i^4=\tilde{A}_i^4/L^4$, we have from (3.17)

$$\begin{aligned}&[(1+r^2)^2-4r^2\cos^2\varepsilon\cos^2\beta]^{3/2}\{\{r[\cos\varepsilon\cos\varepsilon_{v0}\cos(\beta-\beta_{v0})+\sin\varepsilon\sin\varepsilon_{v0}]\\&-\cos\varepsilon_{v0}\cos\beta_{v0}\}\times\{r[\cos\varepsilon\cos\varepsilon_{vi}\cos(\beta-\beta_{vi})+\sin\varepsilon\sin\varepsilon_{vi}]\\&+\cos\varepsilon_{vi}\cos\beta_{vi}\}\}^{-1}\leqslant b_i^4.\end{aligned} \tag{3.18}$$

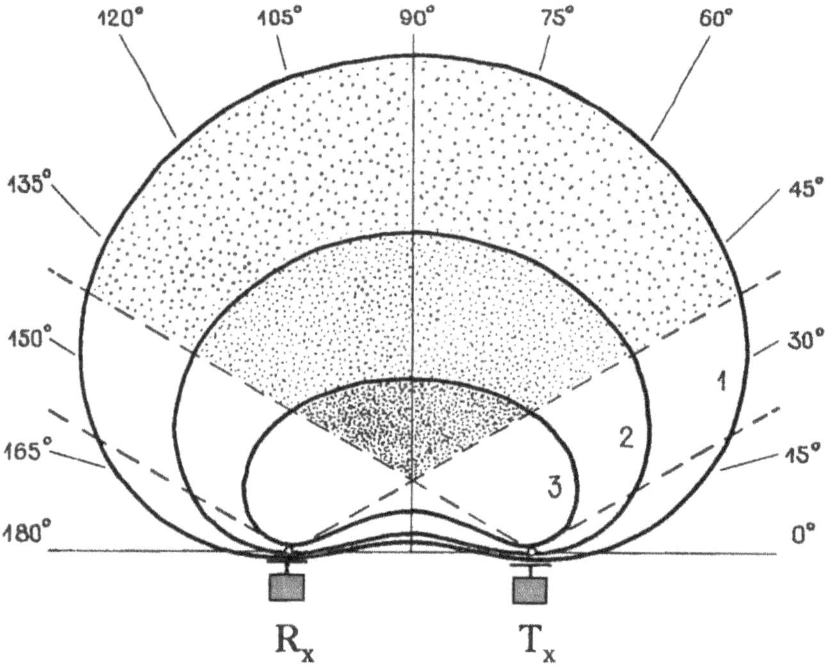

Figure 3.2 Coverages of a bistatic radar with PAAs for different values of b^4 in (3.19). Curve 1: $b^4=307$; curve 2: $b^4=69.1$; curve 3: $b^4=13.2$

In the case considered the boundary of the coverage in a bistatic plane cannot be described by Cassini's ovals.

Assuming for simplicity $\mathbf{n}_0=\mathbf{n}_i$, i.e. $\varepsilon_{v0}=\varepsilon_{vi}$, $\beta_{v0}=\beta_{vi}$ and $\beta_{v0}=\pi/2$. Then we obtain instead of (3.18)

$$\frac{[(1+r^2)^2-4r^2\cos^2\varepsilon\cos^2\beta]^{3/2}}{r^2[\cos\varepsilon\cos\varepsilon_{v0}\sin\beta+\sin\varepsilon\sin\varepsilon_{v0}]^2}\leqslant b^4. \tag{3.19}$$

The section views of the coverages of a bistatic radar with stationary PAAs are shown in Fig. 3.2 relevant to the choice of b_i^4 values at $\varepsilon=\varepsilon_{vi}=15°$. Unlike the case of constant antenna mainbeam gain (Fig. 3.1), we have here not only "far" but also "near" coverage boundaries (for all values of b^4) which are caused by sharp antenna mainlobe gain decreasing at large point direction angles from the PAA normal[3].

Example 3.4. Let us return again to the radar considered in Example 3.1 and assume its transmitting and receiving antennas to be the planar arrays with $G_{0\,\max}=15\times10^3$ and $S_{\mathrm{eff}\,i\,\max}=12\,\mathrm{m}^2$. Then the coverage boundaries in Fig. 3.2 correspond to the following baselengths, $2L$: $b^4=307$, $2L\approx97.2$ km – curve 1; $b^4=69.1$, $2L\approx141.1$ km – curve 2; $b^4=13.2$, $2L\approx213.4$ km – curve 3.

[3] Mainlobe antenna gain dependence on the angular deflection from the normal to the PAA plane of actual planar PAAs (excluding special thinned PAAs) should be approximated by more sharply decreasing functions at the angles greater than 45 60 than at smaller angles. It does not taken into account in Fig. 3.3.

When b^4 is small (low transmitting power and/or receiver sensitivity, large baselength), the real radar coverage may be limited to a small volume. To avoid large variations of antenna gain, the angle over which the PAA can be scanned is usually no more than $\pm 45°$–$60°$ from the array normal. The coverage areas taking into account this limitation ($\pm 60°$) are shaded in Fig. 3.2. Antenna gains vary by no more than 10–15% within these shaded areas, but the "near" boundary moves away which means increasing the minimum range where the target can be detected. Apparently, the mutual turning of the transmitting and receiving PAAs through a certain angle will change the coverage configuration of Fig. 3.2.

When a broader electronically scanned azimuthal sector than in Fig. 3.2 is required, each station may contain several (2–4) PAAs turned about each other or one rotating PAA.

This section has been devoted to the coverage of bistatic radars as "cells" of MSRSs. In accordance with the suggested definition we have considered two most important cases. In the first case antenna gains in the mainbeam directions are constant in a surveillance sector. This is a typical situation for antennas with mechanical scanning. The general relationships (3.2) *and* (3.3) *have been presented. For spherical coordinate systems expressions* (3.6), (3.7) *and* (3.8) *have been obtained for both tracking and search modes. In the second case antenna gains in mainbeam peak directions change essentially within the surveillance sector. Such a situation is typical for modern radar systems with electronically scanning antenna arrays. For this case the general inequality* (3.15) *and* (3.17) (*for the spherical coordinate system*) *have been obtained. Numerical examples* 3.1–3.4 *illustrate practical coverage calculations.*

3.2. MAXIMUM RANGE AND COVERAGE OF ACTIVE MSRSs

The maximum range and coverage of an active MSRS (or an active–passive MSRS in the active mode) depend not only on transmitting and receiving station characteristics, but on the system geometry, the type of the MSRS and the used information fusion techniques.

It was noted in Section 1.2 that using the different number and station arrangement in space, the coverage may be "matched" to the expected environment. However, the same MSRS can have different coverages depending on the information fusion algorithms which, in turn, are determined by the tasks that the MSRS has to carry out. For example, when a MSRS is in a search mode, it is often sufficient to detect a target in one receiving station only. For the accurate measurements in a tracking mode the target must be detected by several spatially separated stations. Clearly, the maximum detection range and coverage in the search mode may be larger than in the tracking one.

We define the coverage of an *active* MSRS as *a space region where a target with a given RCS should be positioned so that for chosen information fusion technique and detection decision rule, the target would be detected with no lower than required detection probability at the fixed false alarm probability.*

Unlike the definition of the coverage for a bistatic radar (see Section 3.1.) we introduce a probabilistic criterion here. For a single monostatic or bistatic radar the signal-to-noise ratio (SNR) determines, as a rule, detection probability characteristics unambiguously, whereas for a MSRS it is not the general case (see Chapters 5 and 6).

The coverage has been defined above for a target with a given RCS assumed equal for all stations of a MSRS. However, it follows from Chapter 2 that RCS σ_b of the same target depends on the bistatic angle β. Hence this RCS, σ_b, may vary with bistatic angle variations when a target position in space is changed. Even for a monostatic radar, RCS of a moving target may vary significantly with aspect angle variations (including averaged RCS over a certain aspect angle). Thus the radar coverage is defined for an artificial "test" target with a given RCS. Such a target cannot always be identified as a real target.

For MSRSs a target with the given RCS is a more artificial term than for both monostatic and bistatic radars. Certain real targets can never have equal RCSs relative to spatially separated stations and at different points of space (because of different aspect and bistatic angles). However, taking into account those differences of a target RCS is possible only if *a priori* information concerning the target is full enough which is usually not the case. Therefore, for the coverage of a MSRS we shall use the definition presented above assuming a given target RCS. Despite its artificial character, it is important for practice. It should be noted that for MSRSs with short baseline distances (see Section 1.1) real targets may often satisfy the condition of equal and constant (averaged) RCSs relative to different stations.

Let us consider coverages of MSRSs of different types containing a single transmitting and m receiving stations. The extension to MSRSs with several transmitting stations is not difficult.

Spatially Coherent MSRSs

As was mentioned in Section 1.1, such MSRSs may be considered as a single radar with sparsely populated antenna array. Hence there is usually an unambiguous relationship between SNR and detection probabilistic characteristics so that the coverage of such MSRSs may be obtained as for a single radar. Spatially coherent MSRSs are designed for receiving and processing spatially coherent echoes with nonfluctuating amplitudes or with completely correlated (at the inputs of different stations) amplitude fluctuations. The information from different stations is integrated at the radio signal level (see Section 1.1.). After this integration the probability distributions of noise and of signal plus noise are the same as in a single receiving station and SNR, $q^2_{\text{out}} = E/N$, is equal to the sum of SNRs in all the receiving stations, $q^2_{\text{out}\,i} = E_i/N_i$, $i=\overline{1,m}$ (see Sections 5.1, 5.4). If antenna gains in mainbeam peak directions are constant within the sector of responsibility we can obtain from (3.1)

$$q^2_{\text{out}} = \sum_{i=1}^{m} q^2_{\text{out}\,i} = \frac{P\tau G_0 \sigma_b \lambda^2}{(4\pi)^3 |\mathbf{R}-\mathbf{L}_0|^2} \sum_{i=1}^{m} \frac{G_i \eta_i}{kT_{\text{eff}\,i} |\mathbf{R}-\mathbf{L}_i|^2}. \tag{3.20}$$

The inequality determining the coverage can be written from (3.20) in the form

$$|\mathbf{R}-\mathbf{L}_0|^2 \left[\sum_{i=1}^{m} \frac{G_i \eta_i}{|\mathbf{R}-\mathbf{L}_i|^2 kT_{\text{eff}\,i}} \right]^{-1} \leqslant C_1^4 \tag{3.21}$$

where

$$C_1^4 = P\tau G_0 \sigma_b \lambda^2 / (4\pi)^3 q^2_{\text{out}}. \tag{3.22}$$

If antenna gains in mainbeam peak directions are changed within a surveillance sector according to (3.10), we have instead of (3.21)

$$\frac{|\mathbf{R}-\mathbf{L}_0|^2}{u_0(\beta_0, \varepsilon_0)} \left[\sum_{i=1}^{m} \frac{G_{i\,\max} u_i(\beta_i, \varepsilon_i) \eta_i}{|\mathbf{R}-\mathbf{L}_i|^2 kT_{\text{eff}\,i}} \right]^{-1} \leqslant C_2^4 \tag{3.23}$$

where the difference between C_2^4 and C_1^4 consists only in replacing G_0 by $G_{0\,\max}$.

Using (3.4), (3.5), (3.14) and (3.16) one can transform (3.21) and (3.23) into the spherical coordinate system as for a bistatic radar in Section 3.1.

The spatially coherent echoes are possible, as a rule, in MSRSs with short baseline distances. The coverage of such a MSRS is close to the coverages of an "equivalent" monostatic radar located at the origin of the coordinate system. The coverage of this monostatic radar may be obtained assuming in (3.21), (3.23) $\mathbf{L}_0 = \mathbf{L}_i = 0$, replacing $\sum_{i=1}^{m} [G_i \eta_i / T_{\text{eff}i}]$ by $G_\Sigma \eta_\Sigma / T_{\text{eff}\Sigma}$ in (3.21) and $\sum_{i=1}^{m} [G_{i\max} u_i(\beta_i, \varepsilon_i) \eta_i / T_{\text{eff}i}]$ by $G_{\Sigma\max} u_\Sigma(\beta_0, \varepsilon_0) \eta_\Sigma / T_{\text{eff}\Sigma}$ in (3.23). Such an approach is useful for approximate estimation.

MSRSs with Short-Term Spatial Coherence and Spatially Incoherent MSRSs with Information Fusion at the Video Signal Level

Because of nonlinear transformation of received signals in such MSRSs, the coverage determination is a more complicated problem. Noise probability distribution after information integration depends on the number of stations and the noise level in each station, whereas signal-plus-noise probability distribution depends besides that on the mutual correlation of signal amplitude fluctuations and the signal energy at each receiving station.

The true coverage of a MSRS may be determined using results presented in Section 5.5. For a chosen information integration algorithm (see Section 5.3) we can obtain output noise probability distribution and an appropriate threshold according to given false alarm probability, P_{fa}. Then for each point $\mathbf{R}$ of space we can calculate SNRs in receiving stations by (3.1) [taking into account (3.10), if necessary], $q^2_{\text{out}i} = E_i / N_i$, $i = \overline{1, m}$. Now, using formulas from Section 5.5, we determine the output signal-plus-noise probability distribution (with allowance for mutual correlation of signal amplitude fluctuations, if it is not zero) and probability of excess of the established threshold, i.e. detection probability, P_d. If for the considered point $\mathbf{R}$ the obtained detection probability, P_d, is no less than the required probability value, $P_\text{d} \geqslant P_{\text{d req}}$, this point belongs to the coverage of the MSRS. A set of points $\mathbf{R}$ for which $P_\text{d} = P_{\text{d req}}$ determines the boundary surface of the coverage. Though the procedure described seems rather awkward, it is quite realizable with the help of modern computers. However, as was mentioned in Section 1.1, information fusion at the video signal level is seldom used in MSRSs.

MSRSs with Short-Term Spatial Coherence and Spatially Incoherent MSRSs with Information Fusion at the Plot Level

Such information fusion is widely used in practice. In this case a decentralized signal detection is implemented (see Chapter 6). A threshold comparison is performed at the output of each receiving station. In each range (or range sum) resolution cell "one" is generated if the threshold is exceeded by the output noise or signal-plus-noise and "zero" is generated if the excess does not occur. Sequences consisting of ones and zeros are then integrated (combined) for the final decision. Decision rules "*k out of m*" ("coincidence criteria") are usually used. They mean that a signal (a target) is considered as a detected one, if thresholds are exceeded at least in *k* receiving stations (out of the total quantity of *m* stations) in resolution cells corresponding to a common "point" in space.

To determine the coverage of such MSRSs, the allowable output false alarm probability, P_{fa}, and the required output detection probability, $P_{\text{d req}}$, should be fixed. As it will be shown in Section 6.2, partial false alarm probabilities (in different receiving stations) are reasonable to be chosen equal, i.e. $P_{\text{fa}i} = P_{\text{fa}0}$, $i = \overline{1, m}$.

The coverage heavily depends on a decision rule. We consider at first the most "soft" rule "1 out of m" ("OR"). In this case

$$P_{fa0}=1-(1-P_{fa})^{1/m}. \tag{3.24}$$

The relationship between P_d and the probability of threshold exceeding by the sum of signal and noise in each station, P_{di}, $i=\overline{1,m}$, is influenced by the statistical dependence of signal fluctuations at different stations. As it will be shown in Section 4.2, strong correlation is typical for MSRSs with small baselengths (compared with usual target ranges). But for such MSRSs, coverages of all "cells" (radars consisting of the transmitting and the ith receiving stations, $i=\overline{1,m}$) nearly coincide in space[4]. Hence the coverage of such a MSRS has the same form. If the transmitting station is located among the receiving ones, the coverage of the MSRS is close to the coverage of an equivalent (in power characteristics) monostatic radar. MSRSs with large baselengths have more complicated form of coverage and the useful capability of its significant reconfiguration (see Section 1.1). But signal fluctuations at the inputs of different receiving stations in such MSRSs are, as a rule, statistically independent. In this case

$$P_d=1-\prod_{i=1}^{m}(1-P_{di}). \tag{3.25}$$

The coverage of such a MSRS may be determined as follows. Using (3.24) the false alarm probability, P_{fa0}, is to be found for a given output false alarm probability, P_{fa}. For each point of space the SNRs $q^2_{out\,i}$, $i=\overline{1,m}$ should be calculated by the equation (3.1) [taking into account (3.10), if necessary] and the corresponding detection probability, P_{di}, at the given P_{fa0}. Substituting P_{di} in (3.25) yields P_d. Those points for which $P_d \geqslant P_{d\,req}$ belong to the coverage of the MSRS.

Approximate estimate of the coverage may be obtained without calculations described (though they are not difficult for a modern computer) using the coverages of all monostatic and bistatic radars ("cells") constituting the MSRS. Let us plot the coverage boundaries of all those radars individually [using (3.1) and (3.10) if necessary] for such a value of q^2_{out} which at given P_{fa0} yields $P_{di}=P_{d\,req}$. Then the disjunction (the sum) of all these coverages is the lower estimate of the MSRS's coverage. As a matter of fact, we assume that in each part of the coverage boundary, where for the kth "cell" $P_{dk}=P_{d\,req}$, for all other "cells" $P_{di}=0, i=\overline{1,m}, i\neq k$. Under this condition (3.25) yields $P_d=P_{dk}$, that is $P_d=P_{d\,req}$. In reality, in those parts of the coverage boundaries $P_{di}>0$ so that $P_d>P_{d\,req}$, i.e. the true coverage is larger. The accuracy of this estimate increases with the difference between the disjunction (the sum) of all coverages of the "cells", on the one hand, and the conjunction (the product, the intersection) of these coverages, on the other hand.

Example 3.5. In Fig 3.3 horizontal projections (views in plane) of coverages are presented for a MSRS containing one transmitting–receiving station (a monostatic radar) and two remote receiving stations. All the three stations are located in the horizontal plane at the points $(0,0,0), (1,45°,0), (1,135°,0)$ in the spherical coordinate system (r,β,ε) where $r=R/L$. Here R is the range, L is the normalized factor which is assumed to be equal to the distances of the remote stations from the

[4] If their power characteristics are not significantly different.

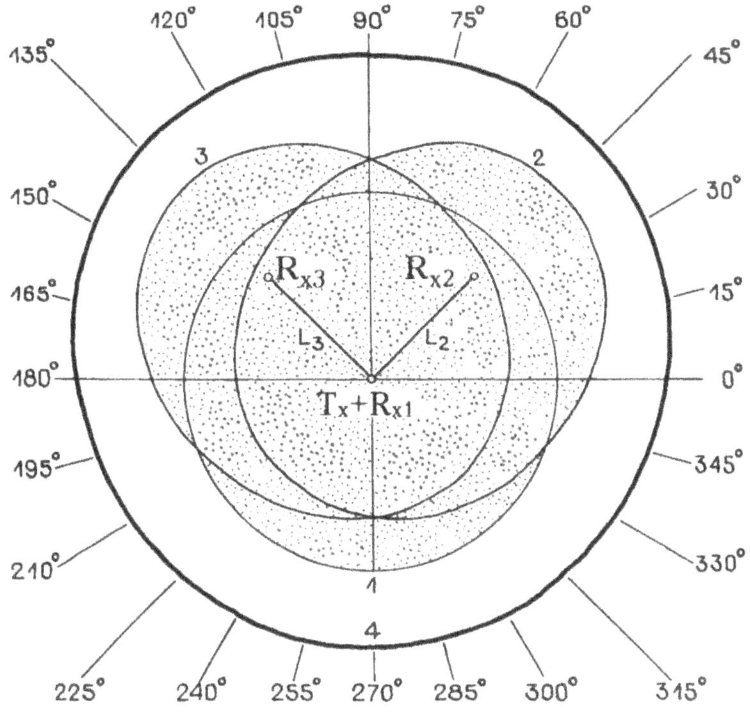

Figure 3.3 Coverages of three radars and the MSRS created on the basis of these radars when the decision rule "1 out of 3" is used. Curve 1: the monostatic radar T_x+R_{x1}; curves 2 and 3: the bistatic radars T_x+R_{x2} and T_x+R_{x3}, respectively; curve 4: MSRS, dotted area corresponds to the lower estimate of the MSRS coverage ($P_d=0.9$, $P_{fa}=10^{-4}$, $P_{d0}=0.9$, $P_{fa0}=3.33\cdot10^{-5}$)

origin. For the radar considered in Example 3.1, $L=158.6$ km. The target elevation angle is equal to 15 . In this figure the boundaries of the individual coverages for the three "cells" of the MSRS – the monostatic and two bistatic radars – are plotted assuming Gaussian complex amplitude fluctuations. The approximate estimate (the sum of all the three coverages) is shown by shaded area. Besides that the bold line is the boundary of the MSRS's true coverage in the case of "1 *out of* 3" decision rule and spatially independent fluctuations of echo complex amplitudes. Calculations have been carried out for $P_{d\,req}=0.9$, $P_{fa}=10^{-4}$ so that $P_{d1}=P_{d2}=P_{d3}=0.9$ and $P_{fa1}=P_{fa2}=P_{fa3}=3.33\times10^{-5}$. The approximate maximum range has turned out to be within 66–82% of the true maximum range.

Let us now consider the most "severe" decision rule "*m out of m*" ("*AND*"). In this case

$$P_{fa0}=(P_{fa})^{1/m};\qquad P_d=\prod_{i=1}^{m}P_{di}. \tag{3.26}$$

The true coverage may be determined as for the decision rule "*OR*" replacing (3.24) and (3.25) by (3.26). Calculating for each point of space $q^2_{out\,i}=E_i/N_i$ by the equation (3.1) [taking into account (3.10), if necessary] and the corresponding probability P_{di} at the fixed probability P_{fa0}, that is determined by (3.26) for the given P_{fa}, we can find P_d using (3.26) again. If $P_d \geq P_{d\,req}$, such a point belongs to the MSRS's coverage.

In this case one may avoid cumbersome computations in a similar manner as for the decision rule "*OR*" if to confine oneself to an approximate lower estimate. The coverage boundaries for all "cells" of a MSRS should be plotted [using (3.1) and (3.10), if necessary] for such q^2_{out} which provides $P_{di} = \sqrt[m]{P_{d\,req}}$ at $P_{fa0} = \sqrt[m]{P_{fa}}$. Then the conjunction (intersection) of all the individual coverages furnishes the approximate lower estimate for the coverage of the MSRS. Here we assume that on each part of the coverage boundary where for the kth "cell" $P_{dk} = \sqrt[m]{P_{d\,req}}$, for the other "cells" $P_{di} = \sqrt[m]{P_{d\,req}}$ too. In reality, these parts of the coverage boundary relating to the kth "cell" lie inside the coverages of other "cells" so that $P_{di} > \sqrt[m]{P_{d\,req}}$, $i = \overline{1, m}$, $i \neq k$. Therefore, the true coverage is larger. The accuracy of the approximate estimate is the higher, the lesser is the difference between the intersection (the product) of individual coverages of all the "cells" of a MSRS, on the one hand, and the sum of those coverages, on the other.

Example 3.6. In Fig. 3.4, for the same MSRS as in Fig. 3.3, horizontal projections of the same individual coverages for all the three radars ("cells") are plotted. The boundary of the true coverage for the MSRS (assuming the "3 *out of* 3" decision rule and Gaussian spatially independent complex amplitude fluctuations) are shown by a bold line. Calculations have been accomplished under the same conditions as for Fig. 3.3: $P_{d\,req} = 0.9$, $P_{fa} = 10^{-4}$ but here $P_{di} = \sqrt[3]{0.9} \approx 0.966$ and $P_{fai} = \sqrt[3]{10^{-4}} \approx 4.64 \times 10^{-2}$, $i = \overline{1, 3}$. The approximate estimate is shown by the shaded area. The

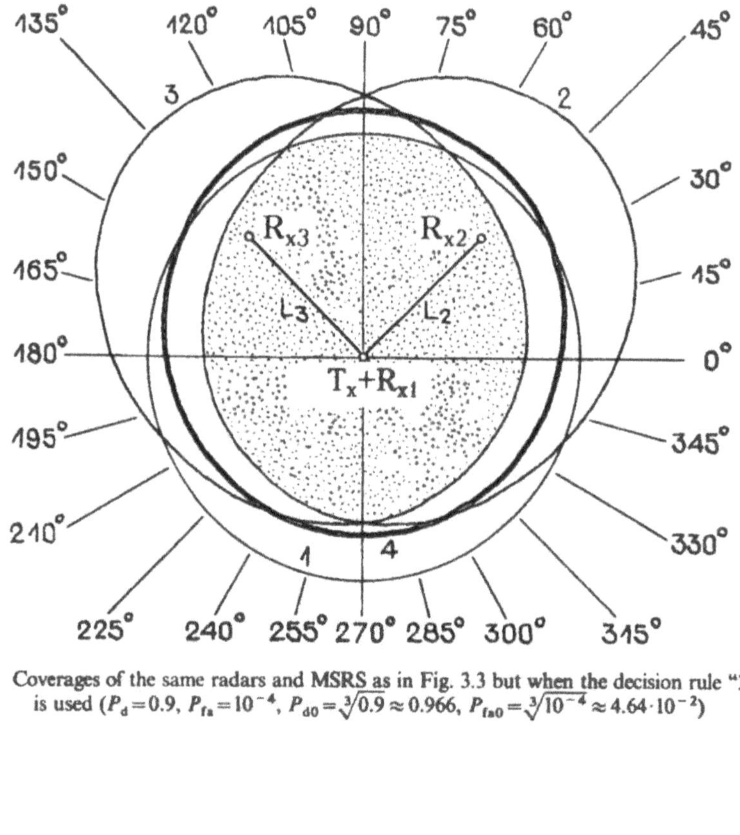

Figure 3.4 Coverages of the same radars and MSRS as in Fig. 3.3 but when the decision rule "3 out of 3" is used ($P_d = 0.9$, $P_{fa} = 10^{-4}$, $P_{d0} = \sqrt[3]{0.9} \approx 0.966$, $P_{fa0} = \sqrt[3]{10^{-4}} \approx 4.64 \cdot 10^{-2}$)

approximate maximum range of the MSRS has turned out to be within 80–92% of the true maximum range.

Comparing Fig. 3.3 with Fig. 3.4 attention should be paid to the fact that when the "severe" decision rule "3 *out of* 3" ("*AND*") is used, the coverage of the same MSRS is less than that for the "soft" decision rule "1 *out of* 3" ("*OR*"). In a similar (though more complicated) manner the MSRS's coverage can be determined for intermediate decision rules of the type "*k out of m*" where $k \neq 1$, $k \neq m$.

Spatially Incoherent MSRSs with Track Fusion

Track fusion is usually used when a MSRS consists of monostatic radars with independent (noncooperative) signal reception (see Section 1.1). Such MSRSs are often called netted radar systems. They may contain heterogeneous information sensors including radars working in different frequency bands.

The coverage of such a system may be divided into two parts. The first part includes regions where a target can be observed by any single radar. The second part contains such regions where each target can be observed at least by two radars. According to the definition of the MSRS (see Section 1.1) only the second part may be considered as a space under responsibility of a MSRS. The system's coverage may be determined by plotting individual coverages for all radars. Boundaries of these coverages should be calculated for such values of q^2_{out} which provide track formation with given confidence. The coverage of the total system is the sum of all individual coverages while the coverage of the MSRS is only a part of that coverage. It includes those regions which are intersections of coverages of at least two radars. It is clear from the above that track fusion is conventionally used for the improvement of tracking accuracy rather than of detection characteristics.

***In this section**, we have introduced the probabilistic definition for the coverage of active MSRSs of different types. For spatially coherent MSRSs we have obtained inequalities* (3.21)–(3.23). *We have described briefly the way of coverage calculations for MSRSs with short-term spatial coherence and for incoherent MSRSs with information fusion at the video signal and the plot levels. For the case of the plot level information fusion, which is widely used in practice, we have considered the "soft" and the "severe" decision rules: "*1 *out of m" ("OR") and "m out of m" ("AND"). We have shown that the same MSRS can have different coverages depending on information fusion decision rules. With the help of numerical Examples* 3.5 *and* 3.6 *(Figs.* 3.3 *and* 3.4, *respectively) we have demonstrated the simplified procedure for approximate coverage calculations.*

3.3. MAXIMUM RANGE AND COVERAGE UNDER NOISE-JAMMING CONDITIONS

As mentioned in Section 1.2, modern radar systems must be able to operate in the presence of external interferences. Military radar systems must also operate in hostile environments, i.e. in the presence of deliberate external interferences called Electronic Countermeasures (ECM). One of the most effective active ECM is the noise-jamming. The main purpose of noise-jamming is to degrade the radar's performance, first of all to reduce its coverage and maximum range. We consider here the influence of noise jammers on the maximum range and coverage of bistatic radars and multistatic radar systems (including multiradar systems).

Maximum Range and Coverage of Bistatic Radars

Let us consider again the ith "cell" of a MSRS: a bistatic radar with the transmitting and the receiving stations located at the points $\mathbf{L}_0$ and $\mathbf{L}_i$, respectively. The total noise-jamming power spectral density (PSD) caused by all jammers at the input of the ith receiver is denoted by N_i^J. The expressions for maximum range and coverage obtained in Section 3.1 remain valid if we replace the thermal noise PSD $N_i = kT_{\text{eff}\,i}$ by the sum $N_i + N_i^J = kT_{\text{eff}\,i} + N_i^J$. For example, instead of (3.1) we can write

$$q_{\text{out}\,i}^2 = \frac{E_i}{N_i + N_i^J} = \frac{P\tau G_0 G_i \sigma_b \lambda^2 \eta_i}{(4\pi)^3 (kT_{\text{eff}\,i} + N_i^J)|\mathbf{R} - \mathbf{L}_0|^2 |\mathbf{R} - \mathbf{L}_i|^2}. \tag{3.27}$$

The increase of the total PSD of the background noise from $kT_{\text{eff}\,i}$ to $kT_{\text{eff}\,i} + N_i^J$ leads to a certain reduction of $q_{\text{out}\,i}^2$ and hence to the decrease of maximum range and coverage ($q_{\text{out}\,i}^2$ must be equal to or greater than q_{out}^2 required for target detection with necessary detection characteristics).

Let M independent "point" jammers operate against the ith receiving station. Let the jammers' positions at the moment t be determined by the radius-vectors $\mathbf{R}_{Jk}$, $k = \overline{1, M}$. Assuming the jamming bandwidths to be much greater than the receiver bandwidth (such jammers are called *barrage jammers* [192]), the kth jammer's transmission in the ith receiving station direction may be characterized by the product of the jamming PSD N_{Jk} and the jammer's antenna gain in this direction G_{Jki} (including all power losses). Then at the input of the ith receiver the total jamming PSD can be written as

$$N_i^J = \sum_{k=1}^{M} \frac{N_{Jk} G_{Jki} G_i g_i(\beta_{Jki}, \varepsilon_{Jki}, \beta_i, \varepsilon_i) \lambda^2 \gamma_i}{(4\pi)^2 |\mathbf{R}_{Jk} - \mathbf{L}_i|^2} \tag{3.28}$$

where $g_i(\beta_{Jki}, \varepsilon_{Jki}, \beta_i, \varepsilon_i)$ is the relative antenna gain factor of the ith receiving station in the kth jammer direction when the mainbeam is pointed at a target; $\beta_{Jki}, \varepsilon_{Jki}, \beta_i, \varepsilon_i$ are the angular coordinates of the kth jammer and of the target, respectively, as they are seen from the ith station; γ_i is the total loss factor ($\gamma_i < 1$) in the ith station for jamming reception.

When all values in (3.28) are known, one can calculate maximum ("burn-through") range and coverage under the noise-jamming condition using the relationships from Section 3.1 (replacing in those relationships $kT_{\text{eff}\,i}$ by $kT_{\text{eff}\,i} + N_i^J$). However, this is not the case in practice. It should be taken into account that since targets as well as jammers usually move in space relative to the radar, the values $\mathbf{R}_{Jk}, \beta_{Jki}, \varepsilon_{Jki}, \beta_i, \varepsilon_i$ and hence g_i change in time. Besides, the product $N_{Jk} G_{Jki}$ may fluctuate. Therefore, $N_i^J = N_i^J(t)$. Apparently, calculations using (3.28), (3.27) for determining maximum range and coverage with the help of equations from Section 3.1 are reasonable only for analysis and simulation of certain situations. For more general estimations, it is desirable to choose some averaged situation. For example, one may assume jammers of approximate equal jamming intensity to be distributed approximately uniformly within the sector $\Delta\beta_{Ji} \times \Delta\varepsilon_{Ji}$ at the range R_i from the ith receiving station. The total PSD of jammers' radiation in the ith station direction is $(N_J G_{Ji})_\Sigma$. Then from (3.28) we have

$$N_i^J = (N_J G_{Ji})_\Sigma G_i \tilde{g}_{iJ} \lambda^2 \gamma_i / (4\pi)^2 R_i^2 \tag{3.29}$$

where

$$\tilde{g}_{iJ} = \frac{1}{\Delta\beta_{Ji}\Delta\varepsilon_{Ji}} \int_{\Delta\beta_{Ji}} \int_{\Delta\varepsilon_{Ji}} g_i(\beta_{Ji}, \varepsilon_{Ji}, \beta_i, \varepsilon_i)\, d\beta_{Ji}\, d\varepsilon_{Ji} \tag{3.30}$$

is the averaged antenna gain factor within the "jamming" sector relative to the gain in mainlobe peak direction.

The receiving antenna gain in the mainbeam direction is usually by 25–40 dB greater than that in sidelobe directions. Hence if jammers operate through the antenna mainbeam, the value of N_i^J in (3.28) and in (3.27) sharply increases. But the situation where many jammers fall simultaneously within the antenna mainbeam is scarcely probable. Therefore, in (3.29) and (3.30) the antenna sidelobe regions are to be considered while the mainbeam sector should be separately analysed.

Using (3.29), (3.30) and the relationships from Section 3.1 one can estimate maximum (burn-through) range and coverage of a bistatic radar (and of a monostatic radar as the particular case) under sidelobe jamming conditions. The effect of mainlobe jamming can be estimated by the direct use of (3.28) and then of relationships from Section 3.1. Expressions for maximum range and coverage calculations of monostatic and bistatic radars under jamming conditions are presented also in [82,192].

Example 3.7. Let $M=5$ jammers with approximately equal jamming power spectral density (PSD) in the ith receiving station direction $N_{Jk}\,G_{Jki}=10$ W/MHz, $k=\overline{1,5}$, operate through antenna sidelobes from the range $R_i=300$ km. The jammers are approximately uniformly distributed in space within the sector $30°\times 5°$ (azimuth × elevation angle). The radar characteristics are assumed to be the same as in Example 3.1 while the averaged relative antenna gain factor within the "jamming" sector [see (3.30)] $\tilde{g}_{iJ}=3.16\times10^{-4}$ (-35 dB) and the total loss factor for jamming reception $\gamma_i=0.3(-5.2\,\text{dB})$. Then substituting $(N_J G_{Ji})_\Sigma=50\times10^{-6}$ W/Hz, $G_i=15\times10^3$, $\tilde{g}_{iJ}=3.16\times10^{-4}$, $\lambda^2=0.01\,\text{m}^2$, $\gamma_i=0.3$ and $R_i=300$ km in (3.29), yields $N_i^J=5\times10^{-20}$ W/Hz. The jamming PSD turns out to be greater than the thermal noise PSD (4×10^{-21} W/Hz, see Example 3.1) by the factor of 12.5. Replacing in (3.3) $kT_{\text{eff}}=4\times10^{-21}$ W/Hz by $(kT_{\text{eff}}+N_i^J)=54\times10^{-21}$ W/Hz, yields $A_i^4=1.27\times10^{20}\,\text{m}^4$. This results in maximum range reduction: $(|\mathbf{R}-\mathbf{L}_0||\mathbf{R}-\mathbf{L}_i|)_{\max}=(R_T R_R)_{\max}=11\,254\,\text{km}^2$ instead of $41\,350\,\text{km}^2$ without jamming. It means that at the point of Cassini's oval where the bisector of the bistatic angle is perpendicular to the baseline ($\theta=0°$) $R_T=R_R\approx106$ km instead of 203 km. It should be noted that decreasing jamming PSD in the receiving station direction by the factor of 10 (i.e. down to 1 W/MHz of each jammer) yields increasing $(R_T R_R)_{\max}$ up to $27\,573\,\text{km}^2$ or R_T, R_R up to 166 km when $R_T=R_R$.

Example 3.8. Let one of 5 jammers from the Example 3.7 (for instance, with the number $k=1$) operate through the receiving antenna mainlobe. The jamming PSD from this single jammer at the receiver may be obtained from (3.28) assuming $M=1$, $g_i(\beta_{J1i}, \varepsilon_{J1i}, \beta_i, \varepsilon_i)\approx1$. Even for $N_{J1}\,G_{J1i}=1$ W/MHz and other conditions being equal, we have $N_i^J=3.16\times10^{-18}$ W/Hz, i.e. greater than kT_{eff} by the factor of 790, so that the maximum detection range becomes too small: $(R_T R_R)_{\max}=1470\,\text{km}^2$ or $R_T, R_R=38.3$ km only (when $R_T=R_R$). Clearly, the remaining 4 jammers operating through sidelobes have not any significance. The jammers cut out narrow wedges in Cassini's ovals when scanning mainlobe is pointed at any jammer.

It can be seen from Examples 3.7 and 3.8 that mainlobe jamming severely constrains radar performance whereas sidelobe jamming has a significant effect on radar performance only if total jamming PSD in the receiving station direction is sufficiently high.

In modern radars adaptive jamming cancellation techniques, based on the space–time processing, are widely used for target detection in a background of jamming (see, e.g. [35,72]). As a rule, these techniques are utilized against sidelobe jamming. For rough estimation of the maximum range and coverage under jamming conditions with adaptive jamming cancellation techniques being used, in the relationships from Section 3.1 after replacing $kT_{\mathrm{eff}\,i}$ by $kT_{\mathrm{eff}\,i}+N_i^J$ the value of N_i^J should be reduced by the expected jamming cancellation coefficient. In the particular case of Example 3.7 sidelobe adaptive jamming cancellers can, under certain conditions, restore nearly the same maximum range and coverage as those of benign environment. For more accurate calculations detailed description and simulation of specific situations are necessary.

Maximum Range and Coverage of MSRSs

For spatially coherent MSRSs (3.20)–(3.23) are valid with $kT_{\mathrm{eff}\,i}+N_i^J$ in place of $kT_{\mathrm{eff}\,i}$ where N_i^J can be obtained from (3.28). As for a single radar, (3.28) presents an "instantaneous" jamming environment and yields the instantaneous value of $N_i^J=N_i^J(t)$, $i=\overline{1,m}$, so that such calculations are reasonable apparently only for analysis and simulation of specific situations. Using (3.29), (3.30) for each receiving station with subsequent substituting N_i^J in (3.20)–(3.23) permits to obtain approximate estimate for the maximum range and coverage of a MSRS under sidelobe jamming conditions. Echoes from targets and interferences from jammers may be considered to be resolved in TDOAs, so that after proper delay equalization and coherent summation of echoes coming from all receiving stations, interfering signals from each jammer are summed incoherently. Adaptive sidelobe jamming cancellation may be taken into account approximately, as for a single radar, by dividing the value of N_i^J obtained from (3.29) by an expected jamming cancellation factor.

When a jammer falls within the antenna mainlobe of any receiving station the noise level in this receiver increases drastically which results in sharp reduction of maximum detection range (see Example 3.8 above). When other receiving stations are not suppressed by mainlobe jamming at that moment, only the "graceful degradation" of the MSRS's performance occurs.

As will be shown in Chapter 8, the joint signal and interference processing in spatially coherent MSRSs and MSRSs with short-term spatial coherence can cancel out external interferences that are correlated at the inputs of receiving stations. This may be used for mainlobe jamming adaptive cancellation without target echo suppression when a jammer is in close proximity to a target or even in the self-screening case where the target and the jammer are colocated [24,65,66]. This important capability is a unique feature of MSRSs. The maximum range under mainlobe jamming conditions with adaptive mainlobe jamming cancellation can be calculated using (3.28) and results from Chapter 8.

For other types of MSRSs where information fusion is implemented at the video signal level, plot level or track level, maximum range and coverage under jamming conditions may be estimated as those for benign environment (see Section 3.2). It is only necessary to replace $kT_{\mathrm{eff}\,i}$ by $kT_{\mathrm{eff}\,i}+N_i^J$ where N_i^J, $i=\overline{1,m}$, can be obtained from (3.28). For sidelobe jamming the averaged value of N_i^J from (3.29), (3.30) may be used. To take approximately into account sidelobe jamming cancellation in each receiving station individually, one should divide the obtained values of N_i^J, $i=\overline{1,m}$, by the expected jamming cancellation factor.

***In this section**, coverages of bistatic radars and of active MSRSs under sidelobe and mainlobe noise-jamming conditions have been considered. We have presented the equations* (3.28) *and* (3.29), (3.30) *for jamming power spectral density (PSD) at the inputs of spatially separated receiving stations. Summation of jamming PSD with thermal noise PSD and then substitution this sum for thermal noise PSD into* (3.27) *and corresponding equations from Section* 3.1 *permits to estimate coverage reduction under jamming conditions. A simple procedure for approximate coverage estimation under sidelobe jamming conditions [with the help of* (3.29) *and* (3.30)] *has been suggested for both bistatic radars and MSRSs. Taking into account much greater antenna gain in mainbeam than in sidelobe directions, we have recommended separate estimation of mainlobe jamming effect on coverage. Numerical examples* 3.7 *and* 3.8 *permit to reveal effects of both sidelobe and mainlobe jamming and illustrate the suggested coverage estimation techniques. These techniques take into account possible sidelobe jamming adaptive cancellation in each receiving station and adaptive mainlobe jamming cancellation by joint coherent jamming processing in MSRSs of certain types (which will be considered in Chapters* 8,9).

3.4. MAXIMUM RANGE AND COVERAGE OF PASSIVE MSRSs

For active MSRSs (and for monostatic radars as well) the most important target characteristic for maximum range and coverage determination is the radar cross section (RCS). For passive systems this role is played by target radiation characteristics. There are variety of forms of radiation sources. They differ in waveform structure, radiation power, bandwidth etc. These may be sources of naturally occurring radiations and sources of deliberate active interferences (active ECM) designed to degrade performance of active radars and MSRSs in active mode. In what follows we shall consider passive MSRSs designed for detection, coordinate measurement and tracking of "point" sources radiating continuous stochastic (noise-like) signals. Mathematical models of such signals will be described in Section 4.1.

The coverage of a *passive* MSRS we define as *the region of space where a radiating source should be positioned so as with chosen signal processing and information fusion algorithms and detection criterion, the source detection probability be no smaller than a required value at a fixed allowable false alarm probability, assuming that the source radiates a stochastic signal with given power spectral density (PSD) in the directions of all receiving stations.*

For active MSRSs we assumed the target bistatic RCS to be equal with respect to all stations (see Section 3.1). Like that we assume the PSD of radiation to be equal in the all stations' directions and constant within the signal bandwidth. Thus, as with active MSRSs a rather artificial target (here a radiation source) is assumed. However, if sufficient *a priori* information is available, one may, of course, take into account characteristics of the actual radiation source including radiation PSD differences in the directions of different receiving stations, variations of PSD within the bandwidth etc.

The most widely practised are passive MSRSs where the triangulation or hyperbolic methods (and their combination) are used for the location of radiation sources. Let us consider maximum ranges and coverages for such MSRSs.

MSRSs Using Triangulation

Three space coordinates of a signal source are determined by the intersection of the source bearings from spatially separated stations so that a "cell" of such a MSRS is a receiving station. As will be shown in Sections 7.1–7.3, detection probability of a stochastic signal in a background of the self-noise of the ith station ($i=\overline{1,m}$) at given false alarm probability is determined by the power SNR at the receiver, q_{Si}^2, and the accumulation (integration) factor, $n_i=\Delta f_{Si}T_{Si}$, where $\Delta f_{Si}, T_{Si}$ are the processed signal bandwidth and duration, respectively. For the receiving station located at the point that is determined by the radius-vector $\mathbf{L}_i$, we have

$$q_{Si}^2=\frac{(NG)G_i\lambda^2\gamma_i}{(4\pi)^2kT_{eff\,i}|\mathbf{R}-\mathbf{L}_i|^2},\quad i=\overline{1,m} \tag{3.31}$$

where (NG) is the signal PSD in the directions of all stations; $\mathbf{R}$ is the radius-vector of the signal source; γ_i is the total loss factor for the stochastic signal reception and processing; $\lambda, k, T_{eff\,i}$ and G_i are the same as in (3.1).

Signal reception from the source of interest often occurs in a background of interferences (noise jamming) from other sources. In this case the signal-to-interference-plus-noise ratio (SINR), q_{SIi}^2, is to be considered:

$$q_{SIi}^2=\frac{(NG)G_i\lambda^2\gamma_i}{(4\pi)^2(kT_{eff\,i}+N_i^J)|\mathbf{R}-\mathbf{L}_i|^2},\quad i=\overline{1,m} \tag{3.32}$$

where the total jamming PSD N_i^J is determined in (3.28) or (3.29), (3.30). If the sidelobe jamming cancellation is used N_i^J should be divided by the expected jamming cancellation factor.

If the antenna gains in the mainbeam peak directions are not constant within the region of responsibility, G_i in (3.31) and (3.32) should be replaced by $G_{i\,max}u_i(\beta_i,\varepsilon_i)$ according to (3.10).

For fixed accumulation factor, n_i, the maximum detection range and coverage for the ith station can be obtained from (3.31), (3.32):

$$|\mathbf{R}-\mathbf{L}_i|^2\leqslant D_{1i}^2,\qquad [|\mathbf{R}-\mathbf{L}_i|^2/u_i(\beta_i,\varepsilon_i)]\leqslant D_{2i}^2. \tag{3.33}$$

If external interferences (jamming) are absent,

$$D_{1i}^2=\frac{(NG)G_i\lambda^2\gamma_i}{(4\pi)^2kT_{eff\,i}q_i^2},\qquad D_{2i}^2=\frac{(NG)G_{i\,max}\lambda^2\gamma_i}{(4\pi)^2kT_{eff\,i}q_i^2}, \tag{3.34}$$

where q_i^2 is the value of q_{Si}^2 which is necessary to obtain the required detection characteristics.

Under jamming conditions with total jamming PSD N_i^J (taking into account jamming cancellation techniques if it is used), $kT_{eff\,i}$ in (3.34) should be replaced by $kT_{eff\,i}+N_i^J$.

It follows from (3.33) that when the antenna mainbeam gains are constant, the coverage turns out to be a sphere with the centre at the station location point. When the antenna mainbeam gains vary within the coverage, its form is no longer a sphere and maximum range $|\mathbf{R}-\mathbf{L}_i|_{max}$ becomes dependent on the direction of signal arrival. In the spherical coordinate system with the origin in a conventional "centre" of the MSRS, dividing both sides of the inequality by L_i^2 and introducing the "relative range" $r_i=R/L_i$, we obtain from (3.33), (3.4), (3.5)

$$1+r_i^2-2r_i[\cos\varepsilon\cos\varepsilon_{ai}\cos(\beta-\beta_{ai})+\sin\varepsilon\sin\varepsilon_{ai}]\leqslant D_{1i}^2/L_i^2. \tag{3.35}$$

For MSRSs with varying antenna mainbeam gains within the coverage region, assuming the approximation of (3.14) is valid and using (3.16) we have

$$\{1+r_i^2-2r_i[\cos\varepsilon\cos\varepsilon_{ai}\cos(\beta-\beta_{ai})+\sin\varepsilon\sin\varepsilon_{ai}]\}^{3/2}$$
$$\times\{r_i[\cos\varepsilon\cos\varepsilon_{vi}\cos(\beta-\beta_{vi})+\sin\varepsilon\sin\varepsilon_{vi}]$$
$$-[\cos\varepsilon_{ai}\cos\varepsilon_{vi}\cos(\beta_{ai}-\beta_{vi})+\sin\varepsilon_{ai}\sin\varepsilon_{vi}]\}^{-1}\leqslant D_{2i}^2/L_i^2. \tag{3.36}$$

When each receiving station can measure both two angle coordinates of a radiation source, then for the source location in space (i.e. for all the three space coordinates determination) it is sufficient to detect signals from this source at least by two stations. If there are several sources, especially closely spaced, the problem of data association arises, i.e. the problem from which radiation source, if any, a particular measurement from a particular station originates (see Section 15.4). For solving this problem, additional measurements by one or two stations may be required. Therefore, as with active MSRSs, the detection decision rules of "*k out of m*" type are used in passive MSRSs. In this case m is the total number of stations and $k=2$–4. Since TDOAs of signals and interferences at the inputs of different stations are usually much greater than the correlation intervals, threshold exceedings in different stations may be considered as statistically independent events. In this case maximum detection range and coverage may be determined as for an active MSRS with information fusion at the plot level (see Section 3.2). SNR or SINR should be obtained from (3.31)–(3.36).

For coverage calculations the value k of a decision rule "*k out of m*" should be chosen, the allowable false alarm probability, P_{fa}, and the required detection probability, $P_{d\,req}$, are to be specified. Assuming partial false alarm probabilities to be equal in all stations $P_{fai}=P_{fa0}$ (see Section 6.2) and estimating SNR value from (3.31) for each point of space [taking into account (3.10), if necessary] one can calculate partial detection probabilities, P_{di}, $i=\overline{1,m}$ (see Sections 7.3, 7.4), and then the output detection probability, P_d, corresponding to the decision rule chosen (see Sections 6.1, 6.2). Those points for which $P_d \geqslant P_{d\,req}$, belong to the coverage of the MSRS. In a similar manner the MSRS's coverage under jamming conditions can be determined if (3.32) is used instead of (3.31).

As with an active MSRS, the approximate estimate of maximum range and coverage for a passive MSRS can be obtained in a form of the combination of individual station coverages.

Example 3.9. In Fig. 3.5 the horizontal projection of the coverage of the passive triangulation MSRS containing three equal receiving stations is shown. These stations are located at the same points as those of the active MSRS in Figs. 3.3 and 3.4 (Section 3.2). The radiation source elevation angle $\varepsilon=15^\circ$. The output probabilities for the "3 *out of* 3" decision rule have been chosen again, $P_{d\,req}=0.9$, $P_{fa}=10^{-4}$. The individual coverages of the three receiving stations are plotted for $P_{di}=\sqrt[3]{0.9}\approx 0.966$ and $P_{fai}=\sqrt[3]{10^{-4}}\approx 4.64\times10^{-2}$, $i=\overline{1,3}$. The boundary of the true coverage is shown by a bold line, the approximate lower estimate is shaded. The approximate maximum range amounts to 84–97% of the true maximum range.

MSRSs Using the Hyperbolic Method

Since TDOAs measurements are based on mutual correlation processing of signals received by pairs of stations (Sections 1.2, 11.1–11.3), the "cell" of such MSRSs is a

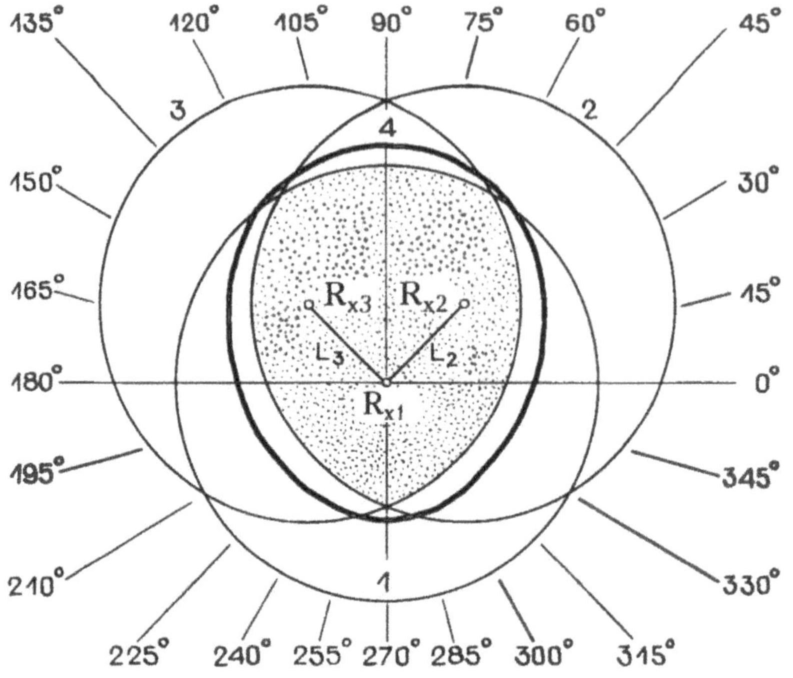

Figure 3.5 Coverages of three receiving stations (curve 1: R_{x1}, curve 2: R_{x2}, curve 3: R_{x3}) and of the triangulation passive MSRS consisting of these stations when the decision rule "3 out of 3" is used (curve 4). The dotted area corresponds to the lower estimate of the MSRS coverage ($P_d=0.9$, $P_{fa}=10^{-4}$, $P_{d0}=\sqrt[3]{0.9}\approx 0.966$, $P_{fa0}=\sqrt[3]{10^{-4}}\approx 4.64\cdot 10^{-2}$)

pair of receiving stations. It will be shown in Sections 7.3, 7.4 that the principal detection parameter for an arbitrary pair of stations with the numbers i,k ($i,k=\overline{1,m}$) is the signal-to-noise ratio (SNR) at the correlator output, $q^2_{\text{out cor }ik}$

$$q^2_{\text{out cor }ik}=q^2_{Si}q^2_{Sk}n/(1+q^2_{Si})(1+q^2_{Sk}) \tag{3.37}$$

where q^2_{Si}, q^2_{Sk} are the power SNR at the receivers (at the correlator inputs); $n=\Delta f_S T_S$ is the signal accumulation factor (assumed equal for all stations). For usual values $n=10^3$–10^4 the required high detection probabilities can be achieved at $q^2_{Si}\ll 1$, $i=\overline{1,m}$ (see Fig. 7.4), so that (3.37) may be simplified

$$q^2_{\text{out cor }ik}\approx q^2_{Si}q^2_{Sk}n,\quad i,k=\overline{1,m},\ i\neq k. \tag{3.38}$$

The quantity q^2_{Si}, $i=\overline{1,m}$, has been determined by (3.31). When antenna mainbeam gains are dependent on mainbeam pointing directions, (3.10) is to be substituted in (3.31). Under jamming conditions q^2_{Si}, q^2_{Sk} in (3.37), (3.38) should be replaced by q^2_{SIi}, q^2_{SIk} in accordance with (3.32). Substituting (3.31) in (3.38) shows that at the fixed value of $n=\Delta f_S T_S$ the maximum range and coverage are determined by the inequality

$$|\mathbf{R}-\mathbf{L}_i|^2|\mathbf{R}-\mathbf{L}_k|^2\leqslant F^4_{1ik} \tag{3.39}$$

or, taking into account (3.10)

$$|\mathbf{R}-\mathbf{L}_i|^2|\mathbf{R}-\mathbf{L}_k|^2/u_i(\beta_i,\varepsilon_i)u_k(\beta_k,\varepsilon_k)\leqslant F^4_{2ik} \tag{3.40}$$

where in benign environments

$$F^4_{1ik}=\frac{(NG)^2G_iG_k\lambda^4\gamma_i\gamma_k}{(4\pi)^4k^2T_{\mathrm{eff}\,i}T_{\mathrm{eff}\,k}q^2_{\mathrm{out\,cor}\,ik}};\qquad F^4_{2ik}=\frac{(NG)^2G_{i\,\mathrm{max}}G_{k\,\mathrm{max}}\lambda^4\gamma_i\gamma_k}{(4\pi)^4k^2T_{\mathrm{eff}\,i}T_{\mathrm{eff}\,k}q^2_{\mathrm{out\,cor}\,ik}} \tag{3.41}$$

and under jamming conditions

$$F^4_{1ik}=\frac{(NG)^2G_iG_k\lambda^4\gamma_i\gamma_k}{(4\pi)^4(kT_{\mathrm{eff}\,i}+N^J_i)(kT_{\mathrm{eff}\,k}+N^J_k)q^2_{\mathrm{out\,cor}\,ik}};\qquad F^4_{2ik}=\frac{(NG)^2G_{i\,\mathrm{max}}G_{k\,\mathrm{max}}\lambda^4\gamma_i\gamma_k}{(4\pi)^4(kT_{\mathrm{eff}\,i}+N^J_i)(kT_{\mathrm{eff}\,k}+N^J_k)q^2_{\mathrm{out\,cor}\,ik}}. \tag{3.42}$$

In (3.41) and (3.42) $q^2_{\mathrm{out\,cor}\,ik}$ is the minimum value of output SNR or SINR affording required detection characteristics.

Excluding constants in the right sides, (3.39) is the same as (3.2) while (3.40) coincides with (3.12) and (3.11). Thus replacing the constants A^4_i in (3.2) by F^4_{1ik} or $\tilde{A}^4_i$ in (3.12) by F^4_{2ik}, we can say that maximum range and coverage of a pair stations of a hyperbolic passive MSRS with correlation signal processing (when input signals are weak, $q^2_{Si}\ll 1$, $i=\overline{1,m}$) are determined by the same inequalities as for an active bistatic radar. Obviously, it is valid for the inequalities (3.6)–(3.8) and (3.17)–(3.19) expressed in the spherical coordinate system. In particular, when antenna mainbeam gains are not dependent on mainbeam pointing directions, then coverage area boundaries in any plane passing through both stations are Cassini's ovals (see Fig. 3.1).

Maximum range and coverage of a MSRS containing several "cells" discussed above (i.e. pairs of stations with mutual correlation signal processing), we shall here estimate for the information fusion at the plot level with the decision rule of "*k out of m*" type. To obtain all the three space coordinates of a radiation source at least three linearly independent TDOA measurements are necessary. When several closely spaced sources are observed, one or two redundant measurements may be required for correct data association. Hence the value of k in the decision rule "*k out of m*" is usually chosen to be equal 3–5.

When the output false alarm probability, P_{fa}, required output detection probability, $P_{\mathrm{d\,req}}$, and chosen decision rule are specified, maximum range and coverage calculations are similar to those for a triangulation MSRS. For each point of space $\mathbf{R}=(R,\beta,\varepsilon)$ one can estimate q^2_{Si} from (3.31) taking into account (3.10), if necessary, and then $q^2_{\mathrm{out\,cor}\,ik}$ from (3.38). Assuming the partial false alarm probabilities at the correlator output of all pair of stations to be the same ($P_{\mathrm{fa}ik}=P_{\mathrm{fa}0}$, see Section 6.1) and using formulas from Section 7.4, one can obtain detection probability, $P_{\mathrm{d}ik}$. As will be shown in Section 7.2, threshold exceedings in different pairs of stations under the condition $q^2_{Si}\ll 1$ may be considered as statistically independent events. The output detection probability, P_{d}, can be calculated according to the decision rule chosen and using the obtained values of $P_{\mathrm{d}ik}$ (see Section 6.1). The set of points $\mathbf{R}=(R,\beta,\varepsilon)$ for which $P_{\mathrm{d}}\geqslant P_{\mathrm{d\,req}}$ form the coverage of the MSRS.

The approximate lower estimate for maximum detection range and coverage may be obtained combining individual coverages of each pair of stations.

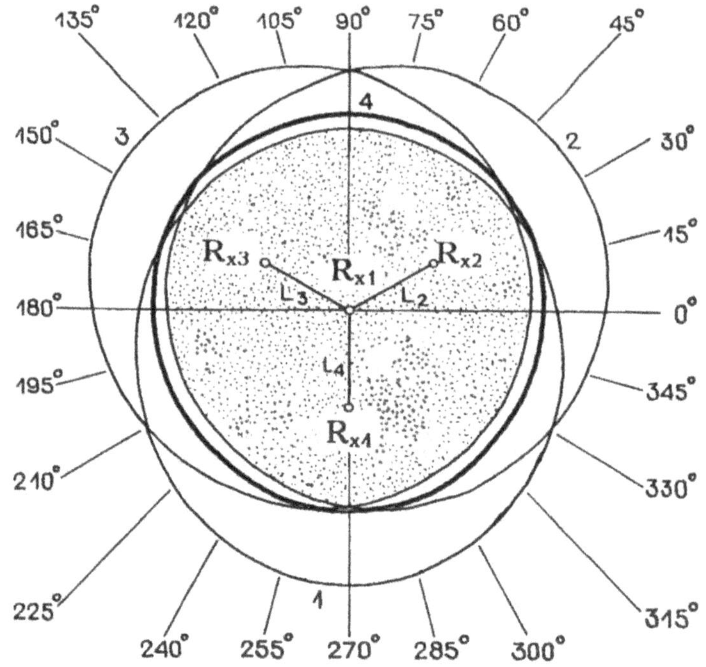

Figure 3.6 Coverages of three pairs of receiving stations using correlation signal processing (curve 1: R_{x1}, R_{x4}; curve 2: R_{x1}, R_{x2}; curve 3: R_{x1}, R_{x3}) and of the hyperbolic MSRS consisting of these pairs of stations when the decision rule "3 out of 3" is used (curve 4). The dotted area corresponds to the lower estimate of the MSRS coverage (P_d and P_{fa} are the same as in Fig. 3.5)

Example 3.10. In Fig. 3.6 the horizontal projection of the coverage of the passive hyperbolic MSRS containing four equal receiving stations is presented. The same spherical coordinate system is used as in Figs. 3.3–3.5. The stations located in the form of a regular three-pointed star. Again the source elevation angle $\varepsilon = 15°$. As with the triangulation MSRS in Fig. 3.5, the "3 *out of* 3" decision rule has been chosen. In Fig. 3.6 the horizontal projections of individual coverages for three pairs of stations and the approximate lower coverage estimate of the MSRS are shown. The approximate maximum range amounts to 92–98% of the true maximum range.

***In this section**, we have defined the coverage of a passive MSRS. The most practically important passive MSRSs using triangulation and hyperbolic methods have been considered. For triangulation passive MSRSs we have presented the general expressions* (3.33) *and* (3.34) *which determine coverage of one receiving station as a "cell" of such MSRSs, and equations* (3.31) *and* (3.32) *for signal-to-noise ratio (SNR) and signal-to-interference-plus-noise ratio (SINR). The general inequalities* (3.33) *have been specified for the spherical coordinate system by* (3.35) *and* (3.36). *In hyperbolic passive MSRSs a pair of receiving stations is played a role of a "cell" of such MSRSs. The general inequalities* (3.39)–(3.42) *determine coverages of an arbitrary pair of stations and the equations* (3.37), (3.38) *determine the principal detection parameter–SNR at the correlator output of a pair of stations. Resultant coverages of MSRSs of both types have been obtained by using decision rules "k out of m". We have presented numerical examples* 3.9, 3.10 *that illustrate application of this approach and of the simplified procedure for approximate coverage estimation.*

PART 2

Target Detection in MSRSs

4. MODELS FOR SIGNALS AND INTERFERENCES. OPTIMIZATION CRITERIA

4.1. PECULIAR FEATURES OF SIGNAL DETECTION IN MSRSs. SIGNAL AND INTERFERENCE MODEL FORMULATION

The signal detection theory for MSRSs may be considered as a part of the general spatial–temporal signal processing theory. However, most practical results of that general theory relate to processing of plane or spherical waves incident upon receiving antennas of given form and size, mainly upon phased antenna arrays (PAAs) [35,42,52,72]. As can be seen from Section 1.1, MSRSs have a number of peculiar features which are to be taken into account when developing the signal detection and parameter estimation theories. The main features are as follows:

1. The number and arrangement of stations may be arbitrary. Therefore the detection theory should be developed without connection with a specific system configuration.
2. The differences of signal propagation time from each target to spatially separated stations (TDOAs) are, as a rule, much greater than the reciprocals of signal bandwidths. In particular, this does not permit to separate the spatial processing of the sum of echoes and external interferences from the temporal processing (which is usually possible for antenna arrays).
3. A target may be positioned at any range from a MSRS, i.e. not necessarily in the far zone and even in the Freshnel's zone with respect to the whole antenna system of the MSRS but, as a rule, in the far zone of each station's antenna.
4. Mutual correlation of scattered signal fluctuations at the inputs of different stations may widely vary – from complete correlation (as at different elements of usual antenna arrays) to zero correlation (when the stations are separated by sufficiently large baselines).
5. There are some peculiar features depending on the type of a MSRS i.e. on the degree of spatial coherence, on the information integration level, cooperative or autonomous signal reception etc. (see Section 1.1).

Of great importance is the choice of mathematical models for signals and interferences. On the one hand, these models should adequately reflect the principal features of actual signals and interferences. On the other hand, they are to be suitable for deriving sufficiently simple and clear results which could be applied to practice. We have chosen generally accepted models for "useful" signals: *the deterministic signals* (i.e. signals with known waveform without unknown or random parameters, completely known signals), *the quasideterministic signals* (i.e. signals with known waveform but containing unknown or random parameters) and *stochastic signals* (random signals with unknown waveform). Deterministic and quasideterministic signals that will be used for target echoes modelling, will also be called *regular signals* (to emphasize their known waveform).

Of course, echoes from real targets cannot be really deterministic signals. Actual echoes always contain unknown parameters. Measurement (estimation) of them is the ultimate goal of any radar. Besides those informative parameters, received signals usually include some interfering unknown or random parameters that do not contain useful information (noninformative parameters). Nevertheless, deterministic signals are widely used in the radar detection theory as a mathematical model. Employing this model is often the simplest way to reveal some principal relationships that hold for actual signals. Thus the deterministic signal model is warranted and practically convenient.

The classical detection theory usually assumes informative signal parameters, such as time and direction (angle) of arrival (TOA and AOA), Doppler shift etc. to be *a priori* known [45–48,54,72]. The same assumption is exploited in the detection theory for MSRSs. In other words, the detection problem is usually formulated as the statistical decision problem of whether a target is present within the considered Space Resolution Cell (SRC) of the MSRS.[1]

Consider a set of wanted signals (echoes) at the inputs of m spatially separated receiving stations of a MSRS when a target is illuminated by a single transmitting station. In a complex form $\mathbf{S}^*(t)=[S_1^*(t),\ldots S_m^*(t)]$. Here and in the sequel the asterisk means complex conjugation for scalars but complex conjugation and transposition (Hermitian conjugation) for vectors and matrices. The wanted signal at the input of the ith station can be written in the form

$$S_i(t)=a_{si}\exp(-j\varphi_{si})s_0(t-t_{si})\exp[j(\omega_0+\Omega_{si})(t-t_{si})] \tag{4.1}$$

where a_{si} is the r.m.s. (effective) value; φ_{si} is the initial phase; t_{si} is the signal propagation delay; ω_0, Ω_{si} are the carrier frequency and the Doppler frequency shift, respectively; $s_0(t)$ is the normalized complex envelope (waveform), that is

$$\frac{1}{2}\int_{-\infty}^{\infty}|S_i(t)|^2\,dt=\frac{a_{si}^2}{2}\int_{-\infty}^{\infty}|s_0(t-t_{si})|^2\,dt=a_{si}^2T_s=E_i \tag{4.2}$$

where T_s is the signal duration; E_i is the signal energy at the input of the ith station.

It is reasonable to synthesize and analyze detectors for MSRSs rather in the frequency domain than in the time domain (see Section 4.3). The spectrum of the signal of (4.1)

$$\Psi_i(\omega)=a_{si}\exp(-j\varphi_{si})\Psi_0(\omega-\omega_0-\Omega_{si})\exp(-j\omega t_{si}) \tag{4.3}$$

where $\Psi_0(\omega)$ is the Fourier transformation of $s_0(t)$ so that

$$(1/2\pi)\int_{-\infty}^{\infty}|\Psi_0(\omega)|^2\,d\omega=\int_{-\infty}^{\infty}|s_0(t)|^2\,dt=2T_s. \tag{4.4}$$

Sometimes it is necessary to take into account the antenna directivity pattern (ADP) of the receiving station. Then instead of (4.3) we have

$$\Psi_i(\omega)=a_{si}\exp(-j\varphi_{si})g_i(\beta_{si},\varepsilon_{si},\omega)\Psi_0(\omega-\omega_0-\Omega_{si})\exp(-j\omega t_{si}) \tag{4.5}$$

where $g_i(\beta_{si},\varepsilon_{si},\omega)$ is the normalized ADP of the ith station $[g_i(\beta_{si},\varepsilon_{si},\omega)\leqslant 1]$ depending, in general, on frequency; $\beta_{si},\varepsilon_{si}$ are the target angle coordinates with respect to the ith station. Equations (4.1)–(4.5) are also valid for quasideterministic

[1] The effect of a target position uncertainty is considered in Sections 5.6 and 6.3.

signals when some or all parameters are unknown or random. At the same time the complex envelope $s_0(t)$ and hence its spectrum $\Psi_0(\omega)$ are assumed to be known.

Signals transmitted by sources of "noise" radiation and received by passive MSRSs (or by active–passive MSRSs in the passive mode) will be modelled as stochastic signals. The term "stochastic" instead of the simpler term "random" is used here in order to distinguish between these signals and quasideterministic signals with random parameters. Sometimes stochastic signals are called "noise signals". However the term "noise" is usually associated with the term "interference", i.e. with some signal contamination. Since in passive MSRSs these signals are "useful" (wanted signals) it will be more appropriate to call them "stochastic" signals. Such signals may be written in the form of (4.1), (4.3) or (4.5), assuming $s_0(t)$ and hence $\Psi_0(\omega)$ to be realizations of random processes. We assume here and in what follows the stochastic signals to be realizations of *complex, zero-mean, narrowband* ($\Delta\omega \ll \omega_0$ where $\Delta\omega$ is the power spectrum bandwidth), *stationary* within the observation interval $(-T/2, T/2)$ *Gaussian stochastic processes.* Note, however, that statistical coupling between processes at the inputs of different stations are *not necessarily stationary*. The Gaussian stochastic processes may be considered as proper characterization for transmissions of many radiation sources observed by MSRSs including noise-like jammers. This model permits exploiting well developed mathematical techniques. As well known, a multivariable, stationary, zero-mean Gaussian process with zero mean can be completely determined by its correlation (covariance) matrix[2] or by the power spectral density (PSD) matrix.

In writing stochastic signal in the form of (4.1), (4.3) or (4.5) we assume the parameters a_{si}, φ_{si}, $i=\overline{1,m}$, to be constant within the observation time interval. At the same time the complex envelope $s_0(t)$ varies fast so that each observation interval contains many correlation intervals of $s_0(t)$. Such a model is appropriate for signals from many radiation sources observed by a MSRS. For stochastic signals of this type, it is reasonable to introduce the so-called *conditional correlation matrices* at the fixed values of parameters characterizing intensity (power), differential phase shift and (or) time delay. When these values are known, we have a stochastic signal with *the deterministic correlation matrix*, when they are unknown or random, it is a stochastic signal with *the quasideterministic correlation matrix*.

An arbitrary element of the correlation matrix (we omit the term "conditional" for the sake of brevity) for complex stochastic signals at the inputs of m receiving stations of a MSRS can be written as follows

$$B_{ik}(t_1, t_2) = 0.5\,\overline{S_i(t_1) S_k^*(t_2)}, \quad i,k=\overline{1,m}. \tag{4.6}$$

The overbar here and in the sequel denotes the expectation, in this case over the set of complex envelopes $s_0(t)$. The factor 0.5 has been introduced for the variance $B_{ii}(0)$ to coincide with the variance (power) of the real part of a signal, $\mathrm{Re}\, S_i(t)$, i.e. $B_{ii}(0) = 0.5\overline{|S_i(t)|^2} = \overline{[\mathrm{Re}\, S_i(t)]^2}$ since $\overline{[\mathrm{Im}\, S_i(t)]^2} = \overline{[\mathrm{Re}\, S_i(t)]^2}$. Substituting (4.1) into (4.6) and averaging yield

$$B_{ik}(t_1, t_2) = \sqrt{P_{si} P_{sk}}\, \rho_s(t_1 - t_2 + \tau_{sik}) \exp\{j[(\omega_0 + \Omega_{si})(t_1 - t_2) - \Delta\Omega_{sik} t_2 + \omega_0 \tau_{sik} + \Delta\varphi_{sik}]\}. \tag{4.7}$$

[2] For considered stochastic processes and variables with zero mean, correlation functions and matrices coincide with covariance functions and matrices, respectively.

In (4.7) we denote

$$\tau_{sik}=t_{sk}-t_{si};\qquad \Delta\Omega_{sik}=\Omega_{sk}-\Omega_{si};\qquad P_i=a_{si}^2; \tag{4.8}$$

and

$$\rho(t_1-t_2)=0.5\overline{s_0(t_1)s_0^*(t_2)} \tag{4.9}$$

which is the correlation function of the signal complex envelope $s_0(t)$. Doppler phase shifts are included into the differential phase $\Delta\varphi_{sik}$. For target echoes ω_0 is the known carrier frequency of a transmitted signal whereas in (4.7) ω_0 is the conventional (usually unknown) carrier frequency of the stochastic signal from a stationary (with respect to the receiving stations) radiation source. When a radiation source moves, then the frequency $\omega_0+\Omega_{si}$, $i=\overline{1,m}$, is perceived as a carrier frequency. Nevertheless, it is convenient to separate ω_0 from Ω_{si}. Note, that in (4.8) t_{si} is the one-way signal propagation delay (from a source to the ith station) and Ω_{si} is the Doppler frequency caused by the one-way source range variations relative to the ith station. In (4.1) t_{si} is the two-way signal propagation delay (from the transmitting station to the target and then to the ith receiving station) and Ω_{si} is the Doppler frequency caused by the target range sum variations.

The autocorrelation functions in (4.7), i.e. at $i=k$, $i,k=\overline{1,m}$, depend on the difference (t_1-t_2) which means that the signal is stationary. At the same time, the mutual correlation functions, i.e. at $i\neq k$, depend also on the time itself, t_2, which indicates the nonstationary coupling between signals at different stations. When the differential phase shifts accumulated during the observation time interval and caused by the Doppler frequency differences may be neglected, i.e. when $|\Delta\Omega_{sik}T|\ll 2\pi$, then signals at all stations are not only stationary but stationary coupled processes.

To the correlation function (4.7) corresponds the "instantaneous" power spectrum [33]

$$\begin{aligned}\Phi_{ik}(\omega,t)&=\int_{-\infty}^{\infty}B_{ik}(t,t-x)\exp(-j\omega x)\,dx\\ &=(\sqrt{P_{si}P_{sk}}/\Delta f_s)F_s(\omega-\omega_0-\Omega_{sk})\exp[j(\omega\tau_{sik}+\Delta\psi_{sik})]\\ &\quad\times\exp(-j\Delta\Omega_{sik}t)\end{aligned} \tag{4.10}$$

where $\Delta\psi_{sik}=\Delta\varphi_{sik}+\Delta\Omega_{sik}t_{si}=\varphi_{sk}-\varphi_{si}+\Delta\Omega_{sik}t_{si}$; Δf_s is the bandwidth of the signal power spectrum; $F_s(\omega)$ is the dimensionless normalized power spectrum of the complex envelope [see (4.9)]:

$$\begin{aligned}F_s(\omega)&=\Delta f_s\int_{-\infty}^{\infty}\rho_s(x)\exp(-j\omega x)\,dx;\\ (1/2\pi\Delta f_s)&\int_{-\infty}^{\infty}F_s(\omega)\,d\omega=1.\end{aligned} \tag{4.11}$$

If the Doppler frequency differences at different stations may be neglected (i.e. if $|\Delta\Omega_{sik}T|\ll 2\pi$, $i,k=\overline{1,m}$), then

$$\Phi_{ik}(\omega,t)=\Phi_{ik}(\omega)=(\sqrt{P_{si}P_{sk}}/\Delta f_s)F_s(\omega-\omega_0)\exp[j(\omega\tau_{sik}+\Delta\psi_{sik})]. \tag{4.12}$$

Here $\Delta\psi_{sik}=\Delta\varphi_{sik}=\varphi_{sk}-\varphi_{si}$. At $i=k$ equations (4.12) with (4.8) yield the signal power spectrum at an arbitrary receiving station.

We shall model *self-noises of receiver* systems as *white, zero-mean, stationary Gaussian processes mutually independent at different stations.* Then

$$B_{ik}(t_1,t_2)=0.5\overline{n_i(t_1)n_k^*(t_2)}=\sqrt{N_iN_k}\,\delta_{ik}\delta(t_1-t_2);$$
$$\Phi_{ik}(\omega)=\sqrt{N_iN_k}\,\delta_{ik},\quad i,k=\overline{1,m} \tag{4.13}$$

where N_i is the one-sided noise power spectral density at the ith station; δ_{ik} is Kronecker's symbol ($\delta_{ik}=1$ if $i=k$, $\delta_{ik}=0$ if $i\neq k$); $\delta(t)$ is Dirac's delta-function.

Apart from self-noises, there may be external interferences at the inputs of receiving stations. We shall consider mainly *noise-like interferences from "point" sources* (first of all, noise-like jamming). Such interferences can be modelled as stochastic signals, i.e. as *complex, zero-mean, narrowband, stationary Gaussian processes.* Each radiation source may appear as a source of wanted stochastic signal (when it is a target of interest) and as a jammer (when it interferes in observing other targets). Once we use the same models, all the considerations and equations concerning stochastic signals are applicable to interferences. In order to distinguish between stochastic signals and interferences (e.g., jamming), we shall omit the subscript "s" in expressions related to the latter case.

For monostatic radars, correlation of echo complex amplitude fluctuations in time (the temporal correlation) is of great significance. For MSRSs apart from temporal correlation, an important role is played by *the interstation (spatial) correlation of complex amplitude fluctuations* (after signal delay equalization). Just as with the temporal correlation, one should consider first of all the extreme situations. Being completely spatially correlated, the complex amplitudes at the inputs of different receiving stations are tightly coupled and fluctuate simultaneously. Since the initial phases are tightly coupled as well, such signals may naturally be called "spatially coherent signals". In particular, deterministic signals and stochastic signals with deterministic correlation matrices are spatially coherent. On the contrary, when there is no correlation between complex amplitudes at the inputs of different stations (or, in the more general case, complex amplitudes are statistically independent), such signals may be called "spatially incoherent signals". Quasideterministic signals and stochastic signals with quasideterministic correlation matrix can be spatially incoherent but in other situations they can be spatially coherent. Of course, the intermediate cases of partial temporal or/and spatial correlation are possible.

It is important to stress that spatial (interstation) correlation of complex amplitudes and their temporal correlation (in each station individually) are independent signal features. This is easy to clarify by way of an example. Let us consider a pulse train, i.e. a sequence of pulses. When an incoherent pulse train is transmitted there is no correlation between echo pulses, i.e. the temporal correlation is absent. At the same time, echo pulses of the same number at different stations may be perfectly spatially correlated in a MSRS with sufficiently short baselines. In the opposite situation, when a coherent pulse train of not too long duration is transmitted, complex amplitude of echo pulses may be completely correlated at each station individually (the perfect temporal correlation). If, however, the baselengths between receiving stations are sufficiently large, the complex amplitudes of echo pulses at any pair of stations are mutually independent (the absence of the spatial correlation).

***In this section**, we have considered peculiar features of the signal detection problem in MSRSs and have introduced some mathematical models for signals, receiving system (thermal) self-noises and external interferences. For wanted signal we use the deterministic, quasideterministic and stochastic signal models. All these signals may be described by expressions* (4.1) *in the time domain and* (4.3), (4.5) *in the frequency domain. Receiver self-noises and external interferences as well as stochastic signals are modelled by complex Gaussian random processes with zero mean. Correlation (covariance) matrix of stochastic signals and PSD matrix are determined by the equations* (4.7) *and* (4.10), *respectively. They take into account that statistical coupling between these signals at the inputs of stations may be nonstationary because of different signal Doppler frequencies with respect to different stations. When these Doppler differences may be neglected, we have stationary multidimensional processes with the PSD matrix* (4.12). *The same equations determine external noise-like interferences (e.g. jamming). Receiver self-noises are assumed to be white with PSD matrix* (4.13). *The introduced models will be exploited in the subsequent Sections and Chapters of the book for synthesis and analysis of different signal processing and information fusion algorithms.*

4.2. SPACE–TIME CORRELATION FUNCTION OF SCATTERED SIGNAL FLUCTUATIONS IN MSRSs

As discussed in Section 2.1, the signal scattered by a target whose dimensions are much larger than the wavelength of illuminating signal, may be considered as the sum of partial signals scattered by FSs ("Flare Spots", scattering centres) of the target. Even small target vibration and rotation about its centre of mass lead to significant changes in distances (as compared with the wavelength $\lambda = 2\pi c/\omega_0$) from different FSs to transmitting and receiving stations, and hence cause sharp phase variations in those partial signals. This results in complex amplitude fluctuations, i.e. in real amplitude and phase fluctuations, of the total echo from the target. These fluctuations are usually modelled as random variables. Most widely used are the Rayleigh probability distribution for real amplitude fluctuations and uniform probability distribution within the limits $(-\pi, \pi)$ for phase fluctuations (Swerling models 1 and 2 [203]) which correspond to the zero-mean Gaussian distribution for complex amplitudes. Such a model describes quite satisfactory signal fluctuations from target that can be represented as a collection of several (the more, the better) FSs of approximately equal intensity.

Complete characterization of the Gaussian complex amplitude fluctuation model can be yielded by the space–time correlation matrix[3]. This matrix may also be useful for some other probability distributions of fluctuations. In this section we shall derive approximate expressions determining the space–time correlation matrix.

Obviously, nonzero interstation correlation is possible only when the angle between directions from the target to receiving stations is small, while nonzero temporal correlation can be maintained during the small time intervals. It should be taken into account in problem formulation.

Let us assume a moving target of large dimensions (as compared with λ) to be perceived by all the stations as the same system of N FSs rigidly fixed relative to the target's centre of mass. Each FS can be characterized by the radius-vector $\boldsymbol{\rho}_i$, velocity vector $\mathbf{v}_i$, describing motion relative to the target centre of mass, and complex amplitude b_i of scattered signal ($i = \overline{1, N}$). Due to the small angle between

[3] See the footnote on page 91.

directions from the target to any pair of receiving stations (see above), the differences in $|b_i|$ with respect to different stations may be neglected.

Let us now consider a transmitting, T_x, and a pair of receiving stations, R_{x1}, R_{x2} (see Fig. 4.1). The unit vectors directed from the target centre of mass (the origin of the coordinate system) to the stations are assumed constant during the considered time interval. We denote these unit vectors by $\mathbf{r}_0, \mathbf{r}_1, \mathbf{r}_2$, and the station distances from the origin at the initial time instant by R_0, R_1, R_2. Let the illuminating signal be a pair of short pulses with the interval T_0 between them. Additionally we assume that different FSs cannot be resolved in range and angle coordinates and that phase shifts accumulated during τ_p (each pulse duration) and changes in the time interval T_0 in scattered signals, both caused by the Doppler effect, may be neglected. Under these conditions the signal received by the kth station ($k=1,2$) can be written in the form

$$S_k(t)=\sum_{m=1}^{2}\sum_{i=1}^{N} b_i s_0[t-(R_0+R_k)/c-(m-1)T_0]\exp\{j\omega_0\{t-(R_0+R_k)/c$$
$$-[\boldsymbol{\rho}_i+\mathbf{v}_i(m-1)T_0](\mathbf{r}_0+\mathbf{r}_k)/c\}\}. \qquad (4.14)$$

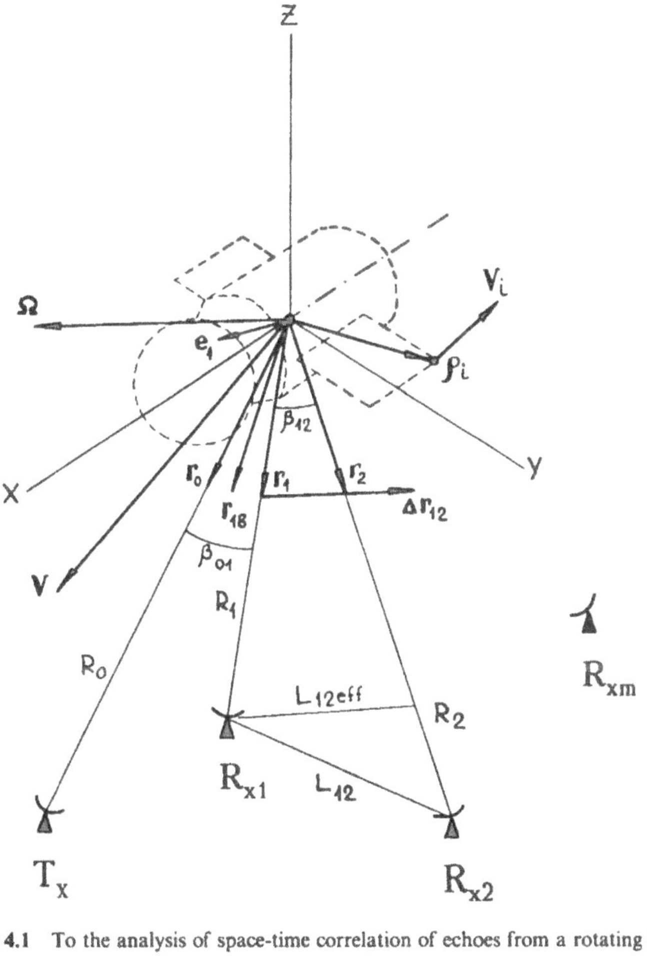

Figure 4.1 To the analysis of space-time correlation of echoes from a rotating target

If the target centre of mass is moving with the velocity $\mathbf{v}$ relative to the some "centre" of the MSRS, the additional factor $\exp[j(\omega_0/c)\mathbf{v}(\mathbf{r}_0+\mathbf{r}_k)(m-1)T_0]$ should be inserted into (4.14). From (4.14) we can write the complex amplitude of the mth pulse at the kth station as

$$A_{km}=A_k[(m-1)T_0]=\sum_{i=1}^{N} b_i \exp\{j(\omega_0/c)[\boldsymbol{\rho}_i+\mathbf{v}_i(m-1)T_0](\mathbf{r}_0+\mathbf{r}_k)\}. \tag{4.15}$$

The space–time correlation matrix of complex amplitudes can be presented in a partitioned form

$$\mathbf{B}=\begin{pmatrix}\mathbf{B}_{11} & \mathbf{B}_{12}\\ \mathbf{B}_{21} & \mathbf{B}_{22}\end{pmatrix} \tag{4.16}$$

where each element (block) is the 2×2 matrix: $\mathbf{B}_{kl}=\|B_{klmn}\|=0.5\,\|\overline{A_{km}A_{ln}^*}\|$, $k,l,m,n=1,2$. Averaging in (4.16) should be carried out over random initial aspects of the target.

In many real situations the target initial random aspect can vary within narrow limits about its known mean value. The key assumption which will permit obtaining the required relationships is that *the initial random aspect of the target is a result of a preceding rotation from the mean position through a small random angle $\delta\theta$*. Under this assumption the position of the ith FS at the initial time instant ($t=0$) may be approximately presented in the form ($i=\overline{1,N}$)

$$\boldsymbol{\rho}_i(0)=\boldsymbol{\rho}_{i0}+\Delta\mathbf{v}_i\tau=\boldsymbol{\rho}_{i0}+(\boldsymbol{\omega}\times\boldsymbol{\rho}_{i0})\tau=\boldsymbol{\rho}_{i0}+\boldsymbol{\delta\theta}\times\boldsymbol{\rho}_{i0} \tag{4.17}$$

where $\boldsymbol{\rho}_{i0}$ determines the mean position of the ith FS with respect to the target centre of mass; $\boldsymbol{\delta\theta}=\boldsymbol{\omega}\tau$ is the random angle by which the ith FS rotates with the angle velocity $\boldsymbol{\omega}$ during the time interval τ. The sign "$\times$" means a vector product. It is convenient to consider this angle as a vector directed along $\boldsymbol{\omega}$ and to express it as the sum of the Cartesian components

$$\boldsymbol{\delta\theta}=\boldsymbol{\delta\theta}_x+\boldsymbol{\delta\theta}_y+\boldsymbol{\delta\theta}_z \tag{4.18}$$

i.e. of the angles of rotation about the fixed axes X,Y and Z with the origin at the target centre of mass. It may often be considered that $|\boldsymbol{\delta\theta}_x|=\delta\theta_x$, $|\boldsymbol{\delta\theta}_y|=\delta\theta_y$, $|\boldsymbol{\delta\theta}_z|=\delta\theta_z$ are mutually independent random variables distributed uniformly within known limits $\pm\Delta\theta_x/2$, $\pm\Delta\theta_y/2$, $\pm\Delta\theta_z/2$, respectively. It can be shown that the relative error of the linear approximation (4.17), i.e. the excess of $|\boldsymbol{\rho}_i(0)|$ over $|\boldsymbol{\rho}_{i0}|$, is of the order of $0.5\delta\theta^2$.

Now let us assume that during the observation time interval T_0 the target rotates about its centre of mass by a small angle ΩT_0 with the constant angle velocity $\boldsymbol{\Omega}$. Then being turned the radius-vector of the ith FS can be approximately presented in the form

$$\boldsymbol{\rho}_i(T_0)=\boldsymbol{\rho}_i+\mathbf{v}_iT_0=\boldsymbol{\rho}_i+(\boldsymbol{\Omega}\times\boldsymbol{\rho}_i)T_0 \tag{4.19}$$

where $\boldsymbol{\rho}_i=\boldsymbol{\rho}_i(0)$ is the random initial position determined in (4.17). The relative error in length of the vector $\boldsymbol{\rho}_i(T_0)$ in (4.19) is of the order of $0.5(\Omega T_0)^2$.

Substituting (4.19) and (4.17) in (4.15) permits to obtain B_{klmn} in (4.16), i.e. all elements of the space–time correlation matrix of interest. After simple but cumbersome transformations, taking into consideration (4.18) and averaging over

$\delta\theta_x$, $\delta\theta_y$, and $\delta\theta_z$ with uniform distributions within $\pm\Delta\theta_x/2, \pm\Delta\theta_y/2, \pm\Delta\theta_z/2$, respectively, we have

$$B_{klmn}=\sum_{i=1}^{N}\sum_{p=1}^{N} 0.5b_i b_p^* \,\mathrm{sinc}(\mathbf{C}_{ip}^{klmn}\Delta\boldsymbol{\theta}_x)\mathrm{sinc}(\mathbf{C}_{ip}^{klmn}\Delta\boldsymbol{\theta}_y)$$
$$\times\,\mathrm{sinc}(\mathbf{C}_{ip}^{klmn}\Delta\boldsymbol{\theta}_z)\exp(jd_{ip}^{klmn}),\quad k,l,m,n=1,2 \tag{4.20}$$

where $\Delta\boldsymbol{\theta}_x, \Delta\boldsymbol{\theta}_y, \Delta\boldsymbol{\theta}_z$ are vectors directed along $\boldsymbol{\omega}_x, \boldsymbol{\omega}_y, \boldsymbol{\omega}_z$, respectively;

$$\begin{aligned}
&\mathrm{sinc}(\alpha)=(\sin\alpha)/\alpha;\\
&\mathbf{C}_{ip}^{klmn}=(\pi/\lambda)\{\boldsymbol{\rho}_{i0}\times[(\mathbf{r}_0+\mathbf{r}_k)+(\mathbf{r}_0+\mathbf{r}_k)\times\boldsymbol{\Omega}(m-1)T_0]\\
&\qquad-\boldsymbol{\rho}_{p0}\times[(\mathbf{r}_0+\mathbf{r}_l)+(\mathbf{r}_0+\mathbf{r}_l)\times\boldsymbol{\Omega}(n-1)T_0]\};\\
&d_{ip}^{klmn}=(2\pi/\lambda)\{[\boldsymbol{\rho}_{i0}+(\boldsymbol{\Omega}\times\boldsymbol{\rho}_{i0})(m-1)T_0](\mathbf{r}_0+\mathbf{r}_k)\\
&\qquad-[\boldsymbol{\rho}_{p0}+(\boldsymbol{\Omega}\times\boldsymbol{\rho}_{p0})(n-1)T_0](\mathbf{r}_0+\mathbf{r}_l)\}.
\end{aligned} \tag{4.21}$$

If the target dimensions are much larger than the wavelength λ, for all i,p, $i\neq p$ in (4.20) at least one of the inequalities $\mathbf{C}_{ip}^{klmn}\Delta\boldsymbol{\theta}_x\geqslant\pi$, $\mathbf{C}_{ip}^{klmn}\Delta\boldsymbol{\theta}_y\geqslant\pi$, $\mathbf{C}_{ip}^{klmn}\Delta\boldsymbol{\theta}_z\geqslant\pi$, holds so that corresponding factor sinc(·) is close to zero. Therefore all terms at $i\neq p$ in (4.20) may be neglected. Besides that, we take into account (see Fig. 4.1) that $\mathbf{r}_0+\mathbf{r}_k=2\cos(\beta_{0k}/2)\mathbf{r}_{kb}$; $\mathbf{r}_0-\mathbf{r}_k=2\sin(\beta_{12}/2)\Delta\mathbf{r}_{12}$ where β_{0k} is the bistatic angle between $\mathbf{r}_0$ and $\mathbf{r}_k$; β_{12} is the angle between $\mathbf{r}_1$ and $\mathbf{r}_2$; $\mathbf{r}_{kb}$ and $\Delta\mathbf{r}_{12}$ are the unit vectors directed along the bisector of β_{0k} and from $\mathbf{r}_1$ to $\mathbf{r}_2$ perpendicular to the bisector of β_{12}, respectively. It is worth reminding that the subscript "0" relates to the transmitting station whereas the subscript "k" denotes the number of the receiving station, so that $k=1$ or $k=2$.

Let us denote $P_i=0.5|b_i|^2$ and omit the second subscripts i in $\mathbf{C}_{ii}^{klmn}$ and d_{ii}^{klmn}. Then (4.20) and (4.21) take the form ($k,l=1,2$; $m,n=1,2$).

$$B_{klmn}=\sum_{i=1}^{N} P_i\,\mathrm{sinc}(\mathbf{C}_i^{klmn}\Delta\boldsymbol{\theta}_x)\mathrm{sinc}(\mathbf{C}_i^{klmn}\Delta\boldsymbol{\theta}_y)\,\mathrm{sinc}(\mathbf{C}_i^{klmn}\Delta\boldsymbol{\theta}_z)\exp(jd_i^{klmn}); \tag{4.22}$$

$$\begin{aligned}
&\mathbf{C}_i^{12mn}=(2\pi/\lambda)[-(\boldsymbol{\rho}_{i0}\times\Delta\mathbf{r}_{12})\sin(\beta_{12}/2)+(m-n)T_0(\boldsymbol{\rho}_{i0}\times\mathbf{r}_{1b}\times\boldsymbol{\Omega})\cos(\beta_{01}/2)\\
&\qquad-(n-1)T_0(\boldsymbol{\rho}_{i0}\times\Delta\mathbf{r}_{12}\times\boldsymbol{\Omega})\sin(\beta_{12}/2)];\\
&d_i^{12mn}=(4\pi/\lambda)[-\boldsymbol{\rho}_{i0}\Delta\mathbf{r}_{12}\sin(\beta_{12}/2)+(m-n)T_0\,\boldsymbol{\rho}_{i0}(\mathbf{r}_{1b}\times\boldsymbol{\Omega})\cos(\beta_{01}/2)\\
&\qquad-(n-1)T_0\boldsymbol{\rho}_{i0}(\Delta\mathbf{r}_{12}\times\boldsymbol{\Omega})\sin(\beta_{12}/2)].
\end{aligned} \tag{4.23}$$

Equations for $\mathbf{C}_i^{11mn}$ and d_i^{11mn} can be obtained from (4.23) at $\beta_{12}=0$, for $\mathbf{C}_i^{22mn}$ and d_i^{22mn} also from (4.23) at $\beta_{12}=0$ and replacing $\mathbf{r}_{1b}$ by $\mathbf{r}_{2b}$ and $\beta_{01}$by β_{02}. To obtain $\mathbf{C}_i^{21mn}$ and d_i^{21mn} one should reverse the sign of the first term and replace n by m in the third term.

It is seen from (4.22) and (4.23) that B_{11mn} and B_{22mn} depend on the time difference $(m-n)T_0$ whereas B_{12mn} and B_{21mn} depend on the current time $(n-1)T_0$ or $(m-1)T_0$. It means that signal complex amplitudes are stationary at each station individually, but are nonstationary coupled at difference stations. However, we are interested in the region of mainlobe of the correlation functions. This region corresponds to small baselengths between stations and small angles of the target rotation. Then we may assume $\beta_{12} \approx L_{12\,\mathrm{eff}}/R_1 \ll 1$ and $\Omega T_0 \ll 1$ where $L_{12\,\mathrm{eff}}$ is the effective baselength between receiving stations, i.e. the projection of L_{12} on $\Delta\mathbf{r}_{12}$ (see Fig. 4.1). Under these conditions the last terms in (4.23) may be neglected and thus nonstationary coupling between signal complex amplitudes at different stations may be ignored. Substituting into (4.23) $2\sin(\beta_{12}/2) \approx L_{12\,\mathrm{eff}}/R_1$; $\mathbf{r}_{1\mathrm{b}} \times \boldsymbol{\Omega} = \Omega \sin\varepsilon_{1\mathrm{b}}\mathbf{e}_1$; $\mathbf{r}_{2\mathrm{b}} \times \boldsymbol{\Omega} = \Omega \sin\varepsilon_{2\mathrm{b}}\mathbf{e}_2$ (where $\varepsilon_{1\mathrm{b}}, \varepsilon_{2\mathrm{b}}$ are the angles between $\mathbf{r}_{1\mathrm{b}}$ and $\boldsymbol{\Omega}$ and between $\mathbf{r}_{2\mathrm{b}}$ and $\boldsymbol{\Omega}$, respectively; $\mathbf{e}_1$ and $\mathbf{e}_2$ are the unit vectors in the directions determined by the vector products $\mathbf{r}_{1\mathrm{b}} \times \boldsymbol{\Omega}$ and $\mathbf{r}_{2\mathrm{b}} \times \boldsymbol{\Omega}$, respectively) yields

$$\begin{aligned}
\mathbf{C}_i^{12mn} &= -\mathbf{C}_i^{21mn} = (\pi/\lambda)[-(L_{12\,\mathrm{eff}}/R_1)(\boldsymbol{\rho}_{i0} \times \Delta\mathbf{r}_{12}) \\
&\quad + 2\Omega \sin\varepsilon_{1\mathrm{b}}(m-n)T_0(\boldsymbol{\rho}_{i0} \times \mathbf{e}_1)\cos(\beta_{01}/2)]; \\
d_i^{12mn} &= -d_i^{21mn} = (2\pi/\lambda)[-(L_{12\,\mathrm{eff}}/R_1)\boldsymbol{\rho}_{i0}\Delta\mathbf{r}_{12} \\
&\quad + 2\Omega \sin\varepsilon_{1\mathrm{b}}(m-n)T_0\boldsymbol{\rho}_{i0}\mathbf{e}_1\cos(\beta_{01}/2)].
\end{aligned} \tag{4.24}$$

Within the frames of the same approximations ($L_{12\,\mathrm{eff}}/R_1 \ll 1$ and $\Omega T_0 \ll 1$)

$$\begin{aligned}
\mathbf{C}_i^{11mn} &= \mathbf{C}_i^{22mn} = (2\pi/\lambda)[\Omega \sin\varepsilon_{1\mathrm{b}}(m-n)T_0(\boldsymbol{\rho}_{i0} \times \mathbf{e}_1)\cos(\beta_{01}/2)] \\
&= (2\pi/\lambda)[\Omega \sin\varepsilon_{2\mathrm{b}}(m-n)T_0(\boldsymbol{\rho}_{i0} \times \mathbf{e}_2)\cos(\beta_{02}/2)]; \\
d_i^{11mn} &= d_i^{22mn} = (4\pi/\lambda)[\Omega \sin\varepsilon_{1\mathrm{b}}(m-n)T_0\boldsymbol{\rho}_{i0}\mathbf{e}_1\cos(\beta_{01}/2)] \\
&= (4\pi/\lambda)[\Omega \sin\varepsilon_{2\mathrm{b}}(m-n)T_0\boldsymbol{\rho}_{i0}\mathbf{e}_2\cos(\beta_{02}/2)].
\end{aligned} \tag{4.25}$$

Equations (4.22), (4.24) determine the interstation *space–time correlation* (*covariance*) *of the signal complex amplitudes* at the inputs of an arbitrary pair of receiving stations, whereas equations (4.22), (4.25) determine the interpulse temporal correlation of these amplitudes at the input of each station. (We recall that complex amplitudes are modelled here as zero-mean Gaussian variables.) It is seen that under conditions considered the matrix $\mathbf{B}$ in (4.16) is the Hermitian one ($\mathbf{B}_{21} = \mathbf{B}_{12}^*$) and, furthermore, $\mathbf{B}_{11} = \mathbf{B}_{22}$. The influence of both receiver spatial separation and target rotation is additive.

Spatial and temporal correlation of signal complex amplitudes may be considered separately. The spatial correlation is the interstation correlation of echoes caused by a single illuminated pulse at the random target aspect. *The spatial correlation* (*covariance*) *function* can be obtained from (4.22), (4.24) at $m=n$ or from (4.22) and more general equation (4.23) at $\Omega=0$. Then

$$\mathbf{C}_i^{12} = -\pi L_{12\,\mathrm{eff}}(\boldsymbol{\rho}_{i0} \times \Delta\mathbf{r}_{12})/\lambda R_1; \qquad d_i^{12} = -2\pi L_{12\,\mathrm{eff}}\boldsymbol{\rho}_{i0}\Delta\mathbf{r}_{12}/\lambda R_1. \tag{4.26}$$

It can be seen from (4.22) and (4.26) that the spatial correlation decreases with the increase of the ratios of the effective baselength, $L_{12\,\mathrm{eff}}$, to the target range, R_1, and of the target dimensions to the wavelength. Here the target dimension along $\Delta\mathbf{r}_{12}$ (in the effective baseline direction) is important. To reveal a role of the ranges of aspect uncertainty, $\Delta\theta_x, \Delta\theta_y, \Delta\theta_z$, let, for instance, the X axis run along $\Delta\mathbf{r}_{12}$. Denoting the unit vectors of such a right-handed Cartesian coordinate system by $\mathbf{x}_0, \mathbf{y}_0, \mathbf{z}_0$, yields instead of (4.22) and (4.26)

$$B_{12} = \sum_{i=1}^{N} P_i \operatorname{sinc}(\pi L_{12\,\mathrm{eff}} \Delta\theta_y \boldsymbol{\rho}_{i0} \mathbf{z}_0 / \lambda R_1) \operatorname{sinc}(\pi L_{12\,\mathrm{eff}} \Delta\theta_z \boldsymbol{\rho}_{i0} \mathbf{y}_0 / \lambda R_1)$$

$$\times \exp(-\mathrm{j}2\pi L_{12\,\mathrm{eff}} \boldsymbol{\rho}_{i0} \mathbf{x}_0 / \lambda R_1). \qquad (4.27)$$

Random rotations about the X axis have no effect on the spatial correlation (B_{12} does not depend on $\Delta\theta_x$), since the projections of all FSs' radius-vectors on $\mathbf{x}_0$ remain constant. Increasing the aspect uncertainty limits $\Delta\theta_y$ and $\Delta\theta_z$ reduces the spatial correlation because of broadening the range of possible variations of those projections.

The *temporal correlation* of signal complex amplitudes should be considered at each station individually. In this case, in general, the condition $\beta_{12} \approx L_{12\,\mathrm{eff}}/R_1 \ll 1$ is not required, the spatial (interstation) correlation may be equal to zero and the target may be perceived by the two receiving stations even as different systems of FSs. Those systems of FSs are required only to be the same for each station individually and do not change during the target rotation by a considered small angle. Under these conditions $\mathbf{B}_{11} \neq \mathbf{B}_{22}$. If we denote by $\boldsymbol{\rho}_{iok}$ the radius-vector of the ith FS at the mean target aspect angle as seen from the kth receiving station ($k = 1, 2$), the signal intensity from this FS by P_{ik} and the total number of FSs by N_k, then

$$\mathbf{B}_{kkmn} = \sum_{i=1}^{N_k} P_{ik} \operatorname{sinc}(\mathbf{C}_i^{kkmn} \Delta\boldsymbol{\theta}_x) \operatorname{sinc}(\mathbf{C}_i^{kkmn} \Delta\boldsymbol{\theta}_y) \operatorname{sinc}(\mathbf{C}_i^{kkmn} \Delta\boldsymbol{\theta}_z) \exp(\mathrm{j} d_i^{kkmn}), \qquad (4.28)$$

where

$$\mathbf{C}_i^{kkmn} = (2\pi/\lambda)[\Omega \sin \varepsilon_{kb} (m-n) T_0 (\boldsymbol{\rho}_{iok} \times \mathbf{e}_k) \cos(\beta_{0k}/2)];$$

$$d_i^{kkmn} = (4\pi/\lambda)[\Omega \sin \varepsilon_{kb} (m-n) T_0 \boldsymbol{\rho}_{iok} \mathbf{e}_k \cos(\beta_{0k}/2)]. \qquad (4.29)$$

It is seen that the temporal correlation is determined, firstly, by the target rotation angle during the time interval T_0 about the axis perpendicular to $\mathbf{r}_{kb}$ and lying in the plane $(\mathbf{r}_{kb}, \boldsymbol{\Omega})$, i.e. $\Omega \sin \varepsilon_{kb} T_0$, and, secondly, by the target dimension to the wavelength ratio. Unlike the spatial correlation, here the target dimension along the unit vector $\mathbf{e}_k$ is important. This unit vector lies in the rotation plane and is perpendicular to $\mathbf{r}_{kb}$. The role of $\mathbf{e}_k$ here is similar to the role of $\Delta\mathbf{r}_{12}$ for spatial correlation. Any rotation about $\mathbf{r}_{kb}$ (when $\mathbf{r}_{kb} \times \boldsymbol{\Omega} = \mathbf{0}$) does not lead to decorrelation because in this case no variations occur in Doppler frequencies of scattered signals.

To reveal the effect of the ranges of target aspect uncertainty, $\Delta\theta_x, \Delta\theta_y, \Delta\theta_z$, on the temporal correlation, let the X axis be directed along the unit vector $\mathbf{e}_k$ so that $\mathbf{x}_0 = \mathbf{e}_k$. Then taking into account that $m, n = 1, 2$, we have in such a Cartesian right-handed coordinate system

$$B_{kk}(0) = \sum_{i=1}^{N_k} P_{ik};$$

$$B_{kk}(\pm T_0) = \sum_{i=1}^{N_k} P_{ik} \operatorname{sinc}[2\pi\Omega \sin \varepsilon_{kb} T_0 \Delta\theta_y \boldsymbol{\rho}_{iok} \mathbf{z}_0 \cos(\beta_{0k}/2)]$$

$$\times \operatorname{sinc}[2\pi\Omega \sin \varepsilon_{kb} T_0 \Delta\theta_z \boldsymbol{\rho}_{iok} \mathbf{y}_0 \cos(\beta_{0k}/2)]$$

$$\times \exp[\pm \mathrm{j}4\pi\Omega \sin \varepsilon_{kb} T_0 \boldsymbol{\rho}_{iok} \mathbf{x}_0 \cos(\beta_{0k}/2)]; \qquad k = 1, 2. \qquad (4.30)$$

It is seen that aspect uncertainties caused by target random rotation about the X axis produce no effect on correlation [$B_{kk}(\pm T_0)$ is independent of $\Delta\theta_x$]. Broadening of uncertainty regions $\Delta\theta_y$ and $\Delta\theta_z$ decreases the correlation because it extends variation ranges for the target projections on the X axis which is parallel to $\mathbf{x}=\mathbf{e}_k$.

Example 4.1. Let us consider a MSRS containing a single transmitting station with a pair of receiving stations, and a target which may be represented as a collection of the seven ($N=7$) scattering centres (flare spots, FSs), see Fig. 4.2. The Cartesian coordinates, x_i, y_i, z_i, of these FSs and amplitudes, $|b_i|$, $(i=\overline{1,7})$ of signals scattered by the FSs are shown in Table 4.1. The unit vectors $\mathbf{r}_1$ and $\mathbf{r}_2$ are directed to the receiving stations Rx_1 and Rx_2 respectively. The curves of the moduli and phases of the spatial and temporal correlation coefficients are plotted in Figs 4.3 and 4.4. These curves have been obtained by echo simulation at random target aspects and by calculations using equations (4.22), (4.26), (4.28), (4.29). It is seen that in the mainlobe regions the approximate equations derived above for the moduli of the correlation coefficients exhibit satisfactory accuracy though present slightly reduced values (by 7–15%). The values of phases are in worse agreement (especially for the temporal correlation coefficient) but all phases in the mainlobe region do not exceed 10–30°.

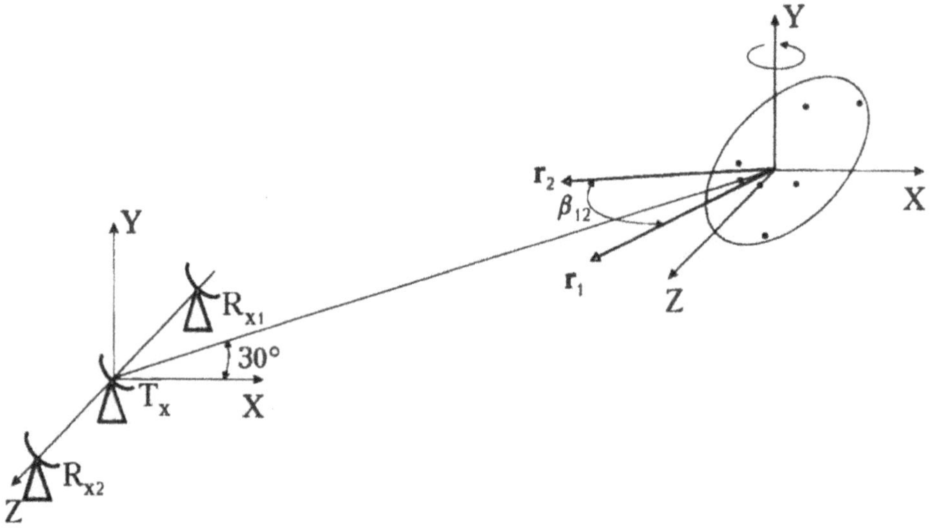

Figure 4.2 The variant of MSRS and target configuration used for calculations and simulation

Table 4.1

Coordinates and intensities of FSs	Number of FSs						
	1	2	3	4	5	6	7
x	-10λ	0	-7λ	7λ	0	-7λ	0
y	0	-10λ	-7λ	7λ	10λ	7λ	-10λ
z	20λ	20λ	0	-20λ	-10λ	10λ	-4λ
$\|b\|$	1.0	1.5	2.0	1.0	1.5	2.0	1.0

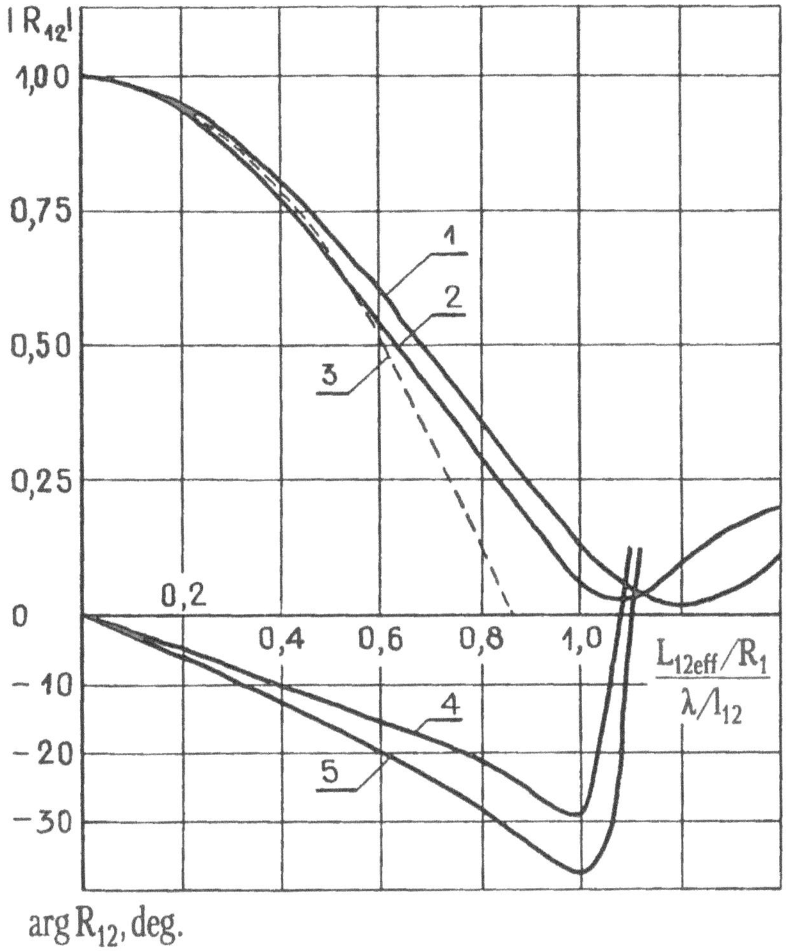

Figure 4.3 Modulus (curves 1, 2, 3) and phase (curves 4, 5) of the target echo spatial correlation coefficient for the conditions of Fig. 4.2. Curves 1, 4: simulation results; curves 2, 5: calculation results according to equations (4.22), (4.26); curve 3: calculation results according to equation (4.38) (l_{12} = the target dimension along the unit vector $\Delta\mathbf{r}_{12}$)

Unfortunately, the number and locations of FSs, amplitudes and phases of scattered signals are usually *a priori* unknown in real situations. Therefore direct exploiting of the obtained relationships is seldom possible. However, in many cases it is sufficient to estimate the boundaries of *the regions of high and low fluctuation correlation*. This may be done having only little target information.

Region of High Correlation

Each factor in (4.22), (4.28) can be expanded into the Taylor series at zero point and truncated including square terms. After some transformations one can obtain approximate equations for modulus and phase of both the spatial (interstation),

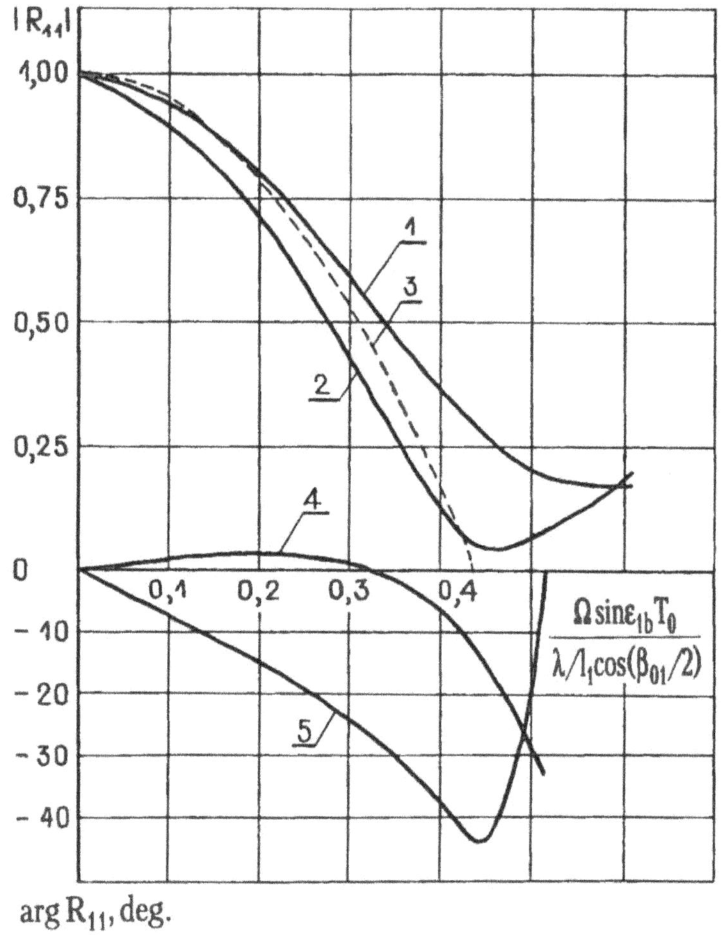

Figure 4.4 Modulus (curves 1, 2, 3) and phase (curves 4, 5) of the target echo temporal correlation coefficient for the conditions of Fig. 4.2. Curves 1, 4: simulation results; curves 2, 5: calculation results according to equations (4.28), (4.29); curve 3: calculation results according to equation (4.39) (l_1 = the target dimension along the unit vector $\mathbf{e}_1$)

R_{12}, and temporal (interpulse), R_{kk}, $k=1,2$, correlation coefficients:

$$|R_{12}| \approx 1-(2\pi^2/\lambda^2)(L_{12\,\mathrm{eff}}^2/R_1^2)(J_{12}-M_{12}^2); \tag{4.31}$$

$$\arg R_{12} \approx -\operatorname{arctg}\frac{(2\pi/\lambda)(L_{12\,\mathrm{eff}}/R_1)M_{12}}{1-(2\pi^2/\lambda^2)(L_{12\,\mathrm{eff}}^2/R_1^2)J_{12}}; \tag{4.32}$$

$$|R_{kk}| \approx 1-(8\pi^2/\lambda^2)\cos^2(\beta_{0k}/2)\sin^2\varepsilon_{kb}\Omega^2T_0^2(J_{kk}-M_{kk}^2); \tag{4.33}$$

$$\arg R_{kk} \approx -\operatorname{arctg}\frac{(4\pi/\lambda)\cos(\beta_{0k}/2)\sin\varepsilon_{kb}\Omega T_0 M_{kk}}{1-(8\pi^2/\lambda^2)\cos^2(\beta_{0k}/2)\sin^2\varepsilon_{kb}\Omega^2T_0^2J_{kk}}. \tag{4.34}$$

Equations (4.31) and (4.32) have been derived for a system comprising two arbitrary receiving stations and a single transmitting station (see Fig. 4.1). For a system consisting of two arbitrary monostatic radars (two transmitting–receiving stations) with independent signal reception (see Section 1.1) and the same working frequency, in the numerator of (4.31) and in the denominator of (4.32) $2\pi^2$ must be replaced by $8\pi^2$ whereas in the numerator of (4.32) 2π must be replaced by 4π. This is a result of two-way signal propagation from each radar.

The quantities M_{12}, M_{11}, M_{22} and J_{12}, J_{11}, J_{22} may be treated as the first and second generalized moments of the target FS distributions in corresponding directions:

$$M_{12}=\left(\sum_{i=1}^{N} P_i \boldsymbol{\rho}_{i0}\Delta\mathbf{r}_{12}\right)\bigg/\sum_{i=1}^{N} P_i;\quad M_{11}=\left(\sum_{i=1}^{N} P_i \boldsymbol{\rho}_{i0}\mathbf{e}_1\right)\bigg/\sum_{i=1}^{N} P_i;$$
$$M_{22}=\left(\sum_{i=1}^{N} P_i \boldsymbol{\rho}_{i0}\mathbf{e}_2\right)\bigg/\sum_{i=1}^{N} P_i; \tag{4.35}$$

$$J_{12}=\sum_{i=1}^{N} P_i\{(\boldsymbol{\rho}_{i0}\Delta\mathbf{r}_{12})^2+\sigma_x^2[\boldsymbol{\rho}_{i0}(\Delta\mathbf{r}_{12}\times\mathbf{x}_0)]^2+\sigma_y^2[\boldsymbol{\rho}_{i0}(\Delta\mathbf{r}_{12}\times\mathbf{y}_0)]^2$$
$$+\sigma_z^2[\boldsymbol{\rho}_{i0}(\Delta\mathbf{r}_{12}\times\mathbf{z}_0)]^2\}\bigg/\sum_{i=1}^{N} P_i. \tag{4.36}$$

To obtain J_{11} and J_{22} one should replace in (4.36) $\Delta\mathbf{r}_{12}$ by $\mathbf{e}_1$ or $\mathbf{e}_2$. For the sake of simplicity we ignore here possible differences in N, P_i and $\boldsymbol{\rho}_{i0}$ with respect to different stations. In (4.36) $\mathbf{x}_0, \mathbf{y}_0, \mathbf{z}_0$ are the unit vectors of the X, Y, Z axes; $\sigma_x^2, \sigma_y^2, \sigma_z^2$ are the variances of the random angles $\delta\theta_x, \delta\theta_y, \delta\theta_z$ determining the random target aspect. Expressions (4.31)–(4.36) are valid for arbitrary probability distribution of mutually independent variables $\delta\theta_x, \delta\theta_y, \delta\theta_z$ with zero mean and limited variance.

It follows from (4.32), (4.34) and (4.35) that the phases of the correlation coefficients are equal to zero, when distributions of FSs in the $\Delta\mathbf{r}_{12}$, $\mathbf{e}_1$ and $\mathbf{e}_2$ directions, respectively, are symmetric about the target centre of mass. Therefore, for $\arg R_{12}$ or $\arg R_{kk}$ $(k=1,2)$ evaluation, the degree of asymmetry of those distributions is to be known. Sometimes certain target features are *a priori* known which permits obtaining at least coarse estimates of the first moments M_{12}, M_{11} and M_{22}. More often *a priori* target information does not suffice to estimate phases even roughly. However, these phases are small in the regions of high correlation (see, for instance, Figs 4.3 and 4.4) and may often be neglected. Moduli of the correlation coefficients are the most significant. For approximate calculations M_{12} and M_{kk} in (4.31) and (4.33) may be neglected too assuming FS distributions to be close to symmetric. Besides that we may take into account that aspect uncertainty regions are usually not large so that $\sigma_x^2 \ll 1, \sigma_y^2 \ll 1, \sigma_z^2 \ll 1$. The main role in J_{12}, J_{11} and J_{22} is played by the first terms in braces which are determined by the target dimensions along $\Delta\mathbf{r}_{12}$, $\mathbf{e}_1$ and $\mathbf{e}_2$, respectively. We denote them l_{12}, l_1 and l_2. Then J_{12}, J_{11} and J_{22} may be written in the form

$$J_{12}=\kappa_{12}^2(l_{12}^2/12);\quad J_{11}=\kappa_{11}^2(l_1^2/12);\quad J_{22}=\kappa_{22}^2(l_2^2/12) \tag{4.37}$$

where κ_{12}, κ_{11} and κ_{22} are the dimensionless coefficients determined by the difference between actual FS distribution on the target and the uniform distribution when

$\kappa_{12}=\kappa_{11}=\kappa_{22}=1$. Substituting (4.37) in (4.31), (4.33) yields

$$|R_{12}|\approx|R_{21}|\approx 1-(\pi^2\kappa_{12}^2 l_{12}^2/6\lambda^2)(L_{12\,\mathrm{eff}}^2/R_1^2); \quad (4.38)$$

$$|R_{kk}|\approx 1-(2\pi^2\kappa_{kk}^2 l_k^2/3\lambda^2)\cos^2(\beta_{0k}/2)\sin^2\varepsilon_{kb}\Omega^2 T_0^2. \quad (4.39)$$

Like (4.31), equation (4.38) is valid for a system containing two arbitrary receiving stations and a single transmitting station. For a system containing two monostatic radars with the same working frequency and independent signal reception instead of $(\pi^2\kappa_{12}^2 l_{12}^2/6\lambda^2)$ should be $(2\pi^2\kappa_{12}^2 l_{12}^2/3\lambda^2)$.

From the last equations we can obtain the maximum allowable station separation and the angle of target rotation during the time interval T_0 which do not lead to correlation reduction below the prescribed level. If, for instance, this level is set at $|R_{12}|\geqslant 0.95$ and $|R_{kk}|\geqslant 0.95$, $(k=1,2)$ we have

$$(L_{12\,\mathrm{eff}}/R_1)\leqslant 0.17\lambda/\kappa_{12}l_{12} \quad \text{(for two receiving stations);}$$

$$(L_{12\,\mathrm{eff}}/R_1)\leqslant 0.085\lambda/\kappa_{12}l_{12} \quad \text{(for two monostatic radars);} \quad (4.40)$$

$$\Omega\sin\varepsilon_{kb}T_0\leqslant 0.085\lambda/\kappa_{kk}l_k\cos(\beta_{0k}/2).$$

As was mentioned above, for the uniform FS distribution $\kappa_{12}=\kappa_{kk}=1$. For the triangle and "reverse triangle" distributions (in the latter case, with the linear increase from the midpoint to the edges) $\kappa_{12}=\kappa_{kk}=0.7$–$1.2$. Assuming these values as the extreme ones and substituting in (4.40) yields

$$(L_{12\,\mathrm{eff}}/R_1)\leqslant(0.14\text{–}0.24)\lambda/l_{12} \quad \text{(for two receiving stations);}$$

$$(L_{12\,\mathrm{eff}}/R_1)\leqslant(0.07\text{–}0.12)\lambda/l_{12} \quad \text{(for two monostatic radars);} \quad (4.41)$$

$$\Omega\sin\varepsilon_{kb}T_0\leqslant(0.07\text{–}0.12)\lambda/l_k\cos(\beta_{0k}/2); \quad (k=1,2).$$

These are practically important *estimates for the boundaries of the high correlation regions*. Apparently, for Gaussian signal complex amplitude fluctuations the values $|R_{12}|=|R_{11}|=|R_{22}|=0.95$ are the minimum acceptable values of correlation coefficients when echoes may still be considered as "strongly correlated" and "coherent". The correlation coefficient of the real amplitudes is reduced in this case down to 0.9 and root-mean-square value of the phase difference increases up to $\sigma_{\Delta\varphi}\approx 30°$. Writing the inequality for two monostatic radars we bear in mind that those radars are working at the same frequency with independent (autonomous) signal reception.

Thus for maintaining high spatial (interstation) correlation of echo fluctuations in a MSRS the angle between the directions from a target to the receiving stations $(L_{12\,\mathrm{eff}}/R_1)$ should not exceed a small fraction (0.14–0.24) of the "averaged" width of the target scattering pattern lobes. This "averaged" width is determined by the target dimension along the effective baseline direction. For a MSRS containing monostatic radars with independent (autonomous) signal reception this fraction is to be even less by the factor of two. To maintain the high temporal correlation of echo fluctuations at each station, the angle of the target rotation about the axis perpendicular to the bistatic plane (passing through the transmitting station, the target and the receiving station) during the considered time interval should not exceed a small fraction (0.07–0.12) of the "averaged" width of the target scattering pattern lobes which, in this case, is determined by the target dimension along the direction perpendicular to both the axis of rotation and the bisector of the bistatic angle β_{01} or β_{02}. Due to the round-trip phase delay (as with the spatial correlation in the case of two monostatic radars) the values of the coefficients in the inequalities for the temporal correlation are equal to half of the values of the coefficients for the spatial correlation in the case of two receiving stations.

Example 4.2. For the same target and under the same conditions as in Example 4.1 the dashed curves are plotted from (4.38), (4.39) in Figs 4.3 and 4.4. A good agreement with the simulation results has been obtained at $\kappa_{12} \approx 0.87$ and $\kappa_{11} \approx 0.95$ which are in the limits assumed above. It is interesting to note that the quadratic approximation in (4.38) and (4.39) turns out to be sufficiently good for $|R_{12}|, |R_{11}|, |R_{22}| \geqslant 0.4$–$0.5$.

Region of Low Correlation

When the correlation coefficient of real amplitudes does not exceed 0.20–0.25, their fluctuations may be considered as practically uncorrelated. In this case the modulus of the correlation coefficient of complex amplitudes is no more than 0.45–0.50, and the probability distribution of the phase difference is close to the uniform distribution ($\sigma_{\Delta\varphi} \approx 80$–$77°$ whereas for the uniform distribution $\sigma_{\Delta\varphi} \approx 104°$). Taking it into account, we may assume the region of low correlation to be limited by the inequality $|R_{12}|, |R_{11}|, |R_{22}| \leqslant 0.45$–$0.50$. [When a target contains only a few scattering centres (FSs), sidelobes of the correlation coefficients (as functions of the station separation or of the angle of target rotation) can exceed the above specified level but these "irregular spikes" are not considered here.]

Consider first the modulus of the spatial correlation coefficient $|R_{12}|$. Let a target be prolate in the direction $\mathbf{x}_0 = \Delta\mathbf{r}_{12}$. Then the last factors, $\exp(-j2\pi L_{12\,\text{eff}} \boldsymbol{\rho}_{i0} \mathbf{x}_0 / \lambda R_1)$, mainly contribute to B_{12} in (4.27). All terms with sinc(·) may be considered as approximately equal to one. Assuming again the uniform, the triangle and the "reverse triangle" distributions of the target FSs we can obtain from the condition

$$|R_{12}| \approx \left[\left| \sum_{i=1}^{N} P_i \exp(-j2\pi L_{12\,\text{eff}} \boldsymbol{\rho}_{i0} \mathbf{x}_0 / \lambda R_1) \right| \Big/ \sum_{i=1}^{N} P_i \right] \leqslant 0.45$$

the inequality $(\pi l_{12} L_{12\,\text{eff}} / \lambda R_1) \geqslant 1.6$–$2.8$. Accepting the round-off "mean" value 2.5 in the right side yields $(L_{12\,\text{eff}}/R_1) \geqslant 0.8\lambda/l_{12}$. When a target is prolate in the directions of $\mathbf{y}_0$ or $\mathbf{z}_0$, the correlation reduction with the increase of $(L_{12\,\text{eff}}/R_1)$ is determined by the second or the first factor, respectively, of each addend in (4.27). Considering, for instance, the sum of the first factors only (when the target is prolate in the $\mathbf{z}_0$ direction) and the same three variants of FS' distributions as earlier, from the condition $|R_{12}| \leqslant 0.45$ we can obtain the inequality $(L_{12\,\text{eff}}/R_1) \geqslant (1.9$–$4.0)\lambda/\Delta\theta_y l_z$. Accepting the round-off "mean" multiplier 3.0 in the right side and combining results for targets which can be prolate in difference directions, yield the *approximate condition of the low spatial (interstation) correlation*

$$\begin{aligned} (L_{12\,\text{eff}}/R_1) &\geqslant \min[0.8\lambda/l_{12}; 3\lambda/\Delta\theta_y l_z; 3\lambda/\Delta\theta_z l_y] \\ (L_{12\,\text{eff}}/R_1) &\geqslant \min[0.4\lambda/l_{12}; 1.5\lambda/\Delta\theta_y l_z; 1.5\lambda/\Delta\theta_z l_y]. \end{aligned} \tag{4.42}$$

The first inequality in (4.42) relates to the case of a pair of receiving stations whereas the second inequality relates to the case of a pair of monostatic radars with independent signal reception working at the same frequency.

Similar considerations based on (4.30) lead to following *approximate condition of low temporal correlation* for the kth station

$$\Omega \sin \varepsilon_{kb} T_0 \geqslant \min[0.4\lambda/l_k \cos(\beta_{0k}/2); 1.5\lambda/\Delta\theta_y l_z \cos(\beta_{0k}/2);$$

$$1.5\lambda/\Delta\theta_z l_y \cos(\beta_{0k}/2)], \quad k=1,2. \tag{4.43}$$

It is seen from (4.42) that if a target is prolate along $\Delta\mathbf{r}_{12}=\mathbf{x}_0$, i.e. along the effective baseline direction, the condition of low spatial correlation requires the angle between the directions from the target to each pair of receiving stations to be no less than 0.8 of the "averaged" width of target scattering pattern lobes determined by the dimension l_{12}. In the case of a pair of monostatic radars 0.4 should be substituted for 0.8. When l_{12} is small, the above mentioned angle should exceed the "averaged" width of lobes determined by any of the two other target dimensions perpendicular to $\Delta\mathbf{r}_{12}=\mathbf{x}_0$, i.e. λ/l_z or λ/l_y. This excess is to be no less than by $3/\Delta\theta_y$ or $3/\Delta\theta_z$ for a pair of receiving stations and by $1.5/\Delta\theta_y$ or $1.5/\Delta\theta_z$ for a pair of monostatic radars with the same working frequency and autonomous signal reception. It is worth to recall that $\Delta\theta_y$ and $\Delta\theta_z$ are the ranges of initial target aspect uncertainty with respect to rotation about the Y and Z axes respectively. When, as usually the case, $\Delta\theta_y<1$ and $\Delta\theta_y<1$, the values of $3/\Delta\theta_y$ and $3/\Delta\theta_z$ (and even $1.5/\Delta\theta_y$ and $1.5/\Delta\theta_z$) may be comparatively large.

The similar meaning is presumed by the requirements of (4.43) to the angle of target rotation during the time interval T_0 about the axis perpendicular to $\mathbf{r}_{kb}$, i.e. to the angle $\Omega \sin \varepsilon_{kb} T_0$, in order for the temporal correlation of echo complex amplitudes at each station to be low. In this case the determining target dimension l_1 (or l_2) is the dimension along the vector $\mathbf{e}_1=\mathbf{x}_0$ (or $\mathbf{e}_2=\mathbf{x}_0$) which is perpendicular to the plane $(\mathbf{r}_{1b}, \Omega)$ or $(\mathbf{r}_{2b}, \Omega)$.

***In this section**, we have demonstrated an approach to the estimation of the spatial–temporal (and separately of both the spatial and the temporal) correlation (covariance) of echo complex amplitude fluctuations at the inputs of receivers of a MSRS. Knowledge of these correlations is necessary for developing optimum and suboptimum algorithms for joint signal processing and information fusion in MSRSs. It is also necessary for the performance evaluation of signal processing and information fusion algorithms. It has been shown that it is the angle between directions from a target to receiving stations (or, which is approximately the same, the effective baselength-to-target range ratio) that determines the spatial (interstation) correlation. The temporal correlation is determined by the angle of the target rotation about its centre of mass during the observation time interval. Since available information concerning specific target features is not usually sufficient for detailed calculations, we have considered how to determine the most important regions of high and low correlation when only little target information is available. Of major practical significance are the inequalities* (4.40), (4.41), (4.42) *and* (4.43) *which determine those regions. Their main advantage is that they may be used in practice when target information is rather poor. The relationships which have been derived in this section are in satisfactory agreement with the simulation results* (*Example* 4.1).

4.3. CRITERIA FOR OPTIMUM SIGNAL DETECTION IN MSRSs. INITIAL RELATIONSHIPS

As well known, for a number of "classical" optimality criteria for statistical decisions (the Bayes criteria, the Neyman–Pearson criterion etc.) the optimum signal processing is reduced to the likelihood ratio test. The specific criteria differ only by threshold levels. It means that we may consider the likelihood ratio criterion as a sufficiently universal optimality criterion which determines the structure of optimum detectors, i.e. signal processing algorithms up to threshold comparison.

In this book we shall use *the likelihood ratio criterion* for the synthesis of optimum detection algorithms. For the performance analysis of optimum and simplified suboptimum algorithms we shall use *the detection characteristics* which present the detection probability as functions of the Signal-to-Noise Ratio (SNR) or Signal-to-Interference plus Noise Ratio (SINR) at the inputs of receivers with the probability of the false alarm as a parameter.

A set of overall signals received by m stations of a MSRS in the time interval $(-T/2,\ T/2)$ represents a vector of overall signals which can be written in a complex form as follows: $\mathbf{X}^*(t)=[X_1^*(t),\ldots,X_m^*(t)]$. This may be a sum of wanted signals, $\mathbf{S}(t)$, external interferences, $\mathbf{J}(t)$, and receiver system self-noises (thermal noises), $\mathbf{N}(t)$, i.e. $\mathbf{X}(t)=\mathbf{S}(t)+\mathbf{J}(t)+\mathbf{N}(t)$, or a sum of external interferences and self-noises only, i.e. $\mathbf{X}(t)=\mathbf{J}(t)+\mathbf{N}(t)$ where $\mathbf{J}^*(t)=[J_1^*(t),\ldots,J_m^*(t)]$ and $\mathbf{N}^*(t)=[N_1^*(t),\ldots,N_m^*(t)]$.[4] In particular cases external interferences may be absent.

Such a representation of overall received signals, wanted signals, external interferences and noises is valid for an arbitrary m-channel receiving system. Peculiar features of MSRSs can be revealed after the substitution of the specific expressions for signals, interferences and self-noises corresponding to mathematical models introduced in Section 4.1. Taking into account the assumed Gaussian probability distribution for noises and external interferences and imposing no constraints on the MSRS' configuration we can write the known general likelihood ratio relationship for regular signals (e.g. [56])

$$\Lambda=\exp\left\{\mathrm{Re}\int_{-T/2}^{T/2}\int_{-T/2}^{T/2}\mathbf{S}^*(t_1,\boldsymbol{\Theta})\mathbf{R}(t_1,t_2)\mathbf{X}(t_2)\,\mathrm{d}t_1\,\mathrm{d}t_2\right.$$
$$\left.-0.5\int_{-T/2}^{T/2}\int_{-T/2}^{T/2}\mathbf{S}^*(t_1,\boldsymbol{\Theta})\mathbf{R}(t_1,t_2)\mathbf{S}(t_2,\boldsymbol{\Theta})\,\mathrm{d}t_1\,\mathrm{d}t_2\right\} \tag{4.44}$$

where $\boldsymbol{\Theta}$ is the vector of signal parameters, and the $m\times m$ matrix $\mathbf{R}(t_1,t_2)$ is the solution of the integral-matrix equation

$$\int_{-T/2}^{T/2}\mathbf{B}(t_1,t)\mathbf{R}(t,t_2)\,\mathrm{d}t=\mathbf{I}\delta(t_1-t_2),\quad -T/2\leqslant(t_1,t_2)\leqslant T/2. \tag{4.45}$$

In (4.45) $\mathbf{I}$ is the identity matrix of the order m; $\delta(t)$ is the Dirac delta-function.

The kernel of the equation (4.45) is the $m\times m$ space–time correlation (covariance) matrix of the sum of external interferences and receiver systems' self-noises at the inputs of spatially separated receiving stations, $\mathbf{B}(t_1,t_2)$. For the synthesis of optimum detection structures we shall, as a rule, assume this matrix to be known, i.e. the sum of self-noises and external interferences to be a *Gaussian zero-mean random process with deterministic correlation (covariance) matrix* (see Section 4.1). Of course, in most practical situations the correlation matrix at least of external interferences contains unknown or random parameters, i.e. it should be considered

[4] We remind here that the asterisk with scalars denotes the complex conjugate but the same asterisk with vectors and matrices denotes complex conjugate and transpose (Hermitian conjugate).

as a quasideterministic matrix. There has been growing interest over the last decades in the application of detection algorithms adapting to unknown interference parameters. In essence, such adaptive algorithms include usually the same procedures as corresponding algorithms obtained for deterministic correlation matrices, but instead of *a priori* unknown matrices they use statistical estimates of those matrices. Performance of "good" adaptive algorithms under steady-state conditions has to be close to that of optimum detection algorithms obtained for completely known matrices. Therefore, optimum detection algorithms derived under assumption of deterministic correlation matrix $\mathbf{B}(t_1, t_2)$ are of great importance. On the one hand, they serve as a base for developing adaptive detection algorithms, and, on the other hand, they determine the potential performance level for any detection algorithms.

It should be noted that in the case of target detection in a background of receiver system thermal self-noises, the correlation matrix of these noises may often be considered as deterministic one. Adaptive detection algorithms using correlation matrix estimation will be presented in Chapter 9.

In signal (target) detection problems, we shall assume the interferences to be stationary and, in the case of their interstation correlation, stationary coupled during the observation time interval $(-T/2, T/2)$. This assumption simplifies detection problems significantly and at the same time often holds in practice. Nonstationarity of interferences has not sufficient time to manifest itself in a short time interval T which is usually limited by the target illuminating signal duration. The motion of interference sources (e.g., jammers) in space leads to nonstationary coupling of spatially correlated interferences (see Section 4.1). However, the strong interstation correlation is possible (after proper delay equalization) if the baselengths between stations are not too large. In these cases differential Doppler frequency shifts of each interference at the inputs of any pair of stations $\Delta\Omega_{ik}$, $i, k = \overline{1, m}$, are often small, i.e. $|\Delta\Omega_{ik} T| \ll 1$, so that the nonstationarity of interference coupling may be neglected. Adaptive detection algorithms can usually eliminate this nonstationary coupling by the proper phase (or frequency) alignment.

For stationary and stationary coupled interferences $\mathbf{B}(t_1, t_2) = \mathbf{B}(t_1 - t_2)$ and $\mathbf{R}(t_1, t_2) = \mathbf{R}(t_1 - t_2)$. Now it is convenient to go over *from the time domain to the frequency domain*. Let us denote the power spectral density (PSD) of the sum of interferences and self-noises by $\mathbf{\Phi}(\omega)$ which is the Fourier transformation of $\mathbf{B}(t_1 - t_2)$. The Fourier transformation of $\mathbf{R}(t_1 - t_2)$ is denoted by $\mathbf{f}(\omega)$. The observation time interval T is usually much longer than the correlation interval of interferences, and all signals fall within the observation interval. Then we may replace the limits of integrals in (4.45) by infinity and solve this system of equations with the help of the Fourier transformation. Instead of (4.45) we have

$$\mathbf{\Phi}(\omega)\mathbf{f}(\omega) = \mathbf{I}. \tag{4.46}$$

The likelihood ratio in the frequency domain takes the form

$$\Lambda = \exp\left\{\operatorname{Re} \frac{1}{2\pi} \int_{-\infty}^{\infty} \mathbf{\Psi}^*(\omega, \mathbf{\Theta})\mathbf{f}(\omega)\boldsymbol{\chi}(\omega)\, d\omega - \frac{1}{4\pi} \int_{-\infty}^{\infty} \mathbf{\Psi}^*(\omega, \mathbf{\Theta})\mathbf{f}(\omega)\mathbf{\Psi}(\omega)\, d\omega\right\} \tag{4.47}$$

where

$$\mathbf{\Psi}^*(\omega)=[\Psi_1^*(\omega,\mathbf{\Theta}),\ldots,\Psi_m^*(\omega,\mathbf{\Theta})];$$
$$\boldsymbol{\chi}^*(\omega)=[\chi_1^*(\omega),\ldots,\chi_m^*(\omega)] \tag{4.48}$$

are the vectors of the Fourier transformations (the spectra) of the wanted signals, $\mathbf{S}^*(t,\mathbf{\Theta})=[S_1^*(t,\mathbf{\Theta}),\ldots,S_m^*(t,\mathbf{\Theta})]$, and of the overall signals, $\mathbf{X}^*(t)=[X_1^*(t),\ldots,X_m^*(t)]$, received during the time interval $(-T/2,T/2)$, respectively. It is seen that *the system of integral equations to solve for the functions of time* $R_{ik}(t_1-t_2)$, $i,k=\overline{1,m}$, *has been transformed into the much simpler system of functional equations to solve for the spectral functions* $f_{ik}(\omega)$, $i,k=\overline{1,m}$. This is one of the most important advantages of the spectral approach. Another advantage which is especially significant for the MSRS investigations is that *the signal envelope delays in the time domain are transformed into "phasors" (phase factors) in the frequency domain.* If, for instance, the wanted signal is $S_i(t)=a_{si}\exp(-j\varphi_{si})s_0(t-t_{si})\exp[j\omega_0(t-t_{si})]$, then its spectrum is $\Psi_i(\omega)=a_{si}\exp(-j\varphi_{si})\Psi_0(\omega-\omega_0)\exp(-j\omega t_{si})$ (see Section 4.1).

It follows from (4.46) that in (4.47) $\mathbf{f}(\omega)=\mathbf{\Phi}^{-1}(\omega)$, i.e. the inverse of the PSD matrix for the external interferences and self-noises, $\mathbf{\Phi}(\omega)$.

When wanted signals are deterministic (i.e. vector $\mathbf{\Theta}$ is known), the second integrals in (4.44) and (4.47) may be discarded since they have no effect on received signal processing. Passing to $\ln\Lambda$ we have from (4.47) *the optimum processing algorithm for deterministic signals*

$$\ln\Lambda=L=\mathrm{Re}\frac{1}{2\pi}\int_{-\infty}^{\infty}\mathbf{\Psi}^*(\omega,\mathbf{\Theta})\mathbf{f}(\omega)\boldsymbol{\chi}(\omega)\,d\omega. \tag{4.49}$$

The decision of whether a target is present in the MSRS resolution cell being probed can be obtained by comparing the decision variable L with a predetermined threshold which, in its turn, is determined by the allowable false alarm probability (the Neyman–Pearson criterion).

When wanted signals contain random parameters $\mathbf{\Theta}$, i.e. the signals are quasideterministic (e.g. fluctuating), the likelihood ratio (4.47) should be considered as a conditional one at a fixed $\mathbf{\Theta}$. The unconditional likelihood ratio can be obtained by averaging over $\mathbf{\Theta}$.

$$\bar{\Lambda}=\int_{\Theta}w(\mathbf{\Theta})\Lambda(\mathbf{\Theta})\,d\mathbf{\Theta}=\int_{\Theta}w(\mathbf{\Theta})\exp[L(\mathbf{\Theta})]$$
$$\times\exp\left[-\frac{1}{4\pi}\int_{-\infty}^{\infty}\mathbf{\Psi}^*(\omega,\mathbf{\Theta})\mathbf{f}(\omega)\mathbf{\Psi}(\omega,\mathbf{\Theta})\,d\omega\right]d\mathbf{\Theta} \tag{4.50}$$

where $w(\mathbf{\Theta})$ is the Probability Density Function (PDF) of $\mathbf{\Theta}$; $L(\mathbf{\Theta})$ is determined by (4.49) and the integrals are taken over the total domain of $w(\mathbf{\Theta})$. The unconditional likelihood ratio determines *the optimum processing algorithm for signals containing random parameters.* The processing result (the decision variable $\bar{\Lambda}$) should be compared with a threshold to make a decision as to the presence of target.

In many cases *the PDF* $w(\mathbf{\Theta})$ *is not known or parameters* $\mathbf{\Theta}$ *are nonrandom though unknown.* In these cases *an adaptive approach* may be used. Following this approach, we utilize the algorithm (4.49) which is optimum for known $\mathbf{\Theta}$, but really *unknown*

5. OPTIMUM TARGET DETECTION IN ACTIVE MSRSs IN A BACKGROUND OF SPATIALLY UNCORRELATED INTERFERENCES

5.1. DETECTION OF DETERMINISTIC SIGNALS

The problem of signal detection in a background of spatially uncorrelated interferences is, first of all, the problem of *signal detection in a background of receiving system self-noises.*

Many MSRSs, especially military systems, must be able to operate under deliberate interference (e.g., jamming) conditions. Nevertheless, operation in a background of receiver system thermal self-noises remains one of the most important regimes for any MSRS.

It should be noted that even under jamming conditions it is not always reasonable to take into account jamming interstation correlation. Such a typical situation takes place when jamming enters through sidelobes of receiving antenna directivity patterns (ADPs). Apparently, one could take into account interstation correlation of jamming from all sources and synthesize an appropriate detection algorithm employing joint processing of all noises, jamming and wanted signals impinging the MSRS. But such an algorithm would be too awkward. At the same time much simpler two-staged detection algorithms may be used for these situations without noticeable performance losses. At the first stage sidelobe jamming is cancelled out in each receiving station (or in each monostatic radar) individually. For sidelobe jamming cancellation, the spatial correlation of jamming at each antenna aperture only should be taken into account. At the second stage residual (not perfectly cancelled) jamming together with self-noises and, possibly, wanted signals are transmitted from all receiving stations via data links to the information fusion centre (FC) for joint processing[1]. In these cases *sums of receiver self-noises and residual weak sidelobe jamming may be treated as spatially uncorrelated (at different stations).* For the sake of simplicity, we shall assume those sums to be white noises as receiver self-noises alone. Effect of "coloured" noises (with nonzero temporal correlation) on detection algorithms will be considered briefly at the end of this section.

We begin the synthesis of optimum detection algorithms from the simplest case of deterministic (completely known) wanted signals. As it was noted in Section 4.3, this signal model is rather idealized and unrealistic but convenient from the methodological point of view. It is the simplest way for revealing some essential features of detection algorithms for more realistic signal models which will be considered in the following sections.

---

[1] Decentralized signal detection where preliminary decisions about the presence or absence of a target are transmitted to the FC instead of residual external interferences, self-noises and signals themselves, is considered in Chapter 6.

Synthesis of Optimum Detection Algorithm

For spatially uncorrelated white noises with power spectral density (PSD) N_i at the ith station, $i=\overline{1,m}$, both the PSD matrix $\mathbf{\Phi}(\omega)$ and its inverse, the matrix $\mathbf{f}(\omega)$ are diagonal:

$$\begin{aligned} \mathbf{\Phi}(\omega) &= \operatorname{diag}(N_1,\ldots,N_m) = \|\delta_{ik}N_i\| = \mathbf{N}, \\ \mathbf{f}(\omega) &= \operatorname{diag}(N_1^{-1},\ldots,N_m^{-1}) = \|\delta_{ik}N_i^{-1}\| = \mathbf{N}^{-1}. \end{aligned} \tag{5.1}$$

Substituting (5.1) in (4.49) yields *the optimum processing algorithm for deterministic signals* (the vector Θ of known parameters is omitted)

$$L_1 = \operatorname{Re}\sum_{i=1}^{m}(1/2\pi N_i)\int_{-\infty}^{\infty}\chi_i(\omega)\Psi_i^*(\omega)\,d\omega. \tag{5.2}$$

The algorithm includes matched filtration in each station and coherent weighted summation of the outputs of the matched filters. To specify the processing procedures let us substitute the expression (4.3) for the wanted signal spectrum, $\Psi(\omega)$, into (5.2)

$$L_1 = \operatorname{Re}\sum_{i=1}^{m}\frac{a_{si}\exp(j\varphi_{si})}{2\pi N_i}\int_{-\infty}^{\infty}\chi_i(\omega)\Psi_0^*(\omega-\omega_0-\Omega_{si})\exp(j\omega t_{si})\,d\omega. \tag{5.3}$$

The matched filter in the ith station takes into account the expected Doppler frequency, Ω_{si}. Output signals of the filters are equalized both in delays, t_{si}, and phase shifts, φ_{si}, and then a weighted sum is formed. The weights are proportional to the ratios of expected signal r.m.s. values to noise PSD. These weights increase the contribution of stations with higher SNRs.

As was noted in Section 4.1, deterministic signals are always spatially coherent. Clearly, the optimum processing (5.2) and (5.3) is feasible in spatially coherent MSRSs only (see Section 1.1). It may be treated as focussing of the whole receiving station system at the point from which a target echo is expected. Doppler frequency shifts should be taken into account in this procedure.

We have paid attention in Section 4.3 to the fact that synthesis and analysis of detection algorithms in the frequency domain does not mean the necessity to perform signal processing just in that domain. The frequency domain has been chosen for simpler presentation of most relationships and solution of principal equations. The choice of the most suitable domain for signal processing implementation depends on specific requirements and conditions. It is always possible to transform results obtained in the frequency domain into the time domain and vice versa. For example, substituting

$$\begin{aligned} \chi_i(\omega) &= \int_{-T/2}^{T/2} X_i(t_1)\exp(-j\omega t_1)\,dt_1; \\ \Psi_0^*(\omega) &= \int_{-T/2}^{T/2} s_0^*(t_2)\exp(j\omega t_2)\,dt_2 \end{aligned} \tag{5.4}$$

into (5.3) we obtain

$$L_1 = \mathrm{Re}\sum_{i=1}^{m} \frac{a_{si}\exp(j\varphi_{si})}{2\pi N_i} \int_{-\infty}^{\infty} \int_{-T/2}^{T/2} X_i(t_1)\exp(-j\omega t_1)\,dt_1$$

$$\times \int_{-T/2}^{T/2} s_0^*(t_2)\exp[j(\omega-\omega_0-\Omega_{si})t_2]\,dt_2 \exp(j\omega t_{si})\,d\omega. \tag{5.5}$$

We can change the integration order and take into account that

$$(1/2\pi)\int_{-\infty}^{\infty} \exp[-j\omega(t_1-t_2-t_{si})]\,d\omega = \delta(t_1-t_2-t_{si}) \tag{5.6}$$

where $\delta(t)$ is Dirac's delta-function. Then substituting (5.6) in (5.5) and integrating over t_2 yields

$$L_1 = \mathrm{Re}\sum_{i=1}^{m} \frac{a_{si}\exp(j\varphi_{si})}{N_i} \int_{-T/2}^{T/2} X_i(t)s_0^*(t-t_{si})\exp[-j(\omega_0+\Omega_{si})(t-t_{si})]\,dt. \tag{5.7}$$

This is *the optimum processing algorithm for deterministic signal detection in the time domain.* The algorithm (5.7) is fully equivalent to (5.3) and as well as (5.3) can be implemented both in analogous and digital form. The functional scheme of the synthesized algorithm is shown in Fig. 5.1.

Performance Analysis of the Optimum Algorithm

To obtain detection characteristics, probability distributions of the statistic (the decision random variable) L_1 are to be known for both the possible cases: when a wanted signal (target echo) is absent (the H_0 hypothesis) and when it is present (the H_1 hypothesis). As can be seen from (5.3) and (5.7), the optimum processing consists of linear procedures only. Since the input process is either a Gaussian noise or a sum of a deterministic signal and a Gaussian noise, the output statistic L_1 is a Gaussian variable too. It is completely determined by its mean value and variance. To obtain the mean value $\overline{L_1}$, the quantity $\chi_i(\omega)$ in (5.3) or $X_i(t)$ in (5.7) should be replaced by its mean values. When a wanted signal is absent, $\overline{\chi_i(\omega)}=\overline{X_i(t)}=0$ at each $i=\overline{1,m}$ so that $\overline{L_{1n}}=0$. When a wanted signal is present, $\chi_i(\omega)$ takes the form

$$\overline{\chi_i(\omega)} = a_{si}\exp(-j\varphi_{si})\Psi_0(\omega-\omega_0-\Omega_{si})\exp(-j\omega t_{si}). \tag{5.8}$$

For this case $\overline{X_i(t)}$ may be obtained in a similar manner.

Substituting (5.8) in (5.3) and taking into account normalizing equation (4.4) yields

$$\overline{L_{1sn}} = L_{1s} = \sum_{i=1}^{m} 2a_{si}^2 T_s \Big/ N_i = \sum_{i=1}^{m} 2E_i/N_i. \tag{5.9}$$

Here $\overline{L_{1sn}}$ is the mean value of L_{1sn}, i.e. of the decision variable L_1 at the presence of the wanted signal; L_{1s} is the signal (nonrandom) component of L_1; E_i is the expected signal energy at the ith station.

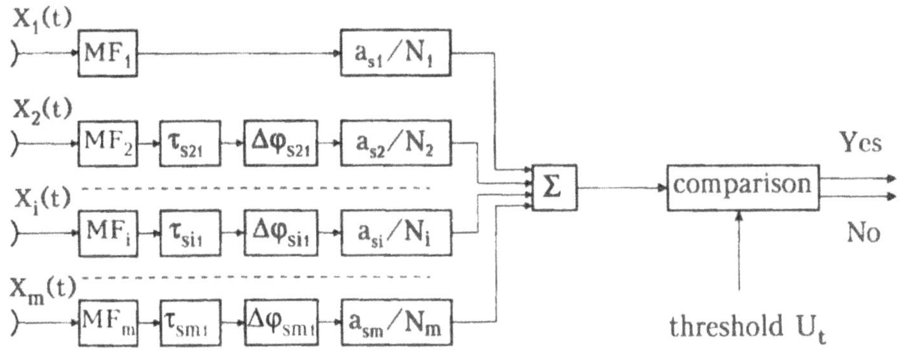

Figure 5.1 Structure of the optimum detector for deterministic signals [algorithm L_1, equations (5.3), (5.7)]. MF = Matched Filter

The variance of L_1 is the same in both cases: when the signal is absent and when it is present

$$\sigma^2(L_1)=\sigma^2(L_{1sn})=\sigma^2(L_{1n})$$

$$=\overline{[\mathrm{Re}\sum_{i=1}^{m}\frac{a_{si}\exp(j\varphi_{si})}{2\pi N_i}\int_{-\infty}^{\infty}\tilde{N}_i(\omega)\Psi_0(\omega-\omega_0-\Omega_{si})\exp(j\omega t_{si})\,d\omega]^2} \quad (5.10)$$

where $\tilde{N}_i(\omega)$ is the spectrum of a random noise realization, $N_i(t)$, at the ith station. [Random functions $\tilde{N}_i(\omega)$ and $N_i(t)$ should not be confused with nonrandom power spectral density, PSD, N_i]. For the narrowband signals considered [i.e. $(\Delta\omega_s \ll \omega_0)$] the following equation holds: $(\mathrm{Re}\,z)^2=0.5\,\mathrm{Re}\,zz^*$. Besides that [see (4.13) and (5.6)]

$$\overline{\tilde{N}_i(\omega_1)\tilde{N}_k^*(\omega_2)}=\int_{-T/2}^{T/2}\int_{-T/2}^{T/2}\overline{N_i(t_1)N_k^*(t_2)}\exp(-j\omega_1t_1+j\omega_2t_2)\,dt_1\,dt_2$$

$$=2\delta_{ik}N_i\int_{-T/2}^{T/2}\int_{-T/2}^{T/2}\delta(t_1-t_2)\exp(-j\omega_1t_1+j\omega_2t_2)\,dt_1\,dt_2$$

$$=4\pi\delta_{ik}N_i\delta(\omega_1-\omega_2). \quad (5.11)$$

Substituting (5.11) in (5.10) and taking into account equation (4.4) we can obtain

$$\sigma^2(L_1)=\sum_{i=1}^{m}2a_{si}^2T_{si}/N_i=\sum_{i=1}^{m}2E_i/N_i. \quad (5.12)$$

Detection characteristics are completely determined by the output SNR. We denote this output power SNR by q_{out}^2. Then from (5.9) and (5.12) we have

$$q_{out}^2=(L_{1s})^2/\sigma^2(L_1)=\sum_{i=1}^{m}2E_i/N_i=\sum_{i=1}^{m}q_{out\,i}^2 \quad (5.13)$$

where $q_{out\,i}^2$ is the "partial" power SNR at the output of the ith station. The false alarm probability, P_{fa}, and the detection probability, P_d, can be calculated from the following equations:

$$P_{fa}=0.5[1-\mathrm{erf}(u_0/\sqrt{2})];$$
$$P_d=0.5\{1-\mathrm{erf}[(u_0-q_{out})/\sqrt{2}]\}. \quad (5.14)$$

In (5.14) $u_0=u_t/\sigma(L_1)$ is the normalized threshold level; erf(x) is the error function

$$\mathrm{erf}(x)=(2/\sqrt{\pi})\int_0^x \exp(-t^2)\,dt. \quad (5.15)$$

Equations (5.14) are the same as for monostatic radars. The difference is that the quantity q^2_{out} is determined according to (5.13) by summing over all the stations of a MSRS.

Thus detection characteristics for the deterministic signal detection in a spatially coherent MSRS are similar to the detection characteristics of a monostatic radar, but the output power signal-to-noise ratio (SNR) is equal to the sum of output power SNRs of all the receiving stations of the MSRS.

When distances and RCSs of a target relative to all stations are not much different, then at $N_i=N_0$, $i=\overline{1,m}$, the summed output SNR is proportional to the area $S_{eff\Sigma}$ of the sum of receiving antenna effective apertures of all stations, S_{effi}:

$$q^2_{out}=(2K/N_0)\sum_{i=1}^{m} S_{effi}=(2K/N_0)S_{eff\Sigma} \quad (5.16)$$

where K is a factor of proportionality. It means that detection characteristics for such a MSRS are the same as for a monostatic radar with the receiving antenna effective aperture $S_{eff\Sigma}$. It follows also from (5.16) that if the receiving antenna aperture of a monostatic radar were cut into m equal parts and spatially separated so that to obtain m spatially coherent stations, then for the same noise PSD, target RCS and (approximately) range, spatially coherent signals and optimum processing, the detection characteristics would not change.

All results obtained above are valid for the case where a target is illuminated by a single transmitting station. These results may easily be extended to the case of n transmitting stations. By means of similar considerations, it may be shown that optimum processing in this case includes additional weighted summation (after delay and phase shift equalization) of echoes caused by all transmitting stations in each receiving station. Detection characteristics are determined by equations (5.14) too, but the output power SNR takes the form

$$q^2_{out}=\sum_{k=1}^{n}\sum_{i=1}^{m} 2E_{ik}/N_i \quad (5.17)$$

where E_{ik} is the energy of the signal caused by the kth transmitting station at the ith receiving station.

Equation (5.17) relates to the case where target echoes caused by different transmitting stations, arrive at each receiving station at different moments of time, i.e. are resolved in time. In this case the optimum processing is applied to mn segments of independent Gaussian processes with mean values equal either to zero or

Θ is replaced by its maximum likelihood estimate $\hat{\Theta}$ [43]. Then instead of (4.49) we have

$$L = \mathrm{Re}\,\frac{1}{2\pi}\int_{-\infty}^{\infty} \boldsymbol{\Psi}^*(\omega, \hat{\Theta})\mathbf{f}(\omega)\boldsymbol{\chi}(\omega)\,d\omega. \tag{4.51}$$

This algorithm is often called "the generalized likelihood ratio algorithm" (see, e.g., [8]).

The stochastic wanted signal, according to models introduced in Section 4.1, does not differ by its features from interferences. The log-likelihood ratio *optimum processing algorithm in the frequency domain for Gaussian signals with deterministic correlation (covariance) matrices* can be written in the form

$$L = \frac{1}{2\pi}\int_{-\infty}^{\infty} \boldsymbol{\chi}^*(\omega)[\mathbf{f}(\omega) - \mathbf{f}_{SI}(\omega)]\boldsymbol{\chi}(\omega)\,d\omega \tag{4.52}$$

where $\mathbf{f}(\omega)$ is the inverse matrix for the PSD matrix of the sum of external interferences and self-noises, $\boldsymbol{\Phi}(\omega)$; $\mathbf{f}_{SI}(\omega)$ is the inverse matrix for the PSD matrix of the sum of wanted signals, external interferences and self-noises, $\boldsymbol{\Phi}_{SI}(\omega) = \boldsymbol{\Phi}_S(\omega) + \boldsymbol{\Phi}(\omega)$.

When mutual phase shifts and (or) intensities of wanted stochastic signals at the inputs of receiving stations are random, correlation (covariance) matrices of such signals are quasideterministic. It means that $\boldsymbol{\Phi}_{SI}(\omega)$ and hence $\mathbf{f}_{SI}(\omega)$ contain random parameters $\boldsymbol{\Theta}$. One can obtain unconditional likelihood ratio by averaging over $\boldsymbol{\Theta}$:

$$\bar{\Lambda} = \int_{\Theta} w(\boldsymbol{\Theta})\frac{\mathfrak{R}_1(\boldsymbol{\Theta})}{\mathfrak{R}_0}\exp\left\{\frac{1}{4\pi}\int_{-\infty}^{\infty} \boldsymbol{\chi}^*(\omega)[\mathbf{f}(\omega) - \mathbf{f}_{SI}(\omega, \boldsymbol{\Theta})]\boldsymbol{\chi}(\omega)\,d\omega\right\}d\boldsymbol{\Theta} \tag{4.53}$$

where, as earlier, $w(\boldsymbol{\Theta})$ is the PDF of $\boldsymbol{\Theta}$; the integral is taken over the domain of $w(\boldsymbol{\Theta})$; $\mathfrak{R}_1(\boldsymbol{\Theta})$ is the likelihood functional coefficient when the wanted signal is present; $\mathfrak{R}_0$ is the similar coefficient when the wanted signal is absent. If random parameters $\boldsymbol{\Theta}$ are not connected with signal energy (for instance, phase shifts) then not only $\mathfrak{R}_0$ but $\mathfrak{R}_1$ does not depend on $\boldsymbol{\Theta}$. In these cases the ratio $\mathfrak{R}_1/\mathfrak{R}_0$ may be factored outside the integral sign and omitted.

Two remarks are necessary in closing. Firstly, using the frequency domain in formulas above does not mean that signal processing must be carried out in that domain. The frequency domain has been chosen here for simpler presentation of most relationships and solution of principal equations. In practice detection algorithms may be implemented in both the frequency and time domains depending on specific conditions. Secondly, here and in the subsequent chapters we consider mainly continuous signals, interferences and noises which is in accordance with their actual physical form at the inputs of spatially separated stations of any MSRS. In conformity with this presentation, the optimum detection algorithms have been written in the analogous form. But exploiting these algorithms, the engineer-designer may always choose the analogous or the digital variant of their implementation using, in the latter case, well-known sampling techniques.

***In this section**, we have chosen the likelihood ratio criterion for the synthesis of optimum signal processing and the Neyman–Pearson criterion for decision making in the target detection problem for MSRSs. We have chosen also the frequency domain for simpler signal processing algorithm presentation. Following this approach we have presented the* *general (initial) equations determining optimum signal processing for deterministic signals* (4.49), *quasideterministic signals* (4.50), (4.51) *and stochastic signals* (4.52), (4.53). *These equations include expressions for target echo spectra, power spectral densities of stochastic signals, external interferences and receiver system self-noises which have been presented in Section* 4.1 [*see* (4.3)–(4.5), (4.10)–(4.13)]. *To obtain practical algorithms, it is necessary to invert the PSD matrix of the sum of external interferences and self-noises,* $\mathbf{\Phi}(\omega)$, *or both the matrix* $\mathbf{\Phi}(\omega)$ *and the matrix of the sum of wanted stochastic signals, external interferences and self-noises,* $\mathbf{\Phi}_{SI}(\omega)$ *in order to specify matrices* $\mathbf{f}(\omega)$ *or both the matrices* $\mathbf{f}(\omega)$ *and* $\mathbf{f}_{SI}(\omega)$ *for different specific conditions of MSRS operation.*

to the expected signals. A similar result may be obtained in the case where deterministic signals are resolved in frequency.

More profitable, from the power point of view, is to sum all illuminating signals in phase at a target. Then at the input of each receiving station only a single echo can be expected but its effective value is equal to the sum of the effective values of echoes caused by all transmitting stations. Only m segments of independent Gaussian processes undergo the optimal processing, but when the signal is present, the mean value of these processes at each receiving station is equal to the phased sum of n signals. The optimal processing does not differ from (5.3) after replacing a_{si} by the sum $a_{si}=\sum_{k=1}^{n} a_{sik}$ where a_{sik} is the effective value of the signal caused by the kth transmitting station at the ith receiving station. As before, the output power SNR is determined by (5.13) if to substitute there

$$E_i=\left(\sum_{k=1}^{n} a_{sik}\right)^2 T_s.$$

In the simplest case, where $N_i=N_0$, $a_{sik}=a_0$, $i=\overline{1,m}$, $k=\overline{1,n}$, when echoes caused by different transmitting stations are resolved in each receiving station, then from (5.17)

$$q_{\text{out}}^2=mnq_{\text{out}\,0}^2. \tag{5.18}$$

In the same case under the condition that illuminating signals from all transmitting stations are summed in phase at a target

$$q_{\text{out}}^2=mn^2q_{\text{out}\,0}^2 \tag{5.19}$$

where $q_{\text{out}\,0}^2$ is the power output SNR at the linear part of a monostatic radar with the same transmitting and receiving stations (possibly, colocated) as each of the corresponding stations of the MSRS.

Thus if echoes caused by all transmitting stations are resolved in receiving stations, the total power gain (as compared with the above mentioned monostatic radar) is equal to mn. If illuminating signals from all transmitting stations are summed in phase at a target, the total power gain of a MSRS is equal to mn^2. The extra gain by n times is obtained owing to the fact that n illuminating signals sum at a target without receiver noises. At the same time the coherent summation of n resolved signals in receivers is accompanied by the incoherent summation of n independent noises.

As was noted in Chapter 4, the deterministic signal is a convenient ideal model not claiming an adequate description of actual signals. Therefore we do not discuss here practical difficulties of the illuminating signal coherent summation at a target (see Section 5.3).

Let us return to the case where a target is illuminated by a single transmitting station and take into account the influence of receiving ADP $g_i(\beta_{si}, \varepsilon_{si}, \omega)$, $i=\overline{1,m}$. In this case one should substitute (4.5) for (4.3) into (5.3). Then (5.3) takes the form

$$L_1=\operatorname{Re}\sum_{i=1}^{m}\frac{a_{si}\exp(j\varphi_{si})}{2\pi N_i}\int_{-\infty}^{\infty}\chi_i(\omega)g_i^*(\beta_{si},\varepsilon_{si},\omega)\Psi_0^*(\omega-\omega_0-\Omega_{si})\exp(j\omega t_{si})\,d\omega. \tag{5.20}$$

The factor $g_i(\beta_{si}, \varepsilon_{si}, \omega)$ should be inserted into (5.8) and (5.10) too. Then the antenna ADP $g_i(\beta_{si}, \varepsilon_{si}, \omega)$ will change the expression for the signal energy, E_i. Taking into

account the normalization (4.4) we have

$$\frac{1}{2\pi}\int_{-\infty}^{\infty}|g_i(\beta_{si},\varepsilon_{si},\omega)|^2|\Psi_0(\omega-\omega_0-\Omega_{si})|^2\,d\omega=2\alpha_i T_s \quad (5.21)$$

where $\alpha_i \leqslant 1$ is the multiplier which accounts for the ADP's effect. Now replacing $2T_s$ by $2\alpha_i T_s$ and $E_i=a_{si}^2 T_s$ by $E_i=a_{si}^2\alpha_i T_s$ we can use equations (5.9), (5.12) and all subsequent expressions.

Up to now we assume noises to be white. This is an adequate and convenient model for receiver system noises and for wideband (with respect to receiver bandwidth) external interferences (e.g., sidelobe jamming). In some cases it is desirable to take into consideration possible temporal correlation, i.e. a limited bandwidth and nonuniformity of noise spectra. In these cases we have the problem of signal detection in MSRSs in a background of spatially uncorrelated but "coloured" noises.

Let the sum of a white self-noise and external interference be at the input of each receiving station so that the total PSD can be written in the form

$$\Phi_{ii}(\omega)=N_i[1+q_{Ii}^2 F_i(\omega-\omega_0)],\quad i=\overline{1,m} \quad (5.22)$$

where q_{Ii}^2 is the power Interference-to-Noise Ratio (INR); $F_i(\omega)$ is the dimensionless normalized PSD of the interference complex envelope [see (4.11)]. The entries of the matrix $\mathbf{f}(\omega)=\mathbf{\Phi}^{-1}(\omega)$ are equal to the reciprocals of (5.22) instead of (5.1). Therefore the optimum processing algorithm (5.3) takes the form

$$L_1=\operatorname{Re}\sum_{i=1}^{m}\frac{a_{si}\exp(j\varphi_{si})}{2\pi N_i}\int_{-\infty}^{\infty}\frac{\chi_i(\omega)\Psi_0^*(\omega-\omega_0-\Omega_{si})\exp(j\omega t_{si})}{1+q_{Ii}^2F_i(\omega-\omega_0)}\,d\omega. \quad (5.23)$$

The matched filter for the expected signal contaminated with a white noise is replaced by a new filter which is also optimal according to the maximum output SNR criterion but for a "coloured" noise.

The equations for the mean value, $\overline{L_1}=L_{1s}$, the variance, $\sigma^2(L_1)$, and output SNR, q_{out}^2, should be changed as follows

$$L_{1s}=\sigma^2(L_1)=q_{out}^2=\sum_{i=1}^{m}\frac{2K_iE_i}{N_i}=\sum_{i=1}^{m}q_{out\,i}^2 \quad (5.24a)$$

where $K_i \leqslant 1$ is "the signal energy utilization factor" [47,73] (in comparison with the case of a white noise where $K_i=1$)

$$K_i=\frac{1}{4\pi T_s}\int_{-\infty}^{\infty}\frac{|\Psi_0(\omega-\omega_0-\Omega_{si})|^2}{1+q_{Ii}^2F_i(\omega-\omega_0)}\,d\omega,\quad i=\overline{1,m}. \quad (5.24b)$$

It is possible, of course, to take into account the effect of antenna ADP in (5.23) and (5.24) as it has been done above for white noises.

***This section** is the first among the several sections devoted to the group of problems concerning signal (target echo) detection in MSRSs in a background of spatially uncorrelated Gaussian noises. First of all, these are problems of signal detection in a background of receiver system self-noises. It has been emphasized, however, that in*

many cases external interferences (e.g. jamming) operated through antenna sidelobes, are reasonable to be treated as spatially uncorrelated noises too (i.e. at different stations). We have considered the problem of deterministic signal detection in spatially coherent MSRSs as the simplest problem of that group. Optimum signal processing algorithms (5.3), (5.7) *and* (5.20) *for white noise contamination (the latter takes into account the effect of antenna directivity patterns, ADPs) and* (5.23) *for "coloured" noises have been synthesized. The corresponding functional scheme (Fig.* 5.1) *has been presented. Performance of optimal algorithms has been analyzed for the cases where a target is illuminated by both a single and several transmitting stations. In the latter case we have considered situations where echoes caused by different transmitting stations are resolved or not resolved in each receiving stations. This analysis has led to equations* (5.13), (5.17)–(5.19), (5.24) *and* (5.14), (5.15) *that permit calculations of detection characteristics and power gain estimation as compared with corresponding monostatic radars. The obtained results will serve as a base for optimum and suboptimum algorithm synthesis and analysis in the subsequent sections when more adequate signal models will be considered.*

5.2. OPTIMUM DETECTION ALGORITHMS FOR FLUCTUATING SIGNALS WHEN A TARGET IS ILLUMINATED BY A SINGLE TRANSMITTING STATION

The model of fluctuating (quasi-deterministic) signals, or of signals with random complex amplitudes, is usually a good, practically adequate model for many actual signals scattered by targets. Therefore, the results obtained in Sections 5.2–5.6 may in many cases be immediately applied to practical problems.

We start the synthesis of optimum processing algorithms from equations (4.50), (4.51) taking into account that the matrix $\mathbf{f}(\omega)$ is diagonal [see (5.1)].

In equation (4.3) for the wanted signal spectrum each random complex amplitude contains a random initial phase, φ_{si}, and a random r.m.s. value, a_{si}, $i=\overline{1,m}$,. For revealing their effect on optimum processing algorithms, let us separate these random parameters from the total signal spectrum. We can present the vector of wanted signals in (4.48)–(4.50) in the following form:

$$\boldsymbol{\Psi}(\omega,\boldsymbol{\Theta})=\boldsymbol{\Psi}(\omega,\boldsymbol{\varphi}_s,\mathbf{a}_s)=\tilde{\boldsymbol{\Psi}}(\omega)\tilde{\mathbf{E}}(\boldsymbol{\varphi}_s)\mathbf{a}_s \tag{5.25}$$

where $\tilde{\boldsymbol{\Psi}}(\omega)=\|\delta_{ik}\Psi_0(\omega-\omega_0-\Omega_{si})\exp(-j\omega t_{si})\|$, $i,k=\overline{1,m}$ is the diagonal matrix of wanted signal spectra containing no random parameters; $\tilde{\mathbf{E}}(\boldsymbol{\varphi}_s)$, $\mathbf{a}_s$ are the diagonal matrix and the vector of random parameters:

$$\tilde{\mathbf{E}}(\boldsymbol{\varphi}_s)=\operatorname{diag}[\exp(-j\varphi_{s1}),\ldots,\exp(-j\varphi_{sm})];$$
$$\mathbf{a}_s^t=(a_{s1},\ldots,a_{sm}). \tag{5.26}$$

The superscript "t" denotes the transposition as before. Now we can substitute (5.25) in (4.50) and take into account that $\boldsymbol{\varphi}_s$ and $\mathbf{a}_s$ are mutually independent so that $w(\boldsymbol{\Theta})=w(\boldsymbol{\varphi}_s,\mathbf{a}_s)=w(\boldsymbol{\varphi}_s)w(\mathbf{a}_s)$ where "$w(\cdot)$" denotes the Probability Density

Function (PDF). Then the unconditional likelihood ratio takes the form

$$\tilde{\Lambda} = \int_{\mathbf{a}_s}\int_{\boldsymbol{\varphi}_s} w(\mathbf{a}_s)w(\boldsymbol{\varphi}_s)\exp\{\mathbf{a}_s^t \operatorname{Re}\left[\tilde{\mathbf{E}}^*(\boldsymbol{\varphi}_s)\frac{1}{2\pi}\int_{-\infty}^{\infty}\tilde{\boldsymbol{\Psi}}^*(\omega)\mathbf{f}(\omega)\boldsymbol{\chi}(\omega)\,d\omega\right]$$

$$-0.5\mathbf{a}_s^t\tilde{\mathbf{E}}^*(\boldsymbol{\varphi}_s)\left[\frac{1}{2\pi}\int_{-\infty}^{\infty}\tilde{\boldsymbol{\Psi}}^*(\omega)\mathbf{f}(\omega)\tilde{\boldsymbol{\Psi}}(\omega)\,d\omega\right]\tilde{\mathbf{E}}(\boldsymbol{\varphi}_s)\mathbf{a}_s\}\,d\boldsymbol{\varphi}_s\,d\mathbf{a}_s. \quad (5.27)$$

For simplifying the equations we denote in (5.27)

$$\mathbf{G} = \frac{1}{2\pi}\int_{-\infty}^{\infty}\tilde{\boldsymbol{\Psi}}^*(\omega)\mathbf{f}(\omega)\boldsymbol{\chi}(\omega)\,d\omega;$$

$$\mathbf{C} = \frac{1}{2\pi}\int_{-\infty}^{\infty}\tilde{\boldsymbol{\Psi}}^*(\omega)\mathbf{f}(\omega)\tilde{\boldsymbol{\Psi}}(\omega)\,d\omega. \quad (5.28)$$

The $m\times 1$ vector $\mathbf{G}$ and the $m\times m$ matrix $\mathbf{C}$ will play an important role in optimum and suboptimum processing algorithms.

Substituting (5.28) into (5.27) yields

$$\tilde{\Lambda} = \int_{\mathbf{a}_s}\int_{\boldsymbol{\varphi}_s} w(\mathbf{a}_s)w(\boldsymbol{\varphi}_s)\exp\{\mathbf{a}_s^t\operatorname{Re}[\tilde{\mathbf{E}}^*(\boldsymbol{\varphi}_s)\mathbf{G}]-0.5\mathbf{a}_s^t\tilde{\mathbf{E}}^*(\boldsymbol{\varphi}_s)\mathbf{C}\tilde{\mathbf{E}}(\varphi_s)\mathbf{a}_s\}\,d\boldsymbol{\varphi}_s\,d\mathbf{a}_s. \quad (5.29)$$

Now let us take into account that $\mathbf{f}(\omega)=\mathbf{N}^{-1}$ [see (5.1)]. Then it is convenient to write the likelihood ratio (5.29) in the form

$$\tilde{\Lambda} = \int_{\mathbf{a}_s}\int_{\boldsymbol{\varphi}_s} w(\mathbf{a}_s)w(\boldsymbol{\varphi}_s)\exp\{\mathbf{a}_s^t\mathbf{N}^{-1}\operatorname{Re}[\tilde{\mathbf{E}}^*(\boldsymbol{\varphi}_s)\tilde{\mathbf{G}}]-T_s\mathbf{a}_s^t\mathbf{N}^{-1}\mathbf{a}_s\}\,d\boldsymbol{\varphi}_s\,d\mathbf{a}_s. \quad (5.30)$$

The vectors $\mathbf{G}$ in (5.29) and $\tilde{\mathbf{G}}$ in (5.30) are connected by the equation: $\mathbf{G}=\mathbf{N}^{-1}\tilde{\mathbf{G}}$. Each element of the vector $\mathbf{G}$ $(i=\overline{1,m})$

$$G_i = \frac{\tilde{G}_i}{N_i} = \frac{1}{2\pi N_i}\int_{-\infty}^{\infty}\chi_i(\omega)\dot{\Psi}_0^*(\omega-\omega_0-\Omega_{si})\exp(j\omega t_{si})\,d\omega \quad (5.31)$$

is a result of the matched filtration of an overall signal received by the ith station (normalized to the noise PSD). Each element of the diagonal matrix $\mathbf{C}=2T_s\mathbf{N}^{-1}$ is a constant

$$C_{ii} = \frac{1}{2\pi N_i}\int_{-\infty}^{\infty}|\Psi_0(\omega-\omega_0-\Omega_{si})|^2\,d\omega = 2T_s/N_i. \quad (5.32)$$

Further specification of the likelihood ratio $\tilde{\Lambda}$ depends on the type of expected signals and on the class of a MSRS.

Spatially Coherent Signals with Completely Mutually Dependent Amplitude Fluctuations in Spatially Coherent MSRSs

The conditions (4.40), (4.41) of strong spatial correlation of signal complex amplitudes are assumed to be satisfied. As noted in Section 1.1, spatial coherence of a MSRS means the long term phase equipment stability. In practice slow phase variations in such MSRSs are to be periodically aligned with the help of some reference reflector.

In the case considered the expected signal phases at all receiving stations are functionally interconnected:

$$\varphi_{si}=\varphi_{s1}-\Delta\varphi_{si1}; \qquad w(\boldsymbol{\varphi}_s)=w(\varphi_{s1})\prod_{i=2}^{m}\delta(\varphi_{si}-\varphi_{s1}+\Delta\varphi_{si1}) \tag{5.33}$$

where φ_{s1} is the unique random initial phase (among initial phases of expected signals at the inputs of all stations) at some (arbitrary) "reference" station which we may refer to as the first station without loss of generality; $\Delta\varphi_{si1}$ is the known signal phase shift at the ith station ($i=\overline{2,m}$) with respect to the reference station; $\delta(\cdot)$ is the Dirac delta-function. The r.m.s. signal values, a_{si}, at different stations are functionally interconnected too, so that all a_{si} may be expressed through a_{s1}:

$$a_{si}=A_{i1}a_{s1}; \qquad w(\mathbf{a}_s)=w(a_{s1})\prod_{i=2}^{m}\delta(a_{si}-A_{i1}a_{s1}) \tag{5.34}$$

where A_{i1}^2 ($i=\overline{1,m}, A_{11}^2=1$) is the known ratio of the second initial moments of probability distributions for a_{si} and a_{s1}, i.e. the ratio of the expected signal power at the ith station to that at the reference station:

$$A_{i1}^2=\int_0^\infty a_{si}^2 w(a_{si})\,da_{si}\Big/\int_0^\infty a_{s1}^2 w(a_{s1})\,da_{s1}. \tag{5.35}$$

Substituting (5.33) and (5.34) in (5.30) yields

$$\tilde{\Lambda}=\int_0^\infty\int_{-\pi}^{\pi} w(\varphi_{s1})w(a_{s1})\exp\{a_{s1}\,\mathrm{Re}[\exp(j\varphi_{s1})\mathbf{A}^t\mathbf{N}^{-1}\mathbf{E}^*\tilde{\mathbf{G}}]$$
$$-a_{s1}^2T_s\mathbf{A}^t\mathbf{N}^{-1}\mathbf{A}\}\,d\varphi_{s1}\,da_{s1} \tag{5.36}$$

where

$$\mathbf{A}^t=(1,A_{21},\ldots,A_{m1});$$
$$\mathbf{E}=\mathrm{diag}[1,\exp(j\Delta\varphi_{s21}),\ldots,\exp(j\Delta\varphi_{sm1})]. \tag{5.37}$$

Assume now (as in Section 4.2) the expected signal complex amplitudes to be Gaussian random variables. Then φ_{s1} is uniformly distributed within the interval $(-\pi,\pi)$ whereas for a_{s1} holds the Rayleigh probability distribution[2]

$$w(\varphi_{s1})=1/2\pi, \quad \varphi_{s1}\in(-\pi,\pi); \tag{5.38a}$$

$$w(a_{s1})=(2a_{s1}/\sigma_1^2)\exp(-a_{s1}^2/\sigma_1^2), \quad a_{s1}\in(0,\infty). \tag{5.38b}$$

Substituting (5.38) into (5.36) and integrating, we take into account that [10]

$$\frac{1}{2\pi}\int_{-\pi}^{\pi}\exp\{a_{s1}\,\mathrm{Re}[\exp(j\varphi_{s1})\mathbf{A}^t\mathbf{N}^{-1}\mathbf{E}^*\tilde{\mathbf{G}}]\}\,d\varphi_{s1}=I_0(a_{s1}|\mathbf{A}^t\mathbf{N}^{-1}\mathbf{E}^*\tilde{\mathbf{G}}|);$$

[2] It has taken into account in (5.38) that a_{s1} is the effective (r.m.s) value, not the amplitude, of a signal so that $\overline{a_{s1}^2}$ is equal to σ_1^2 but not to $2\sigma_1^2$.

$$\int_0^\infty \frac{2a_{s1}}{\sigma_1^2}\exp\left[-a_{s1}^2\left(\frac{1}{\sigma_1^2}+T_s\mathbf{A}'\mathbf{N}^{-1}\mathbf{A}\right)\right]I_0(a_{s1}|\mathbf{A}'\mathbf{N}^{-1}\mathbf{E}^*\tilde{\mathbf{G}}|)\,da_{s1}$$

$$=(1+\sigma_1^2T_s\mathbf{A}'\mathbf{N}^{-1}\mathbf{A})^{-1}\exp\frac{\sigma_1^2|\mathbf{A}'\mathbf{N}^{-1}\mathbf{E}^*\tilde{\mathbf{G}}|^2}{4(1+\sigma_1^2T_s\mathbf{A}'\mathbf{N}^{-1}\mathbf{A})} \tag{5.39}$$

where $I_0(z)$ is the modified Bessel function of zero order. Taking a logarithm, omitting terms independent of $\tilde{\mathbf{G}}$ (since only the vector $\tilde{\mathbf{G}}$ contains received signals that should undergo the optimal processing) and replacing the squared modulus by the modulus, yields *the optimum processing algorithm* in the form

$$L_2=|\mathbf{A}'\mathbf{N}^{-1}\mathbf{E}^*\tilde{\mathbf{G}}|=\left|\sum_{i=1}^{m}(A_{i1}/N_i)\exp(-\mathrm{j}\,\Delta\varphi_{si1})\tilde{\mathbf{G}}_i\right|;\quad A_{11}=1,\ \Delta\varphi_{s11}=0. \tag{5.40}$$

We may substitute $\tilde{\mathbf{G}}_i$ from (5.31) into (5.40). Then

$$L_2=\left|\sum_{i=1}^{m}\frac{A_{i1}\exp(-\mathrm{j}\Delta\varphi_{si1})}{2\pi N_i}\int_{-\infty}^{\infty}\chi_i(\omega)\Psi_0^*(\omega-\omega_0-\Omega_{si})\exp(\mathrm{j}\omega t_{si})\,d\omega\right|. \tag{5.41}$$

The decision variable L_2 is to be compared with a threshold determined by allowable false alarm probability in the MSRS. For weak output signals when $a_{s1}^2|\mathbf{A}'\mathbf{N}^{-1}\mathbf{E}^*\tilde{\mathbf{G}}|^2/16 \ll 1$ with high probability[3], then $I_0(z)\approx 1+z^2/4$, and the algorithm (5.40), (5.41) turns out to be optimal for any probability distribution of a_{s1}.

It is interesting to compare (5.40), (5.41) with the optimum processing algorithm (5.3) for deterministic signals. Two essential distinctions can be seen. Firstly, random phases, φ_{si}, and r.m.s. values, a_{si}, are replaced by the known phase shifts relative to the reference station, $\Delta\varphi_{si1}$, and by the known ratios of the second moments of probability distributions, $\sqrt{\overline{a_{si}^2}/\overline{a_{s1}^2}}$. Secondly, the decision variable L_2 is the modulus of the weighted and phased sum of the matched filter outputs, i.e. the envelope detection procedure is added. A form of the envelope detection characteristic does not matter that is $|\mathbf{A}'\mathbf{N}^{-1}\mathbf{E}^*\tilde{\mathbf{G}}|$ may be subjected to any monotone transformation.

When probability distributions of φ_{s1} and a_{s1} are not available or these quantities are not random[4] (though unknown), then adaptive optimum processing algorithm may be obtained (see Section 4.3). Let us consider the log-likelihood ratio at fixed values of φ_{s1} and a_{s1} [before averaging over φ_{s1} and a_{s1}, see the exponent in (5.36)]

$$\ln\Lambda=a_{s1}\,\mathrm{Re}[\exp(\mathrm{j}\varphi_{s1})\mathbf{A}'\mathbf{N}^{-1}\mathbf{E}^*\tilde{\mathbf{G}}]-a_{s1}^2T_s\mathbf{A}'\mathbf{N}^{-1}\mathbf{A}. \tag{5.42}$$

The maximum likelihood estimate of φ_{s1} from (5.42) is

$$\hat{\varphi}_{s1}=-\arg\mathbf{A}'\mathbf{N}^{-1}\mathbf{E}^*\tilde{\mathbf{G}}. \tag{5.43}$$

Substituting (5.43) in (5.42) yields

$$\ln\Lambda=a_{s1}|\mathbf{A}'\mathbf{N}^{-1}\mathbf{E}^*\tilde{\mathbf{G}}|-a_{s1}^2T_s\mathbf{A}'\mathbf{N}^{-1}\mathbf{A}. \tag{5.44}$$

---

[3] Or else: when $\overline{a_{s1}^4}||\mathbf{A}'\mathbf{N}^{-1}\mathbf{E}^*\tilde{\mathbf{G}}|^2/4-T_s\mathbf{A}'\mathbf{N}^{-1}\mathbf{A}|\ll 2\overline{a_{s1}^2}$ where $\overline{a_{s1}^4}$ and $\overline{a_{s1}^2}$ are the fourth and the second initial moments of the PDF $w(a_{s1})$.

[4] Parameters φ_{s1} and a_{s1} may be nonrandom (though unknown), for instance, when the target is a sphere or some other "point" scatterer.

Solving the likelihood equation $\partial \ln \Lambda / \partial a_{s1} = 0$ for a_{s1}, yields the maximum likelihood estimate for a_{s1}:

$$\hat{a}_{s1} = |\mathbf{A}'\mathbf{N}^{-1}\mathbf{E}^*\tilde{\mathbf{G}}| / 2T_s \mathbf{A}'\mathbf{N}^{-1}\mathbf{A}. \tag{5.45}$$

Let us now substitute (5.45) in (5.44), omit the common factor $(4T_s\mathbf{A}'\mathbf{N}^{-1}\mathbf{A})$ not depending on the vector $\tilde{\mathbf{G}}$ and replace the squared modulus by the modulus itself. Then we obtain the algorithm L_2 (5.40), (5.41) again.

Thus *the optimal adaptive processing algorithm* (according to the generalized likelihood ratio criterion) *coincides with the optimum "in the average" algorithm for Gaussian complex signal amplitudes* (i.e. according to the unconditional likelihood ratio criterion). The functional scheme for the optimum processing algorithm is shown in Fig. 5.2.

Spatially Coherent Signals with Completely Mutually Dependent Amplitude Fluctuations in MSRSs with Short Term Spatial Coherence or in Spatially Incoherent MSRSs

Random and mutually independent phase shifts are added to the received signal phases in receivers of such MSRSs. These random phase shifts may be considered as uniformly distributed within the interval $(-\pi, \pi)$. Therefore, useful information that might be contained in the received signal initial phases, is lost. As a result the expected signal phases at different receiving stations, φ_{si}, $i = \overline{1,m}$, should be considered as random, mutually independent variables, whereas the effective (r.m.s.) signal values, a_{si}, preserve the complete mutual dependence [see (5.34)]. It should be noted that when a_{s1} is Rayleigh distributed and φ_{si}, $i = \overline{1,m}$, are uniformly distributed in $(-\pi, \pi)$, the signal complex amplitudes at different stations remain Gaussian at each station but become jointly non-Gaussian.

Thus let us assume that (5.34) holds for a_{s1}, $i = \overline{1,m}$, but for phases we have instead of (5.33)

$$w(\boldsymbol{\varphi}_s) = \prod_{i=1}^{m} w(\varphi_{si}); \tag{5.46a}$$

$$w(\varphi_{si}) = 1/2\pi, \quad \varphi_{si} \in (-\pi, \pi), \ i = \overline{1,m}. \tag{5.46b}$$

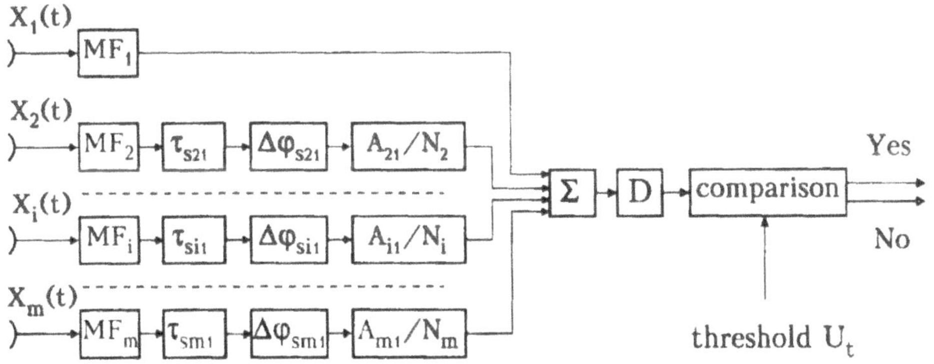

Figure 5.2 Structure of the optimum detector for signals with completely correlated fluctuating complex amplitudes in a spatially coherent MSRS [algorithm L_2, equations (5.40), (5.41)]

Such a mathematical model is similar to the first Swerling model [203]. Substituting (5.34), (5.38) and (5.46) in (5.30) yields

$$\bar{\Lambda}=\int_0^\infty \frac{2a_{s1}}{\sigma_1^2}\exp\left[-a_{s1}^2\left(\frac{1}{\sigma_1^2}+T_s\mathbf{A}'\mathbf{N}^{-1}\mathbf{A}\right)\right]\prod_{i=1}^m I_0(a_{s1}A_{i1}|\tilde{G}_i|/N_i)\,\mathrm{d}a_{s1}. \quad (5.47)$$

The explicit expression for the optimum processing algorithm can be obtained for weak signals at the outputs of matched filters in each station. When $a_{s1}^2A_{i1}^2|\tilde{G}_i|^2/16N_i \ll 1$ with high probability, then $I_0(z)\approx 1+z^2/4$. In this case integrating and omitting terms not depending on received signals yields

$$L_3=\sum_{i=1}^m (A_{i1}^2/N_i^2)|\tilde{G}_i|^2 \quad (5.48)$$

where $\tilde{G}_i$ is determined by (5.31).

Thus *the optimum processing algorithm in the case of weak signals* includes the matched filtration and squared envelope detection in each station and then the weighted summation of all square detector outputs. Though the algorithm L_3 (5.48) has been obtained for Rayleigh distribution of a_{s1} it remains optimal for any probability distribution of a_{s1} under the weak signal condition. It is easy to verify by substituting an arbitrary probability density $w(a_{s1})$ in (5.47) and again replacing $I_0(a_{s1}A_{i1}|\tilde{G}_i|/N_i)$ by $1+a_{s1}^2A_{i1}^2|\tilde{G}_i|^2/4N_i^2$. The only constraint that should be imposed on $w(a_{s1})$ is the convergence of the integral

$$\int_0^\infty w(a_{s1})a_{s1}^2\exp(-a_{s1}^2T_s\mathbf{A}'\mathbf{N}^{-1}\mathbf{A})\,\mathrm{d}a_{s1}.$$

When probability distributions of φ_{si}, $i=\overline{1,m}$, and a_{s1} are not available or φ_{si} and a_{s1} are not random (but unknown), the generalized likelihood ratio criterion may be exploited. From (5.30) we can obtain at the fixed $\boldsymbol{\varphi}_s$ and a_{s1}

$$\ln\Lambda=a_{s1}\,\mathrm{Re}\sum_{i=1}^m (A_{i1}/N_i)\exp(\mathrm{j}\varphi_{si})\tilde{G}_i-a_{s1}^2T_s\sum_{i=1}^m (A_{i1}^2/N_i). \quad (5.49)$$

From (5.49) we can find the maximum likelihood estimate for φ_{si}:

$$\hat{\varphi}_{si}=-\arg\tilde{G}_i,\quad i=\overline{1,m}. \quad (5.50)$$

Substituting (5.50) in (5.49) we can solve the likelihood equation $\partial\ln\Lambda/\partial a_{s1}=0$ for a_{s1} and obtain

$$\hat{a}_{s1}=\left[\sum_{i=1}^m (A_{i1}/N_i)|\tilde{G}_i|\right]\Big/4T_s\sum_{i=1}^m (A_{i1}^2/N_i). \quad (5.51)$$

Substituting the estimate $\hat{a}_{s1}$, discarding the common factor not depending on received signals and replacing the squared sum by the sum itself yields *the optimum adaptive processing algorithm*

$$L_4=\sum_{i=1}^m (A_{i1}/N_i)|\tilde{G}_i|. \quad (5.52)$$

Unlike the algorithm L_3 (5.48) the linear envelope detection is optimal in (5.52). In accordance with this the weights in (5.52) are equal to the square root of the weights in (5.48).

Spatially Incoherent Signals with Mutually Independent Amplitude Fluctuations in MSRSs with Arbitrary Spatial Coherence

In such a MSRS both the initial phases, φ_{si}, and effective values, a_{si}, of wanted signals are mutually independent at the inputs of different stations ($i=\overline{1,m}$) irrespective of the degree of spatial coherence of the MSRS itself. As before, we assume the signal complex amplitudes to be Gaussian random variables. Under this condition the phase probability distribution has been determined in (5.46). For signal effective values we have

$$w(\mathbf{a}_s)=\prod_{i=1}^{m} w(a_{si});$$

$$w(a_{si})=(2a_{si}/\sigma_1^2 A_{i1}^2)\exp(-a_{s1}^2/\sigma_1^2 A_{i1}^2); \quad a_{si}\in(0,\infty). \tag{5.53}$$

Such a model is similar to the second Swerling model [203]. Now we can substitute (5.46) and (5.53) in (5.30). Having integrated firstly with respect to φ_{si} and then to a_{si}, taken the logarithm and omitted terms not depending on $\tilde{G}_i$ we obtain *the optimum processing algorithm* in the form

$$L_5=\sum_{i=1}^{m}\frac{A_{i1}^2|\tilde{G}_i|^2}{N_i^2(1+\sigma_1^2A_{i1}^2T_s/N_i)}=\sum_{i=1}^{m}\frac{A_{i1}^2|\tilde{G}_i|^2}{N_i^2(1+\bar{E}_i/N_i)} \tag{5.54}$$

where $\tilde{G}_i$ is defined in (5.31) and $\bar{E}_i=\sigma_1^2A_{i1}^2T_s=\sigma_i^2T_s$ is the averaged expected signal energy at the ith station so that $\bar{E}_i/N_i=\bar{q}_{\text{out}i}^2$.

Thus for spatially uncorrelated Gaussian signal complex amplitudes the optimum processing algorithm L_5 comes again to the weighted summation of squared signal envelopes from all stations after the matched filtration in each station. For weak signals ($\bar{E}_i/N_i \ll 1$) the algorithm L_5 (5.54) is reduced to L_3 (5.48). Taking into account that L_3 has been obtained for strongly dependent interstation fluctuations, it may be expected that for weak signals L_3 is optimal for arbitrary mutual statistical dependence of a_{si} when φ_{si} are mutually independent ($i=\overline{1,m}$). Indeed, it may be shown by substituting (5.46) into (5.30), averaging over φ_{si}, expanding the exponent and the product of Bessel functions into the Taylor series and discarding all terms containing the third and higher powers of a_{si}. Then averaging over a_{si} with arbitrary $w(a_{s1},\ldots,a_{sm})$ yields (5.48) again [if second moments of $w(a_{s1},\ldots,a_{sm})$ are limited].

In the case of strong signals ($\bar{E}_i/N_i \gg 1$) we have *the optimum processing algorithm* from (5.54)

$$L_6=\sum_{i=1}^{2}|\tilde{G}_i|^2/N_i \tag{5.55}$$

which differs from L_5 by the weights only.

When probability distributions for φ_{si} and a_{si} are not known or these unknown parameters are nonrandom, an adaptive optimum processing algorithm may be constructed as before. The log-likelihood ratio algorithm at fixed values of φ_{si} and a_{si} can be written from (5.30)

$$\ln \Lambda = \sum_{i=1}^{m} (a_{si}/N_i)\,\mathrm{Re}[\exp(j\varphi_{si})\tilde{G}_i] - T_s \sum_{i=1}^{m} a_{si}^2/N_i. \tag{5.56}$$

The maximum likelihood estimates from (5.56) are

$$\hat{\varphi}_{si} = -\arg \tilde{G}_i; \qquad \hat{a}_{si} = |\tilde{G}_i|/2T_s.$$

Substituting these estimates in (5.56) and omitting the common factor $1/4T_s$ yields the algorithm L_6 (5.55) again.

Thus *the optimum adaptive processing algorithm in this case coincides with the optimum algorithm obtained for strong signals and Gaussian spatially independent complex amplitudes* $a_{si}\exp(-j\varphi_{si})$ *by averaging over* φ_{si} *and* a_{si}.

Comparing (5.48), (5.52), (5.54) shows that the weights of all summands in each algorithm become equal under two conditions: when the noise PSDs are equal and the output SNR values (or the signal powers) are equal at all the stations ($N_i = N$ and $\tilde{q}^2_{\text{out}\,i} = \tilde{q}^2_{\text{out}}$ or $A^2_{i1} = 1$ for all $i = \overline{1,m}$). Under these conditions all the three algorithms coincide. When only PSDs are equal, then the algorithms L_3 and L_5 go over into the algorithm L_6. If the output SNR values $\tilde{q}^2_{\text{out}\,i} = \bar{E}_i/N_i = \sigma_1^2 A^2_{i1} T_s/N_i = \sigma_i^2 T_s/N_i$ are not equal at different stations, then $\tilde{q}^2_{\text{out}\,i}$ turn out to be included in the weights of the algorithm L_4 (5.52) for strongly spatially correlated fluctuations and of the algorithms L_3 and L_5 for spatially independent fluctuations but in the latter case only when the signals are not sufficiently strong (the condition $\bar{E}_i/N_i \gg 1$ is not satisfied). The functional scheme of the optimum detector which operates according to the algorithms L_3, L_4, L_5 and L_6 are shown in Fig. 5.3.

For the reader's convenience all the algorithms that have been synthesized in this Section are collected in Table 5.1. It is reasonable to recall that the quantity $\tilde{G}_i$, entering in all optimum algorithms, is a result of the received signal matched filtration. It is defined in the frequency domain by (5.31). As was shown in Section

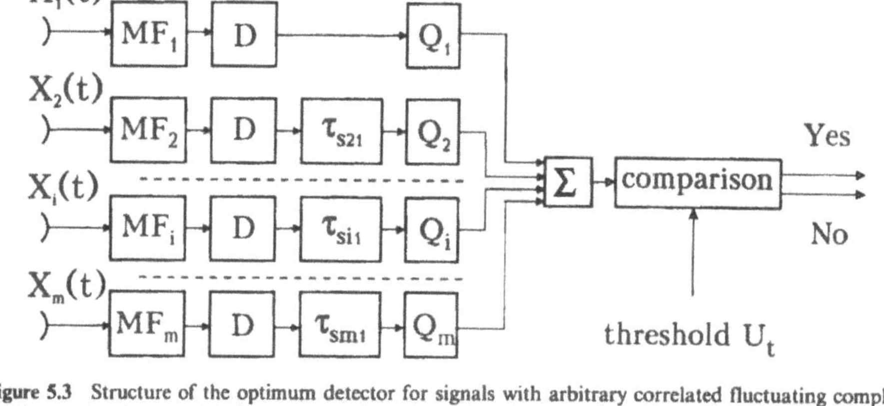

Figure 5.3 Structure of the optimum detector for signals with arbitrary correlated fluctuating complex amplitudes in a MSRS with short-term spatial coherence and in a spatially incoherent MSRS [algorithms $L_3, \ldots, L_6$, equation (5.48), (5.52), (5.54), (5.55); Q_i = weights, $i = \overline{1,m}$]

5.1, it is easy to transform (5.31) into the time domain:

$$\tilde{G}_i = \int_{-T/2}^{T/2} X_i(t) s_0^*(t - t_{si}) \exp[-\mathrm{j}(\omega_0 + \Omega_{si})(t - t_{si})] \, \mathrm{d}t. \tag{5.57}$$

Table 5.1

Signals at the inputs of receiving stations	Spatially coherent MSRS	
	Optimum algorithms averaged over random parameters	Optimum adaptive algorithms
Spatially coherent with completely dependent amplitude fluctuations $w(\varphi_{s1}, \dots, \varphi_{sm}) = w(\varphi_{s1}) \times \prod_{i=1}^{m} \delta(\varphi_{si} - \varphi_{s1} + \Delta\varphi_{si1})$; $w(a_{si} - A_{i1} a_{s1}) = w(a_{s1}) \times \prod_{i=1}^{m} \delta(a_{si} - A_{i1} a_{s1})$;	$L_2 = \left\| \sum_{i=1}^{m} \frac{A_{i1}}{N_1} e^{-\mathrm{j}\Delta\varphi_{si1}} \tilde{G}_i \right\|$. (5.40) *Additional conditions:* $w(\varphi_{s1}) = \frac{1}{2\pi}, \ \varphi_{s1} \in (0, 2\pi)$; $w(a_{s1}) = \frac{2a_{s1}}{\sigma_1^2} \exp\left(-\frac{a_{s1}^2}{\sigma_1^2}\right)$ or any distribution for weak signals	$L_2 = \left\| \sum_{i=1}^{m} \frac{A_{i1}}{N_1} e^{-\mathrm{j}\Delta\varphi_{si1}} \tilde{G}_i \right\|$. (5.40) *Additional conditions:* maximum likelihood estimates $\hat{\varphi}_{s1}$ and $\hat{a}_{s1}$ are used
Spatially incoherent with independent amplitude fluctuations $w(\varphi_{s1}, \dots, \varphi_{s1}) = \prod_{i=1}^{m} w(\varphi_{si})$; $w(a_{s1}, \dots, a_{sm}) = \prod_{i=1}^{m} w(a_{si})$	$L_5 = \sum_{i=1}^{m} \frac{A_{i1}^2}{N_i^2} \|\tilde{G}_i\|^2 \left(1 + \frac{E_i}{N_i}\right)^{-1}$. (5.54) *Additional conditions:* $w(\varphi_{si}) = \frac{1}{2\pi}, \ \varphi_{si} \in (0, 2\pi)$; $w(a_{si}) = \frac{2a_{si}}{\sigma_1^2 A_{i1}^2} \times \exp\left(-\frac{a_{si}^2}{\sigma_1^2 A_{i1}^2}\right); \ i = \overline{1, m}$ ($L_5 \rightarrow L_3$ for weak signals) $L_6 = \sum_{i=1}^{m} \frac{\|\tilde{G}_i\|^2}{N_i}$. (5.55) *Additional conditions:* the same as for L_5 and strong signals	$L_6 = \sum_{i=1}^{m} \frac{\|\tilde{G}_i\|^2}{N_i}$. (5.55) *Additional conditions:* maximum likelihood estimates $\hat{\varphi}_{s1}$ and $\hat{a}_{s1}$ are used, $i = \overline{1, m}$
Spatially incoherent with arbitrary statistical dependence of amplitude fluctuations $w(\varphi_{s1}, \dots, \varphi_{sm}) = \prod_{i=1}^{m} w(\varphi_{si})$;	$L_3 = \sum_{i=1}^{m} \frac{A_{i1}^2}{N_i^2} \|\tilde{G}_i\|^2$. (5.48) *Additional conditions:* $w(\varphi_{si}) = \frac{1}{2\pi}, \varphi_{si} \in (0, 2\pi), i = \overline{1, m}$; any distribution of $a_{s1}, \dots, a_{sm}$ for weak signals	$L_6 = \sum_{i=1}^{m} \frac{\|\tilde{G}_i\|^2}{N_i}$. (5.55) *Additional conditions:* maximum likelihood estimates $\hat{\varphi}_{si}$ and $\hat{a}_{si}$ are used, $i = \overline{1, m}$

Table 5.1 *(continued)*

MSRS with short-term spatial coherence and spatially incoherent MSRS	
Optimum algorithms averaged over random parameters	Optimum adaptive algorithms
$L_3=\sum_{i=1}^{m}\frac{A_{i1}^2}{N_i^2}\lvert\tilde{G}_i\rvert^2.$ (5.48) *Additional conditions:* $w(\varphi_{si})=\frac{1}{2\pi},\ \varphi_{si}\in(0,2\pi),\ i=\overline{1,m};$ any distribution of a_{s1} for weak signals	$L_4=\sum_{i=1}^{m}\frac{A_{i1}}{N_i}\lvert\tilde{G}_i\rvert.$ (5.52) *Additional conditions:* maximum likelihood estimates $\hat{\varphi}_{s1},\ldots,\hat{\varphi}_{sm}$ and $\hat{a}_{s1}$ are used
$L_5=\sum_{i=1}^{m}\frac{A_{i1}^2}{N_i^2}\lvert\tilde{G}_i\rvert^2\left(1+\frac{E_i}{N_i}\right)^{-1}.$ (5.54) *Additional conditions:* $w(\varphi_{si})=\frac{1}{2\pi},\ \varphi_{si}\in(0,2\pi);$ $w(a_{si})=\frac{2a_{si}}{\sigma_1^2A_{i1}^2}\times\exp\left(-\frac{a_{si}^2}{\sigma_1^2A_{i1}^2}\right),\ i=\overline{1,m}$ ($L_5\to L_3$ for weak signals) $L_6=\sum_{i=1}^{m}\frac{\lvert\tilde{G}_i\rvert^2}{N_i}.$ (5.55) *Additional conditions:* the same as for L_5 and strong signals	$L_6=\sum_{i=1}^{m}\frac{\lvert\tilde{G}_i\rvert^2}{N_i}.$ (5.55) *Adiitional conditions:* maximum likelihood estimates $\hat{\varphi}_{si}$ and $\hat{a}_{si}$ are used, $i=\overline{1,m}$
$L_3=\sum_{i=1}^{m}\frac{A_{i1}^2}{N_i^2}\lvert\tilde{G}_i\rvert^2.$ (5.48) *Additional conditions:* $w(\varphi_{si})=\frac{1}{2\pi},\ \varphi_{si}\in(0,2\pi),i=\overline{1,m};$ any distribution of $a_{s1},\ldots,a_{sm}$ for weak signals	$L_6=\sum_{i=1}^{m}\frac{\lvert\tilde{G}_i\rvert^2}{N_i}.$ (5.55) *Additional conditions:* maximum likelihood estimates $\hat{\varphi}_{si}$ and $\hat{a}_{si}$ are used, $i=\overline{1,m}$

When the station ADPs are necessary to be taken into account it may be done as in Section 5.1. The optimum processing algorithms L_2 (5.40), L_3 (48), L_4 (5.52) and L_6 (5.55) remain valid if we replace (5.31) by the following expression [compare with (4.3) and (4.5)]

$$\tilde{G}_i=\frac{1}{2\pi}\int_{-\infty}^{\infty}\chi_i(\omega)g_i^*(\beta_{si},\varepsilon_{si},\omega)\Psi_0^*(\omega-\omega_0-\Omega_{si})\exp(j\omega t_{si})\,d\omega,\quad i=\overline{1,m}. \tag{5.58}$$

Algorithm L_5 (5.54) also remains valid if ADP $g_i^*(\beta_{si}, \varepsilon_{si}, \omega)$ is additionally inserted into the expression for the averaged signal energy $\bar{E}_i$. Using (5.21) yields $C_{ii}=2\alpha_i T_s/N_i$ instead of (5.32) and $\bar{E}_i=\sigma_1^2 A_{i1}^2 \alpha_i T_s$. These expressions should be substituted into (5.54).

If there are "coloured" noises at the inputs of receiving stations (e.g., clutter), it may be taken into consideration as in Section 5.1. The matrix $\mathbf{f}(\omega)=\mathbf{\Phi}^{-1}(\omega)$ in (5.28) takes the form $\mathbf{f}(\omega)=\|\delta_{ik}N_i^{-1}[1+q_{1i}^2 F_i(\omega-\omega_0)]^{-1}\|$. For elements of the vector $\mathbf{G}$ and matrix $\mathbf{C}$ we have instead of (5.31) and (5.32)

$$G_i=\frac{\tilde{G}_i}{N_i}=\frac{1}{2\pi N_i}\int_{-\infty}^{\infty}\frac{\chi_i(\omega)\Psi_0^*(\omega-\omega_0-\Omega_{si})\exp(j\omega t_{si})}{1+q_{1i}^2 F_i(\omega-\omega_0)}\,d\omega;$$

$$C_{ii}=\frac{1}{2\pi N_i}\int_{-\infty}^{\infty}\frac{|\Psi_0(\omega-\omega_0-\Omega_{si})|^2}{1+q_{1i}^2 F_i(\omega-\omega_0)}\,d\omega=2K_i T_s/N_i. \tag{5.59}$$

Here $K_i \leqslant 1$ is the signal energy utilization factor (5.24b) which is equal to one when $q_{1i}^2=0$. If we substitute $\tilde{G}_i$ from (5.59) in the algorithms $L_2, \ldots, L_6$ synthesized above and additionally include in L_5 the factor K_i into the expression for the averaged signal energy $\bar{E}_i=\sigma_1^2 A_{i1}^2 K_i T_s$, all these algorithms will retain their optimality for "coloured" noises at the inputs of receiving stations.

In this section*, we have synthesized a series of optimum processing algorithms for fluctuating signal detection in MSRSs when noises at the inputs of spatially separated receiving stations may be considered as mutually uncorrelated. We have addressed the case where a MSRS consists of a single transmitting and several (at least two) receiving stations. For the reader's convenience all synthesized algorithms are gathered in Table 5.1. All these algorithms contain the matched filtration in each station and then weighted summation either of the matched filters' phased outputs themselves or of their moduli or squared moduli. The character of weighted summation depends on the spatial correlation of the expected signal complex amplitude fluctuations and on the spatial coherence of the MSRS. We have obtained the optimum processing algorithms for Gaussian complex amplitude fluctuations* [L_2 (5.40), (5.41), L_5 (5.54)], *the simplified algorithms which are optimal for weak or strong output signals* [L_3 (5.48) *and* L_6 (5.55)] *and the adaptive algorithms for the cases where probability distributions of complex amplitudes are not available or those amplitudes are not random (though unknown)* [L_2 (5.40), (5.41), L_4 (5.52) *and* L_6 (5.55)]. *The synthesized algorithms presented in the frequency domain, can easily be transformed into the time domain. Clearly, all the obtained algorithms can be implemented in both analogous and digital forms.*

5.3. OPTIMUM DETECTION ALGORITHMS FOR FLUCTUATING SIGNALS WHEN A TARGET IS ILLUMINATED BY SEVERAL TRANSMITTING STATIONS

Let us assume that a MSRS contains m receiving stations as before but a target is illuminated by n transmitting stations. Some or all of these transmitting stations may coincide with receiving stations, i.e. be monostatic radars.

As was discussed in Section 5.1, the two extreme cases are possible. In the first case all echoes from a target caused by different transmitting stations are resolved in each receiving station. In this case there is a detection problem of n echoes in each station and of mn echoes in a MSRS as a whole (in a background of statistically independent noises). In the second case illuminated signals from all transmitting stations are summed at the target in such a manner that echoes caused by different transmitting stations are not resolved in receiving stations. Of course, for such a summation the special synchronization of transmitting stations (with errors much less than the reciprocal of the signal bandwidth) is necessary. When the transmitting waveforms and receivers have the resolution capability in Doppler frequency, then the closeness of carrier frequencies and Doppler frequency shifts at each receiving station are required (the frequency errors must be much less than the reciprocal of the signal duration). For signal coherent (in phase) summation at the target the mutual phase alignment ("phase synchronization") of illuminating signals is required too. Apparently, these strict requirements are not likely to be satisfied in usual MSRSs especially in a surveillance and search mode. However, in some special MSRSs these conditions can be met, for example, by means of the "autofocussing techniques" [127], in particular, in MSRSs designed for radio imaging [114–116].

Obviously, apart from above mentioned extreme cases, intermediate cases are possible where some echoes caused by different transmitting stations can be resolved in receiving stations and some cannot. Corresponding optimum processing algorithms can be derived basing on the results obtained for the extreme cases.

Echoes Caused by Different Transmitting Stations are Not Resolved in All Receiving Stations

In this case the problem of the synthesis of optimum detection algorithms does not differ from the similar problem with a single transmitting station. When a MSRS contains m receiving stations, then there are only m echoes (from each target) which are to be detected in the MSRS regardless of the number of transmitting stations. It is necessary, however, to clarify the statistical features of expected signals (at the inputs of receiving stations) caused by the summed target illumination from all transmitting stations.

Let us assume the expected signal complex amplitudes at the inputs of receiving stations when a target is illuminated by a single arbitrary transmitting station to be Gaussian. Then the complex amplitudes of expected signals caused by the coherent summation of target illuminating signals from several transmitting stations are Gaussian too. The Gaussian distribution of the expected signal complex amplitudes holds also in the case where the Gaussian complex amplitudes of illuminating signals are uncorrelated, i.e. in the case of incoherent summation of illuminating signals from different transmitting stations at the target. It can occur when carrier frequencies of illuminating signals differ by $\delta f > c/2l$ (c is the light velocity, l is the characteristic target dimension) or/and when baselengths between transmitting stations are large enough to destroy the spatial correlation of expected signals caused by different transmitting stations at each receiving station (see Section 4.2). It is assumed as before that even in the case of incoherent summation of the illuminating signals they are so synchronized that echoes caused by different transmitting stations are not resolved in time in all receiving stations and Doppler resolution capability is absent or insufficient.

For Gaussian complex amplitudes of summed expected signals at the inputs of receiving stations algorithms L_2, L_3 and L_5 remain optimal under the same conditions as in the case of a single transmitting station. Algorithm L_3 holds its optimality for weak signals when echoes caused by different transmitting stations are incoherent but their amplitude fluctuations are completely correlated, i.e. when complex amplitudes of the summed signals are non-Gaussian (see Section 8.4). The adaptive algorithms L_2, L_4 and L_6 that are optimal according to the generalized likelihood ratio criterion (when probability distribution of signal complex amplitudes is unknown), remain optimal under the same conditions as in the case of a single transmitting station.

Echoes Caused by Different Transmitting Stations are Resolved in Each Receiving Station

This situation is easier to realize in practice so that we shall consider it in more detail. In this case optimum processing algorithms depend not only on the interstation (spatial) echo fluctuation correlation but on the mutual (temporal) correlation of echoes caused by different transmitting stations and received by each receiving station.

Let us consider the unconditional likelihood ratio (5.30). When at the inputs of m receiving stations mn scattered signals are expected, then each element of the vectors $\tilde{\mathbf{G}}$, $\mathbf{a}_s$ and of the matrices $\tilde{\mathbf{E}}(\boldsymbol{\varphi}_s)$, $\mathbf{N}^{-1}$ becomes a vector $(n\times 1)$ or a matrix $(n\times n)$ itself. Each of the m blocks of the matrix $\mathbf{N}^{-1}$ becomes a diagonal matrix, $\mathbf{N}_i^{-1}$, consisting of n equal elements. The probability density functions $w(\boldsymbol{\varphi}_s)$ and $w(\mathbf{a}_s)$ must take into account mutual statistical properties of initial phases and effective (r.m.s.) values of signals expected not only at different stations but at each station as well. We shall consider here the most typical cases.

When the transmitting stations of a MSRS are spatially coherent, then the MSRS as a whole is usually spatially coherent. On the other hand, trying to get the spatial coherence of a MSRS may be reasonable only in the case where target echoes (expected signals) are spatially coherent too. In practice it means strong mutual dependence not only between phases but between complex amplitudes of expected signals. Therefore, let us consider *the detection problem of mn resolved signals with strong spatial dependence of complex amplitudes in a spatially coherent MSRS containing n transmitting and m receiving stations*. This problem differs only by its dimension from a similar problem for a MSRS with a single transmitting station. In a collection of mn echoes the initial phase and the effective value are random only for a single "reference" signal. Without loss of generality it may be the expected signal with a minimum time of arrival (TOA) at the first receiving station. Repeating the reasoning from Section 5.2 yields *the optimum processing algorithm* in the form [compare with L_2 (5.40)]

$$L_7=|\mathbf{A}^{\mathrm{t}}\mathbf{N}^{-1}\mathbf{E}^{*}\tilde{\mathbf{G}}|=\left|\sum_{k=1}^{n}\sum_{i=1}^{m}(A_{ik1}/N_i)\exp(-\mathrm{j}\Delta\varphi_{ik1})\tilde{G}_{ik}\right| \tag{5.60}$$

where A_{ik1}^2 is the known averaged power ratio of the kth expected signal at the ith receiving station ($k=\overline{1,n}$, $i=\overline{1,m}$) to the "reference" signal, i.e. the ratio of the second moments of the probability distributions for a_{ik} and a_{11} [compare with (5.35)]; $\Delta\varphi_{ik1}$ is the known difference of the initial phases for the same signals; $\tilde{G}_{ik}$ is the result of the kth signal matched filtration in the ith station. For the expected signals resolved in TOA, we have from (5.31)

$$\tilde{G}_{ik}=\frac{1}{2\pi}\int_{-}\chi_{ik}(\omega)\Psi_0^{*}(\omega-\omega_0-\Omega_{sik})\exp(\mathrm{j}\omega t_{sik})\,\mathrm{d}\omega. \tag{5.61}$$

In (5.61) $\chi_{ik}(\omega)$ is the spectrum of the signal segment received by the ith station where the echo caused by the kth transmitting station is expected.

Consider now another possible situation. Let the receiving stations be as before spatially coherent with comparatively short baselines between them so that complex amplitude fluctuations of echoes caused by each transmitting station at different receiving stations are strongly correlated. At the same time either baselines between transmitting stations are large or illuminating signals from different transmitting stations are significantly separated in carrier frequency. Then complex amplitudes of echoes caused by different transmitting stations are mutually independent. In this case the following detection problem arises: *to detect mutually independent signals caused by different transmitting stations at each receiving station together with strongly correlated signals caused by each transmitting station at different spatially coherent receiving stations*. The total collection of mn expected signals can be divided into n mutually independent groups; each group consists of m mutually strongly dependent echoes:

$$w(\boldsymbol{\varphi}_s)=w(\boldsymbol{\varphi})=\prod_{k=1}^{n} w(\varphi_{1k})\prod_{i=2}^{m}\delta(\varphi_{ik}-\varphi_{1k}+\Delta\varphi_{i1k}) \tag{5.62}$$

$$w(\mathbf{a}_s)=w(\mathbf{a})=\prod_{k=1}^{n} w(a_{1k})\prod_{i=2}^{m}\delta(a_{ik}-A_{i1k}a_{1k}). \tag{5.63}$$

Substituting (5.62) and (5.63) in (5.30), taking into account (5.61) and (5.38), we obtain the optimum processing algorithm in the form

$$L_8=\sum_{k=1}^{n}\tilde{A}_{k1}^2\left[\left|\sum_{i=1}^{m}(A_{i1}/N_i)\exp(-j\Delta\varphi_{i1k})\tilde{G}_{ik}\right|^2\Big/\left(1+\sum_{i=1}^{m}\bar{E}_{ik}/N_i\right)\right] \tag{5.64}$$

where $\tilde{A}_{k1}^2$ is the ratio of the averaged powers of signals caused by the kth and "reference" transmitting stations, $k=\overline{1,n}$ (for simplicity this ratio is assumed to be equal for all receiving stations, $i=\overline{1,m}$); A_{i1}^2 is the ratio of the averaged powers of signals caused by a single transmitting station at the ith and the "reference" receiving stations (for simplicity assumed to be the same for all transmitting stations); $\bar{E}_{ik}$ is the averaged energy of the expected signal caused by the kth transmitting station at the ith receiving station. All other notations are the same as in (5.60). When the assumptions concerning independence of $\tilde{A}_{k1}^2$ and A_{i1}^2 on the numbers of receiving and transmitting stations, respectively, are not permissible, then the product $\tilde{A}_{k1}^2\, A_{i1}^2$ should be substituted for the factor A_{ik1}^2 which have been introduced in (5.60).

Comparing L_8 (5.64) with L_7 (5.60) and with L_2 (5.40) and L_5 (5.54) it is easy to notice that the optimum algorithm includes three processing stages. At the first stage the matched filtration is carried out in each receiving station, i.e. the quantities $\tilde{G}_{ik}$, $i=\overline{1,m}$, $k=\overline{1,n}$, are obtained. At the second stage, all $\tilde{G}_{ik}$ for a fixed k, i.e. caused by the kth transmitting station, are summed coherently (over i) with certain weights and undergo the squared envelope detection. Finally, at the third stage, the weighted summation is performed for the squared envelope detection results corresponding to different transmitting stations (over k).

When the probability distributions of a_{1k} and φ_{1k}, $k=\overline{1,n}$, in (5.30) are not available or these parameters are nonrandom (but unknown) their *maximum likelihood estimates and optimum adaptive processing algorithm* can be obtained. As for the case of a single transmitting station [see, for example, (5.50)–(5.52)] we can write

$$\hat{\varphi}_{1k} = -\arg \sum_{i=1}^{m} (A_{1ik}/N_i) \exp(-j\Delta\varphi_{i1k}) \tilde{G}_{ik},$$

$$\hat{a}_{1k} = -\left| \sum_{i=1}^{m} (A_{1ik}/N_i) \exp(-j\Delta\varphi_{i1k}) \tilde{G}_{ik} \right| \Big/ \left(2 \sum_{i=1}^{m} A_{i1k}^2 T_s / N_i \right), \tag{5.65}$$

$$L_9 = \sum_{k=1}^{n} \left| \sum_{i=1}^{m} (A_{i1k}/N_i) \exp(-j\Delta\varphi_{i1k}) \tilde{G}_{ik} \right|^2. \tag{5.66}$$

Algorithm L_9 (5.56) does not differ from algorithm L_8 (which is optimal in average for Gaussian complex amplitudes) under strong signal condition when $\sum_{i=1}^{m} \bar{E}_{ik}/N_i \gg 1$. It is easy to verify taking into account in L_8 that $\bar{E}_{ik} = \sigma_{11}^2 \bar{A}_{k1}^2 A_{i1}^2 T_s$ and omitting the common factor $\sum_{i=1}^{m} \sigma_{11}^2 A_{i1} T_s / N_i$.

Now we can pass to the simplest case (from the technical point of view) which hence most often occurs in practice. It is the case where baselengths between stations are large enough so that the spatial (interstation) statistical coupling of expected signal complex amplitude fluctuations is absent. Besides that carrier transmitting frequencies may be different. Clearly, so far as the expected signals are spatially incoherent, MSRSs may be spatially incoherent too.

Under these conditions the following problem arises: *to detect mn resolved signals with mutually independent complex amplitudes*. Obviously, this problem differs from the similar problem in the case of a single transmitting station by its dimension only. Repeating the considerations which have led to algorithm L_5 (5.54) we can obtain *the optimum (averaged over random parameters) processing algorithm for Gaussian complex amplitudes*

$$L_{10} = \sum_{k=1}^{n} \sum_{i=1}^{m} \bar{A}_{k1}^2 A_{i1}^2 |\tilde{G}_{ik}|^2 / N_i^2 (1 + \bar{E}_{ik}/N_i) \tag{5.67}$$

where all notations are the same as in (5.64) and (5.66). The difference from algorithm L_5 for a single transmitting station is in additional weighted summation of the square envelope detector outputs over all the transmitting stations (over $k = \overline{1, n}$).

If the influence of receiving station ADPs or of "coloured" noises at the receiver inputs is to be taken into account, it may be done as in the case of a single transmitting station (see Section 5.2).

We note in closing that the problem of detection of n resolved echoes from each target in each of the m receiving stations can also arise in a MSRS containing only one transmitting station. This situation occurs when the transmitting station radiates n separated signals and echoes caused by them are resolved in receiving stations in time or in frequency. A typical example is a pulse train signal (see Chapter 13). The optimum algorithms $L_7, \ldots, L_{10}$ are perfectly applicable to such MSRSs.

***In this section**, we have synthesized optimal detection algorithms for MSRSs containing not only several (m) receiving stations but several (n) transmitting stations, which can be colocated or not colocated with the receiving stations. When echoes caused by different transmitting stations cannot be resolved in each receiving station, then all the algorithms* $L_2, \ldots, L_6$ (5.40), (5.41), (5.48), (5.52), (5.54), (5.55) *obtained in*

Section 5.2 for the case of a single transmitting station remain valid under their corresponding conditions. When echoes caused by different transmitting stations are resolved in each receiving station, we have obtained optimum algorithms L_7 (5.60), L_8 (5.64), L_9 (5.66) and L_{10} (5.67) for different degree of interstation dependence of signal fluctuations. These algorithms are the straightforward extensions of the algorithms synthesized in Section 5.2.

5.4. PERFORMANCE ANALYSIS OF OPTIMUM DETECTORS FOR FLUCTUATING SIGNALS. COHERENT SUMMATION ALGORITHMS

The optimal detection algorithms that have been synthesized in Sections 5.2 and 5.3 are similar to corresponding optimal detection algorithms for pulse train signals in monostatic radars. The reason is the similarity or even identity of the likelihood ratio formalism for a MSRS, on the one hand, and for a pulse train signal in a monostatic radar, on the other. This similarity holds also for performance analysis of synthesized detectors. However, all pulses of such a train are usually assumed to be equally mutually correlated and to have equal SNRs at the input of a monostatic radar receiver. It follows from Sections 5.1–5.3 that differences in mutual correlation and SNR of echoes at the inputs of spatially separated stations often exist in MSRSs and are to be taken into account.

To obtain detection characteristics for the optimal algorithms, probability distributions of the statistics L_2–L_{10} should be derived. Algorithms L_2, L_7 including a coherent summation procedure are optimal for spatially coherent MSRSs and strong interstation correlation of echo complex amplitudes. We consider firstly algorithms L_2 (5.40), (5.41). Let us denote the random variable at the input of the envelope detector by $\mathfrak{Z}_2$ (so that $L_2=|\mathfrak{Z}_2|$), the noise and signal components by the subscripts "n", "s", respectively, and the sum of signal plus noise by the subscript "sn" as in Section 5.1. In the absence of a wanted signal $\mathfrak{Z}_{2n}$ is a Gaussian variable with zero mean (as a linear combination of mutually independent Gaussian noises with zero mean) and the variance:

$$\sigma^2(\mathfrak{Z}_{2n})=0.5\sum_{i=1}^{m}\sum_{k=1}^{m}\frac{A_{i1}A_{k1}}{N_iN_k}\exp[-j(\Delta\varphi_{si1}-\Delta\varphi_{sk1})]\overline{\tilde{G}_{ni}\tilde{G}^*_{nk}}.$$

Taking into account (5.31) at $\chi_i(\omega)=\tilde{N}_i(\omega)$ and (5.11) yields

$$\sigma^2(\mathfrak{Z}_{2n})=\sum_{i=1}^{m}2A_{i1}^2T_s/N_i. \tag{5.68}$$

The probability distribution of $\mathfrak{Z}_{2sn}$ depends on the character of echo fluctuations. Let at first a_{s1} and φ_{s1} be nonrandom (though unknown), i.e. expected signals are nonfluctuating. Then $\sigma^2(\mathfrak{Z}_{2sn})=\sigma^2(\mathfrak{Z}_{2n})$ but a nonzero mean value arises [see (5.41)]

$$\mathfrak{Z}_{2s}=\bar{\mathfrak{Z}}_{2sn}=\sum_{i=1}^{m}(A_i/N_i)\exp(-j\Delta\varphi_{si1})\frac{1}{2\pi}\int_{-\infty}^{\infty}\overline{\chi_i(\omega)}\Psi_0^*(\omega-\omega_0-\Omega_{si})\exp(j\omega t_{si})\,d\omega. \tag{5.69}$$

Substituting the signal spectrum (5.8) for $\overline{\chi_i(\omega)}$ and taking into account (5.34) we have from (5.69)

$$\mathfrak{J}_{2s} = a_{s1} \exp(-j\varphi_{s1}) \sum_{i=1}^{m} 2A_{i1}^2 T_s / N_i. \tag{5.70}$$

Thus when a wanted signal (target echo) is absent, then $L_2 = L_{2n}$ is a Rayleigh distributed random variable, so that the false alarm probability is

$$P_{fa} = \exp(-u_0^2/2). \tag{5.71}$$

When a wanted signal is present then $L_2 = L_{2sn}$ is a Rician random variable so that the detection probability can be written as

$$P_d = \int_{u_0}^{\infty} x \exp[-(x^2 + q_{out}^2)/2] I_0(x q_{out})\, dx \tag{5.72}$$

where $u_0 = u_t / \sigma(\mathfrak{J}_{2n})$ is the normalized threshold level u_t, q_{out}^2 is the power SNR at the output of the detection algorithm's linear part. From (5.68) and (5.70) we can find

$$q_{out}^2 = \frac{|\mathfrak{J}_{2s}|^2}{\sigma^2(\mathfrak{J}_{2n})} = a_{s1}^2 \sum_{i=1}^{m} \frac{2A_{i1}^2 T_s}{N_i} = \sum_{i=1}^{m} 2E_i / N_i = \sum_{i=1}^{m} q_{out\, i}^2. \tag{5.73}$$

Now let complex amplitudes of wanted signals be random (but strongly mutually correlated at the inputs of spatially separated stations) and Gaussian. Then $\mathfrak{J}_{2s}$ in (5.70) is a Gaussian variable with zero mean and the following variance:

$$\sigma^2(\mathfrak{J}_{2s}) = 0.5\overline{\mathfrak{J}_{2s}\mathfrak{J}_{2s}^*} = 0.5\overline{a_{s1}^2}\left(\sum_{i=1}^{m} 2A_{i1}^2 T_s / N_i\right)^2 = 0.5\sigma_1^2\left(\sum_{i=1}^{m} 2A_{i1}^2 T_s / N_i\right)^2 \tag{5.74}$$

where $\overline{a_{s1}^2} = \sigma_1^2$ since a_{si}, $i = \overline{1, m}$ are r.m.s. (effective) values. The sum of a wanted signal and noise, $\mathfrak{J}_{2sn}$, is the sum of two Gaussian variables, $\mathfrak{J}_{2sn} = \mathfrak{J}_{2s} + \mathfrak{J}_{2n}$ with zero means and variances (5.74) and (5.68) respectively. Hence $\mathfrak{J}_{2sn}$ is a Gaussian variable too with $\overline{\mathfrak{J}_{2sn}} = 0$ and variance

$$\sigma^2(\mathfrak{J}_{2sn}) = \sum_{i=1}^{m} \frac{2A_{i1}^2 T_s}{N_i}\left(1 + \sigma_1^2 \sum_{l=1}^{m} \frac{A_{l1}^2 T_s}{N_l}\right)$$

$$= \sum_{i=1}^{m} \frac{2A_{i1}^2 T_s}{N_i}\left(1 + \sum_{l=1}^{m} \frac{\bar{E}_l}{N_l}\right) = \sigma^2(\mathfrak{J}_{2n})(1 + \tilde{q}_{out}^2) \tag{5.75}$$

where $\bar{E}_i = \sigma_i^2 T_s = \sigma_1^2 A_{i1}^2 T_s$ is the averaged energy of a random wanted signal at the ith station, and the averaged power output SNR

$$\tilde{q}_{out}^2 = \sigma^2(\mathfrak{J}_{2s}) / \sigma^2(\mathfrak{J}_{2n}) = \sum_{i=1}^{m} \bar{E}_i / N_i = \sum_{i=1}^{m} \tilde{q}_{out\, i}^2 \tag{5.76}$$

is the sum of partial power SNRs of all stations.

The output variable $L_2 = |\mathfrak{Z}_2|$ is a Rayleigh variable both in the absence and in the presence of a wanted signal (target echo). The false alarm probability, P_{fa}, is determined by (5.71) and the detection probability, P_d, is connected with P_{fa} through a known equation

$$P_0 = P_{fa}^{1/(1+\tilde{q}_{out}^2)}. \tag{5.77}$$

Thus detection characteristics of the algorithm L_2 for both nonrandom and random (fluctuating) Gaussian signal complex amplitudes are the same as with a monostatic radar. The only difference is that the detection parameter – output power SNR at the envelope detector input in a MSRS is the sum of partial output power SNRs of all the stations. If we create a spatially coherent MSRS by adding to a monostatic radar of $m-1$ spatially separated receiving stations (with the same antenna and receiver characteristics as of the monostatic radar), then for strong interstation signal correlation and the same signal power we shall obtain the power gain by m times. As in Section 5.1 we may consider that detection characteristics of such a MSRS are the same as those of a monostatic radar whose receiving antenna effective aperture is equal to the sum of effective antenna apertures of all the receiving stations of the MSRS:

$$S_{eff} = \sum_{i=1}^{m} S_{effi}.$$

When illuminating signals from n transmitting stations are summed at a target in phase, the form of detection characteristics for the wanted signals considered is not changed but the SNR increases. As in Section 5.1 for summation in phase we have

$$a_{si} = a_{s1}A_{i1} = A_{i1}\sum_{i=1}^{m} a_{1k};$$

$$E_i = A_{i1}^2\left(\sum_{k=1}^{m} a_{1k}\right)^2 T_s; \quad \bar{E}_i = \sigma_1^2 A_{i1}^2\left(\sum_{k=1}^{m} \bar{A}_{k1}\right)^2 T_s \tag{5.78}$$

where all notations are the same as in (5.64). When signal energy and noise PSDs are equal at all the receiving stations, then as in Section 5.1

$$q_{out}^2 = mn^2 q_{out\,0}^2; \quad \tilde{q}_{out}^2 = mn^2 \tilde{q}_{out\,0}^2 \tag{5.79}$$

where $q_{out\,0}^2$ and $\tilde{q}_{out\,0}^2$ are the doubled ratio of nonfluctuating signal energy to noise PSD and the ratio of averaged fluctuating signal energy to noise PSD, respectively, at the envelope detector input of a monostatic radar whose transmitting and receiving stations (possibly colocated) are the same as each transmitting and each receiving station of the MSRS. As with deterministic signals the power gain of mn^2 is achieved (see Section 5.1).

Let us now go over to algorithm L_7 (5.60). It can be seen from the comparison (5.40) with (5.60) that detection characteristics of L_7 are similar to those of L_2 i.e. are determined by (5.71) and (5.72) or (5.77). However, the coherent summation of mn echoes in a background of independent noises should be taken into account in output SNR. For nonfluctuating signals when detection characteristics are determined by (5.71), (5.72), we can find from (5.60), (5.61):

$$q_{\text{out}}^2 = a_{11}^2 \sum_{k=1}^{n} \sum_{i=1}^{m} 2A_{ik1}^2 T_s / N_i = \sum_{k=1}^{n} \sum_{i=1}^{m} 2E_{ik}/N_i = \sum_{k=1}^{n} \sum_{i=1}^{m} q_{\text{out}\,ik}^2. \tag{5.80}$$

For fluctuating Gaussian signal complex amplitudes when detection characteristics are determined by (5.77) we have

$$\bar{q}_{\text{out}}^2 = \sigma_{11}^2 \sum_{k=1}^{n} \sum_{i=1}^{m} A_{ik1}^2 T_s / N_i = \sum_{k=1}^{n} \sum_{i=1}^{m} \bar{E}_{ik}/N_i = \sum_{k=1}^{n} \sum_{i=1}^{m} \bar{q}_{\text{out}\,ik}^2. \tag{5.81}$$

In (5.80), (5.81) $q_{\text{out}\,ik}^2$ is the power SNR for a nonfluctuating signal caused by the kth transmitting station at the ith receiving station, and $\bar{q}_{\text{out}\,ik}^2$ is the averaged SNR for the same stations and fluctuating signals. When $q_{\text{out}\,ik}^2 = q_{\text{out}\,0}^2$, $\bar{q}_{\text{out}\,ik}^2 = \bar{q}_{\text{out}\,0}^2$, $i = \overline{1,m}$, $k = \overline{1,n}$, we obtain from (5.80) and (5.81)

$$q_{\text{out}}^2 = mn q_{\text{out}\,0}^2; \quad \bar{q}_{\text{out}}^2 = mn \bar{q}_{\text{out}\,0}^2. \tag{5.82}$$

The power gain is equal to mn as compared with a monostatic radar but it is by n times lesser than in the case where illuminating signals from n transmitting stations are summed in phase at the target.

Having calculated q_{out}^2 or $\bar{q}_{\text{out}}^2$ one can obtain detection probability P_d using detection characteristics of monostatic radars. For a prescribed value of P_{fa} we can find the normalized threshold level u_0 from (5.71). If expected signals are nonfluctuating, the probability P_d is determined by q_{out}^2 and u_0 with the help of (5.72). The integral in (5.72) cannot be expressed through elementary functions but it can be expressed through the incomplete Toronto function. Graphs of that function are given[5] in [26,45,202]. The values of P_d versus q_{out}^2 at $P_{fa} = \text{const.}$ can be obtained immediately by computer simulation. Detection characteristics for the most typical values of P_{fa} are shown in Fig. 5.4 (dashed lines).

When complex amplitudes of expected signals are Gaussian (fluctuating echoes), then detection characteristics are determined by (5.77). They are shown in Fig. 5.4 for the same values of P_{fa} (solid lines). The SNR values $\bar{q}_{\text{out}}^2$ for these curves can be calculated from (5.76) or (5.81). It is seen that as the detection probability, P_d, increases, energy losses caused by signal amplitude fluctuations sharply increase too.

***In this section**, we have analyzed performance characteristics of the optimal detection algorithms* L_2 (5.40), (5.41) *for MSRSs with a single transmitting station and* L_7 (5.60) *for MSRSs with several transmitting stations. These algorithms may be called "algorithms of coherent summation". We have considered two cases: the case of nonfluctuating expected signals (target echoes) and the case of fluctuating expected signals with Gaussian complex amplitudes. For both cases we have presented expressions for practical calculations of detection characteristics:* (5.71), (5.72), (5.73), (5.80) *for nonfluctuating signals and* (5.76), (5.77), (5.81) *for fluctuating signals. We have estimated the power gain in such MSRSs as compared with monostatic radars* [*see* (5.79), (5.82)]. *Detection characteristics are plotted in Fig. 5.4 for typical values of false alarm probability.*

---

[5] To use the graphs of $T_{\sqrt{v}}(1, 0, \sqrt{q})$ presented in [26,45,202] one should substitute $v = u_0^2$ and $q = q_{\text{out}}^2/2$.

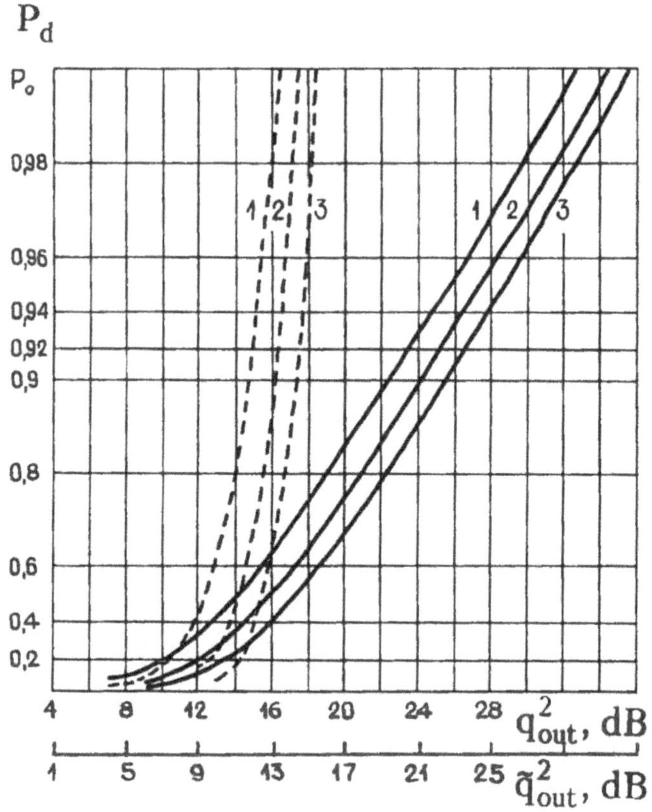

Figure 5.4 Detection characteristics of optimum coherent summation algorithms L_2, L_7 for nonfluctuating signals (dashed lines) and for signals with completely correlated complex amplitude fluctuations (solid lines). Curve 1: $P_{fa}=10^{-4}$; curve 2: $P_{fa}=10^{-6}$; curve 3: $P_{fa}=10^{-8}$

5.5. PERFORMANCE ANALYSIS OF OPTIMUM DETECTORS FOR FLUCTUATING SIGNALS. INCOHERENT SUMMATION ALGORITHMS

Let us consider the algorithms L_3, L_5, L_6. All of them include weighted sums of random variables $|\tilde{G}_i|^2$ which are square-law envelope detector outputs after matched filtration of received signals in each station. Taking into account (5.11) and (4.4) we can see that the variance of the noise component of $\tilde{G}_i$ is equal to $2N_iT_s$. Hence it is convenient to transform $\tilde{G}_i$ as follows [see (5.31)]:

$$\tilde{G}_i=\sqrt{2N_iT_s}\,\frac{1}{2\pi}\int_{-\infty}^{\infty}\left[\chi_i(\omega)/\sqrt{2N_iT_s}\right]\Psi_0^*(\omega-\omega_0-\Omega_{si})\exp(j\omega t_{si})\,d\omega.$$

Substituting the sum of signal and noise spectra for $\chi_i(\omega)$ yields

$$\tilde{G}_i=\sqrt{2N_iT_s}\left[\sqrt{2\tilde{q}_{out}}a_{0i}\exp(-j\varphi_{si})+n_{0i}\right] \tag{5.83}$$

where a_{0i} are the normalized random effective signal values $(\overline{a_{0i}^2}=1,\ i=\overline{1,m})$; n_{0i} are the normalized noise components with unit variance which are determined by the relationship

$$n_{0i}=\frac{1}{2\pi\sqrt{2N_iT_s}}\int_{-\infty}^{\infty}\tilde{N}_i(\omega)\Psi_0^*(\omega-\omega_0-\Omega_{si})\exp(j\omega t_{si})\,d\omega.$$

Now the statistics L_3, L_5, L_6 can be presented in a generalized form for the cases of both the absence and the presence of expected signals as follows:

$$L_n=\sum_{i=1}^{m}|Q_i n_{0i}|^2;$$
$$L_{sn}=\sum_{i=1}^{m}\left|Q_i\left[\sqrt{2}\tilde{q}_{\text{out}\,i}a_{0i}\exp(-j\varphi_{si})+n_{0i}\right]\right|^2. \qquad (5.84)$$

The weights Q_i can be expressed from (5.48), (5.54) and (5.55) with (5.83): $Q_i=\tilde{q}_{\text{out}}$ for the algorithm L_3; $Q_i=\tilde{q}_{\text{out}\,i}/\sqrt{1+\tilde{q}^2_{\text{out}\,i}}$ for the algorithm L_5; $Q_i=1$ for the algorithm L_6. We have taken into account that $A_{i1}^2T_s=\tilde{q}^2_{\text{out}\,i}/\sigma_1^2$ where σ_1^2 is the averaged wanted signal power at the first ("reference") station. It follows from (5.84) that detection characteristics are completely determined by SNRs $\tilde{q}^2_{\text{out}\,i}$, $i=\overline{1,m}$. When the SNRs, $\tilde{q}^2_{\text{out}\,i}$, are equal at all the stations ($\tilde{q}^2_{\text{out}\,i}=\tilde{q}^2_{\text{out}}$; $Q_i=Q$) then detection characteristics of all the three algorithms are the same. It should be noted that the algorithms themselves depend on both $\tilde{q}^2_{\text{out}\,i}$ and PSDs N_i at all the stations.

False Alarm Probability

To calculate *the false alarm probability*, P_{fa}, let us consider L_n in (5.84). This random variable is described by a chi-square probability distribution with $2m$ degrees of freedom for the algorithm L_6 and for the algorithms L_3, L_5 under the additional condition of the same $\tilde{q}^2_{\text{out}\,i}$ at all the receiving stations. In this case we can write [49]

$$P_{fa}=\exp(-u_0/2)\sum_{k=0}^{m-1}\frac{(u_0/2)^k}{k!} \qquad (5.85)$$

where $u_0=u_t/Q^2$ is the normalized threshold level. The tables permitting to calculate $u_0/2$ for given values of P_{fa} from 10^{-1} to 10^{-12} and for m from 1 to 150 are presented in [45].

When there are distinct SNRs, $\tilde{q}^2_{\text{out}\,i}$, at spatially separated stations, then for calculations of the false alarm probability, P_{fa}, for algorithms L_3 and L_5 a generalized chi-square probability distribution may be used. If all the Q_i in (5.84) are different and arranged in decreasing order, then [14]

$$P_{fa}=\sum_{i=1}^{m}\alpha_i\exp(-u_t/2Q_i^2) \qquad (5.86a)$$

where

$$\alpha_i=-(-1)^{i+1}Q_i^{2(m-1)}\left[\prod_{k=1}^{i-1}(Q_k^2-Q_i^2)\prod_{l=i+1}^{m}(Q_l^2-Q_i^2)\right]^{-1}, \qquad (5.86b)$$

$$Q_1^2>Q_2^2>\cdots>Q_m^2.$$

A general expression for P_{fa} may be obtained in terms of Moment Generating Function (MGF). It is valid when SNRs, $\tilde{q}^2_{\text{out}\,i}$, are equal at certain stations but different at other stations. A MGF $M(v)$ and a probability density function (PDF) $w(x)$ are connected by the following pair of transformations[6]

$$M(v)=\int_{-\infty}^{\infty} w(x)\exp(-vx)\,dx; \qquad w(x)=\frac{1}{2\pi j}\int_{-\infty}^{\infty} M(v)\exp(vx)\,dx. \tag{5.87}$$

The PDF $w(x)$ for the random variable $|Q_i n_{0i}|^2$ where $Q_i n_{0i}$ is a Gaussian random variable with zero mean and the variance Q_i^2 is

$$w_i(x)=(1/2Q_i^2)\exp(-x/2Q_i^2), \quad x\geqslant 0.$$

Hence in accordance with (5.87)

$$M_i(v)=(1+2Q_i^2 v)^{-1}.$$

Consider a MSRS with m receiving stations. Let SNR values, $\tilde{q}^2_{\text{out}\,i}$, be distinct at $N\leqslant m$ stations, i.e. at n_1 stations $\tilde{q}^2_{\text{out}}=\tilde{q}^2_{\text{out}\,1}$, at n_2 stations $\tilde{q}^2_{\text{out}}=\tilde{q}^2_{\text{out}\,2},\ldots,$ at n_N stations $\tilde{q}^2_{\text{out}}=\tilde{q}^2_{\text{out}\,N}$. It is clear that $n_1+n_2+\cdots+n_N=m$. By the independence of noises at different stations the MGF of the statistic L_n takes the form

$$M(v)=\prod_{l=1}^{N}(1+2Q_l^2 v)^{-n_l}. \tag{5.88}$$

It should be noted that we have from (5.87) for the case considered

$$(1/2Q_l^2)\int_{-\infty}^{\infty}\exp(-x/2Q_l^2)\exp(-vx)\,dx=(1+2Q_l^2 v)^{-1}.$$

Differentiating this equation n_l-1 times with respect to v yields the PDF which corresponds to the MGFs $(1+2Q_l^2 v)^{-n_l}$ from (5.88)

$$(1+2Q_l^2 v)^{-n_l}\leftrightarrow\frac{1}{2Q_l^2}\frac{(x/2Q_l^2)^{n_l-1}}{(n_l-1)!}\exp(-x/2Q_l^2). \tag{5.89}$$

Using the partial fraction expansion method for $M(v)$ from (5.88), taking into account (5.89) and integrating yields[7]

$$P_{fa}=\sum_{l=1}^{N}\sum_{k=0}^{n_l-1}C_{lk}\exp(-u_t/2Q_l^2)[1+(u_t/2Q_l^2)+\cdots+(u_t/2Q_l^2)^k/k!] \tag{5.90a}$$

where

$$C_{lk}=\frac{1}{(2Q_l^2)^{n_l-k-1}(n_l-k-1)!}\frac{d^{n_l-k-1}}{dv^{n_l-k-1}}[(1+2Q_l^2 v)^{n_l}M(v)]_{v=-1/2Q_l^2} \tag{5.90b}$$

and $M(v)$ is taken from (5.88). The expressions (5.85) and (5.86) can be derived from (5.90) as particular cases. Using (5.90) or (5.85), (5.86) one can obtain the threshold level u_t for given false alarm probability P_{fa} and weighting coefficients Q_i from (5.84).

[6] Sometimes another definition of the MGF is used: with the opposite signs at the indexes of the exponential functions [25].

[7] Similar formulas are presented in [126] but they contain several misprints.

Detection Probability

Let us pass to *detection probability*, P_d, calculations. Probability distribution of L_{sn} in (5.84) depends on the joint probability distribution of complex amplitudes $a_{0i}\exp(-j\varphi_{si})$. Consider firstly the case of strong coupling between a_{0i} but mutually independent phases φ_{si}, i.e. the case of *strongly connected amplitude fluctuations of spatially incoherent signals*. This situation takes place, for instance, when complex amplitudes of target echoes are strongly correlated at the inputs of stations of a spatially incoherent MSRS. The relationships (5.34), (5.38b) for the effective values $a_{si}=\sigma_i a_{0i}$ and the relationship (5.46) for phases φ_{si} are valid. In this case m random variables a_{0i} are reduced to a single random variable a_{01}. The variables in brackets in (5.84) are Gaussian but not jointly Gaussian (for different i).

When *SNRs at all stations are equal to the same value*, $\tilde{q}^2_{out}$, one may utilize the expression for detection probability, P_d, through the tabulated Pearson incomplete gamma-function [45]:

$$I(c, M)=(1/M!)\int_0^{c\sqrt{M+1}} e^{-z}z^M\,dz. \tag{5.91}$$

Integrating by parts and neglecting small terms we can obtain

$$P_d \approx (1+1/m\tilde{q}^2_{out})^{m-1}\exp\left[-\frac{(u_0/2)}{1+m\tilde{q}^2_{out}}\right]. \tag{5.92}$$

Calculations performed for $m=2$–10 and $P_{fa}\leqslant 10^{-4}$ have shown that the difference between P_d values obtained from (5.91) and (5.92) is of the order of 10^{-4}–10^{-5} so that the precision of (5.92) may be considered as quite sufficient for practical calculations.

Detection characteristics (the detection probability P_d versus the SNR $\tilde{q}^2_{out}$) calculated with the help of (5.85) and (5.92) for the algorithms L_3, L_5, L_6 are plotted in Fig. 5.5. It is assumed that signal effective values, a_{si}, are strongly mutually connected at different stations, initial phases, φ_{si}, are mutually independent and SNRs are equal at all stations. It can be seen that the power gain as a function of the number of receiving stations, m, is practically the same for different values of detection probability P_d (for $P_d>0.4$) and weakly increases with the decrease of false alarm probability P_{fa}. Unlike for spatially coherent MSRSs, the power gain increases more slowly with the number of stations, m, than the number m itself. It reflects energy losses caused by the mutual independence of initial phases at different stations.

When a_{si} are strongly mutually connected at different stations and φ_{si} are mutually independent but their probability distributions are unknown (or these parameters are nonrandom though unknown) the optimum algorithm L_4 (5.52) includes a linear incoherent summation. It is known (see, e.g. [47,48,54] that for a monostatic radar ($m=1$) and a single pulse echo, detection characteristics do not depend on the form of envelope detector characteristic. Hence for $m=1$ efficiencies of the algorithms L_3, L_5, L_6 and L_4 are strictly equal. It may be expected that in a MSRS at least with few stations, m, the difference of the algorithm efficiency between linear and square-law summation is negligible.

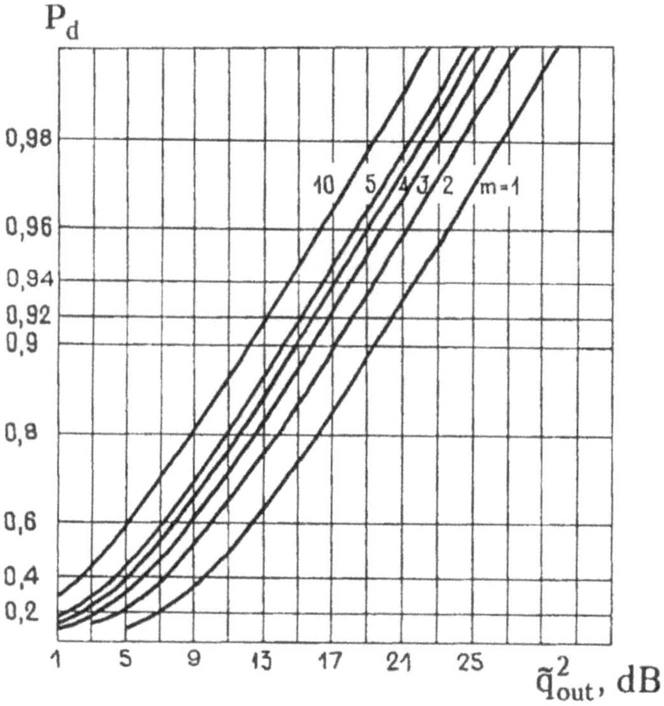

Figure 5.5(a)

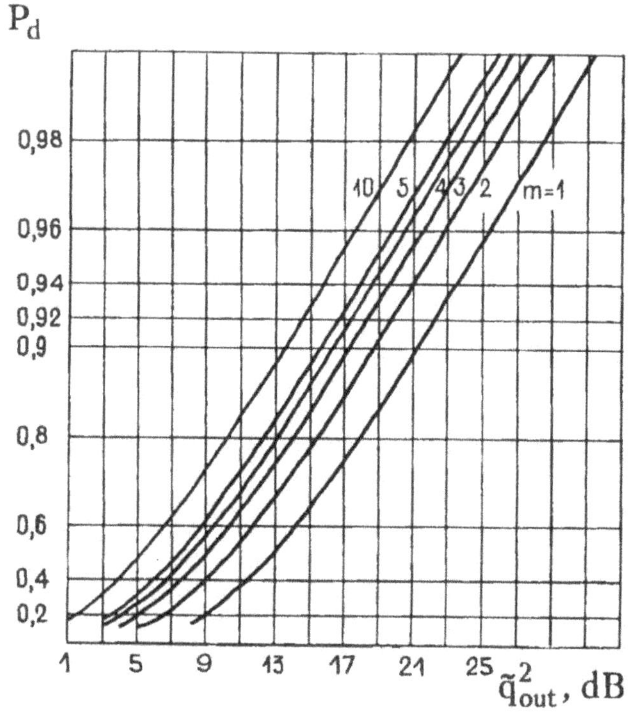

Figure 5.5(b)

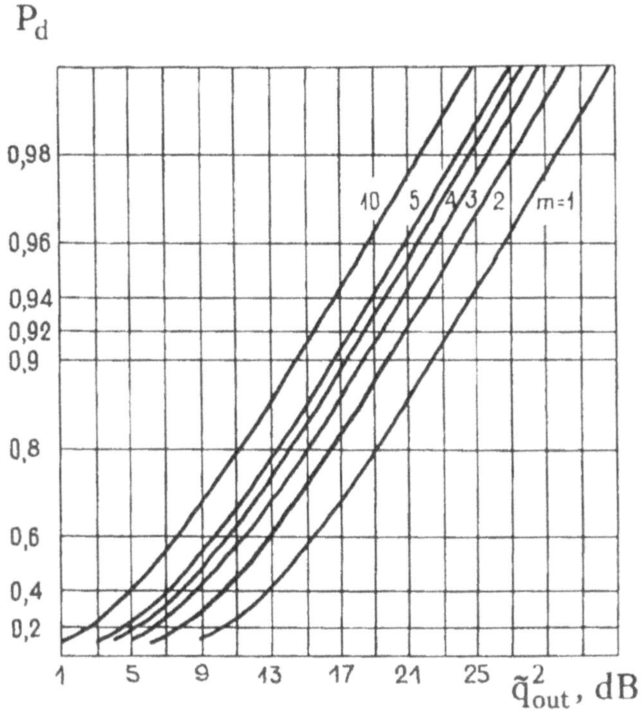

Figure 5.5(c)

Figure 5.5 Detection characteristics of square-law summation algorithms L_3, L_5, L_6 for spatially incoherent signals with completely correlated real amplitude fluctuations: (a) $P_{f_a}=10^{-4}$; (b) $P_{f_a}=10^{-6}$; (c) $P_{f_a}=10^{-8}$ ($m=$ the number of receiving stations)

For the algorithm L_4 there are no exact analytical expressions permitting P_{f_a} and P_d calculations[8]. The detection probability P_d should be obtained by computer simulation of the algorithm at the presence of wanted signal and by estimating the probability $P_d=P\{L_{4sn}\geqslant u_t\}$. The threshold level u_t may also be estimated by simulation of L_4 in the absence of wanted signals for a given value of $P_{f_a}=P\{L_{4n}\geqslant u_t\}$. However, when P_{f_a} is small, as is usually the practice, to obtain statistically reliable estimates, a great many statistical experiments are necessary. Therefore it is often more reasonable to determine u_t by approximate formulas representing the PDF of L_{4n} as a sum of several first terms of its expansion into a series of the Hermite or Laguerre orthogonal polynomials [33].

Simulation of the algorithm L_4 for $m=2$–10, equal values of SNRs, $\tilde{q}^2_{\text{out}\,i}$, $i=\overline{1,m}$ and $P_{f_a}=10^{-4}$ has shown that its performance is practically the same as that of the algorithms L_3, L_5, L_6. The difference in values of $\tilde{q}^2_{\text{out}}$ corresponding to a certain detection probability P_d does not exceed 0.2–0.4 dB even for $m=10$ and almost is independent of P_d. Therefore one may successfully use in practice both linear and square-law envelope detectors without noticeable energy losses.

[8] An exclusion is the case of $m=2$. In this case [33]

$$P_{f_a}=\exp(-u_t^2/2\sigma_n^2)+0.5\sqrt{\pi}(u_t/\sigma_n)\exp(-u_t^2/4\sigma_n^2)\,\mathrm{erf}(u_t/2\sigma_n)$$

where σ_n^2 is the variance of Gaussian noises at each station; erf(x) is the error function (5.15).

When *SNRs* $\tilde{q}^2_{\text{out}\,i}$ *at different stations are unequal*, there is no exact analytical expression for P_d even for algorithms of square-law summation (L_3, L_5, L_6). In this case computer simulation of the statistic L_{sn} for probability estimation $P_d = P\{L_{sn} \geq u_t\}$ is necessary too. The threshold level u_t should be preliminary calculated from (5.90) or (5.86) for a given value of P_{fa}.

There may be many different combinations of $\tilde{q}^2_{\text{out}\,i}$, $i = \overline{1,m}$, when they are unequal. This makes it difficult to present a general quantitative performance analysis. For practical purposes it is important to reveal at what maximum difference between $\tilde{q}^2_{\text{out}\,i}$, $i=\overline{1,m}$, information contribution from "weak" stations (with small $\tilde{q}^2_{\text{out}\,i}$) to joint processing can lead to a noticeable target detection enhancement. This is important because information fusion and joint processing require certain expense (see Section 1.3). Great differences between $\tilde{q}^2_{\text{out}\,i}$, $i=\overline{1,m}$, can occur, for instance, when a MSRS is organized on the basis of several stations (including monostatic radars) with significantly different antenna gains or (and) receiver sensitivity.

For *coherent summation* (see Section 5.4, algorithms L_2, L_7) the power gain as a function of the number m of receiving stations is completely determined by the increase of the total SNR $\tilde{q}^2_{\text{out}}$. Hence the contribution of each station can be easily calculated using (5.76). When, $\tilde{q}^2_{\text{out}\,i}$, $i=\overline{1,m}$, are equal, adding one extra station to m stations leads to the extra power gain $\Delta K_p = 10\lg[(m+1)/m]$ dB. When the SNR, $\tilde{q}^2_{\text{out}(m+1)}$, at a new station (number $m+1$) is by v dB less than at each of m stations, then $\Delta K_p = 10\lg[(m+10^{-0.1v})/m]$ dB. For example, when we go from $m=1$ (a monostatic radar) to $m=2$ and $\tilde{q}^2_{\text{out}\,1} = \tilde{q}^2_{\text{out}\,2}$ then $\Delta K_p \approx 3$ dB. Reduction of the SNR at one of two stations by 4–6 dB leads to the decrease of the power gain down to $\Delta K_p \approx 1.5$–1.0 dB. To achieve $\Delta K_p \approx 3$ dB in this situation, 3–4 such "weak" stations should be added, i.e. the total number of receiving stations should be increased up to $m=4$–5.

Analysis of the influence of differences in SNRs on detection characteristics for *incoherent summation* (algorithms $L_3, \ldots, L_6$) and strongly connected signal intensity fluctuations at spatially separated stations requires computer simulation. Simulation results of the simplest algorithm L_6 for MSRSs with two and three receiving stations are depicted in Fig. 5.6. Without loss of generality, number one may be assigned to the station with maximum $\tilde{q}^2_{\text{out}\,i}$ so that $\tilde{q}^2_{\text{out}\,1} = \max(\tilde{q}^2_{\text{out}\,i})$, $i=\overline{1,m}$. The curves in Fig. 5.6 show the rise of $\tilde{q}^2_{\text{out}\,1}$ required for maintaining a certain value of P_d (at a given value of P_{fa}) when the difference $\Delta\tilde{q}^2_{\text{out}}$ between SNRs at the first and other stations increases. For $m=3$ two curves are plotted corresponding to two extreme cases (variants). In the first variant SNRs at the second and third stations are equal, $\tilde{q}^2_{\text{out}\,2} = \tilde{q}^2_{\text{out}\,3}$, and lower than at the first station, $\tilde{q}^2_{\text{out}\,1}$, by $\Delta\tilde{q}^2_{\text{out}}$ dB. In the second variant SNRs are equal at the first and second stations, i.e. $\tilde{q}^2_{\text{out}\,1} = \tilde{q}^2_{\text{out}\,2}$, but at the third station, $\tilde{q}^2_{\text{out}\,3}$, is lower by $\Delta\tilde{q}^2_{\text{out}}$ dB. Taking into account arbitrary numeration of stations we can see that all other combinations of $\tilde{q}^2_{\text{out}\,1}$, $\tilde{q}^2_{\text{out}\,2}$ and $\tilde{q}^2_{\text{out}\,3}$ yield intermediate results. Besides, in Fig. 5.6 the values of SNR $\tilde{q}^2_{\text{out}\,1}$ for a monostatic radar ($m=1$) and for a MSRS with two receiving stations ($m=2$) and equal SNRs ($\tilde{q}^2_{\text{out}\,1} = \tilde{q}^2_{\text{out}\,2}$) are shown by dashed straight lines. It is seen that for $m=2$ when the difference between $\tilde{q}^2_{\text{out}\,1}$ and $\tilde{q}^2_{\text{out}\,2}$ is equal to 4 dB then the use of information from the "weak" station leads to the power gain[9] of less than 1 dB (as compared to a monostatic radar) instead of approximately 2.5 dB when $\tilde{q}^2_{\text{out}\,1} = \tilde{q}^2_{\text{out}\,2}$. When this difference increases up to 8 dB, the power gain falls off to zero and changes its sign so that instead of gain we obtain energy losses[10] (because of noise addition from the

---

[9] The power gain here is the reduction of a required value of $\tilde{q}^2_{\text{out}\,1}$ as compared to a monostatic radar or to a system with a single transmitting station and two receiving stations where $\tilde{q}^2_{\text{out}\,1} = \tilde{q}^2_{\text{out}\,2}$.

[10] It should be recalled that the algorithm L_6 is optimal for strong expected signals with spatially independent fluctuations.

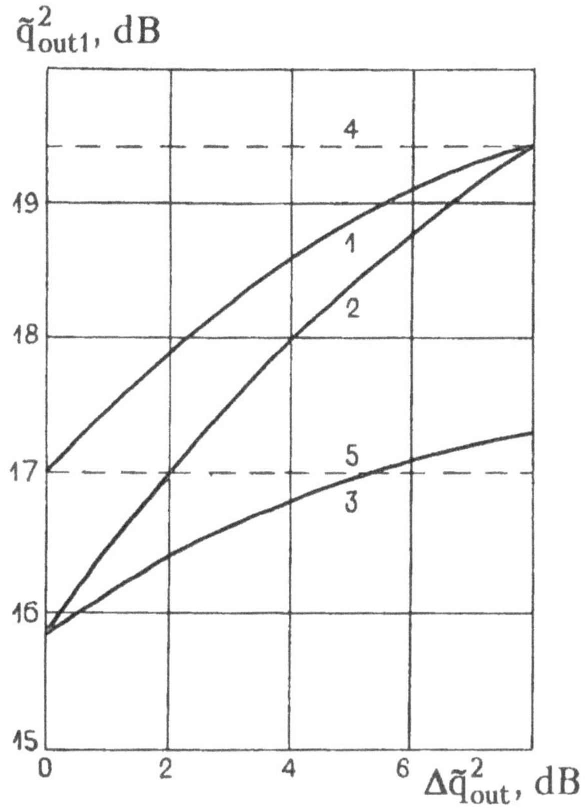

Figure 5.6 Required values of SNR $\tilde{q}^2_{out\,1}$ for maintaining $P_d = 0.9$ at $P_{fa} = 10^{-4}$ when $\tilde{q}^2_{out\,2}$ and $\tilde{q}^2_{out\,3}$ decrease by $\Delta\tilde{q}^2_{out}$ (in dB). Algorithm L_6, completely correlated real amplitude fluctuations. Curve 1: $m = 2$, $\tilde{q}^2_{out\,2} = \tilde{q}^2_{out\,1} - \Delta\tilde{q}^2_{out}$; curve 2: $m = 3$, $\tilde{q}^2_{out\,2} = \tilde{q}^2_{out\,3} = \tilde{q}^2_{out\,1} - \Delta\tilde{q}^2_{out}$; curve 3: $m = 3$, $\tilde{q}^2_{out\,2} = \tilde{q}^2_{out\,1}$; $\tilde{q}^2_{out\,3} = \tilde{q}^2_{out\,1} - \Delta\tilde{q}^2_{out}$; curve 4: $m = 1$; curve 5: $m = 2$, $\tilde{q}^2_{out\,2} = \tilde{q}^2_{out\,1}$

"weak" station). For $m = 3$ when the SNR at one station, $\tilde{q}^2_{out\,3}$, decreases by 4 dB, the contribution of this station to power gain (as compared to a system with two receiving stations, $m = 2$, and $\tilde{q}^2_{out\,1} = \tilde{q}^2_{out\,2}$) is reduced to approximately 0.2 dB instead of approximately 1.2 dB when $\tilde{q}^2_{out\,1} = \tilde{q}^2_{out\,2} = \tilde{q}^2_{out\,3}$. If for $m = 3$ the SNR at each of two stations, $\tilde{q}^2_{out\,2}, \tilde{q}^2_{out\,3}$, decreases by more than 2 dB (relative to $\tilde{q}^2_{out\,1}$), required values of $\tilde{q}^2_{out\,1}$ turn out to be higher than for a system containing two receiving stations ($m = 2$) when $\tilde{q}^2_{out\,1} = \tilde{q}^2_{out\,2}$ so that in this case the power gain changes into energy losses too. All these relationships are practically the same for different values of P_d, linear (L_4) and square-law (L_6) summation.

Thus when echo amplitude fluctuations at the inputs of stations are strongly mutually dependent, information addition from "weak" stations with SNRs by 4–6 dB lower than at the "strong" stations does not lead to a noticeable power gain. Therefore it is practically unreasonable from the viewpoint of target detection enhancement. It does not mean, however, that information fusion from stations with such differences in SNRs is absolutely senseless. Information fusion in such MSRSs may enjoy other advantages, e.g. higher target coordinate measurement accuracy, jamming resistance increase etc. (see Section 1.2).

Let us now go over to the analysis of detection characteristics of the algorithms L_3, L_5, L_6 when there is no interstation correlation of target echo complex amplitudes, i.e. in the case of *independent signal intensity fluctuations in spatially incoherent MSRSs*. The r.m.s. values $a_{si}=\sigma_i a_{0i}$ and phases φ_{si} of wanted signals are described by the probability distributions (5.53) and (5.46), respectively. The statistic L_{sn} in (5.84), as in the absence of wanted signals, represents a sum of squared moduli of mutually independent Gaussian variables $\tilde{G}_i$. Probabilities P_{fa} and P_d are to be calculated using the same formulas. One should only take into account that the presence of wanted signals leads to the increase of variances of the variables $\tilde{G}_i$. If *SNR values are identical at all stations* ($\tilde{q}^2_{\text{out}\,i}=\tilde{q}^2_{\text{out}}$, $i=\overline{1,m}$)

$$P_d=\exp\left[-\frac{(u_0/2)}{1+\tilde{q}^2_{\text{out}}}\right]\sum_{k=0}^{m-1}\frac{(u_0/2)^k}{k!(1+\tilde{q}^2_{\text{out}})^k}. \tag{5.93}$$

The only difference from (5.85) is that in accordance with (5.84) the signal variance $Q^2\tilde{q}^2_{\text{out}}$ is added to the noise variance Q^2.

Detection characteristics (which are the same for the algorithms L_3, L_5 and L_6 when SNRs are equal at all stations) are calculated from the formulas (5.85), (5.93) and plotted in Fig. 5.7. These curves permit estimation of the power gain achieved by optimum target echo detection in a MSRS containing a single transmitting and m receiving stations. Comparing them with the curves in Fig. 5.6 one can easily notice an important difference: when wanted signal fluctuations are spatially independent, the power gain rises with the increase of the detection probability P_d. For visual comparison of optimum detection efficiency in a MSRS with a single transmitting and m receiving stations and wanted signals of different types the power gain K_p(in decibels) is plotted in Fig. 5.8 versus the number m of receiving stations

$$K_p=10\lg(\tilde{q}^2_{\text{out}\,0}/\tilde{q}^2_{\text{out}}) \tag{5.94}$$

where $\tilde{q}^2_{\text{out}}$, $\tilde{q}^2_{\text{out}\,0}$ are the power SNRs corresponding to the required detection probability, P_d, at the given false alarm probability, P_{fa}, at each of the m receiving stations of the MSRS and at the monostatic radar, respectively. It can be seen from Fig. 5.8[11] that when required detection probability is sufficiently high, the power gain for spatially independent fluctuating signals may be much greater than for signals with strongly spatially correlated fluctuations in a spatially coherent MSRS. In the latter case $K_p=10\lg m$ dB. For example, if $P_{fa}=10^{-4}$ the "extra" power gain for $m=2$–10 is 3.1–4.3 dB for $P_d=0.9$ and 7.7–12.3 dB for $P_d=0.99$. This effect is a result of smoothing of intensity fluctuations while the summation of independently fluctuating signals. It is well known for monostatic radars, for example if a transmitting waveform contains several frequencies [47,123], and permits to provide high detection probability with lower transmitting energy in each illuminating pulse while target tracking. In MSRSs such an effect is achieved with arbitrary transmitting waveforms if target echo fluctuations are mutually independent at the inputs of spatially separated stations. For small detection probability this smoothing of intensity fluctuations leads to certain energy losses. Hence the curve for $P_d=0.5$ in Fig. 5.8 runs slightly below than the curve for coherent summation. When signal intensity fluctuations are strongly mutually dependent at different stations then there is no smoothing while summation. In this case, as was mentioned above, the power

[11] Though in Fig. 5.8 continuous curves are plotted for the sake of clarity, the quantity K_p is obviously defined only for integer values of m.

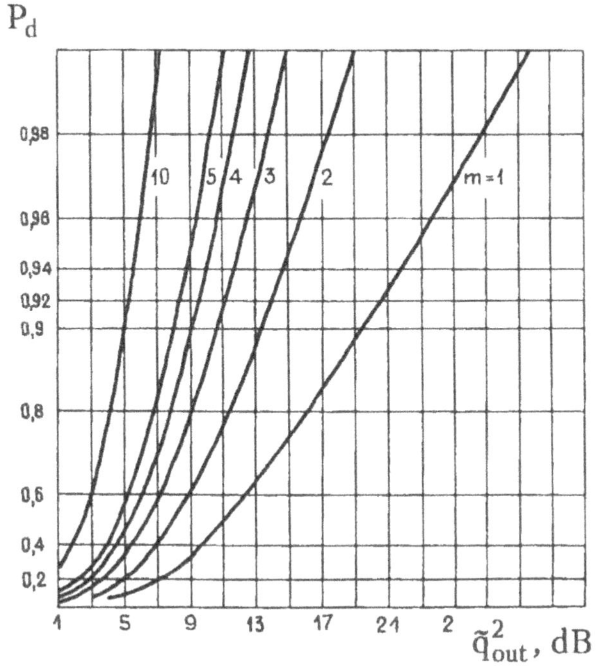

Figure 5.7(a)

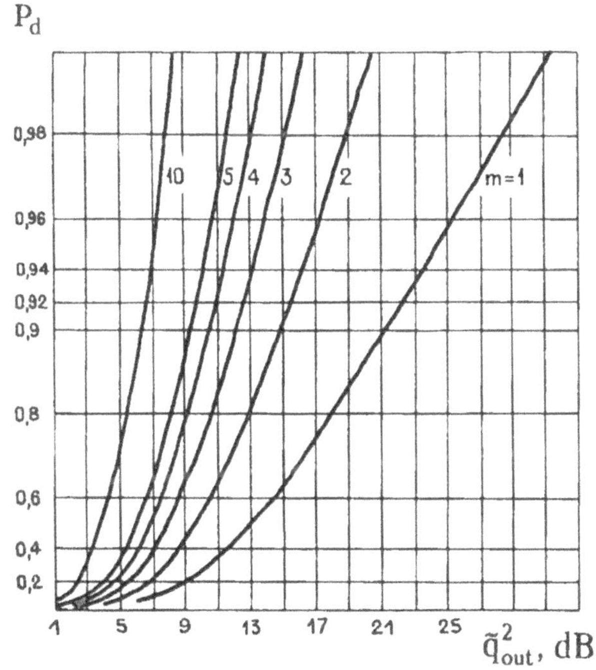

Figure 5.7(b)

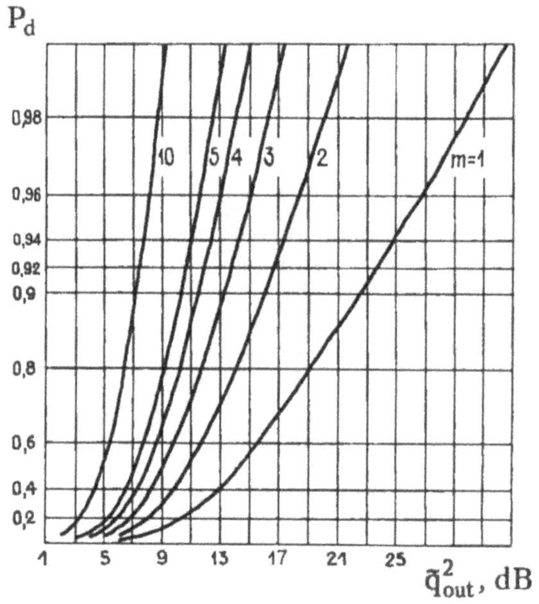

Figure 5.7(c)

Figure 5.7 Detection characteristics of square-law summation algorithms L_3, L_5, L_6 for spatially incoherent signals with independent amplitude fluctuations: (a) $P_{fa}=10^{-4}$; (b) $P_{fa}=10^{-6}$; (c) $P_{fa}=10^{-8}$ (m = the number of receiving stations)

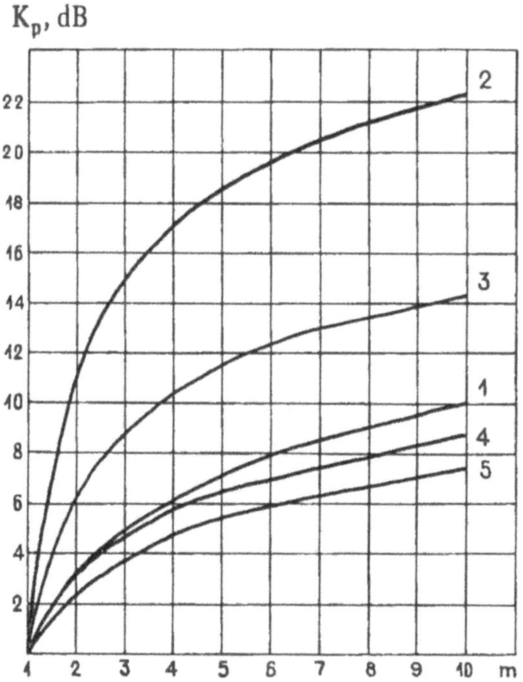

Figure 5.8 Dependences of the energy gain, K_p, on the number of receiving stations, m, for a MSRS with a single transmitting station, $P_{f_a}=10^{-4}$. Curve 1: coherent summation ($K_p = 10\lg m$); curve 2: square-law summation, independent amplitude fluctuations, $P_d=0.99$; curve 3: the same as curve 2 but for $P_d=0.9$; curve 4: the same as curve 2 but for $P_d=0.5$; curve 5: square-law summation, completely correlated fluctuations (K_p does not depend on P_d)

gain rises with the increase of the number of stations m not as fast as for coherent summation.

It is clear that power advantages of a MSRS with m receiving stations and arbitrary waveforms as compared to a monostatic or bistatic radar are to a great extent caused by the fact that the summed effective area of antenna apertures of the MSRS is by m times greater that that of the monostatic radar with the same transmitting and receiving stations. In many cases it is interesting to compare a MSRS with such a monostatic radar whose effective receiving antenna aperture is equal to the sum of effective antenna apertures of all receiving stations of the MSRS. In other words, it is interesting to reveal what would be changed power characteristics of a monostatic radar if, maintaining unchanged the transmitting part of the radar, its receiving antenna were "cut" into m parts (in particular case, equal parts), spatially separated, equipped with the same receivers and in such a MSRS optimal echo processing were performed? Such an approach takes into account the influence of stations' spatial separation only with power potentials of the monostatic radar and the MSRS being equal. This approach is applicable to MSRSs with short baselengths (as compared with target ranges) when the averaged target RCS is approximately equal with respect to all receiving stations.

The analysis carried out in Section 5.4 has shown that in the case of coherent summation antenna "cutting" and separation in space does not change detection characteristics if the summed effective antenna aperture remains constant. Therefore curve 1 corresponding to algorithm L_2 in Fig. 5.8 characterizes a monostatic radar whose receiving antenna aperture has been increased by m times. For each value of m the difference in K_p of any other curve in Fig. 5.8 from the curve for algorithm L_2 shows power gains or loss caused by spatial separation of receiving stations. These gains and losses are depicted in Fig. 5.9. This figure demonstrates an essential power gain for high detection probabilities, P_d, and small energy losses for low P_d when echo fluctuations are mutually independent at the inputs of receiving stations. It is also seen that when echo fluctuations are strongly mutually dependent, antenna spatial separation leads to certain energy losses.

The rise of the power gain gradually decelerates with the increase of the number of stations m (see Fig. 5.8). It can be explained, firstly, by the reduction of relative increase of summed antenna aperture and, secondly, by the decrease of influence on smoothing of echo fluctuations (when they are spatially independent) with adding of each new receiving station. Therefore, maximum increment of K_p is achieved when we go over from a monostatic radar ($m=1$) to the simplest MSRS with two receiving stations ($m=2$).

It should be borne in mind that specific values of $\tilde{q}^2_{\text{out}\,i}$ and of power gain K_p, ΔK_p for very high detection probabilities are heavily dependent on the probability distribution of target echo fluctuations. The presented calculation results have been obtained under the assumption of Gaussian complex amplitudes, i.e. Rayleigh fluctuations of a_{0i} in (5.84). Such a mathematical model is the same as the second Swerling model [203] and is adequate for targets which may be represented as collections of independent fluctuating scattering centres ("flare spots – FSs") of approximately equal intensity. If besides such FSs there is a dominant FS (e.g. a fragment of the target surface with specular signal reflection), more adequate is the second Nakagami distribution [72] (or the fourth Swerling model [203]): $w(a_{0i}) = 8a^3_{0i}\exp(-2a^2_{0i})$, $a_{0i} \geqslant 0$. In this case when P_d is 0.99 a MSRS with $m=2$ provides the power gain (compared to a monostatic radar) only 6.5 dB instead of 10.5 dB for Rayleigh amplitude distribution. When the receiving antenna aperture of a monostatic radar is equal to the summed antenna receiving aperture of a MSRS ($m=2$) the power gain is 3.5 dB instead of 7.7 dB for Rayleigh fluctuations.

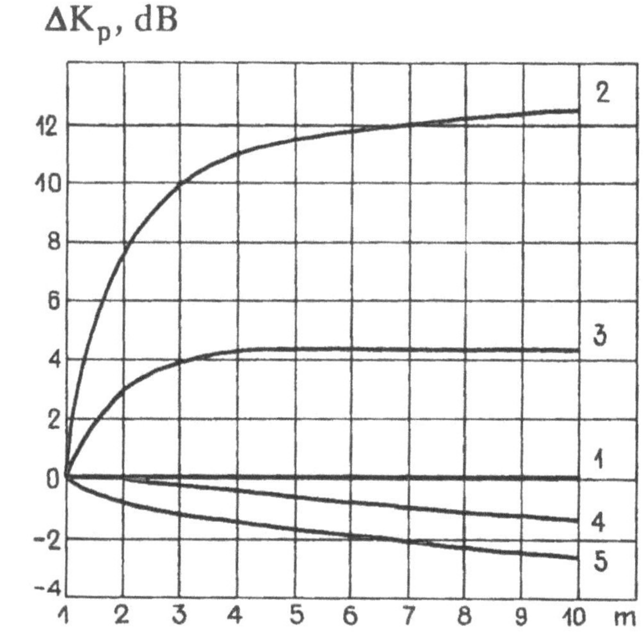

Figure 5.9 Energy gains and losses caused by spatial separation of stations (as compared to a monostatic radar whose receiving antenna aperture is equal to the sum of antenna apertures of m spatially separated receiving stations). Notation is the same as in Fig. 5.8

Let us consider in more detail performance characteristics of the synthesized detection algorithms in *a MSRS with nonidentical SNR values, $\tilde{q}^2_{\text{out}\,i}$, at different receiving stations*. The corresponding curves are plotted in Fig. 5.10. These curves are similar to those of Fig. 5.6 but they are obtained from (5.90) for mutually independent Gaussian echo complex amplitudes at the inputs of receiving stations. Because of strong dependence of the power gain on detection probability, P_d, the curves are plotted for three most interesting values of P_d: $P_d=0.5$, 0.9 and 0.99. For $P_d=0.5$ the curve from Fig. 5.10 is similar in its behavior to that from Fig. 5.6 though absolute values of the power gain are somewhat greater in the former case. However, for $P_d=0.9$ and especially for $P_d=0.99$ there is an essential difference between Figs. 5.10 and 5.6. It can be seen that for $m=2$ the joint processing provides a noticeable power gain when the difference between $\tilde{q}^2_{\text{out}\,2}$ and $\tilde{q}^2_{\text{out}\,1}$ ($\tilde{q}^2_{\text{out}\,2}<\tilde{q}^2_{\text{out}\,1}$) reaches 8–10 dB for $P_d=0.9$ and even is up to 16–18 dB for $P_d=0.99$. Adding of weak but independently fluctuating signal smoothes signal fluctuations and permits a significant reduction of signal energy required for signal detection with high probability. When $m=3$ this effect is not so pronounced but the power gain maintains for much greater differences in SNRs than for strongly mutually dependent fluctuations at the inputs of stations. As with identical SNRs, it should be taken into account here that required SNR values and power gains at high detection probabilities depend strongly on the specific probability distribution of signal fluctuations.

In accordance with (5.84) performance characteristics of the algorithms L_3, L_5, and L_6 are to be different when $\tilde{q}^2_{\text{out}\,i}$, $i=\overline{1,m}$, are unequal. If, however, this difference is not too large one might use the simplest algorithm L_6 which does not require to

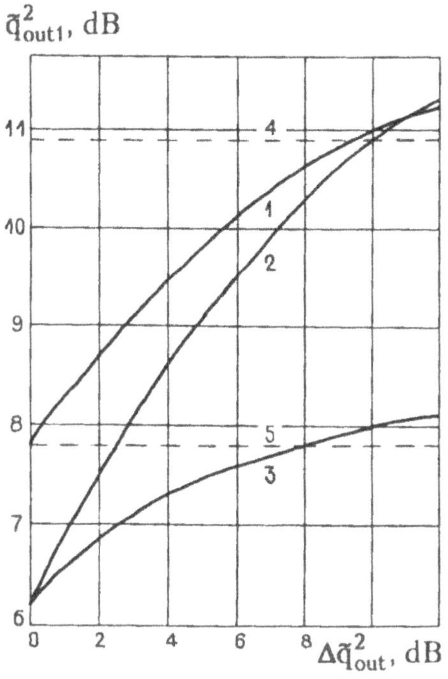

Figure 5.10(a)

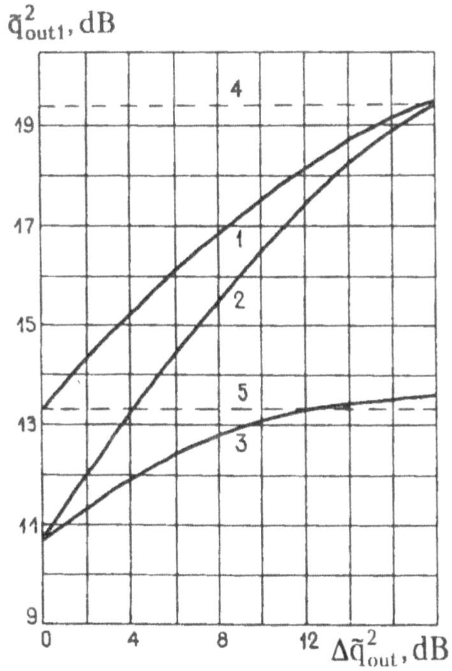

Figure 5.10(b)

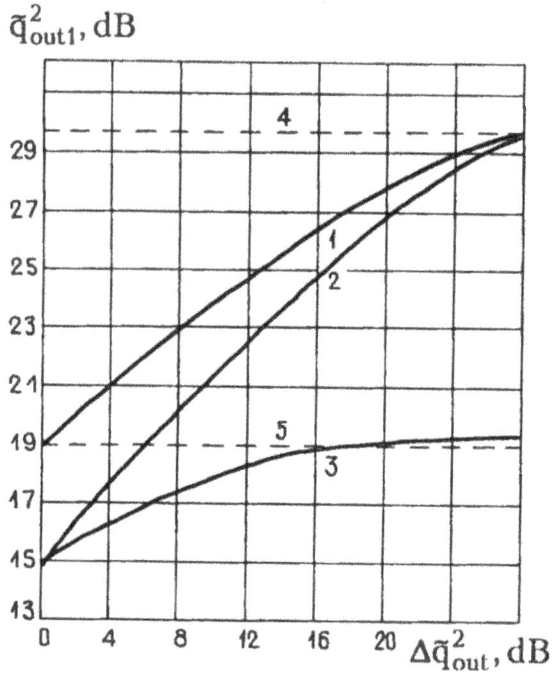

Figure 5.10(c)

Figure 5.10 The same characteristics as in Fig. 5.6 but for independent amplitude fluctuations. (a) $P_d=0.5$; (b) $P_d=0.9$; (c) $P_d=0.99$

know SNR values at spatially separated receiving stations. The comparison of two "extreme" algorithms L_3 and L_6 (optimal for weak and strong signals, respectively) for a MSRS with a few stations ($m=2$ and 3) has shown that L_6 is not worse than L_3 nearly always. When required detection probabilities are sufficiently high L_6 is better than L_3. Typical curves of energy losses for the algorithm L_3 in comparison with L_6 as functions of the difference $\Delta\tilde{q}^2_{out}$ of SNRs at "weak" and "strong" stations for $m=3$ are shown in Fig. 5.11. Here $\Delta Q=\tilde{q}^2_{out}(L_3)[\text{dB}]\ -\tilde{q}^2_{out}(L_6)\,[\text{dB}]$. It is seen that only for $P_d=0.5$ and great differences in SNRs (when including information from "weak" stations into joint processing is rather ineffective) L_6 is slightly worse than L_3 but for $P_d=0.9$ and especially for $P_d=0.99$ L_6 turns out to be better than L_3. As m increases required values of P_d are achieved with lower SNR (weaker signal) at each station so that the power gain of L_3 rises somewhat faster than that of L_6. However even for $m=10$ these differences in performance characteristics are not significant.

Thus it may be considered that the algorithm L_6 is not only simpler but not worse than the algorithm L_3. When the number of stations m is small and required detection probabilities are high, the algorithm L_6 is better than L_3 since it is nearer to L_5 which is optimal for signals of arbitrary intensity.

The algorithm L_6 takes the simplest form in the case where noise PSDs N_i are identical at all receivers. Then L_6 (5.55) is reduced to a simple (unweighted) summation of the outputs of signal processors of all stations. The processor in each station contains a matched filter followed by a square-law envelope detector.

On the basis of the obtained results it is not difficult to evaluate performance characteristics of the algorithms L_8, L_9 and L_{10} optimal for MSRSs with m receiving

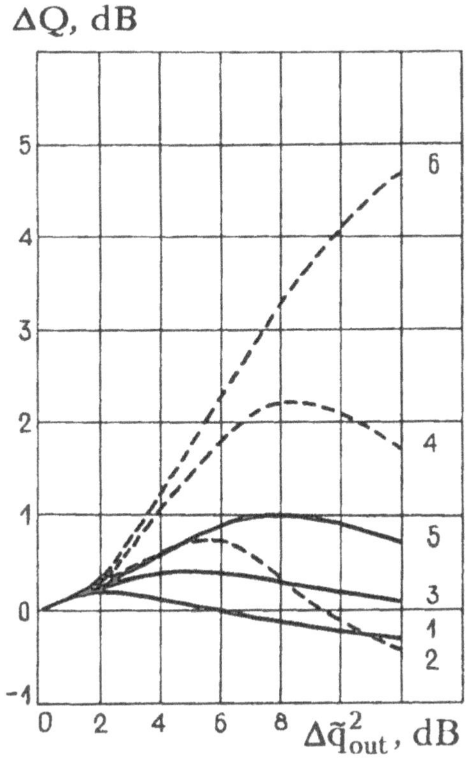

Figure 5.11 Energy loss and gain of algorithm L_3 in comparison with algorithm L_6 for signals with independent amplitude fluctuations, $P_{fa}=10^{-4}$, $m=3$. Curve 1: $P_d=0.5$, $\tilde{q}^2_{out\,2}=\tilde{q}^2_{out\,1}$; $\tilde{q}^2_{out\,3}=\tilde{q}^2_{out\,1}-\Delta\tilde{q}^2_{out}$ (variant 1); curve 2: $P_d=0.5$, $\tilde{q}^2_{out\,2}=\tilde{q}^2_{out\,3}=\tilde{q}^2_{out\,1}-\Delta\tilde{q}^2_{out}$ (variant 2); curve 3: $P_d=0.9$, variant 1; curve 4: $P_d=0.9$, variant 2; curve 5: $P_d=0.99$, variant 1; curve 6: $P_d=0.99$, variant 2

and n transmitting stations when all echoes caused by each transmitting station can be resolved at each receiving station. Obviously, L_8 (5.64) is a combination of L_2 (5.40), (5.41) and L_5 (5.54); L_9 (5.66) is a combination of L_2 (5.40), (5.41) and L_6 (5.55) whereas L_{10} (5.67) differs from L_5 (5.54) only by a number of terms (summands). The statistics (decision variables) L_8 and L_9 may be presented in the absence (L_n) and in the presence (L_{sn}) of wanted signals similar to (5.84) in a generalized form

$$L_n=\sum_{k=1}^{n}|Q_k n_0|^2;\ L_{sn}=\sum_{k=1}^{n}|Q_k[\tilde{q}_{out\,k}\sqrt{2}a_{0k}\exp(-j\varphi_{sk})+n_0]|^2 \qquad (5.95)$$

where the coefficients $Q_k=\tilde{q}_{out\,k}/\sqrt{1+\tilde{q}^2_{out\,k}}$ for the algorithm L_8, $Q_k=1$ for the algorithm L_9, and $\tilde{q}^2_{out\,k}$ is the SNR after the coherent summation of echoes caused by the kth transmitting station at each receiving station:

$$\tilde{q}^2_{out\,k}=\sum_{i=1}^{m}\tilde{q}^2_{out\,ik}=\sum_{i=1}^{m}A^2_{i1}\sigma^2_{ik}T_s/N_i=\tilde{A}^2_{k1}\sigma^2_{11}\sum_{i=1}^{m}A^2_{i1}T_s/N_i. \qquad (5.96)$$

Using (5.95), (5.96), formulas and curves for the algorithms L_3, L_5 and L_6 one can obtain detection characteristics for the algorithms L_8, L_9 and L_{10}. In the latter case the increase of the number of summed echoes up to mn decreases the required SNR of each echo (for given P_{fa} and P_d). Specific values of the power gain can be obtained from these detection characteristics. The same detection characteristics can be utilized for performance analysis when pulse trains or multifrequency signals are used in a MSRS (see Section 5.3).

In this section, *we have analyzed performance characteristics of a series of optimum detection algorithms that were synthesized in Sections* 5.2 *and* 5.3. *These algorithms are optimum for signals (target echoes) which are mutually incoherent at spatially separated receiving stations and therefore include the procedure of incoherent summation. For MSRSs with a single transmitting and m receiving stations we have presented the decision variables of all the synthesized algorithms* (L_3, L_5 and L_6) *in a common generalized form* (5.84) *both for the case where the wanted signal is absent* (L_n) *and for the case where it is present* (L_{sn}). *Such a presentation allows us to consider all the three algorithms together (and algorithm* L_4 *as well).*

For false alarm probability, P_{fa}, *calculations, equations* (5.85), (5.86) *and* (5.90) *are given. The first of them relates to the algorithm* L_6 *and to* L_3, L_5 *if SNRs at all stations are identical. When all SNRs are different expression* (5.85) *is valid. Finally, if there are both equal and unequal SNRs at different stations the general expression* (5.90) *is to be used.*

Detection probability, P_d, *is essentially affected by the degree of statistical dependence between target echo amplitude fluctuations at the inputs of receiving stations. For strongly mutually dependent fluctuations and identical SNR values at all stations, equation* (5.92) *has been presented and detection characteristics have been plotted in Fig.* 5.5. *The case where SNR values are unequal at different stations has been considered in detail for MSRSs with two and three receiving stations (Fig.* 5.6). *These results allow the reader to reveal the condition under which including information from "weak" stations leads (or does not lead) to a noticeable power gain.*

For mutually independent echo amplitude fluctuations we have presented the expression (5.93) *and detection characteristics in Fig.* 5.7 *that are valid when SNR values are identical at all stations. We also have presented the useful curves in Figs* 5.8 *and* 5.9 *which permit to compare the power gain or energy losses inherent in algorithms considered, as functions of the number of receiving stations. The case of unequal SNR values have been discussed also in detail and corresponding curves are given in Fig.* 5.10. *It has been shown that unlike the case of strongly dependent amplitude fluctuations, including information from significantly "weaker" stations with independent fluctuating signals in joint processing can lead to noticeable power gain. It has been also shown that for independent signal fluctuations, the performance characteristics of the simplest algorithm* L_6 *are, as a rule, not worse than those of the algorithms* L_3 *and* L_5 *(see Fig.* 5.11) *so that* L_6 *may be used nearly always without energy losses.*

The results obtained for MSRSs with a single transmitting station have been extended to MSRSs with n transmitting stations.

5.6. ADDITIONAL ENERGY LOSSES CAUSED BY THE IGNORANCE OF TARGET POSITION. "COST OF RESOLUTION"

As mentioned in Section 4.1, target detection problems considered in Sections 5.1–5.5 were reduced to the problems of whether a target is present in the probed spatial resolution cell (SRC). However, a possible target position is known at any radar (including MSRSs) before each illuminating signal transmission within some uncertainty region (Target Position Uncertainty Region – TPUR). This region may be large (in target search mode) or small (in target tracking mode) but it is always greater than one SRC. In each transmission several resolution cells from the TPUR are simultaneously illuminated and analyzed. It is important that during each recurrent period the total number of false alarms does not exceed an allowable value. In fact, it is the number of false alarms, N_{fa}, (not the probability of false alarm, P_{fa}, in one resolution cell) that is an output characteristic of radar signal processing. For monostatic radars there is a simple connection between N_{fa} and P_{fa}. Since self-noises may be considered stationary and statistically independent in different resolution cells, the number of false alarms N_{fa} in M resolution cells is binomially distributed [33] with the mean value $\overline{N_{fa}}=MP_{fa}$. It is convenient to use this mean value as an output characteristic of the number of false alarms[12]. Then the maximum allowable false alarm probability in each resolution cell can be written in the form

$$P_{fa\,max}=\alpha_0=\bar{N}_{fa\,max}/M. \tag{5.97}$$

Just this value of false alarm probability is to be taken into account in detection characteristics from Sections 5.4, 5.5. The more resolution cells M are probed in each radar transmission, the less $P_{fa\,max}$ for given $\bar{N}_{fa\,max}$ and hence the higher threshold detection level. Therefore, the increase of radar resolution capability (with the TPUR being constant) leads to energy losses. This is a well-known price that has to be paid for high resolution capability and may be called "cost of resolution". To reduce these energy losses, both monostatic radars and MSRSs usually use narrowband signals for target search and detection whereas for target tracking, when the uncertainty region of a target position decreases, signals with broader frequency band are used which provide greater resolution capability in range.

The "cost of resolution" for MSRSs has special features since a TPUR is "looked at" from different directions by spatially separated stations. These special features depend on the type of a MSRS and information fusion.

Spatially Coherent MSRSs and Spatially Coherent Target Echoes

In this case there are no significant differences from monostatic radars. As was mentioned in Sections 1.2 and 5.1, such a MSRS may be considered as a "monostatic" radar with a very large (in comparison with the wavelength) and sparsely populated antenna array. Scanning a TPUR, this array is focused at the point of space where a target is expected. Around this point a "focal spot" is made up which in fact is a Space Resolution Cell (SRC) of the MSRS. Spatially coherent MSRSs have usually "short baselines" (see Section 1.1) so that it is reasonable to consider the dimensions of a SRC "in down-range" direction (i.e. along the direction to the conventional centre of a MSRS) and "in cross-range" directions (i.e. in the plane perpendicular to "down-range" direction). Then as with monostatic radar, the "down-range" dimension of a SRC is usually determined by the bandwidth Δf_s of signal used: $c/2\Delta f_s$ where c is the velocity of light; the "cross-range" dimension of a SRC is determined by the quantity of the order of $R\lambda/L_{eff}=Rc/f_0L_{eff}$ where R is the range, L_{eff} is the effective baselength and f_0 is the carrier frequency of a MSRS.

[12] Let a maximum allowable number of false alarms in M resolution cells, $N_{fa\,max}$, be given ($0<N_{fa\,max}\leqslant M$). Then as an averaged value $\bar{N}_{fa\,max}$ may be used the solution of the following equation: $\bar{N}_{fa}+3\sigma(N_{fa})=N_{fa\,max}$ where $\sigma(N_{fa})$ is the r.m.s. value of N_{fa}. Taking into account that for the binomial distribution $\sigma(N_{fa})=\sqrt{MP_{fa}(1-P_{fa})}$ yields for $P_{fa}\ll 1:\sigma(N_{fa})\approx\sqrt{MP_{fa}}=\sqrt{\bar{N}_{fa}}$. Substituting $\sqrt{\bar{N}_{fa}}$ for $\sigma(\bar{N}_{fa})$ and solving the equation for $\bar{N}_{fa}$ we obtain: $\bar{N}_{fa\,max}=(-1.5+\sqrt{2.25+N_{fa\,max}})^2$.

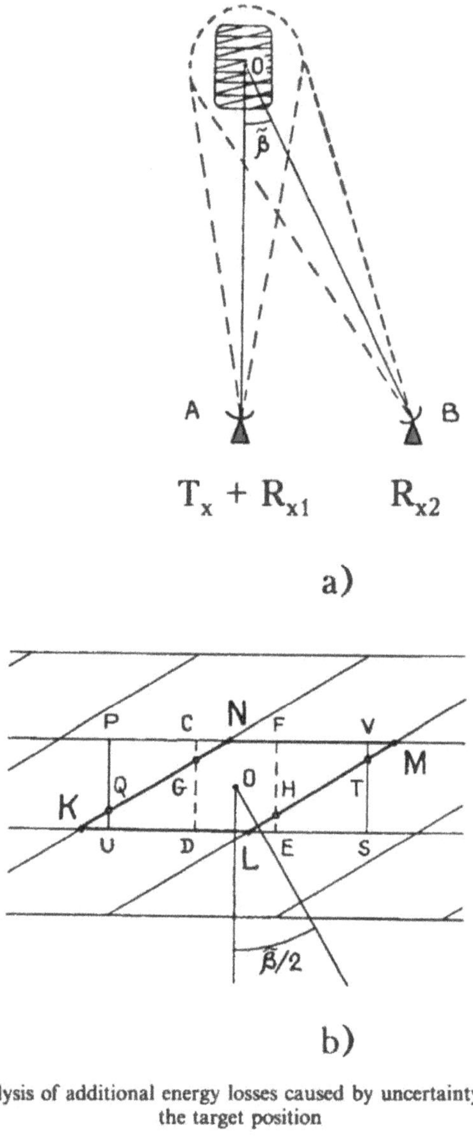

Figure 5.12 To the analysis of additional energy losses caused by uncertainty in *a priori* knowledge of the target position

Consider the simplest example of a system with one transmitting–receiving station (a monostatic radar) and one receiving station (see Fig. 5.12(a)). In this figure a TPUR and the boundaries of range resolution cells are shown. For the first station these boundaries are the fragments of spherical surfaces with the centre at the point A. For the second station these boundaries are the fragments of surfaces of prolate spheroids with their foci at the points A and B. Within the antenna main beams these fragments of surfaces may be replaced by fragments of planes perpendicular to OA and to the bisector of the angle $\tilde{\beta}$ between OA and OB, respectively.

In Fig. 5.12(b) a fragment of the intersection of several resolution cells in the plane of figure is shown in a closed-up view. The boundaries of a single focal spot are shown by the dashed lines CD and EF. It is clear that differences in receiving directions are negligible within any SRC $CDEF$ if $CG \ll CD/2$ and $EH \ll EF/2$. But $CD \approx c/2\Delta f_s$ is the dimension of a range resolution cell. It is seen from Fig. 5.12(b) that $CG = EH \approx (CF/2)\,\mathrm{tg}(\bar{\beta}/2) \approx CF(\bar{\beta}/4)$ since $\bar{\beta} \ll 1$. On the other hand, $CF \approx R\lambda/L_{eff} \approx Rc/f_0 L_{eff}$ is the cross-range dimension of the focal spot. The condition $CG \ll CD/2$ means that $Rc\bar{\beta}/4f_0\, L_{eff} \ll c/4\Delta f_s$ and since $\bar{\beta} \approx L_{eff}/R$ we obtain $f_0 \gg \Delta f_s$. The last inequality holds nearly always in practice. Therefore, for a spatially coherent MSRS the difference in directions of looking at a TPUR from spatially separated stations may be neglected. This means that the "cost of resolution" can be evaluated by the same technique as for a monostatic radar. For example, when a TPUR is wholly illuminated by a single probe from the transmitting station, then the number M in (5.97) is equal to the number of independent focal spots falling within this region. It is important to take into account only independent focal spots. In practice their number is equal to the product of the number of receiving stations by the number of range resolution cells within the uncertainty region. In the case of the interferometer with two stations considered in Fig. 5.12, only two independent systems of periodically repeating focal spots can be formed. These two systems are shifted with respect to each other by the width of a focal spot.

MSRSs without Long-Term Spatial Coherence or (and) with Incoherent Target Echoes

As a result of optimum signal processing (see Sections 5.2, 5.3) a SRC of a MSRS is formed as an intersection of resolution cells ("pulse volumes") of each pair of the transmitting and receiving stations. Such a SRC (its projection on plane) is shown in Fig. 5.12(b) and denoted by $KLMN$. If the MSRS of interest is a "short baseline" one we have again the situation which can be quite similar to that of a monostatic radar. It can occur if the following condition is satisfied (see Fig. 5.12(b): $PQ \ll PU/2$, $ST \ll SV/2$. The difference from a spatially coherent MSRS is that the width of a focal spot $CF = DE$ is replaced here by the width PV = US of the TPUR in the cross-range direction. If we denote this quantity by $R\Theta$ (where R is the range and Θ is the angle width of the uncertainty region) and substitute again $c/2\Delta f_s$ for $PU = SV$, we obtain $PQ = (R\Theta/2)\,\mathrm{tg}(\bar{\beta}/2) \ll c/4\Delta f_s$. Replacing once more $\mathrm{tg}(\bar{\beta}/2) \approx (\bar{\beta}/2) \approx L_{eff}/2R$, we obtain the above condition in the form

$$\Theta \ll c/L_{eff}\Delta f_s. \tag{5.98}$$

The expression in the right part of this inequality is the width of a "resultant directivity pattern" (RDP, see Section 1.2). It determines the resolution capability of a MSRS in the angle coordinate in the baseline plane (passing through the baseline and the target). Thus the difference between directions from the target to spatially separated receiving stations may be neglected if the angle width of a TPUR is much less than the RDP of the MSRS. When the condition (5.98) is satisfied, then resolution cells of different stations practically coincide in space (almost completely overlap) so that the "cost of resolution" may be evaluated as for a monostatic radar, i.e. with the help of (5.97).

The condition (5.98) is usually satisfied in target tracking process when target coordinates are measured by the same MSRS. In target search mode or when a target is tracked by measurement facilities with lower accuracy this inequality may have the opposite sign. In this case the angle width of a TPUR can be covered by several SRCs of the MSRS (see Fig. 5.12). Then receiving stations must "look" at all SRCs of the TPUR which are illuminated by the transmitting station in each recurrent period. If the optimum detection algorithms $L_3, \ldots, L_6$ (see Section 5.2) are used, a multichannel system of weighted summation is required with the number of channels equal to the number of illuminated SRCs. Delays (and Doppler shifts, if necessary) of wanted signals coming from each SRC to spatially separated stations are to be equalized in front of each summator. However, since noises in different SRCs are not statistically independent, the total number of SRCs within the uncertainty region illuminated by the transmitting station is not equal to M which determines false alarm probability, P_{fa}, at a given value of average number of false alarms, N_{fa}. It is seen from Fig. 5.12 that because of the difference in directions of echo reception by spatially separated stations, each range[13] resolution cell of one station is covered by several SRCs which correspond to several range resolution cells of the other station. Noises in these SRCs are the sums of mutually independent noises from different range resolution cells of the second station with the same noise from the range resolution cell of the first station.

To calculate the total number of false alarms from the whole illuminated part of the uncertainty region, the joint probability distribution of threshold exceedings by mutually dependent noises from all SRCs of interest is required. For each specific situation computer simulation may be used to obtain statistical characteristics of the number of false alarms. Apparently, the synthesized algorithms $L_3, \ldots, L_6$ are reasonable to use in practice when a TPUR covers only a few SRCs of a MSRS. Otherwise multichannel weighted summation with delay (and possibly, Doppler) equalization for each SRC turns out to be too complex and awkward. However, when a number of SRCs in a TPUR is not too large, necessary reduction of P_{fa} (in comparison to the case where the TPUR is equal to a single SRC) is small and leads to an insignificant energy losses. The following inequality may be used for evaluation of M

$$n_R < M < n_R(\Theta_1 \Theta_2 / \Delta\Theta_1 \Delta\Theta_2) \tag{5.99}$$

where n_R is the number of range resolution cells within the TPUR (for the station where this number is the largest); Θ_1, Θ_2 are the widths of the TPUR in angle coordinates; $\Delta\Theta_1, \Delta\Theta_2$ are the RDPs of the MSRS in these two angle coordinates. The RDP's width is described by the right part of (5.98) where the largest baselength determining the resolution capability in corresponding coordinate should be substituted. The left inequality of (5.99) relates to the case where the condition (5.98) is satisfied. In this case we have nearly complete overlapping of resolution cells of different stations. The right inequality gives the upper estimate. It corresponds to the total number of SRCs illuminated by the transmitting station in a TPUR and does not take into account the noise mutual dependence of adjacent SRCs.

Substituting M from (5.99) in (5.97) yields $P_{fa\,max}$. Now using detection characteristics from Sections 5.4 and 5.5 we can evaluate energy losses caused by *a priori* uncertainty of the target position in comparison to the case where this uncertainty is absent i.e. $M = 1$.

***In this section**, we have taken into account that before each radar transmission the possible position of a target is in practice known with some uncertainty. This Target Position Uncertainty Region (TPUR) may be large (in a search regime) or small (in a tracking regime) but is always larger than a single resolution cell. This leads to extra energy losses as compared with detection characteristics obtained in the previous sections. The higher the resolution capability of a radar, the greater these energy losses,*

[13] For the sake of brevity we shall not distinguish between "range" and "range-sum" (for a bistatic radar). In general, a range resolution cell may be treated as a resolution cell in signal time of arrival.

since it is the number of false alarms per a recurrent period (or per a second) that is a practical output radar characteristic. The increase of the number of resolution cells in a TPUR requires the higher threshold levels and hence results in extra energy losses. This is a well-known "cost of resolution". We have considered the special features of MSRSs caused by the difference in directions from a TPUR to spatially separated stations. We have shown that for spatially coherent MSRSs with short baselines a simple relationship (5.97) *between the average number of false alarms and the false alarm probability may be used. For other types of MSRSs we have derived the condition* (5.98) *under which the relationship* (5.97) *may also be exploited. However, in most cases the condition* (5.98) *is not satisfied. For these cases we have presented the inequality* (5.99) *permitting to estimate the number of resolution cells of a TPUR which in its turn allows to take into account the uncertainty of a target position when using the detection characteristics obtained in Sections* 5.4 *and* 5.5.

6. DECENTRALIZED TARGET DETECTION IN A BACKGROUND OF SPATIALLY UNCORRELATED INTERFERENCE

6.1. OPTIMIZATION OF DECENTRALIZED TARGET DETECTION

All algorithms that have been considered in Sections 5.1–5.6 may be called algorithms of *centralized* detection. In fact, received signals (i.e. sums of wanted signals and receiver self-noises or self-noises alone) are transferred from all receiving stations of a MSRS (after matched filtration and, possibly, envelope detection) to a Fusion Centre (FC). Decisions regarding the presence or absence of a target (a wanted signal) are made only in the FC by the comparison of weighted sums of received signals coming from all stations with a predetermined threshold.

As was mentioned in Section 1.1 *decentralized* (*distributed*) information processing is possible in MSRSs. Specifically, preliminary (local) decisions concerning target detection may be made in each receiving station. Only these preliminary decisions are then communicated to the FC. If in an arbitrary range resolution cell of the ith station ($i=\overline{1,m}$) a predetermined threshold is exceeded, a unity is generated and transferred to the FC. If the threshold is not exceeded, a zero is assumed to be in that resolution cell. Hence preliminary decisions are sequences of zeros and ones. A final (output, global) decision is made in the FC as a result of joint processing (combining) of those sequences (preliminary decisions) coming from all receiving stations.

Thus the decentralized target detection defined above is in reality partially decentralized. Fully decentralized detection is also possible in MSRSs. In this case final decisions are made in each station independently of other stations. Joint information processing (which is one of the main attributes of MSRSs) is, for instance, reduced to measurement or track fusion. It is clear that fully decentralized detection does not differ from monostatic detection and hence is not considered here. About measurement and track fusion see Chapters 14 and 15.

Practical implementation of decentralized detection algorithms is much simpler than that of centralized ones. Besides, only limited information of signals exceeded the predetermined threshold is transferred from each receiving station via Data Transmission Lines (DTLs) to FC. Hence requirements to the handling capacity of DTLs are drastically reduced. Under certain conditions decentralized processing provides the enhancement of reliability and survivability of systems. All these advantages are of great practical significance so that decentralized detection is often preferable to use in actual MSRSs. Therefore it is important to optimize algorithms of decentralized detection and to evaluate their performance characteristics.

A structural scheme of the decentralized detection for a MSRS with a single transmitting and m receiving stations is depicted in Fig. 6.1. This is so called parallel topology which is most widely used. The same scheme is valid for a MSRS consisting of m monostatic radars illuminating simultaneously one and the same space region. Signal processing in each receiving station includes optimum linear filtration (matched filtration if noises are white), envelope detection and threshold comparison. If the

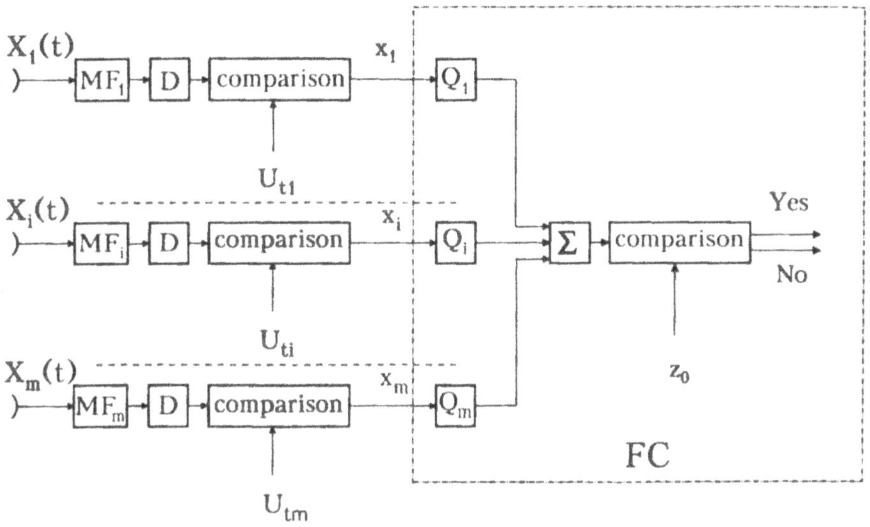

Figure 6.1 Structure of a decentralized (distributed) detector

threshold is exceeded in the ith station ($i=\overline{1,m}$) in a range (or time of arrival) resolution cell corresponding to the analyzed SRC of the MSRS, $x_i=1$ is communicated via the DTL to the FC. Otherwise $x_i=0$ is assumed to be in this resolution cell. For final decision of whether a target is present in the analyzed SRC, sequences of zeros and ones, $x_1, x_2, \ldots, x_m$, undergo joint processing. Evidently, x_i is a random variable with the Bernoulli probability distribution

$$P(x_i)=P_i^{x_i}(1-P_i)^{(1-x_i)} \tag{6.1}$$

where P_i is the probability of excess of a threshold in the ith receiving station. When a wanted signal (target echo) is absent, P_i means the (local) false alarm probability, $P_i=P_{fai}$. When it is present, P_i is the (local) detection probability, $P_i=P_{di}$.

The problem of optimum processing will be considered firstly in a "narrow sense". Let a train $x_1, \ldots, x_m$ come to the input of the FC. Joint probability distributions for all possible combinations of x_i, $i=\overline{1,m}$, both in the absence and in the presence of a wanted signal (hypotheses H_0 and H_1, respectively), i.e. $P(x_1, \ldots, x_m|H_0)$ and $P(x_1, \ldots, x_m|H_1)$, are assumed to be arbitrary but completely known. For each specific sequence $x_1, \ldots, x_m$ we can write the likelihood ratio

$$\Lambda=P(x_1, \ldots, x_m|H_1)/P(x_1, \ldots, x_m|H_0). \tag{6.2}$$

The processing of $x_1, \ldots, x_m$ described in (6.2) is optimum in accordance with the likelihood ratio criterion. Comparison of Λ with a threshold determined by the allowable false alarm probability provides the Neyman–Pearson optimal decision about the presence or absence of a wanted signal (target echo).

By the independence of noises at different receiving stations [see (6.1)]

$$P(x_1, \ldots, x_m|H_0)=\prod_{i=1}^{m} P(x_i|H_0)=\prod_{i=1}^{m} P_{fai}^{x_i}(1-P_{fai})^{(1-x_i)}. \tag{6.3}$$

Considering the probability $P(x_1,\ldots,x_m|H_1)$ conditioned by the assumption that a wanted signal is present, one should take into account the mutual statistical dependence of target echo fluctuations at the inputs of spatially separated stations. The modulus of mutual correlation coefficient of the sums of wanted signals and noises in the ith and kth stations before envelope detectors is

$$|r_{ik}|=\tilde{q}_{\text{out}\,i}\tilde{q}_{\text{out}\,k}|R_{ik}|/\sqrt{(1+\tilde{q}^2_{\text{out}\,i})(1+\tilde{q}^2_{\text{out}\,k})} \qquad (6.4)$$

where $|R_{ik}|$ is the modulus of the interstation correlation coefficient of signal complex amplitude fluctuations (see Section 4.2); $\tilde{q}^2_{\text{out}\,i}$, $\tilde{q}^2_{\text{out}\,k}$ are the power SNRs at the inputs of envelope detectors in the ith and kth stations. When fluctuations are spatially strongly dependent, i.e. when $|R_{ik}|\approx 1$, and $\tilde{q}^2_{\text{out}\,i}\gg 1$, $\tilde{q}^2_{\text{out}\,k}\gg 1$, the mutual dependence of threshold exceeding (or nonexceeding) in different stations is to be taken into account. In this case the probability $P(x_1,\ldots,x_m|H_1)$ can be written in the form

$$P(x_1,\ldots,x_m|H_1)$$

$$=\lim_{A\to\infty}\int_{x_1u_1}^{(1-x_1)u_1+x_1A}\cdots\int_{x_mu_m}^{(1-x_m)u_m+x_mA} w(y_1,\ldots,y_m|H_1)\mathrm{d}y_1,\ldots,\mathrm{d}y_m \qquad (6.5)$$

where u_i is the threshold level in the ith station, $i=\overline{1,m}$; $w(y_1,\ldots,y_m|H_1)$ is the joint m-dimensional PDF of signal-plus-noise envelopes in front of thresholding in all receiving stations. For Gaussian signal complex amplitudes and linear envelope detectors in each station $w(y_1,\ldots,y_m|H_1)$ is the joint PDF of m mutually dependent Rayleigh variables. Having obtained a sequence of zeros and ones we can substitute it in the right part of (6.5) to evaluate the likelihood function $P(x_1,\ldots,x_m|H_1)$, then substitute this value in (6.2) taking into account (6.3) and then compare the result with a threshold. However the use of such an algorithm in practice is rather difficult and, as a rule, is not a success. Even for $m=2$ the likelihood function $P(x_1,x_2|H_1)$ as a function of two mutually dependent Rayleigh variables can be expressed only in the form of an infinite series. For $m>2$ the problem becomes more complex.

At the same time, when target dimensions are much larger than the wavelength, the spatial (interstation) correlation of echo complex amplitude fluctuations in actual MSRSs is most likely low (see Section 4.2). If $|R_{ik}|<0.5\ldots0.6$ in (6.4) we may, as a rule, consider threshold exceedings in different stations as approximately mutually independent events since the degree of their mutual dependence is determined by $|r_{ik}|^2$ from (6.4). In this case

$$P(x_1,\ldots,x_m|H_1)=\prod_{i=1}^{m}P(x_i|H_1)=\prod_{i=1}^{m}P_{\mathrm{d}i}^{x_i}(1-P_{\mathrm{d}i})^{(1-x_i)}. \qquad (6.6)$$

Substituting (6.6) and (6.3) in (6.2) and taking the logarithm yields

$$L=\ln\Lambda=\sum_{i=1}^{m}x_i(\ln P_{\mathrm{d}i}-P_{\mathrm{fa}i})+(1-x_i)[\ln(1-P_{\mathrm{d}i})-\ln(1-P_{\mathrm{fa}i})].$$

We denote the multipliers at x_i by

$$Q_i=\ln P_{\mathrm{d}i}-\ln P_{\mathrm{fa}i}-\ln(1-P_{\mathrm{d}i})+\ln(1-P_{\mathrm{fa}i})=\ln\left(\frac{P_{\mathrm{d}i}}{1-P_{\mathrm{d}i}}\,\frac{1-P_{\mathrm{fa}i}}{P_{\mathrm{fa}i}}\right) \qquad (6.7)$$

and omit the summands not depending on x_i. Besides, we denote by z_0 the threshold level determined by the given output false alarm probability P_{fa}. Then the optimum detection algorithm according to the Neyman–Pearson criterion takes the form

$$L = \sum_{i=1}^{m} Q_i x_i \gtrless z_0. \tag{6.8}$$

If $L \geqslant z_0$ the FC decides a target is present (the hypothesis H_1 is accepted). If $L < z_0$ the FC decides a target is absent (the hypothesis H_0 is accepted).

Thus the optimum joint processing is reduced to a weighted summation of zeros and ones, x_i, $i=\overline{1,m}$, which represent preliminary decisions made in all receiving stations. From the definition of weights Q_i [see (6.7)] it is evident that they increase the contribution of those stations where preliminary decisions are more reliable, i.e. where detection probability, P_{di}, is higher at lower false alarm probability, P_{fai}. Hence, as for the centralized detection (see Sections 5.2, 5.3) weights Q_i depend on both SNRs and noise levels at receiving stations.

Since x_i, $i=\overline{1,m}$, is equal to 0 or to 1 only, the sum in (6.8) turns out to be a sum of $n \leqslant m$ weights Q_i and hence can be equal only to certain discrete values. The threshold level z_0 must lie within the interval $(0, \sum_{i=1}^{m} Q_i)$, i.e. $0 < z_0 \leqslant \sum_{i=1}^{m} Q_i$, in order to exclude trivial decisions that a target is present even when $x_1 = x_2 = \cdots = x_m = 0$, on the one hand, and that a target is absent even when $x_1 = x_2 = \cdots = x_m = 1$, on the other. When all Q_i, $i=\overline{1,m}$, are different and a sum of any group of Q_i is not equal to a sum of any other group of Q_i, then $2^m - 1$ different values of $L > 0$ are possible for different combination of x_i, $i=\overline{1,m}$. Choosing the threshold z_0 between Q_i and their different sums we may form $2^m - 1$ different detection criteria (decision rules).

Example 6.1. The case of $m=3$ is shown in Fig. 6.2. Along the axis $2^3 - 1 = 7$ values of $L > 0$ are marked in accordance with all possible combinations of zeros and ones in the sequence x_1, x_2, x_3. Choosing the threshold level z_0 within the interval $(0, Q_1)$, i.e. $0 < z_0 \leqslant Q_1$, we have the known rule "at least one out of m" ("1 out of m", "OR"). It means that a target is decided to be present if at least in one (arbitrary) station the predetermined threshold has been exceeded. As the threshold z_0 rises the decision rules become more "severe". When $Q_1 < z_0 \leqslant Q_2$ a target echo is considered to be detected if a preliminary detection has been declared in at least one out of m stations but not in the first station only. Since $Q_1 < Q_2 < Q_3$ in Fig. 6.2, decisions coming from the first station are the least reliable, i.e. the first station is the "weakest" one [see (6.7)]. In a similar manner, if $Q_2 < z_0 \leqslant Q_3$ the presence of a target is not decided if preliminary detections are obtained only in the first or only in the second station

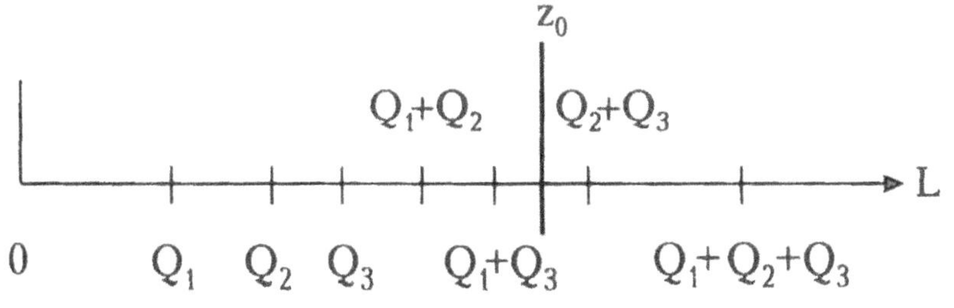

Figure 6.2 Feasible values of the decision variable L in equation (6.8) for $m=3$

and so on. When $Q_2+Q_3<z_0\leqslant Q_1+Q_2+Q_3$ preliminary detections in all the three stations are necessary for a final decision that a target is present. It is a well-known decision rule of the type "*m* out of *m*" ("AND").

When $P_{\mathrm{f}ai}=P_{\mathrm{f}a0}, P_{\mathrm{d}i}=P_{\mathrm{d}0}, i=\overline{1,m}$, then all weights are identical: $Q_1=Q_2=\cdots=Q_m=Q$. In this case we may take Q outside the sum sign in (6.8) and divide both parts of (6.8) by Q. We obtain instead of (6.8)

$$L=\sum_{i=1}^{m} x_i \gtrless z_0'. \tag{6.9}$$

For all possible x_i the random variable $L>0$ can be equal only to m different integers from 1 to m. Choosing z_0' so that $k-1<z_0'\leqslant k$ where $k=1,2,\ldots,m$, yields known decision rules (detection criteria) of the type "*k* out of *m*". According to these rules the decision "a target is present" is made if preliminary detections have been obtained at least in *k* arbitrary receiving stations out of total number of *m* stations (in corresponding range resolution cells).

It should be noted that 2^m-1 decision rules determined by (6.8) can include all the m decision rules "*k* out of *m*" followed from (6.9).

Example 6.2. It can be seen from Fig. 6.2 that for $0<z_0\leqslant Q_1$, $Q_3<z_0\leqslant Q_1+Q_2$ and $Q_2+Q_3<z_0\leqslant Q_1+Q_2+Q_3$ we obtain the decision rules "1 out of 3" ("OR"), "2 out of 3" and "3 out of 3" ("AND"), respectively.

Thus the difference between weights Q_i leads to greater variety of possible decision rules but does not exclude the rules "*k* out of *m*" corresponding to identical weights, Q_i, $i=\overline{1,m}$.

However, this is true only for "natural" relationships between weights Q_i in (6.8) when a sum of arbitrary two weights is greater that any one weight, a sum of arbitrary three weights is greater than any sum of arbitrary two weights and so on. Such an order corresponds to not too large differences between Q_i. Just this case is shown in Fig. 6.2 for $m=3$. When the differences between Q_i, $i=\overline{1,m}$, are large [which in its turn is a consequence of large differences in stations' power characteristics – see (6.7)] a sum of several weights can be less than a sum of a lesser number of other weights (for instance, $Q_1+Q_2<Q_3$ for $m=3$). Then 2^m-1 decision rules followed from (6.8) do not include all the rules "*k* out of *m*" (for instance, if $Q_1+Q_2<Q_3$, the decision rule "2 out of 3" is excluded). Furthermore, according to (6.8) if a sum of several arbitrary weights $\sum_{i=1}^{n} Q_i$ is less than arbitrary one remaining weight Q_j and a threshold level z_0 is chosen so that $\sum_{i=1}^{n} Q_i<z_0\leqslant Q_j$, information from stations with numbers from 1 to *n* should not be taken into account in detection algorithm. In particular, under conditions of Example 6.1 (Fig. 6.2) if $Q_1+Q_2<Q_3$ for $m=3$ and z_0 is chosen so that $Q_1+Q_2<z_0\leqslant Q_3$, the optimum algorithm requires to ignore information from the first and second stations, i.e. to go over to a monostatic detection by the third "strong" station. Apparently, it may be reasonable when differences between power characteristics (sensitivities) of receiving stations are so large that including of "weak" stations in joint processing cannot enhance detection characteristics of MSRSs. However, as was mentioned in Section 5.5, such cases are not of practical importance. Therefore, we shall, as a rule, consider "natural" relationships between weights Q_i in (6.8).

For fixed probabilities of preliminary decisions, $P_{\mathrm{f}ai}$ and $P_{\mathrm{d}i}$, $i=\overline{1,m}$, different decision rules (detection criteria) in a FC provide different output probabilities P_{fa} and P_{d}. To choose an optimal rule (according to the Neyman–Pearson criterion), i.e. to choose the threshold z_0 in (6.8), let us derive expressions for output (final) probabilities P_{fa} and P_d corresponding to the optimum processing (6.8). Since x_i is described by the Bernoulli distribution (6.1) with the PDF

$$w(x_i)=P_i\delta(x_i-1)+(1-P_i)\delta(x_i)$$

the random variable $z_i=Q_ix_i$ is described by the PDF

$$w(z_i)=P_i\delta(z_i-Q_i)+(1-P_i)\delta(z_i)$$

and by the characteristic function

$$\Theta_i(u)=\int_{-\infty}^{\infty} w(z_i)\exp(juz_i)dz_i=P_i\exp(juQ_i)+(1-P_i).$$

The characteristic function of L in (6.8) which is a sum of independent random variables

$$\Theta_L(u)=\prod_{i=1}^{m}\Theta_i(u)=\prod_{i=1}^{m}[P_i\exp(juQ_i)+(1-P_i)]. \tag{6.10}$$

The inverse Fourier transformation yields the PDF of L

$$w_L(z)=\prod_{i=1}^{m}(1-P_i)\delta(z)+\sum_{k=1}^{m}\sum_{i_1=1}^{m-k+1}\sum_{i_2=i_1+1}^{m-k+2}\cdots\sum_{i_k=i_{k-1}+1}^{m}P_{i_1}P_{i_2}\dots P_{i_k}$$

$$\times\delta\left(z-\sum_{r=1}^{k}Q_{i_r}\right)\prod_{j=1}^{m}(1-P_j),\quad j\neq i_1,i_2,\dots,i_k. \tag{6.11}$$

As k changes from 1 to m the number of sums in (6.11) also changes from 1 to m: for $k=1$ in (6.11) is a single sum over i_1; for $k=2$ – a double sum over i_1 and i_2 and so on.

The false alarm probability, P_{fa}, and the detection probability, P_d, can be obtained by substituting P_{fai} or P_{di}, respectively, for P_i, $i=\overline{1,m}$, in (6.11) and by integrating between the limits from z_0 to infinity. Since $z_0>0$ the first term of (6.11) does not contribute to the integral. The same is true for all terms where arguments of Dirac's delta-functions contain $\sum_{r=1}^{k}Q_{i_r}<z_0$. For instance, if the nearest to z_0 but greater than z_0 sum of weights contains n summands ($n=\overline{1,m}$) and is equal to $\sum_{r=1}^{n}Q_{l_r}$, then the probability of excess of the threshold level z_0 can be written as follows:

$$P=\sum_{k=n}^{m}\sum_{i_1=1}^{m-k+1}\sum_{i_2=i_1+1}^{m-k+2}\cdots\sum_{i_k=i_{k-1}+1}^{m}P_{i_1}P_{i_2}\dots P_{i_k}\prod_{j=1,\,j\neq i_1,i_2,\dots,i_k}^{m}(1-P_j) \tag{6.12}$$

and when $k=n$ then $i_1\geqslant l_1$, $i_2\geqslant l_2,\dots,i_n\geqslant l_n$.

Example 6.3. Let a number of receiving stations m be again equal to 3. Then from (6.11) we have the PDF

$$w_L(z)=\prod_{i=1}^{3}(1-P_i)\delta(z)+\sum_{i_1=1}^{3}P_{i_1}\delta(z-Q_{i_1})\prod_{j=1,j\neq i_1}^{3}(1-P_j)$$
$$+\sum_{i_1=1}^{2}\sum_{i_2=i_1+1}^{3}P_{i_1}P_{i_2}\delta[z-(Q_{i_1}+Q_{i_2})]\prod_{j=1,j\neq i_1,i_2}^{3}(1-P_j)$$
$$+P_1P_2P_3\delta[z-(Q_1+Q_2+Q_3)]. \tag{6.13}$$

When $Q_1+Q_3<z_0<Q_2+Q_3$ from (6.12) we have $n=2$, $l_1=2$, $l_2=3$. It means that only two summands remain in the right side of (6.12): for $k=2$, $i_1=2$, $i_2=3$ and for $k=3$, $i_1=1$, $i_2=2$, $i_3=3$ so that $P=P_2P_3(1-P_1)+P_1P_2P_3$. If a target echo is absent, $P_{fa}=P_{fa2}P_{fa3}(1-P_{fa1})+P_{fa1}P_{fa2}P_{fa3}$. When a target echo is present, then $P_d=P_{d2}P_{d3}(1-P_{d1})+P_{d1}P_{d2}P_{d3}$.

When false alarm and detection probabilities are identical in all receiving stations, i.e. $P_{fa1}=\cdots=P_{fam}=P_{fa0}$; $P_{d1}=\cdots=P_{dm}=P_{d0}$, then $Q_1=\cdots=Q_m$. In this case for the algorithm (6.9) and $k-1<z_0'\leqslant k$ (decision rules of "k out of m" type) we have instead of (6.12)

$$P=\sum_{n=k}^{m}C_m^n\tilde{P}^n(1-\tilde{P})^{m-n}. \tag{6.14}$$

Here $P=P_{fa}$ if $\tilde{P}=P_{fa0}$ and $P=P_d$ if $\tilde{P}=P_{d0}$;

$$C_m^n=\binom{m}{n}=\frac{m!}{n!(m-n)!}$$

are the combinations of m elements taken as groups of n elements.

Using equations (6.12) and (6.14) we can calculate the output probabilities P_{fa} and P_d for given values of P_{fai} and P_{di}, $i=\overline{1,m}$, an arbitrary threshold z_0 and a decision rule corresponding to the chosen threshold. The higher z_0 (i.e. the more "severe" a decision rule), the lower P_{fa} and P_d. If $P_{fa}\leqslant\alpha_0$ is required, the optimum decision rule (in accordance with the Neyman–Pearson criterion) is such a rule which provides a pair of maximal values of P_{fa} and P_d under the condition $P_{fa}\leqslant\alpha_0$.

Example 6.4. Let for $m=3$ the probabilities of preliminary (local) decisions be: $P_{fa1}=10^{-2}$, $P_{d1}=0.9$; $P_{fa2}=5\cdot10^{-3}$, $P_{d2}=0.85$ and $P_{fa3}=3\cdot10^{-3}$, $P_{d3}=0.95$. According to (6.7) and (6.8) we have an optimum algorithm of joint processing in the form

$$L=6.79x_1+7.03x_2+8.75x_3\gtrless z_0.$$

For different values of z_0 we obtain seven different possible decision rules. Using (6.12) yields seven corresponding pairs of output probabilities P_{fa} and P_d. The results are shown in Table 6.1.

For instance, if $a_0=3\cdot10^{-5}$ the optimum rule is "the second and the third stations or all the three stations". It leads to $P_{fa}=1.5\cdot10^{-5}$ and $P_d=0.808$. The optimum threshold z_0 is $15.54<z_0\leqslant15.78$. If $\alpha_0\leqslant10^{-4}$ the optimum decision rule is "2 out of 3" with $P_{fa}=9.5\cdot10^{-5}$ and $P_d=0.974$. The optimum threshold z_0 must lie

Table 6.1

	Threshold level	Decision rule	P_{fa}	P_d
1	$0<z_0\leqslant(Q_1=6.79)$	"1 out of 3"	$1.8\cdot10^{-2}$	0.999
2	$(Q_1=6.79<z_0\leqslant(Q_2=7.03)$	"1 out of 3 but not only the first"	$8.0\cdot10^{-3}$	0.993
3	$(Q_2=7.03)<z_0\leqslant(Q_3=8.75)$	"1 out of 3 but not only the first or the second"	$3.1\cdot10^{-3}$	0.988
4	$(Q_3=8.75)<z_0\leqslant(Q_1+Q_2=13.82)$	"2 out of 3"	$9.5\cdot10^{-5}$	0.974
5	$(Q_1+Q_2=13.82)<z_0 \leqslant(Q_1+Q_3=15.54)$	"2 out of 3 but not only the first and the second"	$4.5\cdot10^{-5}$	0.936
6	$(Q_1+Q_3=15.54)<z_0 \leqslant(Q_2+Q_3=15.78)$	"The second and the third or all the three"	$1.5\cdot10^{-5}$	0.808
7	$(Q_2+Q_3=15.78)<z_0 \leqslant(Q_1+Q_2+Q_3=22.57)$	"3 out of 3"	$1.5\cdot10^{-7}$	0.727

within the interval (8.75, 13.82], i.e. $8.75<z_0\leqslant13.82$. When $\alpha_0<1.5\cdot10^{-7}$ or $\alpha_0>1.8\cdot10^{-2}$ there are no optimum decision rules for given probabilities P_{fai} and P_{di}, $i=1,2,3$.

The algorithm (6.9) optimal for $Q_1=\cdots=Q_m$ is much simpler than (6.8) since it does not require to estimate local probabilities P_{fai} and P_{di} in each station, to transfer these estimate via DTLs to a FC and to calculate weights Q_i, $i=\overline{1,m}$. Therefore, it is desirable to employ (6.9) not only for identical but also for different weights Q_i. Let us evaluate possible energy losses in this case.

While discussing Fig. 6.2 we noted that in most practically important case of "natural" relationships between weights Q_i, decision rules of the algorithm (6.8) include all decision rules of the algorithm (6.9). Hence if we go over from (6.8) to (6.9) the total number of possible decision rules is simply reduced from 2^m-1 (in the general case) to m. It means that the total number of possible pairs of P_{fa} and P_d is reduced too. If at a given allowable $P_{fa}=\alpha_0$ the optimum rule from (6.8) turns out to be one of the m rules from (6.9), then choosing this rule of "k out of m" type does not lead to any energy losses. In Example 6.4 the rules 1, 4 and 7 from (6.8) are the same as from (6.9). Therefore at $\alpha_0=10^{-4}$ the rule "2 out of 3" is optimum for both (6.8) and (6.9). Performance characteristics of the simple algorithm (6.9) do not differ from those of (6.8). However if $\alpha_0=3\cdot10^{-5}$ the algorithm (6.8) provides $P_d=0.808$ (the rule 6 is optimal) whereas the algorithm (6.9) provides $P_d=0.727$ (the rule 7 is optimal).

Thus when for a given value of $P_{fa\,max}=\alpha_0$ the nearest but lower value of P_{fa} is provided by one of the rules from (6.9), then it is reasonable to use the simple algorithm (6.9) not requiring weights calculation. No energy losses occur in this case.

Up to now we consider the decentralized detection optimization in a "narrow" sense when (arbitrary) probabilities P_{fai} and P_{di} of local preliminary decisions coming from each receiving station are given. In actual MSRSs these probabilities may be controlled by changing threshold levels. Therefore it is important to reveal possibilities of the decentralized detection optimization in a "wide" sense. It means a joint optimization of both the fusion of preliminary (local) decisions and preliminary decisions themselves in each receiving station.

It can be shown [112, 113, 206] that such a joint optimization results in a combination of the algorithm (6.8) at a FC with the likelihood ratio test at each station. This likelihood ratio should be derived from envelope detector outputs y_i under both assumptions: a target echo is absent (hypothesis H_0) and a target echo is present (hypothesis H_1). Then the likelihood ratio should be compared to a threshold. For Rayleigh signal amplitude fluctuations we have

$$\Lambda_i = \frac{w(y_i|H_1)}{w(y_i|H_0)}$$

$$= \frac{[y_i/\sigma_i^2(1+\tilde{q}_{\mathrm{out}\,i}^2)]\exp[-y_i^2/2\sigma_i^2(1+\tilde{q}_{\mathrm{out}\,i}^2)]}{(y_i/\sigma_i^2)\exp(-y_i^2/2\sigma_i^2)}, \quad i=\overline{1,m}.$$

Omitting common multipliers, taking a logarithm and discarding summands independent of y_i we obtain that the optimum algorithm of the preliminary detection is reduced to comparing the variable y_i with a threshold, i.e.

$$L_i = y_i \gtrless \sigma_i u_{0i}, \quad i=\overline{1,m}.$$

Thus usual processing is to be performed in each receiving station: the optimum linear filtration according to the maximum SNR criterion (matched filtration if noises are white) envelope detection and thresholding.

However, the principal difficulty is that the optimum threshold level u_{0i} depends not only on the output (final) false alarm probability, P_{fa}, and the chosen decision rule in a FC, but on SNRs at all the m receiving stations. In principle, for each of 2^m-1 possible decision rules followed from the algorithm (6.8), given probability P_{fa} and measured values of SNRs $\tilde{q}_{\mathrm{out}\,i}^2$, it is possible to compute optimal probabilities $P_{\mathrm{fa}i}$, corresponding threshold levels u_{0i} and detection probability $P_{\mathrm{d}i}$, $i=\overline{1,m}$, which maximize the output (final) detection probability P_{d}.

It can be done by maximizing the objective function (see, e.g. [113])

$$J(P_{\mathrm{fa}1},\ldots,P_{\mathrm{fa}m}) = P_{\mathrm{d}}(P_{\mathrm{fa}1},\ldots,P_{\mathrm{fa}m}) + \beta[P_{\mathrm{fa}}(P_{\mathrm{fa}1},\ldots,P_{\mathrm{fa}m}) - \alpha_0] \tag{6.15}$$

where α_0 is the desired value for P_{fa} and β is the Lagrange multiplier. Differentiating the objective function with respect to $m+1$ unknowns ($P_{\mathrm{fa}1},\ldots,P_{\mathrm{fa}m}$ and β) yields a system of $m+1$ nonlinear equations

$$\frac{\partial J(P_{\mathrm{fa}1},\ldots,P_{\mathrm{fa}m})}{\partial P_{\mathrm{fa}i}} = 0, \ i=\overline{1,m}; \quad P_{\mathrm{fa}}(P_{\mathrm{fa}1},\ldots,P_{\mathrm{fa}m}) = \alpha_0. \tag{6.16}$$

Using (6.11) and (6.12) we can express P_{d} and P_{fa} through $P_{\mathrm{fa}1},\ldots,P_{\mathrm{fa}m}$ for any chosen decision rule and then solve the equations (6.16). Having obtained $P_{\mathrm{fa}1},\ldots,P_{\mathrm{fa}m}$, we can determine the thresholds u_{0i} from (5.71) and detection probabilities, $P_{\mathrm{d}i}$, from (5.77) for given SNR values $\tilde{q}_{\mathrm{out}\,i}^2$. An application of this technique has been illustrated by a simple example in [113].

It is clear that the described technique results in achieving an "optimum distribution" of a given output false alarm probability, P_{fa}, between all receiving stations of a MSRS. It turns out that the optimum threshold should be lowered (i.e. greater probabilities $P_{\mathrm{fa}i}$ may be allowed) in those stations where SNR values are higher (i.e. probabilities of correct preliminary decisions are higher). However, using this technique leads to a complicated and rather cumbersome procedure. It requires storing envelope detector outputs and estimating of SNRs at all receiving stations, solving nonlinear equations to determine an optimum threshold level for each station and for each decision rule. These nonlinear equations should be solved each time when a SNR is changed at any station.

An alternative approach is the simplest technique of a "uniform distribution" of P_{fa} between all receiving stations, i.e. equalization of $P_{\mathrm{fa}i}$ (and hence equalization of relative threshold levels u_{0i}) at all stations. It may be expected that such a simple technique does not lead to noticeable performance losses as compared to the complex optimum technique. Indeed, it is clear from symmetry considerations that for identical SNRs $\tilde{q}^2_{out\,i}$ and any decision rule at a FC, optimal false alarm probabilities, P_{fai}, and relative threshold levels u_{0i} are identical too ($i=\overline{1,m}$). Hence for small differences in SNRs $\tilde{q}^2_{out\,i}$ the optimal probabilities P_{fai} are to be quite close to each other. However, when the differences in SNRs $\tilde{q}^2_{out\,i}$ are large, contributions of "weak" stations to output detection characteristics of a MSRS are not significant even in the case of optimum centralized detection (see Section 5.5). Therefore optimization of local false alarm probabilities, P_{fai}, in decentralized detection under the condition of large differences in SNRs is unlikely to influence markedly final detection characteristics.

Example 6.5. To verify the above considerations comparative computations were carried out for a MSRS with a single transmitting station and two ($m=2$) receiving stations (when a role played by each station is maximal) and for the "mildest" decision rule "1 out of 2" ("OR"). The difference in SNRs between $\tilde{q}^2_{out\,1}$ and $\tilde{q}^2_{out\,2}$ was chosen to be equal to 10 dB which is near the extreme value when a "weak" station could yet contribute to detection characteristics of a MSRS. Two algorithms were compared: the optimum algorithm (with "optimal distribution" of $P_{fa}=10^{-4}$ between P_{fa1} and P_{fa2} according to [113]) and the suboptimum algorithm with $P_{fa1}=P_{fa2}=5\cdot 10^{-5}$. The calculation results are presented in Table 6.2.

In Table 6.2 $\delta\tilde{q}^2_{out\,1}$ denotes energy losses of the suboptimum algorithm in comparison to the optimal one, i.e. $\delta\tilde{q}^2_{out\,1}$ is the increase of $\tilde{q}^2_{out\,1}$ required to achieve the same value of detection probability P_d as the optimum algorithm provides. It is seen that for most important region $P_d>0.7$ the energy losses are extremely low, so that elimination of those losses does not warrant the use of the complex optimum algorithm.

The results of Example 6.5 mean that we may use in practice the simplest technique of uniform distribution of false alarm probabilities, i.e. of identical relative threshold levels $u_{0i}=u_{00}$ and identical local false alarm probabilities $P_{fai}=P_{fa0}$ at all receiving stations of a MSRS.

Thus the decentralized detection optimization is practically reduced, firstly, to the choice (by a threshold level z_0) of one of the 2^m-1 possible decision rules corresponding to the algorithm (6.8) of joint processing at a FC, and, secondly, to setting identical relative thresholds u_{00} providing such identical local false alarm probabilities P_{fa0} at each station that at the chosen decision rule result in required output (final, global) false alarm probability P_{fa}.

Unlike the above considered optimization in a "narrow" sense, the allowable false alarm probability $P_{fa}\leqslant\alpha_0$ does not determine now a threshold level z_0 in (6.8) or z_0' in (6.9), i.e. does not permit to choose an optimum decision rule. Varying the

Table 6.2

Algorithm	$\tilde{q}^2_{out\,1}/\tilde{q}^2_{out\,2}$ (dB)	P_{fa1} $\times 10^{-5}$	P_{fa2} $\times 10^{-5}$	P_{d1}	P_{d2}	P_d	$\delta\tilde{q}^2_{out\,1}$ (dB)
Optimum	10/0	9.917	0.083	0.433	0.001	0.433	
Suboptimum	10/0	5.000	5.000	0.406	0.007	0.411	0.3
Optimum	15/5	8.100	1.900	0.749	0.073	0.768	
Suboptimum	15/5	5.000	5.000	0.738	0.093	0.762	<0.1
Optimum	20/10	6.108	3.892	0.908	0.397	0.945	
Suboptimum	20/10	5.000	5.000	0.906	0.406	0.945	<0.1

threshold level u_{00} and hence the probability P_{fa0} at each station we can satisfy the condition $P_{fa}=\alpha_0$ for any decision rule. It means that an additional technique is necessary for choosing an optimum decision rule out of 2^m-1 possible decision rules followed from (6.8) or out of m rules followed from (6.9).

Let us consider the simpler algorithm (6.9) optimal for identical SNRs at all receiving stations. The choice of optimal threshold level z_0' means here the choice of $k=k_{opt}$ in decision rules "k out of m" ($1\leqslant k\leqslant m$). To obtain k_{opt} we may utilize (6.14) and solve the following optimization problem: to find the value of k maximizing the following objective function

$$P_d=\sum_{n=k}^{m} C_m^n P_{d0}^n(1-P_{d0})^{m-n}\to\max(k)$$

with the constraint in the form of equality

$$P_{fa}=\sum_{n=k}^{m} C_m^n P_{fa0}^n(1-P_{fa0})^{m-n}=\alpha_0$$

and under additional condition (5.77) for P_{fa0}, P_{d0}. If m is not too large the simplest way to solve this problem is to exhaust all possible values of k. This is reduced to the analysis and performance comparison of all decision rules of "k out of m" type (see Section 6.2). In a similar manner optimum decision rules out of 2^m-1 rules followed from the algorithm (6.8) can be found for the general case of different SNRs at receiving stations.

The choice of a decision rule in a MSRS may be determined not only by the desire for the best possible target detection. For instance, precise "multistatic" (or "multiradar") target coordinate measurements are possible only when the target is detected by several spatially separated stations (see Chapter 11). In this case more "severe" decision rules may be required than for optimum target detection.

In recent years much attention has been paid to decentralized detection in literature. Not only the Neyman–Pearson but the Bayesian approach and the Wald sequential probability test have been used [204–208]. Many papers were devoted to different extensions of the parallel topology considered above. These extensions include serial topology [209,210], detection systems with feedback [211–213], and systems with arbitrary topology [214]. More complicated schemes of decentralized detection permit in some cases a certain enhancement of detection characteristics. But in practice it is important to take into account the price that has to be paid for possible enhancement.

In this section, *we have considered principles of optimal processing algorithms for the practically important decentralized (distributed) detection in MSRSs. The decentralized detection optimization problem has been discussed in a "narrow" sense and in a "wide" sense. In the former case probabilities of preliminary (local) decisions at all receiving stations,* P_{fai} and P_{di}, $i=\overline{1,m}$, *are assumed to be known. An optimal algorithm* (6.8) *for joint processing has been presented and discussed. A general expression* (6.12) *for output (final) false alarm and detection probabilities,* P_{fa} *and* P_d, *has been derived. We have considered the conditions under which the general algorithm* (6.8) *providing up to* 2^m-1 *different decision rules can be reduced to the significantly simpler algorithm* (6.9) *determining up to m decision rules. The algorithm* (6.9) *is a well-known algorithm of "k out of m" type. It may often be used in practice instead of the markedly more complicated algorithm* (6.8). *Corresponding performance losses have been evaluated with the help of numerical Example* 6.3.

The optimization problem in a "wide" sense takes into account that probabilities of preliminary (local) decisions are not known in practice and can be easily changed by the choice of threshold levels in each receiving station (including monostatic radars). For this case we have described the optimization technique permitting the optimal "distribution" of a given output false alarm probability between all stations which maximizes the output detection probability given SNR values at all stations and a chosen decision rule at the FC. However, using this technique leads to a complicated and cumbersome procedure including storing envelope detector outputs and estimating of SNRs at all receiving stations, solving nonlinear equations to determine an optimum threshold level for each station and for each decision rule. These nonlinear equations should be solved each time when a SNR is changed at any station. It has been shown that keeping identical local false alarm probabilities (and hence relative threshold levels) at all receiving stations is a reasonable though not optimal alternative approach (Example 6.4). *An optimization problem has been formulated as the problem of choosing an optimal decision rule in the case where the simple algorithm* (6.9) *of "k out of m" type is used. In a similar manner optimal decision rules may be found in the more general case where algorithm* (6.8) *is preferable to use.*

6.2. PERFORMANCE ANALYSIS OF DECENTRALIZED DETECTION ALGORITHMS

The algorithms obtained in Section 6.1 have been optimized for most frequently occurring situation when target echo fluctuations are mutually independent at the inputs of spatially separated stations. Nevertheless we shall analyze the performance of these algorithms for both independent and strongly mutually dependent Rayleigh amplitude fluctuations.

At first (as in Section 5.5) we shall consider the case of *identical SNR values*, $\tilde{q}^2_{\mathrm{out}\,i}$, $i=\overline{1,m}$, when "multistatic" detection is most effective. In this case identical relative threshold levels $u_{0i}=u_{00}$, providing identical false alarm probabilities $P_{\mathrm{fa}i}=P_{\mathrm{fa}0}$ should be at all receiving stations. Detection probabilities are identical too, $P_{\mathrm{d}i}=P_{\mathrm{d}0}$, $i=\overline{1,m}$. Under these conditions decision rules of the "k out of m" type are optimal for joint data processing at FCs (see Section 6.1). The output (final) false alarm probability, P_{fa}, is determined by (6.14) where $\tilde{P}=P_{\mathrm{fa}0}$ and $P=P_{\mathrm{fa}}$ should be substituted. For a given allowable P_{fa} and a chosen decision rule we can calculate $P_{\mathrm{fa}0}$ at each station from (6.14).

The output (final) detection probability P_{d} for decision rules of the "k out of m" type and *strongly mutually dependent amplitude fluctuations* can be obtained by computer simulation. It follows from the simulation results that for each number of receiving stations m an optimal $k=k_{\mathrm{opt}}$ exists. The decision rule "k_{opt} out of m" provides maximum output detection probability, P_{d}, at a given output false alarm probability, P_{fa}. The optimum is achieved as a result of the influence of two conflicting factors. On the one hand, the increase of k makes a decision rule more "severe" which leads to the decrease of P_{d}. On the other hand, for preserving the output probability P_{fa}, when more severe decision rule is used, the increase of $P_{\mathrm{fa}0}$, i.e. lowering of threshold levels at each station is necessary, which in its turn leads to higher $P_{\mathrm{d}0}$ and hence to higher output P_{d}. Optimal values of k ($k=k_{\mathrm{opt}}$) are presented in Table 6.3 for $m=2,3,4,5,10$ and completely mutually dependent Rayleigh amplitude fluctuations at the inputs of spatially separated stations. For $m=10$ the decision rules "5 out of 10" and "6 out of 10" at $P_{\mathrm{fa}}=10^{-4}$ as well as "6 out of 10" and "7 out of 10" at $P_{\mathrm{fa}}=10^{-8}$ lead to nearly equal results. Detection characteristics for optimal decision rules "k_{opt} out of m" are shown in Fig. 6.3. It is seen that the general behaviour of the curves is the same as in Fig. 5.5 for centralized detection.

However the decentralization of detection results in certain energy losses. These losses are plotted in Fig. 6.4 versus the total number of receiving stations m for the decision rules "k_{opt} out of m". For comparison similar curves are shown for the decision rules "m out of m" ("AND") and "1 out of m" ("OR"). Though all curves are plotted for $P_d=0.9$ and $P_{fa}=10^{-4}, 10^{-8}$ the losses are practically constant within the interval $0.9 \geqslant P_d \geqslant 0.5$. Only when P_d increases from 0.9 to 0.99 the energy losses are reduced by 0.2–0.6 dB. It is clear from Fig. 6.4 that when the decision rules "k_{opt} out of m" are used, the energy losses as a price for the decentralization of target echo detection are of the order of 1 dB. Slightly greater are energy losses for the decision rules "m out of m" whereas the largest losses occur for the most "soft" rules "1 out of m". It can be explained by the strong mutual dependence of target echoes at the inputs of receiving stations whereas noises are mutually independent. When a threshold is exceeded (or not exceeded) in one station, then the same event occurs in other stations with high probability (of course, if SNRs are not too small). Hence the choice of "softer" decision rules, i.e. dropping the requirement of "simultaneous" threshold exceeding in several stations, does not compensate for losses caused by the

Table 6.3

m	2	3	4	5	10	P_{fa}
k_{opt}	2	2	3	3	5;6	10^{-4}
k_{opt}	2	3	3	4	6;7	10^{-8}

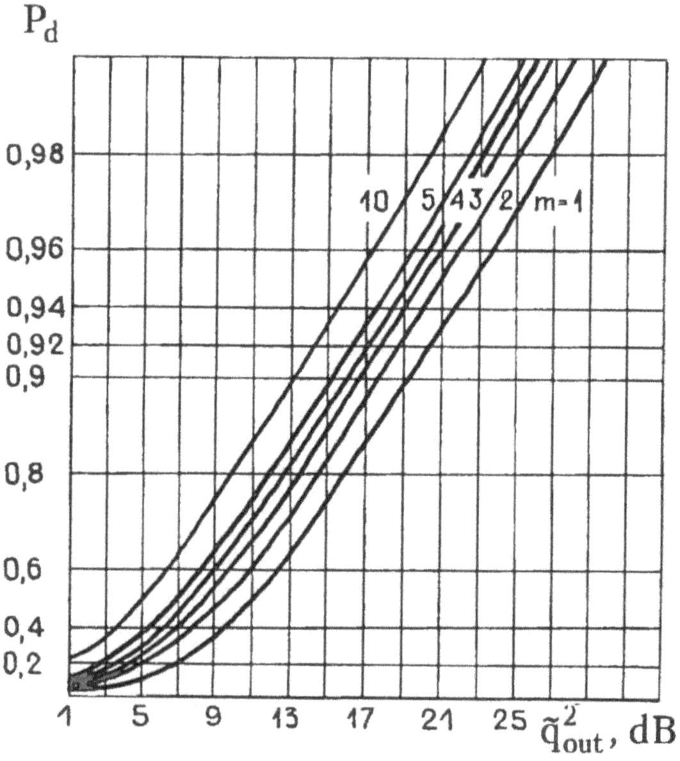

Figure 6.3(a)

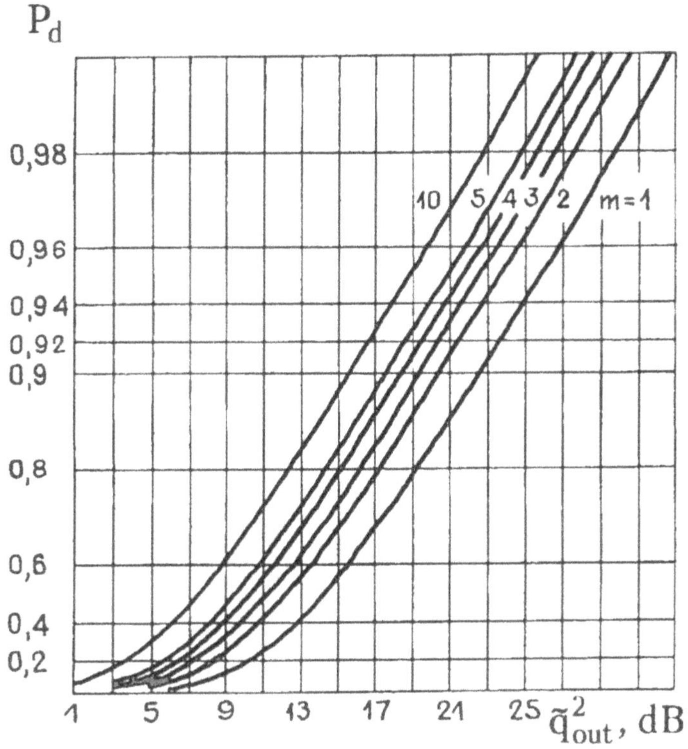

Figure 6.3(b)

Figure 6.3 Characteristics of the decentralized detector for signals with completely correlated amplitude fluctuations when the decision rules "*k* out of *m*" are used: (a) $P_{f_a} = 10^{-4}$; (b) $P_{f_a} = 10^{-8}$

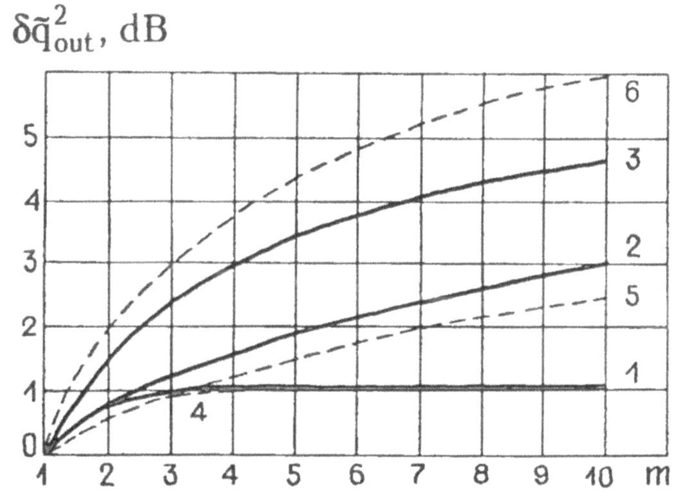

Figure 6.4 Energy losses of decentralized detection in comparison with centralized square-law summation for spatially completely correlated signal amplitude fluctuations. Curves 1, 2, 3: $P_{f_a} = 10^{-4}$; curves 4, 5, 6: $P_{f_a} = 10^{-8}$; curves 1, 4: decision rules "k_{opt} out of *m*"; curves 2, 5: decision rules "*m* out of *m*"; curves 3, 6: decision rules "1 out of *m*"

necessary increase of threshold levels at each station to preserve the predetermined output false alarm probability, P_{fa}.

Let us now consider performance characteristics of algorithms of the "k out of m" type for *mutually independent signal (target echo) amplitude fluctuations* at the inputs of spatially separated receiving stations. Having used (6.14) for computing P_{fa0} given arbitrary output probability P_{fa} and chosen decision rule, we can obtain P_{d0} from (5.77) and then find output detection probability P_d again from (6.14). Calculations show that for independent amplitude fluctuations and for each m there also exists an optimal decision rule "k_{opt} out of m". However, the optimal values k_{opt} for independent and completely dependent fluctuations are different. The optimal values $k=k_{opt}$ for independent fluctuations (equal for $P_{fa}=10^{-4}\ldots10^{-8}$) are presented in Table 6.4.

When the number of receiving stations is small, the rule "1 out of m" ("OR") is optimal. For $m=5$ the rules "1 out of 5" and "2 out of 5" lead to approximately equal results. As m further increases k_{opt} increases too, i.e. more severe decision rules become optimal. The reason is that threshold lowering caused by the necessity to preserve a given output false alarm probability, P_{fa}, overcomes the direct effect of more severe rules on decreasing the output detection probability P_d.

It should be noted that detection characteristics for the decision rules "m out of m" are the same as for a monostatic radar regardless of the number m of receiving stations. In this case [see (5.77)]

$$P_{fa0}=P_{fa}^{1/m};$$

$$P_d=P_{d0}^m=[P_{fa0}^{1/(1+\bar{q}_{out}^2)}]^m=[P_{fa}^{1/m(1+\bar{q}_{out}^2)}]^m=P_{fa}^{1/(1+\bar{q}_{out}^2)}]. \tag{6.17}$$

Therefore, when the decision rule "m out of m" ("AND") is employed for decentralized detection of target echoes with spatially independent Rayleigh amplitude fluctuations we have not obtained any power gain as compared to a monostatic radar.

This is approximately true for amplitude fluctuations modelled by the second Nakagami law (Swerling 4 model) when [72,203]

$$P_{d0}=P_{fa0}^{1/(1+0.5\bar{q}_{out}^2)}\left[1-\frac{0.5\bar{q}_{out}^2}{(1+0.5\bar{q}_{out}^2)^2}\ln P_{fa0}\right].$$

Substituting $P_{fa0}=P_{fa}^{1/m}$ and taking into account that $P_d=P_{d0}^m$ yields

$$P_d=P_{fa}^{1/(1+0.5\bar{q}_{out}^2)}\left[1-\frac{0.5\bar{q}_{out}^2}{m(1+0.5\bar{q}_{out}^2)^2}\ln P_{fa}\right]^m.$$

For large SNR $\bar{q}_{out}^2$ we may expand the binomial expression in the brackets and truncate this expansion retaining only the first term.

Detection characteristics for the decision rules "k_{opt} out of m" are presented in Fig. 6.5 with the number of stations m as a parameter. In Fig. 6.6 energy losses are

Table 6.4

m	2	3	4	5	10	P_{fa}
k_{opt}	1	1	1	1;2	3	$10^{-4}\ldots10^{-8}$

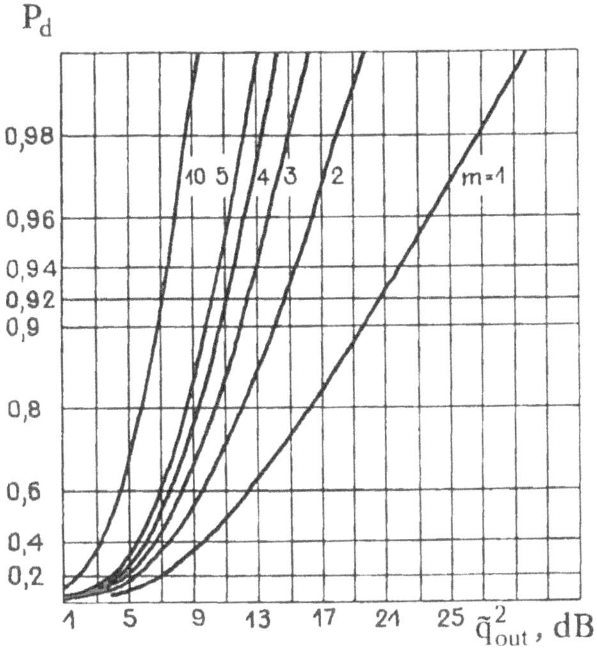

Figure 6.5(a)

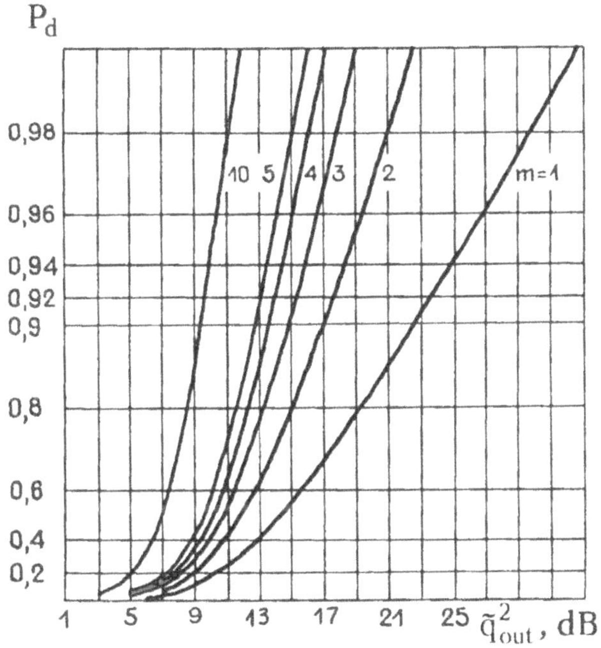

Figure 6.5(b)

Figure 6.5 The same characteristics as in Fig. 6.3 for decision rules "k_{opt} out of m" but for signals with spatially independent amplitude fluctuations: (a) $P_{fa} = 10^{-4}$; (b) $P_{fa} = 10^{-8}$

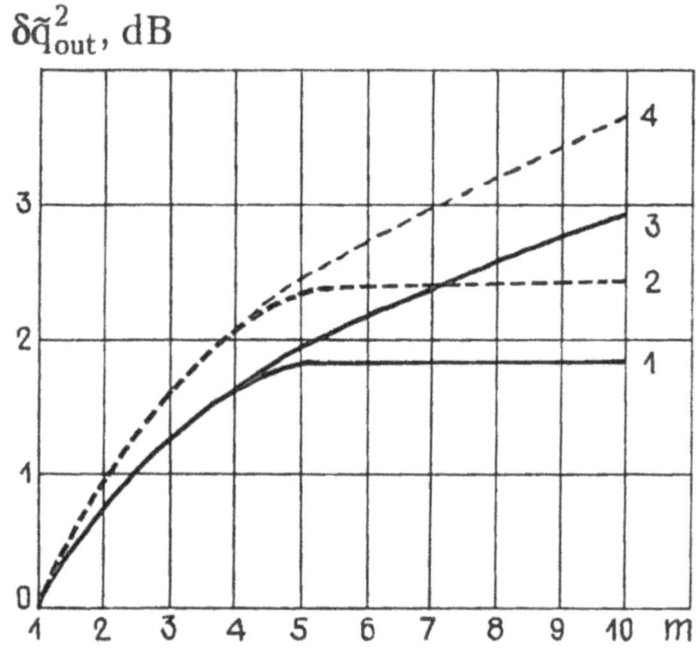

Figure 6.6 The same characteristics as in Fig. 6.4 but for signals with spatially independent amplitude fluctuations. Curves 1, 3: $P_{f_a}=10^{-4}$; curves 2, 4: $P_{f_a}=10^{-8}$; curves 1, 2: "k_{opt} out of m"; curves 3, 4: "1 out of m"

plotted versus the number m of receiving stations. As in Fig. 6.4 we assume $P_d=0.9$, but the dependence of losses on P_d is weak here too. In going from $P_d=0.9$–0.99 the losses increase by 0.02–0.2 dB and from $P_d=0.9$ to $P_d=0.5$ the losses decrease by 0.04–0.4 dB. Only for $m=10$ the differences are up to ± 0.5 dB (they increase as m increases). Curves for the decision rules "m out of m" are not shown. In accordance with (6.17) these energy losses are equal to the power gain for optimum centralized detection of independently fluctuating signals shown in Fig. 5.8.

It can be seen from Figs. 6.4 and 6.6 that energy losses are slightly greater for independent than for completely dependent fluctuations. In general, however, replacing centralized detection by decentralized detection, when SNRs at all stations are equal to each other and optimal decision rules are used, results in small energy losses, especially if a MSRS contains only few stations.

Let us now pass to the situation in which SNRs $\tilde{q}^2_{\text{out}\,i}$, $i=\overline{1,m}$, *are different*. We confine ourselves to the most practically important case of *mutually independent target echo amplitudes* at the inputs of spatially separated receiving stations. Under these conditions the optimum processing is reduced to a weighted summation of preliminary (local) decisions coming to a FC from all receiving stations [see (6.8)]. Weighting coefficients take into account the reliability of those preliminary decisions [see (6.7)]. In accordance with the results of Section 6.1 we assume here that relative threshold levels, u_{0i}, and local false alarm probabilities, P_{f_ai}, are identical at all the m stations ($u_{0i}=u_{00}$, $P_{f_ai}=P_{f_a0}$, $i=\overline{1,m}$).

Output false alarm and detection probabilities, P_{f_a} and P_d, for each decision rule out of 2^m-1 possible rules are determined by equation (6.12). For each decision rule and given false alarm probability P_{f_a} we can solve (6.12) for P_{f_a0}. Then we can calculate P_{di} from (5.77) for any combination of $\tilde{q}^2_{\text{out}\,i}$, $i=\overline{1,m}$ and at last obtain P_d using (6.12) again.

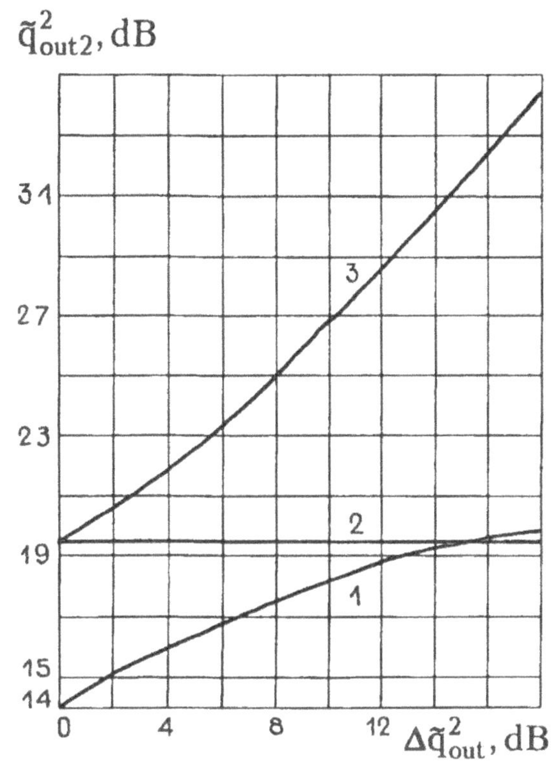

Figure 6.7 Required values of SNR $\tilde{q}^2_{\text{out }2}$, dB, at a "strong" station for maintaining $P_{\text{d}}=0.9$ at $P_{\text{fa}}=10^{-4}$ when SNR $\tilde{q}^2_{\text{out }1}$ dB at a "weak" station decreases by $\Delta\tilde{q}^2_{\text{out}}$ (with respect to the "strong" station). Spatially independent amplitude fluctuations. Curve 1: "1 out of 2": curve 2: "at least one station but not the "weak" one"; curve 3: "2 out of 2"

For rules of the "m out of m" type like in (6.17) we have

$$P_{\text{fa}0}=P_{\text{fa}}^{1/m};$$

$$P_{\text{d}}=\prod_{i=1}^{m} P_{\text{d}i}=\prod_{i=1}^{m} P_{\text{fa}0}^{1/(1+\tilde{q}^2_{\text{out }i})}=P_{\text{fa}}^{(1/m)\sum_{i=1}^{m}[1/(1+\tilde{q}^2_{\text{out }i})]}. \tag{6.18}$$

The detection probability, P_{d}, for any m is the same as for a monostatic radar given equal false alarm probability, P_{fa}, and if the SNR is

$$\tilde{q}^2_{\text{out eq}}=\left(\frac{1}{m}\sum_{i=1}^{m}\frac{1}{1+\tilde{q}^2_{\text{out }i}}\right)^{-1}-1.$$

If SNRs are identical for all $i=\overline{1,m}$ (6.18) is reduced to (6.17).

The specific analysis of decision rules at different $\tilde{q}^2_{\text{out }i}$, $i=\overline{1,m}$, will be performed for the simplest but most important practical cases: $m=2$ and $m=3$. We shall reveal a role of weights Q_i in (6.8) and evaluate performance losses in the case of replacing the weighted summation (6.8) by the significantly simpler algorithm (6.9), i.e. by decision rules of the "k out of m" type.

In Fig. 6.7 curves for $m=2$ similar to those of Fig. 5.10 are presented[1]. They show what a SNR $\tilde{q}^2_{out\,2}$ must be at a "strong" station to provide $P_d=0.9$ at $P_{fa}=10^{-4}$ if $\tilde{q}^2_{out\,1}$ at a "weak" station is by $\Delta\tilde{q}^2_{out}$ dB less than at the "strong" one.

For $m=2$ the total quantity of possible different decision rules is $2^2-1=3$. The rule "1 out of 2" (curve 1) leads to the best results. The losses compared with centralized detection (see the corresponding curve for $m=2$ in Fig. 5.10(b)) are changed from 0.74 dB at $\tilde{q}^2_{out\,1}=\tilde{q}^2_{out\,2}$ to 0.1–0.2 dB as $\Delta\tilde{q}^2_{out}$ increases. The rule "at least one but not the "weak" one" in this case means excluding information from the "weak" station from joint processing: $P_d=P_{d2}$. In Fig. 6.7 the horizontal straight line 2 at the level of $\tilde{q}^2_{out\,2}=19.4$ dB corresponds to this rule (this is the monostatic detection for $P_d=0.9$ at $P_{fa}=10^{-4}$, see Fig. 5.4 or 5.5(a) for $m=1$). This decision rule becomes better than the rule "1 out of 2" only if $\Delta\tilde{q}^2_{out}>15$ dB when detection probability in the "weak" station $P_{d1}<0.1$. In this case it is reasonable to exclude the "weak" station so that to permit threshold reducing in the "strong" station at a constant output $P_{fa}=10^{-4}$.

Thus when the difference between SNRs $\Delta\tilde{q}^2_{out}$ exceeds 15 dB, the considered joint processing becomes senseless since instead of enhancing it leads to a certain deterioration of detection characteristics (as compared to the monostatic detection). At the same time excluding the "weak" station leads to an insignificant gain. Therefore, when the difference between SNRs $\Delta\tilde{q}^2_{out}$ is unknown or can vary over a wide range, the rule "1 out of 2" should be used.

The rule "2 out of 2" (curve 3) is the least effective. The output detection probability P_d is determined mostly by the probability P_{d1} of the "weak" station. Hence $\tilde{q}^2_{out\,2}$ is to be increased nearly by $\Delta\tilde{q}^2_{out}$ in order to provide a sufficiently high value of $\tilde{q}^2_{out\,1}=\tilde{q}^2_{out\,2}-\Delta\tilde{q}^2_{out}$ in the "weak" station. Nevertheless the use of the rule "2 out of 2" may be necessary, for instance, for multistatic target coordinate measurement.

For $m=3$ we shall consider two extreme versions (variants) as in Section 5.5[2]. In the first version two stations are equally "strong" and the third station is "weak": $\tilde{q}^2_{out\,2}=\tilde{q}^2_{out\,3}$; $\tilde{q}^2_{out\,1}=\tilde{q}^2_{out\,3}-\Delta\tilde{q}^2_{out}$, dB. In the second version one station is a "strong" one and two others are equally "weak": $\tilde{q}^2_{out\,1}=\tilde{q}^2_{out\,2}=\tilde{q}^2_{out\,3}-\Delta\tilde{q}^2_{out}$, dB. Taking into account arbitrary numeration of stations all other combinations of $\tilde{q}^2_{out\,1}$, $\tilde{q}^2_{out\,2}$ and $\tilde{q}^2_{out\,3}$ lead to intermediate results. For both versions only five different decision rules are possible since SNRs $\tilde{q}^2_{out\,i}$ are equal at two stations. When $P_{fai}=P_{fa0}$, $i=\overline{1,3}$, it means that in accordance with (6.7) two of three weights Q_i in the algorithm (6.8) are equal which results in the "confluence" of two pairs of rules into two rules. In Fig. 6.8 corresponding curves are shown for the first version and in Fig. 6.9 for the second version. Besides that curves for the rule "3 out of 3" are presented in Fig. 6.9 for both extreme versions considered.

Let us consider the first version (Fig. 6.8). It is seen that when $m=3$ the rule "1 out of 3" ($0<z_0\leqslant Q_1$, curve 1) is most effective too. The performance losses (as compared to the centralized detection algorithm L_6) are changed from 1.2 dB at $\Delta\tilde{q}^2_{out}=0$ dB to 0.4–0.3 dB as $\Delta\tilde{q}^2_{out}$ increases. The rule "at least one station but not a "weak" one" ($Q_1<z_0\leqslant Q_2=Q_3$, curve 2) is reduced in the version 1 to the rule "1 out of 2" for two "strong" stations. The "weak" station is excluded from joint processing. Hence variations of $\Delta\tilde{q}^2_{out}$, i.e. $\tilde{q}^2_{out\,1}$, have no effect on $\tilde{q}^2_{out\,2}=\tilde{q}^2_{out\,3}$ which

---

[1] Since receiving stations may be numbered arbitrarily, for remaining the similarity with the material of Section 6.1, we denote here $\tilde{q}^2_{out\,1}\leqslant\tilde{q}^2_{out\,2}$, so that $\tilde{q}^2_{out\,2}$ is laid off as the ordinate instead of $\tilde{q}^2_{out\,1}$ as in Fig. 5.10.

[2] Here the first variant (version) corresponds to the second variant (version) in Section 5.5 and vice versa.

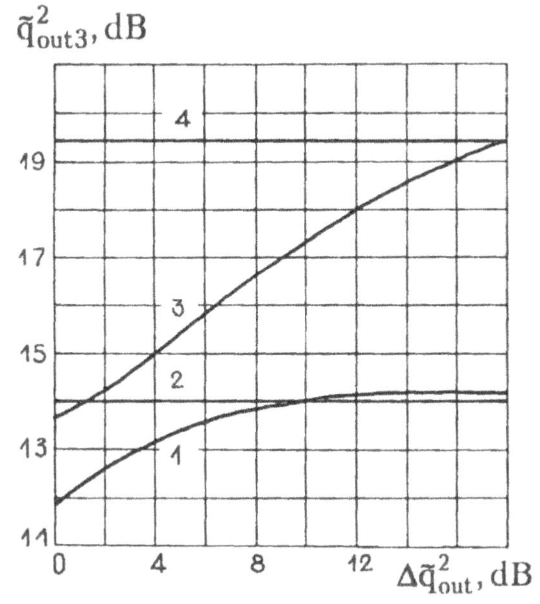

Figure 6.8 The same characteristics as in Fig. 6.7 but for $m = 3$, variant 1. Curve 1: "1 out of 3"; curve 2: "at least one station but not the "weak" one"; curve 3: "2 out of 3"; curve 4: "at least two stations but excluding the "weak" one"

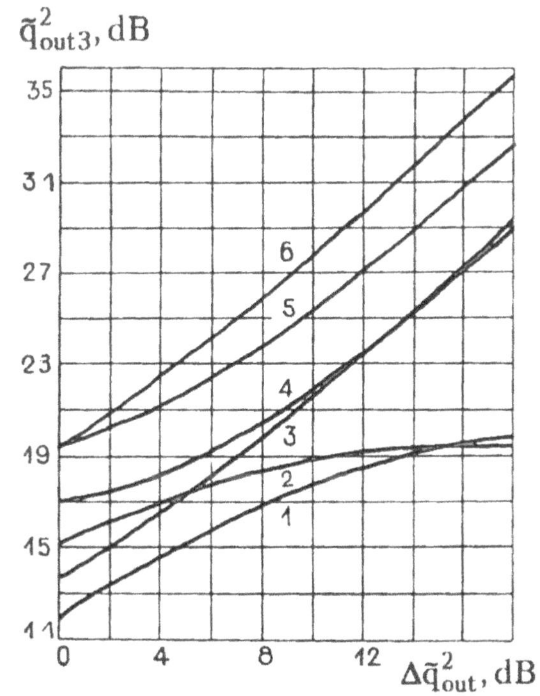

Figure 6.9 The same characteristics as in Fig. 6.8 but for variant 2. Curves 1, 2, 3, 4 for the same decision rules as in Fig. 6.8. Curve 5: "3 out of 3", variant 1; curve 6: "3 out of 3", variant 2

are necessary to achieve $P_d=0.9$ at $P_{fa}=10^{-4}$. This rule is better than "1 out of 3" if $\Delta\tilde{q}^2_{out}>9$ dB but the advantage is insignificant. Thus when $\Delta\tilde{q}^2_{out}>8\ldots9$ dB, then from the point of view of target detection it is not reasonable to complicate joint processing by including information from the "weak" station. However, when $\Delta\tilde{q}^2_{out}$ is unknown the rule "1 out of 3" may be used without noticeable losses.

The rule "2 out of 3" ($Q_2=Q_3<z_0\leqslant Q_1+Q_2=Q_1+Q_3$, curve 3) requires significantly larger SNR values $\tilde{q}^2_{out\,2}=\tilde{q}^2_{out\,3}$ which increase more rapidly as $\Delta\tilde{q}^2_{out}$ increases. This rule like the rule "2 out of 2" should be used in the cases where it is necessary, for instance, to measure target coordinates by no less than two spatially separated stations.

The rule "at least two stations but without the "weak" one" ($Q_1+Q_2=Q_1+Q_3<z_0\leqslant 2Q_2=2Q_3$, curve 4) in the first version is reduced to the rule "2 out of 2" for two "strong" stations. The performance characteristics are the same as for a monostatic detector [see (6.17)] for $P_d=0.9$ at $P_{fa}=10^{-4}$. When $\Delta\tilde{q}^2_{out}>18$ dB this rule becomes better than the rule "2 out of 3". It means that under this condition inclusion of the third "weak" station into joint processing leads to the deterioration of detection characteristics. However, when $\Delta\tilde{q}^2_{out}$ is not large the advantages of the rule "2 out of 3" are evident.

The rule "3 out of 3" ("AND") is the least effective rule (see Fig. 6.9, curve 5). Therefore it is to be used only if it is necessary (for instance, when target coordinates are simultaneously measured with respect to three spatially separated stations).

For the second version (variant 2) the general character of the curves is similar to that of the first version (Fig. 6.9). However, for constant $P_d=0.9$ at $P_{fa}=10^{-4}$ the values of $\tilde{q}^2_{out\,3}$ are changed more sharply with the increase of $\Delta\tilde{q}^2_{out}$. The rule "1 out of 3" ($0<z_0\leqslant Q_1=Q_2$, curve 1) is in this case also the best rule whose performance is the nearest to that of the centralized detection. Energy losses are the same as for the first version (see curve 2 in Fig. 5.10(b)). When $\Delta\tilde{q}^2_{out}>15$ dB the rule "1 out of 3" compares unfavorably with monostatic radar detection so that for $\Delta\tilde{q}^2_{out}>14\ldots15$ dB multistatic (multiradar) target detection in this version becomes senseless.

The rule "at least one station but not the "weak" one" ($Q_1=Q_2<z_0\leqslant Q_3$, curve 2) is actually reduced to the requirement of threshold exceeding either in the "strong" station (with arbitrary situation in the two "weak" stations) or in both "weak" stations: $P_d=P_{d3}+P^2_{d1}(1-P_{d3})=P_{d3}+P^2_{d2}(1-P_{d3})$. As $\Delta\tilde{q}^2_{out}$ increases the effectiveness of this rule approaches near that of the monostatic detection in the "strong" station so that when $\Delta\tilde{q}^2_{out}>10$–$11$ dB, then using information from "weak" stations is not profitable. However, this rule is worse than the rule "1 out of 3" and does not guarantee target echoes detection in both two stations (for target coordinate measurement with respect to these two stations). Therefore using the rule "at least one station but not the "weak" one" in the second version (variant 2) is rather unreasonable.

Performance characteristics of the rule "2 out of 3" ($Q_3<z_0\leqslant 2Q_1=2Q_2$, curve 3) in the second version are strongly dependent on $\Delta\tilde{q}^2_{out}$. As $\Delta\tilde{q}^2_{out}$ increases, $\tilde{q}^2_{out\,3}$ at the "strong" station is to be risen so that $\tilde{q}^2_{out\,1}=\tilde{q}^2_{out\,2}=\tilde{q}^2_{out\,3}-\Delta\tilde{q}^2_{out}$ at the "weak" stations are sufficiently high. Therefore, as $\Delta\tilde{q}^2_{out}$ increases this rule in the second version becomes more and more unprofitable.

The rule "at least two stations but not two "weak" ones" ($2Q_1=2Q_2<z_0\leqslant Q_1+Q_3=Q_2+Q_3$) means that for declaring a hit ("a target echo is present") threshold exceeding must occur in the "strong" station and at least in one of the "weak" stations: $P_d=P_{d3}\,[1-(1-P_{d1})(1-P_{d2})]$. Performance characteristics of this rule are the same that those of the rule "2 out of 3" when $\Delta\tilde{q}^2_{out}>10$–$12$ dB while for small $\Delta\tilde{q}^2_{out}$ the rule "2 out of 3" is better. When the "strong" station has no other advantages (for instance, significantly higher coordinate measurement accuracy) employment of the rule "at least two stations but not two "weak" ones" is unlikely reasonable.

At last the rule "3 out of 3" ($Q_1+Q_3=Q_2+Q_3<z_0\leqslant Q_1+Q_2+Q_3$, curve 6) is the least effective rule as in the first version (Fig. 6.8). To avoid large energy losses it should be used (if necessary) when SNRs at different stations $\Delta\bar{q}^2_{\text{out}}$ are not large.

The above analysis has shown that for decentralized detection of target echoes with independent fluctuations (at the inputs of few different receiving stations m) it is apparently possible to give up the weighted summation according to the algorithm (6.8) but to employ significantly simpler decision rules of the "k out of m" type according to (6.9).

Performance characteristics which have been considered above do not cover wide range of conditions. But the described approach permits deriving numerical results for arbitrary combinations of the number of stations m, output probabilities P_{fa} and P_{d} and SNR values $\bar{q}^2_{\text{out}\,i}$, $i=\overline{1,m}$. It may be used also for other probability distributions of spatially independent target echo amplitude fluctuations if corresponding formulas for $P_{\text{fa}i}$ and $P_{\text{d}i}$ at each station are employed.

In this section, *we have considered performance characteristics of the decentralized detection algorithms synthesized in the previous section.*

For the case of equal SNRs at all receiving stations where decision rules of "k out of m" type are optimal, we have derived optimal values of k for different values of m and mutually both completely dependent and independent target echo amplitude fluctuations at the inputs of spatially separated receiving stations (Tables 6.3 *and* 6.4, *respectively). We have presented detection characteristics for the rules "*k_{opt} *out of m": Fig.* 6.3 *for completely dependent and Fig.* 6.5 *for independent fluctuations. Much attention has been paid to the comparative analysis of energy losses with respect to the centralized detection. It has been shown (Figs.* 6.4 *and* 6.6) *that replacing the centralized by the decentralized detection, when SNRs at all stations are equal to each other and optimal decision rules are used, results in small additional energy losses, especially if a MSRS contains only few stations.*

The case of different SNRs at receiving stations has been considered in detail for MSRSs with two and three receiving stations ($m=2$ *and* $m=3$) *and spatially independent target echo amplitude fluctuations. For all possible decision rules we have analyzed required value of SNR in a "strong" station to maintain output detection and false alarm probabilities* ($P_{\text{d}}=0.9$, $P_{\text{fa}}=10^{-4}$) *when SNRs at "weak" stations are lowered by certain value (Figs.* 6.7, 6.8 *and* 6.9). *The practically important conclusion is that even if SNRs at all stations are unequal, instead of the complicated algorithm* (6.8) *with weighted summation much simpler algorithm* (6.9) *of "k out of m" type may be used without significant extra energy losses. The approach that has been demonstrated allows us to evaluate performance characteristics for more general situations* ($m>3$, *different* P_{fa} *and* P_{d}, *different relationships between SNRs).*

6.3. ADDITIONAL ENERGY LOSSES OF DECENTRALIZED DETECTION CAUSED BY IGNORANCE OF TARGET POSITION. "COST OF RESOLUTION" FOR DECENTRALIZED DETECTION

According to the results of Sections 6.1 and 6.2 let us consider decision rules of the "k out of m" type. If the condition (5.98) is satisfied, then *a single range resolution cell of one arbitrary station corresponds in space to a single range resolution cell of each other receiving station.* The resolution capability of such a MSRS in angle coordinates is not higher than that of each combination of one transmitting and one receiving stations (i.e. of a monostatic or bistatic radar). For each combination of m range resolution cells corresponding to each other (one cell of each station) and creating a single space resolution cell (SRC) of the MSRS, all results of Sections 6.1 and 6.2 are valid. Connection between the number of false alarms in an uncertainty region and the false alarm probability in a single SRC at the output of the fusion centre (FC) is determined by the relationship (5.97) where M is equal to the number of range resolution cells at each station covering the target position uncertainty region (TPUR)[3].

The situation is drastically complicated when *the difference in directions from each station to a target may not be neglected*, i.e. when the condition (5.98) is not satisfied and even the inequality (5.98) has an opposite sign. Such a situation is the most frequently occurring in practice. Just in this case high resolution capability and measurement accuracy in range of each station provide markedly higher space resolution capability and measurement accuracy of the MSRS.

Let us consider decision rules of "1 out of m" ("OR") type. These rules are to be used at the stage of target searching. Then after reducing the target position uncertainty region more severe decision rules should be conventionally used to realize measurement advantages of a MSRS. It is not reasonable to employ the concept of SRC for rules of "1 out of m" type since each SRC represents an intersection (multiplication, conjunction) of resolution cells of different stations whereas these rules assume combination (summation, disjunction) of those resolution cells. When a TPUR illuminated by the transmitting station covers n_i range resolution cells of the ith receiving station (in each angle resolution cell) and false alarm probabilities are the same at all receiving stations, $P_{\mathrm{fa}i}=P_{\mathrm{fa}0}$, $i=\overline{1,m}$, then we have at the output of joint processing

$$P_{\mathrm{fa}}=1-(1-P_{\mathrm{fa}0})^{n_\Sigma};\quad \bar{N}_{\mathrm{fa}}=n_\Sigma P_{\mathrm{fa}0};\ n_\Sigma=\sum_{i=1}^{m} n_i. \tag{6.19}$$

Here we have taken into account that the number of false alarms N_{fa} is binomially distributed.

When the uncertainty region is covered by only a single range resolution cell of each station, i.e. $n_i=1$, $i=\overline{1,m}$, then

$$P_{\mathrm{fa}}=1-(1-P_{\mathrm{fa}0})^{m};\qquad \bar{N}_{\mathrm{fa}}=mP_{\mathrm{fa}0}. \tag{6.20}$$

Comparing (6.20) with (6.19) we can see that the increase of range resolution capability (when the uncertainty region is covered by $n_i>1$ range resolution cells of the ith station, $i=\overline{1,m}$) is equivalent to the increase of the number of stations with $n_i=1$ by n_0 times where n_0 is the average range resolution cells in each station

$$n_0=n_\Sigma/m=(1/m)\sum_{i=1}^{m} n_i. \tag{6.21}$$

This means that for the predetermined average number of false alarms, $\bar{N}_{\mathrm{fa}}$, increasing the resolution capability ($n_i>1$ range resolution cells of the ith station, $i=\overline{1,m}$) we must reduce the false alarm probability at each station, $P_{\mathrm{fa}0}$, by n_0 times. Using (5.77), (6.14) and (6.20) it is easy to evaluate energy losses – "cost of

[3] The number of receiving antenna beams covering the uncertainty region and the corresponding number of receiving channels are assumed to be the same for all stations.

resolution"

$$\delta\tilde{q}^2_{\text{out}} = 10\lg\left[1 + \frac{\lg n_0}{\lg(1-\sqrt[m]{1-P_d}) + \lg m - \lg \bar{N}_{\text{fa}}}\right], \text{dB}. \quad (6.22)$$

where $\bar{N}_{\text{fa}}$ and P_d are the average number of false alarms and detection probability, respectively, at the output after employing a decision rule "1 out of m".

If the TPUR of a monostatic radar covers n_0 range resolution cells the "cost of resolution" is

$$\delta\tilde{q}^2_{\text{out}\,1} = 10\lg\left[1 + \frac{\lg n_0}{\lg P_d - \lg \bar{N}_{\text{fa}}}\right], \text{dB}. \quad (6.23)$$

Example 6.6. Let a MSRS contain one transmitting and three receiving stations, i.e. $m=3$. Besides that let $n_0=10$, $\bar{N}_{\text{fa}}=10^{-4}$, $P_d=0.9$. Then energy losses ("cost of resolution") we obtain from (6.22): $\delta\tilde{q}^2_{\text{out}} \approx 0.93\,\text{dB}$. For a monostatic radar ($m=1$) under similar conditions, $n_0=10$, $\bar{N}_{\text{fa}}=10^{-4}$, $P_d=0.9$, we have from (6.23): $\delta\tilde{q}^2_{\text{out}\,1} \approx 0.98\,\text{dB}$. In fact, when decision rules of "1 out of m" ("OR") type are used the resolution capability of a MSRS is not higher than that of a corresponding monostatic radar. Therefore, energy losses ("cost of resolution") are approximately equal.

When a TPUR can be covered by more than one antenna mainlobe of some or all receiving stations, then the necessary number of parallel receiving channels should be formed with displaced mainlobe ADPs from each other so that to cover the part of the uncertainty region illuminated by the transmitting station during one probe. The remaining part of the uncertainty region is to be examined sequentially in time. Parallel receiving channels of each station may be combined by different techniques including optimal algorithms of linear or squared summation (see Section 5.2). If the rules "1 out of m" are used for combining parallel receiving channels, the value of n_i in (6.19) should be equal to the total number of range resolution cells in all parallel channels of the ith station covering an illuminated part of the uncertainty region.

Let us pass to the *general decision rules of "k out of m" type where* $1 \leqslant k \leqslant m$. When $k \geqslant 3$ then threshold exceedings in any k stations in range resolution cells corresponding to each other, determines an SRC where a target may be positioned. The size of this SRC depends on the arrangement of the stations (geometry of the MSRS) with respect to the TPUR. In practice a large quantity of different combinations of sizes, forms and orientations of SRCs are possible.

The problem considered can be solved applying directly the decision rules "k out of m" to the analysis of an uncertainty region. Usually, after each illumination of the uncertainty region by the transmitting station, the corresponding range resolution cells of "extreme" stations, having the largest baselengths of a MSRS, are analyzed. Often for a specific number of receiving stations m and a value of k the condition "k out of m" can be satisfied only when thresholds are exceeded in those "extreme" stations. In this case if thresholds are not exceeded in "extreme" stations we may conclude that there is no target in the uncertainty region. Only when m is large and (or) k is small, threshold exceeding in "extreme" stations is not necessary for declaring a hit ("a target is present"). In these cases when thresholds are not exceeded in "extreme" stations one should analyze corresponding range resolution cells in remaining stations beginning again from the most remote station (with respect to the "centre" of the MSRS). Let a threshold be exceeded in the ith station just once in the analyzed region. Then the uncertainty region is immediately reduced to the intersection of the initial uncertainty region with the resolution volume of the ith station (where the threshold is exceeded). Then the corresponding range resolution cells within this new reduced uncertainty region are to be analyzed at the $(i+1)$th station having the largest effective baselength relative to the ith station. If there is no threshold exceeding one should pass to the next station. If a threshold is exceeded, the uncertainty region is further reduced to the intersection of the previous uncertainty region with the corresponding resolution volume of the $(i+1)$th station [or, otherwise, to the intersection of the initial uncertainty region with the corresponding resolution volumes of the ith and $(i+1)$th stations]. Then one should go over to $(i+2)$th station and analyze range resolution cells corresponding to the new reduced uncertainty region. If there is no threshold exceeding the next station should be considered. If a threshold is exceeded in the $(i+2)$th station, the possible target position is localized in a SRC which has been formed by the three certain stations. When $k>3$, then, as a rule, only a single range resolution cell corresponding to this SRC is to be analyzed at each of remaining stations.

The described procedure is to be continued up to the moment when the required condition "k out of m" either is satisfied or is not satisfied after the analysis of all remaining stations. When the condition is satisfied earlier than all stations are examined, one should fix that a target has been detected and analyze whether the condition "k out of m" is satisfied for remaining stations.

When $k<3$ then after threshold exceeding in k stations the decision "a target is present" is made and other combinations of similar events are to be searched for. If several threshold exceedings within the illuminated region occur in the ith or other stations, the described procedure should be repeated for each of them.

To evaluate the averaged number of false alarms, $\bar{N}_{fa}$, let us consider the general expression

$$\bar{N}_{fa} = \sum_{j=1}^{j_{max}} jP(j) \tag{6.24}$$

where $P(j)$ is the probability of exactly j false alarms in an uncertainty region. When P_{fa0} is small (as is usually the practice) then the values of $P(j)$ sharply decrease as j increases. Therefore if the number of SRCs within the illuminated region for each probe is not too large (<100–300) it is possible to use only the first term in the sum of (6.24) $\bar{N}_{fa} \approx P(1)$ with sufficient accuracy (with the error of the order of 10–20%). If we take into account probabilities of different combinations of possible events in detection procedure described above, we can obtain the following approximate expression for $P(1)$

$$P(1) \approx \sum_{p=0}^{m-k} P_{fa0}^{k+p}(1-P_{fa0})^{n_{\Sigma}-k-p} \times \sum_{i_1=1}^{m-k-p+1} \sum_{i_2=i_1+1}^{m-k-p+2} \sum_{i_3=i_2+1}^{m-k-p+3} C_{m-i_3}^{k+p-3} n_{i_1} v_{i_1 i_2} v_{i_1 i_2 i_3} \tag{6.25}$$

where C_a^b is the binomial coefficient (the number of combinations of a taken b in a time); n_{i_1} is the number of range resolution cells of the (i_1)th station covering the uncertainty region; $v_{i_1 i_2}$ is the number of range resolution cells of the (i_2)th station corresponding to each range resolution cell of the (i_1)th station within the uncertainty region (i.e. intersecting this resolution cell of the (i_1)th station); $v_{i_1 i_2 i_3}$ is the number of range resolution cells of the (i_3)th station corresponding to each intersection of the range resolution cells of the (i_1)th station and the (i_2)th station within the uncertainty region; P_{fa0} is the false alarm probability in a resolution cell of each station $(P_{fai}=P_{fa0},\ i=\overline{1,m})$; $n_\Sigma=\sum_{i=1}^{m} n_i$.

The expression (6.25) is valid for $k\geqslant 3$ in the decision rule "k out of m". If $k<3$ summation in (6.25) should be made not over i_1, i_2, i_3 but over $i_1,\ldots,i_s$ where $s=\min(k+p,3)$. If $k+p=2$, the last sum vanishes and $C_{m-i_s}^{k+p-3}=1$, $v_{i_1i_2i_3}=1$; If $k+p=1$, the two last sums vanish and $C_{m-i_1}^{k+p-3}=1$, $v_{i_1i_2}=1$, $v_{i_1i_2i_3}=1$.

We have assumed here that within the analyzed target position uncertainty region a value of $v_{i_1i_2}$ is the same for all range resolutions cells of the (i_1)th station, a value of $v_{i_1i_2i_3}$ is the same for all intersections of range resolution cells of the (i_1)th and the (i_2)th stations, i.e. boundary effects have been neglected. We also have not taken into account that when a false alarm occurs there may be threshold exceedings in different stations which however do not correspond to any SRC i.e. do not form additional false alarms for the MSRS.

In (6.25) for the rules of "k out of m" type we have taken into account probabilities of threshold exceeding not only in k but also in $(k+1),\ldots,(k+p),\ldots,m$ stations. However for small P_{fa0} these probabilities sharply decrease as p increases. Since all summands in (6.24) for $j\geqslant 2$ are neglected there is no sense in using in (6.25) all terms with powers of P_{fa0} higher than the minimal power of P_{fa0} in the expression for $P(2)$. It means that one should set in (6.25) $p\leqslant k-3$ (but $p\geqslant 0$). Then (6.25) can be written in a final form

$$\bar{N}_{fa}\approx P(1)\approx\sum_{p=0}^{\min(m-k,\,k-3)} P_{fa0}^{k+p}(1-P_{fa0})^{n_\Sigma-k-p}$$

$$\times\sum_{i_1=1}^{m-k-p+1}\ \sum_{i_2=i_1+1}^{m-k-p+2}\ \sum_{i_3=i_2+1}^{m-k-p+3} C_{m-i_3}^{k+p-3}\, n_{i_1} v_{i_1i_2} v_{i_1i_2i_3} \tag{6.26}$$

where if $\min(m-k,k-3)<0$ then $p_{max}=0$.

Using (6.26) one can calculate the allowable false alarm probability, P_{fa0}, i.e. determine the threshold level for each station if a value of $\bar{N}_{fa}$ and the parameters of an uncertainty region n_{i_1}, $v_{i_1i_2}$ and $v_{i_1i_2i_3}$ are given. Then with the help of (6.14) for the required output detection probability P_d and accepted specific decision rule of "k out of m" type the detection probability in each station, P_{d0}, can be found. Having P_{fa0} and P_{d0} one can evaluate the required SNR $\tilde{q}_{out}^2$ from (5.77). To estimate energy losses ("cost of resolution") one should compare the obtained value of $\tilde{q}_{out}^2$ with the value of SNR that is required to achieve the same value of P_d when there is no uncertainty in a target position (i.e. when a single SRC is analyzed where a target can be present or absent). This value of SNR can be obtained from (5.77) with P_{fa0} calculated before from (6.14) where P in the left side should be replaced by the given value of $\bar{N}_{fa}$. The expression for energy losses through the output characteristics $\bar{N}_{fa}$ and P_d in an explicit form is difficult to obtain since $\bar{N}_{fa}$ and P_d are connected with P_{fa0} and P_{d0} in (6.26) and (6.14) by equations of high power (in the general case). Therefore, we express these losses ("cost of resolution") of a MSRS in the following form

$$\delta\tilde{q}_{out}^2=10\lg\frac{\lg P_{fa0}-\lg P_{d0}}{\lg P_{fa0}^{(1)}-\lg P_{d0}},\ \mathrm{dB} \tag{6.27}$$

where P_{fa0} and $P_{fa0}^{(1)}$ are the false alarm probabilities at each station corresponding to required value of $\bar{N}_{fa}$ for a given uncertainty region and without uncertainty, respectively; P_{d0} is the required detection probability at each station.

Example 6.7. Let us consider again the same MSRS as in Example 6.6. with $m=3$ under the same conditions ($\bar{N}_{fa}=10^{-4}$, $P_d=0.9, n_1=n_2=n_3=10$) but instead of the rule "1 out of 3" the rules "2 out of 3" and "3 out of 3" are used. We assume that $v_{12}=v_{13}=v_{23}=v_{123}=5$. Just these values show the increase of space resolution capability of the MSRS in comparison with a corresponding monostatic radar.
For the rules "2 out of m" we can write from (6.26)

$$\bar{N}_{fa} \approx P_{fa0}^2(1-P_{fa0})^{n_\Sigma-2}\sum_{i_1=1}^{m-1}\sum_{i_2=i_1+1}^{m} n_{i_1}v_{i_1i_2}. \tag{6.28}$$

Hence for the rule "2 out of 3" ($m=3$)

$$\bar{N}_{fa} \approx P_{fa0}^2(1-P_{fa0})^{(n_1+n_2+n_3-2)}(n_1v_{12}+n_1v_{13}+n_2v_{23}).$$

For given numerical values we have $\bar{N}_{fa}\approx 150\,P_{fa0}^2(1-P_{fa0})^{28}$. From this for $\bar{N}_{fa}=10^{-4}$ we have $P_{fa0}\approx 8.3\cdot 10^{-4}$. From (6.14) for $m=3$ and $k=2$ we obtain $P_{fa0}^{(1)}\approx 5.8\cdot 10^{-3}$ and $P_{d0}=0.804$ [$P_{fa}\approx \bar{N}_{fa}=10^{-4}$ and $P_d=0.9$, respectively, should be substituted for P into the left side of (6.14)]. The substitution of the obtained values in (6.27) gives $\delta\tilde{q}_{out}^2\approx 1.4\,\text{dB}$.

For the rules "m out of m" from (6.26) it can be written

$$\bar{N}_{fa} \approx P_{fa}^m(1-P_{fa0})^{n_\Sigma-m}n_1v_{12}v_{123}. \tag{6.29}$$

For given parameter values $\bar{N}_{fa}\approx 250\;P_{fa0}^3(1-P_{fa0})^{27}$. From this for $\bar{N}_{fa}=10^{-4}$ we have $P_{fa0}\approx 7.9\cdot 10^{-3}$. From (6.14) for $m=3$ and $k=3$ we obtain $P_{fa0}^{(1)}\approx 4.6\cdot 10^{-2}$ and $P_{d0}=0.936$. Substituting the obtained values in (6.27) yields $\delta\tilde{q}_{out}^2\approx 2\,\text{dB}$.

Since the rule "3 out of 3" provides higher space resolution than the rule "2 out of 3" energy losses are larger. This is the "cost of resolution". It is useful to remind that for the corresponding monostatic radar and for the same MSRS when the rule "1 out of 3" is used the "cost of resolution" under the same conditions is $\delta q_{out\,1}^2\approx 1\,\text{dB}$ (see Example 6.6). Thus in our example the additional energy losses caused by the increase of space resolution capability of the MSRS for the rule "2 out of 3" $\delta\tilde{q}_{out}^2-\delta\tilde{q}_{out\,1}^2\approx 0.4\,\text{dB}$ and for the rule "3 out of 3" $\delta q_{out}^2-\delta q_{out\,1}^2\approx 1\,\text{dB}$.

Usually $P_{fa0}\ll 1$ so that $(1-P_{fa0})^{n_\Sigma-m}\approx 1$. In this case it is easy to express the "cost of resolution" for the rules "m out of m" explicitly through the output characteristics $\bar{N}_{fa}$ and P_d as in the case of the rule "1 out of m" in (6.22)

$$\delta\tilde{q}_{out}^2=10\lg\left[1+\frac{(\lg n_1+\lg v_{12}+\lg v_{123})}{\lg P_d-\lg\bar{N}_{fa}}\right], \text{dB}. \tag{6.30}$$

It is reasonable to repeat that all stations are assumed to be numbered from the "extreme" stations with the largest effective baselengths, i.e. from the stations which provide the highest space resolution (the largest values of $v_{12}, v_{13}, v_{23}, v_{123}$).

In this section, for the decentralized (distributed) detection with decision rules of the "k out of m" type we have obtained useful expressions permitting evaluation of additional energy losses caused by the uncertainty of a target position and especially by the increase of resolution capability of a MSRS as compared with a monostatic radar (the "cost of resolution"). If the difference between directions from receiving stations to a target may be neglected, i.e. the condition (5.98) *is satisfied, the resolution capability of a MSRS is not better than that of a corresponding monostatic radar. In this case energy losses are the same as those of a monostatic radar* [*see* (6.23)]. *Of more practical importance are situations where the difference between directions from stations to a target may not be neglected. For these cases we have derived equation* (6.22) *for energy losses when decision rules "1 out of m"* ("*OR*") *are employed, the general approximate equations* (6.26) *and* (6.27) *for rules of the "k out of m" type and equation* (6.30) *for rules "m out of m"* ("*AND*"). *As has been shown in Example* 6.6, *the rules* "1 *out of m" do not lead to additional energy losses with respect to a monostatic radar since they do not enhance the space resolution capability of a MSRS. On the contrary, when the rules* "2 *out of m"* (*in the particular case* "2 *out of* 3") *and especially "m out of m"* (*in the particular case* "3 *out of* 3") *are used the space resolution capability of the MSRS increases giving rise to additional energy losses* (*Example* 6.7).

7. DETECTION OF STOCHASTIC SIGNALS IN PASSIVE MSRSs IN A BACKGROUND OF SPATIALLY UNCORRELATED INTERFERENCES

7.1. DETECTION OF STOCHASTIC SIGNALS WITH KNOWN CORRELATION MATRICES

As was mentioned in Section 4.1, detection of stochastic signals is the task of passive MSRSs (or active–passive MSRSs in a passive mode) which receive signals from sources of "noise" radiation. The assumption that a correlation matrix of stochastic signals is completely known (deterministic correlation matrix) is a methodically convenient idealization similar to the assumption of the deterministic character of regular signals (Sections 4.1, 5.1).

Synthesis of Optimum Detectors

The general form of the *optimum processing algorithm* (in the frequency domain) for realizations of stationary and stationary coupled (at the inputs of different stations) Gaussian stochastic processes with zero means is presented in Section 4.3. For stochastic signal detection in a background of spatially uncorrelated interferences equation (4.52) takes the form

$$L=(1/2\pi)\int_{-\infty}^{\infty}\chi^*(\omega)[\mathbf{f}_n(\omega)-\mathbf{f}_{sn}(\omega)]\chi(\omega)\,d\omega \tag{7.1}$$

where $\chi(\omega)$ is the vector $(m\times 1)$ of the Fourier transformations of received signals (realizations of sums of wanted signals plus noises or noises alone) at the inputs of m stations within the observation time interval $(-T/2, T/2)$; $\mathbf{f}_n(\omega)$ and $\mathbf{f}_{sn}(\omega)$ are the inverse of the PSD noise and signal plus noise matrices, $\mathbf{\Phi}_n(\omega)$ and $\mathbf{\Phi}_{sn}(\omega)$, respectively. To specify the algorithm L we should reveal the structure of the matrices $\mathbf{f}_n(\omega)$ and $\mathbf{f}_{sn}(\omega)$. By the spatial independence of noises, $\mathbf{\Phi}_n(\omega)=\mathbf{N}$ and $\mathbf{f}_n(\omega)=\mathbf{N}^{-1}$ are diagonal matrices [see (5.1)]. As was discussed in Section 4.1, stationary within the interval $(-T/2, T/2)$ Gaussian stochastic signals are stationary coupled when the differences between Doppler frequency shifts at the inputs of all stations may be neglected: $|\Delta\Omega_{sik}T|\ll 2\pi$, $i,k=\overline{1,m}$. Then a Doppler frequency shift common for all stations may be included into the carrier frequency ω_0. In this case we may write $\mathbf{\Phi}_{sn}(\omega)=\mathbf{N}+\mathbf{\Phi}_s(\omega)$ where an arbitrary element of the PSD matrix of the wanted signals at the inputs of m stations, $\mathbf{\Phi}_s(\omega)$, is determined by (4.12). To obtain $\mathbf{f}_{sn}(\omega)=\mathbf{\Phi}_{sn}^{-1}(\omega)$ we make use of the fact that from (4.12)

$$\mathbf{\Phi}_{sn}(\omega)=\mathbf{N}+\mathbf{\Phi}_s(\omega)=\mathbf{N}+F_s(\omega-\omega_0)\mathbf{U}(\omega)\mathbf{U}^*(\omega) \tag{7.2}$$

where $\mathbf{U}(\omega)$ is the m-dimensional vector with the following elements

$$U_i(\omega)=\sqrt{P_{si}/\Delta f_s}\exp[-j(\omega t_{si}+\varphi_{si})],\quad i=\overline{1,m}. \tag{7.3}$$

For matrices having the form of (7.2) the simple inversion rule is known [72,215]

$$\mathbf{f}_{sn}(\omega)=\mathbf{\Phi}_{sn}^{-1}(\omega)=\mathbf{N}^{-1}-\gamma F_s(\omega-\omega_0)\mathbf{N}^{-1}\mathbf{U}(\omega)\mathbf{U}^*(\omega)\mathbf{N}^{-1};$$
$$\gamma=[1+F_s(\omega-\omega_0)\mathbf{U}^*(\omega)\mathbf{N}^{-1}\mathbf{U}(\omega)]^{-1}. \tag{7.4}$$

Substituting (7.3) into (7.4) yields the expression for an arbitrary element of the matrix $\mathbf{f}_{sn}(\omega)$:

$$f_{snik}(\omega)=\frac{1}{\sqrt{N_iN_k}}\left\{\delta_{ik}-\frac{q_{si}q_{sk}F_s(\omega-\omega_0)}{1+F_s(\omega-\omega_0)q_{s\Sigma}^2}\exp[\mathrm{j}(\omega\tau_{sik}+\Delta\varphi_{sik})]\right\} \tag{7.5}$$

where as in Section 4.1,

$$\tau_{sik}=t_{sk}-t_{si};\qquad \Delta\varphi_{sik}=\varphi_{sk}-\varphi_{si} \tag{7.6}$$

are the TDOA and phase difference of expected signals at the inputs of the ith and kth stations;

$$q_{s\Sigma}^2=\sum_{i=1}^{m}q_{si}^2 \tag{7.7}$$

and $q_{si}^2=P_{si}/N_i\Delta f_s$ is the SNR at the input of the ith station; δ_{ik} is Kronecker's symbol. It follows from (7.6) that

$$\tau_{sik}=-\tau_{ski};\quad \Delta\varphi_{sik}=-\Delta\varphi_{ski},\qquad i,k=\overline{1,m};$$
$$\tau_{sik}+\tau_{skl}+\tau_{sli}=0;\quad \Delta\varphi_{sik}+\Delta\varphi_{skl}+\Delta\varphi_{sli}=0,\qquad i,k,l=\overline{1,m}. \tag{7.8}$$

Assigning (without loss of generality) the first station to be the "reference" one we can write

$$\tau_{sik}=\tau_{si1}-\tau_{sk1};\quad \Delta\varphi_{sik}=\Delta\varphi_{si1}-\Delta\varphi_{sk1}. \tag{7.9}$$

Now we take into account that according to (5.1) $f_{nik}(\omega)=\delta_{ik}/\sqrt{N_iN_k}$. Substituting this relationship together with (7.5) and (7.9) in (7.1) yields *the optimal algorithm* in the form

$$L=\frac{1}{2\pi}\int_{-\infty}^{\infty}\left|H(\omega)\sum_{i=1}^{m}\frac{q_{si}\exp(-\mathrm{j}\Delta\varphi_{si1})}{\sqrt{N_i}}\chi_i(\omega)\exp(-\mathrm{j}\omega\tau_{si1})\right|^2\mathrm{d}\omega \tag{7.10}$$

where

$$|H(\omega)|^2=F_s(\omega-\omega_0)/[1+F_s(\omega-\omega_0)q_{s\Sigma}^2]. \tag{7.11}$$

When the condition $|\Delta\Omega_{sik}T|\ll 2\pi$, $i,k=\overline{1,m}$ is not satisfied then *stationary Gaussian signals at the inputs of receiving stations turn out to be nonstationary coupled*. An arbitrary element of the correlation matrix takes the form of (4.7) and an arbitrary element of power spectral density (PSD) matrix takes the form of (4.10). We have to consider the more complicated expression for the optimal signal processing instead of (7.1).

$$L=\sum_{i=1}^{m}\sum_{k=1}^{m}\frac{1}{4\pi^2}\int_{-\infty}^{\infty}\int_{-\infty}^{\infty}\chi_i^*(\omega_1)\chi_k(\omega_2)$$

$$\times\int_{-\infty}^{\infty}[\tilde{f}_{nik}(\omega_1,t)-\tilde{f}_{snik}(\omega_1,t)]\exp(j\omega_2 t)dt d\omega_1 d\omega_2. \qquad (7.12)$$

The functions $\tilde{f}_{nik}(\omega,t)$ and $\tilde{f}_{snik}(\omega,t)$ are Fourier transformations (with respect to the argument t_1) of the elements $R_{nik}(t_1,t_2)$ and $R_{snik}(t_1,t_2)$ of the matrices $\mathbf{R}_n(t_1,t_2)$ and $\mathbf{R}_{sn}(t_1,t_2)$, respectively. The last matrices are solutions of the equation (4.45). From (4.45) we can immediately obtain a system of equations for $\tilde{f}_{nik}(\omega,t)$ and $\tilde{f}_{snik}(\omega,t)$ integrating over infinite limits, applying Fourier transformation to both sides with respect to the argument t_1-t_2, carrying out the change of variables and employing (4.10). After these manipulations we have

$$\tilde{f}_{nik}(\omega,t)=(\delta_{ik}/N_i)\exp(-j\omega t),\quad i,k=\overline{1,m}. \qquad (7.13)$$

The functions $\tilde{f}_{snik}(\omega,t)$ can be found as solutions of the following system of equations

$$\sum_{k=1}^{m}\sqrt{N_i N_k}\,[\delta_{ik}+q_{si}q_{sk}F_s(\omega-\omega_0-\Omega_{si})]$$

$$\times\exp\{j[(\omega-\Omega_{si})\tau_{sik}+\Delta\psi_{sik}]\}\tilde{f}_{snkl}(\omega+\Delta\Omega_{sik},t)\exp(j\omega t)=\delta_{il},\quad i,l=\overline{1,m}. \qquad (7.14)$$

As can be verified by substitution these solutions take the form

$$\tilde{f}_{snik}(\omega,t)=\frac{\exp[-j(\omega+\Delta\Omega_{sik})t]}{\sqrt{N_i N_k}}\left[\delta_{ik}-\frac{q_{si}q_{sk}F_s(\omega-\omega_0-\Omega_{si})}{1+F_s(\omega-\omega_0-\Omega_{si})q_{s\Sigma}^2}\right]$$

$$\times\exp\{j[(\omega-\Omega_{si})\tau_{sik}+\Delta\psi_{sik}]\}. \qquad (7.15)$$

Using (7.13), (7.15) and (7.9) and assuming the first station to be the reference one yields the *optimal processing algorithm* (7.12) in the following form

$$L=\frac{1}{2\pi}\int_{-\infty}^{\infty}\left|H_1(\omega)\sum_{i=1}^{m}\frac{q_{si}\exp(-j\Delta\psi_{si1})}{\sqrt{N_i}}\chi_i(\omega-\Delta\Omega_{si1})\exp[-j(\omega-\Omega_{s1})\tau_{si1}]\right|^2 d\omega \qquad (7.16)$$

where

$$|H_1(\omega)|^2=F_s(\omega-\omega-\Omega_{s1})/[1+F_s(\omega-\omega-\Omega_{s1})q_{s\Sigma}^2]. \qquad (7.17)$$

Comparing (7.16) with (7.10) shows that the difference of expected signal Doppler frequencies at the inputs of receiving stations requires equalization of the signal carrier frequencies before summation.

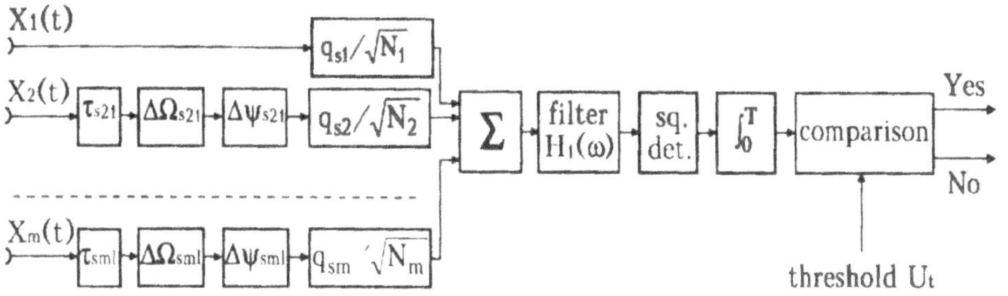

Figure 7.1 Structure of the optimum detector for stochastic signals with a known correlation (covariance) matrix

The structural scheme of the obtained optimum detector is presented in Fig. 7.1. Received signals pass through delay lines, phase shifters and (if necessary) frequency shifters which provides equalization in time delays, phase and central (carrier) frequencies with respect to the reference station. Then weighted summation, filtration, square envelope detection and integration are carried out. It may be said that the scheme of Fig. 7.1 provides an estimate of the power (energy) of coherent (for wanted signals) weighted and filtered sum of received signals. When input wanted signals are strong ($q_{s\Sigma}^2 \gg 1$) the filter in Fig. 7.1 may be dropped. On the contrary, when input wanted signals are weak ($q_{s\Sigma}^2 \ll 1$) which more often occurs in practice, this filter is necessary and the squared modulus of its transfer function is equal to PSD of the wanted signal at the reference station. If wanted signals are stationary coupled at the inputs of spatially separated stations, frequency shifters $\Delta\Omega_{si1}$, $i=\overline{2,m}$, should be excluded from the scheme in Fig. 7.1.

When the influence of antenna ADPs is to be taken into account, then additional multipliers–complex ADPs $g_i(\beta_{si}, \varepsilon_{si}, \omega)$ should be included into the sums (7.10) and (7.16) as it was made in Section 5.1. It should be noted that the scheme in Fig. 7.1 after the summator does not differ from the so called "energy receiver" which is the optimal monostatic detector for Gaussian stochastic signals.

Performance Analysis of Optimum Detectors

As usual, it is necessary to obtain probability distribution of the variable L from (7.10) and (7.16) when wanted signals are absent and when they are present. It is seen from Fig. 7.1 that the optimum processing includes a nonlinear operation: the square-law envelope detection of the sum of Gaussian stochastic processes. Therefore the variable L is in general non-Gaussian. In the particular case if discrete processing is employed and the signal PSD is rectangular with the bandwidth Δf_s the variable L is chi-square distributed with $2\Delta f_s T$ degrees of freedom ($\Delta f_s T$ independent complex samples and each of them consists of two quadratures). However, usually $\Delta f_s T \gg 1$ (e.g. $\Delta f_s T \approx 10^3 - 10^4$). In this case it is possible to consider that the integration provides the effective normalization of the output process from the square envelope detector so that the variable L may be assumed to be Gaussian with practically sufficient accuracy. Then it is only necessary to derive mean values and variances of the variable L in the absence and in the presence of expected signals.

Let us consider more general expression (7.16). The mean value can be written as

$$\bar{L}=\frac{1}{2\pi}\int_{-\infty}^{\infty}|H_1(\omega)|^2\sum_{i=1}^{m}\sum_{k=1}^{m}\frac{q_{si}q_{sk}\exp[-j(\Delta\psi_{si1}-\Delta\psi_{sk1})]}{\sqrt{N_iN_k}}$$

$$\times\overline{\chi_i(\omega-\Delta\Omega_{si1})\chi_k^*(\omega-\Delta\Omega_{sk1})}\exp[-j(\omega-\Omega_{s1})(\tau_{si1}-\tau_{sk1})]\,d\omega. \qquad (7.18)$$

The spectra $\chi_i(\omega)$ and $\chi_k(\omega)$ are to be shifted by $\Delta\Omega_{si1}$ and $\Delta\Omega_{sk1}$, $i,k=\overline{2,m}$, respectively. Then signal central frequencies at all stations become equal to $\omega_0+\Omega_{s1}$ [since according to (4.8) $\Omega_{si}+\Delta\Omega_{si1}=\Omega_{s1}$, $i=\overline{2,m}$]. It means that all received signals turn out to be stationary coupled. Using (4.12) it can be shown that when wanted signals are present and $\Delta f_sT\gg 1$

$$\overline{\chi_i(\omega-\Delta\Omega_{si1})\chi_k^*(\omega-\Delta\Omega_{sk1})}=2T\Phi_{sn\,ik}(\omega-\Omega_{s1})$$

$$=2T\sqrt{N_iN_k}\{\delta_{ik}+q_{si}q_{sk}F_s(\omega-\omega_0-\Omega_{s1})$$

$$\times\exp\{j[(\omega-\Omega_{s1})\tau_{sik}+\Delta\psi_{sik}]\}\}. \qquad (7.19)$$

If wanted signals are absent, $q_{si}=q_{sk}=0$ in (7.19) and the right side of (7.19) is equal to $2TN_i\delta_{ik}$. Substituting (7.19) in (7.18) yields mean values of L in the absence and in the presence of wanted signals

$$m_{10}=m_1\{L_0\}=\bar{L}_0=\frac{q_{s\Sigma}^2T}{\pi}\int_{-\infty}^{\infty}\frac{F_s(\omega-\omega_0-\Omega_{s1})\,d\omega}{1+F_s(\omega-\omega_0-\Omega_{s1})q_{s\Sigma}^2}; \qquad (7.20)$$

$$m_{11}=m_1\{L_1\}=\bar{L}_1=\frac{q_{s\Sigma}^2T}{\pi}\int_{-\infty}^{\infty}F_s(\omega-\omega_0-\Omega_{s1})\,d\omega. \qquad (7.21)$$

To calculate variances of L the fourth moment of $\chi_i(\omega-\Delta\Omega_{si1})$, $i=\overline{1,m}$ should be obtained. For jointly Gaussian variables the following expression of the fourth moment through the second moments is known [49]

$$\overline{\chi_i(\omega_1-\Delta\Omega_{si1})\chi_k^*(\omega_1-\Delta\Omega_{sk1})\chi_p(\omega_2-\Delta\Omega_{sp1})\chi_q^*(\omega_2-\Delta\Omega_{sq1})}$$

$$=\overline{\chi_i(\omega_1-\Delta\Omega_{si1})\chi_k^*(\omega_1-\Delta\Omega_{sk1})}\;\overline{\chi_p(\omega_2-\Delta\Omega_{sp1})\chi_q^*(\omega_2-\Delta\Omega_{sq1})}$$

$$+\overline{\chi_i(\omega_1-\Delta\Omega_{si1})\chi_q^*(\omega_2-\Delta\Omega_{sq1})}\;\overline{\chi_k^*(\omega_1-\Delta\Omega_{sk1})\chi_p(\omega_2-\Delta\Omega_{sp1})}. \qquad (7.22)$$

Besides that,

$$\overline{\chi_i(\omega_1-\Delta\Omega_{si1})\chi_q^*(\omega_2-\Delta\Omega_{sq1})}=4\pi\Phi_{iq}(\omega_1-\Omega_{s1})\delta(\omega_1-\omega_2). \qquad (7.23)$$

Taking into account (7.22) and (7.23) we can obtain expressions for variances of L when wanted signals are absent (M_{20}) and they are present (M_{21})

$$M_{20}=M_2\{L_0\}=\overline{L_0^2}-(\overline{L}_0)^2=\frac{2(q_{s\Sigma}^2)^2T}{\pi}\int_{-\infty}^{\infty}\frac{F_s^2(\omega-\omega_0-\Omega_{s1})\,d\omega}{[1+F_s(\omega-\omega_0-\Omega_{s1})q_{s\Sigma}^2]^2}; \tag{7.24}$$

$$M_{21}=M_2\{L_1\}=\overline{L_1^2}-(\overline{L}_1)^2=\frac{2(q_{s\Sigma}^2)^2T}{\pi}\int_{-\infty}^{\infty}F_s^2(\omega-\omega_0-\Omega_{s1})\,d\omega. \tag{7.25}$$

It is seen that the main energy parameter is the summed signal-to-noise ratio (SNR) at the inputs of all receiving stations $q_{s\Sigma}^2$. It should be noted that the number of stations m has only an indirect effect through the sum $q_{s\Sigma}^2=\sum_{i=1}^m q_{si}^2$. Besides, the mean values and variances of L depend on the duration of received signal realization (observation interval) T and on certain integral parameters of signal power spectra. Detection characteristics are determined by the expressions (7.20), (7.21) and (7.24), (7.25)

$$\begin{aligned}P_{fa}&=0.5[1-\mathrm{erf}(y_0/\sqrt{2})];\\ P_d&=0.5[1-\mathrm{erf}(y_0/\sqrt{M_{20}/2M_{21}}-(m_{11}-m_{10})/\sqrt{2M_{21}})]\end{aligned} \tag{7.26}$$

where erf(x) is the error function [see (5.15)]. For a given false alarm probability, P_{fa}, we can calculate y_0 from the first equation of (7.26). Substituting y_0 in the second equation of (7.26) yields detection probability P_d.

Evidently, the obtained results are valid for the case where wanted signals are stationary coupled at the inputs of spatially separated receiving stations, and the optimal signal processing is determined by (7.10). For this case in (7.20), (7.21), (7.24) and (7.25) Ω_{s1} should be omitted.

Example 7.1. Let a signal PSD have the rectangular form

$$F_s(\omega-\omega_0-\Omega_{s1})=\begin{cases}1, & |\omega-\omega_0-\Omega_{s1}|\leqslant\Delta\omega_s/2;\\ 0, & |\omega-\omega_0-\Omega_{s1}|>\Delta\omega_s/2.\end{cases} \tag{7.27}$$

Then from (7.20), (7.21), (7.24)–(7.27) we have

$$\begin{aligned}m_{10}&=\frac{2q_{s\Sigma}^2n}{1+q_{s\Sigma}^2};\quad m_{11}=2q_{s\Sigma}^2n;\\ M_{20}&=\frac{4(q_{s\Sigma}^2)^2n}{(1+q_{s\Sigma}^2)^2};\quad M_{21}=4(q_{s\Sigma}^2)^2n;\end{aligned} \tag{7.28a}$$

$$\begin{aligned}P_{fa}&=0.5[1-\mathrm{erf}(y_0/\sqrt{2})];\\ P_d&=0.5\left\{1-\mathrm{erf}\left[\frac{y_0-q_{s\Sigma}^2\sqrt{n}}{\sqrt{2}(1+q_{s\Sigma}^2)}\right]\right\}.\end{aligned} \tag{7.28b}$$

The signal accumulation (integration) factor, $n=\Delta f_sT$, is determined by the number of independent samples of a stochastic process accumulated by the integrator in Fig. 7.1.

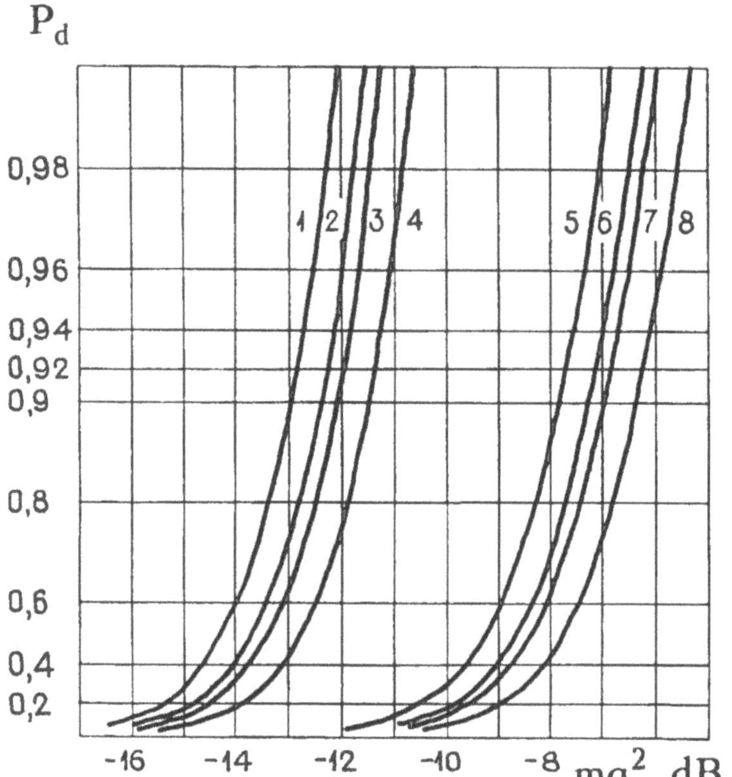

Figure 7.2 Detection characteristics of the optimum and correlation detectors for stochastic signals with a known correlation matrix, $P_{f_a}=10^{-4}$. Curves 1, 5: optimum detector; curves 2, 3, 4, 6, 7, 8: correlation detector; curves 2, 6: $m=4$; curves 3, 7: $m=3$; curves 4, 8: $m=2$; curves 1, 2, 3, 4: $n=10^4$, curves 5, 6, 7, 8: $n=10^3$

Detection characteristics calculated from (7.28) are plotted in Fig. 7.2 for $P_{f_a}=10^{-4}$ and two values of n: 10^3 and 10^4. As can be seen from Fig. 7.2, when the accumulation factor n is large enough, very weak input signals ($q_{s\Sigma}^2 \ll 1$) can be detected with high confidence.

In this section, *we have synthesized optimum processing algorithms for detection of stochastic signals with completely known correlation (covariance) matrices in a passive MSRS with m spatially separated receiving stations when noises and other interferences at all stations are Gaussian and mutually independent. The assumption that a signal correlation matrix is completely known is a useful idealization permitting to reveal some principal features of stochastic signal detection algorithms. For the case where stochastic signals at the inputs of receiving stations are stationary coupled we have derived the optimum algorithm* (7.10). *We have also considered the more general case where stationary signals are nonstationary coupled. It may take place, for instance, when the differences of Doppler frequency shifts of the signals from a moving source (target) at the inputs of all stations may not be neglected. For this case we have obtained the optimum processing algorithm* (7.16). *The structure of this algorithm is shown in Fig.* 7.1.

We have also analyzed performance characteristics of the synthesized algorithms. We have assumed that for usually employed large signal accumulation factors, n, the output variable L (which is to be compared with a predetermined threshold) may be considered to be Gaussian. Under this assumption we have derived the expressions for the mean values and variances of L in the absence and in the presence of wanted signals: (7.20), (7.24) *and* (7.21), (7.25), *respectively. These expressions allow us to calculate detection characteristics using* (7.26). *The employment of the described technique has been illustrated by the Example* 7.1 *and Fig.* 7.2.

7.2. DETECTION OF STOCHASTIC SIGNALS WITH CORRELATION MATRICES CONTAINING RANDOM PARAMETERS

In actual MSRSs designed for detection and location of stochastic ("noise") radiation sources, mutual phase shifts of wanted signals, and hence initial phases of mutual correlation (covariance) matrices, at the inputs of spatially separated receiving stations are, as a rule, unknown and randomly vary from one observation interval T to others. These random variations may be caused by the short term spatial coherence of the MSRS itself (see Section 1.1), by random variations of the source position with respect to the MSRS, by fluctuations of propagation medium and so on. Besides mutual phase shifts, signal intensities (powers), and hence variances of correlation matrices, may vary too. All these variations are slow so that signal phase and amplitude relationships may be considered to be constant during each observation interval $(-T/2, T/2)$. Under this assumption we have a detection problem formulated in Section 4.3 for stochastic signals with quasi-deterministic correlation matrices containing random parameters: initial phases and possibly variances (powers).

Synthesis of Optimal Detectors

The general expression for the unconditional likelihood ratio (in the frequency domain) averaged over a collection of random parameters Θ of a correlation matrix of stationary and stationary coupled stochastic signals has been presented in (4.53). For the problem considered here it may be written in the form

$$\bar{\Lambda} = \int_{\Theta} w(\Theta) \frac{\mathfrak{R}_1(\Theta)}{\mathfrak{R}_0} \exp[0.5L(\Theta)]\, \mathrm{d}\Theta \tag{7.29}$$

where notations are the same as in (4.53) and $L(\Theta)$ is the result of optimal processing either (7.10) or (7.16) which is considered here as conditional for a given Θ.

Let *only initial phases be random* in a signal correlation matrix. Then $\mathfrak{R}_1(\Theta)$ does not depend on Θ so that the ratio $\mathfrak{R}_1(\Theta)/\mathfrak{R}_0$ may be dropped. The variable $L(\Theta)$ we take from more general algorithm (7.16) taking into account possible differences of signal Doppler frequency shifts at the inputs of receiving stations. It can be presented in the form

$$L = \sum_{i=1}^{m} \frac{q_{si}}{N_i} \hat{B}_{ii} + 2\,\mathrm{Re} \sum_{i=1}^{m-1} \sum_{k=i+1}^{m} \frac{q_{si} q_{sk}}{\sqrt{N_i N_k}} \exp(-\mathrm{j}\Delta\psi_{si1} + \mathrm{j}\Delta\psi_{sk1}) \hat{B}_{ik} \tag{7.30}$$

where

$$\hat{B}_{ik}=\frac{1}{2\pi}\int_{-\infty}^{\infty}|H_1(\omega)|^2\chi_i(\omega-\Delta\Omega_{si1})$$

$$\times\chi_k^*(\omega-\Delta\Omega_{sk1})\exp[-j\omega(\tau_{si1}-\tau_{sk1})]\,d\omega \tag{7.31}$$

The phase shifts $\Omega_{s1}\tau_{si1},\Omega_{s1}\tau_{sk1}$ are included in the random phases $\Delta\psi_{si1},\Delta\psi_{sk1}$ When $|\Delta\Omega_{sik}T|\ll 2\pi$, i.e. differences of Doppler frequency shifts may be neglected so that wanted signals at the input of stations are stationary coupled and (7.10) instead of (7.16) is valid, then (7.30) is not changed but it should be taken into account in (7.31) that $\Delta\Omega_{si1}=\Delta\Omega_{sk1}=0$ and $|H_1(\omega)|^2=|H(\omega)|^2$ (Ω_{s1} is included in ω_0). It is clear that $\Delta\psi_{s11}=0$.

The random variables $\hat{B}_{ii}$ and $\hat{B}_{ik}$ are the estimates of the auto- and mutual correlation function, respectively, of complex received signals (realizations of input stochastic processes) after delay and Doppler equalization and passing through the filter with the frequency response $H_1(\omega)$.

The most natural approach is to consider phase shifts at all stations to be mutually statistically independent and uniformly distributed in $(-\pi,\pi)$ with respect to the "reference" station. Then Θ is the vector with $m-1$ elements $\Delta\psi_{si1}$, $i=\overline{2,m}$. However, following this approach the integral in (7.29) can be presented in a closed form only either for $m=2$, $m=3$ or for weak signals at the output. In real MSRSs a high detection probability is usually required so that output signals may not be considered to be weak. For $m\geqslant 4$ and for signals of arbitrary intensity one can obtain the optimal processing algorithm in a closed form if signal phase shifts at the inputs of any two stations may be considered to be statistically independent. In other words, if statistically independent are initial phases of signal correlation matrix (excluding, of course, complex conjugate phases). This condition can be satisfied, for instance, in the cases where mutually independent signal phase shifts occur in data links from one station to other stations, in receiving channels and processing devices after division between distinct paths corresponding to different stations and so on. Assuming in (7.30) all $\Delta\psi_{sik}$, $i,k=\overline{1,m}$, $i<k$, to be mutually statistically independent, replacing $\Delta\psi_{si1}-\Delta\psi_{sk1}$ by $\Delta\psi_{sik}$ and substituting in (7.29) yields

$$\tilde{\Lambda}=\exp\left(\sum_{i=1}^{m}\frac{q_{si}^2}{2N_i}\hat{B}_{ii}\right)(2\pi)^{-m(m-1)/2}\int_{-\pi}^{\pi}\cdots\int_{-\pi}^{\pi}\exp\left\{\sum_{i=1}^{m-1}\sum_{k=i+1}^{m}\frac{q_{si}q_{sk}}{\sqrt{N_iN_k}}\right.$$

$$\left.\times|\hat{B}_{ik}|\cos(\Delta\psi_{sik}+\gamma_{ik})\right\}d\Delta\psi_{s12}\cdots d\Delta\psi_{s(m-1)m}. \tag{7.32}$$

There are $m(m-1)/2$ integrals from $-\pi$ to π in (7.32). The integrand in (7.32) can be presented as a product of $m(m-1)/2$ exponential functions. Each exponential function can be expanded into a well-known series: $\exp(A\cos\alpha)=I_0(A)+\sum_{p=1}^{\infty}I_p(A)\cos p\alpha$. Multiplying together the obtained series and integrating we can transform the multiple integral into a product of the single integrals with separated variables and all terms containing cosines are equal to zero so that

$$\tilde{L}=\ln\tilde{\Lambda}=\sum_{i=1}^{m}\frac{q_{si}^2}{2N_i}\hat{B}_{ii}+\sum_{i=1}^{m-1}\sum_{k=i+1}^{m}\ln I_0\left(\frac{q_{si}q_{sk}}{\sqrt{N_iN_k}}|\hat{B}_{ik}|\right). \tag{7.33}$$

Approximating $\ln I_0(x) \approx x$ for *strong output signals* we have from (7.33)

$$\tilde{L} = \ln \tilde{\Lambda} = \sum_{i=1}^{m} \frac{q_{si}^2}{2N_i} \hat{B}_{ii} + \sum_{i=1}^{m-1} \sum_{k=i+1}^{m} \frac{q_{si} q_{sk}}{\sqrt{N_i N_k}} |\hat{B}_{ik}|. \tag{7.34}$$

In the particular cases where $m=2$ and $m=3$ the same optimum processing algorithm can be obtained when signal phase shifts at all stations with respect to a "reference" station are statistically independent. When *output signals may be considered to be weak*, then the approximation $\ln I_0(x) \approx x^2/4$ is valid and (7.33) takes the form (the common multiplier 0.5 is omitted)

$$\tilde{L} = \ln \tilde{\Lambda} = \sum_{i=1}^{m} \frac{q_{si}^2}{N_i} \hat{B}_{ii} + 0.5 \sum_{i=1}^{m-1} \sum_{k=i+1}^{m} \frac{q_{si}^2 q_{sk}^2}{N_i N_k} |\hat{B}_{ik}|^2. \tag{7.35}$$

This algorithm also can be obtained from (7.29) for statistically independent signal phase shifts at all stations with respect to a "reference" station ($\Delta\psi_{si1}$, $i=\overline{2,m}$) if the exponential function in (7.29) is expanded into a Taylor series and all terms of the third and higher powers are neglected. It should be noted that for the optimal processing according to (7.34) and (7.35) only relationships between (q_{si}^2/N_i) at different stations are to be known since a common factor, e.g. (q_{s1}^2/N_1), may be taken outside the sign of summation and dropped.

Let now *not only initial phases but also variances of signal correlation matrix (signal powers) be random* at the inputs of receiving stations. Usually it is caused by power fluctuations of radiation sources. Therefore after delay equalization signal power fluctuations at the input of receiving stations may be considered to be completely mutually correlated. In this case signal powers at all m stations can be expressed through the signal power at an arbitrary, for instance, the first station. When noise PSDs N_i and signal bandwidths Δf_{si}, $i=\overline{1,m}$ (which is usually determined by a receiver bandwidth) are known, we can write

$$q_{si}^2 = A_{i1}^2 q_{s1}^2; \qquad w(q_{s1}^2, \ldots, q_{sm}^2) = w(q_{s1}^2) \prod_{i=2}^{m} \delta(q_{si}^2 - A_{i1}^2 q_{s1}^2) \tag{7.36}$$

where

$$A_{i1}^2 = \int_0^\infty q_{si}^2 w(q_{si}^2)\, dq_{si}^2 \Big/ \int_0^\infty q_{s1}^2 w(q_{s1}^2)\, dq_{s1}^2 \tag{7.37}$$

is the known ratio of the averaged power SNRs at the ith and the first stations. These ratios usually describe the *a priori* known relationships between antenna apertures, receiver sensitivities and so on.

We utilize equation (7.34), corresponding to the practically most important case of *strong output signals*, as a conditional log-likelihood ratio (for a fixed value of q_{s1}^2). Substituting it in (7.29) yields the unconditional likelihood ratio (averaged over q_{s1}^2)

in the form

$$\bar{\Lambda}=\int_0^\infty w(q_{s1}^2)\frac{\Re_1(q_{s1}^2)}{\Re_0}\Lambda(q_{s1}^2)\,dq_{s1}^2$$

$$=\int_0^\infty w(q_{s1}^2)\frac{\Re_1(q_{s1}^2)}{\Re_0}\exp\left\{q_{s1}^2\left[\sum_{i=1}^{m}\frac{A_{i1}^2}{2N_i}\hat{B}_{ii}+\sum_{i=1}^{m-1}\sum_{k=i+1}^{m}\frac{A_{i1}A_{k1}}{\sqrt{N_iN_k}}|\hat{B}_{ik}|\right]\right\}dq_{s1}^2. \quad (7.38)$$

Note that random variables $\hat{B}_{ik}$ depend implicitly on q_{s1}^2 since the expressions for $|H_1(\omega)|^2$ and $|H(\omega)|^2$ include $q_{s\Sigma}^2=q_{s1}^2\sum_{i=1}^{m}A_{i1}^2$ [see (7.31), (7.17) and (7.11)]. This fact does not allow us to obtain an optimum processing algorithm in the practically convenient explicit form. However, for the most interesting case of *weak input signals* ($q_{s\Sigma}^2\ll 1$) $|H_1(\omega)|^2$ and $|H(\omega)|^2$ do not depend on q_{s1}^2. Under this condition it is possible to obtain an explicit expression for the optimal processing algorithm without carrying out the integration in (7.29). Let us denote the expression in brackets by a and take into account that the integral $\int_0^\infty w(x)[\Re_1(x)/\Re_0]\exp(ax)\,dx=f(a)$ is a monotonous function of a. In fact, assuming the possibility of differentiation with respect to parameter under the integration sign, we have: $df(a)/da=\int_0^\infty w(x)[\Re_1(x)/\Re_0]x\exp(ax)\,dx>0$ since $w(x)\geqslant 0$ [and $w(x)=0$ not in all points $x\in(0,\infty)$], $[\Re_1(x)/\Re_0]>0$ and $\exp(ax)>0$ for $x>0$ and arbitrary real and limited a. But if $f(a)$ is a monotonous function of a, then the argument of this function, i.e. a, may be considered to be the *optimal processing algorithm*. Hence

$$\tilde{L}=\sum_{i=1}^{m}\frac{A_{i1}^2}{2N_i}\hat{B}_{ii}+\sum_{i=1}^{m-1}\sum_{k=i+1}^{m}\frac{A_{i1}A_{k1}}{\sqrt{N_iN_k}}|\hat{B}_{ik}|. \quad (7.39)$$

Thus the optimal processing algorithm (7.39) for random completely correlated signal powers at the inputs of spatially separated receiving stations and weak input signals differs from the optimal processing algorithm for known signal powers (7.34) only by weights (A_{i1} are substituted for random q_{si}, $i=\overline{1,m}$).

The structure of the optimum detector for Gaussian stochastic signals with quasi-deterministic correlation matrix and strong output signals is shown in Fig. 7.3. After delay and Doppler frequency equalization as well as linear filtration [the filter frequency response is $H_1(\omega)$], estimates of signal powers and moduli of estimates of mutual correlation[1] for signals received by all stations are formed. Then weighted summation of these estimates is carried out. Weights depend on the noise level in each receiver and either on SNRs themselves (if they are known) or on relationships between averaged SNRs at all stations and at an arbitrary "reference" one.

In Fig. 7.3 only one of possible versions for obtaining the modulus of an estimate of mutual correlation is presented. This is the analogous technique that is known as "the scheme with frequency shift". At one of the two inputs of each correlator an additional frequency shift δf is included (the value of δf should be larger than the

[1] That is envelopes of output correlator signals.

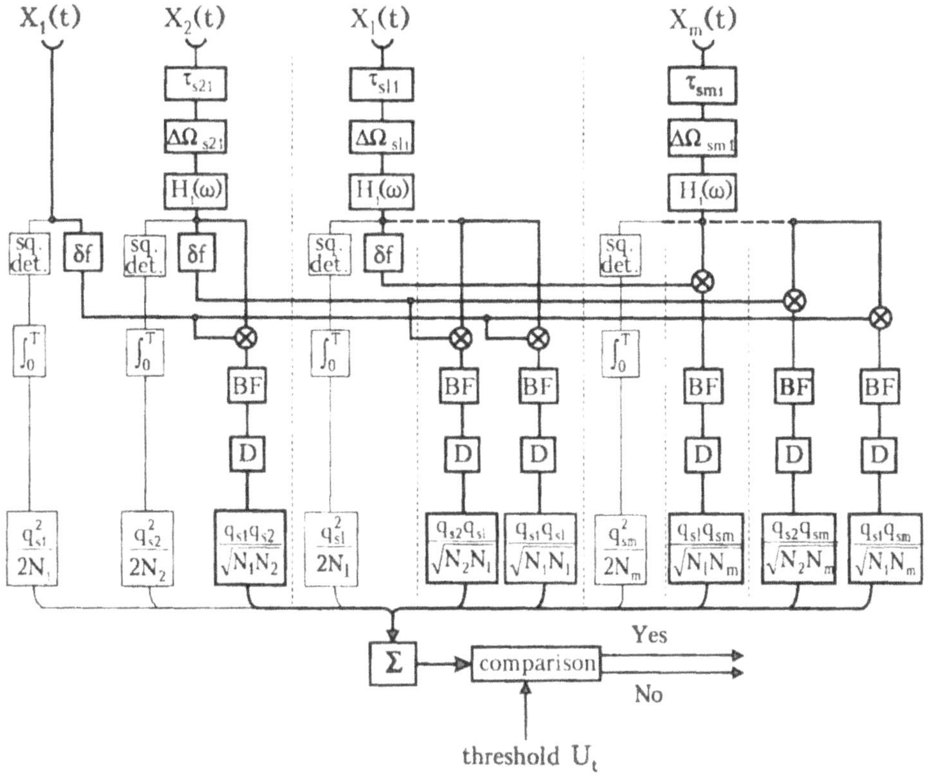

Figure 7.3 Structures of the optimum and correlation detectors for stochastic signals with quasideterministic correlation matrix. Correlation detector is highlighted by bold lines. sq. det. = square-law detector; δf = device for frequency shift by δf; BF = (narrow) bandpass filter tuned to the frequency δf; $\int_0^T$ = integrator; D = linear envelope detector; Σ = summator

receiver bandwidth Δf_s). The integrator of each correlator is a narrowband filter tuned to the frequency δf followed by an envelope detector.

Other versions of technical realization for the optimum algorithms (7.34) and (7.39) are, of course, possible. In the particular cases of digital processing all operations are carried out directly according to (7.34) and (7.39) where $\hat{B}_{ii}$ and $\hat{B}_{ik}$ are presented in the digital form.

For weak signals at correlator outputs linear envelope detectors should be replaced by square-law envelope detectors and weights should be changed in accordance with (7.35). When SNRs are random and completely spatially correlated, then A_{i1}^2 and A_{k1}^2 are to be substituted for q_{si}^2 and q_{sk}^2 respectively, in (7.35).

For the arbitrary statistical dependence between signal powers at the inputs of different stations an optimal processing algorithm may be obtained only for weak signals at the outputs of correlators and power estimators. The corresponding processing turns out to be rather cumbersome [57][2].

[2] Equations (19)–(21) in [57] are valid not for mutally independent phases φ_i, $i=\overline{1,m}$, but for mutally independent phase differences $\Delta\varphi_{ik}$, $i, k=\overline{1,m}$, $i<k$.

In spatially coherent MSRSs initial phases of expected signal correlation matrix may sometimes be known whereas signal powers (variances of that matrix) are random and completely spatially correlated at the inputs of receiving stations. In this case using $\Lambda=\exp[0.5L(q_{s1}^2)]$ as a conditional likelihood ratio where $L(q_{s1}^2)$ is determined by (7.16), replacing q_{si} by $A_{i1}q_{s1}$ and repeating the derivations which resulted in (7.39) yields (instead of (7.16))

$$\tilde{L}=\frac{1}{2\pi}\int_{-\infty}^{\infty}\left|\tilde{H}_1(\omega)\sum_{i=1}^{m}\frac{A_{i1}\exp(-j\Delta\psi_{si1})}{\sqrt{N_i}}\chi_i(\omega-\Delta\Omega_{si1})\exp[-j(\omega-\Omega_{s1})\tau_{si1}]\right|^2 d\omega \tag{7.40}$$

where A_{i1} as above is determined by (7.37) and it is assumed that $q_{s\Sigma}^2\ll 1$ so that $|\tilde{H}_1(\omega)|^2=F_s(\omega-\omega_0-\Omega_{s1})$.

Performance Analysis of the Synthesized Detectors

Let us consider the practically most important case of strong output signals [the algorithms (7.34) and (7.39)]. For fixed wanted signal powers and large values of signal accumulation factor $n=\Delta f_s T$ it is possible to consider the decision variable $\tilde{L}$ in (7.34) *when there is no signal* (i.e. when $\tilde{L}=\tilde{L}_0$) to be a sum of m Gaussian and $m(m-1)/2$ Rayleigh random variables. Obviously, the probability distribution of such a sum is nearly Gaussian, especially for large values of m. Expressing the probability distribution function of $\tilde{L}_0$ through the Edgeworth series [27] an approximate equation for the false alarm probability, P_{fa}, can be written in the form

$$P_{fa}\approx 1-F(y_0)+\frac{1}{3!}\frac{\kappa_3}{\kappa_2^{3/2}}F^{(3)}(y_0)-\frac{1}{4!}\frac{\kappa_4}{\kappa_2^2}F^{(4)}(y_0)-\frac{10}{6!}\frac{\kappa_3^2}{\kappa_2^3}F^{(6)}(y_0) \tag{7.41}$$

where

$$F(y_0)=\frac{1}{\sqrt{2\pi}}\int_{-\infty}^{y_0}\exp(-t^2/2)\,dt=0.5[1+\text{erf}(y_0/\sqrt{2})] \tag{7.42}$$

is the Laplace function; $F^{(k)}(y_0)$ is its kth derivative at the point y_0 and y_0 is the centred and normalized threshold level: $y_0=(u_t-m_{10})/\sqrt{M_{20}}$ where u_t is the threshold level, m_{10} and M_{20} are the mean value and the variance of L_0, respectively; k_2, k_3, k_4 are the second, third and fourth cumulants (semiinvariants) of the distribution of $\tilde{L}_0$. To obtain y_0 for a given value of P_{fa} one should calculate k_2, k_3 and k_4. It is not difficult to show that when a wanted signal is absent all terms in (7.34) are statistically mutually independent. Then assuming for simplicity the values of N_i and q_{si} to be equal at all stations ($N_i=N$, $q_{si}=q_s$, $i=\overline{1,m}$) and omitting the common factor q_s^2/N in (7.34) we can obtain the characteristic function of $\tilde{L}_0$ in the form

$$\Theta_\Sigma(v)=\Theta_1^m(v)\Theta_2^{m(m-1)/2}(v) \tag{7.43}$$

where $\Theta_1(v)$ and $\Theta_2(v)$ are the characteristic functions of the Gaussian variable $\hat{B}_{ii}/2$ and of the Rayleigh variable $|\hat{B}_{ik}|$, respectively. Let us denote the mean value of each $\hat{B}_{ii}/2$ by a_1, the variance of each $\hat{B}_{ii}/2$ by σ_1^2 and the variance of each Gaussian variable $\hat{B}_{ik}$ (in the absence of wanted signal) by σ_2^2. Substituting into (7.43) the known expressions for the characteristic functions of Gaussian and Rayleigh random variables [33] we have

$$\Theta_\Sigma(v)=\exp(jma_1 v-m\sigma_1^2 v^2/2)[1+j\sigma_2\sqrt{2\pi}vF(j\sigma_2 v)$$
$$\times\exp(-\sigma_2^2 v^2/2)]^{m(m-1)/2}. \tag{7.44}$$

Expanding $\Theta_\Sigma(v)$ into the Taylor series at the point $v=0$ yields

$$\begin{aligned}
k_1&=m_{10}=ma_1+[m(m-1)/2]\sigma_2\sqrt{\pi/2};\\
k_2&=M_{20}=m\sigma_1^2+m(m-1)\sigma_2^2(1-\pi/4);\\
k_3&=m(m-1)\sigma_2^3(\pi-3)\sqrt{\pi}/2\sqrt{2};\\
k_4&=m(m-1)\sigma_2^4[3\pi(1-\pi/4)-2].
\end{aligned} \tag{7.45}$$

Using (7.19), (7.22) and (7.23) the quantities a_1, σ_1^2 and σ_2^2 can be written from (7.31) as follows

$$\begin{aligned}
a_1&=\frac{NT}{2\pi}\int_{-\infty}^{\infty}\frac{F_s(\omega-\omega_0-\Omega_{s1})\,d\omega}{1+F_s(\omega-\omega_0-\Omega_{s1})mq_s^2};\\
\sigma_1^2&=\frac{N^2T}{2\pi}\int_{-\infty}^{\infty}\frac{F_s^2(\omega-\omega_0-\Omega_{s1})\,d\omega}{[1+F_s(\omega-\omega_0-\Omega_{s1})mq_s^2]^2}; \qquad \sigma_2^2=2\sigma_1^2.
\end{aligned} \tag{7.46}$$

For any known signal PSD $F_s(\omega-\omega_0-\Omega_{s1})$, one can find cumulants of $\tilde{L}_0$ from (7.45) and (7.46), then solve[3] the equation (7.41) and calculate the threshold level y_0. Note that the ratios of cumulants in (7.41) depend only on the total number of stations m. It can be verified taking into account the equality $\sigma_2^2=2\sigma_1^2$.

Let us now pass to the case where *a wanted signal is present*. In this case the decision variable $\tilde{L}=\tilde{L}_1$ is nearer to a Gaussian random variable than $\tilde{L}_0$ since $\tilde{L}_1$ includes $m(m-1)/2$ Rician variables[4] instead of Rayleigh ones. It is well known that the Rician distribution differs from the Gaussian distribution much less than the Rayleigh distribution, especially when SNR at the correlator output is large.

---

[3] The explicit expression for y_0 as a function of P_{fa}, i.e. the solution of the equation (7.41), is given in [129].

[4] The random variable $|\hat{B}_{ik}|$, $i\neq k$, is Rician one if the Gaussian variables $\mathrm{Re}\,\hat{B}_{ik}$ and $\mathrm{Im}\,\hat{B}_{ik}$ are mutually uncorrelated and have equal variances. When a signal is present, $\mathrm{Re}\,\hat{B}_{ik}$ and $\mathrm{Im}\,\hat{B}_{ik}$ though remain uncorrelated have, in general, different variances. It can be shown that, for instance, for the rectangular PSD within the frequency band Δf_s [see (7.27)]

$$M_{21}\{\mathrm{Re}\,\hat{B}_{ik}\}=\frac{2N^2n}{(1+mq_s^2)^2}[(1+q_s^2)^2+q_s^4];$$

$$M_{21}\{\mathrm{Im}\,\hat{B}_{ik}\}=\frac{2N^2n}{(1+mq_s^2)^2}[(1+q_s^2)^2-q_s^4].$$

However for the most important case of weak signals at the inputs of stations, where $q_s^2\ll 1$, the difference of variances may be neglected so that $|\hat{B}_{ik}|$, $i\neq k$, may be considered as Rician variables.

Therefore for the detection probability, P_d, computation we assume $\tilde{L}_1$ to be a Gaussian random variable. Let us derive expressions for the mean value and variance of $\tilde{L}_1$.

Employing known equations for the first and second moments of the Rice probability distribution [33] yields

$$m_{11}=m_1\{\tilde{L}_1\}=ma_2+\frac{m(m-1)}{2}\sigma_3\sqrt{\frac{\pi}{2}}\left[\left(1+\frac{a_3^2}{2\sigma_3^2}\right)I_0\left(\frac{a_3^2}{4\sigma_3^2}\right)\right.$$

$$\left.+\frac{a_3^2}{2\sigma_3^2}I_1\left(\frac{a_3^2}{4\sigma_3^2}\right)\right]\exp\left(-\frac{a_3^2}{4\sigma_3^2}\right). \tag{7.47}$$

In (7.47) $I_0(\cdot)$ and $I_1(\cdot)$ are the modified Bessel functions of zero and first order, respectively; a_2 is the mean value of the Gaussian variable $\hat{B}_{ii}/2$; a_3 is the mean value modulus of the Gaussian variable $\hat{B}_{ik}$, $i\neq k$, and σ_3^2 is the variance of $\hat{B}_{ik}$ (in the presence of a wanted signal). These distribution parameters for $\hat{B}_{ii}/2$ and $\hat{B}_{ik}$, like the moments a_1, σ_1^2 and σ_2^2 considered above for the same variables but in the absence of a wanted signal, can be obtained from (7.31), using (7.19), (7.22), (7.23):

$$a_2=\frac{NT}{2\pi}\int_{-\infty}^{\infty}\frac{F_s(\omega-\omega_0-\Omega_{s1})[1+q_s^2F_s(\omega-\omega_0-\Omega_{s1})]\,d\omega}{1+F_s(\omega-\omega_0-\Omega_{s1})mq_s^2};$$

$$a_3=\frac{2NTq_s^2}{2\pi}\int_{-\infty}^{\infty}\frac{F_s^2(\omega-\omega_0-\Omega_{s1})\,d\omega}{1+F_s(\omega-\omega_0-\Omega_{s1})mq_s^2}; \tag{7.48}$$

$$\sigma_3^2=\frac{2N^2T}{2\pi}\int_{-\infty}^{\infty}\frac{F_s^2(\omega-\omega_0-\Omega_{s1})[1+q_s^2F_s(\omega-\omega_0-\Omega_{s1})]^2\,d\omega}{[1+F_s(\omega-\omega_0-\Omega_{s1})mq_s^2]^2}.$$

Comparing (7.48) with (7.46) we can see that $a_2=a_1+a_3/2$.

To calculate the variance $M_{21}=M_2\{\tilde{L}_1\}$ it is necessary to take into account the possible correlation between summands in (7.34) but in this case the problem becomes significantly complicated. However, when *input* wanted signals are weak ($q_s^2 \ll 1$) this correlation may be neglected. Indeed, the mutual correlation coefficients for $\hat{B}_{ii}$ and $\hat{B}_{ik}$, $i\neq k$, are evaluated as $q_s^4/(1+q_s^2)^2$ and $q_s^2/(1+q_s^2)$, respectively. As was mentioned above the condition of *weak input wanted signals* is the most interesting for practice. Just under this condition the algorithm (7.39) is optimal. We shall see that for $n=\Delta f_s T \geqslant 10^3$ high detection probabilities can be achieved if $q_s^2<0.1$. When the correlation in (7.34) may be neglected, then we can write

$$M_{21}=\frac{m\sigma_3^2}{2}+\frac{m(m-1)}{2}\left\{2\sigma_3^2+a_3^2-\frac{\pi\sigma_3^2}{2}\right.$$

$$\left.\times\left[\left(1+\frac{a_3^2}{2\sigma_3^2}\right)I_0\left(\frac{a_3^2}{4\sigma_3^2}\right)+\frac{a_3^2}{2\sigma_3^2}I_1\left(\frac{a_3^2}{4\sigma_3^2}\right)\right]^2\exp\left(-\frac{a_3^2}{2\sigma_3^2}\right)\right\}. \tag{7.49}$$

Now detection probability can be calculated from (7.26) where y_0 is the threshold level which has been computed with the help of (7.41). We shall explain the described technique by a specific example.

Example 7.2. Let the signal PSD be rectangular as in Example 7.1 [see (7.27)]. Then denoting $n=\Delta f_s T$ we can obtain from (7.46) and (7.45)

$$a_1=\frac{Nn}{1+mq_s^2};\quad \sigma_1^2=\frac{N^2n}{(1+mq_s^2)^2};\quad \sigma_2^2=\frac{2N^2n}{(1+mq_s^2)^2};$$

$$\kappa_1=m_{10}=\frac{mN\sqrt{n}}{1+mq_s^2}\left(\sqrt{n}+\frac{m-1}{2}\sqrt{\pi}\right);$$

$$\kappa_2=M_{20}=\frac{mN^2n}{(1+mq_s^2)^2}[1+2(m-1)(1-\pi/4)]; \tag{7.50}$$

$$\kappa_3=\frac{m(m-1)N^3n\sqrt{n}(\pi-3)\sqrt{\pi}}{(1+mq_s^2)^3};$$

$$\kappa_4=\frac{4m(m-1)N^4n}{(1+mq_s^2)^4}[3\pi(1-\pi/4)-2].$$

Values of the coefficient of asymmetry $k_a=\kappa_3/\kappa_2^{3/2}$ and the coefficient of excess $\gamma_e=\kappa_4/\kappa_2^2$ calculated with the help of (7.50) for different number of stations are presented in Table 7.1. Comparing it with the corresponding values for the Rayleigh distribution ($k_a\approx 0.6311$, $\gamma_e\approx 0.2451$) we can see that the probability distribution of $\tilde{Z}_0$ is much nearer to Gaussian ($k_a=\gamma_e=0$) than to Rayleigh distribution. As can be expected, approximation to Gaussian distribution improves as m increases.

Using Table 7.1 one can evaluate the convergence rate of the Edgeworth series (7.41). However, the number of terms which are to be used for calculating y_0 should be determined by the increments in detection probability, P_d. In many cases the Gaussian approximation may be sufficient so that only two first terms from (7.41) are to be taken into account: $P_{f_a}=1-F(y_0)$.

To calculate the detection probability we can find from (7.48) for the rectangular signal PSD (7.27)

$$a_2=\frac{N(1+q_s^2)n}{1+mq_s^2};\qquad a_3=\frac{2Nq_s^2n}{1+mq_s^2};\qquad \sigma_3^2=\frac{2N^2(1+q_s^2)^2n}{(1+mq_s^2)^2}. \tag{7.51}$$

Substituting (7.51) in (7.47) and (7.49) we can calculate m_{11} and M_{21} for any number of stations m and given low SNR at the input of each station, q_s^2. Then for the threshold level y_0 obtained earlier with the help of the Edgeworth series (7.41) or of the explicit solution from [129] detection probability P_d can be calculated from (7.26). Though the noise PSD N at each station is incorporated in (7.50) and (7.51) it is eliminated in the computation process so that when N is unknown it does not make difficult P_d calculations.

Table 7.1

m	2	3	4	5	6	10
k_a	0.1039	0.1144	0.1019	0.1003	0.0918	0.0666
γ_e	0.0221	0.0174	0.0129	0.0098	0.0076	0.0034

The described technique has been applied to detection characteristic calculations for the optimal (under the condition of strong output signals) algorithm (7.34). The obtained characteristics are presented in Fig. 7.4. As distinct from detectors for stochastic signals with deterministic correlation matrix, these detection characteristics depend on the SNR value, q_{si}^2, at the input of each station individually. Therefore for the considered case of equal SNRs ($q_{si}^2 = q_s^2$, $i=\overline{1,m}$) q_s^2 is laid off as abscissa. To evaluate energy losses caused by random phases of the signal correlation matrix one should compare the values of SNR from Fig. 7.2 and mq_s^2 from Fig. 7.4, i.e. $q_s^2\,(\mathrm{dB})+10\lg m$, corresponding to the same values of P_{f_a} and P_d. It can be seen from Fig. 7.4 that required values of SNR q_s^2 for fixed probabilities of false alarm, P_{f_a}, and detection, P_d, decrease as the number of stations increases from $m=2$ to $m=4$. However, the required q_s^2 decreases more slowly than m increases so that the summed SNR $q_{s\Sigma}^2 = mq_s^2$ rises. Therefore, energy losses with respect to the optimal detector for stochastic signals with deterministic correlation matrices increase slightly with the increase of the number of stations m. For instance, when $P_{f_a}=10^{-4}$, $n=\Delta f_s T=10^3$, then for $P_d=0.6$ these energy losses change from 0.35 dB for $m=2$ to 1.13 dB for $m=4$ whereas for $P_d=0.95$ they change from 0.27 dB for $m=2$ to 0.85 dB for $m=4$.

Performance analysis of the optimal processing algorithm (7.34) for different SNR values q_{si}^2 at the inputs of receiving stations can be performed in a similar manner as for equal $q_{si}^2 = q_s^2$, but all manipulations are more cumbersome.

Let us consider now the optimal algorithm (7.39) for random signal power. To obtain detection characteristics one should replace in (7.34) q_{si}^2 by $A_{i1}^2 q_{s1}^2$ and average over q_{s1}^2. Power fluctuations from stochastic signal sources at the inputs of receiving stations are usually not too intensive (as compared, for instance, with the fluctuations of echoes from a complex target whose dimensions are much greater than the wavelength). For many artificial sources (e.g. jammers) intensive power fluctuations are, as a rule, undesirable. We assume for the analysis that: (a) the quantity q_{s1}^2 is distributed uniformly (in the logarithmic scale) within ± 3 dB and within ± 5 dB; (b) all those conditions under which the optimum algorithms (7.34) and (7.39) have been synthesized are satisfied, and (c) (for simplicity) mean signal powers at the inputs of all stations are equal to each other ($A_{i1}^2 = 1$, $i=\overline{2,m}$). Then for signals with rectangular PSD (7.27) we obtain detection characteristics presented in Fig. 7.5. The comparison with characteristics for fixed (known) power (Fig. 7.4) shows that, as may be expected, signal power fluctuations lead to a certain energy gain for low detection probabilities and to significant energy losses for high detection probabilities. Obviously, this effect can be seen more markedly for more intensive fluctuations. Physical reasons for this effect are the same as for a similar effect in detection of quasi-deterministic echoes with amplitude fluctuations (see Section 5.4).

***In this section**, we have synthesized and analyzed optimum detection algorithms for stochastic signals with more realistic features than in Section* 7.1, *i.e. with random phase differences and random powers at the inputs of spatially separated receiving stations. For a signal with random and mutually independent phases of all elements of the signal correlation matrix (excluding complex conjugate elements) we have derived the optimal processing algorithm* (7.33). *We have presented the approximate optimal algorithm* (7.34) *for the most practically interesting cases of strong output signals (i.e. for the cases of high detection probability) and the approximate optimal algorithm* (7.35) *for weak output signals. For the case where not only phases but also variances of signal correlation matrix (i.e. signal powers) are random and these variances are completely correlated (as is usually the practice) the processing algorithm* (7.39) *has been obtained which is optimal for weak input and strong output signals. The structural scheme of the*

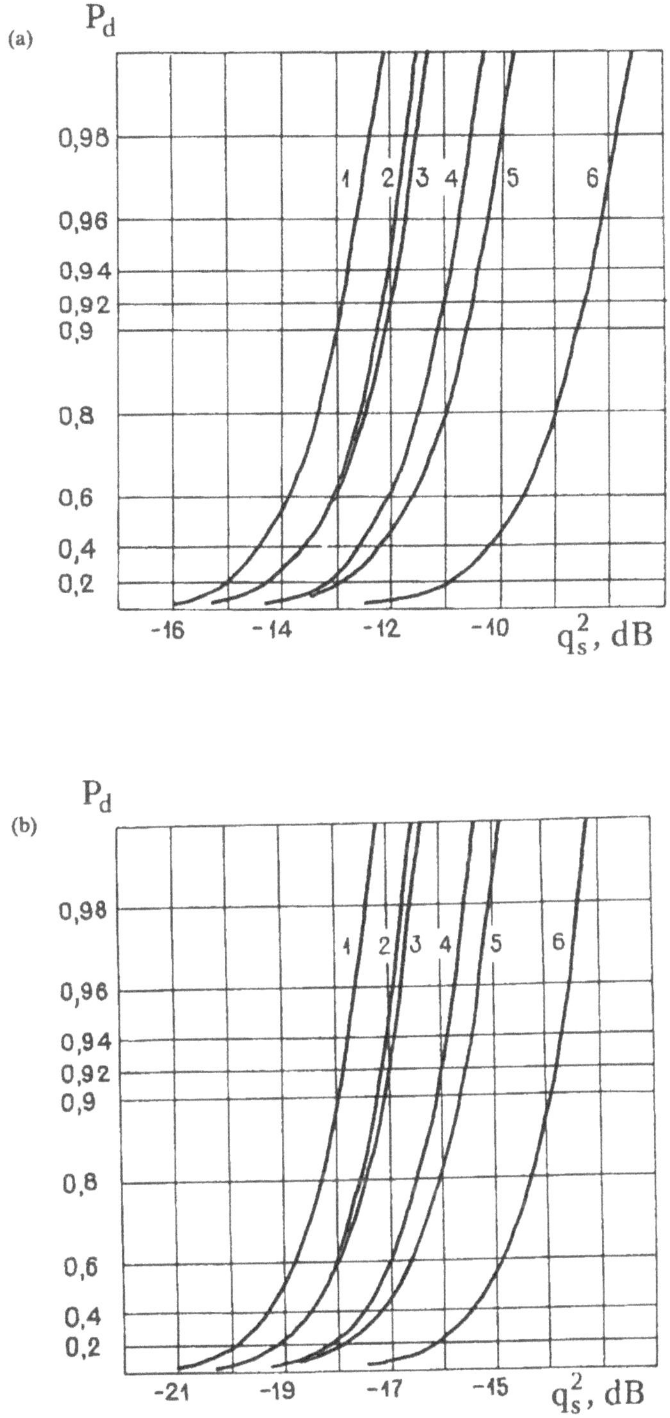

Figure 7.4 Detection characteristics of the optimum and correlation detectors for stochastic signals with random phases of the correlation matrix, $P_{fa} = 10^{-4}$. (a) $n = 10^3$; (b) $n = 10^4$; curves 1, 3, 5: optimum detector; curves 2, 4, 6: correlation detector; curves 1, 2: $m = 4$; curves 3, 4: $m = 3$; curves 5, 6: $m = 2$

(a)

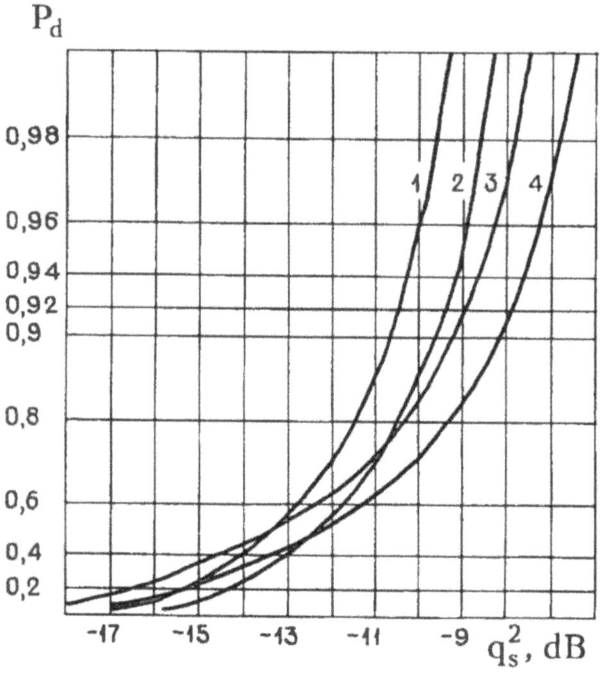

(b)

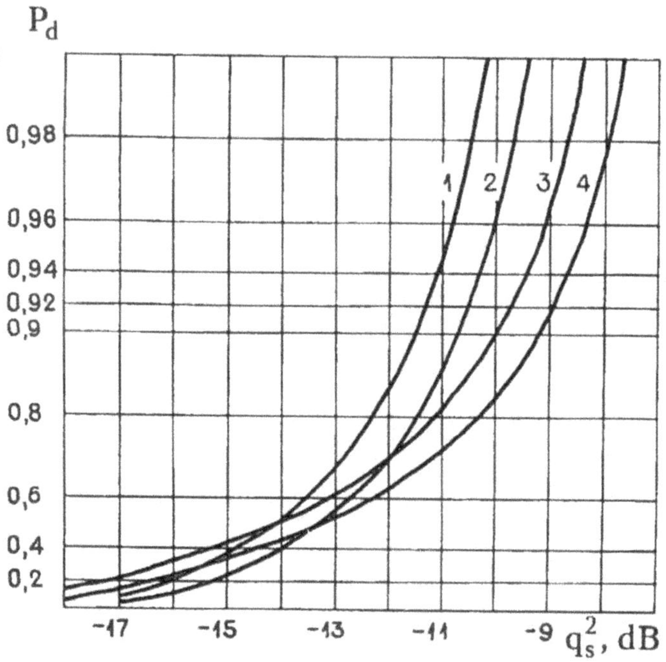

Figure 7.5 Detection characteristics of the optimum and correlation detectors for stochastic signals with random phases and variances of a correlation matrix (signal powers, SNRs), $P_{\mathrm{fa}} = 10^{-4}$, $n = 10^3$. (a) $m = 3$; (b) $m = 4$; curves 1, 3: optimum detector; curves 2, 4: correlation detector; curves 1, 2: SNRs are uniformly distributed within ± 3 dB; curves 3, 4: SNRs are uniformly distributed within ± 5 dB

optimal detector has been shown (Fig. 7.3). The decision variable (which is to be compared with a predetermined threshold) is a result of weighted summation of received signal power estimates from all stations with moduli of all possible mutual correlation estimates.

Performance analysis has been carried out for the practically most important cases of strong output signals, i.e. for the optimal algorithms (7.34) and (7.39). We have described the practical technique for calculating both false alarm and detection probabilities. The false alarm probability should be calculated with the help of the Edgeworth series (7.41), (7.42) since the output decision variable in the absence of a wanted signal cannot always be approximated by a Gaussian variable. The presented expressions (7.43)–(7.46) are sufficient for false alarm probability calculations. In the presence of wanted signals the decision variable may be usually approximated by a Gaussian variable. Therefore detection probabilities can be calculated with the help of expressions (7.47)–(7.49). We have considered in detail an important Example 7.2 for a rectangular signal PSD. The corresponding detection characteristics have been plotted in Figs. 7.4 and 7.5.

7.3. SUBOPTIMUM DETECTORS FOR STOCHASTIC SIGNALS

As for regular signal detection, of great practical importance are suboptimum detectors whose performance characteristics are close to those of more complicated optimum detectors.

Correlation Detector for Stochastic Signal with Deterministic Correlation Matrix

It follows from (7.10) and (7.16) that the estimate of power of the weighted sum of signals received by all stations can be split into the sum of estimates of received signal power at each station and the sum of estimates of mutual correlation of these signals. Mutual correlation processing is used for the coordinate measurement of a radiation source (see Section 11.1). Therefore, it is interesting to consider a correlation detector estimating only mutual correlation of received signals without estimating their power.

A processing algorithm takes the form [see (7.16)]

$$L=\mathrm{Re}\sum_{i=1}^{m-1}\sum_{k=i+1}^{m}\frac{q_{si}q_{sk}\exp(-\mathrm{j}\Delta\psi_{sik})}{\pi\sqrt{N_iN_k}}\int_{-}|H_1(\omega)|^2$$
$$\times x_i(\omega-\Delta\Omega_{si1})\chi_k^*(\omega-\Delta\Omega_{sk1})\exp[-\mathrm{j}(\omega-\Omega_{s1})\tau_{sik}]\,\mathrm{d}\omega. \qquad (7.52)$$

For the performance analysis let us assume (as in Section 7.1) that when the signal accumulation factor $n=\Delta f_s T$ is large enough, the decision variable L may be considered to be Gaussian in both the absence and the presence of wanted signals. Then it is sufficient to derive expressions for mean values and variances. Using the same relationships as in Section 7.1 for the performance analysis of optimum detector we can write

$$m_{10}=m_1\{L_0\}=0;$$
$$M_{20}=M_2\{L_0\}=\frac{2T}{\pi}\left[(q_{s\Sigma}^2)^2-\sum_{i=1}^{m}q_{si}^4\right]\int_{-\infty}^{\infty}\frac{F_s^2(\omega-\omega_0-\Omega_{s1})\,\mathrm{d}\omega}{[1+F_s(\omega-\omega_0-\Omega_{s1})q_{s\Sigma}^2]^2}; \qquad (7.53)$$

$$m_{11}=m_1\{L_1\}=\frac{T}{\pi}\left[(q_{s\Sigma}^2)^2-\sum_{i=1}^{m}q_{si}^4\right]\int_{-\infty}^{\infty}\frac{F_s^2(\omega-\omega_0-\Omega_{s1})\,d\omega}{1+F_s(\omega-\omega_0-\Omega_{s1})q_{s\Sigma}^2};$$

$$M_{21}=M_2\{L_1\}=\frac{2T}{\pi}\left[(q_{s\Sigma}^2)^2-\sum_{i=1}^{m}q_{si}^4\right]\int_{-\infty}^{\infty}\frac{F_s^2(\omega-\omega_0-\Omega_{s1})\,d\omega}{[1+F_s(\omega-\omega_0-\Omega_{s1})q_{s\Sigma}^2]^2}$$

$$+\frac{4T}{\pi}\sum_{i=1}^{m}\sum_{\substack{k=1\\k\neq i}}^{m}\sum_{\substack{l=1\\l\neq i}}^{m}q_{si}^2q_{sk}^2q_{sl}^2\int_{-\infty}^{\infty}\frac{F_s^3(\omega-\omega_0-\Omega_{s1})\,d\omega}{[1+F_s(\omega-\omega_0-\Omega_{s1})q_{s\Sigma}^2]^2} \qquad (7.54)$$

$$+\frac{2T}{\pi}\left[(q_{s\Sigma}^2)^2-\sum_{i=1}^{m}q_{si}^4\right]^2\int_{-\infty}^{\infty}\frac{F_s^4(\omega-\omega_0-\Omega_{s1})\,d\omega}{[1+F_s(\omega-\omega_0-\Omega_{s1})q_{s\Sigma}^2]^2}.$$

The false alarm and detection probabilities, P_{fa} and P_d, are determined by (7.26) where (7.53) and (7.54) should be substituted.

Example 7.3. Consider again the case of a rectangular signal PSD (7.27) and assume additionally that $q_{si}^2=q_s^2$, $i=\overline{1,m}$. Then we can obtain from (7.53) and (7.54):

$$m_{10}=m_1\{L_0\}=0; \qquad M_{20}=M_2\{L_0\}=4m(m-1)q_s^4n/(1+mq_s^2)^2;$$

$$m_{11}=m_1\{L_1\}=2m(m-1)q_s^4n/(1+mq_s^2);$$

$$M_{21}=M_2\{L_1\}=[4m(m-1)q_s^4n/(1+mq_s^2)^2][1+2(m-1)q_s^2+m(m-1)q_s^4]. \qquad (7.55)$$

Detection characteristics of the considered correlation detector for the rectangular signal PSD are shown in the same Fig. 7.2 together with detection characteristics of optimal detectors (7.10) and (7.16). If the number of stations $m\geqslant 3$ energy losses are no more than 1 dB.

Correlation Detector with Envelope Detection at the Correlator Output

This detector distinguishes from the optimum one for signals with quasi-deterministic correlation matrices (see Fig. 7.3) by excluding signal power estimates. Only elements drawn by bold lines remain in Fig. 7.3. The decision variable which is to be compared with a threshold is a weighted sum of the moduli of estimates of received signals' mutual correlation [see (7.34)]

$$L=\sum_{i=1}^{m-1}\sum_{k=i+1}^{m}\frac{q_{si}q_{sk}}{\sqrt{N_iN_k}}|\hat{B}_{ik}|. \qquad (7.56)$$

For random and completely spatially correlated signal powers [see (7.39)]

$$L=\sum_{i=1}^{m-1}\sum_{k=i+1}^{m}\frac{A_{i1}A_{k1}}{\sqrt{N_iN_k}}|\hat{B}_{ik}| \qquad (7.57)$$

where $\hat{B}_{ik}$ is determined by (7.31) as above. Assuming again $\hat{B}_{ik}$ to be a Gaussian variable (for large values of $n=\Delta f_s T$) we may conclude that the decision variable L in (7.56) for $m=2$ is a Rayleigh variable in the absence of a signal and a Rice variable when a signal is present[5]. Detection characteristics of a passive correlation system with two receiving stations (such systems are of great practical importance) are the same as detection characteristics in Fig. 5.4 of an optimum detector for spatially coherent quasi-deterministic nonfluctuating signals, if q_{out}^2 in Fig. 5.4 is assumed to be the SNR at the correlator output [see (7.49)]:

$$q_{\text{out}}^2 = 2q_{\text{out cor}}^2 = a_3^2/\sigma_3^2.$$

Example 7.4. For the rectangular signal PSD (7.27) we obtain in accordance with (7.48)

$$q_{\text{out cor}}^2 = q_s^4 n/(1+q_s^2)^2.$$

In the case of different SNR values at the inputs of receiving stations

$$q_{\text{out cor}\,12}^2 = q_{s1}^2 q_{s2}^2 n/(1+q_{s1}^2)(1+q_{s2}^2). \tag{7.58}$$

If $m>2$ the normalized threshold level y_0 can be calculated with the help of the Edgeworth series at a given false alarm probability, P_{fa}, as for the optimum detector [see (7.41), (7.42)]. Then for calculations of detection probability, P_d, a Gaussian approximation of the decision variable L may be used (in the presence of a weak input signal $L=L_1$ is a sum of $m(m-1)/2$ Rician random variables). As in Section 7.2, let us assume for simplicity that $q_{si}=q_s$, $N_i=N$, $A_{i1}=1$, $i=\overline{1,m}$. Under this assumption common factors in (7.56), (7.57) may be taken outside the signs of summation and dropped. The first and second cumulants of the probability distribution of $L=L_0$, that is of the sum of $m(m-1)/2$ Rayleigh variables, are determined by the second summands in the right side of (7.45) [in the case of the rectangular signal PSD (7.27) these cumulants are determined by the second summands of the right side of (7.50)]. The third and fourth cumulants are the same as in (7.45) [or in (7.50) for the rectangular PSD (7.27)] since for m Gaussian variables which are taken into account in (7.45) and (7.50) these cumulants are equal to zero.

Having y_0 we can calculate the detection probability, P_d, from (7.26) where m_{10}, m_{11}, M_{20} and M_{21} are now the second summands of equations (7.45)–(7.49).

Example 7.5. Let us consider again the case of the rectangular signal PSD (7.27). In this case we can obtain from (7.47)

$$m_{10} = \kappa_1 = \frac{m(m-1)\sqrt{\pi n}N}{2(1+mq_s^2)}; \quad M_{20} = \kappa_2 = \frac{2m(m-1)n(1-\pi/4)N^2}{(1+mq_s^2)};$$
$$\kappa_3 = \frac{m(m-1)n\sqrt{\pi n}(\pi-3)N^3}{(1+mq_s^2)^3}; \quad \kappa_4 = \frac{4m(m-1)n^2N^4}{(1+mq_s^2)^4}[3\pi(1-\pi/4)-2]. \tag{7.59}$$

[5] See footnote 4 on page 201.

Table 7.2

m	2	3	4	5	6	10
k_a	0.6311	0.3644	0.2576	0.1996	0.1630	0.0941
γ_e	0.2451	0.0817	0.0408	0.0245	0.0163	0.0054

Using (7.59), the coefficient of asymmetry $k_a = \kappa_3/\kappa_2^{3/2}$ and the coefficient of excess $\gamma_e = \kappa_4/\kappa_2^2$ of the probability distribution of L_0 have been calculated for different number m of receiving stations. The calculated values are presented in Table 7.2.

Looking up the data in Table 7.2 and using (7.41), (7.42) we can calculate a normalized threshold level y_0 at a given false alarm probability P_{fa}. Comparing Table 7.2 with Table 7.1 shows that the values of the asymmetry and excess coefficients in Table 7.2 are greater than those in Table 7.1. Obviously, it can be expected since the decision variable L_0 of the optimum detector is the sum of not only $m(m-1)/2$ Rayleigh variables but m Gaussian variables as well. Therefore, if a more precise estimate of the threshold level y_0 is required for the considered suboptimum detector, we can retain terms of higher order in the Edgeworth series than in (7.41) [27]. However, it is seldom required in practice.

Having y_0 the detection probability, P_d, can be calculated from (7.26) where m_{10}, and M_{20} are to be taken from (7.59), whereas m_{11} and M_{21} are now the second summands of equations (7.47) and (7.49) taking into account (7.51).

Detection characteristics for the considered correlation detector (7.56) [and the rectangular signal PSD (7.27)] are presented in the same Fig. 7.4 as for the optimum detector (7.34). Taking into account the practical importance of the particular case where $m=2$, detection characteristics for this case are presented too though, as was mentioned above, they do not differ from corresponding detection characteristics in Fig. 5.4.

Comparing detection characteristics of the optimum and correlation detectors shows that excluding power estimates of signals coming from all stations leads to small energy losses which are reduced as the number m of stations increases. In particular, these losses are of the order of 1 dB for $m=3$ and 0.8 dB for $m=4$ if $P_{fa} = 10^{-4}$ and $n = \Delta f_s T = 10^3$.

Detection characteristics for completely correlated fluctuating signal powers at the inputs of receiving stations when the algorithm (7.57) is used (at $A_{i1}=1$ and $N_i = N, \overline{1,m}$) and signal PSD has a rectangular form, are presented in Fig. 7.5 together with the detection characteristics of the optimum algorithm (7.39) under similar conditions. The general behaviour of corresponding curves is the same but, as for the known signal input powers, the correlation detector is inferior to the optimum one approximately by 0.8–1.0 dB.

Detector Using only Signal Power Estimates

From the optimum detector of Fig. 7.3 it is possible to exclude estimates of mutual signal correlation instead of signal power estimates. Such detectors are essentially simpler to implement since there is no need in wideband (with high handling capacity) DTLs which are necessary for mutual correlation processing of signals coming from all stations when optimum or correlation detectors are used. Only signal power estimates undergo joint processing but these estimates may be transferred to a fusion centre via narrowband (with drastically lower handling capacity) DTLs.

The decision variable for the case where relationships between nonrandom signal powers at the different stations are known, can be written as follows

$$L=\sum_{i=1}^{m}\frac{q_{si}^2}{N_i}\hat{B}_{ii} \tag{7.60}$$

where $\hat{B}_{ii}$ is determined in (7.31). As in Section 7.2 we assume the quantities $\hat{B}_{ii}$, $i=\overline{1,m}$, to be Gaussian and mutually independent. Under this assumption detection characteristics can be calculated from (7.26). Mean values and variances of L in (7.60) (for $q_{si}=q_s$, $N_i=N$, $i=\overline{1,m}$ when the common factor q_s^2/N may be dropped) both in the absence and in the presence of wanted signals we can find from the first summands of the corresponding expressions in (7.45), (7.47) and (7.49):

$$m_{10}=m_1\{L_0\}=2ma_1;\quad M_{20}=M_2\{L_0\}=4m\sigma_1^2;$$

$$m_{11}=m_1\{L_1\}=2ma_2;\quad M_{21}=M_2\{L_1\}=2m\sigma_3^2.$$

For a_1, a_2, σ_1^2 and σ_3^2 equations (7.46) and (7.48) [or (7.50) and (7.51) for the rectangular signal PSD (7.27)] are valid.

Calculations show that detection characteristics of the algorithm (7.60) are worse than those of the correlation algorithm (7.56).

Example 7.6. Let us return to the case of the signal rectangular PSD (7.27). Using (7.50) and (7.51) yields that for $P_{fa}=10^{-4}$, $n=\Delta f_s T=10^3$, $q_{si}^2/N_i=q_s^2/N$, $\overline{1,m}$, additional energy losses as compared with correlation detector vary from 0.5 to 1.0 dB for $m=3$ and from 1.0 to 1.5 dB for $m=4$ with P_d variation from 0.5 to 0.99. As the number of receiving stations increase, additional energy losses increase too.

However, excluding mutual correlation processing of signals coming from different receiving stations make it impossible to measure time differences of arrival (TDOAs) of signals from radiation sources to spatially separated stations. It means that the hyperbolic method of location of radiation sources cannot be used. This is a principal disadvantage of such a detector. The mutual correlation of signals from each source at the inputs of different stations cannot also be utilized for signal resolution in TDOA and identification of sources. Therefore, the algorithm (7.60) considered here is reasonable to use in the cases where radiation sources are located by the triangulation method and there is not necessary to employ mutual signal correlation processing either for TDOAs measurements or for other ("auxiliary") purposes.

In this section*, we have considered three practically most interesting simplified suboptimal detection algorithms for stochastic signals. These are: the correlation detector for a signal with a deterministic correlation matrix and two detectors for signals with correlation matrices containing random parameters – the correlation detector with envelope detection at the correlator output (deriving moduli of signal mutual correlation function estimates) and the detector using only signal power estimates. For each detector we have presented expressions determining received signal processing* [(7.52) *for the first,* (7.56), (7.57) *for the second and* (7.60) *for the third detector*]. *We have obtained equations* (7.53) *and* (7.54) *permitting calculations of detection characteristics for the first detector. We also have indicated the technique for calculating detection characteristics for the two other detectors using equations* (7.41), (7.42), (7.45)–(7.49) *from Section* 7.2. *All these equations relate to the general case of arbitrary signal PSD. For particular case of the rectangular signal PSD* (7.27) *we have considered Examples* 7.3–7.6 *including numerical calculations. On the basis of these* *calculations detection characteristics have been plotted for all the three suboptimum detectors. The principal conclusion from the analysis of obtained detection characteristics is that energy losses of considered suboptimum algorithms are not large as compared with the more complex optimum algorithm from Sections 7.1 and 7.2. Among all suboptimal algorithms discussed in this Section the correlation algorithm deriving moduli of signal correlation function estimates is of key practical importance.*

7.4. DECENTRALIZED SUBOPTIMUM DETECTORS FOR STOCHASTIC SIGNALS

All stochastic signal detectors considered in Sections 7.1–7.3 are centralized ones since decisions of whether a wanted signal is present are made in a fusion centre (FC) only. However, as in the case of regular signals, decentralized detection of stochastic signals is possible.

Decentralized Correlation Detector Using Moduli of Mutual Correlation Function Estimates

Preliminary decisions are made at the output of each correlator by threshold comparison of mutual correlation estimate moduli for signals received by spatially separated stations. After each excess of the threshold a "one" is sent to the FC. In those resolution cells in TDOA (delay) where there are no threshold exceeding, zeros are generated. Final decisions of signal detection by the MSRS are made in the FC as a result of joint processing (combination) of binary sequences coming from all correlators. Decision rules of the "k out of m" type (coincidence criteria) are conventionally used.

As with nonradiating target detection the radiation source position is *a priori* known only approximately. A source position uncertainty region covers, as a rule, several (sometimes, many) space resolution cells (SRCs) of a MSRS. To this space uncertainty region correspond certain ranges of values of the signal TDOA (delay) at the inputs of correlators. Hence centralized detectors require complicated multichannel summation structures. These structures must sum signals from correlator outputs with different delay combinations at their inputs so that each combination corresponds to a certain SRC of the source position uncertainty region.

Joint processing at the FC is essentially simplified in a decentralized detector. First, only binary information – ones and zeros – is to be processed. Second, a coincidence criterion test should be applied only if at least a single one occurs at the output of any correlator. In this case preliminary decisions should be jointly processed from only those delay resolution cells at the correlator outputs which together with the cell where the above mentioned one has occurred, correspond to the same SRC. Though for information fusion wideband DTLs (with high handling capacity) are not required, such DTLs cannot be excluded at all (unlike the case of decentralized target echo detection) since they are necessary to transfer signals and interferences received by different stations to the inputs of correlators.

The choice of a specific decision rule from rules of the "k out of m" type depends on the number of receiving stations, method of radiation source coordinate measurement and environmental conditions. In a MSRS containing m receiving stations there are $m(m-1)/2$ pairs of stations where signal TDOAs from each radiation source, τ_{ik}, $i,k=\overline{1,m}$, $i<k$, can be measured. If source coordinates are determined by the

hyperbolic method (see Section 11.1), then, as it follows from (7.8), only $m-1$ TDOAs are linearly independent among those $m(m-1)/2$ TDOAs. These are signal TDOAs from a source to such stations where baselines between them do not form a closed figure. A certain combination is often chosen beforehand among all possible combinations of such baselines so that to provide, for instance, the most accurate source coordinate measurement for expected source positions. Baselines of this combination we shall refer to as measuring baselines. It is not necessary to detect signals at the outputs of all correlators corresponding to measuring baselines. When $m>4$ signal detection at the outputs of any p these correlators, where $m-1 \geqslant p \geqslant 3$, is sufficient since TDOAs measurement with respect to such three baselines permits the radiation source location in space.

When several signal sources are acting simultaneously and these sources are positioned closely to each other (including the case where they are not resolved in angle coordinates by the antennas of receiving stations) the data association problem arises. It means association between TDOA measurements obtained with respect to different baselines, on the one hand, and radiation sources, on the other. To solve this problem, measurements of linearly dependent TDOAs are often utilized, i.e. TDOA measurements with respect to "closing" baselines (see Section 15.4). In this case a signal source may be decided as a detected one by a MSRS if, for instance, the signal is detected at the correlator outputs corresponding to at least p measuring baselines and to at least $p-1$ other baselines "closing" those measuring baselines ($m-1 \geqslant p \geqslant 3$). As earlier we assume that all the quantities $\hat{B}_{ik}$ in (7.56) are mutually statistically independent (for weak signals at the stations' inputs). If false alarm probabilities, P_{fa0}, are equal to each other after threshold comparison at the outputs of all correlators and the same are detection probabilities, P_{d0}, then the algorithm described above leads to the following equations

$$P_{\mathrm{fa}}=\sum_{k=p}^{m-1} C_{m-1}^{k} P_{\mathrm{fa0}}^{k}(1-P_{\mathrm{fa0}})^{m-1-k} \sum_{l=k-1}^{k(k-1)/2} C_{k(k-1)/2}^{l} P_{\mathrm{fa0}}^{l}(1-P_{\mathrm{fa0}})^{k(k-1)/2-l},$$
$$P_{\mathrm{d}}=\sum_{k=p}^{m-1} C_{m-1}^{k} P_{\mathrm{d0}}^{k}(1-P_{\mathrm{d0}})^{m-1-k} \sum_{l=k-1}^{k(k-1)/2} C_{k(k-1)/2}^{l} P_{\mathrm{d0}}^{l}(1-P_{\mathrm{d0}})^{k(k-1)/2-l}. \tag{7.61}$$

For strong signals at the output of each correlator

$$P_{\mathrm{fa0}}=\exp(-y_0^2/2);$$
$$P_{\mathrm{d0}}=\int_{\sigma_2 y_0/\sigma_3}^{\infty} z\exp[-(z^2/2+q_{\mathrm{out\,cor}}^2)]I_0(q_{\mathrm{out\,cor}}\sqrt{2}z)\,\mathrm{d}z. \tag{7.62}$$

For weak signals at correlator outputs

$$P_{\mathrm{fa0}}=\exp(-x_0/2);$$
$$P_{\mathrm{d0}}=\int_{\sigma_2^2 x_0/\sigma_3^2}^{\infty} \exp[-(z/2+q_{\mathrm{out\,cor}}^2)]I_0(q_{\mathrm{out\,cor}}\sqrt{2z})\,\mathrm{d}z. \tag{7.63}$$

In (7.62) and (7.63) σ_2^2 and σ_3^2 are the variances of the Gaussian variables $\hat{B}_{ik}$ in the absence and presence of wanted signals which are determined by (7.46), (7.48), respectively, and by (7.50), (7.51) for rectangular signal PSD; $q^2_{\text{out cor}}$ is the SNR at the correlator output [see (7.58)].

It was shown in Sections 6.1 and 6.2 that decision rules of the "k out of m" type are optimal decentralized detection rules when false alarm probabilities as well as detection probabilities are equal at all stations. They do not lead to large energy losses when these probabilities are different if differences between stations in SNRs and noise levels are not too large. The same is valid for pairs of stations in passive MSRSs. On the other hand, if those differences are large, joint processing of information coming from "weak" and "strong" pairs of stations is unreasonable (from the viewpoint of signal detection in the MSRS).

Example 7.7. Consider a MSRS with $m=4$, $p=3$. Measuring bases are between one station and the three remaining stations. It is assumed that a wanted signal (a radiation source) is detected by the MSRS if it is detected at correlator outputs corresponding to all the three measuring bases and to any two "closing" bases. From (7.61) we have

$$P_{\text{fa}} = P_{\text{fa0}}^5(3-2P_{\text{fa0}}); \qquad P_{\text{d}} = P_{\text{d0}}^5(3-2P_{\text{d0}}). \tag{7.64}$$

Similar though more complicated expressions can be obtained for the case where false alarm probabilities as well as detection probabilities are not equal at the correlator outputs corresponding to different bases (see Section 6.1).

Of course, other decision rules are possible too. For instance, we may require threshold exceeding at the correlator outputs corresponding at least to five (of the total number six) bases. Then instead of (7.64) we can write

$$P_{\text{fa}} = P_{\text{fa0}}^5(6-5P_{\text{fa0}}); \qquad P_{\text{d}} = P_{\text{d0}}^5(6-5P_{\text{d0}}). \tag{7.65}$$

In Fig. 7.6 a detection characteristic is plotted using (7.64) for $m=4$, $n=\Delta f_s T=10^3$, $P_{\text{fa}}=10^{-4}$ and strong signals at the correlator outputs [P_{fa0} and P_{d0} are determined by (7.62)]. For the comparison the corresponding detection characteristic of the centralized correlation detector using moduli of mutual signal correlation function is shown (see Fig. 7.4). It can be seen that energy losses caused by the decentralization of the decision procedure is about 1.4–1.5 dB.

Decentralized Detector Using Only Power Estimates

When coordinates of radiation sources are determined by triangulation, preliminary decisions may be made at each station individually by comparing received signal power (energy) estimates with a threshold. Final decisions are made at a FC by combination of preliminary decisions using decision rules of the "k out of m" type. The most important advantage of such a detection structure is that it does not require DTLs with high handling capacity (wideband DTLs). Only information of signals exceeded the threshold are to be transferred to FC. However, this detector has all the disadvantages of the similar centralized detector and besides some additional energy losses caused by the decentralization of decision making.

If each receiving station can measure two angle coordinates of a radiation source, it is sufficient for signals from this source to be detected at least by two stations. When there are simultaneously several sources resolved in angle coordinates, then

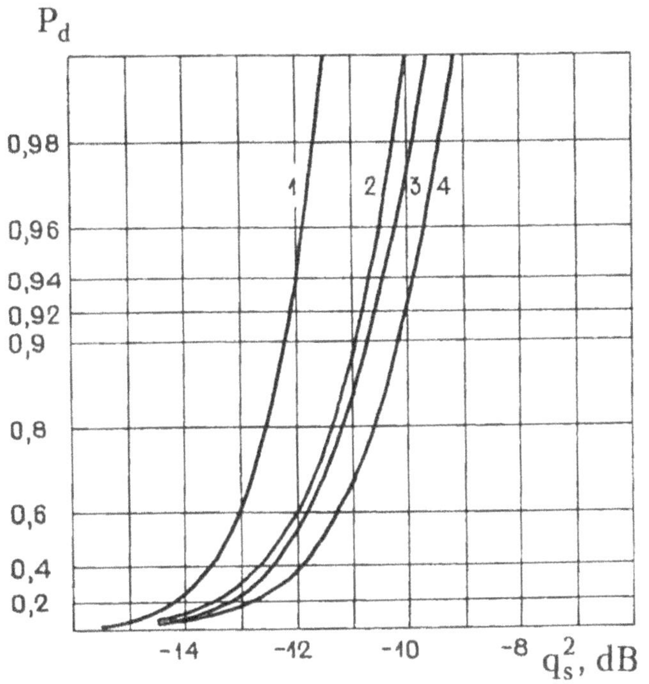

Figure 7.6 Detection characteristics for stochastic signals with random phases of a correlation matrix, $P_{fa}=10^{-4}$, $n=10^3$. Curve 1, 3: centralized and decentralized correlation detectors, respectively; curves 2, 4: centralized and decentralized detectors, respectively combining only signal power estimates

signals from each source are desirable to be detected additionally by one–two stations in order to help the data association problem solution.

Example 7.8. Let us employ, for instance, the decision rule "3 out of m", i.e. we consider a source as the detected one if its signals are detected at least by any three stations. Under the condition of mutual independence of the quantities $\hat{B}_{ii}$ in (7.60), and equal probabilities P_{fa0}, P_{d0} at all stations we have

$$P_{fa}=\sum_{k=3}^{m} C_m^k P_{fa0}^k(1-P_{fa0})^{m-k}; \qquad P_d=\sum_{k=3}^{m} C_m^k P_{d0}^k(1-P_{d0})^{m-k}. \tag{7.66}$$

For usual large values of $n=\Delta f_s T$ the quantities $\hat{B}_{ii}$ may be assumed to be Gaussian variables. The probabilities P_{fa0} and P_{d0} can be calculated using (7.26) where mean values and variances of $\hat{B}_{ii}$ are determined as for the corresponding centralized algorithm (7.60): $m_{10}=2a_1$, $M_{20}=4\sigma_1^2$; $m_{11}=2a_2$, $M_{21}=2\sigma_3^2$. Expressions (7.46), (7.48) [or (7.50), (7.51) for the rectangular signal PSD] are valid for $a_1, a_2, \sigma_1^2, \sigma_3^2$. The corresponding detection characteristic ($P_{fa}=10^{-4}$, $n=10^3$) calculated from (7.66) for $m=4$ according to

$$P_{fa}=P_{fa0}^3(4-3P_{fa0}); \qquad P_d=P_{d0}^3(4-3P_{d0}) \tag{7.67}$$

is shown in Fig. 7.6. The detection characteristic of the similar centralized detector for the same conditions is depicted in Fig. 7.6 too. Energy losses caused by the decentralization are 0.7–0.9 dB whereas energy losses as compared with decentralized correlation detector are about 0.5 dB.

Decentralized Correlation Detector with Amplitude Limiters in Receiver Channels

If radiation source coordinates are determined by the hyperbolic method using TDOA measurement, amplitude limiters may be included in receiver channels of each receiving station. As a result the required dynamic range of receivers and analogous DTLs as well as the handling capacity of digital DTLs may be reduced. Besides it provides a constant false alarm rate at the correlator outputs (including the case of external interferences).

Decision rules (detection criteria) of the "k out of m" type are to be chosen in the same manner as for the decentralized correlation detector without limiters. Probabilities P_{fa0} and P_{d0} at the correlator outputs are determined by (7.62) and (7.63) if $q^2_{\text{out cor}}$ is replaced by q^2_{lim}, i.e. by the SNR at the correlator output when amplitude limiters are included at its inputs. On account of the elimination of amplitude information $q^2_{\text{lim}} \leqslant q^2_{\text{out cor}}$. If limiters are ideal, i.e. hard (output amplitude is constant, phase structure of a random process is not disturbed), and input signals are weak $(q^2_{s1}q^2_{s2}/(1+q^2_{s1})(1+q^2_{s2}) \ll 1)$, then [71]

$$q^2_{\text{lim}} \approx (\pi/4)^2\, q^2_{\text{out cor}}. \tag{7.68}$$

It means that losses in SNR at the correlator output are 2.1 dB. According to the simulation results from [101] these losses for $q^2_{s1}=q^2_{s2}=1$ are 1.65 dB. It follows from Figs. 7.5 and 7.6 that for large values of $n=\Delta f_s T$ the most interesting are the cases where $q^2_s \leqslant 0.1$. Therefore we shall employ (7.68).

When $q^2_{si}=q^2_s$, $i=\overline{1,m}$, then

$$q^2_{\text{lim}} \approx (\pi/4)^2 q^4_s n/(1+q^2_s)^2. \tag{7.69}$$

Setting the right side of (7.69) equal to that of (7.58) we can evaluate losses in the input SNR q^2_s at each station. When q^2_s varies from −15 to −9 dB these losses vary from 1.1 to 1.2 dB. Such an increase of the SNR q^2_s at the input of each station is necessary in order to obtain the same detection characteristics as without amplitude limiters. This is valid both for any algorithms of joint processing of preliminary decisions in FC and for centralized correlation detector. Small energy losses accompanying the complete elimination of amplitude information show that the main role in the correlation processing of narrowband stochastic processes is played by their phase structure.

Decentralized Detectors Based on Digital Polarity Coincidence Correlators

Comparatively high performance characteristics of correlators with amplitude limiters in input channels allows us to go further on the way of correlation detector simplification. In particular it would be important from this standpoint not to use all the phase information. The simplest type of such correlators is the polarity coincidence correlator (PCC). The most convenient implementation of PCC is in a digital form [11,71]. Narrowband processes at each receiving station are splitted into two quadrature channels. These quadrature processes are sampled after hard amplitude limiting with the sampling period which is equal or less than $1/\Delta f_s$. When the polarity of the sampled process is positive at the sampling moment, then a positive one (+1) is generated. When this polarity is negative, a negative one (−1) is generated. Thus hard amplitude limiters destruct all the amplitude information, and phases of those amplitude limited processes are binary quantized. As a result the handling capacity requirements to DTLs connecting spatially separated stations and processing correlators can be significantly reduced. Correlation processing itself is drastically simplified. It is reduced to delay and multiplying of binary digits and to product accumulation. Both in-phase and quadrature components at correlator inputs undergone such a processing. Then accumulator outputs of both quadrature channels are squared, summed and the square root is calculated thus obtaining a modulus of the estimate of input processes' mutual correlation. For weak signals at the inputs of correlators [when $q_{s1}^2 q_{s2}^2/(1+q_{s1}^2)(1+q_{s2}^2) \ll 1$] the power SNR at the output of a polarity coincidence correlator is given by [71]

$$q_{pc}^2 \approx (2/\pi)^2 q_{out\ cor}^2. \tag{7.70}$$

Losses in the output SNR are 3.9 dB with respect to the correlator without amplitude limiters so that additional losses caused by the binary phase quantization (after the amplitude limiters) are 1.8 dB.

When SNR values q_s^2 at the correlator inputs are equal to each other, then

$$q_{pc}^2 \approx (2/\pi)^2 q_s^4 n/(1+q_s^2)^2. \tag{7.71}$$

Setting the right side of (7.71) to the right side of (7.69) one can evaluate losses in the input SNRs. For $q_s^2 = -(9\text{–}15)$ dB additional losses caused by binary phase quantization is 1.1–1.0 dB at the input of each station.

Decentralized Detectors Based on Digital Relay Correlators

In some cases it is important that amplitude information containing in the sum of signal and noise at least at one of the correlator input to be preserved at the correlator output. In these cases a relay correlator can be employed. In a digital form the difference between a relay and a polarity coincidence correlators is that a binary sequence of plus ones and minus ones enters only one of correlator inputs. To the other input a discrete sequence is applied which is an amplitude quantized corresponding quadrature component of the input process (of course, without amplitude limiting). The number of bits of the amplitude quantization is usually chosen so that "quantization noise" does not affect the output SNR. It is sufficient to choose the amplitude quantization step by 2–3 times less than the r.m.s value of receiver noise. This results in the noise variance increase no more than by 1–2%. The power SNR at the output of a relay correlator for weak signals at its inputs is given by [11]

$$q_r^2 \approx (2/\pi) q_{out\ cor}^2. \tag{7.72}$$

Losses in output SNR are about 2 dB in comparison with a correlator without amplitude limiters in all channels [see (7.58)], i.e. losses are the same as in a correlator with amplitude limiters in all input channels [see (7.68)] and by 2 dB less than those of polarity coincidence correlator [see (7.70)]. Using (7.72) with (7.61) and (7.62) or (7.63) one can obtain detection characteristics of decentralized detectors based on relay correlators.

An overview of correlators of different types has been presented in [239].

In conclusion of this Chapter it is important to note that when sources to be detected radiate continuous signals or signals of sufficiently long duration (and of sufficiently wide spectrum), then *energy losses of suboptimum detectors may often be* *offset by the increase of processed signal duration (observation time interval) T and/or processed signal bandwidth Δf_s that is by the increase of the accumulation factor* $n=\Delta f_s T$. This is one of the advantages of passive radar systems which utilize the energy of radiation sources but not the "own" energy of transmitting stations as usual active radar systems do.

In this section, *we have considered several decentralized suboptimum algorithms for stochastic signal detection in passive MSRSs. We have noted that decentralized detection structures are significantly simpler than centralized ones. Taking into account that this simplification leads to energy losses we have presented corresponding expressions permitting comparative energy losses evaluation of all the discussed detectors. Particular examples 7.7, 7.8 and detection characteristics plotted in Fig. 7.6 make such a comparison more easy to grasp. Explicit expressions (7.68)–(7.72) for power output SNR and numerical examples are given for detectors with amplitude limiters in receiving channels as well as detectors based on polarity coincidence and relay correlators. These data help the reader to make a deliberate choice of the practically most suitable detector scheme. It has been emphasized that passive radar systems can in certain cases offset energy losses of suboptimum detectors by the increase of the signal accumulation factor,* $n=\Delta f_s T$, *since unlike usual active radar systems they utilize energy of radiation sources – not the energy of "own" transmitting stations.*

8. TARGET DETECTION IN ACTIVE MSRSs IN A BACKGROUND OF EXTERNAL NOISE-LIKE SPATIALLY CORRELATED INTERFERENCES

8.1. SYNTHESIS OF OPTIMUM DETECTORS FOR DETERMINISTIC SIGNALS

Noise-like interference from a point-like source (e.g., a jammer) is *strongly correlated at the inputs of spatially separated receiving stations* (after TDOA elimination and Doppler frequency equalization, if necessary). Optimum detection algorithms have to take into account this *spatial correlation of interferences* [60].

We assume in this chapter that *a PSD matrix or a correlation* (*covariance*) *matrix of external interferences plus self-noises at the inputs of all receiving stations are known*. Of course, in real situations these matrices are, as a rule, unknown and may vary in time. Therefore, an adaptive approach is necessary. It will be considered in Sections 9.1 and 9.2. However, the assumption of a known PSD matrix in this chapter is useful from two different standpoints. Firstly, optimum detection algorithms synthesized under this assumption can serve as a basis for constructing corresponding adaptive algorithms. Secondly, performance analysis of those optimum algorithms gives detection characteristics to which corresponding characteristics of good adaptive algorithms should be sufficiently close in a steady-state condition.

As in chapter 5, we begin with the problem of deterministic (completely known) signal detection. Solution of this problem will allow us to reveal some salient features of signal detection in a background of spatially correlated interferences. In this chapter we assume for simplicity the interference from each external source (jammer) to be not only stationary (during the observation time interval T) but stationary coupled (at the receiver inputs) stochastic processes. It means that Doppler frequency differences of interferences from moving sources at the inputs of receiving stations may be neglected.

Under this condition a likelihood ratio logarithm in the frequency domain can be obtained from (4.49) in the form

$$L = \mathrm{Re}\,\frac{1}{2\pi}\int_{-\infty}^{\infty} \mathbf{\Psi}^*(\omega)\mathbf{f}(\omega)\boldsymbol{\chi}(\omega)\,d\omega. \tag{8.1}$$

It is reasonable to remind that $\mathbf{\Psi}^*(\omega)$ and $\boldsymbol{\chi}(\omega)$ are the Fourier transformed vectors of expected target echoes and overall received signals, respectively; $\mathbf{f}(\omega)$ is the inverse of the power PSD matrix $\mathbf{\Phi}(\omega)$ of interferences plus self-noises at the receiving stations' inputs [see (4.46), (4.48)] which is assumed to be known. Thus the synthesis of optimum detectors reduces to specifying the matrices $\mathbf{f}(\omega) = \mathbf{\Phi}^{-1}(\omega)$.

Let a MSRS with m receiving stations (some or all of them may be transmitting–receiving) be subjected to interferences from M independent point-like sources.

Using the signal and interference models defined in 4.1, arbitrary elements of $\mathbf{\Phi}(\omega)$ and $\mathbf{\Psi}(\omega)$ can be written as follows:

$$\Phi_{ik}^{(M)}(\omega)=\sqrt{N_iN_k}\{\delta_{ik}+\sum_{p=1}^{M}q_{pi}q_{pk}g_i(\beta_{pi},\varepsilon_{pi},\omega)g_k^*(\beta_{pk},\varepsilon_{pk},\omega)$$

$$\times F_p(\omega-\omega_p)\exp[\mathrm{j}(\omega\tau_{pik}+\Delta\varphi_{pik})]\},\quad i,k=\overline{1,m} \tag{8.2}$$

$$\Psi_i(\omega)=a_{si}\exp(-\mathrm{j}\varphi_{si})g_i(\beta_{si},\varepsilon_{si},\omega)\Psi_0(\omega-\omega_0)\exp(-\mathrm{j}\omega_{si}t_{si}). \tag{8.3}$$

Here (8.3) is similar to (4.5) and (8.2) is the extension of (4.12) to the case of M interference sources. Besides, (8.2) takes into account receiver self-noises and antenna directivity patterns (ADPs). In (8.2) and (8.3) N_i is the one-sided power spectral density (PSD) of the ith receiving station's self-noise; q_{pi}^2 is the input interference-to-noise ratio (INR) for the pth interference source when it is at the ith station mainlobe pointing direction; $g_i(\beta,\varepsilon,\omega)$ is the normalized ADP of the ith station; $\beta_{si},\varepsilon_{si},\beta_{pi},\varepsilon_{pi}$ are the angular coordinates (azimuth and elevation) of the target and the pth interference source with respect to the ith station; ω_0,ω_p are the carrier frequencies of the target echo and interference from the pth source; $\tau_{pik},\Delta\varphi_{pik}$ are the time difference of arrival (TDOA) and the phase difference of the interference from the pth source at the inputs of the ith and the kth stations; $a_{si},\varphi_{si},t_{si}$ are the r.m.s. value, initial phase and time of arrival (TOA) of the target echo at the ith station input; $\Psi_0(\omega),F_p(\omega)$ are the normalized spectrum of the signal complex envelope and the normalized PSD of the interference complex envelope from the pth source [see (4.4), (4.9), (4.11)]. The superscript (M) in (8.2) denotes simultaneous effect of M interference sources. It is assumed in (8.2) that spatial decorrelation of interferences in the propagation medium may be neglected.

The matrix $\mathbf{\Phi}^{(M)}(\omega)$ with elements (8.2) can be written in the form

$$\mathbf{\Phi}^{(M)}(\omega)=\mathbf{N}+\sum_{p=1}^{M}F_p(\omega-\omega_p)\mathbf{U}_p(\omega)\mathbf{U}_p^*(\omega);\quad \mathbf{\Phi}^{(0)}(\omega)=\mathbf{N} \tag{8.4}$$

where $\mathbf{N}=\mathrm{diag}(N_1,\ldots,N_m)$; $\mathbf{U}_p^*(\omega)=\{q_{p1}\sqrt{N_1}g_1^*(\beta_{p1},\varepsilon_{p1},\omega)\exp[-\mathrm{j}(\omega t_{p1}+\varphi_{p1})],\ldots,q_{pm}\sqrt{N_m}g_m^*(\beta_{pm},\varepsilon_{pm},\omega)\exp[-\mathrm{j}(\omega t_{pm}+\varphi_{pm})]\}$. It was taken into account that as for stochastic signals [see (7.6)]

$$\tau_{pik}=t_{pk}-t_{pi};\qquad \Delta\varphi_{pik}=\varphi_{pk}-\varphi_{pi}. \tag{8.5}$$

Equation (8.4) may be rewritten in the recurrent form

$$\mathbf{\Phi}^{(p)}(\omega)=\mathbf{\Phi}^{(p-1)}(\omega)+F_p(\omega-\omega_p)\mathbf{U}_p(\omega)\mathbf{U}_p^*(\omega). \tag{8.6}$$

Using the known inversion rule for matrices which can be represented in the form of (8.6) [see (7.4)], we can obtain the recurrent equation for the matrix $\mathbf{f}^{(P)}(\omega)$ from the optimum algorithm (8.1)

$$\mathbf{f}^{(p)}(\omega)=\mathbf{f}^{(p-1)}(\omega)-\gamma_p(\omega)F_p(\omega-\omega_p)\mathbf{f}^{(p-1)}(\omega)\mathbf{U}_p(\omega)\mathbf{U}_p^*(\omega)\mathbf{f}^{(p-1)}(\omega);$$

$$\gamma_p(\omega)=[1+F_p(\omega-\omega_p)\mathbf{U}_p^*(\omega)\mathbf{f}^{(p-1)}(\omega)\mathbf{U}_p(\omega)]^{-1},\quad p=\overline{1,M}; \tag{8.7}$$

$$\mathbf{f}^{(0)}(\omega)=\mathbf{N}^{-1}.$$

In accordance with (8.7) an element of the matrix $\mathbf{f}^{(p)}(\omega)$ can be written as follows:

$$f_{ik}^{(p)}(\omega)=f_{ik}^{(p-1)}(\omega)-|H^{(p)}(\omega)|^2 D_i^{(p)}(\omega)D_k^{(p)*}(\omega),$$
$$f_{ik}^{(0)}(\omega)=\delta_{ik}/N_i;\quad i,k=\overline{1,m};\quad p=\overline{1,M} \tag{8.8}$$

where

$$|H^{(p)}(\omega)|^2=F_p(\omega-\omega_p)\{1+F_p(\omega-\omega_p)\sum_{l=1}^{m}\sum_{n=1}^{m}q_{pl}q_{pn}\sqrt{N_lN_n}\,g_l(\beta_{pl},\varepsilon_{pl},\omega)$$
$$\times g_n^*(\beta_{pn},\varepsilon_{pn},\omega)f_{ln}^{(p-1)}(\omega)\exp[-j(\omega\tau_{pln}+\Delta\varphi_{pln})]\}^{-1}; \tag{8.9}$$

$$D_i^{(p)}(\omega)=\sum_{l=1}^{m}q_{pl}\sqrt{N_l}\,g_l(\beta_{pl},\varepsilon_{pl},\omega)f_{il}^{(p-1)}(\omega)\exp[j(\omega\tau_{pls}+\Delta\varphi_{pls})]. \tag{8.10}$$

In (8.8)–(8.10): δ_{ik} is Kronecker's symbol; s is the number of an arbitrary station with respect to which TDOAs and phase differences are determined. Note, that $\Phi_{ik}(\omega)=\Phi_{ki}^*(\omega)$; $f_{ik}(\omega)=f_{ki}^*(\omega)$ and equations (7.8) hold for any pth interference source (if s is replaced by $p=\overline{1,M}$).

Expressions (8.1) and (8.7) or (8.8)–(8.10) determine completely the optimum algorithm for deterministic signal detection in a MSRS with m receiving stations in a background of self-noises plus noise-like external Gaussian interferences from M independent point-like sources. The result of the optimum processing L in (8.1) has to be compared with a threshold selected to obtain prescribed false alarm probability.

Let us substitute scalars for vectors and matrices in (8.1) taking into account M interference sources

$$L=\mathrm{Re}\sum_{i=1}^{m}\frac{1}{2\pi}\int_{-\infty}^{\infty}\Psi_i^*(\omega)\sum_{k=1}^{m}f_{ik}^{(M)}(\omega)\chi_{ik}(\omega)\,d\omega. \tag{8.11}$$

The corresponding schematic diagram is shown in Fig. 8.1. As in a background of spatially uncorrelated interferences (see Section 5.1), the optimum detector represents a linear filter. But there is the mutual interstation processing of overall received signals before linear filtering matched to expected echo in each receiving station. If $M=0$, (8.11) reduces to (5.2), (5.3).

The algorithms (8.1), (8.11) together with (8.8)–(8.10) are optimal for arbitrary positions of receiving stations (antennas) including the case of multichannel antenna systems in monostatic radars (e.g., phased antenna arrays – PAA). However, optimum space–time detectors in monostatic radars may often be simplified if it is possible to separate spatial processing from temporal one [35,42,60]. Then $\Psi_i^*(\omega)$ and $f_{ik}(\omega)$ in (8.11) may be represented as products of two factors. One of them depends only on the numbers $i,k=\overline{1,m}$ of antenna array elements whereas the other depends only on the frequency (or time): $\Psi_i^*(\omega)=\Psi_i^*\Psi^*(\omega)$; $f_{ik}(\omega)=f_{ik}f(\omega)$. Thus, instead of (8.11) we have

$$L=\mathrm{Re}\frac{1}{2\pi}\int_{-\infty}^{\infty}\Psi^*(\omega)f(\omega)\chi(\omega)\,d\omega \tag{8.12}$$

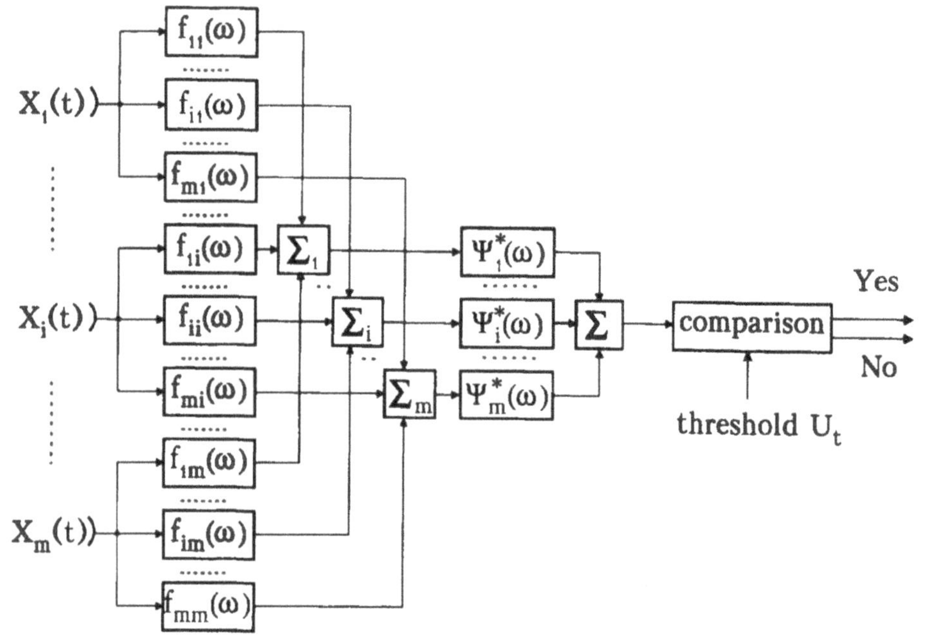

Figure 8.1 Structure of the optimum detector for deterministic signals in a background of spatially correlated interferences (e.g., jammers)

where

$$\chi(\omega) = \sum_{i=1}^{m} \sum_{k=1}^{m} \Psi_i^* f_{ik} \chi_k(\omega). \tag{8.13}$$

It is seen that (8.13) describes the optimum spatial processing while (8.12) determines the optimum temporal processing. From (8.8)–(8.10) and (8.3) the following conditions may be obtained under which the space–time processing separation is possible [60].

1. TDOAs of echoes and interferences from all sources at different elements of a receiving antenna system are substantially less than the reciprocal value of the signal bandwidth: $|\tau_{sik}| \ll 1/\Delta f_s$; $|\tau_{pik}| \ll 1/\Delta f_s$; $i,k=\overline{1,m}$; $p=\overline{1,M}$.
2. ADP of any receiving antenna system's element do not depend upon frequency (within the signal frequency band) in the target and all interference sources directions: $g_i(\beta_{si}, \varepsilon_{si}, \omega) = g_i(\beta_{si}, \varepsilon_{si}, \omega_0) = g_i(\beta_{si}, \varepsilon_{si})$; $g_i(\beta_{pi}, \varepsilon_{pi}, \omega) = g_i(\beta_{pi}, \varepsilon_{pi}, \omega_0) = g_i(\beta_{pi}, \varepsilon_{pi})$, $\omega \in (\omega_0 - \Delta\omega_s/2, \omega_0 + \Delta\omega_s/2)$, $i=\overline{1,m}$, $p=\overline{1,M}$.
3. Normalized PSDs of interferences from all sources are equal to each other and to those of self-noises: $F_p(\omega-\omega_p) = F_n(\omega-\omega_0)$, $p=\overline{1,M}$. In particular, if PSDs of self-noises are constant within the signal frequency band so are PSDs of interferences from all sources: if $F_n(\omega-\omega_0)=1$ then $F_p(\omega-\omega_p)=1$, $\omega \in (\omega_0 - \Delta\omega_s/2, \omega_0 + \Delta\omega_s/2)$, $p=\overline{1,M}$. At high INR this condition may be excluded.

If all the three conditions (or the first two at high INR) are satisfied, the function $f_{ik}^{(M)}(\omega)$ does not depend on frequency ω (at least within the signal frequency band) for all M, that is $f(\omega)=1$ in (8.12).

This can be proved by induction. It follows from (8.8) that in the absence of external interferences $f_{ik}^{(0)}(\omega)=f_{ik}^{(0)}$, i.e. does not depend on frequency. Let now in the presence of $M-1$ interference sources $f_{ik}^{(M-1)}(\omega)=f_{ik}^{(M-1)}$ under conditions 1–3. Consider $f_{ik}^{(M)}(\omega)$ in the presence of M interference sources. In accordance with (8.8)–(8.10)

$$f_{ik}^{(M)}(\omega)=f_{ik}^{(M-1)}-|H^{(M)}|^2 D_i^{(M)} D_k^{(M)*} \tag{8.14}$$

where

$$|H^{(M)}|^2=\left[1+\sum_{l=1}^{m}\sum_{n=1}^{m} q_{Ml}q_{Mn}\sqrt{N_l N_n}\, g_l(\beta_{Ml},\varepsilon_{Ml})\, g_n^*(\beta_{Mn},\varepsilon_{Mn})\, f_{ln}^{(M-1)}\exp(-j\Delta\varphi_{Mln})\right]^{-1};$$

$$D_i^{(M)} D_k^{(M)*}=\sum_{l=1}^{m}\sum_{n=1}^{m} q_{Ml}q_{Mn}\sqrt{N_l N_n}\, g_l(\beta_{Ml},\varepsilon_{Ml})\, g_n^*(\beta_{Mn},\varepsilon_{Mn})\, f_{il}^{(M-1)} f_{kn}^{(M-1)*}\exp(j\Delta\varphi_{Mln}). \tag{8.15}$$

It can be seen that $f_{ik}^{(M)}(\omega)=f_{ik}^{(M)}$. Consequently, independence of frequency keeps validity for arbitrary number of interference source M. In the case of strong interferences (a high INR) a unity in equations for $|H^{(p)}(\omega)|^2, |H^{(p)}|^2$ may be neglected.

Thus, if spatial and temporal processing can be separated, the optimal temporal processing in (8.12) becomes

$$L=\mathrm{Re}\frac{1}{2\pi}\int_{-} \Psi^*(\omega)\chi(\omega)\,d\omega \tag{8.16}$$

where $\chi(\omega)$ is defined in (8.13). In the structure shown in Fig. 8.1 filters with complex transfer functions $f_{ik}(\omega)$ and $\Psi_i^*(\omega)$ have to be replaced by inertialess amplifiers with complex "gain" f_{ik} and Ψ_i^*, respectively (they can be implemented, for example, by two quadrature channels). A common filter with the transfer function $\Psi^*(\omega)$ is included after the second summator. The spatial processing is accomplished in the antenna system and is reduced to the weighted summation of received signals. The temporal processing is carried out in the common receiver channel.

The spatial and temporal processing separation is, as a rule, impossible in MSRSs primarily because the condition 1 is not satisfied. However, in the cases of closely spaced interference sources, small differences of TDOAs may occur for interferences from all sources at any pair of receiving stations (while TDOAs themselves are not small), i.e.

$$|\tau_{pik}-\tau_{1ik}|=|\Delta\tau_{pik}|\ll 1/\Delta f_s,\quad p=\overline{2,M},\ i,k=\overline{1,m}. \tag{8.17}$$

Hence, within the signal bandwidth Δf_s we have $\omega\Delta\tau_{pik}\approx\omega_0\Delta\tau_{pik}$ and these phases may be included into phase terms of the spatial processing algorithm. If the condition (8.17) and the above indicated conditions 2 and 3 (or only the condition 2 at high INR) are satisfied, the optimum detector may be simplified as compared with (8.11) but not so drastically as in the case of the space–time processing separation.

Under those conditions it follows from (8.8)–(8.10) that $f_{ik}^{(1)}(\omega)=f_{ik}^{(1)}\exp(j\omega\tau_{1ik})$. Let us show by induction that a similar expression is valid for $f_{ik}^{(M)}(\omega)$ at any M.

Assume that conditions 2, 3 and (8.17) are satisfied and for $M-1$ interference sources $f_{ik}^{(M-1)}(\omega)=f_{ik}^{(M-1)}\exp(j\omega\tau_{1ik})$, $i,k=\overline{1,m}$. Consider $f_{ik}^{(M)}(\omega)$ for M sources. Substituting $\omega\tau_{pik}\approx\omega\tau_{1ik}+\omega_0\Delta\tau_{pik}$ for $p=M-1$ and $p=M$ in (8.8)–(8.10), including phases $\omega_0\Delta\tau_{pik}$ into phase terms and taking into account that $\tau_{1il}-\tau_{1kn}+\tau_{1ln}=\tau_{1ik}$ we obtain

$$f_{ik}^{(M)}(\omega)=f_{ik}^{(M-1)}\exp(j\omega\tau_{1ik})-|H^{(M)}|^2 D_i^{(M)}D_k^{(M)*}\exp(j\omega\tau_{1ik})=f_{ik}^{(M)}\exp(j\omega\tau_{1ik}). \tag{8.18}$$

It means that the representation $f_{ik}^{(M)}(\omega)=f_{ik}^{(M)}\exp(j\omega\tau_{1ik})$, $i,k=\overline{1,m}$, is valid for arbitrary number of interference sources M.

The optimal processing (8.11) takes the form

$$L=\operatorname{Re}\sum_{i=1}^{m}\frac{1}{2\pi}\int_{-\infty}^{\infty}\Psi_1^*(\omega)\sum_{k=1}^{m}f_{ik}^{(M)}\exp(j\omega_{1ik})\chi_k(\omega)\,d\omega. \tag{8.19}$$

The series connection of simple inertialess amplifiers with the $f_{ik}^{(M)}$ and delay lines with the delays $\tau_{\max}-\tau_{1ik}$ have to be substituted in the diagram of Fig. 8.1 for complicated filters with the transfer functions $f_{ik}^{(M)}(\omega)$. Here $\tau_{\max}$ is an arbitrary time delay (common for all channels) exceeding τ_{1ik} for all $i,k=\overline{1,m}$. The role played by this delay is to exclude "negative" delays.

As can be seen from (8.8)–(8.10), the complexity of filters $f_{ik}^{(M)}(\omega)$ grows rapidly with the increase of the number of interference sources M. The key cause is that in the general case substantially different TDOA values of interferences from all sources have to be eliminated simultaneously at the input of each summator. Under the condition (8.17) these TDOA values are practically equal for any pair of receiving stations. Therefore they can easily be eliminated by only one delay value.

In this section, *we have considered the general problem of optimum signal detection by a MSRS with m receiving stations when that signal is embedded not only in stations' self-noises but in external interferences from M independent point-like sources (e.g., jammers). Both receiving stations and interference sources are arbitrarily positioned in space. We have synthesized an optimum processing algorithm* [see (8.1), (8.11) *together with* (8.8)–(8.10) *and Fig.* 8.1] *for a known PSD matrix of interferences plus self-noises at the inputs of stations and for an idealized signal model – the deterministic signal. This algorithm, however, permits to reveal a common structure and some salient features of optimum detection algorithms for more realistic signal models. We have shown that the optimum algorithm includes at each station the joint processing (a certain linear filtration) of interferences received by all stations and then matched filtration of echoes expected at the input of this station.*

Since the synthesized algorithm can be applied to systems with arbitrary spatial separation between stations, antennas or antenna elements including PAAs of monostatic radars, we have formulated conditions under which this algorithm can be drastically simplified [see (8.13) *and* (8.12), (8.16)]. *These are conditions of spatial and temporal processing separation. As a rule, they are not satisfied in MSRSs but we have considered the case where significant processing simplification is possible in MSRSs* [see (8.19)]. *This is the case where interference sources are close to each other in space so that TDOAs of different interferences at the inputs of stations can be eliminated by a common value of delay.*

8.2. PERFORMANCE ANALYSIS OF OPTIMUM DETECTORS FOR DETERMINISTIC SIGNALS. EXTERNAL INTERFERENCE CANCELLATION

The most important feature of the optimum detection algorithm synthesized in Section 8.1 [see (8.1), (8.11) together with (8.8)–(8.10)] is that it includes cancellation (coherent mutual suppression) of spatially correlated external interferences (e.g., jamming). This can be shown without specifying the characteristics of MSRSs or interference sources [60].

Let us modify (8.11) separating a "main" channel in each receiving station

$$L = \operatorname{Re} \sum_{i=1}^{m} \frac{1}{2\pi} \int_{-\infty}^{\infty} \Psi_i^*(\omega) f_{ii}(\omega) \xi_i(\omega)\, d\omega \tag{8.20}$$

where

$$\xi_i(\omega) = \chi_i(\omega) + \sum_{k=1,\, k\neq i}^{m} \frac{f_{ik}(\omega)}{f_{ii}(\omega)} \chi_k(\omega). \tag{8.21}$$

A schematic diagram corresponding to (8.20) and (8.21) is shown in Fig. 8.2. The linear filters in the left section of the diagram inside the dashed lines provide minimum interference plus self-noise power at the output of each receiving station's summator. To prove this, let us substitute arbitrary transfer functions $K_{ik}(\omega)$ for the functions $f_{ik}(\omega)/f_{ii}(\omega)$ in (8.21) and derive the expression for $K_{ik}(\omega)$ under the condition of minimum interference plus self-noise variance (power) at the output

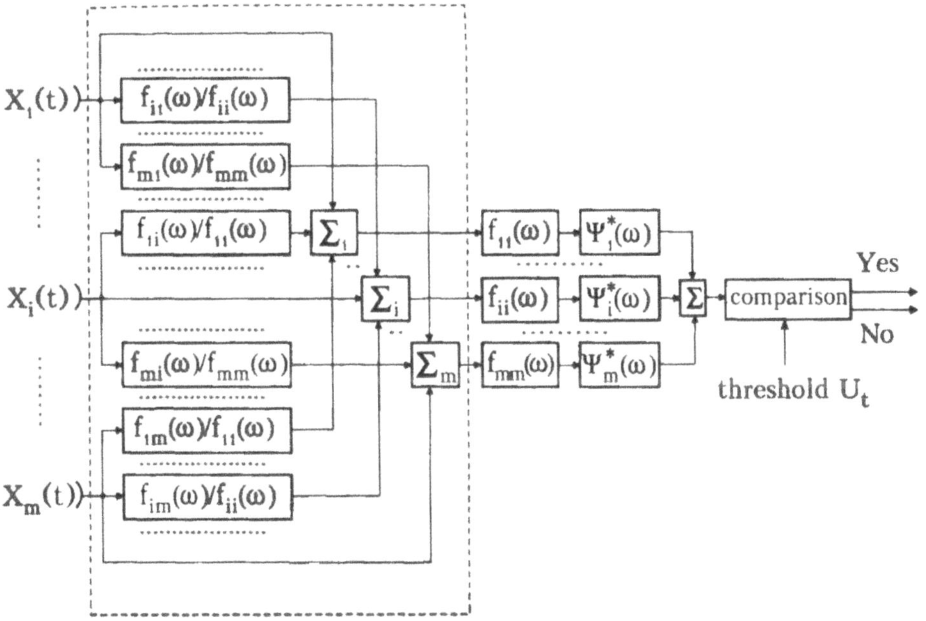

Figure 8.2 The same detector as in Fig. 8.1 after transformation according to (8.20) and (8.21)

of the ith summator ($i,k=\overline{1,m}$). Solving the corresponding variational problem yields

$$\mathbf{K}_i(\omega)=-[\tilde{\mathbf{\Phi}}^{(i)t}(\omega)]^{-1}\tilde{\mathbf{\Phi}}_i(\omega)=\tilde{\mathbf{f}}_i(\omega)/f_{ii}(\omega), \tag{8.22}$$

that is

$$K_{ik}(\omega)=f_{ik}(\omega)/f_{ii}(\omega), \quad i,k=\overline{1,m}. \tag{8.23}$$

In (8.22) $\tilde{\mathbf{\Phi}}^{(i)}(\omega)$ is the interference plus self-noise PSD matrix at the inputs of all receiving stations except the ith station [this matrix may be derived from the matrix $\mathbf{\Phi}(\omega)$ (8.4) by deleting the ith row and the ith column]; $\tilde{\mathbf{\Phi}}_i(\omega)$ is the column vector of the interference plus self-noise mutual PSDs at the inputs of the ith and other stations [this is the ith column without the ith element of the matrix $\mathbf{\Phi}^t(\omega)$]; $\tilde{\mathbf{f}}_i(\omega)$ is the ith column without the ith element of the matrix $\mathbf{f}(\omega)=\mathbf{\Phi}^{-1}(\omega)$; The superscript "t" denotes transposition as before.

The left identity of (8.22) is the known Wiener–Hopf equation solution (e.g., [35]). Hence the multichannel filter with transfer functions of (8.23) is the Wiener filter. It provides the best r.m.s. approximation (with the opposite sign) of the sum of interferences and self-noises from the inputs of all stations, except the ith one, to the ith station's input interferences plus self-noise. Adding this sum to the input interferences plus self-noise of the ith station means subtracting from them their best r.m.s. approximation. The result is a possible minimum of the total interference plus self-noise variance (power) at the ith summator output. The fact, that the right identity of (8.22), i.e. (8.23), is valid as well, proves that the optimum detector has the property to cancel spatially correlated interferences (e.g., jamming). It should be emphasized, that the interference cancellation does not depend on wanted signals (target echoes).

The linear filtering of signals embedded in self-noise and interference residues is accomplished in the right section of Fig. 8.2. This filtering is optimum according to the maximum signal-to-interferences-plus-self-noise ratio (SINR) criterion. The PSD of the interference and self-noise sum at the ith summator output (after interference cancellation) can be obtained from (8.21):

$$\Phi_{\Sigma i}(\omega)=\lim_{T\to\infty}\frac{1}{2T}\overline{\xi_i(\omega)\xi_i^*(\omega)}=1/f_{ii}(\omega). \tag{8.24}$$

The echo amplitude spectrum at the same point in Fig. 8.2

$$\Psi_{\Sigma i}(\omega)=\Psi_i(\omega)+[1/f_{ii}(\omega)]\sum_{k=1,k\neq i}^{m} f_{ik}(\omega)\Psi_k(\omega). \tag{8.25}$$

It follows from (8.24), (8.25) that a filter with the transfer function $f_{ii}(\omega)\Psi_i^*(\omega)$ really provides maximum SINR for a target echo expected at the ith station. Echoes received by other stations are not taken into account. However, those echoes [$\Psi_k(\omega)$, $k\neq i$, see (8.25)] come to the ith summator too ($i=\overline{1,m}$). A result is that the wanted signals $\Psi_k(\omega)$ and $\Psi_i(\omega)$ interfere when they overlap in time of arrival (TOA) at the output of the filter with the transfer function $f_{ii}(\omega)\Psi_i^*(\omega)$. It may occur, for instance, when a target and a jammer are closely spaced. Such signal interfering may lead to subtraction of overlapping signals, i.e. to the signal mutual suppression together with the external interference (jamming) cancellation. Nevertheless, the algorithm (8.20),

(8.21) is optimal, i.e. provides the best possible signal detection (within the frames of assumed models and optimality criteria).

A capability to cancel spatially correlated interferences from external sources (e.g., jammers) is a fundamental and practically important feature of the optimum detectors in MSRSs. As will be shown below, this feature permits in certain cases to solve one of the most difficult problem: *cancelling out mainlobe jamming without target echo suppression.*

However, effective cancellation is not always possible. We define the effective cancellation as perfect (to zero) suppression of spatially correlated interferences from all external sources with the unlimited increase of INR: $\Phi_{\Sigma i}(\omega) \to 0$ in (8.24) when $q_{pi}^2 \to \infty$, $p=\overline{1,M}$, $i=\overline{1,m}$. In view of the identity $\mathbf{f}(\omega)=\mathbf{\Phi}^{-1}(\omega)$ we have $f_{ii}(\omega)=\operatorname{adj}\Phi_{ii}(\omega)/\det\mathbf{\Phi}(\omega)$ where $\operatorname{adj}\Phi_{ii}(\omega)$ is the adjunction to the element $\Phi_{ii}(\omega)$ of the matrix $\mathbf{\Phi}(\omega)$ and $\det\mathbf{\Phi}(\omega)$ denotes the determinant of $\mathbf{\Phi}(\omega)$. For satisfying the condition $\Phi_{\Sigma i}(\omega) \to 0$ at the output of each summator ($i=\overline{1,m}$) it is necessary to have $\det\mathbf{\Phi}(\omega) \to 0$. As it is seen from (8.4), when $q_{pi}^2 \to \infty$, i.e. $N_i \to 0$, $i=\overline{1,m}$, the matrix $\mathbf{\Phi}^{(M)}(\omega)$ tends to be a sum of M matrices, each of rank one. It is known [34] that the rank of such a matrix is equal to M. Therefore, $\det\mathbf{\Phi}(\omega) \to 0$ as $\mathbf{N} \to \mathbf{0}$ only if the dimension of $\mathbf{\Phi}^{(M)}(\omega)$, which is equal to the number of receiving stations of a MSRS, m, exceeds the rank of $\mathbf{\Phi}^{(M)}(\omega)$, i.e. M. Thus, the effective cancellation of external interferences from M arbitrary positioned in space independent sources (e.g., jammers) is possible only under the constraint $m \geqslant M+1$.

For multichannel antenna systems of monostatic radars (e.g., PAAs) with separate spatial and temporal processing it is well known the similar constraint for the effective interference cancellation by the spatial processing: $m \geqslant M+1$. Here, m is the number of receiving antenna channels (e.g., antenna array elements) [35]. So the result derived above extends that constraint to the general space–time processing.

In the case of a MSRS the baselengths between all stations are not necessary to be of the same order of magnitude. But it should be kept in mind that the spatial resolution capability of a MSRS depends on its baselengths.

As it was shown, the optimum detector for deterministic signals is a linear spatial–temporal filter. Owing to its linearity, the output probability distribution is Gaussian since probability distributions of input interferences and self-noises are assumed to be Gaussian too. When a target echo comes, it does not change the variance of output probability distribution, but a nonzero mean appears. As in the case of spatially uncorrelated interferences, detection characteristics may be calculated using (5.14) and are determined by the output SINR. From (8.11) at $\chi_k=\Psi_k$ we find the output signal component[1]

$$L_s=\sum_{i=1}^{m}\sum_{k=1}^{m}\frac{1}{2\pi}\int_{-\infty}^{\infty}\Psi_i^*(\omega)f_{ik}(\omega)\Psi_k(\omega)\,d\omega. \tag{8.26}$$

The output variance of interferences plus self-noises $M_2[L_n]=\overline{L_n^2}$ can be obtained from (8.11) too, assuming that $\chi_k(\omega)$ does not contain a target echo and taking into account that $\overline{\chi_k(\omega_1)\chi_n^*(\omega_2)}=4\pi\delta(\omega_1-\omega_2)\Phi_{kn}(\omega_2)$ [see (7.23)]. After some straightforward manipulations we find that $M_2[L_n]=\overline{L_n^2}=L_s$ and

$$q_{out}^2=L_s^2/M_2[L_n]=\sum_{i=1}^{m}\sum_{k=1}^{m}\frac{1}{2\pi}\int_{-}\Psi_i^*(\omega)f_{ik}(\omega)\Psi_k(\omega)\,d\omega. \tag{8.27}$$

[1] The symbol "Re" is omitted since $f_{ik}(\omega)=f_{ki}^*(\omega)$ and hence the right part of (8.26) is a real quantity.

Using (8.27), (8.3), (8.8)–(8.10) and (5.14) one may calculate q^2_{out} and detection characteristics for any particular situation.

For signal detection in a background of external interferences (e.g., jamming) *the detector spatial resolution capability* is of prime importance. It may be characterized by the dependence of output SINR (or of detection probability given false alarm probability) on the distance between a target and an interference source. We confine ourselves to one simple example for revealing some important features of the optimum detector spatial resolution capability (other examples using more realistic signal models see in Sections 8.4, 8.5).

Example 8.1. Let us consider a MSRS with one transmitting and two receiving stations (one of the latter may coincide with the transmitting station, i.e. may be a monostatic radar). The MSRS is subjected to a single source of interference (for instance, to a jammer). We assume for simplicity that in (8.2) (at $M=1$) and (8.3) ADP does not depend on frequency over the angles of interest and besides that

$$a_{s1}=a_{s2}=a_s; \qquad N_1=N_2=N; \qquad q_{11}^2=q_{12}^2=q^2; \qquad \omega_1=\omega_0; \tag{8.28}$$

$$\begin{aligned} &g_1(\beta_{s1},\varepsilon_{s1})=g_2(\beta_{s2},\varepsilon_{s2})=g(\beta_s,\varepsilon_s);\\ &g_1(\beta_{11},\varepsilon_{11})=g_2(\beta_{12},\varepsilon_{12})=g(\beta,\varepsilon); \end{aligned} \tag{8.29}$$

$$|\Psi_0(\omega-\omega_0)|^2=\begin{cases}2T_s/\Delta f_s, & |\omega-\omega_0|\leqslant\pi\Delta f_s,\\ 0, & |\omega-\omega_0|>\pi\Delta f_s;\end{cases}$$

$$F_1(\omega-\omega_0)=\begin{cases}1, & |\omega-\omega_0|\leqslant\pi\Delta f_s,\\ 0, & |\omega-\omega_0|>\pi\Delta f_s.\end{cases} \tag{8.30}$$

Then we obtain from (8.8)–(8.10) at $M=1$ omitting the subscript 1 on τ_{1ik} and $\Delta\varphi_{1ik}$

$$f_{ik}^{(1)}(\omega)=\frac{1}{N}\left[\delta_{ik}-\frac{q^2|g(\beta,\varepsilon)|^2}{1+2q^2|g(\beta,\varepsilon)|^2}\right]\exp(j\omega\tau_{ik}+\Delta\varphi_{ik}),\quad i,k=1,2. \tag{8.31}$$

Substituting (8.31) and (8.30) into (8.27), taking into account (8.28), (8.29) yields

$$q_{out}^2=\frac{4E|g(\beta_s,\varepsilon_s)|^2}{N[1+2q^2|g(\beta,\varepsilon)|^2]}\{1+q^2|g(\beta,\varepsilon)|^2[1-\mathrm{sinc}(\pi\Delta f_s\delta\tau_{12})\cos(\omega_0\delta\tau_{12}+\delta\varphi_{12})]\} \tag{8.32}$$

where $E=a_s^2T_s$ is the signal (echo) energy at each receiving station when $g(\beta_s,\varepsilon_s)=1$; $\mathrm{sinc}(x)=(\sin x)/x$; $\delta\tau_{12}=\tau_{12}-(t_{s2}-t_{s1})=\tau_{12}-\tau_{s12}$; $\delta\varphi_{12}=\Delta\varphi_{12}-(\varphi_{s2}-\varphi_{s1})=\Delta\varphi_{12}-\Delta\varphi_{s12}$ are the differences in TDOAs and phase differences between the interference and the target echoes at the receiving station inputs.

A typical plot of $q^2_{out}/(4E/N)$ as a function of the dimensionless argument $\Delta f_s|\delta\tau_{12}|$ calculated from (8.32) is presented in Fig. 8.3. It is assumed that $\delta\varphi_{12}=0$ and a target lies in the maximum antenna pointing direction $[g(\beta_s,\varepsilon_s)=1]$. The value $4E/N$ is the signal-to-noise ratio [SNR, see (5.13)] under the same conditions but in the absence of spatially correlated external interference ($q^2=0$). Thus the ratio

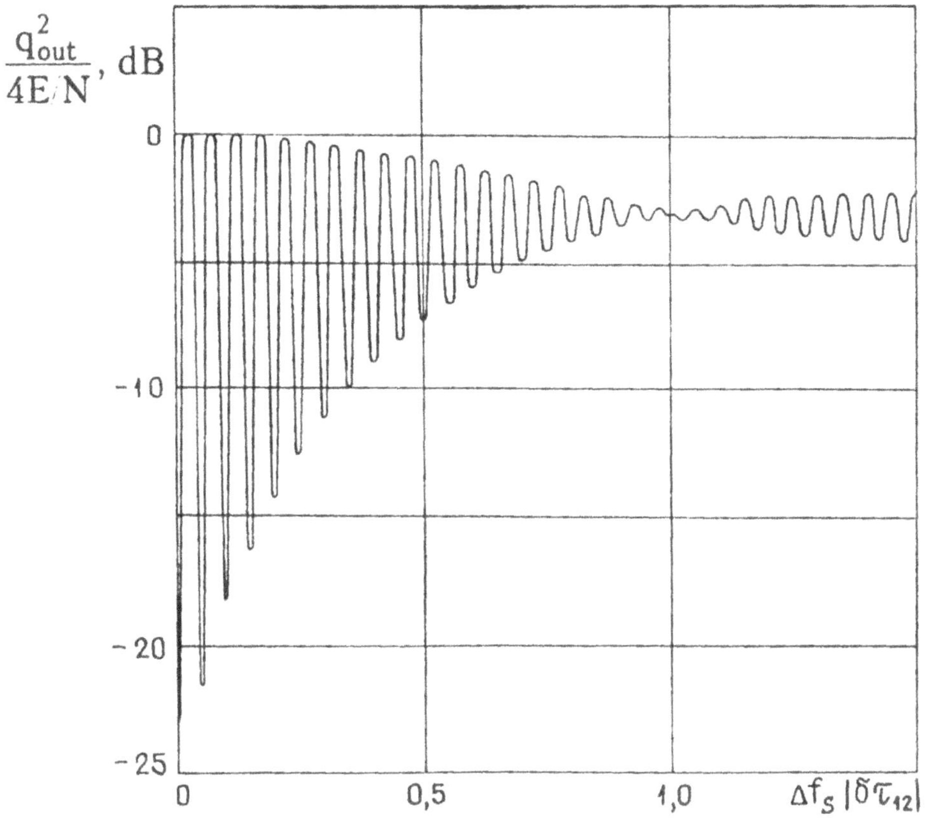

Figure 8.3 Dependence of the output (normalized) SINR on the displacement in TDOA of an interference source with respect to the target [see (8.32)]: $q^2 = 20\,\text{dB}$, $\Delta f/f_0 = 5\%$

$q^2_{out}/(4E/N)$ shows energy losses in comparison with the case where self-noises only are at the inputs of both stations. The curve in Fig. 8.3 may be called "the Resultant Signal Reception Pattern (RSRP)" of a MSRS. The argument is proportional to the shift in TDOA of an interference with respect to that of a target echo. If ranges of both the target and the interference source are greater by several times than the effective baselength L_{eff} between receiving stations, this shift in TDOA primarily depends on target-interference source angular displacement in the baseline plane. In this case the RSRP becomes the "Resultant Directivity Pattern (RDP)" of a MSRS which is similar to the antenna directivity pattern (ADP) of a monostatic radar. It is seen that the RSRP oscillates fast with the increments of $\delta\tau_{12}$. When a target and an interference source coincide in space, the target echo is suppressed together with the interference so that $q^2_{out} = 4E/N(1+2q^2)$. However, a slight shift in the interference source TDOA ($|\delta\tau_{12}| = \pi/\omega_0 = 1/2f_0$) yields $q^2_{out} = 4E/N$, i.e. the coherent echo summation from both stations at the perfect interference cancellation. Such shift in TDOA corresponds to the target-interference source angular shift in the baseline plane of the order of $\lambda/2L_{eff}$ where λ is the wavelength. Usually L_{eff} is by several orders of magnitude greater than a receiving station antenna linear size, L_A. Therefore, $\lambda/2L_{eff}$ is, as a rule, a very small fraction of antenna mainlobe beamwidth. For instance, if $\lambda = 0.1\,\text{m}$, $L_{eff} = 10\,\text{km}$ and $L_A = 10\,\text{m}$, the coherent echo summation with the perfect interference cancellation takes place when the angular distance between a target and an interference source is of the order of 10^{-5} rad, i.e. 2″. At the range $R \approx 400$ km the corresponding linear cross-range distance is of the order of 4 m.

Thus, the spatial (angular) resolution capability of the optimum detector is very high. In fact, for the differences in TDOAs of an echo and an interference $|\delta\tau_{12}| = 1/f_0, |\delta\tau_{12}| = 2/f_0$ and so on the echo is suppressed again together with the interference. However, as it often happens, a target moves relative to an interference source, so the target echo falls within different segments of RSRP. Such a target can be detected quite consistent during several target illuminations. As $|\delta\tau_{12}|$ increases further, oscillations in Fig. 8.3 decay. When $|\delta\tau_{12}| > 1/\Delta f_s$ the ratio $q_{out}^2/(4E/N)$ approaches $(1+q^2)/(1+2q^2)$. When $q^2 \gg 1$ then $q_{out}^2 \to 2E/N$, i.e. q_{out}^2 approaches the SNR for a monostatic radar in the absence of external interference. In this case signals (target echoes) received by both stations are resolved in TOA (after interference suppression) at the outputs of the matched filters (see Figs. 8.1 and 8.2). Therefore, those signals can be processed independently in a common receiver channel. On the contrary, when $|\delta\tau_{12}| \ll 1/\Delta f_s$, the signals almost coincide in time and interfere at the matched filter outputs. That is why the RSRP has the oscillating character in the region of signal interference.

If the range of both a target and an external interference source (e.g., a jammer) is sufficiently large ($R \gg L_{eff}$), the width of a minimum in RDP in an interference source direction is (by level of -3 dB)

$$\Delta\Theta_0 \approx \lambda/2L_{eff} \tag{8.33}$$

and the width of the signal interference region (where the oscillation swing is more than 6 dB) is

$$\Delta\Theta \approx (\lambda/L_{eff})(f_0/\Delta f_s) = 2(f_0/\Delta f_s)\Delta\Theta_0. \tag{8.34}$$

In Fig. 8.3 at $\Delta f_s = 0.05 f_0$ and $L_{eff}/L_A = 10^3$ the role of the receiving station's ADP is not yet revealed for $\Delta f_s |\delta\tau_{12}| \leqslant 1.5$. The signal interference region occupies a small fraction (nearly 0.02) of the antenna beamwidth. However, if, for instance, $\Delta f_s = 0.001 f_0$ and $L_{eff}/L_A = 5 \cdot 10^2$, the signal interference region exceeds the antenna beamwidth. In this case calculations using (8.32) require to take into account the factors $|g(\beta, \varepsilon)|^2$.

It follows from (8.8)–(8.10) that as the number of interference sources (e.g., jammers) M increases, the complexity of expressions for $f_{ik}^{(M)}(\omega)$ and consequently for q_{out}^2 (8.27) is growing up (except for the special cases, which have been considered in Section 8.1). However, in practice only few interference sources (jammers) should be taken into account for the synthesis and analysis of optimum (and suboptimum) detectors for MSRSs. Even when a MSRS is subjected to interferences from many sources, most of them can be cancelled out in each receiving station individually by using known sidelobe interference (jamming) cancellation techniques (see, e.g., [35,72,73]). The multistatic detection optimization should account only for spatially correlated interferences entering mainlobe of ADP [24], when the antenna angular resolution capability is not sufficient for interference cancellation without target echo suppression. In these cases calculating $f_{ik}^{(M)}(\omega)$ using (8.8)–(8.10) is significantly less cumbersome than by the direct (numerical) inversion of the matrix $\mathbf{\Phi}^{(M)}(\omega)$.

***In this section**, we have revealed one of the most important feature of optimum MSRS' signal detectors: a capability to cancel external interferences correlated at the inputs* *of spatially separated stations. Such spatial correlation (after delay and, if necessary, Doppler shift equalization) is typical for interferences from point-like sources, e.g., jammers. As it has been shown (without specifying characteristics of the MSRS and interference sources), interference cancellation (mutual coherent suppression) is accomplished with joint linear filtering of interferences received by all stations.*

We have obtained a necessary condition of the effective interference (i.e. jamming) cancellation from M independent sources in a MSRS with m receiving stations: $m \geqslant M+1$. *It has been emphasized, however, that such joint processing in MSRSs with interference cancellation is reasonable to use only against mainlobe interference sources when the angular resolution capability of each receiving station is not sufficient for external interference cancellation without target echo suppression. The expression* (8.27) *together with* (8.3) *and* (8.8)–(8.10) *allows us to calculate output SINR and detection characteristics for any specific situation of the deterministic signal detection by a MSRS.*

A simple example 8.1 *of a MSRS with two receiving stations* ($m=2$) *and a single interference source* ($M=1$) *has been considered in detail. The output SINR has been plotted in Fig.* 8.3 *as a function of the target and the interference source mutual displacement in space. It can be seen from Fig.* 8.3 *that a moving target can be detected with confidence even when it is in the vicinity of an interference source.*

8.3. DETECTORS FOR FLUCTUATING SIGNALS

In Sections 8.1 and 8.2, we have not specified the type of MSRSs, since when we use an idealized model of the deterministic signal, the spatial coherence of MSRSs is assumed. Moving on to more realistic signal models, namely fluctuating signals, it is important to preserve the main feature of synthesized detectors: a capability to cancel spatially correlated external interferences. This cancellation is possible not only in spatially coherent MSRSs but in MSRSs with short-term spatial coherence as well (see Section 1.1). Therefore here and in the sequel (Sections 8.3–8.5) we shall consider MSRSs of these two types.

To synthesize optimum detection algorithms [65] let us employ the expression (8.3) representing a wanted signal at the input of the ith station ($i=\overline{1,m}$) where the effective (r.m.s.) value a_{si} and the initial phase φ_{si} are assumed to be random or unknown variables.

We use (5.29) as an initial expression for the synthesis. This is an averaged likelihood ratio in the frequency domain. The vector $\mathbf{a}_s$ and the diagonal matrix $\tilde{\mathbf{E}}(\varphi_s)$ are defined by (5.26), the vector $\mathbf{G}$ and the matrix $\mathbf{C}$ are defined by (5.28). The diagonal matrix $\tilde{\mathbf{\Psi}}(\omega)$ is defined by (5.25). In Section 5.2 ADPs of receiving antennas have not been taken into account in signal spectra $\Psi_i(\omega)$ but here we may consider these factors to be included into each element of the matrix $\tilde{\mathbf{\Psi}}(\omega)$. Then arbitrary elements of $\mathbf{G}$ and $\mathbf{C}$ take the form ($i,k=\overline{1,m}$)

$$G_i=\frac{1}{2\pi}\int_{-\infty}^{\infty} g_i^*(\beta_{si},\varepsilon_{si},\omega)\Psi_0^*(\omega-\omega_0)\exp(j\omega t_{si})\sum_{k=1}^{m} f_{ik}(\omega)\chi_k(\omega)\,d\omega$$

$$C_{ik}=\frac{1}{2\pi}\int_{-\infty}^{\infty} g_i^*(\beta_{si},\varepsilon_{si},\omega)g_k(\beta_{sk},\varepsilon_{sk},\omega)|\Psi_0(\omega-\omega_0)|^2$$

$$\times\exp(-j\omega\tau_{sik})f_{ik}(\omega)\,d\omega. \tag{8.35}$$

Comparing G_i in (8.35) with (8.11) and taking into consideration Fig. 8.1, we can note that G_i are the outputs of the matched filters. They differ from the input signals of the last summator in Fig. 8.1 by the absence of the complex amplitudes $a_{si}\exp(j\varphi_{si})$ only. Hence each G_i is the result of the joint coherent space–time processing (linear filtration) of signals coming to the ith station from the inputs of all the m stations. As has been shown in Section 8.2, such processing results in cancellation of spatially correlated external interferences (e.g., jamming) and extraction the wanted signal which is expected at the ith station. We have noted in Section 8.2 that external interference cancellation does not depend on wanted signals at all. Therefore, the random character of the complex amplitudes $a_{si}\exp(j\varphi_{si})$, $i=\overline{1,m}$, have no influence on the interference cancellation. Specific detection algorithms depend on the features of a MSRS and wanted signal fluctuations at the inputs of receiving stations.

Spatially Coherent Signals with Completely Mutually Dependent Amplitude Fluctuations in Spatially Coherent MSRSs

In this case equations (5.33) for random phases $\boldsymbol{\varphi}_s^t=(\varphi_{s1},\ldots,\varphi_{sm})$ and (5.34), (5.35) for real amplitudes $\mathbf{a}_s^t=(a_{s1},\ldots,a_{sm})$ are valid. Substituting probability density functions (PDFs) $w(\boldsymbol{\varphi}_s)$ from (5.33) and $w(\mathbf{a}_s)$ from (5.34) into (5.29) yields

$$\bar{\Lambda}=\int_0^{\infty}\int_{-\pi}^{\pi} w(a_{s1})w(\varphi_{s1})\exp\{a_{s1}\mathrm{Re}[\exp(j\varphi_{s1})\mathbf{A}^t\mathbf{E}^*\mathbf{G}] \\ -0.5a_{s1}^2\mathbf{A}^t\mathbf{E}^*\mathbf{CEA}\}d\varphi_{s1}\,da_{s1}. \tag{8.36}$$

where $\mathbf{A}$ and $\mathbf{E}$ are defined in (5.35), (5.37). When complex amplitudes are Gaussian random variables, then φ_{s1} and a_{s1} are uniformly [within $(-\pi,\pi)$] and Rayleigh distributed, respectively. Now we can substitute (5.38) in (8.36), omit terms which do not depend on $\mathbf{G}$ (i.e. on received signals), take a logarithm of $\bar{\Lambda}$ and replace the squared modulus by the modulus itself (since this is a monotonous transformation). As a result we have an optimum processing algorithm in the form

$$L_2=|\mathbf{A}^t\mathbf{E}^*\mathbf{G}|=\left|\sum_{i=1}^{m} A_{i1}\exp(-j\Delta\varphi_{si1})G_i\right|. \tag{8.37}$$

The decision variable L_2 is to be compared with a predetermined threshold. Under the additional condition of weak output signals (when $a_{s1}^2|\mathbf{A}^t\mathbf{E}^*\mathbf{G}|^2/16\ll 1$ with high probability) $I_0(z)\approx 1+z^2/4$ and the algorithm (8.37) is optimal for any PDF $w(a_{s1})$ with limited second order moments.

When PDFs $w(\varphi_{s1})$ and $w(a_{s1})$ are not known, or φ_{s1} and a_{s1} are nonrandom (though unknown) variables, an adaptive detection algorithm can be derived as in Section 5.2. This algorithm is optimal according to the generalized likelihood ratio criterion. A logarithm of the conditional likelihood ratio at fixed values of φ_{s1} and a_{s1} can be written in the form [see (8.36)]

$$\ln\Lambda=a_{s1}\,\mathrm{Re}[\exp(j\varphi_{s1})\mathbf{A}^t\mathbf{E}^*\mathbf{G}]-0.5a_{s1}^2\mathbf{A}^t\mathbf{E}^*\mathbf{CEA}. \tag{8.38}$$

Solving the equations $\partial\ln\Lambda/\partial\varphi_{s1}=0$ and $\partial\ln\Lambda/\partial a_{s1}=0$ we find the maximum likelihood estimates

$$\hat{\varphi}_{s1}=-\arg(\mathbf{A}^t\mathbf{E}^*\mathbf{G});\qquad \hat{a}_{s1}=|\mathbf{A}^t\mathbf{E}^*\mathbf{G}|/\mathbf{A}^t\mathbf{E}^*\mathbf{CEA}. \tag{8.39}$$

Substituting (8.39) in (8.38), omitting the denominator (independent of **G**) and replacing the squared modulus by the modulus itself yield L_2 (8.37) again. Thus, as in the absence of spatially correlated external interferences, the optimum adaptive algorithm does not differ from the optimum algorithm averaged over random Gaussian signal complex amplitudes.

A comparison of (8.37) with (8.11) [taking into account (8.3) and (8.35)] shows that the random character of φ_{s1} and a_{s1} leads, as in Section 5.2, to the replacement of the real part of a phased and weighted sum of matched filter outputs by the modulus of this sum which can be obtained, for instance, by the envelope detection procedure. On the other hand, comparing (8.37) with (5.40) and (5.41) permits to conclude that the presence of spatially correlated interferences has an effect only on the quantities G_i, $i=\overline{1,m}$. The independent in each station temporal filtration matched to the target echo expected at this station, is replaced by the joint spatial–temporal filtration of signals received by all stations [compare (8.35) with (5.31)].

Spatially Coherent Signals with Completely Mutually Dependent Amplitude Fluctuations in MSRSs with Short-Term Spatial Coherence

Random and mutually independent (at different stations) phase shifts are added to input signal and external interference phases in receiver facilities of such MSRSs (see Section 1.1). The phase shifts may assumed to be uniformly distributed within the interval $(-\pi, \pi)$. Due to the condition of short-term spatial coherence, these phase shifts may be considered as constant values during each observation interval T but random and mutually independent variables in different observation intervals.

For spatially coherent MSRSs, we used (5.29) as an initial equation for the optimum detector synthesis. The equation (5.29) has been derived under the condition that the interference correlation matrix is a deterministic one. However, in MSRSs with short-term spatial coherence this matrix becomes quasideterministic because of random phase shifts in receivers (see Section 4.1).

Nevertheless, if those phase shifts are constant during a sufficiently large time interval T (as compared to the reciprocal value of the wanted signal bandwidth Δf_s), one can employ algorithms adapting to unknown interference phase shifts at the beginning of each observation interval T. To make it possible in the general case, the time duration of the spatial coherence of a MSRS is to be much greater than the duration of adaptation processes. Then an adaptive algorithm can "track" slow phase and gain (amplitude) fluctuations in receiver equipment. Let us assume that this condition is satisfied so that interference phase shifts are eliminated (before mutual interference coherent suppression) at each observation interval by adaptation to interferences. Then, taking into account that we are interested in the role played by random signal parameters, we may consider (5.29) to be valid, i.e. synthesize optimum target echo detection algorithms as for the deterministic interference correlation matrix[2].

If interference phase shifts caused by receiver equipment are eliminated in the interference adaptation process, the same phase shifts of received target echoes are eliminated too (at least phase shifts in common parts of receiver channels for target echoes and interferences). However, even for a single interference source (e.g., a jammer) after interference phase elimination, an unknown signal phase difference

[2] Adaptive cancellation (coherent mutual suppression) of spatially correlated external interferences is considered in Chapter 9.

remains between any pair of receiving stations, which is determined by relative positions in space of the target and the interference source. The situation becomes more complex in the case of several interference sources. Besides, there can be additional unknown (random) target echo phase shifts caused by phase differences of signals scattered by the target in the directions of spatially separated stations, by different contribution of propagation medium and so on. Thus to use an external interference as a "pilot signal" for wanted signal equipment phasing of a MSRS is, as a rule, impossible. The most realistic signal model for MSRSs with short-term spatial coherence is the model with mutually independent (at different stations) initial phases of target echoes in each observation time interval T. It means that spatially coherent wanted signals at the inputs of stations become spatially incoherent in receivers. At the same time amplitude fluctuations can remain completely mutually dependent.

So let the PDFs of signal effective values a_{si} be determined by (5.34) where $w(a_{s1})$ is not necessary to be specified. Signal initial phases φ_{si} are assumed to be mutually independent and uniformly distributed within the interval $(-\pi, \pi)$ [see (5.46)]. As in Section 5.2, the synthesis of optimum detection algorithms turns out to be successful under the condition of weak signals. In this case we can expand the exponential function in (5.29) into a Taylor series retaining terms up to second order. Integrating first over $\varphi_{s1}, \ldots, \varphi_{sm}$ and then over a_{s1} [taking into account (5.46) and (5.34)], dropping all terms independent of received signals (i.e. of G_i, $i=\overline{1,m}$) yields

$$L_3 = \sum_{i=1}^{m} A_{i1}^2 |G_i|^2. \tag{8.40}$$

This optimum processing algorithm (8.40) differs from the similar algorithm (5.48) for the case of spatially uncorrelated interferences by the quantities G_i, $i=\overline{1,m}$, which are determined in (8.35) instead of (5.31). These quantities are here the results of joint space–time processing of received signals with spatially correlated interference cancellation. The weights A_{i1}^2, $i=\overline{1,m}$, $A_{11}^2=1$, describe as before the relationship of averaged wanted signal powers at different stations [see (5.35)].

The condition of weak signals under which (8.40) is optimal, is given by $(\overline{a_{s1}^2}=\sigma_1^2)$

$$0.5\sigma_1^2 \sum_{i=1}^{m} A_{i1}^2 C_{ii} \ll 1. \tag{8.41}$$

The left side of (8.41) corresponds (by the order of magnitude) to the power SNR after the coherent summation of matched filter outputs from all receiving stations. The processing algorithm (8.40) is optimal for arbitrary PDF $w(a_{s1})$ with limited second order moments.

If the condition of weak signals is not satisfied, an optimum detection algorithm can be synthesized for unknown PDFs $w(a_{s1})$ and $w(\varphi_{si}), i=\overline{1,m}$, under the additional condition that the matrix **C** in (5.29) is a diagonal one. Analyzing (8.35), taking into account (8.8)–(8.10), we note that $C_{ik}=0$ at $i \neq k$ not only in the absence of spatially correlated external interferences (when $f_{ik}(\omega)=\delta_{ik}/\sqrt{N_i N_k}$) but (at least approximately) when the integrand in (8.35) oscillates relative to zero with sufficiently high frequency. For each pair of the ith and kth stations, the frequency of oscillations is determined by the differences in a target echo TDOA with respect to this pair of stations, on the one hand, and in algebraic sums of interference TDOAs with respect to different pairs of stations, on the other. If these differences exceed the reciprocal value of the signal bandwidth Δf_s, then $C_{ik} \approx 0$ for $i \neq k$. The least differences correspond, as a rule, to target echo and interference TDOAs with respect to the same pair of stations, i.e. τ_{sik} and τ_{pik}, $i,k=\overline{1,m}$, $p=\overline{1,M}$. Therefore, to satisfy the condition

$$C_{ik}\approx 0, \quad i\neq k, \quad M>0 \tag{8.42}$$

it is usually sufficient to have $|\tau_{sik}-\tau_{pik}|>1/\Delta f_s$, $p=\overline{1,M}$. In this case signals (target echoes) received from the same target by different stations turn out to be separated in time after the optimal space–time processing with interference cancellation. The time interval between these signals is no less than their duration after matched filtration at the level of the order of 0.5 with respect to the signal maximum. For instance, for a rectangular signal spectrum when $|\tau_{sik}-\tau_{pik}|=1/\Delta f_s$, then $C_{ik}\approx 0$, $i\neq k$, and the envelopes of those signals intercross at the level 0.64. For a Gaussian signal spectrum when $|\tau_{sik}-\tau_{pik}|=1/\Delta f_s$ (Δf_s is here the energy bandwidth), then $C_{ik}\approx 0.04$, $i\neq k$, and the signal envelopes intercross at the level 0.46. When $|\tau_{sik}-\tau_{pik}|>1/\Delta f_s$ the signals overlap only with their sidelobes, so that "multiplication" of signals takes place[3]. Thus it may be considered that the condition (8.42) is satisfied in the case of quasiorthogonality of signals. In other words, (8.42) is valid when wanted signals (target echoes) and interferences are resolved in TDOA at any pair of receiving stations. If the condition $|\tau_{sik}-\tau_{pik}|>1/\Delta f_s$ is not satisfied at an arbitrary pair of stations, the wanted signals overlap essentially one another after external interference cancellation and matched filtration, so that they interfere, and (8.42) is not valid.

Let us consider a logarithm of the conditional likelihood ratio for fixed values of a_{s1} and $\varphi_{s1},\ldots,\varphi_{sm}$ [see (5.29) and (5.34)]:

$$\ln\Lambda=a_{s1}\mathbf{A}'\mathrm{Re}[\tilde{\mathbf{E}}^*(\boldsymbol{\varphi}_s)\mathbf{G}]-0.5a_{s1}^2\mathbf{A}'\tilde{\mathbf{E}}^*(\boldsymbol{\varphi}_s)\mathbf{C}\tilde{\mathbf{E}}(\boldsymbol{\varphi}_s)\mathbf{A}. \tag{8.43}$$

If the condition (8.42) is satisfied, then taking into account (5.26) we have

$$\ln\Lambda=a_{s1}\sum_{i=1}^{m}A_{i1}\exp(j\varphi_{s1})G_i-0.5a_{s1}^2\sum_{i=1}^{m}A_{i1}^2C_{ii}. \tag{8.44}$$

Maximum likelihood estimates of φ_{si} and a_{s1} can be obtained from (8.44):

$$\hat{\varphi}_{si}=-\arg G_i; \qquad \hat{a}_{s1}=\sum_{i=1}^{m}A_{i1}|G_i| \quad \sum_{i=1}^{m}A_{i1}^2C_{ii}. \tag{8.45}$$

Substituting the estimates (8.45) in (8.44) instead of the unknowns a_{s1} and φ_{si}, $i=\overline{1,m}$, omitting the multipliers which do not depend on the quantities G_i and taking a square root (a monotonous transformation) yields

$$L_4=\sum_{i=1}^{m}A_{i1}|G_i|. \tag{8.46}$$

Unlike the algorithm (8.40) a linear weighted summation is optimal here as in the absence of spatially correlated interferences [see L_4 in (5.52)]. However, the difference from (5.52) [as in the case of (8.40)] is that G_i are determined here by (8.35) instead of (5.31).

[3] The influence of this "signal multiplication" on errors of the signal TOA measurement is considered in Section 11.3.

Spatially Incoherent Signals with Mutually Independent Amplitude Fluctuations in MSRSs with Long-Term or Short-Term Spatial Coherence

Obviously, when signal complex amplitudes are mutually independent at the inputs of receiving stations, they remain mutually independent regardless of the degree of MSRS spatial coherence. However, we shall consider above mentioned types of MSRSs since coherent mutual suppression of interferences (external interference cancellation) is possible only in spatially coherent MSRSs (MSRSs with long-term spatial coherence) and in MSRSs with short-term spatial coherence.

For mutually independent a_{si} and φ_{si} $(i=\overline{1,m})$ at different stations and for signals of arbitrary intensity, an optimum detection algorithm can be synthesized in the case of Gaussian signal complex amplitudes under the additional condition (8.42), i.e. the condition that wanted signals and external interferences are resolved in TDOA.

From (5.29) taking into account (5.26), (5.46) and (5.53) we can obtain a logarithm of the likelihood ratio averaged over a_{si} and φ_{si}, $i=\overline{1,m}$, in the form

$$\ln\bar{\Lambda}=\ln\prod_{i=1}^{m}\int_{0}^{\infty}\frac{2a_{si}}{\sigma_1^2A_{i1}^2}\exp\left[-a_{si}^2\left(\frac{1}{\sigma_1^2A_{i1}^2}+\frac{C_{ii}}{2}\right)\right]$$

$$\times\frac{1}{2\pi}\int_{-\pi}^{\pi}\exp\{a_{si}\,\mathrm{Re}[G_i\exp(\mathrm{j}\varphi_{si})]\}\,\mathrm{d}\varphi_{si}\,\mathrm{d}a_{si}. \tag{8.47}$$

Integrating over φ_{si} yields $I_0(a_{si}|G_i|)$. Then after integrating over a_{si} [10], taking a logarithm and dropping the terms which do not depend on G_i we obtain

$$L_5=\sum_{i=1}^{m}\frac{A_{i1}^2|G_i|^2}{1+0.5\sigma_1^2A_{i1}^2C_{ii}}. \tag{8.48}$$

For weak signals (when $0.5\sigma_1^2A_{i1}^2C_{ii}\ll 1$) L_5 is transformed into L_3 (8.40), whereas for strong signals (when $0.5\sigma_1^2A_{i1}^2C_{ii}\gg 1$) L_5 is transformed into

$$L_6=\sum_{i=1}^{m}\frac{|G_i|^2}{C_{ii}}. \tag{8.49}$$

In the absence of spatially correlated interferences (8.48) and (8.49) coincide with (5.54) and (5.55), respectively.

If probability distributions of φ_{si} and a_{si}, $i=\overline{1,m}$, are unknown (or these parameters are nonrandom though unknown) an adaptive optimum algorithm can be synthesized according to the generalized likelihood ratio criterion. Let $b_{si}=a_{si}\exp(-\mathrm{j}\varphi_{si})$ be the signal complex amplitude at the ith station. Then a logarithm of the conditional likelihood ratio for fixed b_{si}, $i=\overline{1,m}$, can be obtained from (5.29) in the form

$$\ln\Lambda=0.5\mathbf{B}^*\mathbf{G}+0.5\mathbf{G}^*\mathbf{B}-0.5\mathbf{B}^*\mathbf{C}\mathbf{B} \tag{8.50}$$

where $\mathbf{B}^*=(b_{s1}^*,\ldots,b_{sm}^*)$. The maximum likelihood estimate of the vector $\mathbf{B}$ from (8.50) is $\hat{\mathbf{B}}=\mathbf{C}^{-1}\mathbf{G}$. Substituting this estimate in (8.50) and taking into account that the matrix **C** is a Hermitian one, yields

$$L_7 = \mathbf{G}^*\mathbf{C}^{-1}\mathbf{G} = \sum_{i=1}^{m}\sum_{k=1}^{m} \alpha_{ik} G_i^* G_k \tag{8.51}$$

where α_{ik} are the elements of the matrix $\mathbf{C}^{-1}$. When wanted signals and interferences are resolved in TDOA, i.e. the condition (8.42) is valid, L_7 is transformed into L_6 (8.49). It means that in this case the optimum adaptive algorithm coincides with the optimum algorithm derived by averaging over random parameters under the condition of strong signals.

Signals with Arbitrary Mutual Statistical Dependence of Amplitude Fluctuations: Spatially Coherent Signals in MSRSs with Short-Term Spatial Coherence or Spatially Incoherent Signals in MSRSs with Long-Term and Short-Term Spatial Coherence

Here we consider the detection problem for signals (target echoes) with mutually independent initial phases φ_{si} and arbitrarily statistically coupled r.m.s. values a_{si}, $i=\overline{1,m}$. For weak signals the same algorithm L_3 (8.40) can be obtained for arbitrary probability distributions (with limited second order moments) and arbitrary mutual statistical dependence of amplitude fluctuations by expanding the exponential in (5.29) into a Taylor series and keeping only linear and quadratic terms (with respect to a_{si}). If expected signals are not weak, the optimum adaptive algorithms in the situation considered are L_7 (8.51) and L_6 (8.49) [the latter under the additional condition (8.42)].

For the reader's convenience all the optimum algorithms that have been synthesized above, are collected in Table 8.1. Decisions of whether a target is present are made by the comparison of the decision variables $L_2, \ldots, L_7$ with a threshold.

All optimum detection algorithms obtained in this Section can be divided into two stages (Fig. 8.4). At the first stage coherent mutual interference suppression (interference cancellation) and wanted signal coherent integration (accumulation) are performed. This stage results in deriving the quantities G_i (8.35), their moduli $|G_i|$ or squared moduli $|G_i|^2$, $i=\overline{1,m}$. At the second stage corresponding quantities are summed with weights. The summed quantities and values of weights depend on the spatial coherence of expected signals and MSRSs as well as the character and mutual statistical dependence of signal amplitude fluctuations at the inputs of different stations.

The obtained results relate mainly to multistatic radars, i.e. radar systems with a single transmitting and m spatially separated receiving stations. However, these results can be easily generalized to MSRSs with several transmitting stations including multiradar systems consisting of several monostatic radars. Apparently, each of these radars must be capable to receive and process echoes from targets illuminated by all other radars and external interferences in corresponding frequency bands, so that such MSRSs are to be MSRSs with cooperative signal reception (see Section 1.1). It should be noted that optimum signal detection by MSRSs in a background of spatially correlated interference requires wideband DTLs to transfer signals and interferences received by spatially separated stations for joint processing with external interference cancellation. However, as was mentioned in Section 8.2 and will be shown in next Sections, such MSRSs can cope with the most dangerous mainlobe external interferences (e.g., mainlobe jamming) without wanted signal suppression.

Of great practical importance are *suboptimum detectors with simplified signal processing*. When external interferences are intensive, as is usually the practice, and

Table 8.1

Signals at the inputs of receiving stations	Spatially coherent MSRS	
	Optimum algorithms averaged over random parameters	Optimum adaptive algorithms
Spatially coherent with completely dependent amplitude fluctuations $w(\varphi_{s1},\dots,\varphi_{sm})=w(\varphi_{s1})$ $\times\prod_{i=2}^{m}\delta(\varphi_{si}-\varphi_{s1}+\Delta\varphi_{si1})$; $w(a_{s1},\dots,a_{sm})=w(a_{s1})$ $\times\prod_{i=2}^{m}\delta(a_{si}-A_{i1}a_{s1})$;	$L_2=\left\lvert\sum_{i=1}^{m}A_{i1}e^{-j\Delta\varphi_{si1}}G_i\right\rvert$. (8.37) *Additional conditions:* $w(\varphi_{s1})=\frac{1}{2\pi},\ \varphi_{s1}\in(0,2\pi)$; $w(a_{s1})=\frac{2a_{s1}}{\sigma_1^2}\exp\left(-\frac{a_{s1}^2}{\sigma_1^2}\right)$ or any distribution of a_{s1} for weak signals	$L_2=\left\lvert\sum_{i=1}^{m}A_{i1}e^{-j\Delta\varphi_{si1}}G_i\right\rvert$. (8.37) *Additional conditions:* maximum likelihood estimates $\hat{\varphi}_{s1}$ and $\hat{a}_{s1}$ are used
Spatially incoherent with independent amplitude fluctuations $w(\varphi_{s1},\dots,\varphi_{sm})=\prod_{i=1}^{m}w(\varphi_{si})$; $w(a_{s1},\dots,a_{sm})=\prod_{i=1}^{m}w(a_{si})$	$L_5=\sum_{i=1}^{m}A_{i1}^2\lvert G_i\rvert^2$ $\times(1+0.5\sigma_1^2A_{i1}^2C_{ii})^{-1}$. (8.48) *Additional conditions:* $w(\varphi_{si})=\frac{1}{2\pi},\ \varphi_{si}\in(0,2\pi)$; $w(a_{si})=\frac{2a_{si}}{\sigma_1^2A_{i1}^2}$ $\times\exp\left(-\frac{a_{si}^2}{\sigma_1^2A_{i1}^2}\right);\ i=\overline{1,m}$ ($L_5\to L_3$ for weak signals) $L_6=\sum_{i=1}^{m}C_{ii}^{-1}\lvert G_i\rvert^2$. (8.49) *Additional conditions:* the same as for L_5 and strong signals	$L_7=\sum_{i=1}^{m}\sum_{k=1}^{m}\alpha_{ik}G_i^*G_k$. (8.51) *Additional conditions:* maximum likelihood estimates $\widehat{a_{si}\exp(-j\varphi_{si})}$ are used, $i=\overline{1,m}$ $L_6=\sum_{i=1}^{m}C_{ii}^{-1}\lvert G_i\rvert^2$. (8.49) *Additional conditions:* the same as for L_7 and $C_{ik}=0$ if $i\neq k$
Spatially incoherent with arbitrary statistical dependence of amplitude fluctuations $w(\varphi_{s1},\dots,\varphi_{sm})=\prod_{i=1}^{m}w(\varphi_{si})$;	$L_3=\sum_{i=1}^{m}A_{i1}^2\lvert G_i\rvert^2$. (8.40) *Additional conditions:* $w(\varphi_{si})=\frac{1}{2\pi},\ \varphi_{si}\in(0,2\pi),\ i=\overline{1,m}$; any distribution of $a_{s1},\dots,a_{sm}$ for weak signals	$L_7=\sum_{i=1}^{m}\sum_{k=1}^{m}\alpha_{ik}G_i^*G_k$. (8.51) $L_6=\sum_{i=1}^{m}C_{ii}^{-1}\lvert G_i\rvert^2$. (8.49) *Additional conditions:* maximum likelihood estimates $\widehat{a_{si}\exp(-j\varphi_{si})}$ are used, $i=\overline{1,m}$; L_6 is valid for $C_{ik}=0$ if $i\neq k$

Table 8.1 (continued)

MSRS with short-term spatial coherence	
Optimum algorithms averaged over random parameters	Optimum adaptive algorithms
$L_3=\sum_{i=1}^{m} A_{i1}^2 \lvert G_i\rvert^2$. (8.40) *Additional conditions:* $w(\varphi_{si})=\frac{1}{2\pi}$, $\varphi_{si}\in(0,2\pi)$, $i=\overline{1,m}$; any distribution of a_{s1} for weak signals	$L_4=\sum_{i=1}^{m} A_{i1}\lvert G_i\rvert$. (8.46) *Additional conditions:* $C_{ik}=0$ if $i\neq k$; maximum likelihood estimates $\hat{\varphi}_{s1},\ldots,\hat{\varphi}_{sm}$ and $\hat{a}_{s1}$ are used
$L_5=\sum_{i=1}^{m} A_{i1}^2 \lvert G_i\rvert^2 \times(1+0.5\sigma_1^2 A_{i1}^2 C_{ii})^{-1}$. (8.48) *Additional conditions:* $w(\varphi_{si})=\frac{1}{2\pi}$, $\varphi_{si}\in(0,2\pi)$; $w(a_{si})=\frac{2a_{si}}{\sigma_1^2 A_{i1}^2}\times\exp\left(-\frac{a_{si}^2}{\sigma_1^2 A_{i1}^2}\right)$, $i=\overline{1,m}$ ($L_5\to L_3$ for weak signals) $L_6=\sum_{i=1}^{m} C_{ii}^{-1}\lvert G_i\rvert^2$. (8.49) *Additional conditions:* the same as for L_5 and strong signals	$L_7=\sum_{i=1}^{m}\sum_{k=1}^{m}\alpha_{ik} G_i^* G_k$. (8.51) *Additional conditions:* maximum likelihood estimates $\widehat{a_{si}\exp(-j\varphi_{si})}$ are used, $i=\overline{1,m}$ $L_6=\sum_{i=1}^{m} C_{ii}^{-1}\lvert G_i\rvert^2$. (8.49) *Additional conditions:* the same as for L_7 and $C_{ik}=0$ if $i\neq k$
$L_3=\sum_{i=1}^{m} A_{i1}^2 \lvert G_i\rvert^2$. (8.40) *Additional conditions:* $w(\varphi_{si})=\frac{1}{2\pi}$, $\varphi_{si}\in(0,2\pi)$, $i=\overline{1,m}$; any distribution of $a_{s1},\ldots,a_{sm}$ for weak signals	$L_7=\sum_{i=1}^{m}\sum_{k=1}^{m}\alpha_{ik} G_i^* G_k$. (8.51) $L_6=\sum_{i=1}^{m} C_{ii}^{-1}\lvert G_i\rvert^2$. (8.49) *Additional conditions:* maximum likelihood estimates $\widehat{a_{si}\exp(-j\varphi_{si})}$ are used, $i=\overline{1,m}$; L_6 is valid for $C_{ik}=0$ if $i\neq k$

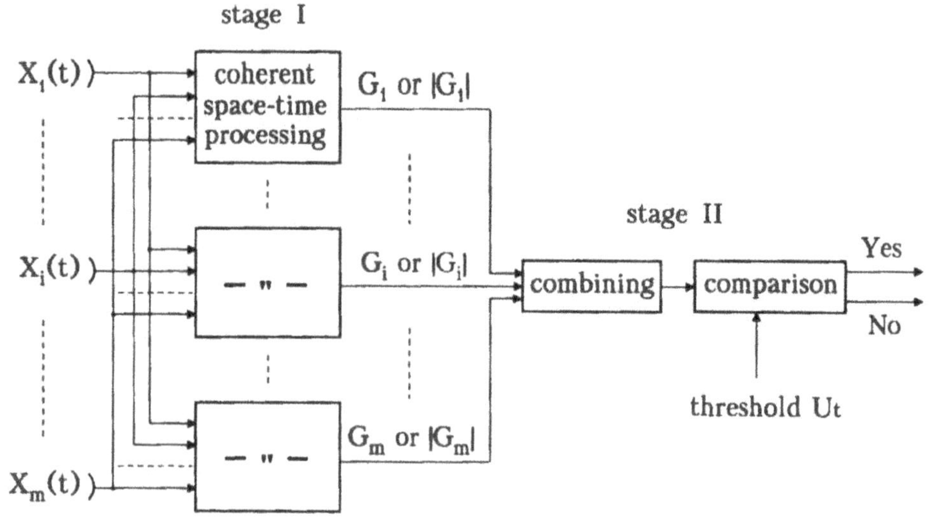

Figure 8.4 General structure of optimal detection algorithms for regular signals in a background of spatially correlated interferences

a MSRS consists of a few stations, the main role in signal detection is played by interference cancellation at the first stage of detection algorithms (see Fig. 8.4). Therefore the simplest suboptimum detector may contain only the first stage of a total detection algorithm. It means that joint processing of signals and interferences (received by all spatially separated stations) with external interference cancellation is performed in a single station only. Then the result of this joint processing is compared with a threshold to make a decision about the presence or the absence of a target. An example of such a MSRS will be considered in Section 9.1.

A more complex and effective suboptimum detector is a structure including all the two stages of detection algorithms but with decentralized decision making at the second stage. Unfortunately, unlike the case of spatially uncorrelated interferences (see Chapter 6), wideband DTLs cannot be excluded from such MSRSs since they are necessary to transfer signals and interferences for joint processing. However, decentralization simplifies significantly the implementation of the second stage of detection algorithms. As in Section 6.1, the most reasonable processing is to compare quantities $|G_i|$ with a threshold in each station and then to combine the preliminary decisions using decision rules of the "k out of m" type.

***In this section**, we have synthesized several optimum algorithms for fluctuating signal (target echo) detection in a background of spatially correlated interferences, e.g., jamming. Since it is practically important to preserve the capability to cancel spatially correlated interferences, we have considered only those MSRSs which have this feature, namely spatially coherent MSRSs and MSRSs with short-term spatial coherence. When expected signals are spatially coherent with completely mutually dependent amplitude fluctuations, algorithm L_2 (8.37) is optimal for spatially coherent MSRSs whereas algorithm L_3 (8.40) or L_4 (8.46) is optimum for MSRSs with short-term spatial coherence. When expected signals are spatially incoherent with mutually independent amplitude fluctuations, algorithm L_5 (8.48), L_3 (8.40) or L_6 (8.49) is optimum. For the case where probability distributions of a_{si} and φ_{si} are unknown (or a_{si} and φ_{si} are nonrandom unknown parameters) adaptive algorithms L_7 (8.51) and L_6(8.49) have been* *derived. The obtained results are similar to those of Section 5.2. The main difference is in deriving the quantities G_i and C_{ik} from* (8.35) *instead of* (5.31) *and* (5.32). *Here these quantities are results of joint spatial–temporal processing of signals and external interferences received by all stations. For the reader's convenience all synthesized algorithms are collected in Table* 8.1.

We have shown (Fig. 8.4) *that all derived algorithms can be divided into two stages. When external interferences are intensive the most important role is played by the first stage where interference cancellation is performed. We have briefly considered simplified suboptimum algorithms with excluded second stage and with decentralized processing at the second stage.*

8.4. PERFORMANCE ANALYSIS OF DETECTION ALGORITHMS FOR FLUCTUATING SIGNALS. EFFICIENCY OF SPACE–TIME PROCESSING

All the optimum and suboptimum detection algorithms considered in Section 8.3 include joint space–time linear filtration of signals and interferences received by spatially separated stations. This filtration results in coherent mutual suppression (cancellation) of spatially correlated interferences. At the output of such processing (of the first stage of detection algorithms – see Fig. 8.4) we have the random variables G_i, $i=\overline{1,m}$, (8.35). Therefore, statistical characteristics of G_i are to be obtained for performance analysis of the synthesized algorithms [66].

It is seen from (8.35) that G_i are the results of linear transformations of additive external interferences, self-noises and, possibly, wanted signals. Hence we can separate a signal component, G_{si}, and a noise component (including external interferences), G_{ni}: $G_i=G_{si}+G_{ni}$, $i=\overline{1,m}$. A correlation (covariance) matrix for the variables G_{ni} with zero mean can be obtained from (8.35)

$$\Gamma_n=\|\Gamma_{nik}\|=\|0.5\overline{G_{ni}G^*_{nk}}\|=\|C_{ik}\|=\mathbf{C},\quad i,k=\overline{1,m}. \tag{8.52}$$

where elements C_{ik} of the matrix $\mathbf{C}$ are defined in (8.35).

For performance analysis of the synthesized detectors, probability distributions of the initial phases φ_{si} and signal r.m.s. values a_{si} should now be specified. As in Section 5.4, we assume for φ_{si} and a_{si} typical the uniform [within $(-\pi,\pi)$] and Rayleigh distributions, respectively. Then $b_{si}=a_{si}\exp(-j\varphi_{si})$ are Gaussian random variables with zero mean. However, *being Gaussian they are not always jointly Gaussian variables.* For instance, when spatially coherent target echoes enter a MSRS with short-term spatial coherence, these signals become spatially incoherent (mutually incoherent at different stations) because of random and mutually independent phase shifts in receiver channels. At the same time amplitude fluctuations may remain completely mutually dependent. The probability distributions of φ_{si} and a_{si} are determined by (5.46) and (5.34), (5.38b). In this case any sum $b_{si}+b_{sk}=a_{s1}[A_{i1}\exp(-j\varphi_{si})+A_{k1}\exp(-j\varphi_{sk})]$, $i,k=\overline{1,m}$, $i\neq k$, is a non-Gaussian variable. The similar situation takes place in another (though rather exotic case) when expected signals in a spatially coherent MSRS are spatially coherent whereas amplitude fluctuations are mutually independent. The probability distributions of φ_{si} and a_{si} are determined by (5.33) and (5.53). Any sum $b_{si}+b_{sk}=\exp(-j\varphi_{s1})[a_{si}\exp(j\Delta\varphi_{si1})+a_{sk}\exp(j\Delta\varphi_{sk1})]$, $i,k=\overline{1,m}, i\neq k$, is a non-Gaussian variable too. Thus the signal components, G_{si}, $i=\overline{1,m}$, which according to (8.35) contain the sums of b_{sk} [included in overall received signal spectra $\chi_k(\omega)$] may not be considered as being Gaussian in the general case. Performance of the synthesized detection algorithms for non-Gaussian G_{si} is to be analyzed with the help of computer simulation.

Let us consider here the most practically important extreme cases where G_{si} are Gaussian variables. When *expected signals are spatially coherent with completely mutually dependent amplitude fluctuations at a spatially coherent MSRS* then φ_{si} and a_{si} are determined by (5.33), (5.34) and (5.38). In this case $b_{si}=A_{i1}\exp(j\Delta\varphi_{si1})$ $a_{s1}\exp(-j\varphi_{s1})$, and G_{si} are jointly Gaussian variables with zero mean. The correlation matrix of G_{si} can be obtained from (8.35) setting $\chi_k(\omega)=\Psi_k(\omega)$ where $\Psi_k(\omega)$, $k=\overline{1,m}$, are defined in (8.3):

$$\Gamma_s=\|\Gamma_{sik}\|=\|0.5\overline{G_{si}G_{sk}^*}\|=\left\|0.5\sigma_1^2\sum_{l=1}^{m}\sum_{n=1}^{m}A_{l1}A_{n1}\exp(j\Delta\varphi_{sln})C_{il}C_{kn}^*\right\|. \quad (8.53)$$

Consider now the other extreme case of *spatially incoherent expected signals with mutually independent amplitude fluctuations* at a MSRS with long-term or short-term spatial coherence. The random parameters φ_{si} and a_{si} are determined by (5.46) and (5.53). In this case $b_{si}=a_{si}\exp(-j\varphi_{si})$ and $b_{sk}=a_{sk}\exp(-j\varphi_{sk})$ are mutually independent Gaussian variables at $i\neq k$. Then G_{si} are jointly Gaussian variables with zero mean and the correlation matrix

$$\Gamma_s=\|\Gamma_{sik}\|=\|0.5\overline{G_{si}G_{sk}^*}\|=\left\|0.5\sigma_1^2\sum_{l=1}^{m}A_{l1}^2C_{il}C_{kl}^*\right\|. \quad (8.54)$$

It follows from (8.54) that though expected signals themselves are mutually independent at the inputs of different stations, the signal components G_{si} and G_{sk}, $i,k=\overline{1,m}$, $i\neq k$, are in general mutually correlated.

When G_{si} as well as G_{ni} are jointly Gaussian variables, then so are their sums G_i. Denoting the correlation matrix of G_i by Γ, we have in the absence of expected signals $\Gamma=\Gamma_n$ whereas in the presence of expected signals $\Gamma=\Gamma_n+\Gamma_s$.

If expected signals (target echoes) and external interferences are resolved in TDOA, i.e. if (8.42) is valid, we have instead of (8.53) and (8.54)

$$\Gamma_s=\|0.5\sigma_1^2A_{i1}A_{k1}\exp(j\Delta\varphi_{sik})C_{ii}C_{kk}\|; \quad (8.55)$$

$$\Gamma_s=\|0.5\sigma_1^2\delta_{ik}A_{i1}^2C_{ii}^2\|. \quad (8.56)$$

It is seen from (8.52) that under the condition (8.42) the interference plus self-noise components, G_{ni}, G_{nk}, are mutually independent if $i\neq k$. It follows from (8.56) that the same is valid for the signal components G_{si}, G_{sk}, in the case of spatially incoherent expected signals with mutually independent amplitude fluctuations. For spatially coherent signals with completely mutually dependent fluctuations the coefficient of mutual correlation between G_{si} and G_{sk} is equal to $\exp(j\Delta\varphi_{sik})$, i.e. its modulus is equal to one [see (8.55)]. When the condition (8.42) is not satisfied, both the signal components, G_{si}, G_{sk}, and interference components, G_{ni}, G_{nk}, at different stations are mutually correlated.

Now the effectiveness of the first stage of optimum processing can be evaluated (see Fig. 8.4). Since G_i are Gaussian random variables with zero mean, it is sufficient to calculate the signal-to-interference-plus-self-noise ratio (SINR) for them. For the case of spatially coherent wanted signals at a spatially coherent MSRS we obtain from (8.52) and (8.53)

$$\tilde{q}_{\text{out}\,i}^2=\Gamma_{sii}/\Gamma_{nii}=(0.5\sigma_1^2/C_{ii})\left|\sum_{l=1}^{m}A_{l1}\exp(-j\Delta\varphi_{sil})C_{il}\right|^2. \quad (8.57)$$

For independently fluctuating spatially incoherent expected signals in a spatially coherent MSRS or in a MSRS with short-term spatial coherence we have from (8.52) and (8.54)

$$\tilde{q}^2_{\text{out}\,i}=\Gamma_{sii}/\Gamma_{nii}=(0.5\sigma_1^2/C_{ii})\sum_{l=1}^{m}A_{l1}^2|C_{il}|^2. \tag{8.58}$$

Under the condition (8.42) both (8.57) and (8.58) take the simpler form

$$\tilde{q}^2_{\text{out}\,i}=\Gamma_{sii}/\Gamma_{nii}=0.5\sigma_1^2 A_{i1}^2 C_{ii}. \tag{8.59}$$

Using the derived expressions (8.57)–(8.59) and taking into account (8.35), (8.8)–(8.10) one can obtain SINR values after the spatially correlated interference cancellation and expected signal matched filtration in each station for arbitrary number of stations, m, and arbitrary number of interference sources (e.g., jammers), M. Relevant mathematical expressions rapidly become more and more complicated with the increase of M. However, in practice, as was noted in Section 8.2, interferences from only few sources fell within mainbeams of receiving ADPs should undergo joint processing in a MSRS. Interferences entering sidelobes of receiving ADPs are to be cancelled in each station individually. Besides, as will be shown in Section 9.1, adaptive versions of the interference cancellation algorithm can be implemented without excessive complication.

In order to put the obtained general results in a more clear and instructive form let us consider the case for which $M=1$. We assume for simplicity that the signal spectrum $|\Psi_0(\omega-\omega_0)|^2$ and interference PSD $F(\omega-\omega_0)$ are rectangular [see (8.30)], the antenna ADPs $g_i(\beta,\varepsilon,\omega)=g_i(\beta,\varepsilon)$ are real and do not depend on frequency ω. We consider that the corresponding factors $g_i(\beta_{si},\varepsilon_{si})$ and $g_i(\beta_{1i},\varepsilon_{1i})$ are included into the signal and interference r.m.s. values a_{si} and $q_{1i}\sqrt{N_i}$, respectively. Besides, we make following simplifying assumptions:

$$A_{i1}=1,\quad \text{i.e. } \sigma_i^2=\sigma^2;\quad N_i=N;\quad q_{1i}=q,\quad i=\overline{1,m};\quad \omega_1=\omega_0. \tag{8.60}$$

Then we have from (8.8)–(8.10) omitting the subscript "1"

$$f_{ik}(\omega)=\frac{1}{N}\left\{\delta_{ik}-\frac{q^2}{1+mq^2}\exp[\,j(\omega\tau_{ik}+\Delta\varphi_{ik})]\right\} \tag{8.61}$$

and from (8.35)

$$C_{ik}=\frac{2T_s}{N}\left\{\delta_{ik}-\frac{q^2}{1+mq^2}\operatorname{sinc}[(\Delta\omega_s/2)\delta\tau_{ik}]\exp[\,j(\omega_0\delta\tau_{ik}+\Delta\varphi_{ik})]\right\} \tag{8.62}$$

where $\operatorname{sinc}(x)=(\sin x)/x$; $\delta\tau_{ik}=\tau_{ik}-\tau_{sik}$ is the difference in TDOA (i.e. the second difference in time of arrival – TOA) between the interference and the expected signal at the inputs of the ith and kth stations. Substituting (8.62) in (8.57) yields the output power SINR in the case of spatially coherent expected signals with completely dependent amplitude fluctuations in a spatially coherent MSRS

$$\tilde{q}^2_{\text{out}\,i}=\frac{\bar{E}}{N(1+mq^2)[1+(m-1)q^2]}\times\left|1+(m-1)q^2-q^2\sum_{l=1,\,l\neq i}^{m}\operatorname{sinc}^2[(\Delta\omega_s/2)\delta\tau_{il}]\exp[\,j(\omega_0\delta\tau_{il}+\delta\varphi_{il})]\right|^2. \tag{8.63}$$

The output power SINR in the case of spatially incoherent signals with mutually independent amplitude fluctuations in a MSRS with long-term or short-term spatial coherence takes the form

$$\tilde{q}^2_{\text{out}\,i}=\frac{\bar{E}}{N(1+mq^2)[1+(m-1)q^2]}\times\left\{[1+(m-1)q^2]^2+q^4\sum_{l=1,\,l\neq i}^{m}\operatorname{sinc}^2[(\Delta\omega_s/2)\delta\tau_{il}]\right\}. \quad (8.64)$$

In (8.63) and (8.64) $\bar{E}=\sigma^2 T_s$ is the mean wanted signal energy at the input of each station; $\delta\varphi_{il}=\Delta\varphi_{il}-\Delta\varphi_{sil}$ is the difference between the interference and wanted signal phase differences (i.e. the second phase difference) at the inputs of the ith and lth stations. Under the condition of quasiorthogonality of wanted signals (received by different stations) after the interference cancellation, when $|\delta\tau_{ik}|>1/\Delta f_s$ and (8.42) is valid, $\operatorname{sinc}[(\Delta\omega_s/2)\delta\tau_{ik}]\approx 0$, and we obtain from both (8.63) and (8.64)

$$\tilde{q}^2_{\text{out}\,i}=\bar{E}[1+(m-1)q^2]/N(1+mq^2). \quad (8.65)$$

It is seen from (8.63) that, as in the case of deterministic signal [see (8.32)], variations of $\delta\tau_{il}=\tau_{il}-\tau_{sil}$ result in fast oscillations of $\tilde{q}^2_{\text{out}\,i}$. When $\delta\varphi_{il}=0$ and a target coincides with an interference source (e.g., a jammer) in space, so that $\delta\tau_{il}=0$, then in the case of intensive interference $\tilde{q}^2_{\text{out}\,i}\approx 0$, i.e. the wanted signal is suppressed together with the interference cancellation. However, for small mutual displacement of the target and the interference source, when $|\delta\tau_{il}|\approx 1/2f_0$ the output SINR value reaches its maximum

$$\tilde{q}^2_{\text{out}\,i\,\max}=\frac{\bar{E}}{N}\frac{[1+2(m-1)q^2]^2}{(1+mq^2)[1+(m-1)q^2]}. \quad (8.66)$$

These events occur when $|\omega_0\delta\tau_{il}+\delta\varphi_{il}|=(2n+1)\pi$ for all $l=\overline{1,m}$, $l\neq i$, and $|\delta\tau_{il}|\ll 1/\Delta f_s$. If $m=2$ the maximum (8.66) takes place at both stations simultaneously since $\delta\tau_{12}=-\delta\tau_{21}$. If $m>2$, $\tilde{q}^2_{\text{out}\,i}$ reaches its maximum at no more than one station. Variations of input interference-to-self-noise ratio (INR) q^2 lead to variations of $\tilde{q}^2_{\text{out}\,i\,\max}$ from $\bar{E}/N$ to $(\bar{E}/N)[4(m-1)/m]$ but at any q^2 interference cancellation occurs without wanted signal suppression. Further increase of $|\delta\tau_{il}|$ leads to reducing of oscillations so that when $|\delta\tau_{il}|>1/\Delta f_s$, the SINR $\tilde{q}^2_{\text{out}\,i}$ is determined by (8.65).

Quite different is the behaviour of $\tilde{q}^2_{\text{out}\,i}$ for spatially incoherent expected signals with mutually independent amplitude fluctuations in a spatially coherent MSRS (i.e. a MSRS with long-term spatial coherence) or in a MSRS with short-term spatial coherence. As can be seen from (8.64), when a target and an interference source (e.g., a jammer) coincide in space ($\delta\tau_{il}=0$), the output power SINR reaches its maximum:

$$\tilde{q}^2_{\text{out}\,i}=\tilde{q}^2_{\text{out}\,i\,\max}=\frac{\bar{E}}{N}\frac{[1+(m-1)q^2]^2+(m-1)q^4}{(1+mq^2)[1+(m-1)q^2]},\quad i=\overline{1,m}. \quad (8.67)$$

As $|\delta\tau_{il}|$ increases, $\tilde{q}^2_{\text{out}\,i}$ slightly monotonously decreases and when $|\delta\tau_{il}|>1/\Delta f_s$ the SINR $\tilde{q}^2_{\text{out}\,i}$ is determined by (8.65). It follows from (8.67) that for intensive interference, when $(m-1)q^2\gg 1$, $\tilde{q}^2_{\text{out}\,i\,\max}\approx\bar{E}/N$ which is equal to the power SNR in a monostatic radar in the absence of external interferences.

The output SINR $\tilde{q}^2_{\text{out}\,i}$ from (8.63) and (8.64) as a function of the difference in TDOA between interference and wanted signal, $\delta\tau_{il}, i,l=\overline{1,m}, i \neq l$, determines the power "resultant signal reception pattern, RSRP" (introduced in Section 8.2) after the first stage of processing (see Fig. 8.4) at the output of the ith station when a target is observed in the presence of self-noises and spatially correlated interference from a single source.

The same function determines completely performance of the simple suboptimum detector which does not contain the second stage of processing (see Section 8.3). Overall signals received by all stations undergo the joint coherent processing (with external interference cancellation) at a single, for instance, the lth station. Then the variable $|G_l|$ is compared with a threshold. The false alarm and detection probabilities (for Gaussian complex amplitudes) are connected with each other by the relationship (5.77).

Example 8.2. Let us consider a numerical example for the case discussed above where there is a single interference source ($M=1$), the signal spectrum and interference PSD are determined by (8.30), the antenna ADPs are real, not depending on frequency and are included into the r.m.s. values of wanted signals and interference, the relationships (8.60) are valid. Let a MSRS contain only two receiving stations ($m=2$), the input interference-to-self-noise ratio (INR) $q^2=20\,\text{dB}$, the input mean signal-to-self-noise ratio (SNR) $\bar{E}/N=15\,\text{dB}$ and the relative signal frequency bandwidth is equal to 5% ($\Delta f_s/f_0=5\%$). The curves of detection probability, P_0, as functions of the normalized difference in TDOA $\Delta f_s|\delta\tau_{12}|$ are plotted in Fig. 8.5 for the false alarm probability $P_{fa}=10^{-4}$. The curves 1 and 2 are calculated using (8.63) and (8.64), respectively. These curves may also be considered as RSRPs of a MSRS but in detection probability instead of in output power SINR. It is important to note that if we had not taken into account the interference spatial correlation, i.e. if we had rejected the joint processing with external interference cancellation, then $\tilde{q}^2_{\text{out}\,i}$ would be equal to $\bar{E}/N(1+q^2) \approx 0.31$ and hence P_0 would be equal approximately to $9\cdot 10^{-4}$. Apparently, in this case the wanted signal could not be detected.

It is seen from Fig. 8.5 that the RSRP for *spatially coherent signals with completely mutually dependent amplitude fluctuations* is similar to the RSRP for deterministic signals (see Fig. 8.3). Oscillations of the RSRP can be explained here as in Fig. 8.3, by the interference of expected signals received by spatially separated stations after the external interference (e.g., mainlobe jamming) cancellation. Though signal complex amplitudes are random, the relationships between signal initial phases at the inputs of different stations as well as between signal r.m.s. values are nonrandom.

For *spatially incoherent expected signals with mutually independent amplitude fluctuations* and $|\delta\tau_{12}| \ll 1/\Delta f_s$, received signals after each target illumination and external interference cancellation are also summed with a certain relationship between their initial phases and effective values. However, these relationships are random since those phases and r.m.s. values are mutually statistically independent at different stations. The phase difference of summed signals is uniformly distributed within the interval $(-\pi, \pi)$. Therefore, oscillations of the summed signal amplitude averaged over many target illuminations turn out to be "smoothed", so that the Gaussian complex amplitude of the summed signal weakly depends on $|\delta\tau_{12}|$.

It should be emphasized that such a character of the RSRP obtained in the Example 8.2 for spatially incoherent signals with mutually independent amplitude fluctuations means that *target detection does not require the target to be spatially separated from an interference source*. When external interference cancellation is sufficiently effective, target echoes not only from the masked target but even from the interference source (the jammer) itself may be detected by a MSRS.

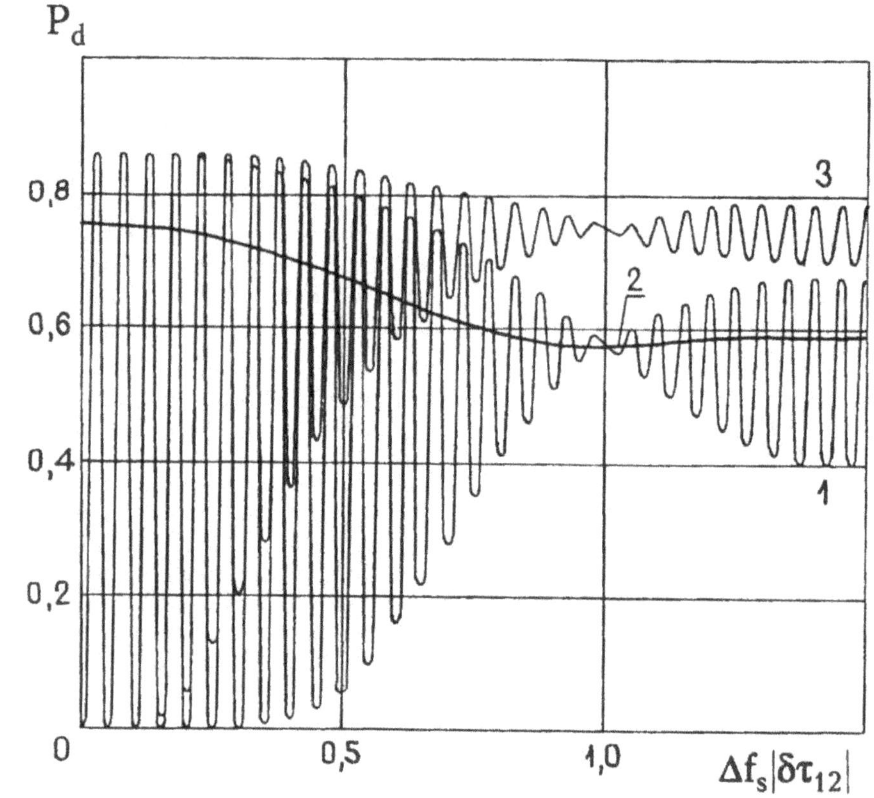

Figure 8.5 Dependences of signal detection probability on the displacement in TDOA of an interference source with respect to the target. Curves 1, 2: suboptimum algorithms without the second processing stage; curve 3: optimum algorithm; curves 1, 3: for spatially coherent signals with completely correlated amplitude fluctuations; curve 2: for spatially incoherent signals with independenr amplitude fluctuations; $m=2$, $P_{f_a}=10^{-4}$, $E/N=15\,\mathrm{dB}$, $q^2=20\,\mathrm{dB}$, $\Delta f/f_0=5\%$

For the sake of clarity we have neglected possible differences in the input interference-to-self-noise ratios (INRs), q_i^2, mean expected signal energies, $\bar{E}_i$, and self-noise PSDs, N_i, as well as in the antenna ADPs in the target direction $g_i(\beta_{si}, \varepsilon_{si}, \omega)$ and in the interference source direction $g_i(\beta_i, \varepsilon_i, \omega)$, $i=\overline{1,m}$. All these differences may be taken into account with the help of the general expressions (8.52)–(8.59), (8.35) and (8.8)–(8.10).

In this section, *we have presented a performance analysis of the first stage of the detection algorithms synthesized in Section* 8.3. *At this stage external interferences, self-noises and, possibly, wanted signals received by all stations undergo the joint processing at each station – the linear filtration with external interference cancellation and wanted signal integration. We have shown that the outputs of such processing,* G_i, *are not always Gaussian variables even for Gaussian external interferences and self-noises. We have considered the most important for practice extreme cases where* G_i *are Gaussian: the case of spatially coherent expected signals with completely mutually dependent amplitude fluctuations and the case of spatially incoherent signals with mutually independent amplitude fluctuations. For both cases we have obtained the correlation (covariance) matrix* Γ_s *of signal components* G_{si}, *and the matrix* Γ_n *for* *noise-plus-interference components,* G_{ni}, $i=\overline{1,m}$: (8.52), (8.53), (8.54) *for the general case and* (8.55), (8.56) *if the condition* (8.42) *is valid. This has made it possible to obtain the general expressions* (8.57), (8.58) *for the output power SINR* $\tilde{q}^2_{\text{out}\,i}$ [(8.59) *under the condition* (8.42)] *after external interference cancellation and wanted signal accumulation in each station. We have analyzed in detail the output power SINR in the particular case of a single interference source within antenna mainlobes of spatially separated stations* [(8.63)–(8.67)]. *A numerical Example* 8.2 *with Fig.* 8.5 *has allowed us to reveal, firstly, high efficiency of the joint linear space–time processing with external interference cancellation, secondly, an oscillating character of the "resultant signal reception pattern (RSRP)" for spatially coherent signals with completely mutually dependent amplitude fluctuations, as for deterministic signals, and, thirdly, a significant difference in the behaviour of the RSRP for spatially coherent signals with completely mutually dependent amplitude fluctuations and for incoherent signals with mutually independent amplitude fluctuations. It has been shown that in the latter case (which is typical in practice) there is no need in a target and an interference source space separation for the target detection, since the maximum detection probability takes place when the target and the interference source coincide in space.*

8.5. PERFORMANCE ANALYSIS OF DETECTION ALGORITHMS FOR FLUCTUATING SIGNALS. RESULTANT EFFICIENCY

Now we pass to the analysis of resultant performance characteristics of the optimum detection algorithms synthesized in Section 8.3 [66].

Detection of Spatially Coherent Signals with Completely Mutually Dependent Amplitude Fluctuations in a Spatially Coherent MSRS

Both the detector which is optimal in average for random Gaussian signal complex amplitudes, and the adaptive optimum detector utilizing maximum likelihood estimates of unknown (random) parameters, "work" according to the same algorithm L_2 (8.37). Since G_i, $i=\overline{1,m}$, are jointly Gaussian random variables with zero mean in the case considered (see Section 8.4), the sum

$$G_0=\sum_{i=1}^{m} A_{i1}\exp(-j\Delta\varphi_{si1})G_i$$

is Gaussian too, and $L_2=|G_0|$ is a Rayleigh variable. Detection characteristics are completely determined by the output power SINR

$$\tilde{q}^2_{\text{out}}=\overline{|G_{s0}|^2}/\overline{|G_{n0}|^2}$$

where G_{s0} and G_{n0} are signal and noise (together with external interference) components of G_0. Using (8.52) and (8.53) we can find from (8.37)

$$\tilde{q}^2_{\text{out}}=0.5\sigma_1^2\sum_{i=1}^{m}\sum_{k=1}^{m} A_{i1}A_{k1}\exp(-j\Delta\varphi_{sik})C_{ik}. \tag{8.68}$$

As in Section 8.4, we consider in more detail the case for which $M=1$ (a single interference source), the signal spectrum and interference PSD are rectangular (8.30) and the assumptions (8.60) are valid. Then $f^{(1)}_{ik}(\omega)$ and C_{ik} are determined by (8.61) and (8.62) and we have from (8.68)

$$\tilde{q}^2_{\text{out}} = \frac{\bar{E}}{N(1+mq^2)}\left\{m[1+(m-1)q^2] - 2q^2 \sum_{i=1}^{m-1} \sum_{k=i+1}^{m} \operatorname{sinc}[(\Delta\omega_s/2)\delta\tau_{ik}]\cos(\omega_0\delta\tau_{ik}+\delta\varphi_{ik})\right\}. \tag{8.69}$$

It is seen that the quantity $\tilde{q}^2_{\text{out}}$ in (8.69), as $\tilde{q}^2_{\text{out}\,i}$ in (8.63), rapidly oscillates as a function of $|\delta\tau_{ik}|$ with the period of $2\pi/\omega_0$. When $\delta\varphi_{ik}=0$ and $\delta\tau_{ik}=0$ (a target coincides with the interference source, e.g., with a jammer), then $\tilde{q}^2_{\text{out}}=(\bar{E}/N)m/(1+mq^2)$, i.e. $\tilde{q}^2_{\text{out}} \ll 1$ for the case of intensive interference ($q^2 \gg 1$). However, when $|\delta\tau_{ik}|$ is changed by π/ω_0, the output SINR $\tilde{q}^2_{\text{out}}$ reaches its maximum. It can be shown that at $|\delta\tau_{ik}| \ll 1/\Delta f_s$ maximum values of $\tilde{q}^2_{\text{out}}$ occur for such mutual positions of a target and an interference source when

$$\sum_{i=1}^{m} \sin(\omega_0\delta\tau_{1i}+\delta\varphi_{1i})=0; \qquad \sum_{i=1}^{m} \cos(\omega_0\delta\tau_{1i}+\delta\varphi_{1i})=0. \tag{8.70}$$

In (8.70) we have taken into account that $\delta\tau_{ik}=\delta\tau_{1k}-\delta\tau_{1i}$. (The first station is chosen as the reference one without loss of generality.) If the conditions (8.70) are satisfied

$$\tilde{q}^2_{\text{out}} = \tilde{q}^2_{\text{out max}} = (\bar{E}/N)m. \tag{8.71}$$

Apparently, spatially correlated interferences are cancelled out. The output power SINR (before the envelope detection deriving $L_2=|G_0|$) is equal to that of the optimum detector for the same type of signals and the same type of a MSRS with a single transmitting and m receiving stations but in a background of station self-noises only, i.e. without spatially correlated external interference [see (5.76)]. As $|\delta\tau_{ik}|$ increases, oscillations of $\tilde{q}^2_{\text{out}}$ decay. When $|\delta\tau_{ik}|>1/\Delta f_s$ then $\operatorname{sinc}[(\Delta\omega_s/2)\delta\tau_{ik}]\approx 0$ and

$$\tilde{q}^2_{\text{out}} = \frac{\bar{E}}{N}\,\frac{m[1+(m-1)q^2]}{1+mq^2}. \tag{8.72}$$

For intensive interference $[(m-1)q^2 \gg 1]$ it follows from (8.72) that

$$\tilde{q}^2_{\text{out}} = (\bar{E}/N)(m-1). \tag{8.73}$$

Thus together with the complete cancellation of spatially correlated external interference one of m stations is "spent" for this procedure. Note that $\tilde{q}^2_{\text{out}}$ from (8.72) is by m times greater than $\tilde{q}^2_{\text{out}\,i}$ (8.65) in each station. Indeed, according to (8.37), (8.52) and (8.55) under the condition (8.42) m completely correlated signals are coherently summed in a background of mutually independent noises.

Example 8.3. Let us consider the detection probability, P_d, as a function of $\Delta f_s|\delta\tau_{12}|$ under the same conditions as in Example 8.2 from Section 8.4 but for the optimum algorithm L_2 (8.37). In Example 8.2 we have analyzed the detection probability for the suboptimum algorithm including only the optimum space–time linear processing at one station, envelope detection and threshold comparison of the quantity $|G_1|$ or $|G_2|$. According to (8.37) the optimum algorithm L_2 includes weighted and phased summation of G_1 and G_2.

The values of P_d for L_2 are calculated using (8.69) and (5.77). The corresponding curve (3) is plotted in the same Fig. 8.5 where the results of Example 8.2 are shown. Comparing the suboptimum with optimum algorithms (curves 1 and 3) we can evaluate the energy advantage of the latter one. It is seen that the energy gain is especially noticeable in minima of P_d. For the most important values of P_d when $P_d > 0.5$ this gain varies from 0 to 6.5 dB.

Example 8.4. The results of P_d calculations using (8.63), (8.69) and (5.77) are shown in Fig. 8.6 for a spatially coherent MSRS containing a single arbitrary positioned transmitting station and four equal receiving stations ($m=4$) positioned in the horizontal plane at the vertices of a square, i.e. at the points with following spherical coordinates: $(l_0, 0, 0)$, $(l_0, \pi/2, 0)$, $(l_0, \pi, 0)$ and $(l_0, 3\pi/2, 0)$. A single interference source (e.g., a mainlobe jammer) is located at the point with spherical coordinates $(R_0, \beta, \varepsilon)$ where azimuth $\beta = 60°$, elevation angle is arbitrary and $R_0 \gg l_0$. A target coincides with the interference source at the initial moment and then is displaced from the interference source along the azimuth direction by $\Delta\beta$. The normalized

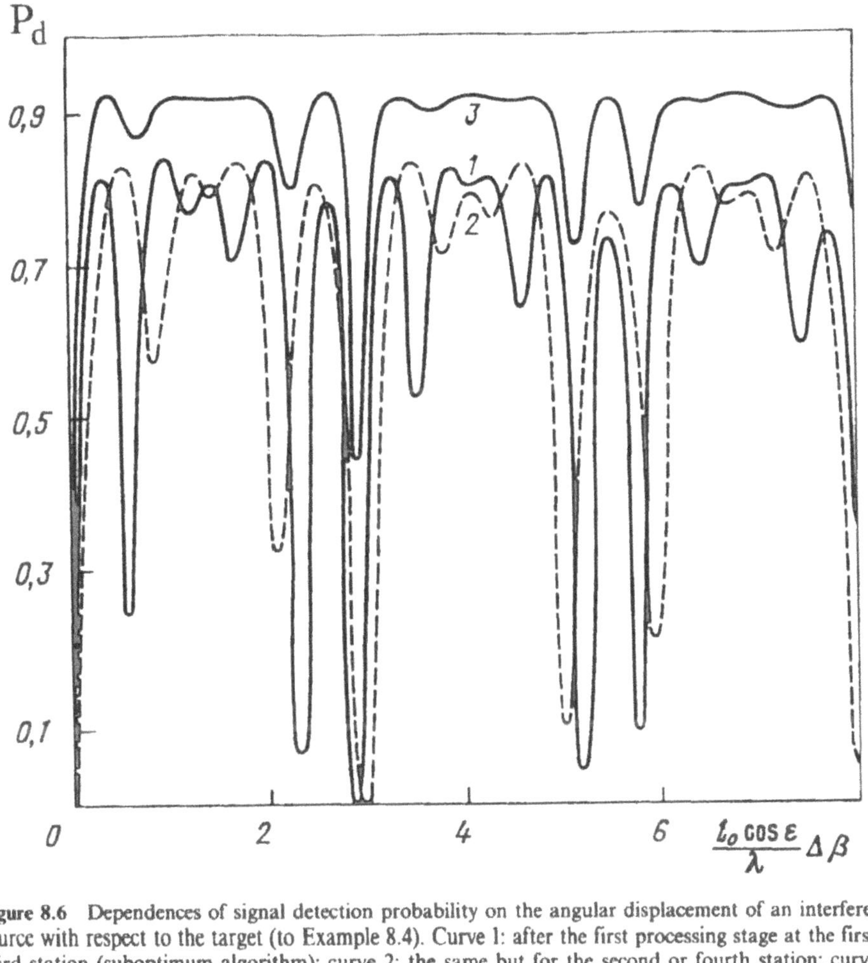

Figure 8.6 Dependences of signal detection probability on the angular displacement of an interference source with respect to the target (to Example 8.4). Curve 1: after the first processing stage at the first or third station (suboptimum algorithm); curve 2: the same but for the second or fourth station; curve 3: optimum algorithm; $m=4$, $P_{f_a}=10^{-4}$, $E/N=15\,\text{dB}$, $q^2=20\,\text{dB}$, $\Delta f/f_0=5\%$

target azimuth displacement $(l_0 \cos \varepsilon / \lambda)\Delta\beta$ is laid as abscissa. In Fig. 8.6 an initial segment of the signal interference region is shown and scaled up $[(l_0 \cos \varepsilon/\lambda)\Delta\beta \leqslant 8$ which corresponds to $|\delta\tau_{12}| \leqslant 0.4/\Delta f_s$ for $\Delta f_s/f_0 = 0.05]$. It can be seen that oscillations of detection probability, P_d, are smoothed as compared to the system with two receiving stations, $m=2$ (see Fig. 8.5) but violent variations remain. When the target and the interference source coincide in space $(|\delta\tau_{12}| = \Delta\beta = 0)$ the wanted signal is suppressed together with the interference. The optimum detector has the energy advantage over the suboptimum one in average of about 5 dB.

Example 8.5. Let us estimate the output SINR when several interference sources $(M=1,2,3,4)$ fall within the receiving antenna mainlobes of the MSRS considered in Example 8.4. In Fig. 8.7 the estimates of the cumulative probability distribution function of the SINR are presented. These results have been obtained by statistical simulation of random, uniformly distributed and mutually independent positions of interference sources within the sector $|\Delta\beta| \leqslant 10\lambda/l_0 \cos\varepsilon$ around the target direction $(\beta = 60°)$ and at a fixed value of ε. In each random experiment the SINR value was calculated using (8.57), (8.58) and (8.68). In Fig. 8.7(a) the results of the first stage of the optimum processing (or, which is the same, the results of the suboptimum processing considered above) are shown. It is seen that for spatially coherent signals with completely mutually dependent amplitude fluctuations the increase of the number of interference sources from $M=1$ to $M=3$ leads to the decrease of $\tilde{q}^2_{out\,1,3}$ which can be provided with high probability and to a certain increase of $\tilde{q}^2_{out\,1,3}$

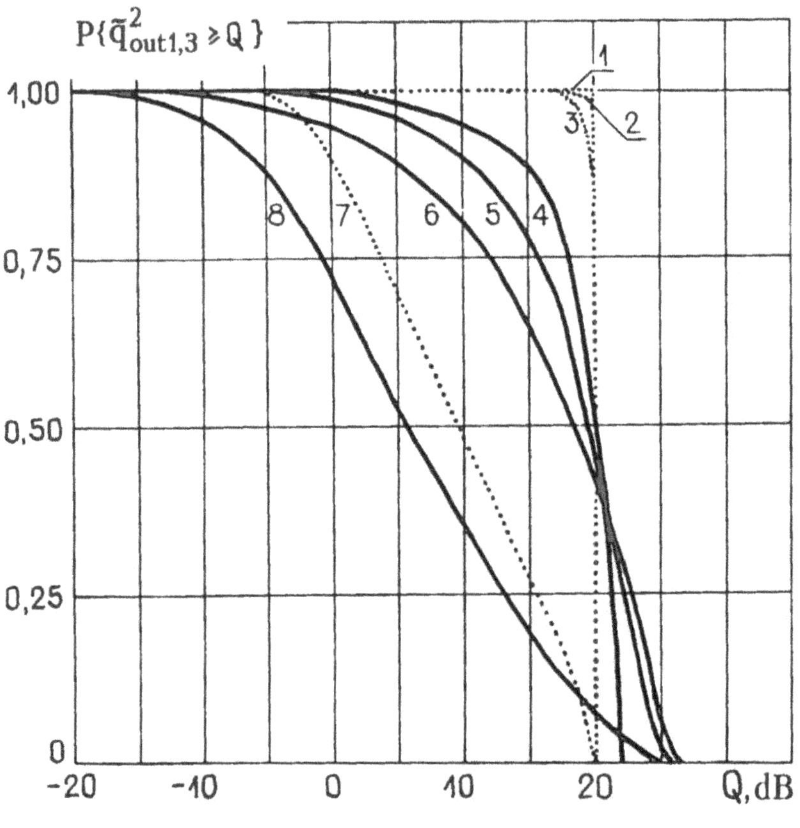

(a)

(b)

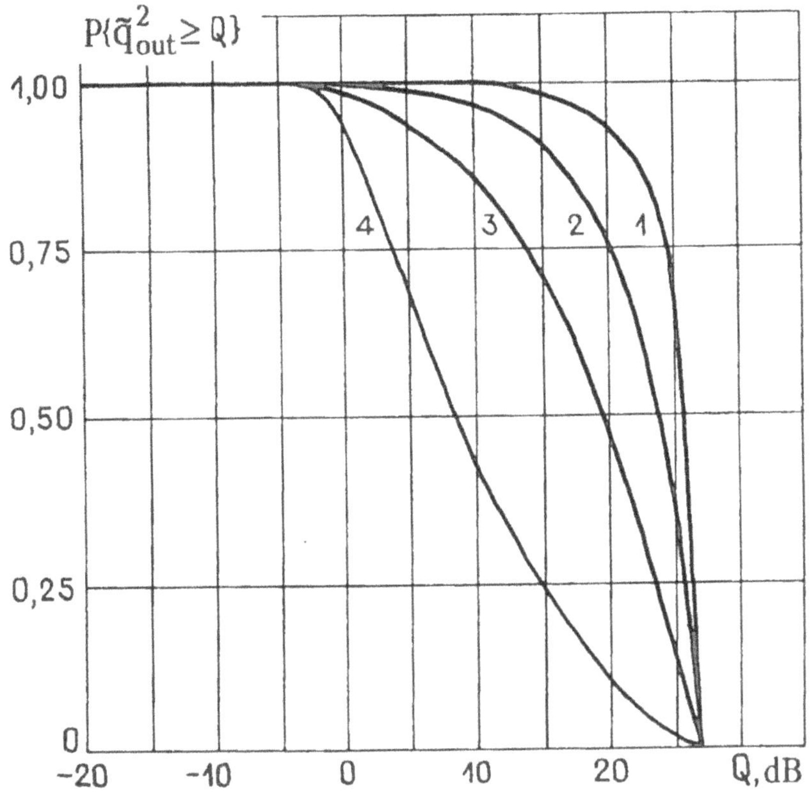

Figure 8.7 Probability distribution of the output SINR under interference condition from M randomly positioned independent sources (to Example 8.5), $q^2 = 20\,dB$, $\bar{E}/N = 20\,dB$. (a) after the first processing stage at any station; curves 1, 2, 3, 7: spatially incoherent signals with independent amplitude fluctuations; curves 4, 5, 6, 8: spatially coherent signals with completely correlated amplitude fluctuations; curves 1, 4: $M=1$, curves 2, 5: $M=2$, curves 3, 6: $M=3$, curves 7, 8: $M=4$; (b) after the second processing stage (optimum combination) for spatially coherent signals with completely correlated amplitude fluctuations; curve 1: $M=1$, curve 2: $M=2$, curve 3: $M=3$, curve 4: $M=4$

which are assured with the probability $P<0.4$. A sharp deterioration takes place for $M=4$ when the condition of effective external interference cancellation $m \geqslant M+1$ turns out to be violated (see Section 8.2).

For spatially incoherent signals with mutually independent amplitude fluctuations the performance is much better. The values of $\tilde{q}^2_{out\,1,3}$ remain the same as in the absence of external interferences with the probability close to one for $1 \leqslant M \leqslant 3$ until the condition $m \geqslant M+1$ is satisfied. However, the effectiveness of interference cancellation sharply declines when $M=4$.

In Fig. 8.7(b) the similar curves are plotted for the optimum algorithm L_2 (8.37). It can be seen that including the second stage of optimum processing (i.e. optimum summation) results in extra power gain as compared to the first stage only. This extra gain, however, decreases with the increase of the number of interference sources, M.

As a whole, Fig. 8.7 shows that until $m \geqslant M+1$ the output SINR is significantly enhanced due to the coherent external interference suppression and nears to the output SNR in the absence of external interferences. It is especially true for the case of spatially incoherent signals with mutually independent (at different stations) amplitude fluctuations. It should be noted that the "matched" processing, ignoring the spatial correlation of external interferences, provides (for assumed in Fig. 8.7 values of $\bar{E}/N = q^2 = 20\,\text{dB}$) the output SINR in each station $\tilde{q}^2_{\text{out }1,3} \approx -10\lg M\,\text{dB}$, and after coherent summation of outputs of four stations $\tilde{q}^2_{\text{out}} \approx (6 - 10\lg M)\,\text{dB}$.

Optimum Detectors for Spatially Incoherent Signals with Mutually Independent Amplitude Fluctuations in MSRSs with Long-Term or Short-Term Spatial Coherence

Several optimum algorithms derived in Section 8.3 (see Table 8.1) include (as their second stage) weighted summation of the squared moduli of the outputs of the first stage, G_i (see Fig. 8.4). These are L_3 (8.40), L_5 (8.48), L_6 (8.49). Similar is the algorithm L_7 (8.51) which includes calculating of a Hermite form of G_i, $i = \overline{1,m}$.

In order to analyze L_3, L_5 and L_6 it is sufficient to consider the following algorithm:

$$L = \sum_{i=1}^{m} |\tilde{G}_i|^2 = \tilde{\mathbf{G}}^* \tilde{\mathbf{G}}. \tag{8.74}$$

The algorithms L_3, L_5 and L_6 can be reduced to L by changing the variables. The PDF of the complex m-dimensional Gaussian vector $\tilde{\mathbf{G}}$ can be written in the form

$$w(\tilde{\mathbf{G}}) = (2\pi)^{-m} (\det \tilde{\mathbf{\Gamma}})^{-1} \exp(-0.5 \tilde{\mathbf{G}}^* \tilde{\mathbf{\Gamma}}^{-1} \tilde{\mathbf{G}}) \tag{8.75}$$

where $\tilde{\mathbf{\Gamma}} = \tilde{\mathbf{\Gamma}}_n$ or $\tilde{\mathbf{\Gamma}} = \tilde{\mathbf{\Gamma}}_s + \tilde{\mathbf{\Gamma}}_n$. Here $\tilde{\mathbf{\Gamma}}_n = \mathbf{H}\mathbf{\Gamma}_n\mathbf{H}$, $\tilde{\mathbf{\Gamma}}_s = \mathbf{H}\mathbf{\Gamma}_s\mathbf{H}$; $\mathbf{\Gamma}_n$ and $\mathbf{\Gamma}_s$ are determined in (8.52) and (8.54); $\mathbf{H} = \text{diag}(h_1, \ldots, h_m)$ is the diagonal matrix of the variable conversion coefficients from L_3, L_5 or L_6 into L. For L_3 (8.40) $h_i = A_{i1}$; for L_5 (8.48) $h_i = A_{i1}/\sqrt{1 + 0.5\sigma_1^2 A_{i1}^2 C_{ii}}$; for L_6 (8.49) $h_i = 1/\sqrt{C_{ii}}$.

If the random variables $\tilde{G}_i$, $i = \overline{1,m}$, were mutually independent (i.e. if the matrix $\tilde{\mathbf{\Gamma}}$ in (8.75) were a diagonal one) detection characteristics of the algorithm L (8.74) could be obtained from (5.86) and (5.90) representing the generalization of the chi-square probability distribution to the case where variances of $\tilde{G}_i$ are different. Therefore, it is reasonable to transform $\tilde{G}_i$ into mutually independent variables. It is known [25,34] that for a positive definite correlation (covariance) matrix $\tilde{\mathbf{\Gamma}}$ there exists a unitary matrix $\mathbf{U}$ so that the transformation $\tilde{\mathbf{G}} = \mathbf{U}\mathbf{X}$ provides diagonalization of $\tilde{\mathbf{\Gamma}}$:

$$0.5\overline{\mathbf{X}\mathbf{X}^*} = 0.5\mathbf{U}^*\overline{\tilde{\mathbf{G}}\tilde{\mathbf{G}}^*}\mathbf{U} = \mathbf{U}^*\tilde{\mathbf{\Gamma}}\mathbf{U} = \mathbf{\Lambda} = \text{diag}(\lambda_1, \ldots, \lambda_m) \tag{8.76}$$

where $\lambda_1, \ldots, \lambda_m$ are the eigenvalues of the matrix $\tilde{\mathbf{\Gamma}}$. Then instead of (8.75) we have the probability distribution of mutually independent Gaussian complex variables X_i, $i = \overline{1,m}$

$$w(\mathbf{X}) = (2\pi)^{-m} \left(\prod_{k=1}^{m} \lambda_k^{-1} \right) \exp\left(-0.5 \sum_{i=1}^{m} \lambda_i^{-1} |X_i|^2 \right). \tag{8.77}$$

The random variable L (8.74) does not change since $\mathbf{U}$ is a unitary matrix:

$$L = \sum_{i=1}^{m} |\tilde{G}_i|^2 = \tilde{\mathbf{G}}^*\tilde{\mathbf{G}} = \mathbf{X}^*\mathbf{U}^*\mathbf{U}\mathbf{X} = \mathbf{X}^*\mathbf{X} = \sum_{i=1}^{m} |X_i|^2. \tag{8.78}$$

It is sufficient in practice to find eigenvalues of the initial matrix $\tilde{\Gamma}$ in order to obtain immediately the probability distribution (8.77).

Now we can employ (5.86) and (5.90) for the analysis of the algorithm (8.78). Let the variances of $N \leqslant m$ variables $X_i, i=\overline{1,m}$, be different, i.e. the matrix $\tilde{\Gamma}$ has $N \leqslant m$ different eigenvalues λ_i. Let n_1 variables X_i have the variance equal to Q_1^2, n_2 variables X_i have the variance equal to Q_2^2, and so on. At last let n_N variables X_i have the variance equal to Q_N^2. It is clear that $n_1+n_2+\cdots+n_N=m$. Then the probability that L from (8.78) exceeds a threshold, is determined by (5.90). Thus having obtained the eigenvalues λ_i of the matrix $\tilde{\Gamma}_n=\mathbf{HCH}$ (which are equal to the variances of X_i in the absence of wanted signals) and using (5.90) we can calculate the threshold u_t corresponding to the allowed false alarm probability P_{fa}. Then having found the eigenvalues μ_i of the matrix $\tilde{\Gamma}=\mathbf{H}(\Gamma_s+\mathbf{C})\mathbf{H}$ (i.e. the variances of X_i in the presence of wanted signals) and using (5.90) again with the known threshold u_t we obtain the detection probability P_d.

Example 8.6. Let us consider again a system with a single transmitting and two receiving stations ($m=2$). One of receiving stations may coincide with the transmitting one. Let now the wanted signals be spatially incoherent with mutually independent amplitude fluctuations. The remaining conditions are the same as in Example 8.2 from Section 8.4: a single interference source ($M=1$), rectangular signal spectrum and interference PSD according to (8.30), effect of antenna ADPs are taken into account in signal and interference r.m.s. values, the condition (8.60) is valid. Then C_{ik} is determined by (8.62), $L_3=L_5=L_6=L$ and the eigenvalues of $\tilde{\Gamma}=\tilde{\Gamma}_n=\mathbf{C}$ are as follows:

$$\lambda_{1,2}=[2T_s/N(1+2q^2)](1+q^2\pm q^2|\operatorname{sinc}\Theta|); \quad \Theta=(\Delta\omega_s/2)\delta\tau_{12}. \tag{8.79}$$

The eigenvalues of $\tilde{\Gamma}=\tilde{\Gamma}_s+\mathbf{C}$ are given by

$$\mu_{1,2}=\frac{2T_s}{N(1+2q^2)}\left\{\left[1+q^2+\frac{\bar{E}}{N}\cdot\frac{(1+q^2)^2+q^4\operatorname{sinc}^2\Theta}{1+2q^2}\right]\right.$$
$$\left.\pm q^2\left(1+\frac{2\bar{E}}{N}\cdot\frac{1+q^2}{1+2q^2}\right)|\operatorname{sinc}\Theta|\right\}. \tag{8.80}$$

If the values of λ_1 and λ_2, μ_1 and μ_2 are different, (5.86) is convenient to be employed. Then the false alarm and detection probabilities, P_{fa} and P_d, can be obtained in the form

$$\left.\begin{aligned} P_{fa}&=(\lambda_1-\lambda_2)^{-1}[\lambda_1\exp(-u_t/2\lambda_1)-\lambda_2\exp(-u_t/2\lambda_2)], \quad \lambda_1>\lambda_2;\\ P_d&=(\mu_1-\mu_2)^{-1}[\mu_1\exp(-u_t/2\mu_1)-\mu_2\exp(-u_t/2\mu_2)], \quad \mu_1>\mu_2.\end{aligned}\right\} \tag{8.81}$$

Under the condition (8.42) $\lambda_1=\lambda_2=\lambda$ and $\mu_1=\mu_2=\mu$. Then [see (5.85)]

$$P_{fa}=(1+u_t/2\lambda)\exp(-u_t/2\lambda); \qquad P_d=(1+u_t/2\mu)\exp(-u_t/2\mu). \tag{8.82}$$

In Fig. 8.8 (curve 1) P_d as a function of $\Delta f_s|\delta\tau_{12}|$ is plotted for the same conditions as in Fig. 8.5. The curve corresponding to the suboptimum algorithm considered in Section 8.4, i.e. to the first stage of the optimum algorithms (curve 2 from Fig. 8.5), is also shown in Fig. 8.8 (curve 3). We referred to such curves as to RSRPs (see Example 8.2 in Section 8.4). It can be seen that performance of the optimum

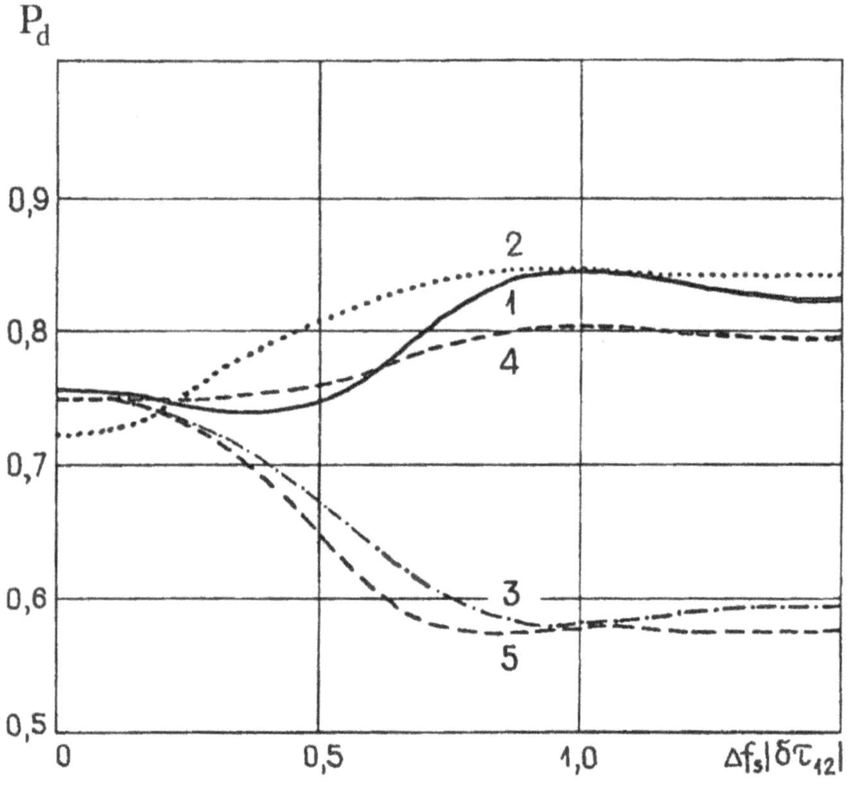

Figure 8.8 The same dependences as in Fig. 8.5 but for spatially incoherent signals with mutually independent amplitude fluctuations, $P_{fa}=10^{-4}$, $\bar{E}/N=15\,\text{dB}$, $q^2=20\,\text{dB}$. Curve 1: optimum algorithm L_3, L_5 or L_6; curve 2: optimum adaptive algorithm L_7; curve 3: after the first processing stage (without output combining); curves 4, 5: two stage algorithms with output combining using decision rules "1 out of 2" and "2 out of 2", respectively

algorithm $L_3=L_5=L_6=L$ including weighted summation of $|G_1|^2$ and $|G_2|^2$ does not differ from that of the suboptimum one in the region $|\delta\tau_{12}|\ll 1/\Delta f_s$. This is a result of high correlation between interference-plus-noise components, G_{n1} and G_{n2}. It follows from (8.52) and (8.62) that when $|\delta\tau_{12}|\ll 1/\Delta f_s$ the modulus of the mutual correlation coefficient, $|\rho_{n12}|$, is equal to $q^2/(1+q^2)$. For $q^2=20\,\text{dB}$ we have $|\rho_{n12}|\approx 0.99$. As $|\delta\tau_{12}|$ increases, the value of $|\rho_{n12}|$ decreases as $\text{sinc}(\pi\Delta f_s\,\delta\tau_{12})$, and the energy gain of the optimum algorithm rises. When $\Delta f_s|\delta\tau_{12}|>0.9$ then $\text{sinc}(\pi\Delta f_s\,\delta\tau_{12})\approx 0$ and the advantage of the optimum algorithm becomes significant. In particular, to obtain the same value of P_d for $\Delta f_s|\delta\tau_{12}|=1$ ($P_d\approx 0.84$) as with optimum algorithm, the SNR $\bar{E}/N$ for the suboptimum algorithm is to be increased up to 20 dB, i.e. by 5 dB. At the second stage of the optimum algorithms (see Fig. 8.4) mutually independent signal amplitude fluctuations are at least partially "smoothed" as a result of the weighted summation (as in the absence of spatially correlated interferences, see Section 5.5). Therefore, the higher the required detection probability P_d, the larger the energy gain at the second stage of the optimum detection algorithms.

Let us now consider the optimum adaptive algorithm L_7 (8.51) containing the initial vectors $\mathbf{G}$ instead of their transformations $\tilde{\mathbf{G}}=\mathbf{HG}$ as in (8.74). The probability distribution of $\mathbf{G}$ is given by (8.75) if $\tilde{\mathbf{G}}$ and $\tilde{\mathbf{\Gamma}}$ are replaced by $\mathbf{G}$ and $\mathbf{\Gamma}$, respectively. In order to employ the generalized chi-square distribution for calculating the false alarm and detection probabilities, such a linear transformation of $\mathbf{G}$ should be found which could diagonalize not only the matrix $\mathbf{\Gamma}$ in (8.75) but the matrix $\mathbf{C}$ in the algorithm (8.51) as well. In the absence of wanted signals $\mathbf{\Gamma}=\mathbf{\Gamma}_n=\mathbf{C}$ [see (8.52)] but in the presence of signals $\mathbf{\Gamma}=\mathbf{\Gamma}_n+\mathbf{\Gamma}_s=\mathbf{C}+\mathbf{\Gamma}_s\neq\mathbf{C}$.

It is known [34] that if at least one out of two matrices $\mathbf{\Gamma}$ and $\mathbf{C}$ is positive definite, a required transformation exists and the positive definite matrix can be brought to the identity matrix. The above condition is satisfied for the matrices under consideration [25,27].

Apparently, the simplest procedure is as follows. Using the known matrix $\mathbf{C}$ we can transform it into an equivalent upper triangle matrix $\mathbf{T}$ by elementary transformations [34]. Having inverted $\mathbf{T}$ we can obtain the diagonal matrix $(\mathbf{T}^{-1})^*\mathbf{C}\mathbf{T}^{-1}=\operatorname{diag}(\gamma_1,\ldots,\gamma_m)$ where γ_i, $i=\overline{1,m}$, are real positive numbers. Let us introduce a normalizing diagonal matrix $\mathbf{V}=\operatorname{diag}(1/\sqrt{\gamma_1},\ldots,1/\sqrt{\gamma_m})$ so that $\mathbf{V}(\mathbf{T}^{-1})^*\mathbf{C}\mathbf{T}^{-1}\mathbf{V}=\mathbf{I}$ where $\mathbf{I}$ is the identity matrix. Hence its inverse is the identity matrix too: $\mathbf{V}^{-1}\mathbf{T}\mathbf{C}^{-1}\mathbf{T}^*\mathbf{V}^{-1}=\mathbf{I}$. It follows from the last equation that we can reduce the Hermite form (8.51) to its canonical form and transform (8.75) to the probability distribution of mutually independent complex Gaussian random variables X_i, $i=\overline{1,m}$, with unit variance by the variable conversion $\mathbf{G}=\mathbf{T}^*\mathbf{V}^{-1}\mathbf{X}$.

$$\begin{aligned} L_7&=\mathbf{G}^*\mathbf{C}^{-1}\mathbf{G}=\mathbf{X}^*\mathbf{V}^{-1}\mathbf{T}\mathbf{C}^{-1}\mathbf{T}^*\mathbf{V}^{-1}\mathbf{X}=\mathbf{X}^*\mathbf{X}=\sum_{i=1}^{m}|X_i|^2;\\ w_n(\mathbf{X})&=(2\pi)^{-m}\exp(-0.5\mathbf{X}^*\mathbf{X})=(2\pi)^{-m}\exp\left(-0.5\sum_{i=1}^{m}|X_i|^2\right). \end{aligned} \tag{8.83}$$

For the case where wanted signals are present, the same variable conversion $\mathbf{G}=\mathbf{T}^*\mathbf{V}^{-1}\mathbf{X}$ transforms the Hermite form in (8.75) (after replacing $\tilde{\mathbf{G}}$ by $\mathbf{G}$ and $\tilde{\mathbf{\Gamma}}$ by $\mathbf{\Gamma}=\mathbf{C}+\mathbf{\Gamma}_s$) into

$$\mathbf{G}^*\mathbf{\Gamma}^{-1}\mathbf{G}=\mathbf{X}^*\mathbf{V}^{-1}\mathbf{T}\mathbf{\Gamma}^{-1}\mathbf{T}^*\mathbf{V}^{-1}\mathbf{X}=\mathbf{X}^*\mathbf{D}^{-1}\mathbf{X}. \tag{8.84}$$

There exists such a unitary transformation $\mathbf{X}=\mathbf{U}_1\mathbf{Y}$ which brings (8.84) to its canonical form (the sum of squared moduli), i.e. the matrix $\mathbf{D}$ to its diagonal form, without changing (8.83). Since $\mathbf{U}_1^*\mathbf{D}\mathbf{U}_1=\boldsymbol{\xi}=\operatorname{diag}(\xi_1,\ldots,\xi_m)$ where $\xi_1,\ldots,\xi_m$ are the eigenvalues of the matrix $\mathbf{D}$, we can obtain instead of (8.84)

$$\mathbf{G}^*\mathbf{\Gamma}^{-1}\mathbf{G}=\mathbf{X}^*\mathbf{D}^{-1}\mathbf{X}=\mathbf{Y}^*\mathbf{U}_1^*\mathbf{D}^{-1}\mathbf{U}_1\mathbf{Y}=\mathbf{Y}^*\boldsymbol{\xi}^{-1}\mathbf{Y}=\sum_{i=1}^{m}\xi_i^{-1}|Y_i|^2. \tag{8.85}$$

Substituting $\mathbf{X}=\mathbf{U}_1\mathbf{Y}$ in (8.83) and (8.85) in (8.75) yields the final expressions

$$\begin{aligned} L_7&=\mathbf{Y}^*\mathbf{Y}=\sum_{i=1}^{m}|Y_i|^2;\\ w_n(\mathbf{Y})&=(2\pi)^{-m}\exp\left(-0.5\sum_{i=1}^{m}|Y_i|^2\right);\\ w_{sn}(\mathbf{Y})&=(2\pi)^{-m}\left(\prod_{k=1}^{m}\xi_k^{-1}\right)\exp\left(-0.5\sum_{i=1}^{m}\xi_i^{-1}|Y_i|^2\right). \end{aligned} \tag{8.86}$$

Thus in order to obtain (8.86) from (8.51) and (8.75) following transformations should be performed step-by-step:

(1) derive $\mathbf{T}$ using the known matrix $\mathbf{C}$;
(2) obtain the inverse $\mathbf{T}^{-1}$;
(3) calculate $(\mathbf{T}^{-1})^*\mathbf{C}\mathbf{T}^{-1} = \mathrm{diag}(\gamma_1, \ldots, \gamma_m)$;
(4) write the normalized matrix $\mathbf{V} = \mathrm{diag}(1/\sqrt{\gamma_1}, \ldots, 1/\sqrt{\gamma_m})$;
(5) calculate the matrix $\mathbf{D} = \mathbf{V}(\mathbf{T}^{-1})^*\mathbf{\Gamma}\mathbf{T}^{-1}\mathbf{V}$ for the known matrix $\mathbf{\Gamma} = \mathbf{C} + \mathbf{\Gamma}_s$;
(6) find the eigenvalues $\xi_1, \ldots, \xi_m$ of the matrix $\mathbf{D}$.

Having performed the described procedures we can employ the usual chi-square probability distribution for calculating P_{fa} and the generalized chi-square distribution for calculating P_d. The normalized threshold level u_0 can be obtained from (5.85) for a given value of the false alarm probability, P_{fa}. Then the detection probability, P_d, can be calculated using (5.90) in the general case. If all eigenvalues $\xi_1, \ldots, \xi_m$ are different, (5.86) can be utilized. For these calculations in (5.90) and (5.86) P_{fa} and Q_i^2 should be replaced by P_d and ξ_i, respectively.

Example 8.7. Let us consider the optimum adaptive algorithm L_7 (8.51) under the same conditions as in Example 8.6 $[m=2, M=1, C_{ik}, i,k=\overline{1,m}$, are determined by (8.62)]. The eigenvalues of the matrix $\mathbf{D}$ can be obtained in the form

$$\xi_{1,2} = 1 + \bar{E}(1 + q^2 \pm q^2|\mathrm{sinc}\,\Theta|)/N(1+2q^2). \tag{8.87}$$

When $|\delta\tau_{12}| > 1/\Delta f_s$ and hence $\mathrm{sinc}\,\Theta = \mathrm{sinc}[(\Delta\omega_s/2)\delta\tau_{12}] \approx 0$, which means that wanted signals and interferences are resolved in TDOA and the condition (8.42) is satisfied, $\xi_1 = \xi_2 = \xi$. In this case P_d can be calculated with the help of (8.82) if μ is replaced by ξ from (8.87).

Using the technique described above we have calculated P_d as a function of $\Delta f_s|\delta\tau_{12}|$, i.e. the RSRP for the algorithm L_7. The results of those calculations are depicted in Fig. 8.8 (curve 2). It is seen that performance of the optimum adaptive algorithm L_7 (8.51), (8.86) in the region where $|\delta\tau_{12}| \ll 1/\Delta f_s$ is nearly the same as that of the optimum algorithm $L_3 = L_5 = L_6 = L$ (8.74) and of the suboptimum algorithm containing only the first stage of the optimum algorithms. As it was explained in Example 8.6, it is a result of high correlation between interference plus noise components of the outputs of both stations, G_{n1} and G_{n2}. When $\Delta f_s|\delta\tau_{12}| > 0.9$, then $\mathrm{sinc}(\pi\Delta f_s\delta\tau_{12}) \approx 0$ and the advantage of the optimum algorithm becomes significant. As for the optimum algorithm $L_3 = L_5 = L_6 = L$ (8.74), in order to obtain the same value of $P_d \approx 0.84$ for $\Delta f_s|\delta\tau_{12}| = 1$ with optimum adaptive algorithm L_7 (8.51), (8.86), the SNR $\bar{E}/N$ for the suboptimum algorithm is to be increased up to 20 dB, i.e. by 5 dB. It is seen from Fig. 8.8 that for the assumed values of parameters the optimum algorithm $L_3 = L_5 = L_6 = L$ (8.74) is slightly better than the optimum adaptive algorithm L_7 (8.51), (8.86) for $|\delta\tau_{12}| < 0.2/\Delta f_s$ and slightly worse for the larger values of $|\delta\tau_{12}|$. However, these differences are small.

Example 8.8. Now we return to the MSRS which was considered in Example 8.4 (with four receiving stations, $m=4$, positioned in the horizontal plane at the vertices of a square) but for spatially incoherent signals with mutually independent amplitude fluctuations. The similar RSRPs as in Fig. 8.8 (curves 1, 2, 3) are plotted in Fig. 8.9 for the same conditions as in Fig. 8.6 (excluding the value of SNR $\bar{E}/N = 10$ dB instead of 15 dB). It can be derived from (8.52) and (8.62) that the modulus of the output interference plus noise components' correlation coefficient for

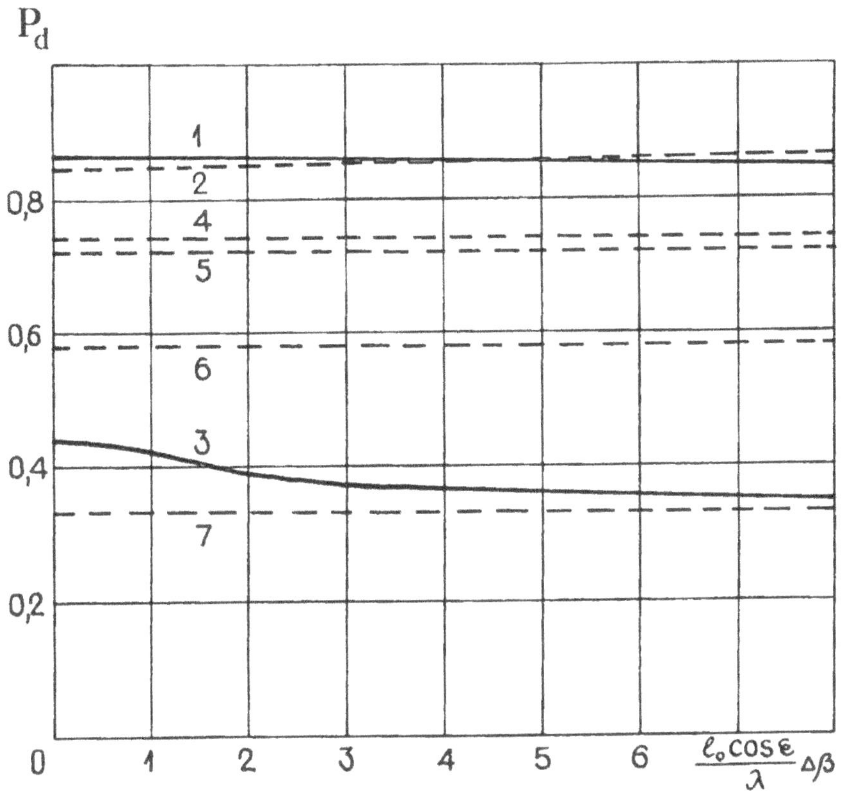

Figure 8.9 The same dependences as in Fig. 8.6 but for spatially incoherent signals with mutually independent amplitude fluctuations, $P_{f_n} = 10^{-4}$, $E/N = 10\,\text{dB}$, $q^2 = 20\,\text{dB}$. Curves 1, 2, 3: the same algorithms as in Fig. 8.8, respectively; curves 4, 5, 6, 7: combining with the decision rules "1 out of 4", "2 out of 4", "3 out of 4", "4 out of 4", respectively

any m and $|\delta\tau_{12}| \ll 1/\Delta f_s$ can be expressed as $|\rho_{nik}| = q^2/[1+(m-1)q^2]$. Therefore, $|\rho_{nik}| \approx 0.33$ for $m=4$ and $q^2 = 20\,\text{dB}$. Owing to correlation reduction (compare with $|\rho_{nik}| \approx 0.99$ in the case of two receiving stations in Examples 8.6 and 8.7) the optimum algorithms have a significant energy advantage over the suboptimum one even when $|\delta\tau_{12}| = \Delta\beta = 0$. For the same reason P_d for the optimum algorithms is nearly independent of $\Delta\beta$.

Example 8.9. In Fig. 8.10 detection characteristics of the adaptive optimum algorithm L_7 (8.51), (8.86) (curve 1) and of the suboptimum algorithm under consideration (curve 2) are plotted for the same MSRS as in Example 8.8 and under the same conditions. These characteristics permit graphic evaluation of the energy advantage of weighted summation of stations' outputs. Both characteristics correspond to the most interesting and difficult case where the target and the interference source coincide in space ($\Delta\beta = 0$). It can be seen that the weighted summation of the outputs from four receiving stations provides a significant energy gain. This gain increases with the increase of detection probability, P_d, as a result of "smoothing" of mutually independent signal amplitude fluctuations.

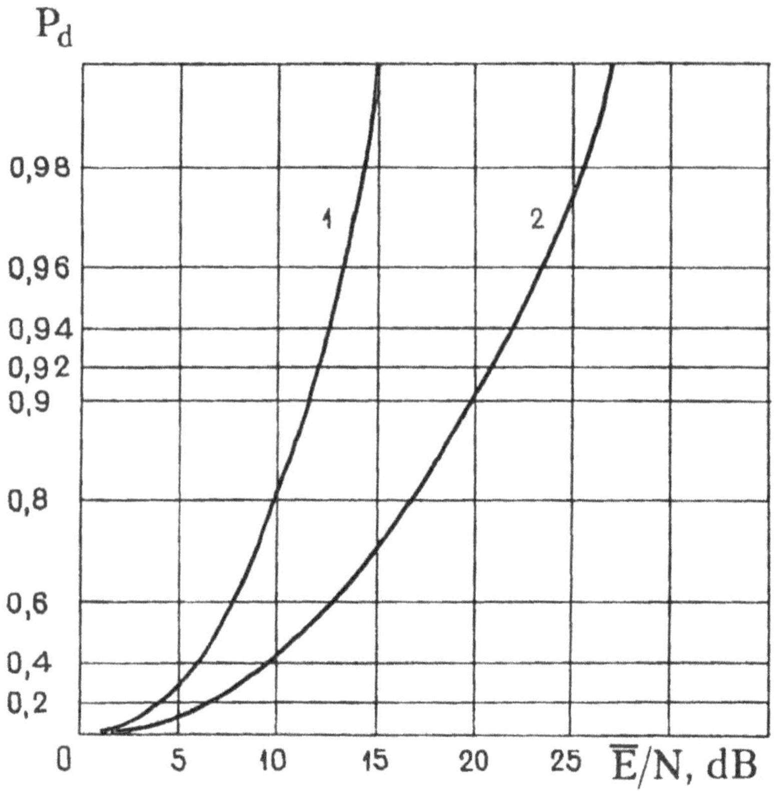

Figure 8.10 Detection characteristics of the optimum adaptive algorithm L_7 (curve 1) and of the suboptimum algorithm without the second processing stage (curve 2). Signals are the same as in Fig. 8.9, $P_{fa}=10^{-4}$, $q^2=20$ dB; The interference source coincides with the target in space

Suboptimum Detectors Combining Preliminary Decisions from Different Stations According to the "k out of m" Decision Rules

As was mentioned in Section 8.3, the variables $|G_i|$ or $|G_i|^2$, $i=\overline{1,m}$, obtained after cancellation of spatially correlated interferences at the first stage of joint processing in such a MSRS, are compared with a threshold in each station individually. The final decision "a target is present" is made at the fusion centre (FC) when the thresholds are exceeded in no less than k stations out of the total quantity of m stations in corresponding range resolution cells. These are the "k out of m" decision rules (the coincidence criteria). Unlike the case of decentralized detection in the absence of spatially correlated interferences (see Sections 6.1, 6.2) wideband DTLs are necessary in the case considered for the spatially correlated interference cancellation at the first stage of processing. Nevertheless, decentralization of the decision making simplifies the required processing and hence is of practical importance.

Simple analytic expressions for detection characteristics can be obtained only for mutually independent G_i, $i=\overline{1,m}$. The interference plus noise components, G_{ni}, are mutually independent when wanted signals and interferences are resolved in TDOA, i.e. when the condition (8.42) is satisfied [see (8.52)]. The signal components, G_{si}, are mutually independent when the condition (8.42) is satisfied and besides expected signals (target echoes) are spatially incoherent with mutually independent amplitude fluctuations [see (8.53)–(8.56)]. It follows from Section 6.2 that the most practically important is the case where probabilities of threshold exceeding are equal at all receiving stations in both the absence and the presence of wanted signals, i.e. $P_{fa\,i}=P_{fa0}$ and $P_{di}=P_{d0}$, $i=\overline{1,m}$. In this case output false alarm and detection probabilities, P_{fa} and P_d, are determined by the expression (6.14). For the Rayleigh amplitude fluctuations P_{fa0} and P_{d0} are connected by the relationship (5.77) where $\tilde{q}^2_{out\,i}$, $i=\overline{1,m}$, should be calculated from (8.59).

If the condition (8.42) is not satisfied, the mutual correlation between G_i, $i=\overline{1,m}$, is to be taken into account. Under this condition the performance analysis is much more complicated. However, when $|\delta\tau_{ik}|\ll 1/\Delta f_s$ and for intensive interferences ($q^2\gg 1$) the mutual correlation between interference plus noise components, G_{ni}, decreases rapidly with the increase of the number of receiving stations m (see Example 8.8). This is valid for signal components, G_{si}, as well when expected signals are spatially incoherent with mutually independent amplitude fluctuations. Therefore, when $m\geqslant 3\ldots 4$ and for approximate performance evaluations one may calculate P_{fa} (or the threshold for the given P_{fa}) using (6.14) for arbitrary expected signals. Under the same conditions P_d may be calculated using (6.14) for spatially incoherent expected signals with mutually independent amplitude fluctuations. It should be noted that when $m=2$ and for arbitrary mutual correlation of G_i the required probabilities P_{fa} and P_d can be calculated with the help of the known precise expression for the two-dimensional probability distribution of the envelope of a Gaussian random process [33].

Example 8.10. Let us apply the obtained results to the MSRSs considered in Examples 8.6 and 8.9. In Fig. 8.8 the RSRPs (P_d as functions of $\Delta f_s|\delta\tau_{12}|$) are shown for the decision rules "1 out of 2", "OR" (curve 4) and "2 out of 2". "AND" (curve 5). These curves are obtained with the help of the expressions from [33] under the same conditions as all other curves in this figure. In the region where $\Delta f_s|\delta\tau_{12}|>1$ the values of P_d calculated using the precise expressions from [33] do not differ from the corresponding values of P_d obtained from (6.14). In Fig. 8.9 the similar RSRPs are plotted using (6.14) which are only approximate for $|\Delta\beta|<8\lambda/l_0\cos\varepsilon$ (or $|\delta\tau_{12}|<0.4/\Delta f_s$). For $m=4$ the following decision rules are employed: "1 out of 4", "OR" (curve 4), "2 out of 4" (curve 5), "3 out of 4" (curve 6) and "4 out of 4", "AND" (curve 7). The presented curves allows us to compare the performance of these suboptimum decentralized algorithms with the optimum centralized algorithms as well as with the suboptimum algorithm considered above (see Section 8.4) without integration of outputs of different stations. It is seen that for $m=2$ the performance of a decentralized algorithm with the decision rule "1 out of 2" is nearly the same as that of the optimum algorithm whereas the performance of another algorithm with the decision rule "2 out of 2" is close to that of the suboptimum algorithm without integration of stations' outputs. For $m=4$ the detection characteristics from Fig. 8.10 (Example 8.9) are convenient to be utilized. For instance, the energy losses of the decentralized algorithm with the decision rule "1 out of 4" with respect to the optimum algorithms are less than 1 dB whereas the energy losses of the decentralized algorithm with the decision rule "4 out of 4" is nearly the same as those of the suboptimum algorithm without output integration.

It should be noted that in Figs. 8.8 and 8.9 (Examples 8.6–8.8, 8.10) we assumed that antenna ADPs of the receiving stations are real, they do not depend on frequency within the signal bandwidth Δf_s and do not affect the RSRPs for considered small values of arguments ($|\delta\tau_{12}<1.5/\Delta f_s$ for $m=2$, $|\Delta\beta|<8\lambda/l_0\cos\varepsilon$ for

$m=4$). In the case of very narrow antenna beamwidths the effect of antenna ADPs can be taken into account with the help of the general relationships (8.52), (8.54), (8.56) and (8.35).

In this section, *we have analyzed optimum processing algorithms synthesized in Section 8.3. We pursued mainly two aims: firstly, to derive expressions for calculating detection characteristics and, secondly, to evaluate energy gain due to signal integration from the outputs of receiving stations, i.e. the contribution of the second stage of optimum processing (after external interference cancellation at the first stage which has been analyzed in Section 8.4).*

We have derived the expression (8.68) for the output SINR which completely determines detection characteristics of the algorithm L_2 (8.37). As shown in Section 8.3, this detection algorithm is optimal for spatially coherent expected signals with perfectly mutually dependent amplitude fluctuations in spatially coherent MSRSs.

For the more complex case (which occurs more often in practice) where expected signals are spatially incoherent with mutually independent amplitude fluctuations, we have shown that the synthesized optimum algorithms L_3 (8.40), L_5 (8.48) and L_6 (8.49) can be reduced to the common algorithm L (8.74) by variable conversion. Using this result and the known technique of correlation (covariance) matrix diagonalization by unitary transformations we have replaced (8.74) and (8.75) by (8.78) and (8.77), respectively. This transformation makes it possible to utilize expressions (5.90), (5.86) from Section 5.5 for detection characteristic calculations. By a similar (though a slightly more complex) transformation we have also derived expressions (8.86) permitting to use (5.85), (5.90) or (5.86) for performance analysis of the optimum adaptive algorithm L_7 (8.51).

Besides the centralized optimum algorithms we have analyzed the decentralized suboptimum ones which employ the decision rules of "k out of m" type for combining preliminary decisions (about the presence or absence of expected signals) made in each receiving station. Performance (detection) characteristics can be calculated with the help of (6.14) and (5.77).

We have considered many numerical examples (Examples 8.3–8.10) which illustrate applications of the developed techniques for performance analysis and permit comparative numerical evaluation of performance characteristics of different optimum and suboptimum algorithms in practically important particular cases. Calculation results of these Examples have been presented in an easy-to-grasp form of plots (Figs. 8.6–8.10).

9. ADAPTIVE CANCELLATION OF NOISE-LIKE EXTERNAL INTERFERENCES IN ACTIVE MSRSs. TARGET DETECTION IN PASSIVE MSRSs IN A BACKGROUND OF NOISE-LIKE EXTERNAL INTERFERENCES

9.1. ADAPTIVE CANCELLATION USING SYSTEMS OF ORTHOGONAL FILTERS

It is reasonable to recall that noise-like interference from a point-like source (e.g., a jammer) is strongly correlated at the inputs of spatially separated receiving stations (after TDOA elimination and Doppler frequency equalization, if necessary). Thus we face *the spatially correlated interferences.*

As shown in Sections 8.2 and 8.3, one of the most important procedures of target detection in a background of spatially correlated external interferences (e.g., jamming) is cancellation (coherent mutual suppression) of those interferences. For the synthesis of optimum and suboptimum detection algorithms in Chapter 8 we assumed the matrix $\mathbf{\Phi}(\omega)$ of the sum of external interferences plus self-noises at the inputs of spatially separated stations as being known. This assumption allowed us to derive the structure of interference cancellation linear filters.

In real-world situations the matrix $\mathbf{\Phi}(\omega)$ is usually unknown and may vary in time (slowly, as compared with the reciprocity of signal and interference bandwidths). These variations may be caused, for instance, by the motion of interference sources, variations of their radiation intensities and so on. Therefore, adaptive processing algorithms[1] are necessary which could "follow" variations of $\mathbf{\Phi}(\omega)$ in time. Obviously, performance of such adaptive algorithms under steady-state conditions is to be sufficiently close to that of the optimum ones synthesized for known $\mathbf{\Phi}(\omega)$.

There is a wealth of literature devoted to adaptive antenna arrays where similar problems are solved for monostatic radars with phased antenna arrays (PAAs) (see, e.g., [35]). Such radars usually satisfy the conditions of space–time processing separation into spatial and temporal processing (see Section 8.1). External interference cancellation is carried out with the help of a multichannel *inertialess* spatial filter, i.e. by the linear weighted summation. Adaptation is reduced to the alignment of weights before summation. This alignment is based on estimates of the correlation (covariance) matrix of interference-plus-self-noise complex amplitudes.

As shown in Sections 8.1 and 8.2, the space–time processing cannot usually be separated into spatial and temporal processing in MSRSs, so that a linear *inertial* filter is necessary for external interference cancellation. In other words, not only interference amplitudes and phases are to be adaptively aligned before subtraction

[1] Adaptive detection algorithms with adaptation to the unknown matrix $\mathbf{\Phi}(\omega)$ considered in this chapter should not be confused with the adaptive detection algorithms derived in previous chapters where only wanted signal parameters [initial phases and r.m.s. values but not $\mathbf{\Phi}(\omega)$] were assumed to be unknown, and the generalized likelihood detection criterion was used.

in a MSRS, but interference spectra at the different receiving stations as well. Such alignment is required to eliminate "linear distortions" of spectra caused by the difference in transfer functions of the propagation medium and receiver channels from an interference source to the input of subtractor (including delay differences). Thus an adaptive inertial filter is necessary for external interference cancellation in MSRSs [61].

In general space–time processing may be inseparable in monostatic radars too (for instance, if the signal propagation time along an antenna aperture is of the same order or greater than the reciprocity of the signal bandwidth). However, if such situations are usual for MSRSs they are rather exclusions for monostatic radars. The theory of adaptive spatial filtration (for separable processing) has been developed to a much higher degree than the general theory of adaptive space–time processing.

Optimum Adaptive Detectors

Let us consider the optimum detector for deterministic signals (8.20), (8.21) and Fig. 8.2. We have shown in Section 8.3 that the external interference cancellation algorithm is the same for both deterministic signals and signals with random complex amplitudes (fluctuating signals). Therefore, for the sake of simplicity and clarity of results, we assume here wanted signals to be deterministic.

It follows from Section 8.2 that the total power of interferences-plus-self-noises is minimized at the output of each summator in the left section of the diagram in Fig. 8.2. It is achieved thanks to the fact that in each ith receiving station input random realizations of the sum of interferences and self-noises from other $m-1$ stations pass through a Wiener filter with $m-1$ channels (before summation with a realization received by the ith station). The channel transfer functions of this filter are determined by (8.22) and (8.23).

Adaptive Wiener filters can be implemented by reducing the adaptive alignment of continuous functions $K_{ik}(\omega)$ to that of a finite number of parameters. To make it possible it is necessary to limit a class of functions $K_{ik}(\omega)$. This limitation should not lead to the essential deterioration of filtration quality.

In practice, without loss of generality, it is sufficient that equations (8.22), (8.23) to be valid only within the frequency bandwidth which is equal to or a little greater than the signal bandwidth $\omega \in (\omega_0 - \Delta\omega_s/2, \omega_0 + \Delta\omega_s/2)$. Within this frequency region each $K_{ik}(\omega)$, $i, k = \overline{1, m}$, can be represented by a generalized Fourier series [61]. Let us consider an arbitrary system of complex functions $\{\varphi_l(\omega)\}$ which are orthonormal and complete in the interval $(\omega_0 - \Delta\omega_s/2, \omega_0 + \Delta\omega_s/2)$. Then $K_{ik}(\omega)$ can be written in the form

$$K_{ik}(\omega) = \sum_{l=-\infty}^{\infty} C_{ikl}\varphi_l(\omega) \tag{9.1}$$

where the Fourier coefficients are

$$C_{ikl} = \frac{1}{2\pi}\int_{\omega_0 - \Delta\omega_s/2}^{\omega_0 + \Delta\omega_s/2} K_{ik}(\omega)\varphi_l^*(\omega)\,d\omega, \quad i, k = \overline{1, m} \tag{9.2}$$

and

$$\frac{1}{2\pi}\int_{\omega_0 - \Delta\omega_s/2}^{\omega_0 + \Delta\omega_s/2} \varphi_l(\omega)\varphi_n^*(\omega)\,d\omega = \delta_{ik} = \begin{cases} 1, & l = n, \\ 0, & l \neq n. \end{cases}$$

Thus *the problem of adaptive control of complex transfer functions* $K_{ik}(\omega)$ *of a Wiener filter has been reduced to the problem of adaptive control of a countable set of complex Fourier coefficients* C_{ikl}. Truncating the series (9.1) yields a *suboptimal adaptive filter*

$$\tilde{\mathbf{K}}_i(\omega)=\mathbf{C}_i\boldsymbol{\varphi}(\omega), \quad i=\overline{1,m}, \tag{9.3}$$

where $\mathbf{C}_i$ is the matrix of Fourier coefficients $\|C_{ikl}\|$, $\boldsymbol{\varphi}(\omega)$ is the vector of functions $\|\varphi_l(\omega)\|$, $k=\overline{1,m}$, $l=\overline{-N,N}$. As well known, the Fourier coefficients provide the best r.m.s. approximation of a finite sum of the Fourier series (9.1) to the optimum transfer function $K_{ik}(\omega)$, i.e. $\tilde{\mathbf{K}}_i(\omega)$ to $\mathbf{K}_i(\omega)$.

The Fourier coefficients C_{ikl} in (9.2) can be adaptively generated with the help of different algorithms which are used in adaptive PAAs, for instance, algorithms based on the correlation feedback [35,72].

Consider Fig. 9.1 where the ith summator of the left part of the diagram in Fig. 8.2 is depicted with its all input channels. Each filter with the transfer function $K_{ik}(\omega)=f_{ik}(\omega)/f_{ii}(\omega)$, $k=\overline{1,m}$, is replaced by the multichannel transversal filter with the transfer function

$$H_{ik}(\omega)=\sum_{l=-N}^{N} W_{ikl}\varphi_l(\omega). \tag{9.4}$$

Let us choose the coefficients W_{ikl} to be proportional (under steady-state conditions) to the cross-correlation between the process at the output of the orthogonal filter with the transfer function $\varphi_l(\omega)$ of the kth station (i.e. the process at the input of the element with the gain W_{ikl}), on the one hand, and the process at the output of the ith summator, on the other:

$$W_{ikl}=\frac{\gamma}{2\pi}\int_{\omega_0-\Delta\omega_s/2}^{\omega_0+\Delta\omega_s/2}\Phi_{ikl}(\omega)\,d\omega. \tag{9.5}$$

Here γ is the multiplier that is proportional to the degree of correlation feedback, $\Phi_{ikl}(\omega)$ is the mutual PSD of those processes (see Fig. 9.1):

$$\Phi_{ikl}(\omega)=\lim_{T\to\infty}\frac{1}{2T}\overline{\chi_k^*(\omega)\varphi_l^*(\omega)\left[\chi_i(\omega)+\sum_{p=1,\,p\neq i}^{m}\chi_p(\omega)\sum_{n=-N}^{N}W_{ipn}\varphi_n(\omega)\right]}$$

$$=\varphi_l^*(\omega)\left[\Phi_{ik}(\omega)+\sum_{p=1,\,p\neq i}^{m}\Phi_{pk}(\omega)\sum_{n=-N}^{N}W_{ipn}\varphi_n(\omega)\right]. \tag{9.6}$$

Substituting (9.6) into (9.5) and assuming $\gamma\gg 1$ [when the left part of (9.5) may be neglected in comparison with the term containing W_{ikl} in the right part] yields

$$\frac{1}{2\pi}\int_{\omega_0-\Delta\omega_s/2}^{\omega_0+\Delta\omega_s/2}\varphi_l^*(\omega)\left[\Phi_{ik}(\omega)+\sum_{p=1,\,p\neq i}^{m}\Phi_{pk}(\omega)\sum_{n=-N}^{N}W_{ipn}\varphi_n(\omega)\right]d\omega=0,$$

$$k=\overline{1,m},\quad k\neq i,\quad l=\overline{-N,N}. \tag{9.7}$$

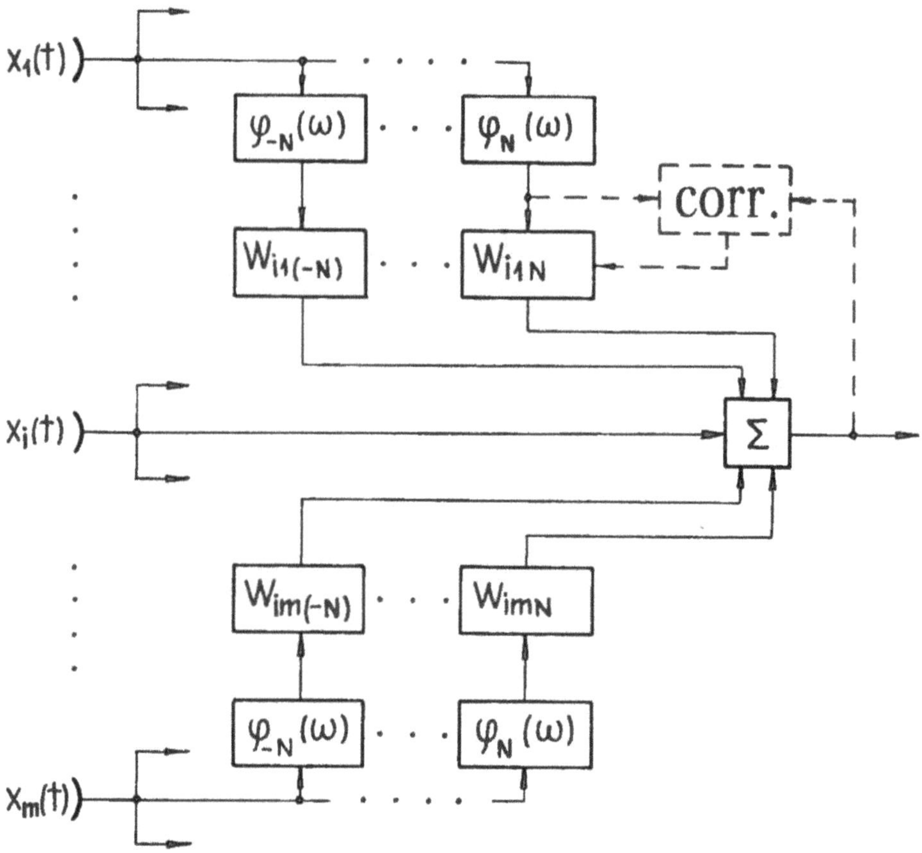

Figure 9.1 A section connected with the ith summator ($i=\overline{1,m}$) of the detector with adaptive interference cancellation. This is a transversal filter in each receiving station with $2N+1$ channel filters and adaptively controlled weights (a correlation feedback circuit is shown by dashed lines)

It follows from the completeness of the $\{\varphi_l(\omega)\}$ that when N is large (strictly speaking, when $N\to\infty$) the equation (9.7) is valid for any $l=\overline{-N,N}$ only if

$$\Phi_{ik}(\omega)+\sum_{p=1,\,p\neq i}^{m}\Phi_{pk}(\omega)\sum_{n=-N}^{N}W_{ipn}\varphi_n(\omega)=0,\quad k=\overline{1,m},\ k\neq i. \tag{9.8}$$

Taking into account (9.4) we obtain from (9.8)

$$\tilde{\mathbf{\Phi}}_i(\omega)+\tilde{\mathbf{\Phi}}^{(i)t}(\omega)\mathbf{H}_i(\omega)=\mathbf{0}. \tag{9.9}$$

Mutually independent receiver self-noises are included in $\tilde{\mathbf{\Phi}}^{(i)}(\omega)$, so that $\det\tilde{\mathbf{\Phi}}^{(i)}(\omega)\neq 0$, and we have from (9.9)

$$\mathbf{H}_i(\omega)=-[\tilde{\mathbf{\Phi}}^{(i)t}(\omega)]^{-1}\tilde{\mathbf{\Phi}}_i(\omega). \tag{9.10}$$

The notation in (9.9) and (9.10) is the same as in (8.22).

Let us compare (9.10), (8.22), (9.4) and (9.1). It follows from this comparison that a transversal filter (see Fig. 9.1) based on a complete system of orthogonal filters with transfer functions $\varphi_l(\omega)$ and containing complex weights W_{ikl} generated with the help of correlation feedback, approximates (under steady-state condition) a Wiener filter if the number of channels (orthogonal filters) is sufficiently large and the correlation feedback is sufficiently deep. As has been shown in Section 8.2, a Wiener filter provides optimum cancellation of spatially correlated interferences. Besides, under the same conditions the weights W_{ikl} coincide with the coefficients C_{ikl} of the generalized Fourier expansion of the Wiener filter's transfer functions $\mathbf{K}_i(\omega)$ by orthonormal functions $\varphi_l(\omega)$. The latter provides the best r.m.s. approximation $\mathbf{H}_i(\omega)$ to $\mathbf{K}_i(\omega)$ for a finite number of channels (orthogonal filters). Thus *adaptive cancellation (coherent mutually suppression) at the first stage of optimum detection algorithms* (see Figs. 8.2 and 8.4) *can be performed with the help of a multichannel transversal filter with orthogonal channels.*

Adaptive transversal filters can be implemented in both analogous and digital versions. In an analogous version the weights W_{ikl} are proportional to the outputs of correlators placed between the summator output and the input of each weight amplifier (such a control circuit is shown in one of channels in Fig. 9.1 by dashed lines). Each complex weight W_{ikl} is generated either with the help of two quadrature channels or in a single channel by additional heterodyning (frequency shifting) [35,72,73]. In a digital version which is widely used in last years, different gradient algorithms may be utilized [35,72,73].

An adaptive transversal filter with a large number of orthogonal channels can eliminate all spectral distortions of interferences on the way from the sources through the spatially separated receivers to interference subtractors. It becomes especially clear if a system $\{\varphi_l(\omega)\}$, $l=\overline{-N,N}$, is chosen in the form of a set of narrowband filters covering the signal frequency bandwidth $(\omega_0-\Delta\omega_s/2,\ \omega_0+\Delta\omega_s/2)$ and having rectangular nonoverlapping amplitude–frequency response with linear phase–frequency response. Obviously, increasing the number of filters (by narrowing their bandwidths) and using proper weights makes it possible to approximate the interference-plus-self-noise PSD of arbitrary form with any degree of accuracy (within the signal bandwidth) and to obtain a required form of mutual interference PSD in the "main" and "auxiliary" ("compensating") receiver channels. Though filters with strictly rectangular amplitude–frequency response are physically nonrealizable, real filters can be constructed with characteristics close to those.

***In this section**, we have analyzed the possibility of adaptive external interference (e.g., jamming) cancellation at the first stage of optimum detection algorithms considered in Chapter* 8. *We have shown that the problem of adaptive control of the complex transfer functions $K_{ik}(\omega)$ of a Wiener filter* (8.22) *can be reduced to the problem of adaptive control of a countable set of the complex coefficients C_{ikl}* (9.2). *These are coefficients of the generalized Fourier expansion* (9.1) *of $K_{ik}(\omega)$ by an arbitrary complete system of orthonormal functions $\{\varphi_l(\omega)\}$ within the signal bandwidth. Truncating the Fourier series yields a suboptimal adaptive filter* (9.3). *Such a filter can be implemented by a multichannel transversal filter* (9.4) *with orthogonal channels having transfer functions $\varphi_l(\omega)$, $l=\overline{-N,N}$, and adaptively controlled weights before summation* (Fig. 9.1). *The required adaptive weights can be obtained, for instance, with the help of correlation feedback* [(9.5), (9.6)]. *We have established* [*by equations* (9.7)–(9.10)] *that when the number of orthogonal channels is sufficiently large and the correlation feedback is deep enough, such an adaptive transversal filter is a close approximation to the Wiener filter* (8.22) *of the optimum detection algorithms.*

9.2. EXTERNAL INTERFERENCE ADAPTIVE CANCELLATION WITH THE HELP OF A SMALL NUMBER OF NONORTHOGONAL FILTERS

Suboptimum Detectors with Nonorthogonal Filters

In practice the number of channels with adaptively controlled weights of the transversal filters considered in Section 9.1 is desirable to be reduced. It is often possible if the matrix of interference-plus-self-noise PSDs at the inputs of spatially separated receiving stations, $\mathbf{\Phi}(\omega)$, varies within *a priori* known limits. For instance, when interference sources move within a certain region in space, only intensities, mutual phase shifts and TDOAs of interferences usually vary in time. Using this *a priori* information about $\mathbf{\Phi}(\omega)$ one can reduce the number of transversal filter channels and choose the most expedient system of channel filter transfer functions $\{\varphi_l(\omega)\}$, $l=\overline{-N,N}$.

However, when N is not large, the coefficients W_{ikl} determined by (9.5) (see Fig. 9.1) differ from the Fourier coefficients, C_{ikl}, determined by (9.2). It means that the transfer functions $\mathbf{H}_i(\omega)=[H_{i1}(\omega),\ldots,H_{im}(\omega)]^t$ (9.4) of the transversal filter differ in general from optimum (9.10). Let us find out the features of W_{ikl} in this situation. To do this we denote in (9.7):

$$\frac{1}{2\pi}\int_{\omega_0-\Delta\omega_s/2}^{\omega_0+\Delta\omega_s/2}\Phi_{ik}(\omega)\varphi_l^*(\omega)\,d\omega=P_{ikl};$$

$$\frac{1}{2\pi}\int_{\omega_0-\Delta\omega_s/2}^{\omega_0+\Delta\omega_s/2}\Phi_{pk}(\omega)\varphi_l^*(\omega)\varphi_n(\omega)\,d\omega=Q_{pknl}. \tag{9.11}$$

Then (9.7) can be rewritten in the form

$$\mathbf{P}_i^t+\mathbf{W}_i^t\mathbf{Q}^{(i)}=\mathbf{0}. \tag{9.12}$$

Here $\mathbf{P}_i^t=(\mathbf{P}_{i1}^t,\ldots,\mathbf{P}_{im}^t)$ is the vector-row consisting of $m-1$ vector-rows $\mathbf{P}_{ik}^t$, $k=\overline{1,m}$, $k\neq i$ where each $\mathbf{P}_{ik}^t$ contains $2N+1$ elements P_{ikl}, $l=\overline{-N,N}$; $\mathbf{W}_i^t=(\mathbf{W}_{i1}^t,\ldots,\mathbf{W}_{im}^t)$ is the similarly constructed vector-row containing $(m-1)\times(2N+1)$ elements W_{ikl}, $k=\overline{1,m}$, $k\neq i$, $l=\overline{-N,N}$; $\mathbf{Q}^{(i)}$ is the Hermitian matrix consisting of $(m-1)\times(m-1)$ square matrices $\mathbf{Q}_{pk}^{(i)}$, $p=\overline{1,m}$, $k=\overline{1,m}$, $p,k\neq i$, each of them containing $(2N+1)\times(2N+1)$ elements Q_{pknl}, $l,n=\overline{-N,N}$; $\mathbf{0}$ is the vector-row consisting of $(m-1)\times(2N+1)$ zeros; t means transposition. It follows from (9.12) that

$$\mathbf{W}_i=-[\mathbf{Q}^{(i)t}]^{-1}\mathbf{P}_i,\quad i=\overline{1,m}. \tag{9.13}$$

Here we have taken into account that $\det\mathbf{Q}^{(i)}\neq 0$ since $\Phi_{pk}(\omega)$ in (9.11) contains mutually independent self-noises.

Let us consider now the variance (power) of the sum of interferences and self-noises at the output of the ith summator, $\sigma^2_{\text{out}\,i}$ in Fig. 9.1. Denoting the PSD of this sum at that point by $\Phi_{\text{out}\,i}(\omega)$ we can obtain from Fig. 9.1:

$$\sigma^2_{\text{out}\,i} = \frac{1}{2\pi}\int_{\omega_0-\Delta\omega_s/2}^{\omega_0+\Delta\omega_s/2} \Phi_{\text{out}\,i}(\omega)\,d\omega$$

$$= \frac{1}{2\pi}\int_{\omega_0-\Delta\omega_s}^{\omega_0+\Delta\omega_s} \lim_{T\to\infty}\frac{1}{2T}\overline{\left|\chi_i(\omega) + \sum_{k=1,k\neq i}^{m}\sum_{l=-N}^{N}\chi_k(\omega)\varphi_l(\omega)W_{ikl}\right|^2}\,d\omega. \qquad (9.14)$$

Using the notation of (9.11) and taking into account that $\lim_{T\to\infty}(1/2T)\times \overline{\chi_i(\omega)\chi_k^*(\omega)} = \Phi_{ik}(\omega)$ permits to reduce (9.14) to the form

$$\sigma^2_{\text{out}\,i} = \sigma_i^2 + \mathbf{W}_i^*\mathbf{P}_i + \mathbf{P}_i^*\mathbf{W}_i + \mathbf{W}_i^*\mathbf{Q}^{(i)t}\mathbf{W}_i \qquad (9.15)$$

where σ_i^2 is the variance (power) of interferences-plus-self-noises in the ith ("main") channel at the input of the ith summator. From (9.15) we can derive an optimum weight vector $\mathbf{W}_i$ minimizing $\sigma^2_{\text{out}\,i}$ for a given number $2N+1$ of channels of the transversal filter, given transfer functions of channel filters $\varphi_l(\omega)$, $l=\overline{-N,N}$, and for each feasible matrix $\mathbf{\Phi}(\omega)$. Differentiating (9.15) with respect to $\mathbf{W}_i$ and putting the derivative equal to zero yields $\mathbf{W}_{i\,\text{opt}}$ that coincides with (9.13).

Thus *if complex weighting coefficients, W_{ikl}, are generated with the help of correlation feedback* [see (9.5)], *they are the best weights in the sense that they provide maximum external interference suppression for any given number $2N+1$ of channels of the transversal filter and for any given system of transfer functions of channel filters, $\{\varphi_l(\omega)\}$, $l=\overline{-N,N}$.* The achievable minimum of interference-plus-self-noise power can be obtained from (9.15) replacing $\mathbf{W}_i$ by the right part of (9.13) and taking into account that the matrix $\mathbf{Q}^{(i)}$ is a Hermitian one:

$$\sigma^2_{\text{out}\,i} = \sigma_i^2 - \mathbf{P}_i^*[\mathbf{Q}^{(i)t}]^{-1}\mathbf{P}_i. \qquad (9.16)$$

Using (9.16) and (9.11) one can choose the minimum necessary number of channels of the transversal filter and the most suitable transfer functions of channel filters, $\varphi_l(\omega)$, $l=\overline{-N,N}$.

It should be noted that *we did not assume the orthogonality of channel filters $\varphi_l(\omega)$* in (9.11)–(9.16). We have to reject orthogonal filters if we want to preserve effective interference cancellation when decreasing the number of transversal filter's channels.

As with orthogonal channels (see Section 9.1) adaptive transversal filters with nonorthogonal channels can be implemented in both analogous and digital versions. An example of the analogous implementation is shown in Fig. 9.1. In a digital version different gradient algorithms may be utilized [35,72,73].

Note that according to (9.11) $\mathbf{Q}^{(i)}$ is the correlation matrix of interferences-plus-self-noises at the outputs of channel filters [with transfer functions $\varphi_l(\omega)$] of the "compensating" channels of the ith summator's circuit (see Fig. 9.1). At the same time $\mathbf{P}_i$ is the vector of the cross-correlation between interferences-plus-self-noises at the outputs of the same channel filters, on the one hand, and those in the "main" channel, on the other. Since the optimum weight vector, $\mathbf{W}_{i\,\text{opt}}$, of the ith summator's circuit is determined by (9.13), an adaptive interference cancellation algorithm may be constructed directly using (9.13) where the matrix $\mathbf{Q}^{(i)}$ and the vector $\mathbf{P}_i$ should be replaced by their estimates, $\hat{\mathbf{Q}}^{(i)}$ and $\hat{\mathbf{P}}_i$ (for instance, by maximum likelihood estimates). Adaptive algorithms of this type are known as *algorithms of correlation matrix direct inversion* [35,72].

Up to now we do not specify transfer functions, $\varphi_l(\omega)$, $l=\overline{-N,N}$, of the channel filters. However, when the number of channels of a transversal filter is small, the choice of $\varphi_l(\omega)$ may have an essential effect on interference cancellation.

As a rule, transversal filters based on delay lines or shift registers are described in many papers and books devoted to the adaptive antenna arrays (see, e.g., [35]). In the case of delay lines $\varphi_l(\omega)=\exp(-j\omega l\tau_0)$ so that each ikth filter contains a tapped delay line where a weight amplifier with complex gain W_{ikl} is included in each of $2N+1$ taps. If we choose $\tau_0=1/\Delta f_s=2\pi/\Delta\omega_s$ then for $l=\overline{-N,N}$ we have a system of orthogonal filters within the frequency band $(\omega_0-\Delta\omega_s/2, \omega_0+\Delta\omega_s/2)$. In order to avoid negative delays a constant delay $\tau>N\tau_0$ is usually placed at the input of the filter. Analogous transversal filters based on delay lines and digital transversal filters based on shift registers are simple for implementation. Such filters are especially expedient when interferences from each source at the inputs of spatially separated stations differ only in intensities, phases and TDOAs within certain limits. In this case more closely spaced taps, i.e. the loss of orthogonality of channel filters, permits reducing of the total number of channels. If the weights W_{ikl} can be adapted sufficiently fast, such a structure can eliminate not too large differential Doppler frequency shifts of interferences at the inputs of stations (or, more exactly, can compensate for varying phase differences).

Example 9.1. Let us illustrate numerically how an adaptive transversal filter operates in a system with two spatially separated stations. An interference from a single source enters through the mainlobes of antenna ADPs. The interference PSD is rectangular within the frequency band $(\omega_0-\Delta\omega_s/2, \omega_0+\Delta\omega_s/2)$, the interference-to-self-noise ratio (INR) values q^2 and self-noise PSD N_0 are the same at the inputs of both stations. Consider a steady-state condition of an adaptive transversal filter based on a tapped delay line (or a shift register) with three and five taps. The spacing between adjacent taps is equal to $\tau_0=1/2\Delta f_s$. The values of Interference Suppression Factor (ISF) calculated from (9.16) and (9.11), $K_s=\sigma_0^2/\sigma_{\text{out}}^2$ [where $\sigma_0^2=\sigma_i^2$ in (9.16) and the subscript "i" of $\sigma_{\text{out}\,i}^2$ is omitted], are plotted in Fig. 9.2 versus the interference TDOA τ_{10} at the inputs of stations. This TDOA varies, for instance, when the interference source moves relative to the radar system. It can be seen that when τ_{10} is equal to the delay corresponding to each tap, i.e. when the interference TDOA τ_{10} is exactly compensated by the delay line, $K_s\approx 0.5q^2$. In this case $\sigma_{\text{out}}^2\approx 2\sigma_0^2/q^2=2N_0(1+q^2)\Delta f_s/q^2\approx 2N_0\Delta f_s$. It means that the external interference is cancelled out. The self-noises of both stations are summed and hence the self-noise power at the output is doubled. The ISF K_s falls sharply beyond the interval of TDOAs covered by the delay line (or by the shift register). When the transversal filter has three adaptively controlled channels, the allowable TDOA interval is equal to $(-\tau_0, \tau_0)=1/\Delta f_s$. When the number of channels is equal to five, the interval of feasible TDOAs increases up to $(-2\tau_0, 2\tau_0)=2/\Delta f_s$. As it can be expected, the least values of ISF K_s inside those intervals take place when values of τ_{10} correspond to the halves of the intertap spacing. The higher INR q^2, i.e. the larger maximum values of K_s, the greater the difference between maxima and minima of K_s. Increasing the number of channels from 3 to 5 yields "pulling" minima of K_s towards its maxima, i.e. "smoothing" of ISF oscillations.

It is interesting to note that the same minimum value of $K_s=27.5$ dB at $q^2=40$ dB as with three-channel filter can be obtained with orthogonal filters [when $\tau_0=1/\Delta f_s$ and the matrix $\mathbf{Q}^{(i)}$ in (9.16) is a diagonal one] if a transversal filter has 256 channels, i.e. delay line has 256 taps or a shift register has 256 elements instead of three. In this case, though, the interval of allowable values of TDOA (to be compensated) greatly increases.

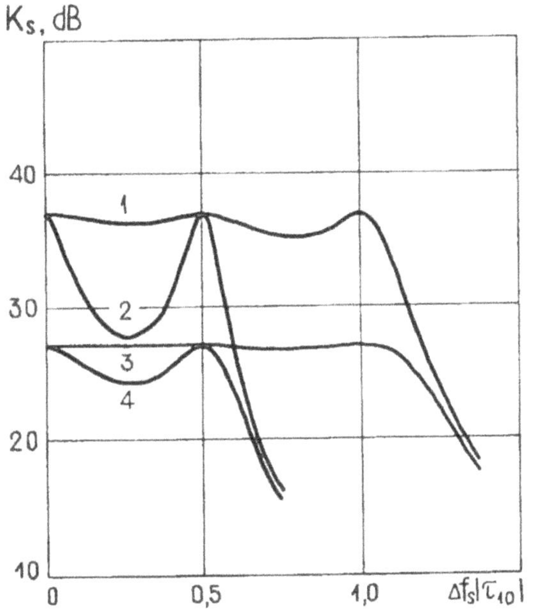

Figure 9.2 Dependences of the interference suppression factor, K_s, on interference TDOA (to Example 9.1). Curves 1, 3: five-channel transversal filter; curves 2, 4: three-channel transversal filter; curves 1, 2: $q^2 = 40$ dB; curves 3, 4: $q^2 = 30$ dB

Two-Stage Adaptive Interference Cancellation Algorithms

In many cases two-stage adaptive algorithms are expedient. Such an algorithm is based on a combination of both the direct inversion technique of a sample correlation matrix and the gradient method. The sample correlation matrix direct inversion provides rapid but not quite accurate adaptation (especially when that matrix varies in time). This is the first stage of adaptation procedure which may be called "preadaptation". At the second stage small residual misalignments can be eliminated by the not so rapid but more accurate gradient method. This stage may be called "final adaptation". The important feature of the preadaptation stage is that, unlike a typical situation with adaptive antenna arrays, the dimension of the correlation matrix is not large, so that its inversion is not difficult. Let us consider one of such algorithms – the Adaptive Mainlobe Jamming Cancellation Algorithm (AMJCA) in more detail [217].

For the sake of clarity we describe here the simplest case. Let a MSRS comprising one transmitting–receiving station (a monostatic radar), T_xR_x, and one remote receiving station, R_x, be subjected to a single intensive mainlobe jammer.

Apart from intensity and phase differences there is usually an essential TDOA between jammer's signals at two spatially separated stations. Therefore, the first (preliminary) step of the AMJCA is a coarse estimation of this TDOA by maximizing the cross-correlation between two sets of samples coming from T_xR_x and R_x stations, respectively.

At the second step of the AMJCA the obtained estimate is used as a delay of the central channel of the transversal filter. These channels are K in number where K

depends on accuracy of the coarse TDOA estimates and on other features of the jamming mutual spectrum, $\Phi_{12}(\omega)$, at the inputs of the stations. The number K may often be limited to 3 or 5 (see Example 9.1). The time delay between adjacent channels is chosen to be equal to half of the reciprocal of the receiver bandwidth (as in Example 9.1). After the second step we have one set of n samples coming from the first station (e.g., T_xR_x) and in certain way delayed K sets coming from the second station, R_x, through delay elements of the transversal filter:

$$u_{1i}=u_1[i];\quad u_{2i}^{(k)}=u_2[i+k-(K+1)/2];\qquad i=\overline{1,n};\ k=\overline{1,K}. \tag{9.17}$$

At the third step the correlation (covariance) matrix $\mathbf{B}$ of those K sample sets, $u_{2i}^{(k)}$, is estimated

$$\hat{\mathbf{B}}=\|\hat{b}_{kl}\|;\quad \hat{b}_{kl}=\frac{1}{I}\sum_{i=1}^{I}u_{2i}^{(k)}u_{2i}^{(l)*},\ l=\overline{1,K}. \tag{9.18}$$

where $I=5K$.

At the fourth step the correlation (covariance) vector $\mathbf{D}$ of the above mentioned K sample sets, on the one hand, and the sample set from the first station, on the other, is estimated

$$\hat{\mathbf{D}}=\|\hat{d}_k\|;\quad \hat{d}_k=\frac{1}{I}\sum_{i=1}^{I}u_{2i}^{(k)}u_{1i}^{*};\ k=\overline{1,K}. \tag{9.19}$$

At the fifth step the linear equation

$$\hat{\mathbf{B}}\mathbf{W}_0=\hat{\mathbf{D}} \tag{9.20}$$

is solved for the vector $\mathbf{W}_0$ of preliminary weights for the transversal filter. As mentioned above, since the transversal filter contains only a few channels, the dimension of the matrix $\mathbf{B}$ is not large, so that inversion of this matrix can be easily performed. This is an essential difference from a usual adaptive antenna array containing a lot of elements. As well known, the inversion of a covariance matrix for such an array may often be a difficult computational problem. The fifth step is the final one of the first stage of the AMJCA. This stage was called the preadaptation. Thus the preadaptation is accomplished by the sample correlation (covariance) matrix inversion technique.

The sixth step of the AMJCA represents the second stage. It is the adaptation itself, or the final adaptation. At this step the final weights before summation in the transversal filter are established with the help of a known gradient method. The preliminary weight vector $\mathbf{W}_0$ is utilized as an initial approximation. The factor μ in equations (9.21) below is usually chosen to be large enough to counter possible jamming differential Doppler and at the same time small enough to ensure the AMJCA stability and convergence.

$$\begin{cases} u_i^{(\mathrm{out})}=u_{1i}-\sum\limits_{k=1}^{K}w_{ik}^{*}u_{2i}^{(k)}; \\ w_{(i+1)k}=w_{ik}+\mu u_{2i}^{(k)}u_i^{(\mathrm{out})*}. \end{cases} \tag{9.21}$$

Example 9.2. Now we can consider a more complex example illustrating the performance of the AMJCA described above and target detection after adaptive mainlobe jamming cancellation when target echoes are spatially coherent with completely dependent amplitude fluctuations and when target echoes are spatially incoherent with independent amplitude fluctuations [217]. The optimum detection algorithms for known interference-plus-self-noise PSD matrix, $\mathbf{\Phi}(\omega)$, at the inputs of receiving stations have been derived and analyzed in Chapter 8.

The system considered for simulation is depicted in Fig. 9.3. A transmitting–receiving station (a monostatic radar) T_xR_x and a spatially separated receiving station R_x are located at the azimuths $\beta = 0°$ and 180°, respectively.

An aircraft moves from the range 300 km, $\beta = 270°$ at a constant height with the constant velocity 0.8 km/s along a straight line towards the radar system (the velocity vector's direction is 90° in azimuth). Phases and amplitudes of target echoes at the receivers are uniformly and Rayleigh distributed, respectively. For completely correlated fluctuations the baselength L has been chosen equal to 1 km, whereas for independent fluctuations the same baseline has been extended to 15 km. At an initial moment the aircraft has a jammer onboard, i.e. they coincide in space (self-screening condition). Then the jammer leaves the aircraft and moves with the same velocity as the aircraft but in a slightly different direction. The difference is 0.1074° in azimuth which provides the increase of a cross-range distance between the aircraft and the jammer of approximately 1.5 m per second. Such a model, rather artificial, is used to reveal the algorithm performance characteristics in the most difficult cases where a jammer and a target coincide in space or are in close proximity to each other. Beamwidths of both stations are 1° in azimuth and elevation so that antenna linear resolution capability (in cross-range directions) at the range 300 km is approximately 5.2 km. The T_xR_x station illuminates the target once every second. Received signals (sums of jamming, wanted signals and self-noises) from R_x station are transmitted to

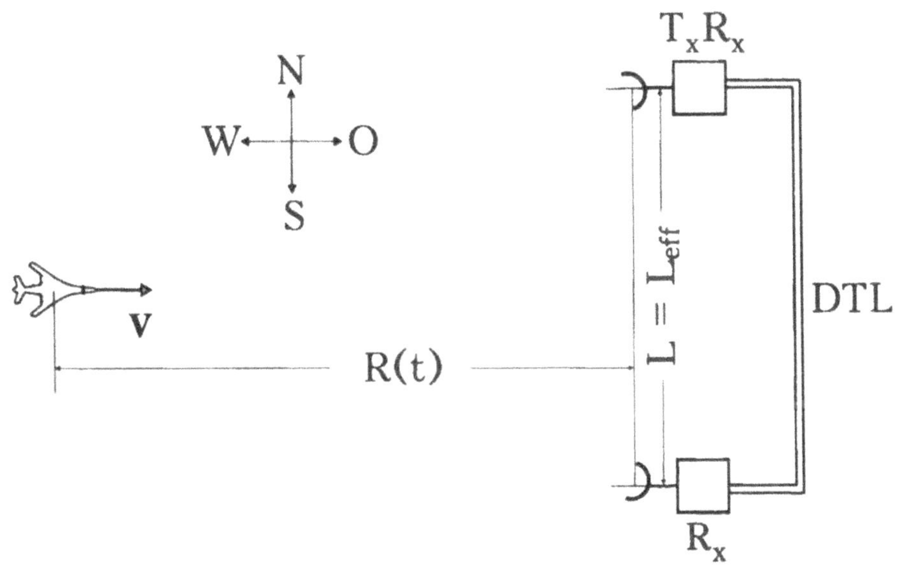

Figure 9.3 System geometry (to Example 9.2). $\mathbf{v}$ = target velocity vector, R(t) = target range, T_xR_x = transmitting receiving station, R_x = receiving station, DTL = data transmission line, $L = L_{eff}$ = baselength between stations equal to the effective baselength between them

the T_xR_x station where they are jointly processed with similar signals received by the T_xR_x station. The output of the AMJCA undergo the matched filtering, envelope detection and thresholding. The signal bandwidth is equal to 5 MHz.

Typical plots of the wanted signal + jamming + noise versus time (sample number) at the output of the envelope detector are shown in Fig. 9.4(a) and (b). In Fig. 9.4(a) the AMJCA is switched off. The target echo is perfectly masked by jamming. In

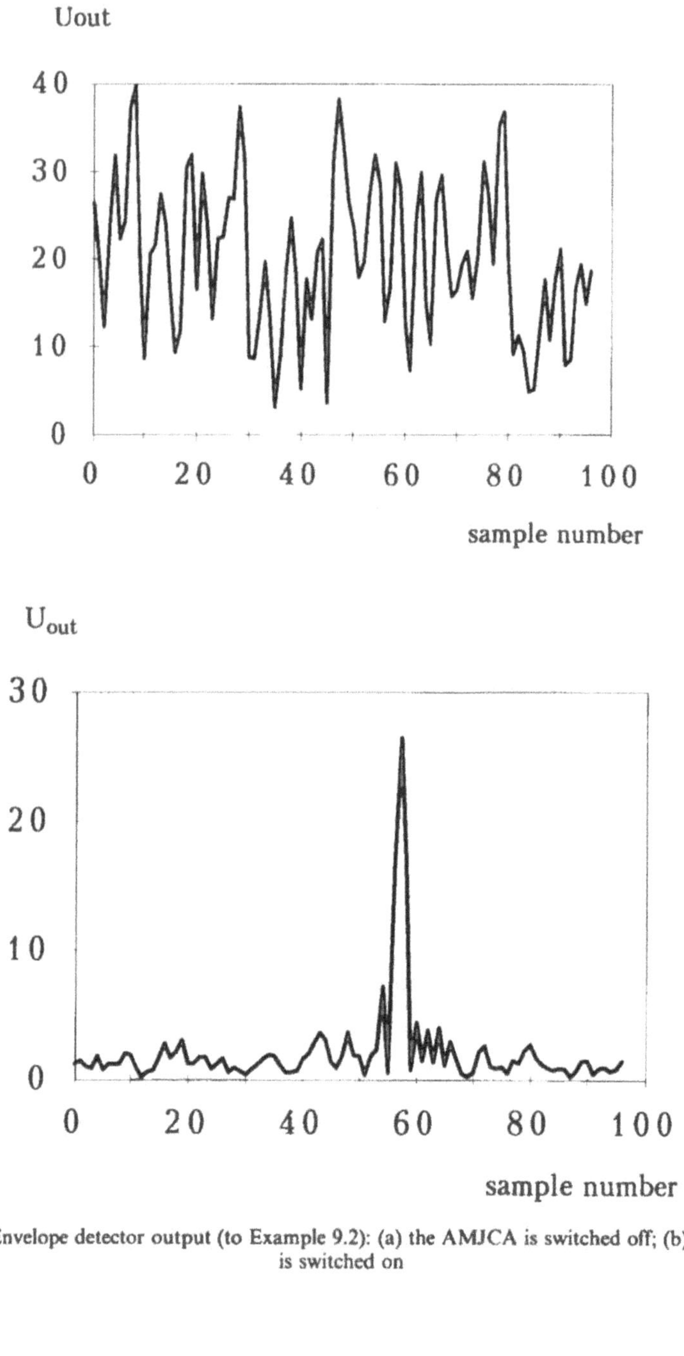

Figure 9.4 Envelope detector output (to Example 9.2): (a) the AMJCA is switched off; (b) the AMJCA is switched on

Fig. 9.4(b) the same time interval is shown but the AMJCA is switched on. The jamming cancellation permits detection of the target echo with confidence. The input and output signal-to-interference-plus-noise ratios (SINRs) are approximately −2 and +23 dB, respectively.

Figure 9.5 relates to the case where signal fluctuations are completely correlated at the stations' inputs. In this figure the input SINR at each receiver and the SINR at the output of the AMJCA before envelope detection are shown. Solid lines represent theoretical values which have been calculated with the help of equation (8.63) from Section 8.4 under the assumption that the matrix $\mathbf{\Phi}(\omega)$ is *a priori* known. The simulation results are shown by dots. Each dot represents a steady-state output SINR value averaged over 100 trials. We display in Fig. 9.5 only the most interesting first segment of the jammer and aircraft paths after their separation. In agreement with the theoretical predictions the simulation results exhibit oscillating, nearly periodic, character of the output SINR. For certain cross-range distances between the target and the jammer (in particular, when they are colocated), target echoes from two spatially separated stations turn out to be in phase (after jamming phase alignment by the AMJCA). Under these conditions target echoes are subtracted together with jamming subtraction. On the contrary, for other cross-range distances when target echoes are in antiphase, then jamming coherent subtraction leads to echo coherent summation. As it can be seen, the output SINR minima are very narrow. Even when the jammer is very close to the target in cross-range direction within the antenna beamwidth, the moving target will be detected nearly always except for short time intervals. Evidently, for assumed input SINR values (of the order of −2 dB) the aircraft cannot be detected without jamming cancellation [see also Fig. 9.4(a)].

Fig. 9.6 demonstrates the case where target echo fluctuations are independent at the stations' inputs. It is a typical situation. All curves are of the same meaning as

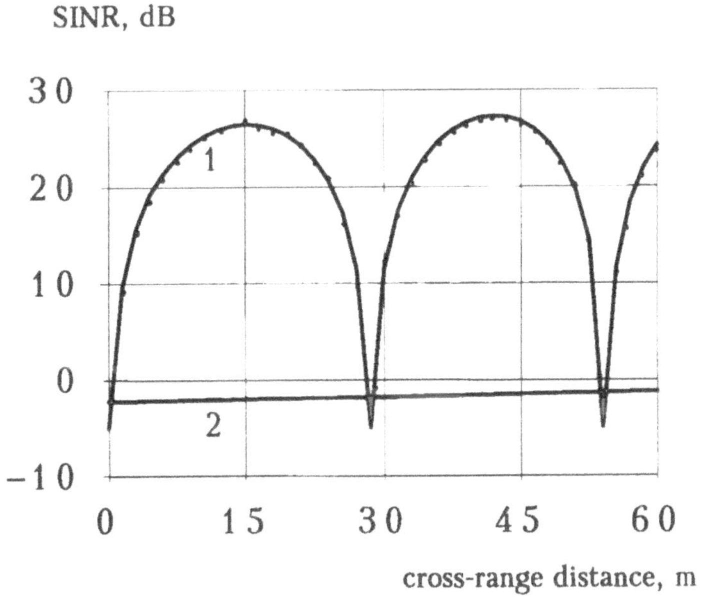

Figure 9.5 Input (curve 2) and output (curve 1) SINR (when the AMJCA is switched on) for spatially coherent signals with completely correlated amplitude fluctuations. Simulation results are shown by dots (to Example 9.2)

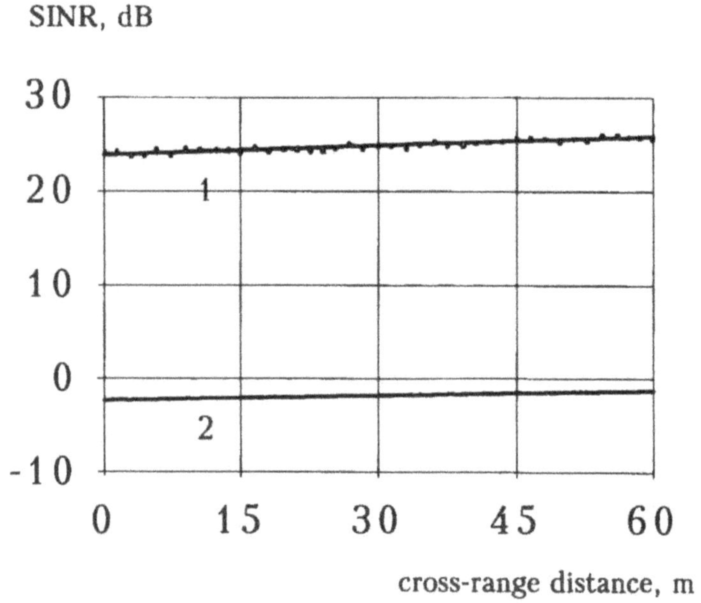

Figure 9.6 The same characteristics as in Fig. 9.5 but for spatially incoherent signals with mutually independent amplitude fluctuations (to Example 9.2)

in Fig. 9.5. The theoretical curves have been obtained from equation (8.64). It is seen that the aircraft can be surely detected even when the jammer is onboard, i.e. when the jammer and the aircraft coincide in space (a self-screening situation). Thanks to the echo fluctuation spatial independence, jamming subtraction is accompanied by incoherent echo summation. The output SINR maintains nearly constant value when the jammer moves apart from the aircraft. A slight increase of the input and output SINRs in Figs. 9.5 and 9.6 has been caused by the slight decrease of the target range. As with correlated target echo fluctuations, the AMJCA performance is practically not worse than that of the theoretical nonadaptive algorithm for *a priori* known matrix $\mathbf{\Phi}(\omega)$. Simulation results are in very good agreement with theoretical predictions from Chapter 8.

According to (9.13) the weights W_{ikl} are to be determined by only interferences and self-noises. Therefore, adaptive generating of these weights should be protected against the influence of expected signals. When signal energy is small (as compared with interference energy which is used for weight generation) it does not distort optimum weights. However, for the case of large signal energy special protection is necessary for the circuits (or algorithms) of adaptation (see, e.g., [72]).

The considered adaptive external interference cancellation techniques using transversal filters with adaptively controlled weights are applicable to signal detectors in a background of spatially correlated interferences when the number of interference sources, M, is arbitrary, provided the number of independent receiver channels, m, exceeds M, i.e. $m > M$ (see Section 8.2). However, as with nonadaptive interference cancellation, detection characteristics depend on the total number, radiation intensities and space positions of interference sources. As has been noted in Section 8.2, only mainlobe external interferences received by spatially separated stations are to be cancelled by joint processing in MSRSs.

***In this section**, we have shown that when there is a* priori *information of the interference-plus-self-noises PSD matrix, $\Phi(\omega)$, and of its variations in time, a significant simplification of adaptive transversal filters for interference cancellation is possible. In these cases dropping the requirement of channel filter orthogonality we can drastically reduce the number of channels with adaptively controlled weights.*

If these weights are generated with the help of a strong correlation feedback (an example of such a circuit is depicted in Fig. 9.1 by dashed lines) they are determined by (9.13), (9.11). *We have established that they are optimum in the sense that provide maximum external interference suppression for any given number of channels of a transversal filter and for any given system of transfer functions of channel filters. We have derived the expression* (9.16) *for the minimum attainable power of interference-plus-self-noise at the output of the transversal filter with optimum weights. These optimum weights can also be obtained immediately according to* (9.13) *(where the matrix* $\mathbf{Q}^{(i)}$ *and the vector* $\mathbf{P}_i$ *should be replaced by their good statistical estimates), i.e. by using the sample correlation matrix direct inversion method.*

We have considered a numerical Example 9.1 *illustrating the operation of an adaptive interference cancellation transversal filter with three and five nonorthogonal channels (Fig.* 9.2*). For instance, the simple filter with three channels may be used instead of a similar filter with* 256 *orthogonal channels without loss of performance if interferences at the inputs of spatially separated stations differ only in intensities, phase and TDOA within the interval less than the reciprocal of the signal bandwidth.*

We have described a two-stage adaptive mainlobe interference cancellation algorithm [AMJCA, (9.17)–(9.21)] *in which a combination of both the sample correlation matrix direct inversion method and the gradient method is used. Such an algorithm has a high speed of the former and a high accuracy of the latter method. Due to the small number of transversal filter channels in a MSRS the dimension of a sample correlation matrix is not large so that it can be inverted without difficulties.*

In Example 9.2 *we have illustrated an application of the two-stage adaptive interference cancellation algorithm to the problem of moving aircraft detection in a background of intensive jamming when a jammer moves in the vicinity of the aircraft or coincides with it in space (a self-screening condition). Simulation results (Figs.* 9.4–9.6) *are in good agreement with theoretical results obtained in Chapter* 8 *for nonadaptive external interference cancellation algorithms.*

9.3. DETECTION OF STOCHASTIC SIGNALS WITH KNOWN CORRELATION MATRICES

Synthesis of the Optimum Detector

The likelihood ratio logarithm is determined by (4.52). When a MSRS is subject to spatially correlated interferences from M sources then (under the conditions of adopted signal and interference models, see Section 4.1) the appearance of a source of wanted stochastic signal is equivalent to adding the $(M+1)$th interference source. Using the notation of Section 8.1 the optimum processing algorithm (4.52) can be written in the form

$$L=\sum_{i=1}^{m}\sum_{k=1}^{m}\frac{1}{2\pi}\int_{-\infty}^{\infty}\chi_i^*(\omega)\chi_k(\omega)[f_{ik}^{(M)}(\omega)-f_{ik}^{(M+1)}(\omega)]\,d\omega. \tag{9.22}$$

Taking into account expressions (8.8)–(8.10) for $f_{ik}^{(p)}(\omega)$, $p=\overline{1,M}$, yields

$$L=\frac{1}{2\pi}\int_{-\infty}^{\infty}\left|H^{(M+1)}(\omega)\sum_{i=1}^{m}\chi_i(\omega)D_i^{*(M+1)}(\omega)\right|^2 d\omega. \tag{9.23}$$

A bloc diagram of the optimum detector is depicted in Fig. 9.7. Comparing (9.23) with (7.10) and Fig. 9.7 with Fig. 7.1 (at $\Delta\Omega_{sik}=0$) we can see that the general structure of the optimum detectors is the same for both the absence ($M=0$) and the presence of spatially correlated interferences. The optimum processing is reduced to the power (energy) estimation of weighted and filtered sum of overall received signals. However, specific transfer functions of filters depend on the number and characteristics of interferences [see (8.8)–(8.10)].

Performance Analysis of the Synthesized Detector [60]

Let Δf_s and T be the signal bandwidth at the input of the square-law detector in Fig. 9.7 and signal duration (or integration time), respectively. As with spatially uncorrelated interferences, we assume (with practically sufficient accuracy) that the decision variable L in (9.23) may be considered as being Gaussian when $\Delta f_s T \gg 1$. Therefore, in order to obtain detection characteristics, expressions for mean values and variances should be derived for both the absence (L_0) and the presence (L_1) of wanted signal.

When there is no wanted signal [see (9.22)]

$$\begin{aligned}\overline{\chi_i^*(\omega)\chi_k(\omega)}&=2T\Phi_{ki}^{(M)}(\omega)=2T\Phi_{ik}^{*(M)}(\omega)\\&=2T[\Phi_{ik}^{*(M+1)}(\omega)-\Phi_{sik}^{*}(\omega)]\end{aligned} \tag{9.24}$$

and when a signal is present

$$\begin{aligned}\overline{\chi_i^*(\omega)\chi_k(\omega)}&=2T\Phi_{ki}^{(M+1)}(\omega)=2T\Phi_{ik}^{*(M+1)}(\omega)\\&=2T[\Phi_{ik}^{*(M)}(\omega)+\Phi_{sik}^{*}(\omega)].\end{aligned} \tag{9.25}$$

In (9.24) and (9.25) $\Phi_{sik}(\omega)$ is the i,kth element of the PSD matrix $\Phi_s(\omega)$ of the wanted signal at the inputs of spatially separated receiving stations. By the definition

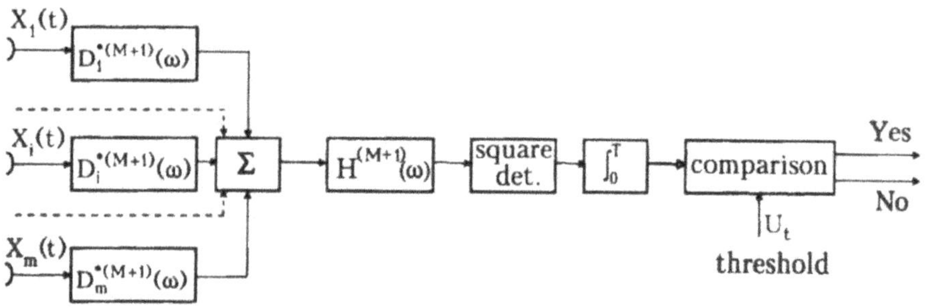

Figure 9.7 Structure of the optimum detector for stochastic signals with a known correlation matrix in a background of spatially correlated interferences

of the matrices $\mathbf{\Phi}(\omega)$ and $\mathbf{f}(\omega)$ [see (4.46)]

$$\sum_{i=1}^{m} \Phi_{li}^{(M)}(\omega) f_{ik}^{(M)}(\omega) = \delta_{kl}; \quad \sum_{i=1}^{m} \Phi_{li}^{(M+1)}(\omega) f_{ik}^{(M+1)}(\omega) = \delta_{kl}. \tag{9.26}$$

Then we obtain from (9.22) the mean values:

$$\begin{aligned} m_{10} &= m_1\{L_0\} = \overline{L_0} = \sum_{i=1}^{m} \sum_{k=1}^{m} \frac{T}{\pi} \int_{-\infty}^{\infty} \Phi_{sik}^{*}(\omega) f_{ik}^{(M+1)}(\omega)\, d\omega, \\ m_{11} &= m_1\{L_1\} = \overline{L_1} = \sum_{i=1}^{m} \sum_{k=1}^{m} \frac{T}{\pi} \int_{-\infty}^{\infty} \Phi_{sik}^{*}(\omega) f_{ik}^{(M)}(\omega)\, d\omega. \end{aligned} \tag{9.27}$$

The variances of L can be obtain from (9.22) taking into account (7.22) and (7.23):

$$\begin{aligned} M_{20} &= M_2\{L_0\} = \overline{L_0^2} - (\overline{L}_0)^2 \\ &= \sum_{i=1}^{m} \sum_{k=1}^{m} \sum_{l=1}^{m} \sum_{n=1}^{m} \frac{2T}{\pi} \int_{-\infty}^{\infty} \Phi_{sil}^{*}(\omega) \Phi_{skn}(\omega) f_{ik}^{(M+1)}(\omega) f_{nl}^{(M+1)}(\omega)\, d\omega; \\ M_{21} &= M_2\{L_1\} = \overline{L_1^2} - (\overline{L}_1)^2 \\ &\leq \sum_{i=1}^{m} \sum_{k=1}^{m} \sum_{l=1}^{m} \sum_{n=1}^{m} \frac{2T}{\pi} \int_{-\infty}^{\infty} \Phi_{sil}^{*}(\omega) \Phi_{skn}(\omega) f_{ik}^{(M)}(\omega) f_{nl}^{(M)}(\omega)\, d\omega. \end{aligned} \tag{9.28}$$

In the particular case when $M=0$, taking into account that $f_{ik}^{(1)}(\omega) = f_{snik}(\omega)$ in (7.5) and $f_{ik}^{(0)}(\omega) = \delta_{ik}/\sqrt{N_i N_k}$, we come from (9.27) and (9.28) to (7.20), (7.21) and (7.24), (7.25) at $\Omega_{s1}=0$.

Having calculated the mean values m_{10}, m_{11} and variances M_{20}, M_{21}, one can obtain detection characteristics with the help of (7.26).

Spatially Correlated Interference Cancellation

As with regular signal detection, spatially correlated external interferences may be expected to be cancelled out (coherently mutually suppressed) in the optimum detectors (9.22), (9.23). To verify this feature let us consider the interference-plus-self-noise PSD at the summator output of the algorithm (9.23). The Fourier transformation of a realization of this stochastic process takes the form

$$\eta_{\text{In}}(\omega) = \sum_{i=1}^{m} \chi_{\text{In}\,i}(\omega) D_i^{*(M+1)}(\omega) \tag{9.29}$$

where $\chi_{\text{In}\,i}(\omega)$ is the Fourier transform of the sum of the interferences and the self-noise at the ith station; $D_i^{*(M+1)}$ is determined by (8.10). From (9.29) we can obtain the PSD [see (9.24)]

$$\begin{aligned} \Phi_{\eta\text{In}}(\omega) &= \lim_{T\to\infty} \frac{1}{2T} \overline{\eta_{\text{In}}(\omega)\eta_{\text{In}}^{*}(\omega)} \\ &= \sum_{i=1}^{m} \sum_{k=1}^{m} \Phi_{ik}^{(M)}(\omega) D_i^{*(M+1)}(\omega) D_k^{(M+1)}(\omega). \end{aligned} \tag{9.30}$$

Note that when the number of interference sources is equal to M then the $(M+1)$th source is a source of a wanted signal. Therefore the subscripts "$M+1$" in the right part of the expression (8.10) for $D_i^{(M+1)}$ may be replaced by the subscripts "s". Besides, for the sake of clarity of results let us simplify (8.10) assuming that

$$q_{sl}=q_s;\quad N_l=N;\quad g_l(\beta_{sl},\varepsilon_{sl},\omega)=1;\quad \Delta\varphi_{sls}=0:\qquad l=\overline{1,m}. \tag{9.31}$$

Substituting (8.10) in (9.30) and taking into account (9.31) and (9.26) yields

$$\Phi_{\eta\,\text{In}}(\omega)=q_s^2 N\sum_{i=1}^{m}\sum_{k=1}^{m} f_{ik}^{(M)}(\omega)\exp(-j\omega\tau_{sik}). \tag{9.32}$$

It is seen that though there are external interferences at the inputs of stations [with the interference-plus-self-noise PSD matrix $\mathbf{\Phi}^{(M)}(\omega)=\|\Phi_{ik}^{(M)}(\omega)\|$, see (8.2)] the PSD, $\Phi_{\eta\,\text{In}}(\omega)$, of the process at the summator output does not contain interference components. It means that spatially correlated external interferences are actually cancelled out by the linear part of the optimum detector (9.23) (see also [41]).

It is interesting to analyze instead of (9.23) the simpler algorithm (7.10) which is optimum in the absence of spatially correlated interferences at the inputs of receiving stations[2]. Substituting $D_i^{*(1)}$ instead of $D_i^{*(M+1)}$ in (9.30) and taking into account that $f_{ik}^{(0)}=\delta_{ik}/N_i$ we obtain at the summator output:

$$\tilde{\Phi}_{\eta\,\text{In}}(\omega)=(q_s^2/N)\sum_{i=1}^{m}\sum_{k=1}^{m}\Phi_{ik}^{(M)}(\omega)\exp(-j\omega\tau_{sik}). \tag{9.33}$$

Unlike (9.32) the PSD $\tilde{\Phi}_{\eta\,\text{In}}(\omega)$ contains the PSD $\Phi_{ik}^{(M)}(\omega)$. It means that external spatially correlated interferences remain at the summator output. Analyzing (8.8)–(8.10) we note that when $N_i=N, q_{si}^2=q_s^2, q_{pi}^2=q_p^2, i=\overline{1,m},\ p=\overline{1,M}$, and for intensive interferences ($q^2\gg 1$), the quantity $f_{ik}^{(M)}(\omega)$ is of the order of $1/N$ whereas $\Phi_{ik}^{(M)}(\omega)$ is of the order of MNq^2. It means that $|\Phi_{\eta\,\text{In}}(\omega)|\ll|\tilde{\Phi}_{\eta\,\text{In}}(\omega)|$. External spatially correlated interferences do not cancelled by "matched" processing (7.10).

Example 9.3. Let us consider the effect of external interference cancellation in a passive system containing two receiving stations ($m=2$) when a single intensive jammer ($M=1$) operates through the antenna mainlobes of spatially separated stations and the conditions (9.31) are satisfied. The PSD $\Phi_\eta(\omega)$ of the process at the summator output (see Fig. 9.7) for the case where the wanted signal is present can be written in the form [see (9.30)]

$$\Phi_\eta(\omega)=\sum_{i=1}^{2}\sum_{k=1}^{2}[\Phi_{ik}^{(1)}(\omega)+\Phi_{sik}(\omega)]D_i^{*(2)}D_k^{(2)}. \tag{9.34}$$

For the sake of simplicity we assume additionally that the dimensionless normalized PSD of complex external interference and signal envelopes, $F_1(\omega)$, $F_s(\omega)$, are determined by (8.30). Then

$$\Phi_{ik}^{(1)}(\omega)=N[\delta_{ik}+q^2\exp(j\omega\tau_{ik})];\quad \Phi_{sik}(\omega)=q_s^2N\exp(j\omega\tau_{sik}),\quad i,k=1,2. \tag{9.35}$$

[2] Such a processing algorithm is often called "the matched processing" which means matching to a wanted signal in the absence of external spatially correlated interferences [28,42].

Now we can obtain the signal, external interference and self-noise components of $\Phi_\eta(\omega)$ from (9.34) taking into account (9.35) and (8.10).

The signal component:

$$\Phi_{\eta s}(\omega)=[4q_s^4/(1+2q^2)^2](1+q^2-q^2\cos\omega\delta\tau_{12})^2; \qquad (9.36)$$

the external interference component:

$$\Phi_{\eta I}(\omega)=[2q_s^2q^2/(1+2q^2)^2](1+\cos\omega\delta\tau_{12}); \qquad (9.37)$$

the self-noise component:

$$\Phi_{\eta n}(\omega)=[2q_s^2/(1+2q^2)^2][(1+q^2)^2+q^4-2q^2(1+q^2)\cos\omega\delta\tau_{12})]. \qquad (9.38)$$

It is seen that the external interference (jamming) is cancelled: $\Phi_{\eta I}(\omega)\to 0$ when $q^2\to\infty$, whereas all other components remain finite (nonzero) values.

The most true notion can be given by analyzing the SINR at the summator output. The variance (power) of each component can be obtained by integrating corresponding PSD, $\Phi_{\eta s}(\omega)$, $\Phi_{\eta I}(\omega)$ or $\Phi_{\eta n}(\omega)$, over ω within the frequency band $|\omega-\omega_0|\leqslant\Delta\omega_s/2$. As a result we have

$$\sigma_{\eta s}^2=\frac{4q_s^4\Delta f_s}{(1+2q^2)^2}\left[(1+q^2)^2+\frac{q^4}{2}+\frac{q^4}{2}\operatorname{sinc}(2\pi\Delta f_s\delta\tau_{12})\cos 2\omega_0\delta\tau_{12}\right.$$
$$\left.-2q^2(1+q^2)\operatorname{sinc}(\pi\Delta f_s\delta\tau_{12})\cos\omega_0\delta\tau_{12}\right]; \qquad (9.39)$$

$$\sigma_{\eta I}^2=\frac{2q_s^2q^2\Delta f_s}{(1+2q^2)^2}[(1+\operatorname{sinc}(\pi\Delta f_s\delta\tau_{12})\cos\omega_0\delta\tau_{12}]; \qquad (9.40)$$

$$\sigma_{\eta n}^2=\frac{2q_s^2\Delta f_s}{(1+2q^2)^2}[(1+q^2)^2+q^4-2q^2(1+q^2)\operatorname{sinc}(\pi\Delta f_s\delta\tau_{12})\cos\omega_0\delta\tau_{12}]. \qquad (9.41)$$

When the jammer and the signal source are positioned on the surface of one and the same hyperboloid of revolution with focuses at the stations' sites (including the particular case, where both the jammer and the signal source coincide in space), then $\delta\tau_{12}=0$ and

$$q_{out}^2=\frac{\sigma_{\eta s}^2}{(\sigma_{\eta I}^2+\sigma_{\eta n}^2)}=\frac{2q_s^2}{(1+2q^2)}. \qquad (9.42)$$

Though the jamming is suppressed the wanted signal is suppressed too. The SINR is the same as for the "matched" processing so that the optimum processing has no advantage. However, at $|\delta\tau_{12}|=1/2f_0$ we have (if $\Delta f_s\ll f_0$)

$$q_{out}^2=2q_s^2. \qquad (9.43)$$

The jamming is cancelled out and the wanted signals coming from both stations are summed coherently. The SINR is the same as in the absence of external interferences when the "matched" processing (7.10) is optimum. As $|\delta\tau_{12}|$ increases further, the values of q^2_{out} oscillate with the period $1/f_0$. When $|\delta\tau_{12}|>1/\Delta f_s$ these oscillations decay, and the output SINR tend to

$$q^2_{out}=2q^2_s\frac{(1+q^2)^2+0.5q^4}{q^2+(1+q^2)^2+q^4}. \tag{9.44}$$

For intensive jamming ($q^2\gg 1$) we obtain from (9.44)

$$q^2_{out}\approx 1.5q^2_s \tag{9.45}$$

which is only slightly worse than in the absence of external interference.

It is clear that the dependence of q^2_{out} on $\delta\tau_{12}$ is similar to the corresponding dependence in the case of deterministic signal detection in a background of spatially correlated interferences from a single source [see Section 8.1, Fig. 8.3). In the case considered here when $|\delta\tau_{12}|\ll 1/\Delta f_s$ the signal source falls into the "signal interference region" of the RSRP, where the SINR rapidly oscillates depending on variations of $\delta\tau_{12}$. When $|\delta\tau_{12}|>1/\Delta f_s$ the condition of signal and interference resolution in TDOA and signal "multiplication" is satisfied. The signal autocorrelation function of the process at the summator output [the Fourier transform of $\Phi_{\eta s}(\omega)$] apart from main peak corresponding to zero delay ($\tau=0$), contains another four peaks corresponding to $\tau=\pm\delta\tau_{12}$ and $\tau=\pm 2\delta\tau_{12}$. The extent of each peak in delay is $\Delta\tau\approx 1/\Delta f_s$ and the period of r.f. carrier is $2\pi/\omega_0$.

In this section*, we have synthesized an optimum detection algorithm* (9.22), (9.23) *for the stochastic wanted signal with a known correlation* (*covariance*) *matrix embedded in external spatially correlated interferences* (*e.g., jamming*). *A general structure of the optimum algorithm is the same as for spatially uncorrelated interferences* (7.10) (*see Figs.* 9.7 *and* 7.1) *but the filter transfer functions in the linear part of detectors depend on the number and characteristics of interferences.*

As in the absence of spatially correlated interferences, the output decision variable L in (9.22), (9.23) *may be considered to be approximately Gaussian if the product of the signal bandwidth Δf_s by the integration time T is large enough* ($\Delta f_s T\gg 1$). *Under this assumption detection characteristics can be obtained with the help of* (7.26) *using the expressions for mean values and variances of L in the absence and in the presence of wanted signals* (9.27), (9.28).

We have shown that the optimum processing algorithm (9.22), (9.23) *includes external interference cancellation in its linear part so that interference (e.g., jamming) component is suppressed at the summator output in Fig.* 9.7 *whereas in the case of the "matched" processing* (7.10) *this component is not suppressed* [*see* (9.32) *and* (9.33)]. *We have considered in more detail the simplest and physically "transparent" case where a system contains two receiving stations, there is a single signal source, and a single jammer operates through the antenna mainlobes. We have derived and analyzed expressions for the PSDs of signal* (9.36), *jamming* (9.37) *and self-noise* (9.38) *components of the process at the summator output and have shown that jamming component is cancelled efficiently. The obtained equations* (9.39), (9.40), (9.41) *for variances of above mentioned components allow us to derive expressions for the summator output's SINR* (9.42)–(9.45). *It has been shown that the dependence of this SINR on the difference between signal and jamming TDOAs is similar to the analogous dependence for the deterministic signal considered in Section* 8.1.

9.4. DETECTION OF STOCHASTIC SIGNALS WITH CORRELATION MATRICES CONTAINING RANDOM PARAMETERS

When a wanted signal correlation matrix contains a vector of random parameters, $\boldsymbol{\Theta}$, the unconditional likelihood ratio determining the structure of the optimum detector is, as earlier, expressed by (7.29). However, $L(\boldsymbol{\Theta})$ in (7.29) is now the random variable (9.22) or (9.23) which is optimum for the case where a MSRS is subject to external spatially correlated interferences (e.g., mainlobe jamming). The variable $L(\boldsymbol{\Theta})$ should be considered here as a conditional variable given $\boldsymbol{\Theta}$.

Substituting $L(\boldsymbol{\Theta})$ from (9.23) in (7.29) yields

$$\bar{\Lambda}=\int_{\boldsymbol{\Theta}} w(\boldsymbol{\Theta})\frac{\mathfrak{R}_1(\boldsymbol{\Theta})}{\mathfrak{R}_0}\exp\left\{\sum_{i=1}^{m}\sum_{k=1}^{m}\frac{1}{4\pi}\int_{-\infty}^{\infty}\chi_i^*(\omega)\chi_k(\omega)\right.$$

$$\left.\times|H^{(M+1)}(\omega,\boldsymbol{\Theta})|^2 D_i^{(M+1)}(\omega,\boldsymbol{\Theta})D_k^{*(M+1)}(\omega,\boldsymbol{\Theta})\,d\omega\right\}d\boldsymbol{\Theta}. \tag{9.46}$$

As in Section 7.2, we assume here the vector of signal random parameters, $\boldsymbol{\Theta}$, as consisting of $m(m-1)/2$ phase differences at the inputs of different pairs of receiving stations, $\Delta\varphi_{sik}$, $i,k=\overline{1,m}$, $i<k$, and of m input SNR values, q_{si}^2, $i=\overline{1,m}$, so that $\boldsymbol{\Theta}^{\text{t}}=(\Delta\varphi_{s12},\ldots,\Delta\varphi_{s(m-1)m},q_{s1}^2,\ldots,q_{sm}^2)$. These parameters are the phases and variances of the signal correlation matrix.

Unlike the case of detection in a background of spatially independent noises (Section 7.2) we cannot obtain an expression for $\bar{\Lambda}$ in a closed form without some additional assumptions, even when only $\Delta\varphi_{sik}$, $i,k=\overline{1,m}$, $i<k$, are random and q_{si}^2, $i=\overline{1,m}$, are known. The cause is that the random phases $\Delta\varphi_{sik}$, $i,k=\overline{1,m}$, $i<k$, are included in the denominator of $|H^{(M+1)}(\omega,\boldsymbol{\Theta})|^2$ (8.9) which makes it difficult to integrate the right part of (9.46) over $\boldsymbol{\Theta}$. Therefore, we confine ourselves to the case of weak input signals which, as mentioned in Chapter 7, is of great practical importance. In this case the double sum in the denominator of (8.9) may be neglected in comparison with unity so that $|H^{(M+1)}(\omega,\boldsymbol{\Theta})|^2=F_s(\omega-\omega_0)$. Substituting (8.10) at $p=M+1$ in (9.46) and changing summation indexes for the sake of convenience yields

$$\bar{\Lambda}=\int_{\mathbf{q}_s}\int_{\Delta\boldsymbol{\varphi}_s} w(\Delta\boldsymbol{\varphi}_s)w(\mathbf{q}_s)\frac{\mathfrak{R}_1(\mathbf{q}_s)}{\mathfrak{R}_0}\exp\left\{0.5\sum_{i=1}^{m}\sum_{k=1}^{m}\frac{q_{si}q_{sk}}{\sqrt{N_iN_k}}\right.$$

$$\left.\times\exp(j\Delta\varphi_{sik})\hat{\bar{B}}_{ik}^{(M)}\right\}d\Delta\boldsymbol{\varphi}_s\,d\mathbf{q}_s \tag{9.47}$$

where $\hat{\bar{B}}_{ik}^{(M)}$ do not depend on random parameters:

$$\hat{\bar{B}}_{ik}^{(M)}=\frac{N_iN_k}{2\pi}\int_{-\infty}^{\infty}g_i(\beta_{si},\varepsilon_{si},\omega)g_k^*(\beta_{sk},\varepsilon_{sk},\omega)F_s(\omega-\omega_0)$$

$$\times\exp(j\omega\tau_{sik})\sum_{l=1}^{m}\sum_{n=1}^{m}\chi_l^*(\omega)\chi_n(\omega)f_{il}^{*(M)}(\omega)f_{kn}^{(M)}(\omega)\,d\omega. \tag{9.48}$$

In many cases we may assume for simplicity (as in Section 7.2) that $g_i(\beta_{si}, \varepsilon_{si}, \omega) = 1$ within the signal bandwidth for all $i = \overline{1, m}$. Then

$$\hat{\mathscr{B}}_{ik}^{(M)} = \frac{N_i N_k}{2\pi} \int_{-\infty}^{\infty} F_s(\omega - \omega_0) \exp(j\omega\tau_{sik}) \times \sum_{l=1}^{m} \sum_{n=1}^{m} \chi_l^*(\omega)\chi_n(\omega) f_{il}^{*(M)}(\omega) f_{kn}^{(M)}(\omega)\, d\omega. \tag{9.49}$$

Let first in (9.47) *only phases, $\Delta\varphi_{sik}$, $i, k = \overline{1, m}$, $i < k$, be random* whereas q_{si}^2, $i = \overline{1, m}$, *let be known*. As in Section 7.2, we consider here an extreme case where $\Delta\varphi_{sik}$ are mutually statistically independent and uniformly distributed within the interval $(-\pi, \pi)$. Then we obtain from (9.47):

$$\tilde{\Lambda} = \exp\left\{\sum_{i=1}^{m} \frac{q_{si}^2}{2N_i} \hat{\mathscr{B}}_{ii}^{(M)}\right\} \frac{1}{(2\pi)^{m(m-1)/2}} \int_{-\pi}^{\pi} \cdots \int_{-\pi}^{\pi} \exp\left\{\sum_{i=1}^{m-1} \sum_{k=i+1}^{m} \frac{q_{si} q_{sk}}{\sqrt{N_i N_k}} |\hat{\mathscr{B}}_{ik}^{(M)}| \cos(\Delta\varphi_{sik} + \gamma_{sik})\right\} d\Delta\varphi_{s12} \cdots d\Delta\varphi_{s(m-1)m}. \tag{9.50}$$

The equation (9.50) is similar to (7.32) but $\hat{\mathscr{B}}_{ik}^{(M)}$ in (9.50) are not the same as $\hat{B}_{ik}$ in (7.32) [compare (9.48), (9.49) and (7.31) at $\Delta\Omega_{si} = 0$, $i = \overline{1, m}$.] In the particular case where external interferences are absent ($M = 0$) and input signals are weak, $\hat{\mathscr{B}}_{ik}^{(0)} = \hat{B}_{ik}$.

As in Section 7.2, integrating (9.50) yields

$$\tilde{\Lambda} = \exp\left\{\sum_{i=1}^{m} \frac{q_{si}^2}{2N_i} \hat{\mathscr{B}}_{ii}^{(M)}\right\} \prod_{i=1}^{m-1} \prod_{k=i+1}^{m} I_0\left(\frac{q_{si} q_{sk}}{\sqrt{N_i N_k}} |\hat{\mathscr{B}}_{ik}^{(M)}|\right) \tag{9.51}$$

where $I_0(x)$ is the modified Bessel function of zero order. In the most practically interesting case of strong output signals (i.e. in the case of reliable target detection) we may employ the approximation $\ln I_0(x) \approx x$. Then the optimum processing algorithm (9.51) takes the form

$$\tilde{L} = \ln \tilde{\Lambda} = \sum_{i=1}^{m} \frac{q_{si}^2}{2N_i} \hat{\mathscr{B}}_{ii}^{(M)} + \sum_{i=1}^{m-1} \sum_{k=i+1}^{m} \frac{q_{si} q_{sk}}{\sqrt{N_i N_k}} |\hat{\mathscr{B}}_{ik}^{(M)}|. \tag{9.52}$$

Let now in (9.47) *both phases, $\Delta\varphi_{sik}$, $i, k = \overline{1, m}$, $i < k$, and SNR values, q_{si}^2, $i = \overline{1, m}$, be random*. Repeating the reasoning from Section 7.2 we can obtain the optimum algorithm for completely mutually dependent signal intensity fluctuations

$$\tilde{\tilde{L}} = \sum_{i=1}^{m} \frac{A_{i1}^2}{2N_i} \hat{\mathscr{B}}_{ii}^{(M)} + \sum_{i=1}^{m-1} \sum_{k=i+1}^{m} \frac{A_{i1} A_{k1}}{\sqrt{N_i N_k}} |\hat{\mathscr{B}}_{ik}^{(M)}| \tag{9.53}$$

where $A_{i1}^2 = \overline{q_{si}^2} / \overline{q_{s1}^2}$ are known quantities [see (7.37)].

Note that the derived expressions for optimum processing algorithms (9.51)–(9.53) are the same as those in the case where external interferences are absent (7.33), (7.34), (7.39). However the quantities $\hat{\mathscr{B}}_{ik}^{(M)}$ are determined by (9.48) or (9.49).

Just at this stage of the processing external spatially correlated interferences are suppressed.

The bloc diagram of the optimum processing is similar to that of Fig. 7.3 excluding the input part where external interferences are cancelled by the joint processing of all received signals.

Performance Analysis of the Synthesized Detectors

This analysis can be performed in the same manner as in Section 7.2. We assume that the product $\Delta f_s T$ is sufficiently large so that the quantities $\hat{B}_{ik}^{(M)}$, $i,k=\overline{1,m}$, may be considered to be Gaussian. To obtain detection characteristics, expressions for mean values and variances of $\hat{B}_{ik}^{(M)}$ should be derived. Using (9.24)–(9.26) under the assumption $\Delta f_s T \gg 1$ yields the mean values of $\hat{B}_{ik}^{(M)}$ in the absence and in the presence of wanted signal in the form

$$m_{10}\{\hat{B}_{ik}^{(M)}\} = \frac{N_i N_k T}{\pi} \int_{-\infty}^{\infty} g_i(\beta_{si}, \varepsilon_{si}, \omega) g_k^*(\beta_{sk}, \varepsilon_{sk}, \omega) \times F_s(\omega-\omega_0) f_{ik}^{*(M)}(\omega) \exp(j\omega\tau_{sik})\, d\omega; \tag{9.54}$$

$$m_{11}\{\hat{B}_{ik}^{(M)}\} = \frac{N_i N_k T}{\pi} \int_{-\infty}^{\infty} g_i(\beta_{si}, \varepsilon_{si}, \omega) g_k^*(\beta_{sk}, \varepsilon_{sk}, \omega) F_s(\omega-\omega_0) \times \exp(j\omega\tau_{sik}) \left[f_{ik}^{*(M)}(\omega) + \sum_{l=1}^{m} \sum_{n=1}^{m} \Phi_{snl}(\omega) f_{il}^{*(M)}(\omega) f_{kn}^{(M)}(\omega) \right] d\omega. \tag{9.55}$$

The variances of $\hat{B}_{ik}^{(M)}$ can be obtained in the same way using additionally the relationships (7.22), (7.23) at $\Delta\Omega_{i1}=0$, $i=\overline{1,m}$:

$$M_{20}\{\hat{B}_{ik}^{(M)}\} = \frac{N_i^2 N_k^2 T}{\pi} \int_{-\infty}^{\infty} |g_i(\beta_{si}, \varepsilon_{si}, \omega)|^2 \, |g_k(\beta_{si}, \varepsilon_{si}, \omega)|^2 \times F_s^2(\omega-\omega_0) f_{ii}^{(M)}(\omega) f_{kk}^{(M)}(\omega)\, d\omega; \tag{9.56}$$

$$M_{21}\{\hat{B}_{ik}^{(M)}\} = \frac{N_i^2 N_k^2 T}{\pi} \int_{-\infty}^{\infty} |g_i(\beta_{si}, \varepsilon_{si}, \omega)|^2 \, |g_k(\beta_{si}, \varepsilon_{si}, \omega)|^2 \times F_s^2(\omega-\omega_0) \left[f_{ii}^{(M)}(\omega) + \sum_{l=1}^{m} \sum_{n=1}^{m} \Phi_{snl}(\omega) f_{il}^{*(M)}(\omega) f_{in}^{(M)}(\omega) \right] \times \left[f_{kk}^{(M)}(\omega) + \sum_{l=1}^{m} \sum_{n=1}^{m} \Phi_{snl}(\omega) f_{kl}^{*(M)}(\omega) f_{kn}^{(M)}(\omega) \right] d\omega. \tag{9.57}$$

The variance of $\hat{B}_{ii}^{(M)}/2$ can be obtained by setting $i=k$ in (9.56) and (9.57) and dividing by two since $\hat{B}_{ii}^{(M)}$ are real quantities so that the multiplier 1/2 is not required for variance calculation.

Comparing (9.54) with (9.55) and (9.56) with (9.57) we can see that the presence of a wanted signal leads to the appearance of the double sums containing elements of the signal PSD matrix $\mathbf{\Phi}_s(\omega)$. It can be shown that these sums differ (by the order of magnitude) from the first terms in brackets of (9.55) and (9.57) by factor q_s^2, i.e. by the input SNR value. Here it should be reminded that the optimum detection algorithms (9.52) and (9.53) have been synthesized under the assumption of weak input signal ($q_s^2 \ll 1$). This assumption is naturally to be taken into account for the analysis too. It means that the wanted signal contribution into the variance of $\hat{B}_{ik}^{(M)}$, $i,k=\overline{1,m}$ as well as into the mean value of $\hat{B}_{ii}^{(M)}$ may be neglected. As far as $m_{11}\{\hat{B}_{ik}^{(M)}\}$, $i \neq k$, is concerned, the contribution of a wanted signal must be taken into account since these mean values differ from zero just because of the presence of a signal [in the absence of signals $m_{10}\{\hat{B}_{ik}^{(M)}\} \approx 0$]. Thus assuming weak input signals following equations may be written (approximately):

$$m_{11}\{\hat{B}_{ii}^{(M)}\}=m_{10}\{\hat{B}_{ii}^{(M)}\}; \quad M_{21}\{\hat{B}_{ik}^{(M)}\}=M_{20}\{\hat{B}_{ik}^{(M)}\}, \qquad i,k=\overline{1,m}. \tag{9.58}$$

For the sake of further simplifying mathematical expressions we assume the ADPs $g_i(\beta_{si}, \varepsilon_{si}, \omega)=1$, $i=\overline{1,m}$, within the signal bandwidth. Then from (9.54)–(9.57) taking into account (9.58), we have ($i,k=\overline{1,m}$):

$$\begin{aligned} m_{10}\{\hat{B}_{ik}^{(M)}\} &= \frac{N_i N_k T}{\pi}\int_{-\infty}^{\infty} F_s(\omega-\omega_0) f_{ik}^{*(M)}(\omega)\exp(j\omega\tau_{sik})\,d\omega; \\ M_{20}\{\hat{B}_{ik}^{(M)}\} &= \frac{N_i^2 N_k^2 T}{\pi}\int_{-\infty}^{\infty} F_s^2(\omega-\omega_0) f_{ii}^{(M)}(\omega) f_{kk}^{(M)}(\omega)\,d\omega. \end{aligned} \tag{9.59}$$

$$m_{11}\{\hat{B}_{ik}^{(M)}\} = \frac{N_i N_k T}{\pi}\int_{-\infty}^{\infty} F_s(\omega-\omega_0)\exp(j\omega\tau_{sik}) \times\left[f_{ik}^{*(M)}(\omega)+\sum_{l=1}^{m}\sum_{n=1}^{m}\Phi_{snl}(\omega) f_{il}^{*(M)}(\omega) f_{kn}^{(M)}(\omega)\right]d\omega, \quad i\neq k. \tag{9.60}$$

Equations (9.58)–(9.60) determine the mean values and variances of $\hat{B}_{ik}^{(M)}$ but their covariances are also necessary in order to obtain detection characteristics. In the same way in which we have derived the equations for variances, we find that in the absence of signals and in the presence of a weak signal at the inputs of receiving stations ($q_s^2 \ll 1$) the correlation coefficient can be written in the form

$$\begin{aligned} \rho(\hat{B}_{ik}^{(M)}, \hat{B}_{ln}^{(M)}) = &\int_{-\infty}^{\infty} F_s^2(\omega-\omega_0)\exp[j\omega(\tau_{sik}-\tau_{sln})] f_{il}^{*(M)}(\omega) f_{kn}^{(M)}(\omega)\,d\omega \\ &\times\left[\int_{-\infty}^{\infty} F_s^2(\omega-\omega_0) f_{ii}^{(M)}(\omega) f_{kk}^{(M)}(\omega)\,d\omega\right]^{-1/2} \\ &\times\left[\int_{-\infty}^{\infty} F_s^2(\omega-\omega_0) f_{ll}^{(M)}(\omega) f_{nn}^{(M)}(\omega)\,d\omega\right]^{-1/2}, \quad i,k,l,n=\overline{1,m}. \end{aligned} \tag{9.61}$$

To employ the detection characteristic calculation technique used in Section 7.2 to the case considered here, we must be sure that the quadrature components of $\hat{B}_{ik}^{(M)}$, $i<k$, in (9.52) and (9.53) are mutually uncorrelated and have the same variances. As mentioned in Section 7.2, only under these conditions $|\hat{B}_{ik}^{(M)}|$ is Rayleigh and Rice distributed in the absence and in the presence of a wanted signal, respectively [33,49].

The correlation (central) moments and variances of $\mathrm{Re}\,\hat{B}_{ik}^{(M)}$, $\mathrm{Im}\,\hat{B}_{ik}^{(M)}$ can be obtained from (9.48) in a similar manner as the variances of $\hat{B}_{ik}^{(M)}$.

$$M_2\{\mathrm{Re}\,\hat{B}_{ik}^{(M)}, \mathrm{Im}\,\hat{B}_{ik}^{(M)}\} = \frac{N_i^2 N_k^2 T}{\pi} \int_{-\infty}^{\infty} F_s^2(\omega-\omega_0) \times \exp(\mathrm{j}2\omega\tau_{sik})[f_{ik}^{*(M)}(\omega)]^2 \, d\omega; \tag{9.62}$$

$$\begin{aligned} M_2\{\mathrm{Re}\,\hat{B}_{ik}^{(M)}\} &= M_2\{\hat{B}_{ik}^{(M)}\} + M_2\{\mathrm{Re}\,\hat{B}_{ik}^{(M)}, \mathrm{Im}\,\hat{B}_{ik}^{(M)}\}; \\ M_2\{\mathrm{Im}\,\hat{B}_{ik}^{(M)}\} &= M_2\{\hat{B}_{ik}^{(M)}\} - M_2\{\mathrm{Re}\,\hat{B}_{ik}^{(M)}, \mathrm{Im}\,\hat{B}_{ik}^{(M)}\}. \end{aligned} \tag{9.63}$$

Here it is also assumed that $q_s^2 \ll 1$ so that the appearance of a wanted signal does not practically influence the second central moments, i.e. $M_{20}\{\cdot\} = M_{21}\{\cdot\}$. It can be shown from (9.62) that when signals and interferences are perfectly resolved in TDOAs at all stations, the correlation moments are equal to zero. Then according to (9.63) the variances of $\mathrm{Re}\,\hat{B}_{ik}^{(M)}$ and $\mathrm{Im}\,\hat{B}_{ik}^{(M)}$ are equal to each other.

Example 9.4. Let us consider again the case of a single interference source (e.g., a mainlobe jammer) where the obtained results can be made easy-to-grasp. Let the conditions (9.31) and (8.30) be satisfied for $F_s(\omega)$ and $F(\omega)$ (i.e. for a wanted signal and interference). Using (8.61) for $f_{ik}^{(1)}(\omega)$ we can obtain from (9.59)

$$m_{10}\{\hat{B}_{ik}^{(1)}\} = \frac{2N\Delta f_s T}{1+mq^2}[(1+mq^2)\delta_{ik} - q^2]\,\mathrm{sinc}(\pi\Delta f_s \delta\tau_{ik}) \times \exp[-\mathrm{j}(\omega_0 \delta\tau_{ik} + \Delta\varphi_{ik})] \tag{9.64}$$

where δ_{ik} is as earlier the Kronecker symbol; $\mathrm{sinc}(x) = (\sin x)/x$; $\delta\tau_{ik} = \tau_{ik} - \tau_{sik}$ is the difference in TDOAs of the interference and the signal at the inputs of the ith and the kth stations ($i, k = \overline{1, m}$). It is seen from (9.64) and (9.58) that

$$m_{10}\{\hat{B}_{ii}^{(1)}\} = m_{11}\{\hat{B}_{ii}^{(1)}\} = 2N\Delta f_s T[1+(m-1)q^2]/(1+mq^2), \quad i = \overline{1, m}. \tag{9.65}$$

When the signal is expected to be reliably resolved in TDOA from the interference, i.e. when $|\delta\tau_{ik}| > (2\ldots3)/\Delta f_s$, then

$$\begin{aligned} &m_{10}\{\hat{B}_{ik}^{(1)}\} \approx 0; \\ &m_{11}\{\hat{B}_{ik}^{(1)}\} = 2N\Delta f_s T q_s^2 \exp(-\mathrm{j}\Delta\varphi_{sik}) \\ &\qquad \times [1+(m-1)q^2]^2/(1+mq^2)^2, \quad i, k = \overline{1, m}, \ i \neq k. \end{aligned} \tag{9.66}$$

Besides,

$$M_{20}\{\hat{B}_{ik}^{(1)}\} = M_{21}\{\hat{B}_{ik}^{(1)}\} = 2N^2\Delta f_s T[1+(m-1)q^2]^2/(1+mq^2)^2, \quad i, k = \overline{1, m}. \tag{9.67}$$

The correlation coefficient [from (9.61)]

$$\begin{aligned} \rho\{\hat{B}_{ik}^{(1)}, \hat{B}_{ln}^{(1)}\} &= [(1+mq^2)\delta_{il} - q^2][(1+mq^2)\delta_{kn} - q^2] \\ &\times \mathrm{sinc}[\pi\Delta f_s(\delta\tau_{ik} - \delta\tau_{ln})]\exp\{-\mathrm{j}[\omega_0(\delta\tau_{ik} - \delta\tau_{ln}) + \Delta\varphi_{ik} - \Delta\varphi_{ln}]\} \\ &\times [1+(m-1)q^2]^{-2}, \quad i, k, l, n = \overline{1, m}. \end{aligned} \tag{9.68}$$

It follows from (9.68) that for the case of perfect resolution in TDOAs when $|\delta\tau_{ik}-\delta\tau_{ln}|>(2\ldots3)/\Delta f_s$, the correlation coefficient between $\hat{B}_{ik}^{(1)}, \hat{B}_{ln}^{(1)}$, for any combination of indexes is nearly zero except for $i=k$, $l=n$. In the latter case

$$\rho\{\hat{B}_{ii}^{(1)}, \hat{B}_{ll}^{(1)}\}=q^4/[1+(m-1)q^2]^2.$$

Under the most practically important conditions of intensive interferences ($q^2 \gg 1$)

$$\rho\{\hat{B}_{ii}^{(1)}, B_{ll}^{(1)}\} \approx 1/(m-1)^2.$$

When a passive MSRS contains only two stations, $m=2$, then $\rho \approx 1$. However, ρ falls down to 0,25–0.09 when there are at least 3–4 stations. Thus the random Gaussian variables $\hat{B}_{ik}^{(1)}$, $i \neq k$, in the optimum processing algorithms (9.52) and (9.53) may be considered as being mutually uncorrelated and uncorrelated with the random variables $\hat{B}_{ii}^{(1)}$ when wanted signals and interferences are resolved in TDOAs in any pair of stations. The variables $\hat{B}_{ii}^{(1)}$ may be considered as being mutually uncorrelated when the number of stations $m \geq 3$–4. When $m=2$ the high correlation between $\hat{B}_{11}^{(1)}$ and $\hat{B}_{22}^{(1)}$ should be taken into account. However, just because of this high correlation the summation of $\hat{B}_{11}^{(1)}$ and $\hat{B}_{22}^{(1)}$ in (9.52) and (9.53) does not enhance detection characteristics so that such a summation may be rejected.

We met with a similar situation in Section 8.5 when we analyzed the detection of regular (nonstochastic) signals in a background of spatially correlated interferences. We saw that the optimum combination of signals coming from two stations did not improve detection characteristics (Fig. 8.8 at $\Delta f_s|\delta\tau_{12}| \approx 0$).

It remains to evaluate the correlation between $\mathrm{Re}\,\hat{B}_{ik}^{(1)}$ and $\mathrm{Im}\,\hat{B}_{ik}^{(1)}$. According to (9.62)

$$M_2\{\mathrm{Re}\,\hat{B}_{ik}^{(1)}, \mathrm{Im}\,\hat{B}_{ik}^{(1)}\} = \frac{2N^2\Delta f_s T q^4}{(1+mq^2)^2}\,\mathrm{sinc}(2\pi\Delta f_s \delta\tau_{ik}) \times \exp[-j2(\omega_0\delta\tau_{ik}+\Delta\varphi_{ik})]. \tag{9.69}$$

It is seen that when $|\delta\tau_{ik}|>2/\Delta f_s$, then $\mathrm{sinc}(2\pi\Delta f_s\delta\tau_{ik})\approx 0$ and $M_2\{\mathrm{Re}\,\hat{B}_{ik}^{(1)}, \mathrm{Im}\,\hat{B}_{ik}^{(1)}\}\approx 0$. Hence, as it follows from (9.63), $M_2\{\mathrm{Re}\,\hat{B}_{ik}^{(1)}\}=M_2\{\mathrm{Im}\,\hat{B}_{ik}^{(1)}\}$.

Now we can employ the detection characteristic calculation technique derived in Section 7.2. To use equations (7.41)–(7.45) and (7.48), (7.50), (7.26) we only have to take into account that the parameters of those equations in the case considered take the form

$$\begin{aligned} &a_1=Nn[1+(m-1)q^2]/(1+mq^2); \quad a_2=a_1;\\ &a_3=2Nq_s^2 n[1+(m-1)q^2]^2/(1+mq^2)^2;\\ &\sigma_1^2=N^2 n[1+(m-1)q^2]^2/(1+mq^2)^2;\\ &\sigma_2^2=2\sigma_1^2; \quad \sigma_3^2=\sigma_2^2. \end{aligned} \tag{9.70}$$

In (9.70) $n=\Delta f_s T$ is the accumulation (integration) factor as before.

Thus the detection characteristic calculation technique developed in Section 7.2 turns out to be appropriate for use when a MSRS is subject to external spatially correlated interferences under the condition where wanted signals and interferences are resolved in TDOAs at all receiving stations. When they are not resolved the corresponding analysis is much more complicated.

In the general case where a MSRS is subject to external interferences from several sources (e.g., from several mainlobe jammers) the mean values and variances of variables $\hat{B}_{ik}^{(M)}$, $i,k=\overline{1,m}$ are different [even when the conditions (9.31) are valid] because of differences in signal and interference TDOAs, interference intensities. For such situations we have to take into account that each term of the (9.52) and (9.53) has its own characteristic function so that it is necessary to replace powers of $\Theta_1(v)$ and $\Theta_2(v)$ in (7.43) by products $\Theta_{1i}(v)$, $i=\overline{1,m}$, and $\Theta_{2ik}(v)$, $i,k=\overline{1,m}$, $i<k$. In Example 9.4 we have confined ourselves to the simplest case ($M=1$) where the technique of Section 7.2 may be employed immediately under the conditions (9.31). Just for this case the expressions (9.70) have been derived.

Simplified Suboptimum Detectors

The most practically important is the *correlation detector with envelope detection at the correlator output*. The corresponding processing algorithm can be obtained from (9.52) and (9.53) by dropping the first sums (containing $\hat{B}_{ii}^{(M)}$). Performance analysis can be made on the basis of the above derived expressions for optimum detectors (9.52) and (9.53) taking into account that $\hat{B}_{ii}^{(M)}$ are excluded. In particular, in the case of a single mainlobe interference source ($M=1$) and under the conditions (9.31) the detection characteristic calculation technique described in Sections 7.2 and 7.3 may be utilized immediately.

Example 9.5. Let us obtain equations for the SINR at the correlator output (before envelope detection). From (9.70) we have

$$q_{\text{out cor}}^2 = a_3^2/2\sigma_3^2 = q_s^4 n[1+(m-1)q^2]^2/(1+mq^2)^2. \tag{9.71}$$

It can be seen that for a weak input wanted signal (when $q_s^2 \ll 1$) the SINR (9.71) nears to (7.58) corresponding to the absence of external spatially correlated interferences and $m=2$. This is a result of interference cancellation (mutual coherent suppression) in the input part of detectors. For the most interesting situation of intensive interferences ($q^2 \gg 1$) (9.71) can be rewritten as follows

$$q_{\text{out cor}}^2 = q_s^4 n(m-1)^2/m^2 \tag{9.72}$$

which is only by $(m-1)^2/m^2$ less than for the case where external interferences are absent [see (7.58) at $q_{s1}^2 = q_{s2}^2 = q_s^2 \ll 1$].

All algorithms for stochastic signal detection considered in Section 9.3 and in this section have been derived under the assumption that the matrix $\mathbf{\Phi}(\omega)$ of interference-plus-self-noise PSDs at the inputs of stations is known. In practice, as with regular signal detection, the matrix $\mathbf{\Phi}(\omega)$ is, as a rule, unknown and may vary in time. Therefore, real stochastic signal detectors which use external interference spatial correlation for cancellation of these interferences, should be capable to adapt themselves to variations of $\mathbf{\Phi}(\omega)$. The obtained results of synthesis and analysis are to be considered as "potential" ones. The actual detectors with adaptive external interference cancellation should "converge" to those results in a steady-state mode.

The technique of adaptive interference cancellation with the help of adaptive Wiener filters based on a transversal filters with orthogonal and nonorthogonal channels considered in Sections 9.1 and 9.2 are appropriate for detection of stochastic signals as well. However, there are some special features here. The main feature is that the structure, duration and other characteristics of wanted signals do not differ from those of interferences. As mentioned in Section 4.1, a stochastic signal source can at one moment be a wanted signal source which is to be detected in a background of interferences from other sources whereas at another moment the same source can be an interference source masking the wanted signal from another source. In order not to distort and the more not to suppress wanted signals, their influence on transversal filter weights is to be eliminated. It is difficult to perform because of similar structure of signals and interferences. However, when a wanted signal and interferences are reliably resolved in TDOA, these differences can be used for the transversal filter weights "protection" against the effect of wanted signals. There is a certain similarity in possible solutions of this problem and of the problem of sidelobe jamming cancellation in adaptive antenna arrays when it is necessary "to protect" the wanted signal of the same structure entering through antenna mainlobe [12,72].

It is important to note that there is an essential distinction between the problem of passive stochastic signal detection in a background of external spatially correlated interferences, on the one hand, and the similar problem of target echo detection in an active MSRS, on the other. It follows from the mentioned in Section 7.4 possibility of accumulation (integration) of long-duration signals in a passive MSRS in many cases. When $n=\Delta f_s T \gg 1$ the accumulated signal energy may be sufficient for reliable signal detection in a background of even intensive external interferences (resolved from the signal in TDOA) without interference cancellation. In such cases we may reject complicated detection schemes with adaptive external interference cancellation and employ much simpler detectors considered in Sections 7.1–7.4.

Example 9.6. Let a MSRS with two receiving stations ($m=2$) be subject to a single mainlobe jammer (apart from a signal source). A correlation detector with envelope detection at the correlator output is used. We consider the possibility of employing the algorithm (7.56), i.e. the possibility of utilizing the much simpler quantities $\hat{B}_{ik}$ from (7.31) at $\Delta\Omega_{sik}=0$, $i,k=\overline{1,m}$, instead of $\hat{B}_{ik}^{(1)}$ from (9.49). It can be shown that when wanted signals and interferences are resolved in TDOA at both the two stations $[|\delta\tau_{12}|>(2\ldots3)/\Delta f_s]$ and under the condition (9.31) the SINR at the correlator output of the simple detector without interference cancellation is determined by (7.58) at $q_{s1}=q_{s2}=q_s$, i.e.

$$q_{\text{out cor}}^2=q_s^4 n_1/(1+q_s^2)^2 \tag{9.73}$$

where

$$q_s^2=q_{s0}^2/(1+q^2). \tag{9.74}$$

In (9.74) q_{s0}^2 and q^2 are the power signal-to-self-noise and interference-to-self-noise ratios, respectively, at the input of each station. Substituting (9.74) in (9.73) yields

$$q_{\text{out cor}}^2=q_{s0}^4 n_1/(1+q^2+q_{s0}^2)^2. \tag{9.75}$$

Now we can evaluate the required accumulation factor n_1 in a simple correlation detector which provides for equal values of the input SNR the same value of the output SINR as in the correlation detector with interference cancellation and the accumulation factor n. Denoting q_s^2 by q_{s0}^2 in (9.71) as in (9.75) and equalizing the right part of (9.75) to the right part of (9.71) at $m=2$ we can obtain the following relationship between required values of the accumulation factor n_1 in (9.75) and n

in (9.71) for weak input signals when $q_{s0}^2 \ll 1$:

$$n_1 = \frac{(1+q^2)^4}{(1+2q^2)^2}\, n. \tag{9.76}$$

For instance, if $q^2 = 10$, then $n_1 \approx 33\, n$; if $q^2 = 100$ then $n_1 \approx 2576\, n$. Thus when the wanted signal duration and bandwidth as well as other operation conditions of a MSRS permit to achieve required values of $n_1 = \Delta f_s T$, simple detectors without interference cancellation (considered in Chapter 7) should be employed. In certain cases a combination of coherent with incoherent signal accumulation can be used.

It is important to emphasize that signal energy accumulation for a wanted signal source detection in a background of external interferences (i.e. by using a "brute force" approach) is performed in a passive mode by more efficient utilization of the energy of signal radiation sources (e.g., jammers). Therefore it does not require additional energy resources of a MSRS. This is an essential advantage of passive MSRSs. Evidently, it is impossible in active MSRSs for target echo detection in a background of external interferences.

***In this section**, we have synthesized the optimum processing algorithms for stochastic signal detection in a passive MSRS under the condition of mainlobe external interferences when the PSD (or correlation) matrix of signals at the inputs of spatially separated receiving stations contains random phases or both random phases and variances (signal intensities) [(9.52) and (9.53), respectively]. The general structure of these algorithms (and corresponding detectors) is the same as that of the algorithms (7.34) and (7.39) which are optimal for detection of similar signals in the absence of external spatially correlated interferences. However, an essential difference is between the random variables $\hat{\hat{B}}_{ik}$ in (9.52), (9.53) and the random variables $\hat{B}_{ik}$ in (7.34), (7.39). In the case where external interferences are present, $\hat{\hat{B}}_{ik}$ are determined by (9.48) whereas otherwise $\hat{B}_{ik}$ is determined by (7.31).*

As in Section 7.2, performance analysis of the synthesized algorithms in this Section is based on the assumption that $\hat{\hat{B}}_{ik}$ are Gaussian variables for large values of the accumulation (integration) factor $n = \Delta f_s T$. We have shown that detection characteristics can be obtained with the help of the same technique as in Section 7.2 though using new equations for main parameters. We have derived equations (9.54)–(9.57) for the mean values and variances of $\hat{\hat{B}}_{ik}$ in both the presence and absence of wanted signals [and the simplified form (9.58)–(9.60) of those equations] as well as equation (9.61) for the mutual correlation coefficient of $\hat{\hat{B}}_{ik}$. The obtained expressions (9.62) and (9.63) for mutual correlation and variances of the quadrature components of $\hat{\hat{B}}_{ik}$ permit to establish that under the conditions of weak input signals ($q_s^2 \ll 1$) and of resolved signal and interferences in TDOA at the inputs of stations, those quadratures are uncorrelated and their variances are equal. Thus all the obtained results allow us to employ the technique of Section 7.2 for numerical performance analysis of the optimum algorithms (9.52) and (9.53).

In Example 9.4 we have shown how to apply this technique in the simple and easy-to-grasp case of a single mainlobe jammer in a passive MSRS with m receiving stations.

We have considered the correlation detector with envelope detection at the correlator output which is the most important among simplified suboptimum detectors for practical purposes. In particular we have shown in Example 9.5 that due to the external interference cancellation the SINR at the correlator output is close to the SNR in the absence of external interferences [see (9.71), (9.72)].

We have pointed out that in real situations where PSD matrices of external interferences and self-noises at the inputs of spatially separated stations are unknown, the adaptive approach from Sections 9.1 and 9.2 can be used.

An important feature of the considered problem has been discussed. This is the possibility of achieving high performance characteristics by using simple algorithms of Chapter 7 without external interference (e.g., jamming) cancellation when signal accumulation (integration) factor $n = \Delta f_s T$ can be chosen sufficiently large. Unlike the case of target echo detection by an active MSRS the realization of such a possibility by a passive MSRS does not require additional energy resources since the passive MSRS utilizes the energy of a radiation source (e.g., a jammer). A numerical illustration of that possibility has been given in Example 9.6.

10. TARGET DETECTION IN CLUTTER BY MSRSs

10.1. EFFICIENCY OF SPATIAL PROCESSING AGAINST CLUTTER

It is well known, that apart from interferences caused by radiation sources (e.g., jamming) and receiver self-noises, wanted signal detection may be hindered by clutter. Radar clutter is defined as unwanted echoes. Natural clutter is typically from the ground and/or sea surfaces, ground objects which are not to be detected, rain and other precipitations, aurora etc. The most important deliberate clutter is caused by chaff [15,47,48,192]. In this chapter we pay main attention to chaff.

The peculiar feature of clutter is that the increase of illumination energy from a radar transmitter leads to the equal increase of the energy of both wanted and unwanted echoes. Therefore, unlike the case of jamming, the "brute force", "burn-through" method for target detection in clutter is useless.

In monostatic radars the time–frequency processing is widely used against clutter. First of all, these are the MTI methods [2,72] based on the difference in radial velocities (with respect to a radar) of targets to be detected, on the one hand, and of Collections (clouds) of Masking Reflectors (CMR), of any nature, on the other. A target radial velocity difference leads, in its turn, to a Doppler frequency shift difference of target echoes which is used for signal detection in clutter. All MTI methods used in monostatic radars are applicable to MSRSs where higher performance of these methods can be achieved due to the possibility of observing both targets and CMRs simultaneously from several different directions and of employing the cooperative signal reception (see Section 10.4).

Apart from the time–frequency processing at each receiving station the space–time (interstation) processing is possible in MSRSs. As shown in Chapters 8 and 9, the space–time processing permits to cancel external interferences from point-like radiation sources, e.g., jammers. The question arises: is it possible to use the space–time processing to cancel clutter?

To answer this question let us consider the efficiency of the space–time processing for signal detection in clutter [64]. We have shown in Chapter 8 that external interference cancellation is performed by the same procedures for both deterministic and quasideterministic (containing unknown or random parameters) signals. Therefore, we may employ here the deterministic signal as a simpler signal model. We assume a MSRS to be spatially coherent or with the short-term spatially coherence (see Section 1.1). To exclude possible influence of Doppler filtration we assume the positions of the target and all elements of both the CMR and the MSRS to be fixed within the observation time interval T_s.

A CMR is Illuminated by a Single Transmitting Station

We consider a volumetric CMR and assume, as usual [53,72,73], that the clutter from this CMR at the inputs of m receiving stations may be modelled by Gaussian stationary (within the time interval T_s) random processes with zero mean. Since the positions of all elements of both the CMR and the MSRS are fixed, these processes are stationary coupled. Apart from clutter we take into account self-noises at the inputs of all receiving stations which we consider to be white mutually independent Gaussian stationary processes with the PSD N_i, $i=\overline{1,m}$. If complex signals with high value of the product of bandwidth by duration are employed ($\Delta f_s T_s \gg 1$), expression (8.1) for the optimum detection algorithm in the frequency domain is valid. Performance characteristics, i.e. the potential efficiency of space–time processing, are completely determined by the output SINR (8.27). To obtain quantitative estimates, the matrix $\mathbf{f}(\omega)=\mathbf{\Phi}^{-1}(\omega)$ in (8.1) and (8.27) is to be expressed in terms of CMR and MSRS parameters.

Let us locate the origin of coordinates O into the geometric centre of a CMR (see Fig. 10.1). The transmitting station is positioned at the point determined by the radius-vector $\mathbf{R}_0$. Consider two arbitrary receiving stations whose positions are determined by the radius-vectors $\mathbf{R}_i$ and $\mathbf{R}_k$, $i,k=\overline{1,m}$. The spectrum (Fourier transform) of the echo from the lth reflector at the input of the ith receiving station, $\Psi_{li}(\omega)$, can be written in the form of (8.3). For simplicity we assume the ADPs to be independent of frequency, i.e. $g_i(\beta_{li}, \varepsilon_{li}, \omega)=g_i(\beta_{li}, \varepsilon_{li})$ within the interval $|\omega-\omega_0| \leqslant \Delta\omega_s/2$. We denote the complex amplitude of the echo by $b_{li}=a_{li}\exp(-j\varphi_{li})$. It is convenient to change the normalization of the complex envelope spectrum used in

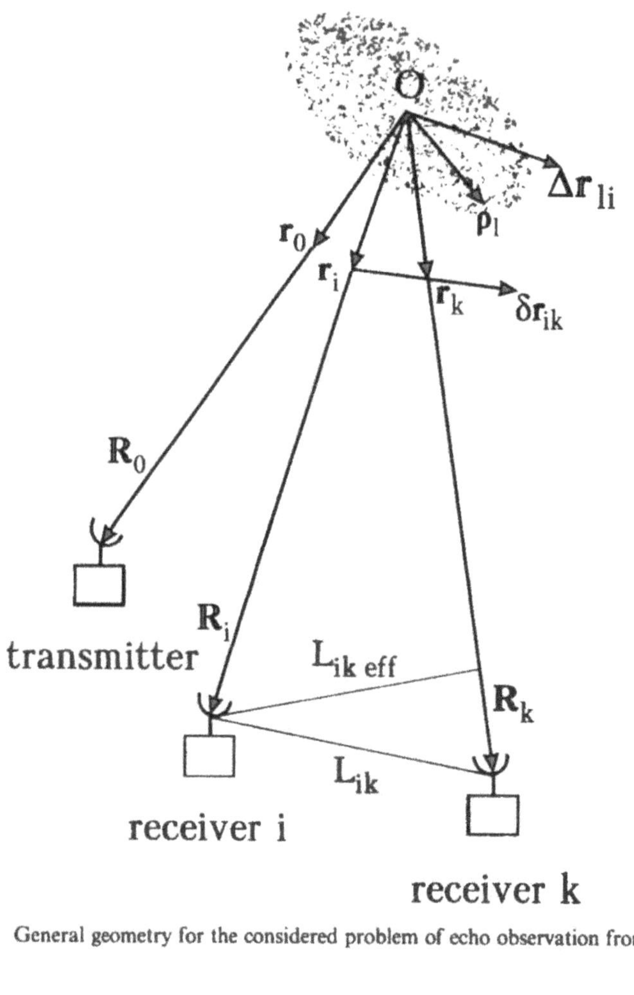

Figure 10.1 General geometry for the considered problem of echo observation from a CMR

previous chapters. Let us introduce

$$\tilde{\Psi}_0(\omega-\omega_0)=\sqrt{\Delta f_s/2T_s}\,\Psi_0(\omega-\omega_0). \tag{10.1}$$

Then we can obtain from (8.3)

$$\Psi_{li}(\omega)=b_{li}g_i(\beta_{li},\varepsilon_{li})\sqrt{2T_s/\Delta f_s}\,\tilde{\Psi}(\omega-\omega_0)\exp[-j(\omega/c)(R_{l0}+R_{li})] \tag{10.2}$$

where R_{l0}, R_{li} are the distances of the lth reflector of the CMR from the transmitting and the ith receiving stations, respectively. Here

$$\frac{1}{2\pi}\int_{-\infty}^{\infty}|\tilde{\Psi}_0(\omega)|^2\,d\omega=\Delta f_s=\Delta\omega_s/2\pi. \tag{10.3}$$

For $\Delta f_s T_s \gg 1$ an arbitrary element of the PSD matrix, $\mathbf{\Phi}(\omega)$, of the clutter-plus-self-noise at the inputs of spatially separated receiving stations is given by

$$\begin{aligned}\Phi_{ik}(\omega)&=\delta_{ik}\sqrt{N_iN_k}+(1/2T_s)\sum_l\sum_n\overline{\Psi_{li}(\omega)\Psi^*_{nk}(\omega)}\\&=\delta_{ik}\sqrt{N_iN_k}+\frac{1}{\Delta f_s}\sum_l\sum_n\overline{b_{li}b^*_{nk}g_i(\beta_{li},\varepsilon_{li})g^*_k(\beta_{nk},\varepsilon_{nk})}\\&\quad\overline{\times|\tilde{\Psi}_0(\omega-\omega_0)|^2\exp[-j(\omega/c)(R_{l0}+R_{li}-R_{n0}-R_{nk})]},\quad i,k=\overline{1,m}.\end{aligned} \tag{10.4}$$

The averaging in (10.4) is performed over the sets of random complex amplitudes and reflector positions in the CMR whereas the summation is carried out over all reflectors contributing to the output SINR (8.27). We assume now that echo complex amplitudes from any arbitrary reflector, b_{li}, are the same at the inputs of all receiving stations, i.e. $b_{li}=b_l$, $i=\overline{1,m}$. Besides, we assume all reflectors of the CMR, taken into consideration in (10.4), to be uniformly illuminated by the transmitting station so that $g_0(\beta_{l0},\varepsilon_{l0})=1$. As usually, we may assume[1] that [47,72]

$$\overline{b_l}=0;\qquad \overline{b_lb^*_n}=\begin{cases}\sigma^2, & l=n;\\ 0, & l\neq n.\end{cases} \tag{10.5}$$

The distance of the lth reflector from the transmitting, the ith and the kth receiving stations can be presented in the form (see Fig. 10.1)

$$\begin{aligned}R_{l0}&\approx R_0-\boldsymbol{\rho}_l\mathbf{r}_0+(\boldsymbol{\rho}_l\Delta\mathbf{r}_{l0})^2/2R_0;\\R_{li}&\approx R_i-\boldsymbol{\rho}_l\mathbf{r}_i+(\boldsymbol{\rho}_l\Delta\mathbf{r}_{li})^2/2R_i;\\R_{lk}&\approx R_k-\boldsymbol{\rho}_l\mathbf{r}_k+(\boldsymbol{\rho}_l\Delta\mathbf{r}_{lk})^2/2R_k\end{aligned} \tag{10.6}$$

where R_0, R_i, R_k are the distances of the transmitting, the ith and the kth receiving stations, respectively, from the centre of the CMR (the coordinate origin); $\boldsymbol{\rho}_l$ is the radius-vector of the lth reflector; $\mathbf{r}_0$, $\mathbf{r}_i$ and $\mathbf{r}_k$ are the unit vectors directed along

[1] It is reasonable to recall that according to (8.3) the quantity a_l in the complex amplitude $b_l=a_l\exp(-j\varphi_l)$ is the r.m.s. value of a target echo.

$\mathbf{R}_0, \mathbf{R}_i$ and $\mathbf{R}_k$, respectively; $\Delta\mathbf{r}_{l0}$, $\Delta\mathbf{r}_{li}$ and $\Delta\mathbf{r}_{lk}$ are the unit vectors lying in the planes perpendicular to $\mathbf{r}_0$, $\mathbf{r}_i$ and $\mathbf{r}_k$, respectively, and directed along the projections of $\boldsymbol{\rho}_l$ on these planes. Let us substitute (10.6) in (10.4) and take into account that for $L_{ik\,\mathrm{eff}} \ll R_i, R_k$

$$\mathbf{r}_k \approx \mathbf{r}_i + (L_{ik\,\mathrm{eff}}/R_{ik})\delta\mathbf{r}_{ik} \tag{10.7}$$

where $\delta\mathbf{r}_{ik}$ is the unit vector lying in the plane $(\mathbf{r}_i, \mathbf{r}_k)$ and perpendicular to the bisector of the angle between $\mathbf{r}_i$ and $\mathbf{r}_k$; $L_{ik\,\mathrm{eff}}$ is the effective baselength between the ith and the kth stations (see the definition in Section 1.1); R_{ik} is the distance from the centre of the CMR to the middle of the effective base. When $L_{ik\,\mathrm{eff}} \ll R_i, R_k$ the difference between the second order terms in (10.6) may be neglected since within the wanted signal bandwidth

$$(\omega/c)[(\boldsymbol{\rho}_l\Delta\mathbf{r}_{li})^2/2R_i - (\boldsymbol{\rho}_l\Delta\mathbf{r}_{lk})^2/2R_k] \ll 2\pi.$$

Under the approximation (10.7) $\beta_{li} = \beta_{lk}$, $\varepsilon_{li} = \varepsilon_{lk}$. Then assuming b_l to be statistically independent of reflector positions, averaging over b_l and taking into account (10.5) we can obtain from (10.4)

$$\Phi_{ik}(\omega) = \delta_{ik}\sqrt{N_i N_k} + \frac{\sigma^2}{\Delta f_s}|\Psi_0(\omega-\omega_0)|^2 \exp[-\mathrm{j}(\omega/c)(R_i - R_k)]$$

$$\times \overline{\sum_l g_i(\beta_l, \varepsilon_l) g_k^*(\beta_l, \varepsilon_l) \exp[-\mathrm{j}(\omega/c)(L_{ik\,\mathrm{eff}}/R_{ik})\,\boldsymbol{\rho}_l\delta\mathbf{r}_{ik}]}. \tag{10.8}$$

Now we have to average the right part of (10.8) over a random position of each reflector which is characterized by the angles β_l, ε_l and by the projection $\boldsymbol{\rho}_l\delta\mathbf{r}_{ik}$ of the radius-vector $\boldsymbol{\rho}_l$ on the effective base direction.

In practice the ADPs $g_i(\beta_l, \varepsilon_l)$, $i = \overline{1,m}$, are slowly varying functions of β_l and ε_l as compared with the exponential functions under the sign of summation in (10.8) since the effective baselength $L_{ik\,\mathrm{eff}}$ is usually much greater than antenna dimensions of each receiving station. Hence ADPs have only a weak effect on the PSD $\Phi_{ik}(\omega)$. It means that for the purpose of this section we may assume $g_i(\beta_l, \varepsilon_l) = 1$, $i = \overline{1,m}$. Taking it into account, random angles β_l and ε_l are convenient to be replaced by random angles $\Theta_{ikl} = \boldsymbol{\rho}_l\delta\mathbf{r}_{ik}/R_{ik}$. For each lth reflector such an angle characterizes the angular shift between the directions from the middle of the effective baseline to the centre of the CMR and to the projection of the lth reflector's position on "the base plane", i.e. the plane passed through the centre of the CMR and the baseline, L_{ik}. Let the angular width of the CMR projection on this plane (as it is seen from the middle of the effective baseline) be equal to $\pm\Delta\Theta_{ik}/2$ and the angles Θ_{ikl} be uniformly distributed within the limits $\pm\Delta\Theta_{ik}/2$. Then averaging the right part of (10.8) over $\Theta_{ikl} = \boldsymbol{\rho}_l\delta\mathbf{r}_{ik}/R_{ik}$ for $g_i(\beta_l, \varepsilon_l) = g_k(\beta_l, \varepsilon_l) = 1$ yields

$$\Phi_{ik}(\omega) = \delta_{ik}\sqrt{N_i N_k} + (\bar{n}\sigma^2/\Delta f_s)|\Psi_0(\omega-\omega_0)|^2$$

$$\times \operatorname{sinc}[\pi(\omega/\omega_0)\alpha_{ik}]\exp[-\mathrm{j}(\omega/c)(R_i - R_k)] \tag{10.9}$$

where $\bar{n}$ is the total number of reflectors contributing to the output SINR (8.27);

$$\alpha_{ik} = (L_{ik\,\mathrm{eff}}\Delta\Theta_{ik})/\lambda_0, \quad \lambda_0 = 2\pi c/\omega_0. \tag{10.10}$$

As can be seen from (10.9) the mutual PSD $\Phi_{ik}(\omega)$ for $i \neq k$ depends on the factor $\text{sinc}[\pi(\omega/\omega_0)\alpha_{ik}]$. For usual narrowband signals $\omega/\omega_0 = 1 + (\omega - \omega_0)/\omega_0 \approx 1$ since $|\omega - \omega_0| \leqslant \Delta\omega_s/2 \ll \omega_0$. Therefore $\text{sinc}[\pi(\omega/\omega_0)\alpha_{ik}] \approx 0$ and hence $\Phi_{ik}(\omega) \approx 0$ for $i \neq k$ when $\alpha_{ik} \geqslant 1$, the more so when $\alpha_{ik} > 2\ldots3$. This means that under the condition

$$\Delta\Theta_{ik} > (1\ldots3)\lambda_0/L_{ik\,\text{eff}} \tag{10.11}$$

mutual correlation of clutter from the CMR at the inputs of receiving stations (i.e. spatial clutter correlation) is practically absent. It follows from (10.9) that the mutual correlation coefficient of clutter at the inputs of the ith and kth stations at $\omega/\omega_0 \approx 1$ after delay equalization can be written in the form

$$\rho_{ik} = \frac{q_{ci}q_{ck}\,\text{sinc}(\pi\alpha_{ik})}{\sqrt{(1+q_{ci}^2)(1+q_{ck}^2)}}, \quad i,k = \overline{1,m}, \; i \neq k \tag{10.12}$$

where $q_{ci}^2 = \tilde{n}\sigma^2/N_i\Delta f_s$ is the power Clutter-to-Noise Ratio (CNR) at the input of the ith station. Under the condition (10.11) $\rho_{ik} \approx 0$. Thus the PSD (or correlation) matrix of the clutter-plus-self-noise at the inputs of receiving stations turns out to be approximately diagonal when the angular width, $\Delta\Theta_{ik}$, of the CMR projection on any "base plane" is greater than the beamwidth of an antenna whose aperture in the same plane is equal to the effective baselength between those stations. In other words, the PSD (or correlation) matrix considered is approximately diagonal when $\Delta\Theta_{ik}$ is greater than the period of the interferometer's ADP (in the CMR direction) which can be created with the same pair of stations. In this case

$$\Phi_{ik}(\omega) \approx 0, \quad i \neq k; \qquad \Phi_{ii}(\omega) = N_i[1 + q_{ci}^2|\Psi_0(\omega - \omega_0)|^2]; \quad i,k = \overline{1,m}. \tag{10.13}$$

The matrix $\boldsymbol{\Phi}(\omega)$ takes the simplest form when $N_i = N$, $q_{ci}^2 = q_c^2$, $i = \overline{1,m}$:

$$\boldsymbol{\Phi}(\omega) \approx N[1 + q_c^2|\Psi_0(\omega - \omega_0)|^2]\mathbf{I} \tag{10.14}$$

where $\mathbf{I}$ is the identity matrix of the order m.

We have come to an important conclusion: *if the distance (baselength) between each pair of receiving stations of a MSRS is such that the angular width of the CMR projection on this "base plane" can be covered with more than one period of the ADP of an interferometer which can be created using the same pair of stations, then clutter from this CMR is spatially uncorrelated, i.e. mutually uncorrelated at the inputs of receiving stations.* This particular situation occurs, as a rule, in practice.

Example 10.1. Let clutter be caused by an illuminated part of a CMR. The angular width $\Delta\Theta_{ik}$ of this part of the CMR (as it is seen from the middle of the effective baseline between the ith and the kth stations) is determined by the transmitting station beamwidth (effect of its sidelobes may be neglected here)

$$\Delta\Theta_{ik} \approx \lambda_0/L_A$$

where L_A is the linear dimension of the transmitting antenna aperture in corresponding direction. It follows from (10.11) that practically zero mutual correlation between clutter at the inputs of the ith and kth receiving stations takes place already when the effective baselength between these stations is greater than the linear dimension of the transmitting antenna aperture, i.e. when

$$L_{ik} > (1\dots3)L_A.$$

Such an inequality is typically valid. The condition (10.11) is, as a rule, satisfied for actual MSRSs even when the angular width of a CMR is smaller than the transmitting beamwidth so that equations (10.13) and (10.14) for the matrix $\mathbf{\Phi}(\omega)$ are usually valid.

Having obtained the equations for $\mathbf{\Phi}(\omega)$ and $\mathbf{f}(\omega)=\mathbf{\Phi}^{-1}(\omega)$ we can utilize (8.27) to evaluate the potential efficiency of spatial processing.

Evidently, in the most typical situations where the condition (10.11) is fulfilled and hence *spatial (interstation) correlation of clutter is nearly zero, spatial processing cannot provide clutter cancellation (mutual compensation)*. In this case the best are optimum processing algorithms for target detection in a background of spatially uncorrelated "coloured" noise. Such algorithms and their performance characteristics have been considered in Chapter 5. In particular, we can express elements of the matrix $\mathbf{f}(\omega)=\mathbf{\Phi}^{-1}(\omega)$ from (10.13) in the form

$$f_{ik}(\omega)=\delta_{ik}\{N_i[1+q_{ci}^2|\Psi_0(\omega-\omega_0)|^2]\}^{-1}, \quad i,k=\overline{1,m}. \tag{10.15}$$

Then substituting (10.15) in (8.1) yields the optimum processing algorithm for a deterministic wanted signal

$$L=\mathrm{Re}\sum_{i=1}^{m}\frac{b_{si}^{*}\sqrt{2T_s}}{2\pi N_i\sqrt{\Delta f_s}}\int_{-\infty}^{\infty}\chi_i(\omega)\frac{\Psi_0^{*}(\omega-\omega_0)\exp(j\omega t_{si})}{1+q_{ci}^2|\Psi_0(\omega-\omega_0)|^2}\,d\omega. \tag{10.16}$$

The output power SINR (SCNR) is given by

$$q_{out}^2=\sum_{i=1}^{m}\frac{2E_i}{N_i}\frac{1}{2\pi\Delta f_s}\int_{-\infty}^{\infty}\frac{|\Psi_0(\omega-\omega_0)|^2\,d\omega}{1+q_{ci}^2|\Psi_0(\omega-\omega_0)|^2} \tag{10.17}$$

where $E_i=|b_{si}|^2T_s$ is the wanted signal energy at the ith station. We employ in (10.16) and (10.17) the same signal presentation (10.2) as for the echo from the lth reflector of the CMR (only the subscript "l" is replaced by the subscript "s"). We also assume $g_i(\beta_{si},\varepsilon_{si})=1$.

As was to be expected, for $E_i=E$, $N_i=N$, $q_{ci}^2=q_c^2$, $i=\overline{1,m}$, the optimum processing in a MSRS with a single transmitting and m receiving stations provides the energy gain equal to m (as compared with a monostatic radar for the same values of E, N and q_c^2). Just the same energy gain has been obtained for signal detection in a background of receiver self-noises (see Section 5.1). Detection characteristics for deterministic and quasideterministic signals are the same as in Chapter 5.

A CMR is Illuminated by Several Transmitting Stations

Let M transmitting stations be positioned at the points $\mathbf{R}_{T1},\dots,\mathbf{R}_{TM}$. The coordinate system origin is located as earlier in the centre of the CMR considered. All transmitting stations radiate the same deterministic signals at times $t_1,\dots,t_m$. We denote the baselength and the effective baselength between the pth and the qth transmitting stations by L_{Tpq} and $L_{Tpq\,eff}$, respectively, and the unit vector similar to $\delta\mathbf{r}_{ik}$ but for the pth and the qth transmitting stations by $\delta\mathbf{r}_{Tpq}$ [see (10.7)]. We assume

b_l in (10.5) to be the same being illuminated by any transmitting station and the quantities $\boldsymbol{\rho}_l \delta \mathbf{r}_{\mathrm{T}pq}/\mathbf{R}_{\mathrm{T}pq}$ to be uniformly distributed within the intervals $\pm\Delta\xi_{pq}/2$ where $\Delta\xi_{pq}$ is the angular width of the CMR projection on the "base plane" (a plane passed through the baseline between the pth and the qth transmitting stations and the centre of the CMR) as it is seen from the middle of this effective baseline. Then taking into account that $\omega/\omega_0 \approx 1$ for $\Delta\omega_s \ll \omega_0$ we can obtain instead of (10.9)

$$\Phi_{ik}(\omega) = \delta_{ik}\sqrt{N_i N_k} + \frac{\tilde{n} M \sigma^2}{\Delta f_s} |\tilde{\Psi}_0(\omega-\omega_0)|^2 \exp\left(-j\omega\frac{R_i - R_k}{c}\right) \operatorname{sinc}(\pi\alpha_{ik})$$
$$\times\left[1+\frac{2}{M}\sum_{p=1}^{M-1}\sum_{q=p+1}^{M}\operatorname{sinc}(\pi\beta_{pq})\cos\frac{\omega(R_{\mathrm{T}p}-R_{\mathrm{T}q})}{c}\right], \quad i,k=\overline{1,m} \quad (10.18)$$

where $R_{\mathrm{T}p}$, $R_{\mathrm{T}q}$ are the distances from the centre of the CMR to the pth and the kth transmitting stations;

$$\beta_{pq} = (L_{\mathrm{T}pq\,\mathrm{eff}}\Delta\xi_{pq})/\lambda_0$$

[see (10.10)]. In particular, under the condition (10.11)

$$\Phi_{ii}(\omega) = N_i\left\{1 + M q_{ci}^2 |\tilde{\Psi}_0(\omega-\omega_0)|^2\right.$$
$$\left.\times\left[1+\frac{2}{M}\sum_{p=1}^{M-1}\sum_{q=p+1}^{M}\operatorname{sinc}(\pi\beta_{pq})\cos\frac{\omega(R_{\mathrm{T}p}-R_{\mathrm{T}q})}{c}\right]\right\}, \quad i=\overline{1,m}, \quad (10.19)$$

and under the similar condition relative to transmitting stations

$$\Delta\xi_{pq} > (1\ldots3)\lambda_0/L_{\mathrm{T}pq\,\mathrm{eff}}, \quad p,q=\overline{1,M}, \; p\neq q \quad (10.20)$$

we can obtain from (10.18) instead of (10.13)

$$\begin{aligned}&\Phi_{ik}(\omega) = 0, \quad i \neq k,\\ &\Phi_{ii}(\omega) = N_i[1 + M q_{ci}^2 |\tilde{\Psi}_0(\omega-\omega_0)|^2], \quad i,k=\overline{1,m}.\end{aligned} \quad (10.21)$$

Under the condition (10.20), clutter from a CMR illuminated by different transmitting stations is mutually uncorrelated at the input of each receiving station. When the conditions (10.11) and (10.20) are valid simultaneously, there is no mutual correlation between total clutter at the inputs of different receiving stations when the CMR is illuminated by all transmitting stations.

The power SCNR at the optimum detector output can be derived from (8.27). To simplify mathematical expressions we assume $N_i = N$, $q_{ci}^2 = q_c^2$, $i=\overline{1,m}$, and the energy of all echoes to be equal, $E_{ip} = E$, $i=\overline{1,m}$, $p=\overline{1,M}$. Then

$$q_{\mathrm{out}}^2 = \frac{2E}{N}\frac{mM}{2\pi\Delta f_s}\int_{-}\frac{|\tilde{\Psi}_0(\omega-\omega_0)|^2}{1+Mq_c^2|\tilde{\Psi}_0(\omega-\omega_0)|^2}$$
$$\times\left[1+\frac{2}{M}\sum_{p=1}^{M-1}\sum_{q=p+1}^{M}\cos\omega\left(t_p - t_q + \frac{R_{\mathrm{Ts}p}-R_{\mathrm{Ts}q}}{c}\right)\right]\mathrm{d}\omega \quad (10.22)$$

where R_{Tsp} is the target range relative to the pth transmitting station, $p=\overline{1,M}$. If the range differences from the target to transmitting stations are compensated by the corresponding choice of transmission times so that $t_p - t_q + (R_{Tsp} - R_{Tsq})/c = 0$, then

$$q^2_{\text{out}} = \frac{2E}{N} \frac{mM^2}{2\pi\Delta f_s} \int_{-\infty}^{\infty} \frac{|\dot{\Psi}_0(\omega-\omega_0)|^2 \, d\omega}{1 + Mq_c^2 |\dot{\Psi}_0(\omega-\omega_0)|^2}. \tag{10.23}$$

In the most important case of intensive clutter, when $q_c^2 \gg 1$, we have from (10.23)

$$q^2_{\text{out}} \approx \frac{2E}{N} \frac{mM}{q_c^2}. \tag{10.24}$$

As was to be expected, CMR illumination by M transmitting stations results in the additional benefit (the increase of the power output SCNR) which is equal to M because of spatially uncorrelated clutter. The total gain in a MSRS with M transmitting and m receiving stations is equal to mM. This result is the same as for deterministic signal detection in a background of spatially uncorrelated noises. This is a maximum possible energy gain for deterministic signals and a target placed near the centre of a CMR. Performance characteristics for fluctuating targets (when echo complex amplitudes are random) also can be obtained from the results of Chapter 5. In particular, when complex amplitude fluctuations of echoes caused by different transmitting stations are mutually independent at the inputs of receiving stations, the attainable energy gain for high probabilities of detection may be much larger than mM thanks to the fluctuation "smoothing" as a result of joint processing. Numerical estimates can be obtained using detection characteristics in Figs. 5.7 and 5.8 assuming mM to be the number of receiving stations with a single transmitting one.

Thus we must give a negative answer to the question which has been formulated at the beginning of this section. *Because of the fact that clutter from an extended CMR is, as a rule, mutually uncorrelated at the inputs of spatially separated receiving stations of a MSRS, clutter cancellation (mutual suppression) in the process of space–time (interstation) processing is impossible* (unlike the case of spatially correlated jamming from a point-like source). As compared with a monostatic radar, such processing can enhance target detection in clutter only as a result of wanted signal accumulation (integration) as in a background of self-noises of receiving stations.

***In this section**, we have analyzed the spatial (interstation) correlation of clutter from an extended Collection of Masking Reflectors (CMR), e.g. a chaff cloud, at the inputs of spatially separated stations of a MSRS. To exclude possible influence of Doppler filtration we assumed the positions of a target and of all the elements of both a CMR and a MSRS to be fixed. At first we have considered the case where a CMR is illuminated by a single transmitting station. Having derived the expression* (10.9) *for mutual clutter PSD at the inputs of different pairs of stations, we have shown that the clutter spatial correlation is nearly zero since the condition* (10.11) *is, as a rule, satisfied. Then the clutter-plus-self-noise PSD matrix at the inputs of stations is approximately diagonal* [(10.13) *and* (10.14)]. *Here is the important difference from the jamming radiated by a point-like source (e.g., jammer). Because of the absence of clutter spatial correlation, clutter cancellation (mutual suppression) as a result of joint interstation (space–time) processing is impossible. Under these conditions optimum detectors and their performance characteristics are the same as for target detection in a background of spatially uncorrelated "coloured" self-noises of receiving stations* [(10.16) *and* (10.17)]. *We have illustrated in Example* 10.1 *that such a situation, for instance, takes* *place when the angular width of a CMR (as it is seen from the middle of the effective baseline between any pair of receiving stations) is determined by the beamwidth of the transmitting station antenna. Clutter from this CMR is practically uncorrelated at the inputs of the receiving stations if the effective baselength between these two stations is greater than the above mentioned linear aperture of the transmitting antenna.*

The obtained results have been extended to the case where a CMR is illuminated by M transmitting stations. Instead of (10.9) *we have derived the more general expression* (10.18) *and the condition* (10.20) *for transmitting stations similar to the condition* (10.11) *for receiving stations. Performance characteristics are now determined by* (10.22)–(10.24).

The detection enhancement due to joint interstation processing for both deterministic and fluctuating wanted signals is caused only by signal accumulation (integration) as in a background of receiver self-noises. Numerical estimates can be obtained from the results of Chapter 5.

10.2. EFFICIENCY OF SPATIAL PROCESSING AGAINST CLUTTER IN THINNED ANTENNA ARRAYS

Antenna Arrays (AAs) including thinned (sparsely populated) ones are a particular case of spatially coherent MSRSs with very small baselengths. Due to small baselengths, target echoes at the inputs of array elements are, as a rule, spatially coherent too. For such AAs the condition (10.11) may be invalid. In this case the PSD matrix $\mathbf{\Phi}(\omega)$ of clutter-plus-self-noise at the inputs of elements is nondiagonal [see (10.9)] so that its inversion turns out to be difficult when the number m of AA elements is large. We recall that the matrix $\mathbf{f}(\omega)=\mathbf{\Phi}^{-1}(\omega)$ determines both the optimum processing algorithms and performance characteristics.

However, the problem may be simplified if we assume an AA to be uniformly spaced and echoes from a CMR at the AA input to be a plane wave. For a linear uniformly spaced AA with the inter-element spacing ΔL we have instead of (10.9):

$$\Phi_{ik}(\omega)=\sqrt{N_i N_k}\,\{\delta_{ik}+q_{ci}q_{ck}|\Psi(\omega-\omega_0)|^2 \times \mathrm{sinc}[\pi(\omega/\omega_0)(i-k)\Delta\alpha]\exp[-j\omega(i-k)\Delta u]\} \tag{10.25}$$

where

$$\Delta\alpha=(\Delta L_{\mathrm{eff}}\Delta\Theta_{ik})/\lambda_0;\quad \Delta u=(\Delta L\sin\beta)/c$$

and $\Delta L_{\mathrm{eff}}=\Delta L\cos\beta$ is the effective inter-element spacing of the AA; β is the bearing of the CMR measured from the normal to the AA. The elements $\Phi_{ik}(\omega)$ depend only on differences of the row and column numbers so that the matrix $\mathbf{\Phi}(\omega)$ is a Toeplitz one. Algorithms for the Toeplitz matrix inversion require significantly less computer operations than those for matrices of general form [9].

An AA Consisting of Two Elements as the Simplest Cell of a Multielement AA

Such a two-element interferometer is interesting for practice and besides it permits to reveal main features of multielement AAs.

It is easy to invert the matrix $\mathbf{\Phi}(\omega)$ with the elements (10.9) for $i,k=1,2$. Substituting its inverse in (8.27), taking into account the expression (10.2) for the wanted signal (replacing the subscript "l" by "s") and assuming for simplicity that

$g_i(\beta_{si}, \varepsilon_{si})=1$, $b_{si}=b_s$, $N_i=N$, $i=1,2$, yields

$$q_{\text{out}}^2=(2E/\pi N\Delta f_s)$$

$$\times\int_{-\infty}^{\infty}\frac{|\Psi_0(\omega-\omega_0)|^2\{1+q_c^2|\Psi_0(\omega-\omega_0)|^2[1-\operatorname{sinc}(\pi\alpha)\cos\pi(\omega/\omega_0)\varepsilon_s]\}}{1+2q_c^2|\Psi_0(\omega-\omega_0)|^2+q_c^4|\Psi_0(\omega-\omega_0)|^4[1-\operatorname{sinc}^2(\pi\alpha)]}. \quad (10.26)$$

where, as before, $\alpha=L_{\text{eff}}\Delta\Theta/\lambda_0$ is the angular width of the CMR projection on the "base plane" normalized to the beamwidth λ_0/L_{eff} of the interferometer under consideration. (The "base plane" is a plane passing through the interferometer baseline and the centre of the CMR). $\varepsilon_s=L_{\text{eff}}\delta\Theta_s/\lambda_0$ is the angular displacement $\delta\Theta_s$ of the CMR centre relative to a target of interest in the same plane also normalized to the interferometer beamwidth. It is assumed that a maximum of a mainbeam is pointed at the target.

Calculation results from (10.26) are depicted in Fig. 10.2. Here plots of the dependence of $q_{\text{out}}^2/(2E/N)$ on ε_s relevant to the different values of α are presented for a narrowband signal with a rectangular spectrum ($\Delta\omega_s/\omega_0)\ll 1$; $|\Psi_0(\omega-\omega_0)|^2=1$ when $|\omega-\omega_0|\leqslant\Delta\omega_s/2$ and $|\Psi_0(\omega-\omega_0)|^2=0$ otherwise), the input power clutter-to-self-noise ratio (CNR) $q_c^2=20\,\text{dB}$. Only the segment $0\leqslant\varepsilon_s\leqslant 0.5$ is shown since all functions are even and periodic with the period $\varepsilon_s=1$. These curves characterize the efficiency of the optimum target detection by the AA in a background of clutter in comparison with that of a single element of the AA in a clear environment when $q_{\text{out}\,0}^2=2E/N$. In this figure corresponding plots for "matched" spatial processing are presented too. This processing is optimal in the absence of spatial correlation of interferences (see Section 5.1). It is reduced to the coherent summation of signals received by both elements of the AA (interferometer) after the optimum temporal processing in each receiver. The output power SCNR in this case is given by

$$q_{\text{out}}^2=\frac{4E}{N}\left\{1+q_c^2\left[\operatorname{sinc}(\pi\alpha)\operatorname{sinc}\frac{\pi\varepsilon_s\Delta\omega_s}{\omega_0}\cos 2\pi\varepsilon_s\right]\right\}^{-1}. \quad (10.27)$$

For $\Delta\omega_s/\omega_0\ll 1$ the term $\operatorname{sinc}(\pi\varepsilon_s\Delta\omega_s/\omega_0)\approx 1$ and hence may be dropped. The dashed line for $\alpha=0$ is the usual power ADP of the interferometer. A dashed-dotted straight line corresponds to a single antenna element. For this case

$$q_{\text{out}}^2=\frac{2E}{N(1+q_c^2)}. \quad (10.28)$$

It can be seen from Fig. 10.2 that when the normalized angular width α of the CMR projection on the "base plane" is small, the output SCNR strongly depends on the displacement of the CMR centre from a target. The optimum processing manifests the essential energy gain (as compared with the "matched" processing) when at small values of α the target is outside the CMR and does not fall into periodic maxima of the interferometer ADP, i.e. when $0.5\alpha+k<|\varepsilon_s|<(1-0.5\alpha)+k$ where $k=0,1,2,\ldots$. The smaller α, the larger the energy gain. This is the result of clutter suppression since at small α the clutter turns out to be spatially correlated at the inputs of receiving antenna elements. As α increases, the spatial correlation of clutter decreases so that efficiency of the optimum processing approaches that of the "matched" one. When $\alpha=1$ the clutter spatial correlation is equal to zero, the "matched" processing

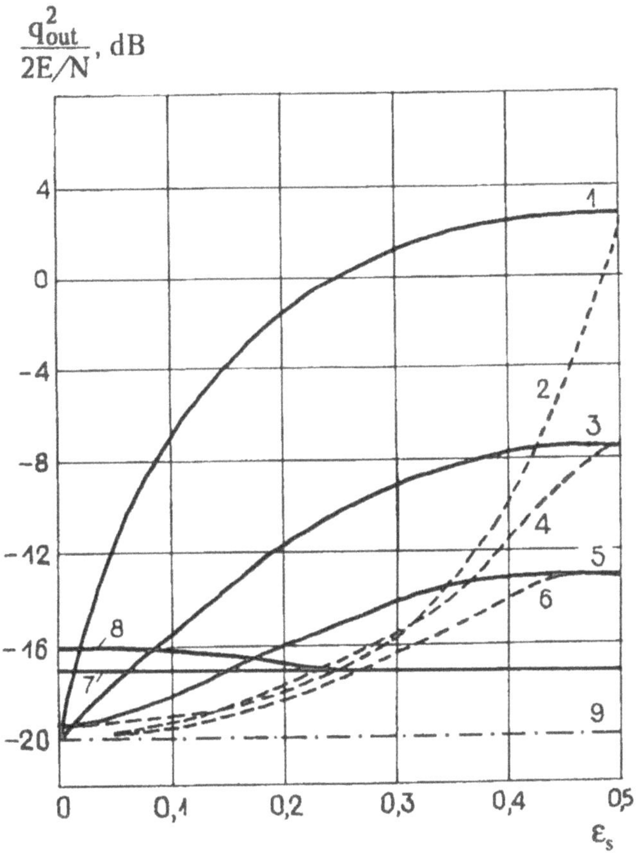

Figure 10.2 Dependences of normalized SCNR for a two-element interferometer on the angular displacement of the CMR centre relative to the target positioned in the direction of a maximum of the interferometer ADP, $q_c^2 = 20$ dB. Curves 1, 3, 5, 7, 8: optimal processing; curves 2, 4, 6: "matched" processing; curve 9: one receiving station; curves 1, 2: $\alpha = 0$, curves 3, 4: $\alpha = 0.25$, curves 5, 6: $\alpha = 0.5$, curve 7: $\alpha = 1.0$, curve 8: $\alpha = 1.5$

is optimal and q^2_{out} does not depend on ε_s (we assume the antenna elements to be omnidirectional).

When a target is positioned near the centre of a CMR ($\varepsilon_s \approx 0$) then for a CMR with small angular width ($\alpha < 0.5$) wanted signals are suppressed together with clutter suppression so that output SCNR values q^2_{out} for both optimum and "matched" processing are approximately the same as for a single receiving antenna element. As α increases further and the total clutter power $\bar{n}\sigma^2$ remains constant, the energy gain of about 3 dB appears with respect to a single antenna element. This energy gain is due to the coherent summation of wanted signals received by both antenna elements in a background of spatially uncorrelated clutter. When $\alpha = 1.5$ the energy gain becomes even slightly higher owing to small negative spatial correlation of clutter [at the first sidelobe maximum of the correlation function of the sinc(·) type]. However, this is evidently of no practical importance (see below). The optimum processing performance is nearly the same as that of the "matched" one.

It follows from the obtained results that when a target is probably positioned near the centre of a CMR ($\delta\Theta_s \ll \Delta\Theta$) the spatial processing with mutual coherent clutter suppression is ineffective since does not lead to the increase of the SCNR. The best we can do is to choose the baselength between receiving antenna elements (or receiving stations) in order to satisfy the condition $\alpha > 1$, i.e. $L_{eff} > \lambda_0/\Delta\Theta$.

A MSRS with m Receiving Stations and Small Baselengths between Them Where the Condition (10.11) *is Not Satisfied*

Let us consider a linear AA with m receiving elements and constant inter-element spacing ΔL. Under the same conditions as above, using (10.25) instead of (10.9) and assuming the wanted signal and clutter spectra to be rectangular within the common bandwidth $[|\Psi_0(\omega-\omega_0)|^2 = 1$ if $|\omega-\omega_0| < \Delta\omega_s/2$ and $|\Psi_0(\omega-\omega_0)|^2 = 0$ otherwise$]$ we obtain from (8.27)

$$q_{out}^2 = 2E\left\{\sum_{i=1}^{m} f_{ii}(\omega) + 2\sum_{i=1}^{m-1}\sum_{k=i+1}^{m} f_{ik}(\omega)\,\mathrm{sinc}\left[\pi\frac{\Delta\omega_s(i-k)}{\omega_0(m-1)}\varepsilon_s\right]\cos\frac{2\pi(i-k)}{m-1}\varepsilon_s\right\}. \quad (10.29)$$

where $\varepsilon_s = (m-1)\Delta L_{eff}\,\delta\Theta_s/\lambda_0$ and ΔL_{eff} is defined in (10.25). The elements $f_{ik}(\omega)$ of the matrix $\mathbf{f}(\omega)$ can be found by the inversion of the Toeplitz matrix $\mathbf{\Phi}(\omega)$ with the elements from (10.25).

The plots of $q_{out}^2/(2E/N)$ versus ε_s calculated from (10.29) for an AA with 10 elements ($m=10$), different values of $\alpha = (m-1)\Delta\alpha = (m-1)\Delta L_{eff}\Delta\Theta/\lambda_0$ and $q_c^2 = 20$ dB are shown in Fig. 10.3. In the same figure similar plots for the "matched" processing (coherent signal summation without clutter suppression) are presented. In the case of "matched" processing

$$q_{out}^2 = \frac{2E}{N}\left\{1 + q_c^2 + \frac{2q_c^2}{m}\sum_{i=1}^{m-1}\sum_{k=i+1}^{m}\mathrm{sinc}\left[\frac{\pi(i-k)}{m-1}\alpha\right] \times \mathrm{sinc}\left[\frac{\pi\Delta\omega_s(i-k)}{\omega_0(m-1)}\varepsilon_s\right]\cos\frac{2\pi(i-k)}{m-1}\varepsilon_s\right\}^{-1}. \quad (10.30)$$

All calculations for Fig. 10.3 are performed with the assumption $\Delta\omega_s/\omega_0 \ll 1$ so that the factors $\mathrm{sinc}[\pi\Delta\omega_s(i-k)\varepsilon_s/\omega_0(m-1)] \approx 1$ are dropped in both (10.29) and (10.30). The dashed curve for $\alpha=0$ is the ADP of the AA. As in Fig. 10.2, the dashed–dotted straight line presents the ADP of a single antenna element.

It is seen from Fig. 10.3 that for a fixed total clutter power, $\tilde{n}\sigma^2$, and when the angular width of the CMR projection on the AA plane (passed through the AA and the CMR centre) is approximately equal to or less than the beamwidth of the AA ($\alpha \approx 1$ or $\alpha < 1$), the optimum processing provides much better angular selectivity than the "matched" one. When the target turns out to be outside the CMR ($|\varepsilon_s| > 0.5\alpha$), the SCNR values q_{out}^2 increase significantly. It can be explained, first, by the effect of clutter mutual correlation at the inputs of most elements of the AA (excluding probably the extreme ones) which makes the clutter suppression effective, and, second, by the difference in bearings of the target and the CMR. When $|\varepsilon_s| > (1+0.5\alpha)$ then clutter operates only through AA sidelobes. It may occur if, for instance, the transmitting station beamwidth is greater than the beamwidth of the AA. As the angular width of the CMR projection increases, clutter mutual correlation

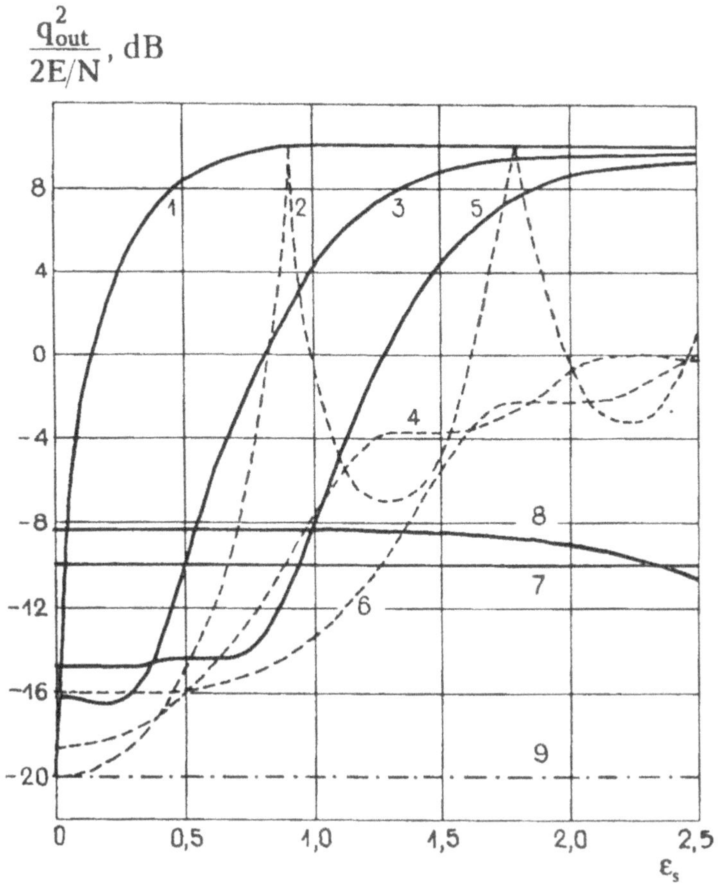

Figure 10.3 The same dependences as in Fig. 10.2 but for a 10-element linear PAA, $q_c^2=20\,\text{dB}$. Curves 1, 2: $\alpha=0$, curves 3, 4: $\alpha=1.0$, curves 5, 6: $\alpha=2.0$, curve 7: $\alpha=9.0$, curve 8: $\alpha=13.6$; curve 9: one-element antenna

at the AA elements (clutter spatial correlation) decreases and the optimum processing nears to the "matched" one. When a target is expected close to the centre of a CMR ($\delta\Theta_s \ll \Delta\Theta$) the only way for enhancing the output SCNR by spatial processing is the wanted signal accumulation (integration) in a background of spatially uncorrelated clutter. This requires the condition $\Delta L_{\text{eff}} > \lambda_0/\Delta\Theta$ to be satisfied.

It has been mentioned in [29] that for certain combinations of ΔL_{eff}, λ_0 and $\Delta\Theta$, spatial correlation of clutter may be negative and hence q_{out}^2 may be additionally increased as compared with the case of zero spatial correlation. This can be seen from Fig. 10.3 too, where q_{out}^2 is approximately by 2 dB higher at $\alpha=13.5$ ($\Delta\alpha=1.5$) than at $\alpha=9$ ($\Delta\alpha=1.0$) when clutter spatial correlation is equal to zero [see (10.25)]. It is clear that when $\Delta\alpha=1.5$ and $|i-k|=1, 5, 9$, then $\text{sinc}[\pi(i-k)\Delta\alpha]<0$. A similar situation was in the case of two-element interferometer (see Fig. 10.2 at $\alpha=1.5$). However, the "additional profit" is small because of low clutter spatial correlation and, what is particularly important, it can unlikely be used in practice. In fact, the angular width $\Delta\Theta$ of a CMR is *a priori* unknown and usually varies in time. The alignment of inter-element (inter-station) spacing in accordance with varying

$\Delta\Theta$ (in order to preserve the required relationship $\Delta\alpha = \Delta L_{\text{eff}}\Delta\Theta/\lambda_0 = 1.5$) is rather unrealistic. Besides, the density of reflectors in a CMR often decreases near its border. Then the sidelobe level of the clutter spatial correlation function will be lower than that of the function sin c(·). This will lead to further decrease of the "additional profit". Therefore we may assert that *if a target is positioned near the centre of a CMR, the maximum gain in the output SCNR is achieved for* $\Delta L_{\text{eff}} \geqslant \Delta L_{\text{effmin}}$ *where* ΔL_{effmin} *is the minimum effective inter-element (inter-station) spacing providing clutter spatial decorrelation for the expected angular width* $\Delta\Theta$ *of the CMR projection on the AA plane*. In particular, it follows from this result that the energy gain in clutter (without taking into account receiver self-noises) of a linear antenna with the aperture L_A in comparison with a single omnidirectional "point-like" antenna element is approximately equal to $L_A/\Delta L_{\text{effmin}}$. It means that *sparsely populated antenna arrays (AAs) with the inter-element spacing* ΔL_{effmin} *are in no way inferior in this sense than solid antennas with the same aperture* L_A.

The increase of the output SCNR by m times in an m-element AA with effective inter-element spacing $\Delta L_{\text{eff}} \geqslant \Delta L_{\text{effmin}}$ may be treated as a result of angular resolution enhancing just by m times (or, in other words, decreasing of the angular width of a "pulse volume" of the signal by m times). When $\Delta L_{\text{eff}} = \Delta L_{\text{effmin}}$ only one AA mainlobe falls within the angular width of the CMR projection on the AA plane. The beamwidth of this AA mainlobe is smaller approximately by a factor of m than the angular width of the CMR projection. Further reducing the effective inter-element spacing ΔL_{eff} with the fixed number of elements m leads to broadening of the AA beamwidth and hence lowers the output SCNR. Further increasing ΔL_{eff} though does not reduce energy gain but does not increase it either. In this case together with one narrower mainlobe, adjacent similar periodic "mainlobes" of the sparsely populated AA fall within the CMR projection so that a total area of the power array ADP within the CMR projection remains approximately constant. Analogous consideration is valid for a two-element interferometer and a MSRS with m receiving stations with deterministic signals.

For the case where a CMR is illuminated by M transmitting stations we may use the expression (10.18) for an arbitrary element of the clutter-plus-self-noise PSD matrix $\mathbf{\Phi}(\omega)$ at the inputs of AA elements. Then the power SCNR at the output of the optimum processor is given by

$$q_{\text{out}}^2 = \frac{2E}{2\pi\Delta f_s}\int_{-\infty}^{\infty} |\Psi_0(\omega-\omega_0)|^2 \sum_{i=1}^{m}\sum_{k=1}^{m} f_{ik}(\omega)\exp\left(j\omega\frac{R_{si}-R_{sk}}{c}\right)$$

$$\times \sum_{p=1}^{M}\sum_{q=1}^{M} \exp\left[j\omega\left(t_p - t_q + \frac{R_{Tsp}-R_{Tsq}}{c}\right)\right]d\omega \tag{10.31}$$

where $f_{ik}(\omega)$ is an arbitrary element of the matrix $\mathbf{f}(\omega)$ which is the inverse of $\mathbf{\Phi}(\omega)$ with the elements from (10.18). All other notations are the same as in (10.18). From (10.18) and (10.31) we can derive simplified expressions for the cases where the "transmitting stations" are the AA elements themselves, i.e. where a fraction of or all AA elements are transceivers.

In this section*, we have considered a particular case of spatially coherent MSRSs: Antenna Arrays (AAs) including sparsely populated ones. To reveal main features of such AAs we have simplified the problem assuming an AA to be a linear uniformly spaced array and echoes from a CMR at the input of the AA to be a plane wave. Under these conditions the PSD matrix* $\mathbf{\Phi}(\omega)$ *of clutter-plus-self-noise at the AA input turns out to be a Toeplitz one with the entries* (10.25). *We have studied first a two-element* *interferometer as a simplest cell of multielement AAs which, however, has an independent practical significance. We have derived the expressions* (10.26) *and* (10.27) *for the power SCNR at the output of both the optimum processing and the "matched" processing, respectively. Calculation results have shown* (*see Fig.* 10.2) *that the less the angular width of the CMR projection on the "AA plane"* (*passed through the AA and the centre of the CMR*) *as compared with the interferometer beamwidth, the greater the advantage of the optimum processing. However, when a target is expected to be placed near the centre of a CMR, the optimum processing with clutter suppression is ineffective because of wanted signal suppression. In this case the effective baselength between antenna elements* (*or receiving stations*) *should be chosen so as to decrease clutter spatial correlation at their inputs close to zero.*

The obtained results have been generalized to m-element AAs. We have also derived the expressions for the power SCNR at the output of the optimum and of the "matched" processing [*see* (10.29), (10.30) *and Fig.* 10.3]. *The optimum processing appears to have a significant advantage in angular target selectivity against clutter when the angular width of the CMR projection on the "AA plane" is of the order of the AA beamwidth and even greater. This advantage disappears only when that angular width approaches to the beamwidth of a pair adjacent antenna elements. However, when a target is positioned in the vicinity of the CMR centre, the best way is in choosing sufficiently large effective baselength between adjacent AA elements in order to achieve approximately zero clutter spatial correlation.*

We have arrived at an important conclusion: spatial processing for target detection in intensive clutter (*when the effect of self-noise may be neglected*) *with a sparsely populated AA may be just as good as with a solid antenna of the same dimensions.*

We have also presented the expression (10.31) *for the power SCNR at the output of the optimum processing in the case where a CMR is illuminated by M transmitting antenna elements* (*or transmitting stations*).

10.3. SIGNALS FROM MOVING TARGETS AND CLUTTER PSD IN MSRSs

To analyze the possibilities of moving target discrimination against clutter in a MSRS we have to consider the received signal (target echo) from a moving target and the clutter PSD at the inputs of spatially separated stations.

Received Signals from a Moving Target

Consider a simplest cell of a MSRS – a single stationary transmitting station (placed at the coordinate system origin) and the ith stationary receiving station (i is an arbitrary number, $i=\overline{1,m}$) whose position in space is determined by the radius-vector $\mathbf{L}_i$ (see Fig. 10.4). Let the target move with respect to the MSRS. Target range variations relative to those stations in the time interval including the signal duration plus propagation time along the path "the transmitting station – the target – the receiving station" may, as a rule, be approximated by a second order power polynomial[2] (Fig. 10.5):

$$R_0(\tau)=R_0+v_{r0}\tau+a_{r0}\tau^2/2; \qquad R_i(\tau)=R_i+v_{ri}\tau+a_{ri}\tau^2/2 \tag{10.32}$$

[2] Diagrams similar to that presented in Fig. 10.5 were originally used in [47,73] for the Doppler effect analysis in monostatic radars.

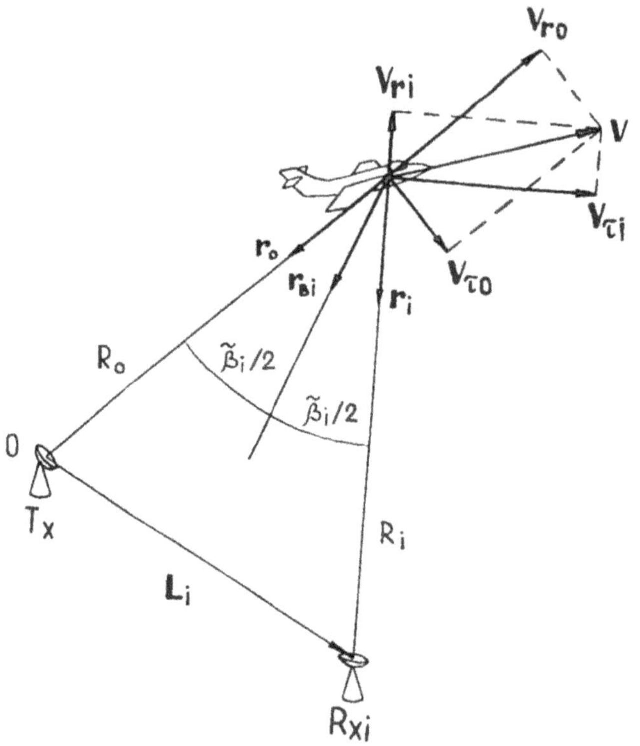

Figure 10.4 Target velocity components with respect to the transmitting station (T_x) and to an arbitrary receiving station (R_{xi})

where R_0, v_{r0} and a_{r0} are the range, the radial velocity and the radial acceleration of the target with respect to the transmitting station at $\tau=0$; R_i, v_{ri} and a_{ri} are the same with respect to the ith receiving station. Radial accelerations may be caused by both velocity vector variations during the observation interval $\mathbf{v}=\mathbf{v}(\tau)$ and, if this vector remains constant, its tangential components $v_{\tau 0}$ and $v_{\tau i}$ when $a_{r0}=v_{\tau 0}^2/R_0$, $a_{ri}=v_{\tau i}^2/R_i$ (just the latter case is presented in Fig. 10.4).

It is seen from Fig. 10.5 that when an initial moment of signal transmission $\tau=0$, then propagating with the light velocity, the initial segment of this signal reaches the target at $\tau=t_1$ and being scattered, arrives at the receiving station at $\tau=t_0$. It means that the received target echo begins at $\tau=t_0$. An arbitrary segment of the illuminating signal being transmitted at time $\tau=t'$ reaches the target at $\tau=t_2$ and comes to the receiving point at $\tau=t$. Thus the target echo at $\tau=t$ is proportional to the signal transmitted at $\tau=t'$:

$$S_i(t)=a_i\exp(j\varphi_i)S(t')=a_i\exp(j\varphi_i)S\{t-[R_0(t_2)+R_i(t_2)]/c\} \qquad (10.33)$$

where a_i, φ_i are the target echo random r.m.s. value and initial phase, respectively, which indirectly account for the fact that the target is not a point-like object, i.e. reflect partial echo interference from several target scattering centres ("flare spots"). It is assumed a_i, φ_i to be constant within the signal duration T_s.

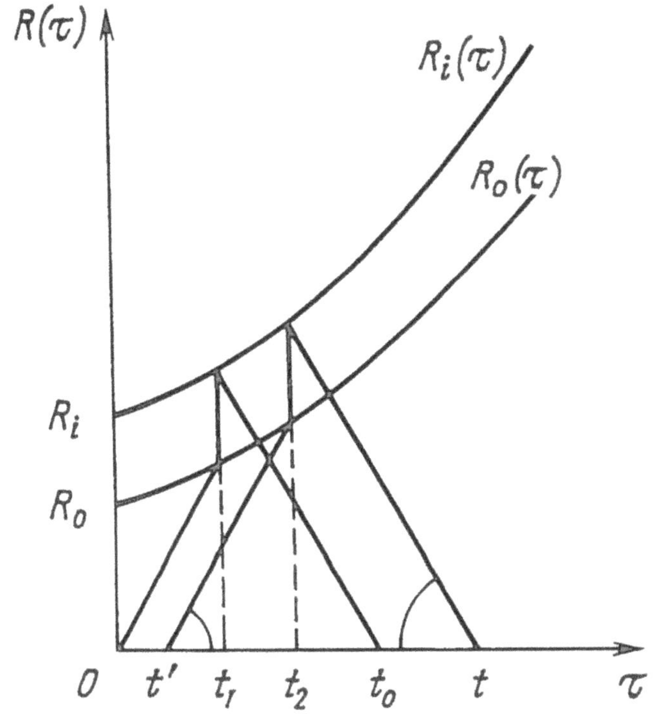

Figure 10.5 To the analysis of the Doppler effect for a pair "a transmitting station and an arbitrary receiving station"

For simplicity of exposition we assume that the transmitting and receiving antennas are pointed in the target direction so that $g_0(\beta_s, \varepsilon_s, \omega) = g_i(\beta_s, \varepsilon_s, \omega) = 1$ within the signal frequency bandwidth.

Now $[R_0(t_2) + R_i(t_2)]/c$ in (10.33) is to be expressed as a function of t. Using (10.32) for $\tau = t_2$ yields

$$[R_0(t_2) + R_i(t_2)]/c = (R_0 + R_i)/c + (v_{r0} + v_{ri})t_2/c + (a_{r0} + a_{ri})t_2^2/2c. \qquad (10.34)$$

It follows from Fig. 10.5 taking into account (10.32) that

$$t_2 = t - R_i(t_2)/c = t - R_i/c - v_{ri}t_2/c - a_{ri}t_2^2/2c. \qquad (10.35)$$

Substituting (10.35) in the right part of (10.34) yields

$$[R_0(t_2) + R_i(t_2)]/c = (R_0 + R_i)/c + [(v_{r0} + v_{ri})/c][t - R_i/c - v_{ri}t_2/c - a_{ri}t_2^2/2c]$$
$$+ [(a_{r0} + a_{ri})/2c][t - R_i/c - v_{ri}t_2/c - a_{ri}t_2^2/2c]^2. \qquad (10.36)$$

Since $v_{r0}/c \ll 1$, $v_{ri}/c \ll 1$, $a_{r0}t/c \ll 1$, $a_{ri}t/c \ll 1$, the remaining terms in (10.36) containing t_2 are small quantities of higher order than in (10.34). If these terms are yet too large to be neglected, we can again substitute (10.35) in (10.36). Such a process of sequential refinement should be stopped when remaining terms containing t_2 may be neglected. On this way we can obtain an approximate expression for

$[R_0(t_2)+R_i(t_2)]/c$ and hence for the target echo $S\{t-[R_0(t_2)+R_i(t_2)]/c\}$ as a function of t with any desired degree of accuracy and for arbitrary parameter combination.

It is important to emphasize that the answer to the question of which term may and which may not be neglected does not depend simply on the magnitude of this term relative to other ones. All terms of the signal envelope arguments must be compared with the reciprocal value of signal bandwidth Δf_s whereas all phase terms must be compared with the value 2π.

Example 10.2. Let $v_{r0\,max}\approx v_{ri\,max}=10^3$ m/s, $R_0\approx R_i=3\cdot 10^5$ m, $a_{r0}\approx a_{ri}=10$ m/s^2. Then for $T_s\approx 0.05$ s, $\lambda_0\approx 0.1$ m and $\Delta f_s\approx 10$–20 MHz we can derive from (10.36) and (10.33)

$$S_i(t)=a_i\exp(j\varphi_i)s_0\left[\left(1-\frac{v_{r0}+v_{ri}}{c}\right)t-\frac{R_0+R_i}{c}\right]$$

$$\times\exp\left\{j\omega_0\left[\left(1-\frac{v_{r0}+v_{ri}}{c}\right)t-\frac{R_0+R_i}{c}+\frac{(v_{r0}+v_{ri})R_i}{c^2}\right]\right\}. \qquad (10.37)$$

We have taken into account in (10.37) that $t_{max}\approx(R_0+R_i)/c+T_s$.

It should be noted that for chosen signal and target motion parameter values square terms in (10.32) have no effect on $S_i(t)$ so that (10.37) can be obtained for a linear model of target range variation in (10.32) as well.

It follows from Fig. 10.4 that $v_{r0}=-\mathbf{v}\mathbf{r}_0$; $v_{ri}=-\mathbf{v}\mathbf{r}_i$ and $\mathbf{r}_0+\mathbf{r}_i=2\cos(\tilde{\beta}_i/2)\mathbf{r}_{bi}$ where $\mathbf{r}_0$, $\mathbf{r}_i$ and $\mathbf{r}_{bi}$ are the unit vectors directed along the vectors $\mathbf{R}_0$, $\mathbf{R}_i$ and the bisector of the bistatic angle $\tilde{\beta}_i$ (between $\mathbf{R}_0$ and $\mathbf{R}_i$), respectively. Then we can transform (10.37) to the form

$$S_i(t)=a_i\exp(-j\psi_i)s_0\left[\left(1+2\frac{\mathbf{v}\mathbf{r}_{bi}}{c}\cos\frac{\tilde{\beta}_i}{2}\right)t-\frac{R_0+R_i}{c}\right]$$

$$\times\exp\left\{j\omega_0\left[\left(1+2\frac{\mathbf{v}\mathbf{r}_{bi}}{c}\cos\frac{\tilde{\beta}_i}{2}\right)t\right]\right\}. \qquad (10.38)$$

The total phase ψ_i is uniformly distributed within $(-\pi,\pi)$.

If, with other conditions being equal, $\Delta f_s<1$ MHz the term $[(2\mathbf{v}\mathbf{r}_{bi}/c)\cos(\tilde{\beta}_i/2)]t$ in the envelope argument in the right part of (10.38) may be neglected. On the contrary, if $\Delta f_s\approx 100$ MHz the term $(2\mathbf{v}\mathbf{r}_{bi}R_i/c^2)\cos(\tilde{\beta}_i/2)$ should be additionally included from (10.36). When $T_s\approx 1$ s, phase terms must include $(a_{r0}+a_{ri})t^2/2c$. The necessity of taking into account other terms from (10.36) depends on corresponding values of v_{r0}, v_{ri}, a_{r0}, a_{ri}, R_0, R_i, Δf_s and λ_0.

From (10.38) we can obtain the signal spectrum at the ith station,

$$\Psi_i(\omega)=a_i\exp(-j\psi_i)\sqrt{\frac{2T_s}{\Delta f_s}}\,\Psi_0\left(\frac{\omega}{1-u_i}-\omega_0\right)\exp\left[-j\left(\frac{\omega}{1-u_i}-\omega_0\right)\frac{R_0+R_i}{c}\right] \qquad (10.39)$$

where $\Psi_0(\omega)$ is the spectrum of the signal complex envelope [see (10.1)–(10.3)]; $u_i=-(2\mathbf{v}\mathbf{r}_{bi}/c)\cos(\tilde{\beta}_i/2)$. For simplicity it is assumed that $a_i/(1-u_i)\approx a_i$.

It can be seen from (10.38) and (10.39) that changes in time and frequency scales are determined not by the target radial velocity (as in a monostatic radar) but by the projection of the velocity vector $\mathbf{v}$ on the bisector of the bistatic angle, $\tilde{\beta}$, between the directions from the target to the transmitting and receiving stations. Increasing $\tilde{\beta}$ leads to reducing the Doppler effect by the factor $\cos(\tilde{\beta}/2)$. When $\tilde{\beta}=0$, then $\mathbf{r}_{bi}=\mathbf{r}_0=\mathbf{r}_i$, $R_0=R_i$ and (10.38), (10.39) are reduced to well-known expressions for a monostatic radar.

The main feature of echoes from moving targets in a MSRS is that *time and frequency scale changes are not the same at different stations*. The above results have been obtained for an arbitrary ith receiving station ($i=\overline{1,m}$) and, as can be seen from (10.36)–(10.39), depend on the arrangement of the transmitting station, receiving station and the target as well as on the target velocity vector direction with respect to the stations of the MSRS. It remains valid for a MSRS with several transmitting stations or several transceivers. Differences in time and frequency scaling, in Doppler frequency shifts are used for both target detection in clutter and target velocity vector measurement (see Section 10.4 and Chapter 13).

For target discrimination against clutter a signal in the form of a coherent pulse train ("coherent burst") consisting of short pulses is often employed. Within the duration of each pulse, τ_p, Doppler phase shifts in target echoes may, as a rule, be neglected [since $|\omega_0 u_i \tau_p| = |\omega_0(2\mathbf{vr}_{bi}/c)\cos(\beta_i/2)\tau_p| \ll 2\pi$] and all the more changes of time scale in the envelope may be neglected. Taking it into account yields instead of (10.38)

$$S_i(t)=a_i\exp(-j\psi_i)\sum_{p=0}^{M-1} s_0\left[t-(1+u_i)pT-\frac{R_0+R_i}{c}\right]$$

$$\times\exp\{j\omega_0[t-(1+u_i)pT]\} \tag{10.40}$$

where T is the time spacing between pulses in the train (the pulse repetition interval); M is the number of pulses; $s_0(t)$ is the normalized complex envelope of each pulse, i.e. $s_0(t)\neq 0$ only when $0<t<\tau_p$, and the normalization (4.4) is valid for $T_s=\tau_p$.

The spectrum of the wanted signal (10.40) at the ith station ($i=\overline{1,m}$) is given by

$$\Psi_i(\omega)=a_i\exp(-j\psi_i)\sqrt{\frac{2\tau_p}{\Delta f_s}}\,\Psi_0(\omega-\omega_0)\exp\left[-j(\omega-\omega_0)\frac{R_0+R_i}{c}\right]$$

$$\times\frac{\sin[\omega(1+u_i)MT/2]}{\sin[\omega(1+u_i)T/2]}\exp[-j\omega(1+u_i)(M-1)T/2] \tag{10.41}$$

where $\sqrt{2\tau_p/\Delta f_s}\,\Psi_0(\omega)$ is the spectrum of $s_0(t)$ normalized according to (10.3). It is seen that the target echo spectrum, like that of the transmitted signal, represents a periodic sequence of narrow peaks from which only a segment near the frequency ω_0, where $\Psi_0(\omega-\omega_0)\neq 0$, is active being cut out by the individual pulse spectrum. Target motion leads to variations of the width of these peaks and of the spacing between them in the frequency scale by $(1+u_i)=1-(2\mathbf{vr}_{bi}/c)\cos(\tilde{\beta}/2)$. Since $|u_i|\ll 1$, these variations themselves are insignificant but in the wide frequency range from $\omega=0$ (where peaks of the transmitted and received signal spectra coincide) to ω near ω_0, a noticeable shift in peak positions of the received signal spectrum with respect to the transmitted one turns out to be accumulated (Fig. 10.6).

A spectrum of the signal received from a stationary target coincides with that of the transmitted (illuminating) signal (excluding possibly an amplitude-phase factor). Thus the whole system of spectral peaks corresponding to the received signal from

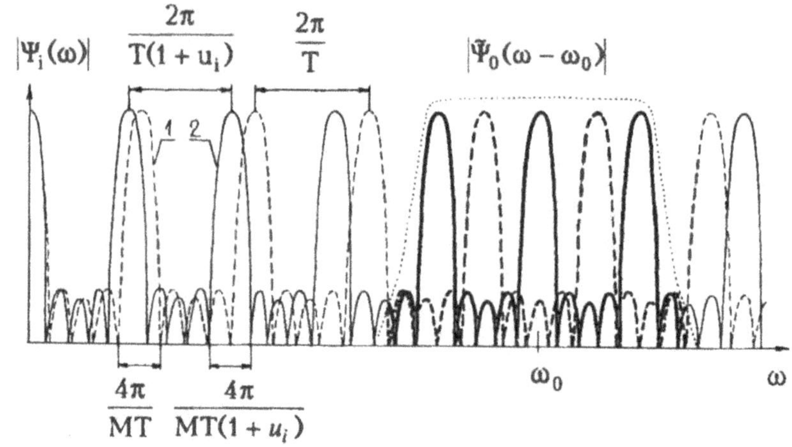

Figure 10.6 Spectral components of a pulse train target echo. Curve 1: for a stationary target; curve 2: for a moving target; the spectrum of one pulse $|\tilde{\Psi}(\omega-\omega_0)|$ is shown by dotted lines, the resultant spectra within the one pulse frequency band are highlighted by bold lines

a moving target is displaced in frequency with respect to the similar system of spectral peaks of the signal received from a stationary target. This displacement is equal to the Doppler shift frequency $\Omega_{si} \approx -\omega_0 u_i = \omega_0(2\mathbf{v}\mathbf{r}_{bi}/c)\cos(\tilde{\beta}_i/2)$. The difference of (10.40) and (10.41) from corresponding expressions for monostatic radars (for which $\tilde{\beta}_i=0$, $\mathbf{r}_{bi}=\mathbf{r}_0=\mathbf{r}_i$, $R_0=R_i$) is the same as for a continuous signal (or a pulse signal of a large duration) considered above. The obtained results may be extended to the case of MSRSs with moving stations.

PSD of Clutter from a CMR

There may be a variety of clutter correlation functions and power spectra depending on the character of CMRs, their motion in space, radar observation conditions and some other factors [53]. We consider volumetric clutter from a CMR (e.g., from a chaff cloud) which may be assumed to be a Gaussian, zero mean, stationary (within the time interval T_s) random process. In general the CMR centre may move relative to a MSRS and besides each reflector may move with respect to the CMR centre.

As shown in Section 10.1, clutter is, as a rule, mutually uncorrelated at the inputs of receiving stations in a MSRS with usual baselengths. Therefore, it is sufficient to consider clutter PSD at the ith station ($i=\overline{1,m}$).

The above obtained results for target echo and its spectrum may be applied to each lth reflector (an element of the CMR). To account for the CMR extension in space we may express the distances R_{l0} and R_{li} (at $t=0$) of the lth reflector from the transmitting and the ith receiving stations, respectively, as in (10.6) in terms of the corresponding distances of the CMR centre and the projection of the radius-vector $\boldsymbol{\rho}_l$ of the lth reflector from the CMR centre. The velocity vector of the lth reflector, $\mathbf{v}_l$, is also convenient to represent in the form: $\mathbf{v}_l=\mathbf{v}_0+\Delta\mathbf{v}_l$ where $\mathbf{v}_0$ is the velocity vector of the CMR centre relative to the MSRS and $\Delta\mathbf{v}_l$ is the velocity vector of the lth reflector with respect to the CMR centre. We must also take into account that the directions from the lth reflector to each of the two stations determined by the unit vectors $\mathbf{r}_{l0}$ and $\mathbf{r}_{li}$, differ from the direction to the same stations from the CMR centre (the unit vectors $\mathbf{r}_0$ and $\mathbf{r}_i$). Hence

$$\mathbf{v}_l\mathbf{r}_{l0}=(\mathbf{v}_0+\Delta\mathbf{v}_l)\{\mathbf{r}_0+[(\boldsymbol{\rho}_l\Delta\mathbf{r}_{l0})/R_0]\Delta\mathbf{r}_{l0}\};$$
$$\mathbf{v}_l\mathbf{r}_{li}=(\mathbf{v}_0+\Delta\mathbf{v}_l)\{\mathbf{r}_i+[(\boldsymbol{\rho}_l\Delta\mathbf{r}_{li})/R_i]\Delta\mathbf{r}_{li}\}. \qquad (10.42)$$

In (10.42) $(\boldsymbol{\rho}_l\Delta\mathbf{r}_{l0})/R_0$ and $(\boldsymbol{\rho}_l\Delta\mathbf{r}_{li})/R_i$ are the tangents of the angles γ_0 and γ_i between $\mathbf{r}_0$ and $\mathbf{r}_{l0}$ and between $\mathbf{r}_i$ and $\mathbf{r}_{li}$, respectively.These angles are small so that $\mathrm{tg}\,\gamma_0\approx\gamma_0$, $\mathrm{tg}\,\gamma_i\approx\gamma_i$ and $\cos\gamma_0\approx\cos\gamma_i\approx 1$.

Using (10.40) and replacing the subscripts "i" by the subscripts "li" yields

$$S_{li}(t)=a_{li}\exp(-j\psi_{li})\sum_{p=0}^{M-1}s_0\left[t-(1+u_{li})pT-\frac{R_{l0}+R_{li}}{c}\right]$$
$$\times\exp\{j\omega_0[t-(1+u_{li})pT]\}. \qquad (10.43)$$

Now we have from (10.42) and (10.6) that

$$u_{li}=-(\mathbf{v}_0+\Delta\mathbf{v}_l)(\mathbf{r}_{l0}+\mathbf{r}_{li})/c\approx-(1/c)\{(2\mathbf{v}_0\mathbf{r}_{bi})\cos(\tilde{\beta}/2)$$
$$+2(\Delta\mathbf{v}_l\mathbf{r}_{bi})\cos(\tilde{\beta}/2)+[(\boldsymbol{\rho}_l\Delta\mathbf{r}_{l0})/R_0]\mathbf{v}_0\Delta\mathbf{r}_{l0}+[(\boldsymbol{\rho}_l\Delta\mathbf{r}_{li})/R_i]\mathbf{v}_0\Delta\mathbf{r}_{li}\};$$
$$(R_{l0}+R_{li})/c\approx(R_0+R_i)/c-(2\boldsymbol{\rho}_l\mathbf{r}_{bi}/c)\cos(\tilde{\beta}/2)$$
$$+(\boldsymbol{\rho}_l\Delta\mathbf{r}_{l0})^2/2cR_0+(\boldsymbol{\rho}_l\Delta\mathbf{r}_{li})^2/2cR_i. \qquad (10.44)$$

We introduce the following notation:

$$\Delta u_{li}=-(2\Delta\mathbf{v}_l\mathbf{r}_{bi}/c)\cos(\tilde{\beta}/2);$$
$$\tau_{li}=(2\boldsymbol{\rho}_l\mathbf{r}_{bi}/c)\cos(\tilde{\beta}/2)-(\boldsymbol{\rho}_l\Delta\mathbf{r}_{l0})^2/2cR_0-(\boldsymbol{\rho}_l\Delta\mathbf{r}_{li})^2/2cR_i; \qquad (10.45)$$
$$h_{l0}=(\boldsymbol{\rho}_l\Delta\mathbf{r}_{l0}/cR_0)\mathbf{v}_0\Delta\mathbf{r}_{l0};\qquad h_{li}=(\boldsymbol{\rho}_l\Delta\mathbf{r}_{li}/cR_i)\mathbf{v}_0\Delta\mathbf{r}_{li}.$$

In general the effect of transmitting and receiving station ADPs should be taken into account. As in Sections 10.1 and 10.2 we assume that all elements of the CMR are illuminated uniformly by the transmitting station, i.e. $g_0(\beta_0,\varepsilon_0)=1$. Besides, we assume the receiving station ADPs $g_i(\beta_i,\varepsilon_i)=\text{const.}$ within the time interval $(0,T_s)$. With these assumptions, substituting (10.44) in (10.43), taking into account (10.45) and the expression for u_i [see (10.39)], an echo at the input of the ith receiving station from the lth reflector illuminated by a coherent M-pulse train can be presented in the form

$$S_{li}(t)=a_{li}\exp(-j\psi_{li})g_i(\beta_{li},\varepsilon_{li})\sum_{p=0}^{M-1}s_0\left[t-(1+u_i)pT-\frac{R_0+R_i}{c}+\tau_{li}\right]$$
$$\times\exp\{j\omega_0[t-(1+u_i+\Delta u_{li}+h_{l0}+h_{li})pT]\}. \qquad (10.46)$$

It is clear that Δu_{li} and τ_{li} account for individual reflector velocities and additional signal delays with respect to the CMR centre, respectively, whereas by h_{l0} and h_{li}

are denoted the contributions to individual reflector radial velocities caused by their cross-range positions relative to the CMR centre.

The expression for h_{l0} can be simplified by introducing the vector $\mathbf{v}_{\tau 0}$ that is the projection of $\mathbf{v}_0$ on the plane which is perpendicular to $\mathbf{r}_0$ (i.e. the tangential component of $\mathbf{v}_0$ with respect to the transmitting station) and the vector $\boldsymbol{\rho}_{l\tau 0}=(\boldsymbol{\rho}_l\,\Delta\mathbf{r}_{l0})\Delta\mathbf{r}_{l0}$ which is the projection of the radius-vector $\boldsymbol{\rho}_l$ of the lth reflector on the same plane. In a similar manner we may transform h_{li} by introducing the vectors $\mathbf{v}_{\tau i}$ and $\boldsymbol{\rho}_{l\tau i}$ which lie in the plane perpendicular to $\mathbf{r}_i$. As a result we have

$$h_{l0}=\mathbf{v}_{\tau 0}\boldsymbol{\rho}_{l\tau 0}/cR_0; \qquad h_{li}=\mathbf{v}_{\tau i}\boldsymbol{\rho}_{l\tau i}/cR_i. \tag{10.47}$$

We recall now that (10.43)–(10.46) are approximate equations. They include only such terms which are significant for typical CMR and signal parameter values: $\lambda\approx 10\,\text{cm}, \Delta f_s<10\text{–}20\,\text{MHz}, T_s<0.05\,\text{s}$. It is assumed that signal delay spread from different reflectors, τ_{li}, is less than the time spacing between pulses T. The spectrum (the Fourier transform) of the signal (10.46) is given by

$$\begin{aligned}\Psi_{li}(\omega)&=a_{li}\exp(-j\psi_{li})g_i(\beta_{li},\varepsilon_{li})\sqrt{2\tau_p/\Delta f_s}\,\Psi_0(\omega-\omega_0)\\ &\times\exp\{-j(\omega-\omega_0)[(R_0+R_i/c)+\tau_{li}]\}\\ &\times\frac{\sin[\omega(1+u_i)+\omega_0(\Delta u_{li}+h_{l0}+h_{li})]MT/2}{\sin[\omega(1+u_i)+\omega_0(\Delta u_{li}+h_{l0}+h_{li})]T/2}\\ &\times\exp\{-j[\omega(1+u_i)+\omega_0(\Delta u_{li}+h_{l0}+h_{li}](M-1)T/2\}.\end{aligned} \tag{10.48}$$

As in (10.41), $\sqrt{2\tau_p/\Delta f_s}\Psi_0(\omega)$ is the complex envelope spectrum of each pulse of the train.

Comparing (10.48) with (10.41) shows that there is a change in frequency scale by $(1+u_i)$ times common for all reflectors. Besides, the spectra of different reflectors have different frequency shifts because of unequal individual Doppler frequencies: $\Delta\Omega_{li}=-\omega_0(\Delta u_{li}+h_{l0}+h_{li})$. It is seen from (10.45)–(10.47) that *the difference between individual Doppler frequencies are caused not only by the projections of individual reflector velocities on the bistatic angle bisector:* $-\Delta u_{li}=(2\Delta\mathbf{v}_l\mathbf{r}_{bi}/c)\cos(\tilde{\beta}/2)$, *but by the cross-range CMR extent with respect to both stations as well.* The more the cross-range distance of a reflector from the CMR centre along the tangential CMR velocity vector direction, the more the angle between this vector and the direction to the corresponding station differs from the right angle. It leads to additional reflector radial velocities and hence to the additional Doppler frequency shifts: $\Delta\tilde{\Omega}_{l0}=-\omega_0 h_{l0}$ and $\Delta\tilde{\Omega}_{li}=-\omega_0 h_{li}$ Thus these contributions to individual reflector echo Doppler frequency are proportional to the cross-range reflector distance from the CMR centre and to the tangential velocity of the CMR as a whole with respect to the transmitting and corresponding receiving stations. This effect is explained in Fig. 10.7 with respect to the transmitting station. A similar situation takes place with respect to any receiving station. In particular, it follows from above results that even when all reflectors have equal velocities ($\Delta\mathbf{v}_l=0$), echoes from different reflectors may have different Doppler frequencies, especially for a CMR of large cross-range dimensions moving with high tangential velocity.

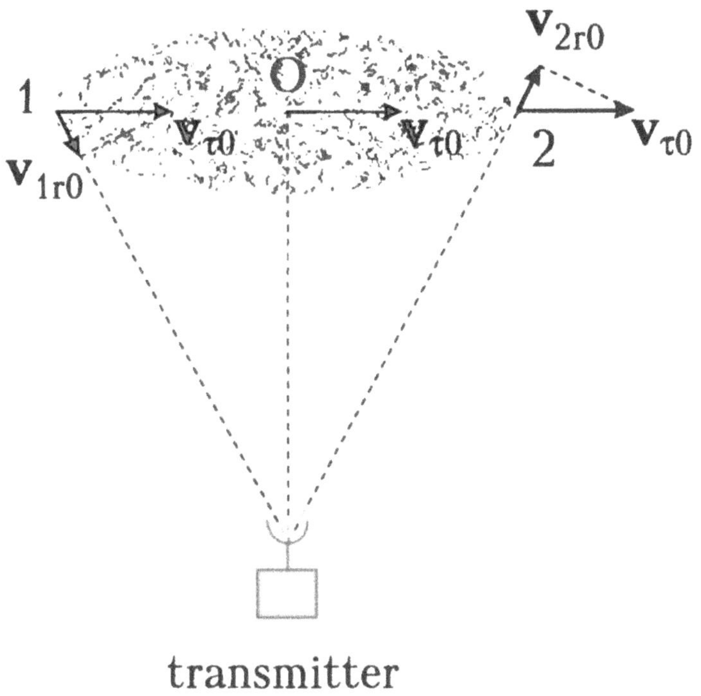

Figure 10.7 Differences in reflector radial velocities caused by tangential velocity of the CMR as a whole

Using (10.48) we can derive an expression for clutter PSD at the ith receiving station ($i=\overline{1,m}$). Under the condition $\Delta f_s \tau_p \gg 1$ we have

$$\Phi_{ii}(\omega)=\frac{1}{2MT_{0i}}\sum_l\sum_n\overline{\Psi_{li}(\omega)\Psi^*_{ni}(\omega)}=\frac{\sigma^2\tau_p}{\Delta f_s MT_{0i}}|\dot{\Psi}_0(\omega-\omega_0)|^2$$

$$\times\overline{\sum_l|g_i(\beta_{li},\varepsilon_{li})|^2\sum_{p=0}^{M-1}\sum_{q=0}^{M-1}\exp\{-j[\omega(1+u_i)+\Delta\Omega_{li}](p-q)T\}}. \qquad (10.49)$$

Here the condition (10.5) is taken into account and the ratio of sines in (10.48) is replaced by the equal sum of exponential functions. T_{0i} is the duration of total echo from the CMR when it is illuminated by a single pulse of the train ($\tau_p \ll T_{0i} < T$). Averaging in (10.49) is performed over reflector random positions and individual velocities. As in Section 10.1, we assume that $|g_i(\beta_{li},\varepsilon_{li})|^2=1$. In the case of the CMR of a small angular extent (as compared with the antenna beamwidth of any receiving station) the effect of nonuniformity of $|g_i(\beta_{li},\varepsilon_{li})|^2$ for different CMR segments may be neglected. For the CMR of a large angular extent we often may consider the receiving station ADPs to have an approximately rectangular form. With this assumption only $\Delta\Omega_{li}$ remain random in (10.49). The simplest solution can be obtained if $\Delta\Omega_{li}$ are mutually independent and equally distributed (for different reflectors). In particular, when they are uniformly distributed within a common interval $(-\Delta\Omega_{i\,max}/2, \Delta\Omega_{i\,max}/2)$ then

$$\Phi_{ii}(\omega)=\frac{\tilde{n}\sigma^2}{\Delta f_s}\frac{\tau_p}{T_{0i}}|\tilde{\Psi}_0(\omega-\omega_0)|^2$$

$$\times\left\{1+\frac{2}{M}\sum_{p=0}^{M-2}\sum_{q=p+1}^{M-1}\operatorname{sinc}\frac{\Delta\Omega_{i\,max}(p-q)T}{2}\cos\omega(1+u_i)(p-q)T\right\} \qquad (10.50)$$

where $\tilde{n}$ is the total number of reflectors illuminated by the transmitting station which fall within the receiving antenna mainbeam. The factor $(\tilde{n}\sigma^2/\Delta f_s)(\tau_p/T_{0i})$ characterizes the maximum clutter PSD averaged over the time interval, T_{0i}, and the frequency band, Δf_s.

The curves corresponding to one period of the PSD, $\Phi_{ii}(\omega)$ [within the interval $\Delta\omega_{per}=2\pi/T(1+u_i)\approx 2\pi/T$], relevant to the different Doppler bandwidth values, $\Delta\Omega_{i\,max}$, of the CMR echo fluctuations are plotted in Fig. 10.8(a) according to (10.48). The power spectrum of each pulse, $|\tilde{\Psi}_0(\omega-\omega_0)|^2$, is assumed to be constant within the frequency band $\Delta\omega_{per}$. As the Doppler spectrum bandwidth, $\Delta\Omega_{i\,max}$, increases, peaks of $\Phi_{ii}(\omega)$ become wider and lower since the same clutter power are distributed over a broader frequency band. When $\Delta\Omega_{i\,max}T\approx 2\pi$, the adjacent peaks are linked up, the periodic structure disappears, and $\Phi_{ii}(\omega)$ becomes the power spectrum of a single pulse of the train.

In Fig. 10.8(b) the same curves are presented as in Fig. 10.8(a) but each curve is normalized to its maximum value. This makes it easier to reveal the effect of echo Doppler spectrum broadening on the form of $\Phi_{ii}(\omega)$. For the peak width, $\delta\Omega_i$, of the spectrum $\Phi_{ii}(\omega)$ (at the -3 dB level) the following approximate equation is valid:

$$\delta\Omega_i=\begin{cases}2\pi/MT, & \Delta\Omega_{i\,max}T<2\pi/M;\\ \Delta\Omega_{i\,max}, & 2\pi>\Delta\Omega_{i\,max}T>2\pi/M.\end{cases} \qquad (10.51)$$

As the product $\Delta\Omega_{i\,max}T$ increases, efficiency of MTI techniques decreases, since the target echo discrimination against clutter in the frequency domain becomes more difficult (the probability for signal spectrum peaks to fall into free intervals between clutter spectrum peaks decreases).

Fourier transformation of (10.50) gives the clutter correlation function

$$B_{ii}(t_1-t_2)=\frac{\tilde{n}\sigma^2}{MT_{0i}}\sum_{p=0}^{M-1}\sum_{q=0}^{M-1}\operatorname{sinc}\frac{\Delta\Omega_{i\,max}(p-q)T}{2}$$

$$\times B_0[t_1-t_2-(1+u_i)(p-q)T]$$

$$\times\exp\{j\omega_0[t_1-t_2-(1+u_i)(p-q)T]\} \qquad (10.52)$$

where

$$B_0(\tau)=0.5\int_{-\infty}^{\infty}s_0(t)s_0^*(t-\tau)\,dt=\frac{\tau_p}{2\pi\Delta f_s}\int_{-\infty}^{\infty}|\tilde{\Psi}_0(\omega)|^2\exp(j\omega\tau)\,d\omega \qquad (10.53)$$

is the correlation function of a single pulse normalized envelope (of the transmitted signal).

Using (10.52) the curves of clutter correlation function envelopes are plotted in Fig. 10.9. These curves are calculated and plotted under the same conditions as the

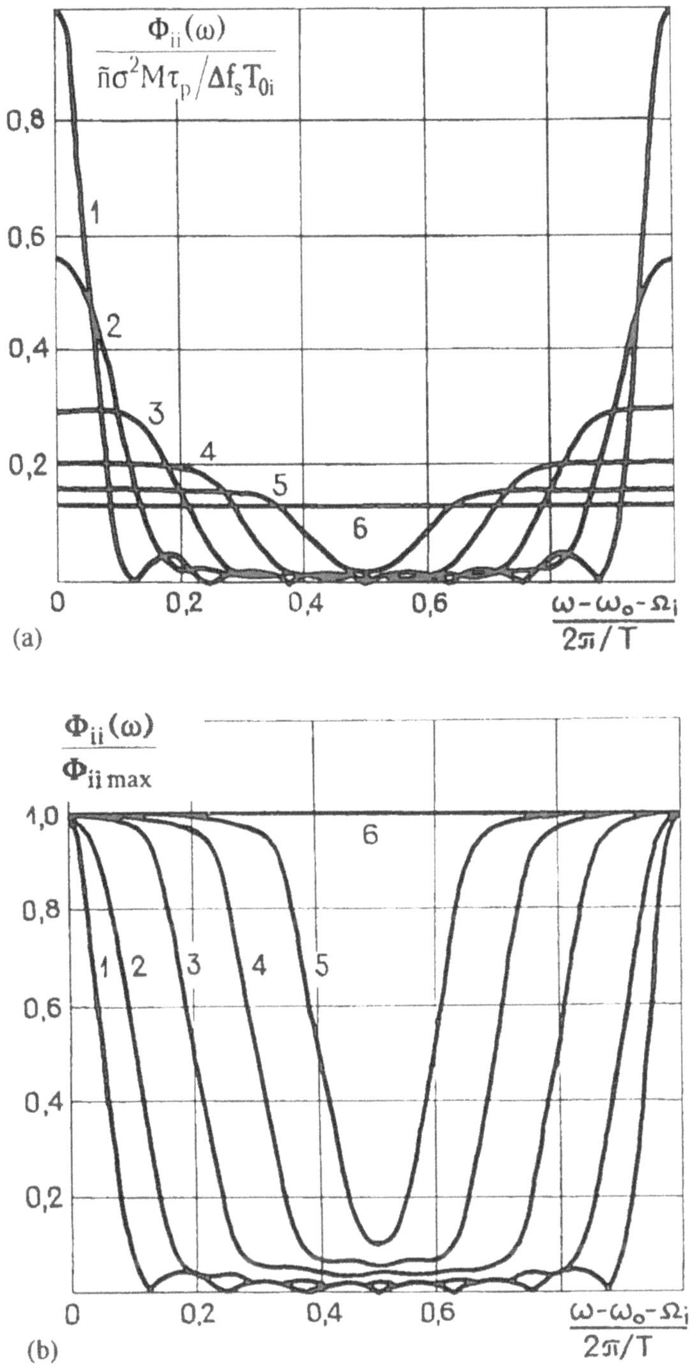

Figure 10.8 One period of clutter PSD at the ith station ($i=\overline{1,m}$) for different normalized clutter Doppler spectrum bandwidth, $\Delta\Omega_{i_{max}}T/2\pi$, when a CMR is illuminated by a pulse train consisting of $M=8$ pulses. (a) all curves are normalized to the spectral maximum for $\Delta\Omega_{i_{max}}=0$; (b) each curve is normalized to its own maximum; $\Delta\Omega_{i_{max}}T/2\pi=0, 0.2, 0.4, 0.6, 0.8, 1.0$ (curve 1, 2, 3, 4, 5, 6, respectively)

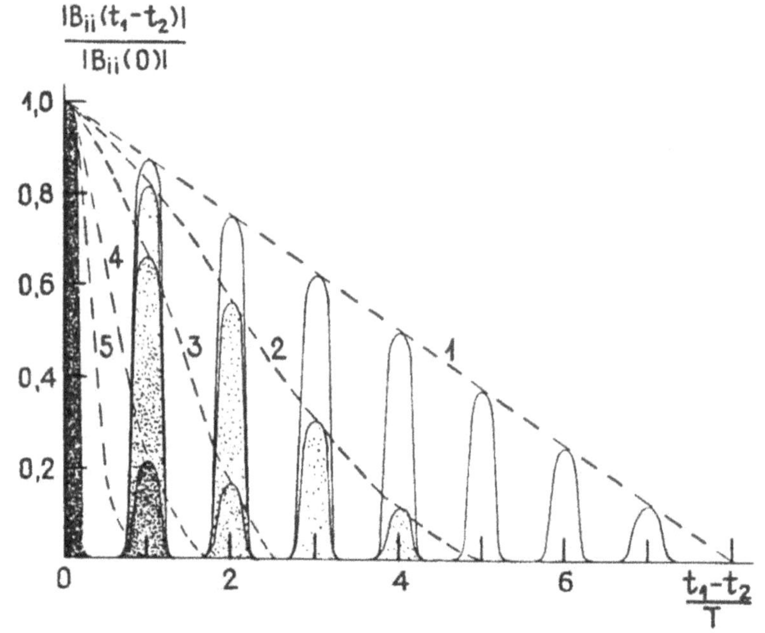

Figure 10.9 Normalized envelope of the clutter correlation function for different clutter spectrum bandwidths (caused by the differences in radial velocities of reflectors). $\Delta\Omega_{i\max}T/2\pi = 0, 0.2, 0.4, 0.6, 1.0$ (curves 1, 2, 3, 4, 5, respectively)

curves of clutter PSD in Fig. 10.8. Only the segment where $t_1 - t_2 \geqslant 0$ is shown in Fig. 10.9 since $B_{ii}(t_1 - t_2)$ is an even function. As the Doppler spectrum bandwidth of CMR echo fluctuations, $\Delta\Omega_{i\max}$, increases, their pulse-to-pulse correlation rapidly degrades. In the extreme case where $\Delta\Omega_{i\max} = 2\pi/T$ this correlation function reduces to only a single peak corresponding to an autocorrelation function of a single pulse. It means that pulse-to-pulse correlation is absent.

Similar results can be obtained with another approach when a clutter average power distribution $\eta(\tau_i, \Delta\Omega_i)$ over delay τ_i and Doppler frequency shift $\Delta\Omega_i$ is given so that

$$\int_{-\infty}^{\infty}\int_{-\infty}^{\infty} \eta(\tau_i, \Delta\Omega_i)\,d\tau_i\,d\Delta\Omega_i = \bar{n}\sigma^2.$$

Then assuming again $|g_i(\beta_i, \varepsilon_i)|^2 = 1$ we obtain

$$\begin{aligned} B_{ii}(t_1, t_2) &= 0.5\overline{S_i(t_1)S_i^*(t_2)} \\ &= \sum_{p=0}^{M-1}\sum_{q=0}^{M-1} 0.5 \int_{-\infty}^{\infty}\int_{-\infty}^{\infty} s_0\left[t_1 - (1+u_i)pT - \frac{R_0 + R_i}{c} + \tau_i\right] \\ &\quad \times s_0^*\left[t_2 - (1+u_i)qT - \frac{R_0 + R_i}{c} + \tau_i\right]\exp\{j\omega_0[t_1 - t_2 - (1+u_i)(p-q)T] \\ &\quad - j\Delta\Omega_i(p-q)T\}\eta(\tau_i, \Delta\Omega_i)\,d\tau_i\,d\Delta\Omega_i. \end{aligned} \tag{10.54}$$

From (10.54) taking into account (10.53) we can obtain (10.52) and corresponding PSD (10.50) for mutually independent and uniform clutter power distribution over τ_i and $\Delta\Omega_i$, i.e. when $\eta(\tau_i, \Delta\omega_i) = \bar{n}\sigma^2/T_{0i}\,\Delta\Omega_{i\,\max}$ for $\tau_i \in (-T_{0i}/2, T_{0i}/2)$, $\Delta\Omega_i \in (-\Delta\Omega_{i\,\max}/2, \Delta\Omega_{i\,\max}/2)$.

For more complex clutter power distributions over signal parameters, approximate calculations methods or computer simulation may be required to obtain $B_{ii}(t_1 - t_2)$ or $\Phi_{ii}(\omega)$. For a detailed analysis the statistical dependence between Doppler frequency shifts and signal delays should be taken into account. However, the general character of the PSD $\Phi_{ii}(\omega)$ and the correlation function $B_{ii}(t_1 - t_2)$ will not change. In particular, this is valid for the dependences of $\Phi_{ii}(\omega)$ and $B_{ii}(t_1 - t_2)$ on the relationship between the clutter Doppler frequency bandwidth $\Delta\Omega_{i\,\max}$ and the pulse time spacing (pulse repetition interval) T of a pulse train.

The main differences from a monostatic radar, as for a single target, consist in Doppler frequency shift decreasing by $\cos(\tilde{\beta}_i/2)$, dependence of this frequency shift on the projections of reflector velocity vectors on the bisector of the bistatic angle $\tilde{\beta}_i$. It is important to take into account the differences between clutter PSDs (and correlation functions) at different receiving stations. These differences are determined by the general geometry of MSRSs, targets and CMRs.

In this section, using square approximation (10.32) *for the target range dependence on time* [*see Fig.*10.5] *we have presented a technique for obtaining expressions* (10.33)–(10.39) *which permit to take into account (with any required degree of accuracy) changes caused by the Doppler effect in moving target echoes received by spatially separated stations of a MSRS. We have shown that time and frequency scale changes caused by the target motion are unequal in different stations and depend on the target velocity vector projections on the bisectors of bistatic angles.*

Since coherent pulse trains are often employed for moving target discrimination against clutter, we have derived the expression (10.40) *and* (10.41) *for the corresponding target echo and its spectrum (the Fourier transform). We have shown how the Doppler effect manifests itself in the spectrum of a moving target echoes with respect to that of a stationary target (Fig.* 10.6).

Taking as a basis the results for a single moving target we have considered clutter PSD from a CMR. Having obtained equation (10.46) *for the echo from an arbitrary element of the CMR illuminated by a coherent pulse train, and equation* (10.48) *for the spectrum of this echo, we have derived the expressions* (10.50) *and* (10.52) *for clutter PSD and correlation function, respectively. The analysis of these PSD and correlation function has revealed the essential dependence of their form on the relationship between the bandwidth of clutter Doppler spectrum and time spacing between pulses in the transmitted pulse train (Figs* 10.8(a,b) *and* 10.9). *The obtained results determine the possibilities of moving target discrimination against volumetric clutter (e.g., chaff).*

10.4. MOVING TARGET DETECTION IN CLUTTER

Using results from Section 10.3 related to the signal and its spectrum from a moving target as well as the PSD of clutter from a CMR, we can synthesize optimum algorithms for wanted signal detection in Gaussian clutter and analyze their performance. Due to the fact that clutter is, as a rule, mutually uncorrelated at the inputs of spatially separated receiving stations (see Section 10.1), this problem is reduced to the problem of signal detection in a background of spatially uncorrelated interferences considered in Chapters 5 and 6. The main results from those chapters are applicable to the problem which is now under consideration. However, it is necessary to take into account that unlike self-noise alone (which we assumed to be "white") the sum of clutter and self-noise at each station should be considered as "coloured" noise. Apparently, this may have an effect on signal processing only in each receiving station individually (see Sections 5.1, 5.2). In this processing the wanted signal from a moving target is detected owing to frequency differences between target echoes and clutter.

Information fusion algorithms, i.e. algorithms combining the quantities $G_i, |G_i|$ $(i=\overline{1,m})$ or preliminary decisions coming from all receiving stations, do not depend on the form of interference PSD at each station. Therefore, the information fusion algorithms which have been obtained and analyzed in Chapters 5 and 6 may be applied without any revision.

As mentioned in Section 10.1, all known MTI methods and techniques are applicable to signal processing in each station of a MSRS. If a MSRS contains only transmitting-receiving stations (transceivers) with independent (autonomous) signal reception (see Section 1.1), i.e. monostatic radars, signal processing in each station is the same as in a usual monostatic radar. When cooperative signal reception is used, specific features of target echoes and clutter PSD considered in Section 10.3 should be taken into account. MTI technique implementation requires some additional measures (translation of reference oscillations or codes determining working frequency from transmitting to receiving stations via DTLs and so on). However, in this case achievements of the MTI theory and technique may be applied as well. Signal and clutter processing in each station of a MSRS has no essential differences as compared with a monostatic radar for which there is a wealth of literature (e.g., [2,72]). Some special features manifest themselves only in the process and as a result of information fusion when certain significant drawbacks of monostatic radars can be eliminated.

The Problem of Zero Target Echo Doppler Frequency from a Moving Target

When a target moves on the surface of a sphere of arbitrary radius with its centre at the position of a monostatic radar, the radial velocity component with respect to this radar is equal to zero. Hence no MTI techniques permit target discrimination against clutter from stationary or slowly moving CMR. A similar situation occurs for a bistatic radar when a target moves on the surface of an ellipsoid of revolution (prolate spheroid) with their foci at the points where the transmitting and receiving stations are positioned.

In a MSRS containing only two transceivers (monostatic radars) with independent (autonomous) signal reception, zero target echo Doppler at the two spatially separated stations simultaneously may be only in the case where a target moves along a circle which is the intersection of two spheres with their centres at the station sites. Obviously, the probability of such a motion is much less than in arbitrary directions on a spherical surface. If the cooperative signal reception is employed in the same MSRS, zero Doppler frequencies with respect to both stations are still less probable since this requires a target to move along a circle which is the intersection of two spheres and an ellipsoid of revolution with their centres and foci at the station sites. When a MSRS contains a single transmitting and two receiving stations the loci of zero Doppler simultaneously at both receiving stations are the intersection curves of two ellipsoids of revolution with one common focus at the transmitting station position and two other foci at the points where the receiving stations are positioned. Adding at least one receiving station to a MSRS consisting of two transceivers or of a single transmitting and two receiving stations leads to a situation where zero Doppler frequency with respect to all receiving stations simultaneously is impossible in principle (excluding cases of degenerate arrangement of stations). Thus for a MSRS with sufficiently large number of stations there are no target trajectories where MTI methods cannot be used.

Let us consider the necessary conditions for efficient application of MTI techniques in the kth receiving station of a MSRS when the Doppler frequency shift is equal to zero at the ith station ($i,k=\overline{1,m}$, $i \neq k$). Let in a MSRS with a single transmitting station be $v_{r0}+v_{ri}=\mathbf{v}(\mathbf{r}_0+\mathbf{r}_i)=0$ [see (10.37)]. To separate the wanted signal from clutter in the kth station with confidence, the modulus of the signal Doppler frequency must be greater than the maximum Doppler frequency modulus value of the clutter fallen within an angle and range resolution cell, i.e. the following inequality has to be valid (see Section 10.3)[3]

$$\omega_0|\mathbf{v}(\mathbf{r}_0+\mathbf{r}_k)|/c>\omega_0|u_k|+(3\ldots4)\omega_0|\Delta u_{lk}+h_{l0}+h_{lk}|_{\max}$$

$$=|\Omega_k|+(1.5\ldots2)\Delta\Omega_{k\,\max}. \tag{10.55}$$

We can express $\mathbf{r}_k$ through $\mathbf{r}_i$ [see (10.7)] and substitute in the left side of (10.55) taking into account that $\mathbf{v}(\mathbf{r}_0+\mathbf{r}_i)=0$. Then

$$\omega_0(L_{ik\,\mathrm{eff}}/R_{ik})|\mathbf{v}\delta\mathbf{r}_{ik}|/c>|\Omega_k|+(1.5\text{–}2)\Delta\Omega_{k\,\max}. \tag{10.56}$$

It follows for (10.56), (10.45), (10.47) and the expression for u_k from (10.39) that

$$L_{ik\,\mathrm{eff}}/R_{ik}>[2|\mathbf{v}_0\mathbf{r}_{bk}\cos(\tilde{\beta}_k/2)|+(3\ldots4)|2\Delta\mathbf{v}_l\mathbf{r}_{bk}\cos(\tilde{\beta}_k/2)$$

$$+\mathbf{v}_{\tau k}\boldsymbol{\rho}_{l\tau0}/R_0+\mathbf{v}_{\tau k}\boldsymbol{\rho}_{l\tau k}/R_k|_{\max(l)}]/|\mathbf{v}\delta\mathbf{r}_{ik}|. \tag{10.57}$$

The numerator of the right part of (10.57) is a maximum summed value of reflector radial velocities (with respect to the transmitting and receiving stations). The denominator contains the "tangential" component of the target velocity: the velocity projection on the effective baseline direction. In the left part of the inequality (10.57) is the ratio of the effective baselength between the ith and kth stations to the range of a target (from the middle of the effective baseline). When the ratio $(L_{ik\,\mathrm{eff}}/R_{ik})$ is not large it is equal to the angle between the directions from the target to the ith and kth stations, i.e. the angle between the unit vectors $\mathbf{r}_i$ and $\mathbf{r}_k$. The relationship (10.57) determines a minimal required effective baselength between the ith and kth stations for reliable target discrimination against clutter in the kth station when wanted signal Doppler frequency is equal to zero at the ith station.

When $\mathbf{v}\delta\mathbf{r}_{ik}\approx 0$, then wanted signal Doppler frequency is close to zero at both the ith and kth stations. It is scarcely probable since this may occur if the target velocity vector, $\mathbf{v}$, is perpendicular simultaneously to $\mathbf{r}_{bi}$ and $\mathbf{r}_{bk}$. Nevertheless, when it occurs the obtained results should be applied to any third, for instance, the pth station (the target velocity vector $\mathbf{v}$ cannot be perpendicular simultaneously to $\mathbf{r}_{bi}$, $\mathbf{r}_{bk}$ and $\mathbf{r}_{bp}$ excluding degenerate cases).

The inequalities (10.56) and (10.57) determine only necessary conditions. In practice actual efficiency of target discrimination against clutter depends on many factors. First of all, the waveform of illuminating signal must have sufficient resolution capability in frequency to use the considered spectral differences between target echoes and clutter.

---

[3] The signs of echo Doppler frequency shifts from the target and from the masking reflectors (clutter) are assumed to be the same.

The Problem of "Blind Velocities"

It is well known that when a coherent pulse train is employed as illuminating signal, the problem of "blind velocities" arises in a monostatic radar. For certain ("blind") target velocities, peaks in wanted signal spectrum coincide with peaks of clutter PSD. In this case frequency discrimination of wanted signal against clutter is impossible. Physically it is clear that such situations take place when the difference in target and CMR range variations during the time interval between adjacent pulses, T, contains an integer number of the halves of the wavelength. As a consequence, phase differences of adjacent pulses for both target echo and clutter are equal to integer numbers of 2π and hence cannot be detected by signal and clutter processing.

An inequality determining blind target radial velocities $v_r = v_{rbl}$ for a monostatic radar taking into account the spread of CMR radial velocities can be obtained from (10.41) and (10.48)

$$0 \leqslant |v_r - v_{rc}| - n\lambda_0/2T < |\Delta v_{lr} + (l_\perp/R_0)v_{\tau c}|_{\max} \qquad (10.58)$$

where $v_r = \mathbf{v}\mathbf{r}_0$, $v_{rc} = \mathbf{v}_0\mathbf{r}_0$ are the radial velocities of a target and of the CMR centre, respectively; Δv_{lr} is the individual radial velocity of the lth reflector; $l_\perp$ is the cross-range extension of the CMR within an angle resolution cell (in the direction of the CMR centre tangential velocity $\mathbf{v}_{\tau c}$); $n = 0, 1, 2, \ldots$

As can be seen from (10.58) a radical method to control target blind velocities consists in reducing the time spacing T between adjacent pulses in an illuminating pulse train. When $\lambda_0/2T > |v_r|_{\max}$ where $|v_r|_{\max}$ is the maximum possible target radial velocity, then only $n = 0$ satisfies the inequality (10.58). It means that frequency target discrimination against clutter is impossible only if the target and CMR centre radial velocities are nearly equal (i.e. their differences do not exceed the spread of CMR radial velocities). However, when we choose a small interpulse spacing T, we face the problem of range measurement ambiguities. For unambiguous target range measurement when the maximum target range is $R_{\max}$, the following inequality must be satisfied: $T > 2R_{\max}/c$. In practice it is usually impossible to satisfy simultaneously two conflicting inequalities: $T > 2R_{\max}/c$ and $T < \lambda_0/2|v_r|_{\max}$ in the centimetre and even decimetre wavelength ranges. To overcome this obstacle, while keeping unambiguity range measurements, variations of interpulse spacing, T, (PRI agility) is often used including sometimes a combination with multifrequency illuminating signals [2]. However, such measures lead to more complicated station structures and to deterioration of MTI performance. All these considerations relate to bistatic radars too. It is only necessary to take into account that Doppler frequency shifts of target echoes and clutter are determined by their velocity projections on the bisectors of bistatic angles (see Section 10.3).

Unlike monostatic and bistatic radars, the problem of blind velocities is easily solved in MSRSs owing to spatial separation of stations since Doppler frequency shifts cannot really be the same at all stations. The inequality determining blind velocities for the ith station can be written in the form

$$0 \leqslant |(\mathbf{v} - \mathbf{v}_0)\mathbf{r}_{bi}\cos(\tilde{\beta}_i/2)| - n\lambda_0/2T < |\Delta\mathbf{v}_l\mathbf{r}_{bi}\cos(\tilde{\beta}_i/2)$$

$$+ (l_{\perp 0}/R_0)\mathbf{v}_0\Delta\mathbf{r}_{l0} + (l_{\perp i}/R_i)\mathbf{v}_0\Delta\mathbf{r}_{li}|_{\max} \qquad (10.59)$$

where $l_{\perp 0}$ and $l_{\perp i}$ are the cross-range extensions of the CMR with respect to the unit vectors $\mathbf{r}_0$ and $\mathbf{r}_i$ (i.e. to the transmitting and receiving station directions) within their beamwidths. To obtain the corresponding inequality for the kth receiving station the subscripts "i" should be replaced by the subscript "k" in (10.59). However, the velocity ambiguity interval, $\lambda_0/2T$, does not depend of the station number, i.e. it is the same with respect to all stations. Therefore, for the case of at least three receiving stations and not too large spread of different masking reflector velocity projections, target and clutter spectral peak coincidence (blind velocities) at all stations is scarcely probable. When a MSRS contains several transmitting stations or/and the co-operative signal reception is used, the total number of target echoes with different Doppler frequencies increases which makes it easier to exclude blind target velocities. In the case of several transmitting stations they may employ pulse trains with unequal interpulse spacing and use different working frequencies. Thus information fusion from the outputs of receiving stations results in reliable moving target detection in clutter if the velocity vectors of a target and a CMR differ sufficiently.

It should be noted that because of blind velocities with respect to certain stations the power SCNRs at the outputs of linear parts of receiving stations may be essentially different. Hence we should take into account the results of detection algorithm performance analysis obtained in Sections 5.4, 5.5 and 6.2 for the case where SNR values are unequal at different stations.

Some additional practical advantages of target discrimination against clutter in MSRSs see in Section 1.2.

In closing of this Section we consider a problem which is important for certain applications.

Moving Target Detection in a Background of Moving CMR When There is No a priori *Difference Between Target and CMR's Element Velocities*

Target velocity may coincide with that of any reflector of a CMR. Such a situation may take place, for instance, for ballistic target observation in space when a CMR is a chaff cloud [218]. Apparently, MTI methods and techniques are ineffective in this case. However, target discrimination against clutter is possible under certain conditions.

We assume that the target RCS is much greater than the RCS of each element of the CMR. If we could divide the CMR by a large number of small cells so that the total RCS of all reflectors fallen into each cell would be essentially less than the target RCS, then the echo from the cell containing the target would be significantly greater than from other cells. This allows the target to be detected if, of course, this target can be detected with confidence in a background of receiving station self-noise. Therefore, for target discrimination against clutter in the case considered, the radar resolution capability should be enhanced in all target parameters measured by a radar: angle coordinates, range (TOA) and velocity (Doppler frequency). This is valid for both monostatic and bistatic radars. However, in MSRSs better results may be achieved owing to joint processing of information coming from spatially separated stations.

It is well known that the angle resolution capability can be increased by narrowing the receiving antenna beamwidth, i.e. by increasing antenna size (at given wavelength). The attainable gain in the output SCNR has been analyzed in Section 10.2. In particular, it has been shown that when a target is positioned near the CMR centre (in angle coordinates) and target echo energy is quite sufficient for target detection in the absence of clutter (in a background of receiver self-noises) there is no need in solid antennas or antenna arrays (AAs) with closely positioned elements. Practically the same value of signal-to-clutter ratio (SCR) as with a solid linear antenna of the aperture L_A can be obtained with a thinned linear AA of the total length L_A and with the inter-element spacing $L_{eff\,min} = \lambda_0/\Delta\Theta$ where $\Delta\Theta$ is the angular width of the CMR projection (as it is seen from the AA) on the plane passed through the AA and the CMR centre. What is the more, if an AA consists of m elements, the beamwidth of each element is greater than the angular width of a CMR and inter-element spacing is no less than $L_{\text{eff min}}$, then the gain in SCR (as compared with a single antenna element) practically does not depend on total array length L_A and spacing between antenna elements. A maximum attainable gain in power SCR is equal to m.

A similar situation can be seen in range resolution which is determined by a signal bandwidth. An illuminating signal with a solid spectrum and the bandwidth Δf_s may be replaced by a coherent multifrequency signal without practical losses in SCR values if the spacing between adjacent frequencies $\delta f_s = \delta f_{s\,\text{min}} = c/2\Delta R$ where ΔR is the CMR extension in range. When a signal spectrum contains n_f frequencies, the inter-frequency spacing δf_s less than $\delta f_{s\,\text{min}}$ is ineffective since it leads to the decrease of the total signal bandwidth, $\Delta f_s \approx (n_f - 1)\delta f_s$, i.e. to the range resolution deterioration. When δf_s exceeds $\delta f_{s\,\text{min}}$, the range resolution capability increases but intervals $c/2\delta f_s$ between peaks of the signal ambiguity function in range become less than the CMR extension in range, ΔR. In this case though each peak is narrower than for $\delta f_s = \delta f_{s\,\text{min}}$, several such peaks fall within ΔR. As a result, the SCR practically does not increase. If transmitting and receiving stations are spatially separated (as in bistatic and multistatic radars) all above consideration relates to range sum "a transmitting station – a target – a receiving station".

Consider now velocity resolution capability. It follows from Section 10.3 that the Doppler frequency spectrum of clutter from a CMR has a certain extent due to differences in both individual reflector velocities, $\Delta \mathbf{v}_l$, and directions from reflectors near the CMR centre and near CMR edges to each station (within an angle and range resolution cell). This leads to reduction of temporal (pulse-to-pulse) clutter correlation (see Figs. 10.8 and 10.9) and hence to losses of MTI technique performance. However, in the problem under consideration extension of clutter Doppler spectrum (Doppler spread) is useful. When the illuminating signal duration, T_s, is greater than the reciprocal value of the Doppler spectrum bandwidth of clutter from reflectors fallen into a resolution cell in range and angular coordinates, then the possibility arises to divide reflectors within this resolution cell into several additional resolution cells in velocity (Doppler frequency) using differences in Doppler frequencies of echoes from different reflectors. The echo from a point-like target will be then in a single frequency resolution cell.

Owing to nonzero extent of the clutter Doppler spectrum, we can "cut away" "unnecessary" segments from the illuminating signal of the required duration, T_s. As a result we obtain a sequence of the same total duration, T_s, consisting of short pulses with spacing between them no less than the reciprocal value of the clutter Doppler spectrum bandwidth. In other words, we can use a coherent pulse train practically without losses in SCR.

Example 10.3. Figure 10.10 shows how variation of the interpulse spacing T in the M-pulse coherent train affects clutter power in a signal Doppler frequency resolution cell for a fixed clutter Doppler spectral bandwidth, $\Delta\Omega_{i\,\text{max}}$. We consider a segment of the width $5\Delta\Omega_{i\,\text{max}}$ of the clutter PSD $\Phi_{ii}(\omega)$. It is assumed that $5\Delta\Omega_{i\,\text{max}} \ll 2\pi\Delta f_s$ so that within this segment $|\Psi_0(\omega-\omega_0)|^2 = \text{const}$. A part of the clutter PSD which falls within a signal frequency resolution cell (taking into account the periodicity of signal spectrum) determines the output clutter power. This part is shaded in Fig. 10.10. When $T = 0.2 \times 2\pi/\Delta\Omega_{i\,\text{max}}$ a significant part of the clutter PSD falls within the signal resolution cell. As T increases, the signal spectral peaks become narrower (the width of each peak is of the order of $2\pi/MT$) but the number of peaks within

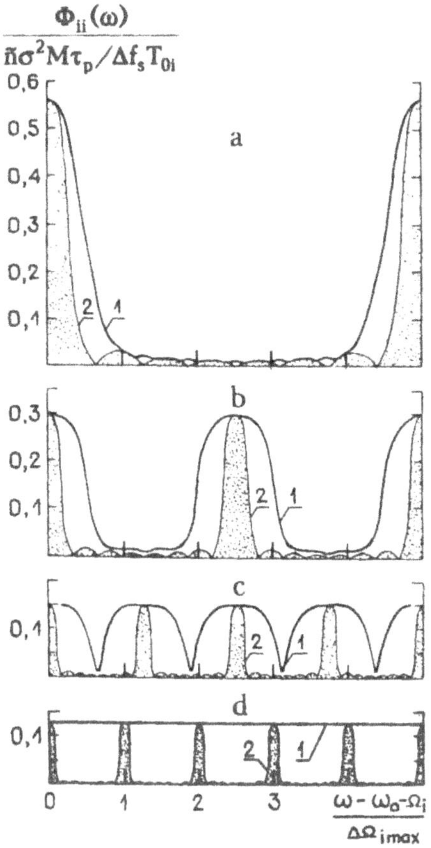

Figure 10.10 Transformations of clutter PSD and clutter power within a frequency resolution cell at the ith receiving station ($i=\overline{1,m}$) caused by the increase of the pulse repetition period of a pulse train, T, when the bandwidth of the clutter Doppler spectrum, $\Delta\Omega_{i\,max}$, remains constant. Curves 1: clutter PSD, $\Phi_{ii}(\omega)$; curves 2: squared filter frequency response corresponding to the target echo spectrum; shaded areas indicate clutter power within a frequency resolution cell; (a) $\Delta\Omega_{i\,max}T/2\pi=0.2$, (b) $\Delta\Omega_{i\,max}T/2\pi=0.4$, (c) $\Delta\Omega_{i\,max}T/2\pi=0.8$, (d) $\Delta\Omega_{i\,max}T/2\pi=1.0$

the clutter PSD increases. Nevertheless, as can be seen from Fig. 10.10, with the increase of T up to $T=2\pi/\Delta\Omega_{i\,max}$ the clutter power in the signal resolution cell is reduced: the shaded area fraction of the total area under the curve $\Phi_{ii}(\omega)$ decreases. When $T=2\pi/\Delta\Omega_{i\,max}$, the clutter PSD coincides with the PSD $|\Psi_0(\omega-\omega_0)|^2$ of a single pulse of the pulse train. The shaded area fraction is by M times less than the total area of the clutter PSD in any frequency band equal to the period $2\pi/T$ of the signal spectrum. This means that an M-pulse coherent pulse train provides the output clutter power decrease in a frequency resolution cell by M times (as compared with a single pulse of the train). Hence the gain in power SCR is equal to M for a point-like target. Further increase of T does not increase the gain in SCR since the clutter PSD remains constant whereas narrowing of signal spectrum peaks is neutralized by the increase of the number of peaks within the clutter PSD. The total shaded area remains practically constant.

Strictly speaking, when $T=2\pi/\Delta\Omega_{i\,max}$ a certain nonuniformity of the clutter PSD appears due to sidelobes of the function $\mathrm{sinc}[\Delta\Omega_{i\,max}(p-q)T/2]$ in (10.48). This leads to some variations of output clutter power in a signal frequency resolution cell. However, it is most likely impossible to take advantage of this phenomenon. A reflector distribution in Doppler frequency may differ from the uniform distribution which was assumed for deriving (10.48). In practice only Doppler spectrum bandwidth $\Delta\Omega_{i\,max}$ is approximately known.

Thus for practical estimates we may consider that an illuminating signal in the form of a coherent pulse train consisting of M pulses can provide a maximum gain in output power signal-to-clutter ratio (SCR) equal to M if the interpulse spacing, T, is chosen so that $T \geqslant 2\pi/\Delta\Omega_{i\,max}$, $i=\overline{1,m}$. It follows from (10.50) and Fig. 10.9 that such a choice of T leads to the absence of pulse-to-pulse clutter correlation. Thus *to achieve maximum values of SCR the interpulse spacing T of a pulse train should be chosen in such a manner so as clutter correlation to be broken down during the time interval T* (because of both mutual displacement of reflectors and CMR motion with respect to a MSRS) (see also [19]). It follows from the obtained results that *a continuous signal of the duration T_s can provide a gain in SCR no greater than $\Delta\Omega_{i\,max}T_s/2\pi$, i.e. the same as a pulse train of the total duration T_s with the interpulse spacing $T \approx 2\pi/\Delta\Omega_{i\,max}$ $i=\overline{1,m}$.*

As shown in Section 10.3, the Doppler frequency spread at the ith station, $i=\overline{1,m}$, is determined not only by differences in individual velocities of reflectors, $\Delta\mathbf{v}_l$, but by the angular width of a CMR within the angle and range resolution cell [the terms h_{l0} and h_{li} in (10.46)–(10.48)]. Therefore high Doppler frequency resolution leads to the angular resolution of CMR elements. This may be treated as a result of the aperture synthesis in consequence of a CMR tangential motion with respect to a MSRS ("inverse aperture synthesis").

Analysis of the resultant SCR and Signal-to-Clutter-plus-self-Noise Ratio, SCNR, at the output of optimal linear processing at each station may be performed as usually with the help of clutter average power distribution over angle coordinates, TOAs and Doppler frequencies as well as resultant ambiguity function. This function may, as a rule, be represented as a product of the antenna system power directivity pattern and signal ambiguity function in TOA and Doppler frequency [2,72].

The obtained results for coherent multifrequency signals and pulse trains as well as for multielement spatially coherent antenna arrays may be treated not only as a result of reducing a resultant resolution cell in all measuring parameters, but as coherent target echo integration in a background of uncorrelated clutter. Incoherent integration is possible too (for instance, when an incoherent illuminating pulse train is used). Losses caused by incoherence may be evaluated by usual techniques. In particular, we may utilize the formulas and curves presented in Chapter 5. Having been obtained for spatial incoherence, they may nonetheless be used in the cases of temporal and frequency incoherence.

It is important to note that the considered methods of enhancing SCR values by the increase of resolution capability (target echo integration) are based on the assumption that CMR spread in the space of radar parameters significantly exceeds that of a target to be detected. A small target spread as compared with that of a CMR is a fundamental limitation on the attainable performance of those methods.

Results of signal processing in a background of clutter at all stations of a MSRS are fused at the FC. This permits further improving target detection in clutter. Even for very high resolution capability it is usually possible to consider clutter in each resolution cell as being approximately Gaussian since sufficient quantity of reflectors falls into each resolution cell (taking into account sidelobes of ADP and signal ambiguity function). In this case the results of Chapters 5 and 6 which relate to performance analysis of different optimum and suboptimum joint signal processing in a background of spatially uncorrelated interferences may be used.

***In this section**, we have noted that all methods and techniques used for MTI in monostatic and bistatic radars are applicable to MSRSs as well. Furthermore, some difficulties that arise in monostatic and even bistatic radars can be removed, or at least markedly reduced. We have considered two such difficulties: the problem of target zero Doppler and the problem of target "blind velocities". Owing to spatial separation of different stations in a MSRS zero Doppler frequencies cannot occur simultaneously with respect to all stations. The same is valid for blind target velocities. We have presented inequality* (10.57) *for the minimum effective baselength between two arbitrary receiving stations which is necessary for reliable Doppler target separation from clutter in one station when the target Doppler frequency at the other station is equal to zero. We have also derived the expressions determining blind velocities for a monostatic radar* (10.58) *and for an arbitrary receiving station of a MSRS* (10.59). *The analysis of those expressions has shown how to overcome such difficulties in MSRSs.*

We have discussed the problem of moving target detection in moving clutter when there are no a priori differences in target and masking reflector velocities. This problem is important for some applications, for instance, for target detection in the exoatmospheric space. If RCS of a target is much greater than RCS of each masking reflector, the target can be detected in clutter in certain cases by the increase of resolution capability in all target parameters measured by radars: angular coordinates, range and velocity. We have revealed an important feature: due to certain spatial extension of a CMR (e.g., a chaff cloud) a solid antenna may be replaced by an antenna array of the same total dimensions but with certain spacing between elements without practically any losses in signal-to-clutter ratio (SCR). In a similar manner an illuminating signal with the continuous spectrum may be replaced by a coherent multifrequency signal and a signal continuous in time may be replaced by a pulse train. Intervals between elements of antenna arrays, signal frequencies and pulse trains should be chosen so as to achieve nearly zero clutter correlation corresponding to each pair of adjacent elements. Example 10.3 *illustrates this feature in the frequency domain.*

10.5. CONSTANT FALSE ALARM RATE (CFAR) DETECTION

Principles of CFAR, Probabilities of False Alarm and Detection

Both monostatic and bistatic radars, especially surveillance ones, often face the problem of target detection in a nonstationary background of clutter and/or jamming plus receiver self-noise. In these cases classical schemes with a fixed threshold cannot be used because the false alarm probability is extremely sensitive to small variations of the received clutter power. For example, when the fixed threshold is set for a false alarm probability $P_{f_a}=10^{-8}$, a 3 dB increase in the total noise power density (clutter plus self-noise) yields a 10 000-fold increase in the false alarm probability [218]. In modern radar with automatic signal and data processing such an increase of the false alarm probability would significantly exceed the radar data-handling rate.

To control false alarm rate in these nonstationary clutter situations the Constant False Alarm Rate (CFAR) processing schemes have been proposed. A CFAR processing sets a detection threshold in each range and/or Doppler frequency resolution cell adaptively using estimates of total noise power by processing a group of reference resolution cells surrounding the cell under investigation.

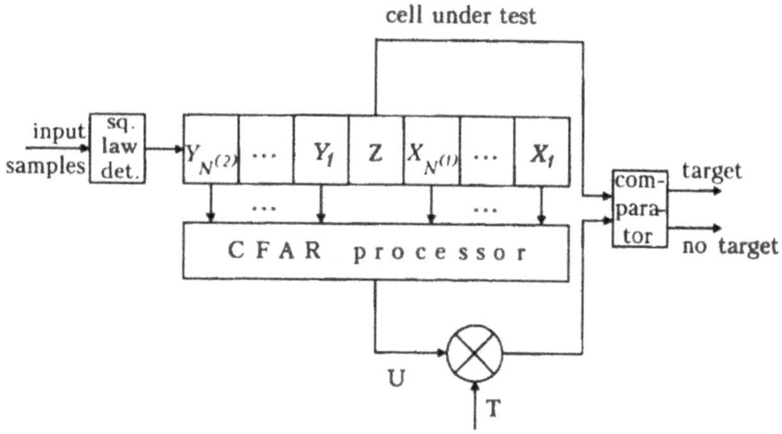

Figure 10.11 General structure of a CFAR detector

A general structure of such a CFAR detector is shown in Fig. 10.11. The sampled square-law envelope detector output is fed into a shift register (or a tapped delay line in the case of analogous processing) of length $N^{(1)}+N^{(2)}+1$ resolution cells. The set of the leading $N^{(1)}$ cells with respect to the cell under test forms one reference window while the set of the $N^{(2)}$ lagging cells forms the other reference window. The outputs of cells from both reference windows, $X_1, \ldots, X_{N^{(1)}}$ and $Y_1, \ldots, Y_{N^{(2)}}$, are jointly processed according to a chosen algorithm. The output random variable U is multiplied by a constant scale factor T which is used to achieve a required false alarm probability for a given window sizes, $N^{(1)}$ and $N^{(2)}$, in the case of homogeneous total background noise. A target is decided to be present if the output Z from the cell under test exceeds the adaptive threshold TU.

The first and probably the simplest processor is the Cell-Averaging CFAR processor (CA-CFAR) [218] which evaluates the sums of X_l and Y_n, $l=\overline{1,N^{(1)}}$, $n=\overline{1,N^{(2)}}$, and then their sum so that

$$U=\sum_{l=1}^{N^{(1)}} X_l+\sum_{n=1}^{N^{(2)}} Y_n=U_1+U_2. \tag{10.60}$$

The CA-CFAR processor is the optimum one (in the sense that it maximizes probability of detection given CFAR) in a homogeneous background when the reference cells contain statistically independent and identically exponentially distributed noise samples. Under these conditions, as the size of the reference windows increases, the performance of CA-CFAR detector approaches that of the optimum one based on a fixed threshold [218,221].

However, the assumption of homogeneous clutter within the reference windows is often violated in practice. The typical inhomogeneities are: regions of clutter power transition and multiple target environments. The first situation occurs when the clutter power changes abruptly within a reference window (for instance, when a CMR edge falls within this reference window). The second situation takes place when there are two or more closely spaced targets. The interfering targets appearing in the reference windows may raise the adaptive threshold which leads to target masking in the cell under test. Both those inhomogeneities lead to serious CA-CFAR performance degradation [143,221,223–226].

A lot of different CFAR processing algorithms have been proposed to overcome the problems associated with above mentioned inhomogeneities. In the Greatest Of selection CFAR (GO-CFAR) algorithm $U=\max(U_1,U_2)$ [226], in Smallest Of selection CFAR (SO-CFAR) algorithm $U=\min(U_1,U_2)$ [227]. Later several algorithms were introduced based on the Ordered Statistics (OS-CFAR schemes). In the OS-CFAR processor, the output samples from all $N^{(1)}+N^{(2)}=N$ reference cells are first ordered in according to their magnitudes. The output quantity U is taken to be the pth largest sample, $X_{(p)}$, i.e. the background power is estimated by selecting the output of the pth largest reference cell [228,229]

$$U=X_{(p)},\quad X_{(1)}\leqslant X_{(2)}\leqslant\cdots\leqslant X_{(p)}\leqslant\cdots\leqslant X_{(N)} \tag{10.61}$$

where $X_{(l)}$ denotes the sample which occupies the lth place among samples from all reference cells ordered in according to their magnitudes.

A scheme has been proposed that combines ordering with arithmetic averaging of reference cell outputs, the so-called Trimmed Mean CFAR (TM-CFAR). In the TM-CFAR processor, the samples of the reference cells are first ordered according to their magnitudes, as in the OS-CFAR processor, and then are trimmed or censored from both the upper and the lower end or only from the upper end. The samples from remaining reference cells are summed, and this sum is used as a background clutter-plus-noise estimate [223,230]. If K_1 and K_2 cells are trimmed from the lower and upper end, respectively, then

$$U=\sum_{l=K_1+1}^{N-K_2} X_l. \tag{10.62}$$

It can be seen that the CA-CFAR algorithm is a special case of the TM-CFAR algorithm with $K_1=K_2=0$. The OS-CFAR is also a special case of TM-CFAR with $K_1=p-1$, $K_2=N-p$.

Another approach to CFAR problem was analyzed in [143]. Adaptive threshold level is set as a result of each resolution cell averaging over several subsequent scans.

A detailed performance analysis of above mentioned CFAR algorithms are presented in [221].

Recently several new CFAR algorithms have been proposed which are intended to overcome CFAR performance degradation under the conditions of clutter power transition and multiple target environments. These algorithms may be considered as different combinations of known schemes. They are: the "mean of order statistics and cell averaging CFAR" (MOSCA-CFAR) [231], several algorithms combining order statistics, cell averaging or trimmed mean with the choice of the greatest or the smallest of corresponding random variables (OSGO-CFAR and OSSO-CFAR [232], GOSCA-CFAR, GOSGO-CFAR, GOSSO-CFAR [233], OSCAGO-CFAR [234], OSTMGO-CFAR [222]), algorithms using the sum of a signal and an ordered statistic (S+OS CFAR [270]) or a linear combination of ordered statistics (L-CFAR [274]). In [235] a special OS-CFAR is used where instead of sliding windows in the time domain a constant window in the frequency domain is employed for obtaining an adaptive threshold. In this case only one adaptive threshold is valid for all test cells so that all cells can be considered the reference window and the cell under test at the same time.

As with any detectors, the main performance characteristics of CFAR detectors are the false alarm and detection probabilities, $P_{\text{fa}i}$ and $P_{\text{d}i}$. The subscript i is introduced to show that these probabilities and all results below may be concerned an arbitrary receiving station (or a monostatic radar) of a MSRS, $i=\overline{1,m}$.

Let us first analyze the CA-CFAR processor (10.60). As mentioned above, we assume that in homogeneous background noise, square-law detected samples for any reference cell of each station are mutually independent and equally exponentially distributed. For the ith station under the hypothesis H_0

$$w_i(x|H_0)=(1/2\sigma_i^2)\exp(-x/2\sigma_i^2),\quad x\geqslant 0 \tag{10.63}$$

where σ_i^2 is the total clutter-plus-self-noise power at the ith station. When the threshold T_iU_i is a random variable (see Fig. 10.11), the false alarm probability is defined as the average quantity

$$P_{\text{fa}i}=\overline{\int_{T_iU_i}^{\infty} w_i(x|H_0)\,dx}=\overline{\int_{T_iU_i}^{\infty}\frac{1}{2\sigma_i^2}\exp\left(-\frac{x}{2\sigma_i^2}\right)dx}=\overline{\exp\left(-\frac{T_iU_i}{2\sigma_i^2}\right)} \tag{10.64}$$

where averaging is performed over possible values of U_i. The last expression in the right part of (10.64) is the moment generating function (MGF) used in Section 5.5 [see (5.87)] so that

$$P_{\text{fa}i}=M(T_i/2\sigma_i^2). \tag{10.65}$$

The MGF of U_i, i.e. of the sum of N_i statistically independent equally exponentially distributed random variables, was derived in Section 5.5. Setting in (5.58) $n_1+n_2+\cdots+n_N=N_i$, $Q_i^2=\sigma_i^2$ and $v=T_i/2\sigma_i^2$ yields

$$P_{\text{fa}i}=1/(1+T_i)^{N_i} \tag{10.66}$$

It is seen that $P_{\text{fa}i}$, as required, depends only on scaling factor, T_i, and the number of reference cells in both windows, N_i, but not on the total clutter-plus-noise power σ_i^2.

To obtain the expression for detection probability at the ith station we assume (as in the previous chapters) the target echo complex amplitude to be Gaussian which means the exponential PDF for signal power (the Swerling 2 model). Under this assumption when a target is present within the cell under test (the hypothesis H_1)

$$w_i(x|H_1)=[1/2\sigma_i^2(1+\bar{q}_i^2)]\exp[-x/2\sigma_i^2(1+\bar{q}_i^2)],\quad x\geqslant 0 \tag{10.67}$$

where σ_i^2 is the same as earlier, and $\bar{q}_i^2$ is the average power SINR (or SCNR) at the ith station. The detection probability $P_{\text{d}i}$ is also defined as an averaged quantity (over the random threshold U_i) and can be obtained from (10.67) in the form

$$P_{\text{d}i}=\overline{\int_{T_iU_i}^{\infty} w_i(x|H_1)\,dx}=\overline{\exp\left(-\frac{T_iU_i}{2\sigma_i^2(1+\bar{q}_i^2)}\right)} \tag{10.68}$$

The last expression in the right part is the MGF (5.87) so that

$$P_{\text{d}i}=M[T_i/2\sigma_i^2(1+\bar{q}_i^2)]. \tag{10.69}$$

Using again (5.58) for $n_1+n_2+\cdots+n_N=N_i$, $Q_i^2=\sigma_i^2$ and $v=T_i/2\sigma_i^2(1+\tilde{q}_i^2)$ yields

$$P_{di}=\frac{1}{[1+T_i/(1+\tilde{q}_i^2)]^{N_i}}=\frac{(1+\tilde{q}_i^2)^{N_i}}{(1+T_i+\tilde{q}_i^2)^{N_i}}. \tag{10.70}$$

It is seen that P_{di} also does not depend on the total clutter-plus-noise power σ_i^2.

Consider now the OS-CFAR processor. To obtain the PDF of U_i [see (10.61) for the ith station] we must take into account the following considerations. Since samples from all the N_i reference cells are mutually statistically independent and equally distributed with the PDF $w_i(x|H_0)$, the PDF of U_i must include a product of $w_i(x|H_0)$ for the pth reference cell by the probability that samples from p_i-1 reference cells are less than $X_{(p_i)}$ while samples from N_i-p_i reference cells are equal to or greater than $X_{(p_i)}$. For each specific arrangement of those samples among physical reference resolution cells the above mentioned product takes the form

$$F_i(x|H_0)^{(p_i-1)}[1-F_i(x|H_0)]^{(N_i-p_i)}w_i(x|H_0)$$

where $F_i(x|H_0)$ is the Cumulative Distribution Function (CDF) corresponding to the PDF $w_i(x|H_0)$, i.e.

$$F_i(x|H_0)=\int_0^x w_i(t|H_0)\,dt; \qquad 1-F_i(x|H_0)=\int_x^\infty w_i(t|H_0)\,dt. \tag{10.71}$$

However, the p_i-1 samples which are less than $X_{(p_i)}$ can be arranged by

$$C_{N_i}^{(p_i-1)}=\binom{N_i}{p_i-1}$$

different ways among all N_i reference cells and for each arrangement of these samples the sample $X_{(p_i)}$ can be in any of N_i-p_i+1 remaining cells. Therefore, the PDF of U_i can be written as follows:

$$w_{U_i}(x)=(N_i-p_i+1)\binom{N_i}{p_i-1}F_i(x|H_0)^{p_i-1}\times[1-F_i(x|H_0)]^{N_i-p_i}w_i(x|H_0). \tag{10.72}$$

Taking into account that

$$(N_i-p_i+1)\binom{N_i}{p_i-1}=p_i\binom{N_i}{p_i}=p_i\frac{N_i!}{p_i!(N_i-p_i)!}$$

expression (10.72) can be presented in a simpler but equivalent form [221]

$$w_{U_i}(x)=p_i\binom{N_i}{p_i}F_i(x|H_0)^{(p_i-1)}[1-F_i(x|H_0)]^{(N_i-p_i)}\,w_i(x|H_0). \tag{10.73}$$

Substituting $w_i(x|H_0)$ from (10.63) and $1-\exp(-x/2\sigma_i^2)$ for $F_i(x|H_0)$ according to (10.71) yields

$$w_{U_i}(x)=p_i\binom{N_i}{p_i}\frac{1}{2\sigma^2}[1-\exp(-x/2\sigma^2)]^{p_i-1}\exp[-x(N_i-p_i+1)/2\sigma^2]. \tag{10.74}$$

The false alarm and detection probabilities for the cell under test, P_{fai} or P_{di}, are defined as for the CA-CFAR by averaging the conditional P_{fai} or P_{di} (given U_i) over possible values of U_i. For instance, to obtain the probability of detection we have to use (10.67)

$$P_{di} = \overline{\int_{T_iU_i}^{\infty} w_i(x|H_1)\,dx} = \overline{\exp[-T_iU_i/2\sigma_i^2(1+\tilde{q}_i^2)]}$$

$$= \int_0^{\infty} \exp[-T_i x/2\sigma_i^2(1+\tilde{q}_i^2)] w_{U_i}(x)\,dx. \tag{10.75}$$

At last, substituting (10.74) in (10.75) and integrating yields

$$P_{di} = p_i \binom{N_i}{p_i} \sum_{l=0}^{p_i-1} (-1)^l \binom{p_i-1}{l} \frac{1}{N_i - p_i + 1 + l + T_i/(1+\tilde{q}_i^2)}. \tag{10.76}$$

After some manipulations (10.76) can be reduced to the form [221]

$$P_{di} = \prod_{l=0}^{p_i-1} \frac{N_i - l}{N_i - l + T_i/(1+\tilde{q}_i^2)}. \tag{10.77}$$

The false alarm probability can be obtained from (10.77) by setting $\tilde{q}_i^2 = 0$:

$$P_{fai} = \prod_{l=0}^{p_i-1} \frac{N_i - l}{N_i - l + T_i}. \tag{10.78}$$

In [238] P_{di} and P_{fai} are presented in another equivalent form:

$$P_{di} = p_i \binom{N_i}{p_i} \frac{(p_i-1)![N_i - p_i + T_i/(1+\tilde{q}_i^2)]!}{[N_i + T_i/(1+\tilde{q}_i^2)]!}; \tag{10.79}$$

$$P_{fai} = p_i \binom{N_i}{p_i} \frac{(p_i-1)!(N_i - p_i + T_i)!}{(N_i + T_i)!}. \tag{10.80}$$

It is seen that neither P_{di} nor P_{fai} depend on the total noise power σ_i^2. Note that unlike the CA-CFAR processor, these probabilities depend not only on T_i and $\tilde{q}_i^2$ but on p_i as well.

In a similar manner one may obtain P_{fai} and P_{di} for CFAR processors of other types. Corresponding expressions and performance characteristics for both clutter transitions and multiple target environments are presented in [219–235].

Applications of CFAR Techniques to MSRSs

As with monostatic radars, methods of coherent clutter and jamming suppression (including those considered in Chapters 8, 9 and Section 10.4) cannot practically provide absolute interference suppression. Therefore, CFAR processing is important for MSRSs for the same reasons and under the same conditions as for monostatic radars. However, there are special features in using CFAR techniques in MSRSs.

In MSRSs with centralized target detection, overall signals received by all spatially separated stations (i.e. sums of receiver self-noises, clutter, jamming with or without wanted signals) are transferred via DTL from all stations of a MSRS (after matched filtration and, possibly, envelope detection) to a fusion centre (FC). Decisions regarding the presence or absence of a target (a wanted signal) are made only at the FC by the comparison of weighted sums of received signals coming from all stations with a threshold. Apparently, after the weighted summation of overall signals coming from all stations, subsequent processing is the same as in a monostatic radar including adaptive thresholding in clutter environments. Thus in this case there are no special features in applying CFAR methods as compared with monostatic radars. All CFAR processing techniques used in monostatic radars can be employed in such MSRSs as well. It is reasonable to recall that the role of the FC may be played by one of the stations of a MSRS.

On the other hand, in MSRSs with measurement or track (trajectory) integration, (for instance, in netted radar systems) target detection processes are, as a rule, completed in each station individually. Since CFAR techniques are associated with the detection process, CFAR processing should be performed in each station independently as in a usual monostatic radar. Obviously, any known CFAR technique (including those outlined above) may be used in spatially separated stations of such a MSRS.

The only case where CFAR processing application to MSRSs has some essential special features is the case of a MSRS with the plot integration level (see Section 1.1) and decentralized target detection. Decentralized detection of a target (a wanted signal) in a background of spatially uncorrelated interferences (i.e. receiver self-noises and, possibly, sidelobe jamming) has been considered in Chapter 6 assuming homogeneous interferences and hence fixed detection thresholds. Now we consider some CFAR processing schemes for such MSRSs when targets are to be detected in clutter. As in Chapter 6, we confine ourselves to the parallel system topology where a FC is connected with all receiving stations and all preliminary decisions regarding the presence or the absence of a target, made in each station individually, are sent directly to the FC via DTLs for a final decision.

To design CFAR processing for a MSRS with decentralized detection, three problems are to be solved for each ith station, $i=\overline{1,m}$: (1) the choice of the type of CFAR processor (CA-CFAR, OS-CFAR, TM-CFAR etc.); (2) the determination of chosen processor's parameters (the numbers of reference cells, N_i, the magnitudes of scaling factors, T_i, the number of trimmed cells, K_{1i}, K_{2i}, in the TM-CFAR etc.); (3) the choice of the decision rule (detection criterion) for preliminary decision combining in the FC.

The first problem should be solved as in a monostatic radar taking into consideration expected clutter and target environments, their possible inhomogeneities, performance requirements, complexity and cost of several devices and so on. It is usually reasonable to have CFAR processors of one and the same type in all stations though it is not necessary.

The way for solving two other problems depends on the general approach to the decentralized detection optimization. As in the case of fixed thresholds at each station considered in Section 6.1, this optimization is possible in a "narrow sense" or in a "wide sense". When the "narrow sense" optimization is performed, all parameters of CFAR processors are chosen for each station as for a monostatic radar (based on the previous experience and preliminary calculations). Optimum decision rules for the FC may be chosen using the results of Section 6.1 without any revision. In fact, having obtained preliminary decisions from all stations, as well as probabilities of false alarm and detection corresponding to these decisions, we may employ optimum decision rules in the FC derived in Section 6.1 regardless of whether adaptive (CFAR) or fixed thresholds were used in each station. Therefore, there is no need in considering the "narrow sense" optimization here.

The "wide sense" optimization has some special features. In this case some of CFAR processors' parameters are to be chosen also as in a monostatic radar. Other parameters should be chosen by an iterative process together with performance analysis. As in Section 6.1, the "wide sense" optimization implies a simultaneous optimum choice of both a decision rule at the FC and thresholds at all stations. It means that for each decision rule at the FC, each type of CFAR processors and their parameters at all stations, given SINR values, scaling factors are to be optimized. Then comparing performance characteristics for each decision rule with the optimized set of scaling factors, i.e. with the best detection thresholds at all stations, we can choose the best decision rule for assumed conditions.

In a monostatic or a bistatic radar the scaling factor T is usually determined from the requirement to achieve a desired constant false alarm probability for a given number of reference cells when the total background noise is homogeneous. In MSRSs the scaling factors at all stations $T_1,\dots,T_m$ can be optimized so as to maximize the output detection probability at the allowed output false alarm probability (at the FC) for given type and other parameters of CFAR processors, the SINR values at all stations and the decision rule at the FC. The optimization can be performed by using an objective function as in Section 6.1 (see (6.15) and also [236–238])

$$J(T_1,\dots,T_m)=P_{\mathrm{d}}(T_1,\dots,T_m)+\beta[P_{\mathrm{fa}}(T_1,\dots,T_m)-\alpha_0] \tag{10.81}$$

where P_{d} and P_{fa} are the probability of detection and false alarm, respectively, at the output of the FC; α_0 is the desired value of P_{fa}; β is the Lagrange multiplier. Differentiating this objective function with respect to $T_1,\dots,T_m$ and β yields the system of $m+1$ nonlinear equations in $m+1$ unknowns:

$$\begin{gathered}\frac{\partial J(T_1,\dots,T_m)}{\partial T_i}=0, \quad i=\overline{1,m},\\ P_{\mathrm{fa}}(T_1,\dots,T_m)=\alpha_0.\end{gathered} \tag{10.82}$$

The relationships between P_{fa} and $P_{\mathrm{fa}i}$ as well as between P_{d} and $P_{\mathrm{d}i}$ are determined by the chosen decision rule at the FC.

The same technique was employed in Section 6.1 [see (6.15), (6.16)] for the "wide sense" optimization. The difference between the objective functions (6.15) and (10.79) is caused by the fact that in (6.15) the false alarm probabilities at all stations are unambiguously connected with corresponding relative thresholds, while for CFAR processors it is not so. In particular, an essential role is played by the number of reference cells. Assuming all other parameters to be chosen before, only scaling factors should be considered as independent arguments to be optimized.

Taking into account the results of Section 6.2, we confine ourselves to the simple and most widely used decision rules of the "k out of m" type (6.9). As shown in Section 10.1, clutter at the inputs of spatially separated stations is, as a rule, mutually statistically independent. It means that total noises (clutter-plus-self-noises) at different stations are mutually independent too. We also assume the condition (4.42) of target echo fluctuations' spatial incoherence (at the inputs of stations) to be fulfilled. Then the output probabilities of false alarm and detection, P_{fa} and P_d, are determined by (6.12) where the indices k should be replaced by l in order not to confuse with k from the decision rule "k out of m". According to these decision rules, n in the first sum of (6.12) should be replaced by k. To obtain P_{fa} or P_d, the corresponding probabilities of each station, P_{fai} or P_{di}, are to be substituted for P_i in (6.13). Specifically, we have the detection probability, P_d, in the form

$$P_d = \sum_{l=k}^{m} \sum_{i_1=1}^{m-l+1} \sum_{i_2=i_1+1}^{m-l+2} \cdots \sum_{i_l=i_{l-1}+1}^{m} P_{di_1} P_{di_2} \cdots P_{di_l} \prod_{j=1, j\neq i_1, i_2, \ldots, i_l}^{m} (1-P_{dj}). \quad (10.83)$$

The false alarm probability, P_{fa}, is expressed by the same equation (10.83) where $P_{di_1}, \ldots, P_{di_l}, P_{dj}$ are to be replaced by $P_{fai_1}, \ldots, P_{fai_l}, P_{faj}$, respectively:

$$P_{fa} = \sum_{l=k}^{m} \sum_{i_1=1}^{m-l+1} \sum_{i_2=i_1+1}^{m-l+2} \cdots \sum_{i_l=i_{l-1}+1}^{m} P_{fai_1} P_{fai_2} \cdots P_{fai_l} \prod_{j=1, j\neq i_1, i_2, \ldots, i_l}^{m} (1-P_{faj}). \quad (10.84)$$

Example 10.4. Consider the decision rule "2 out of 3", i.e. $m=3$, $k=2$. From (10.83) we have

$$P_d = \sum_{l=2}^{3} \sum_{i_1=1}^{4-l} \sum_{i_2=i_1+1}^{5-l} \sum_{i_3=i_2+1}^{3} P_{di_1} P_{di_2} P_{di_3} \prod_{j=1, j\neq i_1, \ldots, i_l}^{3} (1-P_{dj})$$

$$= \sum_{i_1=1}^{2} \sum_{i_2=i_1+1}^{3} P_{di_1} P_{di_2} \prod_{j=1, j\neq i_1, i_2}^{3} (1-P_{dj}) + \sum_{i_1=1}^{1} \sum_{i_2=2}^{2} \sum_{i_3=3}^{3} P_{di_1} P_{di_2} P_{di_3}$$

$$= P_{d1}P_{d2}(1-P_{d3}) + P_{d1}P_{d3}(1-P_{d2}) + P_{d2}P_{d3}(1-P_{d1}) + P_{d1}P_{d2}P_{d3}. \quad (10.85)$$

Replacing P_{d1}, P_{d2}, P_{d3} by $P_{fa1}, P_{fa2}, P_{fa3}$, respectively, yields the output false alarm probability P_{fa}.

To obtain an optimum set of scaling factors $T_1, \ldots, T_m$ one should substitute equations for P_{fai} and P_{di}, $i=\overline{1,m}$, corresponding to a chosen CFAR processor in (10.84) and (10.83), respectively. For the CA-CFAR processor those equations are (10.66) and (10.70). For OS-CFAR processor they are (10.78) or (10.80) and (10.77) or (10.79). As a result we express P_{fa} and P_d in terms of scaling factors T_i and other parameters of CFAR processors which have been chosen before. Then, substituting the obtained expressions in (10.82), we can solve that system of nonlinear equations for $T_1, \ldots, T_m$. Calculation results for the CA-CFAR and OS-CFAR processors under different conditions (the number of stations, the size of reference windows etc.) are presented in [236,238]. Corresponding results for decentralized detection system with S+OS CFAR processor and with L-CFAR processor may be found in [270,274]. The effect of binary integration in each radar and Weibull clutter distribution is considered in [271].

Of course, the optimization described above is not a simple computational problem. It should be emphasized that any changes not only in processor parameters but in SINR (SCNR) values at the input of any station require repeated solutions of those nonlinear equations.

This situation is similar to that considered in Section 6.1. It is reasonable to recall that the scaling factor optimization leads to an "optimal distribution" of the output (final) false alarm probability P_{fa} between all receiving stations (or monostatic radars) of a MSRS. In Section 6.1 we have proposed an essentially simplified suboptimal procedure implying equal false alarm probabilities at all stations, i.e. $P_{fa\,i}=P_{fa\,0}$, $i=\overline{1,m}$ (the "uniform false alarm distribution" between stations).

Unlike the case of fixed thresholds considered in Section 6.1, in the case of CFAR processors equal false alarm probabilities at all stations do not mean equal relative thresholds or equal scaling factors since the numbers of reference cells and even the types of CFAR processors may be different. Nevertheless, as a rule, it is much easier to calculate the scaling factor for a given type of CFAR processor, the number of reference cells and false alarm probability at each station, than to solve the system of nonlinear equations (10.82).

The numerical examples presented in Section 6.1 have shown that in the case of detectors with equal false alarm probability at all stations such a procedure, bypassing the solution of nonlinear equations, does not lead to significant energy losses. Such a result may be expected for similar though more complicated CFAR detectors at all stations. It is clear from the symmetry considerations that in the case of the same type of CFAR processors, equal parameter values (equal sizes of reference windows etc.) and equal SINR values at all stations, optimum false alarm probabilities also must be equal, i.e. $P_{fa\,i}=P_{fa\,0}$, $i=\overline{1,m}$. Furthermore, in this case optimum scaling factors at all stations are equal to each other, i.e. $T_i=T_0$, $i=\overline{1,m}$. For different sizes of reference windows this equality of optimum scaling factors no longer exists but the difference in optimum false alarm probabilities turns out to be not large. The difference in SINR values may have a more significant effect on optimum false alarm probabilities. However, according to results from Section 6.2, when the difference in SINR values is large, "weak" stations do not noticeably contribute to the final detection probability so that including these stations in a MSRS is not expedient (from the point of view of detection process). On the other hand, for moderate differences in SINR values energy losses caused by the equalization of false alarm probabilities are expected to be not too large.

To verify the above arguments we consider a numerical example.

Example 10.5. Let a MSRS contain two receiving stations or two monostatic radars ($m=2$) with the CA-CFAR processors. Just in the case where $m=2$ each station has a most significant effect on the output detection characteristics. We consider the decision rules "1 out of 2" ("OR") and "2 out of 2" ("AND"). It has been shown in Section 6.2 that the decision rule "1 out of 2" ("OR") is optimum for the fixed threshold scheme and spatially statistically independent target echo fluctuations. However, as follows from results presented in [236], for the CA-CFAR detectors with small sizes of reference windows the decision rule "2 out of 2" ("AND") is preferable when the SINRs do not exceed the certain values depending on the number of reference cells. Calculation results are presented in Table 10.1 for the "OR" decision rule and in Table 10.2 for the "AND" decision rule. It can be seen that when the SINR values at both stations are equal and only the sizes of reference windows are different, energy losses caused by the use of the simplified "uniform false alarm distribution" are actually small for both decision rules. When the SINR values are essentially different while the number of reference cells is equal, these losses slightly increase, especially for the "AND" rule. However, for the same differences in the SINR values energy losses decrease for different sizes of reference windows. In general, we may conclude that using the simple algorithm of equal false alarm probabilities at both stations leads to insignificant losses for the considered cases, especially for the "OR" decision rule.

It should be noted that comparing the data from Tables 10.1 and 10.2 with corresponding detection characteristics from Section 6.2, one can see the energy losses of CA-CFAR detectors with respect to detectors with fixed thresholds under

Table 10.1

	$\tilde{q}_1^2/\tilde{q}_2^2$(dB)	10/10	20/20		10/20	
	N_1/N_2	4/8	4/8	4/4	4/8	8/4
Optimum algorithm	T_1/T_2	12.619/2.301	10.474/2.521	21.252/9.105	17.914/2.195	2.894/9.538
	P_{fa1}/P_{fa1}	$2.91\cdot10^{-5}$/$7.09\cdot10^{-5}$	$5.77\cdot10^{-5}$/$4.23\cdot10^{-5}$	$4.08\cdot10^{-6}$/$9.59\cdot10^{-5}$	$7.81\cdot10^{-6}$/$9.22\cdot10^{-5}$	$1.89\cdot10^{-5}$/$8.11\cdot10^{-5}$
	P_{d1}/P_{d2}	0.047/0.219	0.677/0.821	0.0135/0.708	0.021/0.841	0.154/0.697
	P_d	0.256	0.942	0.712	0.845	0.738
Suboptimum algorithm	T_1/T_2	10.892/2.448	10.892/2.448	10.892/10.892	10.892/2.448	2.448/10.892
	P_{fa1}/P_{fa1}	$5\cdot10^{-5}$/$5\cdot10^{-5}$	$5\cdot10^{-5}$/$5\cdot10^{-5}$	$5\cdot10^{-5}$/$5\cdot10^{-5}$	$5\cdot10^{-5}$/$5\cdot10^{-5}$	$5\cdot10^{-5}$/$5\cdot10^{-5}$
	P_{d1}/P_{d2}	0.064/0.200	0.664/0.826	0.064/0.664	0.064/0.826	0.200/0.664
	P_d	0.251	0.941	0.685	0.837	0.731
	Losses (dB)	<0.1	<0.1	0.5	0.3	<0.1

Table 10.2

	$\tilde{q}_1^2/\tilde{q}_2^2$(dB)	10/10	20/20	20/15	
	N_1/N_2	4/8	4/8	4/4	4/8
Optimum algorithm	T_1/T_2	1.154/1.154	1.154/1.154	4.623/0.778	3.642/4.678
	P_{fa1}/P_{fa1}	$4.64\cdot10^{-2}$/$2.15\cdot10^{-3}$	$4.64\cdot10^{-2}$/$2.15\cdot10^{-3}$	0.0010/0.1000	$2.15\cdot10^{-3}$/$4.64\cdot10^{-2}$
	P_{d1}/P_{d2}	0.671/0.450	0.956/0.913	0.836/0.910	0.868/0.892
	P_d	0.302	0.873	0.761	0.774
Suboptimum algorithm	T_1/T_2	2.162/0.778	2.162/0.778	2.162/2.162	0.778/2.162
	P_{fa1}/P_{fa1}	0.010/0.010	0.010/0.010	0.010/0.010	0.010/0.010
	P_{d1}/P_{d2}	0.488/0.578	0.919/0.940	0.919/0.774	0.919/0.828
	P_d	0.282	0.864	0.711	0.761
	Losses (dB)	0.3	0.3	1.0	0.3

the condition of homogeneous noise. It can be seen that for the small number of reference cells these losses are significant. Besides, in agreement with the results from [236], the decision rule "AND" outperforms the decision rule "OR" for small SINR values unlike the fixed threshold detection.

The obtained results in Example 10.5 are by no means comprehensive. Perhaps energy losses caused by using the simple "uniform false alarm distribution" between stations may be large for other types of CFAR processors with certain parameters. In any case, the described approach should be taken into consideration and verified for the specific CFAR processors and other specific conditions.

In this section, *we have considered principles of CFAR detection and some special features of using CFAR processors in MSRSs. We have presented the general structure of a typical CFAR processor* [*Fig.* 10.11] *and briefly described several types of CFAR processors including the most widely used CA-CFAR* (10.60), *OS-CFAR* (10.61), *TM-CFAR* (10.62) *and some new algorithms combining known ones. We have derived the expressions* (10.66) *and* (10.70) *for the false alarm and detection probabilities for the CA-CFAR detector as well as the corresponding expressions for the OS-CFAR detector* [(10.78), (10.80) *and* (10.77), (10.79), *respectively*]. *It has been emphasized that the choice of a specific CFAR processor depends on the expected clutter and target environments. Many CFAR processors proposed in the last years are intended for alleviating the CFAR processor performance degradation under the conditions of clutter power transitions* (*for instance, at the edges of chaff clouds*) *and multiple target environments.*

It has been noted that CFAR applications to MSRSs have special features only in the case where the plot combining with decentralized detection is used. In this case a proper type of a CFAR processor and its certain parameters at each station should be chosen as for a monostatic radar. However, scaling factors determining false alarm probabilities at all stations in a homogeneous total noise background (*clutter-plus-self-noise*) *may be optimized so as to obtain maximum output detection probability at the allowed output probability of false alarm. The situation is similar to that considered in Section 6.1. We have presented the objective function* (10.81) *to be maximized and a system of nonlinear equations* (10.82) *with respect to unknown scaling factors. The formulas* (10.83) *and* (10.84) *as well as Example* 10.4 *permit to express false alarm and detection probabilities at the output of FC by means of scaling factors for decision rules of the "k out of m" type.*

Because such optimization is usually a complicated computational process, we have proposed (*as in Section* 6.1) *a simplified suboptimal procedure of the "uniform false alarm distribution" between stations of a MSRS. The presented numerical data in Example* 10.5 (*Table* 10.1) *show small additional energy losses as compared with the optimum algorithm.*

PART 3

Target Coordinate Estimation and Tracking in MSRSs

11. TARGET POSITION AND VELOCITY ESTIMATION FROM PLOTS. COORDINATE ESTIMATION USING SIGNAL TEMPORAL PARAMETER MEASUREMENT

11.1. TARGET POSITION AND VELOCITY MEASUREMENT METHODS. ONE-STAGE AND TWO-STAGE ALGORITHMS

Methods and techniques for target position and velocity estimation from plots[1] depend on the type of a MSRS.

Spatially Coherent MSRSs

As mentioned in Section 1.1, such a MSRS may be considered a "monostatic radar" with a large thinned antenna array. It can measure parameters of a coherent sum of signals received by all stations. In *the active mode* that "integrated" measurement is then used, as in a conventional monostatic radar, for target range, angular coordinates and radial velocity (or a part of these target parameters) estimation. Target tangential velocity estimation by Doppler method is also possible but its accuracy is, as a rule, low since baselengths in such MSRSs are usually small. In *the passive mode* apart from target angular coordinates, target range can be estimated using signal wavefront curvature since signal sources turn out to be usually in the Freshnel zone of the Antenna System (AS) of a MSRS. Range estimation errors which are determined by the down-range extension of the AS focal spot ("depth of sharpness"), increase in a square law with the increase of the ratio of the signal source range to the linear dimension of the AS. The attainable range accuracy is, as a rule, much less than when a target range is estimated by echo TOA measurements in the MSRS active mode. However, such a method is of great importance for radiation source range estimation in the passive mode of MSRSs [28,36,41,42].

MSRSs with Short-Term Spatial Coherence and Spatially Incoherent MSRSs

Two approaches are possible to the problem of composite measurement formation from plots. We may combine signals received by all stations and estimate target coordinates (and their derivatives) directly from detected plots of this summed signal. Such *one-stage algorithms* are considered in Section 11.2. *Two-stage algorithms* are much simpler and more widely used. At the first stage target coordinates are estimated with respect to each station or each pair of stations as in usual monostatic or bistatic radars. Thus at the first stage we obtain estimates of so-called target local, or "primary" coordinates. At the second stage these local coordinate estimates are combined so as to form a resultant composite measurement, i.e. the estimates of global, or "final" target coordinates and, possibly, their derivatives in a common for

[1] See the footnote on page 6.

a MSRS coordinate reference system and referred to a certain time. The two-stage target measurement is naturally combined with the decentralized target detection in MSRSs (see Chapters 6 and 7). Since local coordinates of only detected targets are sent for joint processing at the FC, requirements to the handling capacity of DTLs are drastically reduced as compared with the one-stage algorithms which are associated with the centralized target detection.

Target coordinate information contained in received signal parameters also depends on the type of a MSRS. For a MSRS consisting of monostatic radars with autonomous (independent) signal reception (see Section 1.1), i.e. for a "netted radar system", a composite target measurement is derived from a set of range, azimuth, elevation angle and radial velocity measurements obtained by all stations with respect to their sites. All measurements are desirable to be made practically simultaneously so as changes in the target state (position and velocity) between different measurements would be much less than expected measurement errors. If it is not so, the measurement extrapolation or interpolation to the same time is necessary (e.g. [85]).

Some or all monostatic radars of a MSRS can often measure only a part of the above mentioned local target coordinates, in particular, the range and the azimuth. Information fusion in MSRSs containing only 2-D radars, allows often to estimate all the three spatial target coordinates and, furthermore, to utilize redundant measurements for the increase of target position estimation accuracy. If three (or more) radars not lying in one plane with a target can measure target echo Doppler frequency shifts, such a MSRS can obtain the target velocity vector using its three projections.

When a MSRS includes bistatic radars, each of these radars measures the range sum "the transmitting station–a target–the receiving station", possibly, the rate of the range sum and the target angular coordinates (usually with respect to the receiving station). Each measurement of the range sum R_Σ determines a surface of a prolate spheroid in space (R_Σ=constant) with their foci at the transmitting and receiving station sites. This surface is a Surface of Position (SOP) in which the target must lie. Let a target be positioned at the point with spherical coordinates R, β, ε and a transmitting station be located at the coordinate system origin. The range sum measured by the ith receiving station at the point $(L_i, \beta_i, \varepsilon_i)$ is given by

$$R_{\Sigma i} = R\left\{1 + \sqrt{1 + \frac{L_i^2}{R^2} - 2\frac{L_i}{R}\left[\cos\varepsilon\cos\varepsilon_i\cos(\beta-\beta_i) + \sin\varepsilon\sin\varepsilon_i\right]}\right\}. \tag{11.1}$$

When $L_i/R \ll 1$, then expanding (11.1) into the Taylor series yields

$$R_{\Sigma i} = R\left\{2 - \frac{L_i}{R}\left[\cos\varepsilon\cos\varepsilon_i\cos(\beta-\beta_i) + \sin\varepsilon\sin\varepsilon_i\right] + O\left(\frac{L_i^2}{R^2}\right)\right\}. \tag{11.2}$$

The approximation (11.2) is equivalent to the assumption that the target is positioned in a far zone of the MSRS (the directions from the target to the transmitting and receiving stations are assumed to be parallel). Such an approximation is often quite sufficient for angular measurement error evaluation. Greater accuracy can be obtained keeping the next term in the Taylor expansion. Then

$$R_{\Sigma i}=R\left\{2-\frac{L_i}{R}[\cos\varepsilon\cos\varepsilon_i\cos(\beta-\beta_i)+\sin\varepsilon\sin\varepsilon_i]\right.$$
$$\left.+\frac{L_i^2}{2R^2}\{1-[\cos\varepsilon\cos\varepsilon_i\cos(\beta-\beta_i)+\sin\varepsilon\sin\varepsilon_i]^2\}+O\left(\frac{L_i^3}{R^3}\right)\right\}. \quad (11.3)$$

In the general case the range sum "the jth transmitting station–the target–the ith receiving station" can be expressed in a Cartesian coordinate system as follows:

$$R_{\Sigma ji}=\sqrt{(x-x_{0j})^2+(y-y_{0j})^2+(z-z_{0j})^2}$$
$$+\sqrt{(x-x_i)^2+(y-y_i)^2+(z-z_i)^2}. \quad (11.4)$$

where (x,y,z), (x_{0j},y_{0j},z_{0j}) and (x_i,y_i,z_i) are the coordinates of the target, the transmitting and receiving station, respectively.

For each bistatic radar of a MSRS it is often convenient to consider a target position in the bistatic plane passing through the target, transmitting and receiving stations (Fig. 11.1). In this case the target ranges from the transmitting station, R_t, and from the receiving station, R_r, can easily be expressed in terms of the range sum $R_\Sigma=R_t+R_r$ and the target bearing with respect to the receiving station, Θ_r [111]

$$R_t=0.5\frac{R_\Sigma^2+L^2+2R_\Sigma L\cos\Theta_r}{R_\Sigma+L\cos\Theta_r}; \qquad R_r=0.5\frac{R_\Sigma^2-L^2}{R_\Sigma+L\cos\Theta_r}. \quad (11.5)$$

Useful expressions for R_t and R_r in terms of other geometrical parameters are presented in [192].

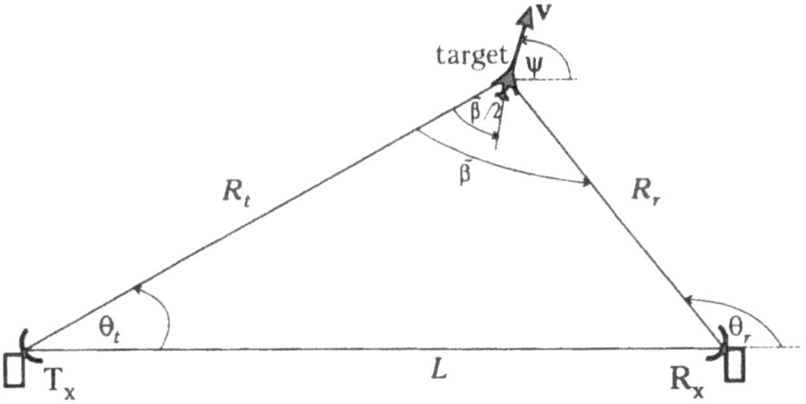

Figure 11.1 Target coordinate measurement in the bistatic plane

For $\Theta_r = \pi$ the prolate spheroid collapses to the straight line. In this case it is clear that a target lies in the baseline but R_t and R_r cannot be measured.

The target echo Doppler frequency (DF) at the receiving station of a bistatic radar is given by

$$F_D = \frac{1}{\lambda}\frac{dR_\Sigma}{dt} = \frac{1}{\lambda}\frac{d(R_t + R_r)}{dt} \tag{11.6}$$

where λ is the wavelength. The dependence of F_D on the target velocity, v, can be expressed in the simplest form when a target moves in the bistatic plane at the angle ψ with the baseline (see Fig. 11.1):

$$F_D = -\frac{2v}{\lambda}\cos(\psi - \Theta_r + \tilde{\beta}/2)\cos(\tilde{\beta}/2). \tag{11.7}$$

The first cosine factor shows that the DF is determined by the target velocity along the bisector of the angle $\tilde{\beta}$. The second cosine factor represents the decrease of the DF caused by the difference in directions from the target to the transmitting and receiving stations[2].

It follows from (11.6) and (11.7) that the Doppler shift in a bistatic radar depends not only on the magnitude and direction of the target velocity vector (as for a monostatic radar) but on the target position in space with respect to the transmitting and receiving stations. The latter dependence becomes stronger as the ratio of the target range to the baselength between stations decreases. Near the baseline, when $\tilde{\beta} \approx \pi$, the Doppler effect is practically imperceptible since the range sum, $R_\Sigma = R_t + R_r$, of a moving target remains nearly constant. It should be noted that the DF dependence on a target position not only influences the target velocity measurements but allows to determine the target coordinates in space by the Doppler method [23,46].

Using (11.4) and (11.6) we can obtain a general expression for the DF at the ith receiving station when a target is illuminated by the jth transmitting station:

$$F_{Dji} = \frac{1}{\lambda}\left[\frac{(x - x_{0j})v_x + (y - y_{0j})v_y + (z - z_{0j})v_z}{\sqrt{(x - x_{0j})^2 + (y - y_{0j})^2 + (z - z_{0j})^2}} + \frac{(x - x_i)v_x + (y - y_i)v_y + (z - z_i)v_z}{\sqrt{(x - x_i)^2 + (y - y_i)^2 + (z - z_i)^2}}\right] \tag{11.8}$$

where v_x, v_y, v_z are the Cartesian components of the target velocity vector $\mathbf{v}$. The remaining notations are the same as in (11.4), (11.6).

Using the cooperative signal reception in a MSRS (see Section 1.1) increases a body of target coordinate information significantly. When a target is illuminated by any transmitting station, several receiving stations can measure the target's angular coordinates, ranges or range sums and range rates or range sum rates (by the Doppler method). Besides, the possibility of measurements of target range differences (and their derivatives) with respect to several pairs of receiving and/or transmitting stations arises. All this leads, as a rule, to a significant redundancy which can be utilized for higher accuracy of the target position and velocity estimation as well as for interstation measurement association (see Section 15.4).

---

[2] It may be treated as the increase of effective wavelength $\lambda_{eff} = \lambda/\cos(\tilde{\beta}/2)$ see Section 2.1.

For target position determination in space based on local coordinate measurements, different methods may be used which are well known in the field of radionavigation [4,23,67]: the elliptic, the hyperbolic, the triangulation methods and their combination.

The elliptic method permits to determine a target position in space using only range sum measurements. It is also called *the range-sum-measurement method, the trilateration* or *multilateration method*. To determine all the three target spatial coordinates when the cooperative signal reception is used, at least three or four stations are required, for example, one transmitting–receiving and two receiving stations or one transmitting and three receiving stations. When the autonomous (independent) signal reception is used, at least three monostatic or bistatic radars are necessary. In the absence of redundant measurements a target position is determined as the intersection point of three SOPs, i.e. of three prolate spheroids, corresponding to the three measured values of range sums. In the case of three monostatic radars prolate spheroids collapse to three spheres.

By *the hyperbolic method* the position of a target is determined using range differences of the target with respect to spatially separated stations. Therefore, it is also called *the range-difference-measurement method*. Each measured value of the target range difference relative to a pair of stations gives rise to the SOP being a hyperboloid of revolution with its foci at these stations' sites. The target position in space is determined by the intersection point of such three hyperboloids. To obtain three linearly independent range differences using echoes scattered by a target, the total number of transmitting and receiving stations has to be no less than five (for instance, one transmitting and four receiving stations). This number is by one station greater than for the elliptic method. Unlike the elliptic method, target range errors rapidly increase (approximately in the square law) with the increase of the target range. The cause of this drawback is that the hyperbolic method does not utilize target range and/or range sum information obtained in MSRSs. Therefore the hyperbolic method in active MSRSs is used only in combination with the elliptic method.

The hyperbolic method is widely used in passive MSRSs or in the passive mode of active–passive MSRSs where the signal propagation time from a target to receiving stations (i.e. the target range or range sum) cannot be measured [18,23,67]. The range difference of a target with respect to each pair of receiving stations is usually evaluated by the TDOA measurement with the help of a correlation technique. In the simplest case, signals received by a pair of receiving stations from a radiation source are conveyed to correlator inputs, from one of stations being passed through a variable delay line. A delay value corresponding to the maximum of the correlator output, cancels out the signal TDOA at the correlator inputs. Having measured this delay value, one can calculate the range difference from the radiation source to the receiving stations. It should be emphasized that this procedure does not require to know *a priori* the signal waveform and the time of signal transmission by the source. Just these properties make the hyperbolic method suitable for radiation source localization when signal waveforms are *a priori* unknown. To obtain three spatial coordinates of a radiation source (e.g. a jammer) by the hyperbolic method, at least four receiving stations are necessary.

Expressions for the range difference from a radiation source to a pair of receiving stations, ΔR_{ik}, depending on the source coordinates in the spherical system used above, can easily be obtained from (11.1)–(11.3) taking into account that $\Delta R_{ik} = R_{\Sigma k} - R_{\Sigma i}$. For evaluation of the angular source coordinates β and ε using measured range differences, ΔR_{ik}, the approximation (11.2) provides usually practically sufficient accuracy. To obtain the source range R, expressions for $R_{\Sigma i}$ and $R_{\Sigma k}$

from (11.3), and in the general case from (11.1), should be substituted in the equation $\Delta R_{ik} = R_{\Sigma k} - R_{\Sigma i}$. In a Cartesian coordinate system the range difference, ΔR_{ik}, can easily be expressed using (11.4) and taking into account that $\Delta R_{ik} = R_{\Sigma jk} - R_{\Sigma ji}$ where j is the number of an arbitrary transmitting station.

When a radiation source moves, Doppler frequencies (DFs) of received signals appear at the inputs of spatially separated stations but, because the central radiation frequency is unknown, these individual DFs cannot be measured. The difference between these DFs (i.e. the differential Doppler frequency, DDF) at each pair of stations is proportional to the source range difference rate with respect to these stations and manifests itself in the central frequency shift of the mutual correlation function of received signals. It can be measured at the correlator output. A DDF depends on both the source position in space and velocity. It is determined by the projection of the source velocity vector on the normal at the point where the source is positioned to the hyperbola which is the intersection of the hyperbolic SOP (containing the source) and the bistatic plane. If the number of receiving stations of a MSRS is no less than four, then in addition to source localization (by determining its three spatial coordinates), source velocity vector can be evaluated by measuring three linearly independent differential Doppler frequencies. In a Cartesian coordinate system the DDF can be expressed immediately from (11.8):

$$\Delta F_{\mathrm{D}ik} = F_{\mathrm{D}jk} - F_{\mathrm{D}ji}$$

for any j.

To determine a target position in space by *the triangulation*[3] *method*, only target bearings from spatially separated stations are used. Therefore this method is sometimes called *the direction finding method*. It is employed in MSRSs with both autonomous and cooperative signal reception. For active MSRSs where target range or range sum measurements are available, triangulation is used only in combination with elliptic method [81,82]. For passive MSRSs or passive mode of active–passive MSRSs, the triangulation method may be employed both individually and together with the hyperbolic method. The minimum number of receiving stations which is necessary to localize a radiation source in space (i.e. to determine its three spatial coordinates) is equal to two, if at least one of these stations can measure both angular coordinates of the source. In this case the target position is determined by the intersection point of a bearing (Line of Position, LOP) and a plane (SOP).

The above mentioned minimal number of stations provides target localization in space and velocity vector determination using only spatial information (local coordinates of a target). If temporal information may be used too, namely several measurements obtained subsequently in time, and there is *a priori* information on the target motion, the required number of stations may be reduced [108,119].

In this section, *we have considered the main features of the target's position and velocity determination process in spatially coherent MSRSs, MSRSs with short-term spatial coherence and spatially incoherent MSRSs. For the last two types of MSRSs one-stage and two-stage algorithms of the target localization in space and velocity evaluation are possible. One-stage algorithms determine the target position directly using redundant signal parameter measurements. Two-stage algorithms are much simpler and widely used. They permit to calculate the target position and velocity by*

[3] Here and in the sequel we use the term "triangulation" though the number of measured target bearings may be arbitrary (more than three). Perhaps, the term "multiangulation" would be more correct.

jointly processing the measurements of local ("primary") target coordinates obtained by spatially separated stations. We have discussed specific target coordinate information included in a measurement of each station or a pair of stations, depending on the type of a MSRS (monostatic or bistatic radars, autonomous or cooperative signal reception). We have presented equation (11.1) *and its approximations* (11.2), (11.3) *for the range sum of a target in the spherical coordinate system as well as* (11.4) *in a Cartesian system. The general expression* (11.8) *determines the signal Doppler frequency at an arbitrary receiving station when a moving target is illuminated by an arbitrary transmitting station. For a bistatic radar as a cell of a MSRS the echo Doppler frequency can be expressed in the simplest form if a target moves in the bistatic plane. Corresponding equation* (11.7) *permits to analyze the special features of the Doppler effect in a bistatic radar.*

Three main methods for the target localization using measured local target coordinates (at the second stage of two-stage algorithms) have been considered. They are elliptic, hyperbolic and triangulation methods. We have briefly discussed possible applications of all these methods (known from radionavigation) to MSRSs and have indicated the minimum number of stations in active and passive MSRSs required to determine three spatial coordinates of a target or a radiation source. We have shown how to obtain expressions for the target range difference from equations (11.1)–(11.4) *and differential Doppler frequency from equation* (11.8).

11.2. ONE-STAGE OPTIMUM COORDINATE MEASUREMENT IN ACTIVE AND PASSIVE MSRSs USING TEMPORAL SIGNAL PARAMETERS

As mentioned in Section 1.2, one of important advantages of MSRSs is the ability to determine three coordinates of a target with high accuracy using only temporal signal parameter measurements. In active MSRSs these parameters are signal TOAs determining target ranges or range sums with respect to spatially separated stations (see *the elliptic method* in Section 11.1). In passive MSRSs such temporal parameters are signal TDOAs at each pair of receiving stations determining range differences for a radiation source (see *hyperbolic method* in Section 11.1).

We consider here one-stage optimum algorithms for the target spatial position determination using temporal signal parameter measurements [58]. One-stage algorithms permit direct target localization in space using signal parameter estimation even with the redundant number of stations. These algorithms may determine the three target spatial coordinates themselves or three linearly independent TOAs, t_{s1}, t_{s2}, t_{s3}, or TDOAs, $\tau_{s12}, \tau_{s13}, \tau_{s14}$, in active and passive MSRSs, respectively (numbering of stations is arbitrary without any loss of generality). For a known geometry of a MSRS the problem of the target three coordinate determination in the latter case is then reduced to the problem of conversion of variables. As well known, such a conversion is mutually single-valued if its Jackobian is non-zero. Owing to one-to-one correspondence, the optimality of signal TOA or TDOA estimates in accordance with any criterion ensures the optimality of coordinate estimates according to the same criterion.

A priori probability distributions of signal parameters to be estimated, are usually much wider than their likelihood functions. Therefore, we adopt *the maximum likelihood criterion* for estimate optimization. The signal and interference models are assumed to be the same as for the detection problems (see Section 4.1).

An Active MSRS Containing One Transmitting and m Receiving Stations

Consider first in more details three linearly independent TOA estimation. We make use of the expression (4.3) for a signal at the input of the ith station (at $\Omega_{si}=0$, $i=\overline{1,m}$) and of the expression (4.50) for the likelihood ratio in the frequency domain. Interferences are assumed to be spatially correlated in the general case.

The vector $\boldsymbol{\Theta}$ in (4.50) contains m useful parameters $(t_{s1},\ldots,t_{sm})$ and $2m$ stray parameters $(a_{s1},\ldots,a_{sm},\varphi_{s1},\ldots,\varphi_{sm})$. Let us separate them as it was done in Section 5.2. Using (5.25)–(5.29), the likelihood ratio logarithm with respect to all unknown parameters can be written as follows:

$$L=\mathbf{a}_s^t\,\mathrm{Re}[\tilde{\mathbf{E}}^*(\boldsymbol{\varphi}_s)\mathbf{G}(t_{s1},\ldots,t_{sm})]-0.5\mathbf{a}_s^t\tilde{\mathbf{E}}^*(\boldsymbol{\varphi}_s)\mathbf{C}(t_{s1},\ldots,t_{sm})\tilde{\mathbf{E}}(\boldsymbol{\varphi}_s)\mathbf{a}_s. \qquad (11.9)$$

The vector $\mathbf{G}$ and the matrix $\mathbf{C}$ [see (5.28)] depend only on the signal TOAs (useful parameters) while $\tilde{\mathbf{E}}(\varphi_s)$ and $\mathbf{a}_s$ are the diagonal matrix and the vector of stray parameters [see (5.26)]. *Optimum estimates of all parameters maximize L.*

Each combination of t_{s1},t_{s2},t_{s3} determines a point in space (singular cases are excluded). This point in its turn determines certain values of the TOAs, t_{si}, at remaining $m-3$ stations. It means that *a priori* dependence exists between true values of TOAs from a target to m stations. All t_{si} for $i>3$ can be expressed by means of t_{s1},t_{s2},t_{s3}:

$$t_{si}=h_i(t_{s1},t_{s2},t_{s3}),\quad i>3 \qquad (11.10)$$

where $h_i(\cdot)$ are the known functions determined by the geometry of a MSRS[4]. Thus only three TOAs, t_{s1},t_{s2},t_{s3}, are to be estimated among the unknown TOAs of signals received by all stations. Hence (11.9) can be rewritten

$$L=\mathbf{a}_s^t\,\mathrm{Re}[\tilde{\mathbf{E}}^*(\varphi_s)\mathbf{G}_0(t_{s1},t_{s2},t_{s3})]-0.5\mathbf{a}_s^t\tilde{\mathbf{E}}^*(\varphi_s)\mathbf{C}_0(t_{s1},t_{s2},t_{s3})\tilde{\mathbf{E}}(\varphi_s)\mathbf{a}_s. \qquad (11.11)$$

By $\mathbf{G}_0$ and $\mathbf{C}_0$ are denoted $\mathbf{G}$ and $\mathbf{C}$, respectively, where t_{si} for $i>3$ are replaced by the functions $h_i(t_{s1},t_{s2},t_{s3})$ in accordance with (11.10).

Further specifying of optimum algorithms depends on the type of MSRSs and the properties of signals. The problem under consideration is most interesting for *MSRSs with short-term spatial coherence and spatially incoherent MSRSs.* As discussed in Section 1.1 a spatial coherent MSRS may be considered as a "monostatic radar" with a large thinned antenna array.

Let *r.m.s. signal values be fully mutually dependent.* Then we assume as in (5.34) and (5.37)

$$a_{si}=A_{i1}a_{s1},\quad i=\overline{1,m};\qquad \mathbf{a}_s=a_{s1}\mathbf{A},\quad \mathbf{A}^t=(1,A_{21},\ldots,A_{m1}) \qquad (11.12)$$

where A_{i1}, $i=\overline{1,m}$, are the known quantities depending on the ratios of receiver sensitivities, antenna gains and other energy parameters of the ith and the first stations. Substituting (11.12) in (11.11) permits to find the maximum likelihood

[4] For example, the equations for calculating three Cartesian coordinates of a target for given range sums with respect to three receiving stations arbitrary positioned on a plane (and hence for given t_{s1},t_{s2},t_{s3}) are presented in [46]. Having calculated the target three coordinates, one can obtain t_{si} using (11.4).

estimate of a_{s1}:

$$\hat{a}_{s1} = \frac{\mathbf{A}'\,\mathrm{Re}[\tilde{\mathbf{E}}^*(\varphi_s)\mathbf{G}_0(t_{s1}, t_{s2}, t_{s3})]}{\mathbf{A}'\tilde{\mathbf{E}}^*(\varphi_s)\mathbf{C}_0(t_{s1}, t_{s2}, t_{s3})\tilde{\mathbf{E}}(\varphi_s)\mathbf{A}}. \tag{11.13}$$

The substitution of this estimate in (11.11) taking into account (11.12) yields

$$L = \frac{0.5\{\mathbf{A}'\,\mathrm{Re}[\tilde{\mathbf{E}}^*(\varphi_s)\mathbf{G}_0(t_{s1}, t_{s2}, t_{s3})]\}^2}{\mathbf{A}'\tilde{\mathbf{E}}^*(\varphi_s)\mathbf{C}_0(t_{s1}, t_{s2}, t_{s3})\tilde{\mathbf{E}}(\varphi_s)\mathbf{A}}. \tag{11.14}$$

An explicit expression for the maximum likelihood estimates of $\varphi_s' = (\varphi_{s1}, \ldots, \varphi_{sm})$ can be derived assuming the matrix $\mathbf{C}_0$ in (11.14) to be diagonal. This assumption is valid when interferences are spatially uncorrelated (see Section 5.2). The conditions under which the matrix $\mathbf{C}$ (and $\mathbf{C}_0$) turns out to be approximately diagonal in the case of spatially correlated interferences are considered in Section 8.3 [see (8.42)]. If $C_{ik} = 0$, $i \neq k$, the denominator in (11.14) does not depend on φ_{si} since the matrix $\tilde{\mathbf{E}}^*(\varphi_s)$ is a diagonal one, and $\tilde{\mathbf{E}}^*(\varphi_s)\tilde{\mathbf{E}}(\varphi_s) = \mathbf{I}$. It follows from (8.35) that C_{ii} does not also depend on TOAs (since $\tau_{sii} = 0$). Then the maximum likelihood estimate of φ_{si} takes the form

$$\hat{\varphi}_{si} = -\arg G_{0i}(t_{s1}, t_{s2}, t_{s3}). \tag{11.15}$$

Substituting (11.15) in (11.14), dropping multipliers independent of t_{si} and omitting the squaring, yields the optimum algorithm

$$L = \mathbf{A}'|\mathbf{G}_0(t_{s1}, t_{s2}, t_{s3})| \to \max(t_{s1}, t_{s2}, t_{s3}). \tag{11.16}$$

"Arguments" of "max" indicate the variables over which a maximum is searched. Using scalar notation we can rewrite (11.16) as follows:

$$L = \sum_{i=1}^{3} A_{i1}|G_i(t_{si})| + \sum_{i=4}^{m} A_{i1}|G_i[h_i(t_{s1}, t_{s2}, t_{s3})]| \to \max(t_{s1}, t_{s2}, t_{s3}). \tag{11.17}$$

In the case of spatially correlated interferences and MSRSs with short-term spatial coherence, $G_i(t_{si})$, $i = \overline{1, m}$, are the results of joint coherent processing in each receiving station of the sums of self-noises, interferences and, possibly, target echoes received by all stations [see (8.35)]. It was shown in Section 8.2, that such joint processing results in cancelling the spatially correlated interferences and target echo accumulation (integration). When either interferences are spatially uncorrelated or MSRSs are spatially incoherent, $G_i(t_{si})$, $i = \overline{1, m}$, are the results of optimum signal filtration (according to the maximum SNR criterion) at each station individually [see (5.31)].

Since there exists one-to-one correspondence between a combination of estimates $\hat{t}_{s1}, \hat{t}_{s2}, \hat{t}_{s3}$ and a point in the three-dimensional space, the optimum algorithm (11.17) requires to find such a point for which the sum (11.17) reaches its maximum (or several such points in the case of several targets). The optimum processing may be implemented as follows. At first, the same processing as in optimum detectors is carried out for corresponding expected signals, and the quantities G_i are generated. The values of moduli, $|G_i|$, in the vicinities of their maxima (i.e. of partial likelihood function peaks) are stored. From the stations number 1, 2 and 3 (arbitrary stations where peaks of $|G_i|$ are sufficiently large), the values of $\hat{t}_{s1}$, $\hat{t}_{s2}$, and $\hat{t}_{s3}$ corresponding to maxima of $|G_1(t_{s1})|$, $|G_2(t_{s2}|$ and $|G_3(t_{s3})|$ are used for calculating $\hat{t}_{si}$ for $i>3$ in accordance with (11.10). This may be done by solving the nonlinear equations from Section 11.1 or using corresponding geometrical relationships presented in [46] and other sources (e.g. [259]). Then the sum (11.17) is evaluated. This procedure is repeated for different combination of $\hat{t}_{s1}$, $\hat{t}_{s2}$ and $\hat{t}_{s3}$ close to maxima of $|G_1(t_{s1})|$, $|G_2(t_{s2})|$ and $|G_3(t_{s3})|$. Such a combination which provides a maximum L in (11.17) is adopted as an optimum estimate. When several targets are within an observed region of space, several optimum combinations will be obtained. To separate "useful" estimates from false alarms, the obtained maximum values of L are to be compared with a detection threshold (see Section 5.2).

Having obtained optimum (maximum likelihood) estimates $\hat{t}_{s1}$, $\hat{t}_{s2}$ and $\hat{t}_{s3}$ one can calculate corresponding optimum (maximum likelihood) estimates of the target spatial coordinates by using the same geometrical relationships.

Owing to one-to-one correspondence between triads $(\hat{t}_{s1}, \hat{t}_{s2}, \hat{t}_{s3})$ and points in space with spatial coordinates (x, y, z), an alternate implementation in the general case (when $m>3$) is possible. An equivalent expression for the optimum algorithm (11.17) is given by

$$L=\sum_{i=1}^{m} A_{i1}|G_i[t_{si}(x, y, z)]| \rightarrow \max(x, y, z). \tag{11.17'}$$

Of course, the algorithm (11.17′) can be derived directly from (11.11) in a similar manner as (11.17) if t_{si}, $i=\overline{1,m}$, in (11.11) are considered as functions of spatial coordinates: $t_{si}=t_{si}(x, y, z)$.

The algorithm (11.17′) implies scanning over the region of space where a target may be *a priori*, estimating $|G_i(t_{si})|$ for all $i=\overline{1,m}$, and evaluating the sum in (11.17′) for each point (x, y, z) within this region [using, for example, (11.4)]. Such a point [with the coordinates $(\hat{x}, \hat{y}, \hat{z})$] for which L in (11.17′) reaches its maximum, is adopted as an optimal estimate of the target spatial coordinates.

Though this approach seems to be simpler in the computational sense, it requires to search over all points within the whole region of space where a target (or targets) may be *a priori*. At the same time the former implementation permits to reduce this search region to one or several small regions in the vicinities of maxima of $|G_1|$, $|G_2|$, $|G_3|$.

Apparently, the best technique combines both approaches described above. At first, the values of $\hat{t}_{s1}$, $\hat{t}_{s2}$ and $\hat{t}_{s3}$ corresponding to maxima of $|G_1(t_{s1})|$, $|G_2(t_{s2})|$ and $|G_3(t_{s3})|$ are determined and used for calculating coordinates of the initial point $(\hat{x}_1, \hat{y}_1, \hat{z}_1)$. Then the sum in (11.17′) is evaluated for each point in the vicinity of $(\hat{x}_1, \hat{y}_1, \hat{z}_1)$. The point $(\hat{x}, \hat{y}, \hat{z})$ for which the sum in (11.17′) reaches its maximum, is declared the optimum (maximum likelihood) estimate of the target position in space. The functions $t_{si}(x, y, z)$ in (11.17′) may be expressed, for example, as follows: $t_{si}(x, y, z)=(1/c)R_{\Sigma ji}$ where c is the light velocity and $R_{\Sigma ji}$ is determined by (11.4).

If the peaks $|G_1(\hat{t}_{s1})|$, $|G_2(\hat{t}_{s2})|$ and $|G_3(\hat{t}_{s3})|$ do not originate from a common target (they may be caused by noise or clutter), the sums in (11.17) and (11.17′) will be small. In this case other combination of $\hat{t}_{s1}$, $\hat{t}_{s2}$ and $\hat{t}_{s3}$ should be tested.

If $m\leqslant 3$ only the first sum remains in (11.17). Since t_{s1}, t_{s2}, t_{s3} are independent parameters, this sum reaches its maximum when each term (summand) has its own maximum. Therefore, the optimum three-dimensional algorithm breaks down into three one-dimensional algorithms

$$L=|G_i(t_{si})| \rightarrow \max(t_{si}), \quad i=1, 2, 3. \tag{11.18}$$

In this case the one-stage target location algorithm coincides with the two-stage algorithm (see Section 11.3). Accordingly, only those maxima of $|G_i(t_{si})|$ are to be taken into account which exceed a detection threshold.

For *mutually independent target echo r.m.s. values at the inputs of receiving stations* we assume again the matrix $\mathbf{C}_0$ in (11.11) to be diagonal. Then the maximum likelihood estimates for a_{si} and φ_{si} can be obtained from (11.11)

$$\hat{\varphi}_{si} = -\arg G_{0i}(t_{s1}, t_{s2}, t_{s3}); \quad \hat{a}_{si} = |G_{0i}(t_{s1}, t_{s2}, t_{s3})|/C_{ii}. \tag{11.19}$$

Substituting (11.19) in (11.11) yields the optimum algorithm:

$$L = \sum_{i=1}^{3} C_{ii}^{-1} |G_i(t_{si})|^2 + \sum_{i=4}^{m} C_{ii}^{-1} |G_i[h_i(t_{s1}, t_{s2}, t_{s3})]|^2 \to \max(t_{s1}, t_{s2}, t_{s3}). \tag{11.20}$$

The algorithm (11.20) is similar to (11.17), but the moduli of G_i are replaced by their squared moduli and the weights A_{i1} are replaced by C_{ii}^{-1}. All above considerations concerning the implementation of (11.17) are valid for (11.20) including the other form of the optimum algorithm

$$L = \sum_{i=1}^{m} C_{ii}^{-1} |G_i[t_{si}(x, y, z)]|^2 \to \max(x, y, z). \tag{11.20'}$$

A Passive MSRS with Short-term Spatial Coherence Containing m Receiving Stations

We assume stochastic signals and interferences to be stationary and stationary coupled Gaussian random processes as before. The likelihood ratio with respect to the vector $\mathbf{\Theta}$ of unknown parameters can be written from (4.53)

$$\Lambda = \frac{\mathfrak{R}_1(\mathbf{\Theta})}{\mathfrak{R}_0} \exp\left\{\frac{1}{4\pi} \int_{-\infty}^{\infty} \boldsymbol{\chi}^*(\omega)[\mathbf{f}(\omega) - \mathbf{f}_{s1}(\omega, \mathbf{\Theta})]\boldsymbol{\chi}(\omega)\, d\omega\right\}. \tag{11.21}$$

It can be shown that in the general case (taking into account possible dependence $\mathfrak{R}_1$ on $\mathbf{\Theta}$) a likelihood equation system determining the optimum estimate of n-dimensional vector $\mathbf{\Theta}$ is given by

$$\sum_{i=1}^{m} \sum_{k=1}^{m} \frac{1}{2\pi} \int_{-\infty}^{\infty} [2T \partial \Phi^*_{slik}(\omega, \mathbf{\Theta})/\partial \Theta_l] f_{slik}(\omega, \mathbf{\Theta})$$

$$+ \chi_i^*(\omega)\chi_k(\omega)[\partial f_{slik}(\omega, \mathbf{\Theta})/\partial \Theta_l]\, d\omega = 0, \quad l = \overline{1, n} \tag{11.22}$$

where $\Phi_{slik}(\omega, \mathbf{\Theta})$, $f_{slik}(\omega, \mathbf{\Theta})$, $\chi_i(\omega)$ are the elements of the matrices $\mathbf{\Phi}_{sl}(\omega, \mathbf{\Theta})$, $\mathbf{f}_{sl}(\omega, \mathbf{\Theta})$ and the vector $\boldsymbol{\chi}(\omega)$. Consider the most practically important case where *interference spatial correlation is absent or may be neglected and the signal correlation matrix is quasideterministic*. The reasons for neglecting interference spatial correlation in passive MSRSs were discussed in Section 9.3.

As for regular signals, consider at first estimation of three linearly independent TDOAs. Though in a MSRS containing m receiving stations there are $m(m-1)/2$ pairs of stations, and $m(m-1)/2$ different TDOAs, $\tau_{sik}, i, k = \overline{1, m}$, $i < k$, may be measured, only $m-1$ TDOAs among them are linearly independent [see (7.8)]. Let us denote these TDOAs by $\tau_{s12}, \ldots, \tau_{s1m}$ (numbering of stations is arbitrary).

In the three-dimensional physical space three linearly independent TDOAs determine all remaining ones. For a known geometry of a MSRS we have the relationships similar to (11.10):

$$\tau_{sik}=\bar{h}_{ik}(\tau_{s12},\tau_{s13},\tau_{s14}),\quad \begin{cases} k\geq 5, & i=1; \\ k\geq i+1, & i>1, \end{cases} \quad i,k=\overline{1,m}. \tag{11.23}$$

As in Section 7.2, we assume a signal model with *mutually independent initial phases of the correlation matrix. Signal intensities (variances) we set first to be known.* Under these assumptions the vector Θ contains $m(m-1)/2$ noninformative phase differences and three informative TDOAs

$$\Theta=(\Delta\varphi_{s12},\ldots,\Delta\varphi_{sm(m-1)},\tau_{s12},\tau_{s13},\tau_{s14}). \tag{11.24}$$

Assuming the noises to be white, we can employ expressions (7.2)–(7.7) for $\Phi_{snik}(\omega,\Theta)$, and $f_{snik}(\omega,\Theta)$ (the subscript "sI" is replaced by "sn"). When the vector Θ of unknown parameters is expressed by (11.24), the first term under the integral in (11.22) is equal to zero since $\Re_1$ in (11.21) does not depend on Θ (all unknown parameters do not relate to signal energy). Substituting (7.5) we can transform (11.22) to the form

$$L=\mathrm{Re}\sum_{k=2}^{4}\frac{q_{s1}q_{sk}\exp(j\Delta\varphi_{s1k})}{\sqrt{N_1N_k}}\frac{1}{2\pi}\int_{-\infty}^{\infty}\chi_1^*(\omega)\chi_k(\omega)|H(\omega)|^2\exp(j\omega\tau_{s1k})\,d\omega$$

$$+\mathrm{Re}\sum_{\substack{i=1\\ \text{for } i=1}}^{m-1}\sum_{\substack{k=i+1\\ k\neq 2,3,4}}^{m}\frac{q_{si}q_{sk}\exp(j\Delta\varphi_{sik})}{\sqrt{N_iN_k}}\frac{1}{2\pi}\int_{-\infty}^{\infty}\chi_i^*(\omega)\chi_k(\omega)|H(\omega)|^2$$

$$\times\exp[j\omega\bar{h}_{ik}(\tau_{s12},\tau_{s13},\tau_{s14})]\,d\omega\rightarrow\max(\Delta\varphi_{s12},\ldots,\Delta\varphi_{s(m-1)m},\tau_{s12},\tau_{s13},\tau_{s14}) \tag{11.25}$$

where the squared modulus of the filter frequency response is given by

$$|H(\omega)|^2=\frac{F_s(\omega-\omega_0)}{1+F_s(\omega-\omega_0)q_{s\Sigma}^2};\quad q_{s\Sigma}^2=\sum_{i=1}^{m}q_{si}^2. \tag{11.26}$$

Here $F_s(\omega-\omega_0)$ is the normalized PSD of the signal complex envelope [see (4.11)]; ω_0 is the carrier frequency. Maximum L in (11.25) for independent $\Delta\varphi_{sik}$ is reached when each term of the sum has its own maxima. Hence a maximum likelihood estimates of $\Delta\varphi_{sik}$ are as follows:

$$\widehat{\Delta\varphi}_{sik}=-\arg\left[\frac{1}{2\pi}\int_{-\infty}^{\infty}\chi_i^*(\omega)\chi_k(\omega)|H(\omega)|^2\exp(j\omega\tau_{sik})\,d\omega\right] \tag{11.27}$$

where τ_{sik} for $i=1$, $k\geq 5$ and for $i>1$, $k\geq i+1$ are determined by (11.23). To obtain an optimum (maximum likelihood) algorithm for estimating τ_{s12}, τ_{s13} and τ_{s14}, we have to substitute (11.27) in (11.25). The optimum estimation algorithm is given by

$$L = \sum_{k=2}^{4} \frac{q_{s1}q_{sk}}{\sqrt{N_1 N_k}} \left| \frac{1}{2\pi} \int_{-\infty}^{\infty} \chi_1^*(\omega)\chi_k(\omega)|H(\omega)|^2 \exp(j\omega\tau_{s1k})\, d\omega \right|$$

$$+ \sum_{\substack{i=1 \\ \text{for } i=1}}^{m-1} \sum_{\substack{k=i+1 \\ k\neq 2,3,4}}^{m} \frac{q_{si}q_{sk}}{\sqrt{N_i N_k}} \left| \frac{1}{2\pi} \int_{-\infty}^{\infty} \chi_i^*(\omega)\chi_k(\omega)|H(\omega)|^2 \right.$$

$$\left. \times \exp[j\omega\bar{h}_{ik}(\tau_{s12}, \tau_{s13}, \tau_{s14})]\, d\omega \right| \rightarrow \max(\tau_{s12}, \tau_{s13}, \tau_{s14}). \qquad (11.28)$$

Thus the one-stage algorithm (11.28) for optimum estimation of the independent signal TDOAs, $\tau_{s12}, \tau_{s13}, \tau_{s14}$, requires performing the following procedures. The input signal at each station undergoes preliminary filtration. From the outputs of the filters these signals from all stations are transferred to a FC and are applied to the inputs of correlators for all $m(m-1)/2$ pairs of stations. Then the values of $\hat{\tau}_{s12}, \hat{\tau}_{s13}, \hat{\tau}_{s14}$ are determined. They correspond to maxima of the moduli of estimates of correlation between signals received by the first and the second, the first and the third and the first and the fourth stations. Numbering of pairs of stations is arbitrary. It is reasonable to use those pairs where maxima of the moduli are large and which have the largest effective baselengths. Having $\hat{\tau}_{s12}, \hat{\tau}_{s13}, \hat{\tau}_{s14}$ the point in space corresponding to this triad is determined and the values of functions $\bar{h}_{ik}(\hat{\tau}_{s12}, \hat{\tau}_{s13}, \hat{\tau}_{s14})$ are calculated. For the known geometry of a MSRS this may be done using relationships from Section 11.1, corresponding expressions from [46] and other publications. Then the quantity L is evaluated according to (11.28). By suitable selection of combinations of $\hat{\tau}_{s12}, \hat{\tau}_{s13}, \hat{\tau}_{s14}$ a maximum of L is achieved. The combination of $\hat{\tau}_{s12}, \hat{\tau}_{s13}, \hat{\tau}_{s14}$ for which L reaches its maximum, is adopted as an optimum estimate. This combination determines an optimum estimate of the radiation source's three spatial coordinates. Of course, only those maxima of L are taken into account which exceed a predetermined detection threshold.

As for regular signals, owing to one-to-one correspondence between the set of triads $\hat{\tau}_{s12}, \hat{\tau}_{s13}, \hat{\tau}_{s14}$ and the set of points with spatial coordinates $(\hat{x}, \hat{y}, \hat{z})$ in the three-dimensional space, the optimum algorithm (11.28) can be written in another equivalent form

$$L = \sum_{i=1}^{m-1} \sum_{k=i+1}^{m} \frac{q_{si}q_{sk}}{\sqrt{N_i N_k}} \left| \frac{1}{2\pi} \int_{-}^{\infty} \chi_i^*(\omega)\chi_k(\omega)|H(\omega)|^2 \exp[j\omega\tau_{sik}(x, y, z)]\, d\omega \right|$$

$$\rightarrow \max(x, y, z). \qquad (11.28')$$

The algorithm (11.28′) can be derived directly from (11.21), (11.22) in a similar manner as (11.28) if τ_{sik}, $i, k = \overline{1, m}$, $i < k$, are expressed as functions of spatial coordinates: $\tau_{sik} = \tau_{sik}(x, y, z)$.

The implementation of (11.28′) implies scanning in spatial coordinates over the region where a radiation source (or several radiation sources) may be positioned. For each point (x, y, z) of this region the corresponding TDOAs $\tau_{sik}(x, y, z)$ are calculated using relationships presented in Section 11.1 $[\tau_{sik}(x, y, z) = (1/c)\Delta R_{ik} = (1/c)(R_{\Sigma jk} - R_{\Sigma ji})$ where $R_{\Sigma ji}$ is determined by (11.4)]. Such a point with the coordinates $(\hat{x}, \hat{y}, \hat{z})$ for which L in (11.28′) reaches its maximum, and this maximum exceeds the threshold, is adopted as an optimum estimate of a source position in space.

This approach requires analyzing all points in the whole region where radiation sources may be. The former approach according to (11.28) permits to reduce the search region in space to one or several small regions in the vicinities of points determined by the values of $\hat{\tau}_{s12}, \hat{\tau}_{s13}, \hat{\tau}_{s14}$ corresponding to maxima of three cross-correlation estimate moduli. In this case the search for cross-correlation envelope maxima is necessary. However the algorithm (11.28) is more complicated in the computational sense than (11.28′). Probably, like the case of regular signals (see above), the best implementation is a combination of both approaches. At first, large correlation peaks should be found, values of $\hat{\tau}_{s12}, \hat{\tau}_{s13}, \hat{\tau}_{s14}$ and the spatial coordinates $(\hat{x}, \hat{y}, \hat{z})$ of corresponding points should be determined. Then the algorithm (11.28′) may be employed in the vicinities of those points.

We now turn to the case where *signal intensities are unknown* (*random*). As in Section 7.2, we consider the most realistic situation for which intensity fluctuations at the inputs of different stations are completely mutually dependent. Then the vector $\boldsymbol{\Theta}$ in (11.24) has an additional parameter which may (without loss of generality) be assumed to be the SNR at the first station, q_{s1}^2. Then $q_{si}^2 = A_{i1}^2 q_{s1}^2$, $i = \overline{2, m}$, where A_{i1}^2 are the known quantities. The substitution of this expression in (11.22), (11.25) shows that q_{s1}^2 is a common factor which does not influence on the estimates of $\Delta\varphi_{sik}$ and $\tau_{s12}, \tau_{s13}, \tau_{s14}$. It means that the algorithm (11.28) and its equivalent (11.28′) are optimal also for unknown completely mutually dependent signal intensities if unknown quantities q_{si} are replaced by A_{i1}, $i = \overline{2, m}$.

When signal differential Doppler frequencies (DDFs) at the correlator inputs cannot be neglected (i.e. when $\Delta\Omega_{sik} T \geqslant 2\pi$ where T is the correlator integration time which should be equal to signal duration), then, repeating the considerations from Section 7.1 which led to (7.16), we obtain instead of (11.28) and (11.28′) following expressions (for instance, when signal intensities are unknown):

$$L = \sum_{k=2}^{4} \frac{A_{k1}}{\sqrt{N_1 N_k}} \left| \frac{1}{2\pi} \int_{-\infty}^{\infty} \chi_1^*(\omega) \chi_k(\omega + \Delta\Omega_{s1k}) |H_1(\omega)|^2 \exp[\mathrm{j}(\omega - \Omega_{s1}) \tau_{s1k}] \, d\omega \right|$$

$$+ \sum_{\substack{i=1 \\ \text{for } i=1}}^{m-1} \sum_{\substack{k=i+1 \\ k \neq 2,3,4}}^{m} \frac{A_{i1} A_{k1}}{\sqrt{N_i N_k}} \left| \frac{1}{2\pi} \int_{-\infty}^{\infty} \chi_i^*(\omega) \chi_k(\omega + \Delta\Omega_{sik}) |H_i(\omega)|^2 \right.$$

$$\left. \times \exp[\mathrm{j}(\omega - \Omega_{si}) \tilde{h}_{ik}(\tau_{s12}, \tau_{s13}, \tau_{s14})] \, d\omega \right| \to \max(\tau_{s12}, \tau_{s13}, \tau_{s14}). \tag{11.29}$$

$$L = \sum_{i=1}^{m-1} \sum_{k=i+1}^{m} \frac{A_{i1} A_{k1}}{\sqrt{N_i N_k}} \left| \frac{1}{2\pi} \int_{-\infty}^{\infty} \chi_i^*(\omega) \chi_k(\omega + \Delta\Omega_{sik}) |H_i(\omega)|^2 \right.$$

$$\left. \times \exp[\mathrm{j}(\omega - \Omega_{si}) \tau_{sik}(x, y, z)] \, d\omega \right| \to \max(x, y, z). \tag{11.29′}$$

It can be seen that additional operations appear: eliminating the signal DDFs at the correlator inputs and tuning the filter in each station according to the frequency of expected input signal [the expression for $|H_i(\omega)|^2$ see in (7.17)].

As mentioned above, the optimal estimate for the triad $(\hat{\tau}_{s12}, \hat{\tau}_{s13}, \hat{\tau}_{s14})$ completely determines the optimal estimates for the three spatial coordinates $(\hat{x}, \hat{y}, \hat{z})$ of a radiation source and vice versa.

In this section, *we have synthesized one-stage algorithms for optimum target localization in space, i.e. maximum likelihood estimation of spatial coordinates of a target in both active and passive MSRSs using signal temporal parameter measurements.*

For active MSRSs and target echoes with random (noninformative) initial phases and random amplitudes (r.m.s. values) we have derived algorithms (11.17) *and* (11.20) *for completely mutually dependent and mutually independent amplitudes, respectively. These algorithms determine a triad of TOA estimates* $(\hat{t}_{s1}, \hat{t}_{s2}, \hat{t}_{s3})$ *having the single-valued relation to the target spatial coordinate estimates (e.g.,* $\hat{x}, \hat{y}, \hat{z}$*). We have presented the equivalent algorithms* (11.17′) *and* (11.20′) *which determine the optimum spatial coordinate estimates* $(\hat{x}, \hat{y}, \hat{z})$ *directly. Advantages and drawbacks of the implementation of those algorithms have been discussed.*

For passive MSRSs we have synthesized algorithms (11.28) *and* (11.28′) *for optimum localization of a radiation source in space. Algorithm* (11.28) *determines a maximum likelihood estimate for the triad of TDOAs* $(\hat{\tau}_{s12}, \hat{\tau}_{s13}, \hat{\tau}_{s14})$ *having single-valued relation to the source spatial coordinate estimates (e.g.,* $\hat{x}, \hat{y}, \hat{z}$*). The equivalent algorithm* (11.28′) *determining the spatial coordinates* $(\hat{x}, \hat{y}, \hat{z})$ *directly, represents another implementation of the optimum source localization. Having utilized the same signal model as in Section* 7.2 *(stochastic signals with quasideterministic correlation matrix and mutually independent phases of that matrix) we have also derived algorithms* (11.29) *and* (11.29′) *for the case where signal intensities at the inputs of receiving stations are unknown (random) but completely mutually dependent.*

11.3. OPTIMUM SIGNAL TOA MEASUREMENT FOR TWO-STAGE TARGET POSITION ESTIMATION IN ACTIVE MSRSs. SIGNAL MULTIPLICATION AFTER SPATIALLY CORRELATED INTERFERENCE CANCELLATION

As noted in Section 11.1, the two-stage target position estimation implies "primary" or "local" target coordinate measurement with respect to spatially separated stations (or pairs of stations) at the first stage, and "final" or "global" target position estimation at the second stage using redundant measurements obtained at the first stage. To achieve high accuracy for a final estimate, precise estimates of local coordinates at the first stage are required. We consider here optimum (maximum likelihood) estimation at the first stage of two-staged target localization algorithms based on signal temporal parameter measurements. For active MSRSs, this means a measurement algorithm for signal TOAs that are proportional to the target ranges or range sums.

Optimum Estimation of Signal TOAs in Active MSRSs

Let *a MSRS with the short-term spatial coherence or a spatially incoherent MSRS* (see Section 1.1) contain a single transmitting and m receiving stations arbitrary positioned. Unlike the one-stage algorithm, target echo TOAs, $t_{s1}, \ldots, t_{sm}$ at all the m receiving stations are to be estimated.

We make use of the expression (11.9) for the likelihood ratio logarithm and assume (as in Section 11.2) the matrix **C** to be diagonal. The second addend in (11.9) does not depend on received signals [see (8.35)], and when $C_{ik}=0$, $i \neq k$, this addend is independent of both φ_{si} and t_{si}, $i=\overline{1,m}$. Hence it may be dropped. As a result we have

$$L = \sum_{i=1}^{m} a_{si} \operatorname{Re}[\exp(j\varphi_{si}) G_i(t_{si})] \tag{11.30}$$

where $G_i(t_{si})$ is defined in (8.35) for the case where external spatially correlated interferences are present and in (5.31) for the case where they are absent. To obtain an optimum algorithm for t_{si} estimation, it is necessary to exclude stray parameters which should be replaced by their maximum likelihood estimates. For MSRSs under consideration, signal initial phases, φ_{si}, $i=\overline{1,m}$, are mutually statistically independent. Clearly, the maximum likelihood estimates for mutually independent phases, φ_{si} [maximizing L in (11.30)], are

$$\hat{\varphi}_{si} = -\arg G_i(t_{si}).$$

Substituting $\hat{\varphi}_{si}$ in (11.30) yields

$$L = \sum_{i=1}^{m} a_{si}|G_i(t_{si})|. \tag{11.31}$$

Unlike the one-stage algorithms (11.16), (11.17), (11.17′), (11.20) and (11.20′), all t_{si}, $i=\overline{1,m}$, are assumed to be independent parameters. Hence maximizing the sum (11.31) means maximizing each term of the sum. Thus an optimum estimation algorithm takes the form

$$L = |G_i(t_{si})| \rightarrow \max(t_{si}), \quad i=\overline{1,m}. \tag{11.32}$$

It is seen that the optimum (maximum likelihood) estimate of the signal TOA at each receiving station (when signal initial phases are unknown or random) is the time when the output signal modulus (envelope) has its maximum. This output signal is obtained after coherent signal integration (matched filtration) in a background of noises and, if necessary, after external spatially correlated interference cancellation [see (5.31) and (8.35)]. In the former case this algorithm is the same as for a monostatic radar. Using algorithm (11.32) yields *m maximum likelihood estimates* of TOAs, $\hat{t}_{si}$, $i=\overline{1,m}$, which are used at the second stage of target localization algorithms. It is seen that these optimum estimates, do not depend on signal r.m.s. values a_{si}.

Signal "Multiplication" after Spatially Correlated Interference Cancellation

When external mainlobe interferences (e.g., jamming) are incident upon an MSRS and spatially correlated interference cancellation techniques are used (see Chapters 8, 9), there may arise gross (anomalous) errors in target localization. Such errors may appear at the outputs of two-stage algorithms employing optimum TOA measurements (11.32) and under certain conditions even at the outputs of one-stage algorithms (11.16), (11.17), (11.17′), (11.20) and (11.20′).

As shown in Sections 8.1–8.3, the quantities $|G_i|$, $i=\overline{1,m}$, used for t_{si} estimation, are obtained after the coherent subtraction of spatially correlated interferences received by all stations. In the process of this subtraction (interference cancellation) target echoes (signals) coming from all stations are summed. Under the condition that signals and interferences are resolved in TDOAs at the inputs of receivers, several signals originated from a single target appear at the output of the interference cancellation filter in each station. Just under this condition the matrix $\mathbf{C}$ in (11.9) is diagonal [see (8.42) in Section 8.2 and corresponding discussion there] and the measurement algorithms (11.16), (11.17), (11.17′), (11.20), (11.20′) and (11.32) have been derived. In this case each quantity $|G_i|$, $i=\overline{1,m}$ has several peaks of approximately equal magnitudes. Among these maxima only one (not necessarily the greatest) is near the true value of t_{si} while others originate from target echoes received by other stations and hence may be significantly displaced in time. This effect can be easier understood with the help of a simple example.

Example 11.1. For clarity of exposition consider the optimum TOA estimation for a two-stage target localization algorithm in a MSRS with the short-term spatial coherence and containing a single transmitting and two receiving stations. A spatially correlated interference from a single jammer enters through antenna mainlobes. The interference PSD (or correlation) matrix at the inputs of stations is assumed to be known. Such an assumption simplifies derivations and does not affect output results. In real situations when this matrix is unknown, adaptive interference cancellation is used (see Section 9.1, 9.2). Let us analyze the optimum TOA estimate according to the algorithm (11.32) in the first station ($i=1$). Substituting $G_1(t_{s1})$ from (8.35) in (11.32) and ignoring the influence of the antenna ADP [i.e. assuming $g_1(\beta_{s1}, \varepsilon_{s1}, \omega)=1$] yields

$$L=\left|\frac{1}{2\pi}\int_{-\infty}^{\infty}\Psi_0^*(\omega-\omega_0)\exp(j\omega t_{s1})\sum_{k=1}^{2}f_{1k}(\omega)\chi_k(\omega)\,d\omega\right|\rightarrow\max(t_{s1}). \quad (11.33)$$

Now we can employ the expression (8.61) for $f_{ik}(\omega)$ setting $i=1$ and $m=2$. For the sake of clarity assume $\Delta\varphi_{12}=0$ and $q^2\gg1$ (large INR). Then

$$f_{ik}(\omega)=\frac{1}{N}[\delta_{1k}-0.5\exp(j\omega\tau_{1k})],\quad k=1,2;\ \omega\in(\omega_0-\Delta\omega_s/2,\omega_0+\Delta\omega_s/2). \quad (11.34)$$

Substituting (11.34) in (11.33) and dropping the multiplier $1/N$ yields

$$L=\left|\frac{1}{2\pi}\int_{\omega_0-\Delta\omega_s/2}^{\omega_0+\Delta\omega_s/2}\Psi_0^*(\omega-\omega_0)\exp[j\omega(t_{s1}+\tau_{12})][\chi_1(\omega)\exp(-j\omega\tau_{12})-\chi_2(\omega)]\,d\omega\right|. \quad (11.35)$$

Replacing $\chi_1(\omega)$ and $\chi_2(\omega)$ by their interference components $\tilde{I}_1(\omega)=\tilde{I}(\omega)$ and $\tilde{I}_2(\omega)=\tilde{I}(\omega)\exp(-j\omega\tau_{12}^0)$ we can see that the TDOA τ_{12}^0 is eliminated by the delay τ_{12} (of course, if $\tau_{12}=\tau_{12}^0$) so that the interference components are cancelled out: $L=L_I=0$ (The superscript "0" denotes the true value of a parameter). The signal component $L=L_s$ we obtain replacing $\chi_1(\omega)$ and $\chi_2(\omega)$ by their signal components [see (4.3) at $\Omega_{si}=0$]

$$\Psi_1(\omega)=a_s^0\exp(-j\varphi_{s1}^0)\Psi_0(\omega-\omega_0)\exp(-j\omega t_{s1}^0);$$

$$\Psi_2(\omega)=a_s^0\exp(-j\varphi_{s2}^0)\Psi_0(\omega-\omega_0)\exp(-j\omega t_{s2}^0). \quad (11.36)$$

It is assumed for simplicity that signal r.m.s. values are equal, i.e. $a_{s1}^0=a_{s2}^0=a_s^0$. Substituting (11.38) in (11.37) and dropping the common multiplier a_s^0 yields the final result

$$L_s=\left|\frac{1}{2\pi}\int_{\omega_0-\Delta\omega_s/2}^{\omega_0+\Delta\omega_s/2}|\Psi_0^*(\omega-\omega_0)|^2\exp[j\omega(t_{s1}-t_{s1}^0)]\,d\omega-\exp(-j\Delta\varphi_{s12}^0)\right.$$
$$\left.\times\frac{1}{2\pi}\int_{\omega_0-\Delta\omega_s/2}^{\omega_0+\Delta\omega_s/2}|\Psi_0(\omega-\omega_0)|^2\exp[j\omega(t_{s1}-t_{s1}^0+\tau_{12})]\,d\omega\right|\rightarrow\max(t_{s1}). \quad (11.37)$$

If the signals and interferences are resolved in TDOA, the quantity L_s reaches its maxima for two values of t_{s1}: $t_{s1}=t_{s1}^0$ and $t_{s1}=t_{s1}^0-\tau_{12}$. When $t_{s1}=t_{s1}^0$, the first integral in (11.37) is equal to $2T_s$, i.e. $L_s=2T_s$ [see (4.4)] while the second integral is nearly zero (since $|t_{s1}^0-t_{s2}^0+\tau_{12}|>2\pi/\Delta\omega_s$). On the contrary, when $t_{s1}=t_{s1}^0-\tau_{12}$, the first integral is nearly zero while the second integral is equal to $2T_s$ so that again $L_s=2T_s$. The first estimate is a "correct" one and permits to obtain a consistent estimate of the target spatial coordinates. The second estimate leads to a gross error, i.e. yields an inconsistent estimate of the target position.

Figure 11.2 illustrates the obtained results. It is seen that after TDOA elimination and interference cancellation, two signal pulses of equal amplitudes appear in the output envelope $|G_1(t_{s1})|$. Clearly, If joint processing of received signals and interferences for interference cancellation is not used (for instance, when interferences are spatially uncorrelated) such gross errors do not appear.

One-stage optimum algorithms for the target localization have, in principle, an inherent "mechanism" for rejecting gross errors (inconsistent target position estimates). In accordance with (11.16) and (11.17), for each triad (t_{s1}, t_{s2}, t_{s3}) the unique point in space $(\hat{x}, \hat{y}, \hat{z})$ is determined and the values of all t_{si}, $i=\overline{4,m}$, are calculated.

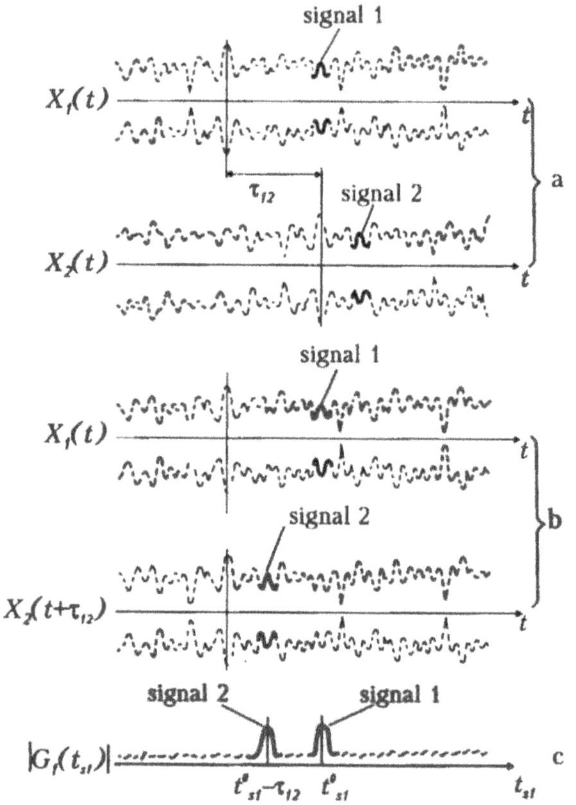

Figure 11.2 Signal multiplication after interference cancellation (to Example 11.1). (a) envelopes of the sum of a signal and an interference at the first station, $X_1(t)$, and at the second station, $X_2(t)$, (r.f. carrier is not shown); (b) the same envelopes after TDOA elimination (with the help of a delay line) before subtraction; (c) signals after interference cancellation and envelope detection; τ_{12}=interference TDOA at the inputs of receiving stations

Let at least a single anomalous estimate be among $\hat{t}_{s1}, \hat{t}_{s2}, \hat{t}_{s3}$. This means that there is no target near the point $(\hat{x}, \hat{y}, \hat{z})$. Hence there are no signal peaks at times $\hat{t}_{si} = h_i(\hat{t}_{s1}, \hat{t}_{s2}, \hat{t}_{s3})$, $i = \overline{4, m}$, and the output quantity L will have a smaller magnitude than for correct combinations of $\hat{t}_{s1}, \hat{t}_{s2}, \hat{t}_{s3}$ (of course, this is true if the output SNR is sufficiently large). The same is valid for the algorithm (11.17′). "Spurious" signal peaks at one or (more seldom) several stations may correspond to a point $\hat{x}, \hat{y}, \hat{z}$ where there is no target. However, there will be no signal peaks corresponding to this point at other stations, so that the output quantity L will have a smaller magnitude than for those points where targets are present. Thus gross errors are rejected due to the redundance of the number of stations in a MSRS. If there are no redundant stations gross (anomalous) errors of target position determination arise with one-stage estimation algorithms as well as with two-stage ones. Specifically, in the MSRS considered in Example 11.1 which contains one transmitting and two receiving stations, the one-stage algorithms for the target localization (11.17) or (11.17′) are not proof against inconsistent target position estimates.

The TOA measurement at the first stage of two-stage target localization algorithms does not contain any "mechanism" for rejecting gross errors in $\hat{t}_{si}$ estimation. However, if the number of stations is redundant as compared with the space dimensionality, these gross errors may be rejected at the second stage of two-stage algorithms. This can be done in the association process, i.e. determining from which target, if any, a particular measurement originates (see Section 15.4).

It can be seen from Example 11.1 that anomalous estimates $\hat{t}_{si}$ at the ith station are determined by target echo TOAs at other stations and interference TDOAs at different pairs of stations. If at least some stations can measure target angular coordinates using the amplitude monopulse method, amplitude differences appear between signals coming from different stations (taking into account the effect of weights in interference cancellation filters). All this information may be utilized for anomalous estimate rejection.

In this section, *we have discussed optimum (maximum likelihood) signal TOA estimation at the first stage of two stage target localization algorithms. At this stage signal TOAs at the inputs of all stations are assumed to be mutually independent parameters. We have considered the most important for practice MSRSs with short-term spatial coherence or spatially incoherent MSRSs and corresponding signals with random mutually independent phases and amplitudes. Maximum likelihood estimates of "useful" parameters – signal TOAs are synthesized by maximizing the likelihood function* (11.30) *with respect to all unknown parameters including phases. We have shown that for such signals an optimum estimate of the set of all the signal TOAs breaks down to the set of individual optimum TOA estimates at each station* (11.32).

When external (mainlobe) interference cancellation techniques are employed in a MSRS (see Chapters 8 *and* 9*) and target echoes (signals) and interferences are resolved in TDOAs at the inputs of stations, the signal "multiplication" effect takes place at the output of interference cancellation filters. This is a result of the summation of signals coming from all stations. These "spurious" signals yield inconsistent estimates of target positions in space. We have analyzed this effect with the help of a simple Example* 11.1 *[expressions* (11.33)–(11.37) *and Fig.* 11.2*]. We have noted that in the case where the number of stations is redundant as compared with the dimensionality of space, one-stage algorithms, considered in Section* 11.2, *have an inherent "mechanism" for rejecting gross errors caused by the signal multiplication effect. Two-stage target localization algorithms under similar conditions can reject those gross errors at the second stage in the process of measurement association, i.e. determining from which target, if any, each measurement at each station originates.*

11.4. SIGNAL TDOA MEASUREMENT FOR TWO-STAGE RADIATION SOURCE LOCALIZATION BY PASSIVE MSRSs

For passive MSRSs, the first stage of two-stage radiation source localization implies measurement algorithms for signal TDOAs which are proportional to the target (radiation source) range differences. We consider the most interesting for practice stochastic signals *with quasideterministic correlation* (*covariance*) *matrices* containing random phases and variances (signal intensities, see Section 4.1). For a passive MSRS with m receiving stations two different approaches are possible for the optimum TDOA estimation. *The first approach* (which may be referred to as a one-step approach) implies direct optimum estimation of $m-1$ linearly independent signal TDOAs, $\tau_{s12}, \ldots, \tau_{s1m}$, utilizing signals received by all the m stations. These $m-1$ optimum TDOA estimates, $\hat{\tau}_{s12}, \ldots, \hat{\tau}_{s1m}$, are used for the optimal estimation of radiation source spatial coordinates, $\hat{x}, \hat{y}, \hat{z}$, at the second stage of a radiation source localization algorithm. *The second approach* (which may be referred to as a two-step approach) implies at the first step optimal estimation of all the $m(m-1)/2$ signal TDOAs, τ_{sik}, $i,k=\overline{1,m}$, $i<k$, individually. Then, at the second step, optimum estimates $\hat{\tau}_{s12}, \ldots, \hat{\tau}_{s1m}$ of $m-1$ linearly independent TDOAs are calculated using the obtained estimates $\hat{\tau}_{s12}, \ldots, \hat{\tau}_{s1m}, \hat{\tau}_{s23}, \ldots, \hat{\tau}_{s2m}, \ldots, \hat{\tau}_{s(m-1)m}$. The second stage of a two-stage source localization algorithm is the same as for the first approach (see, e.g., [242]).

An optimum algorithm for TDOAs estimation according to *the first approach* can be derived like in Section 11.2. Using (11.25) but without (11.23) we have an optimum algorithm in the form

$$L=\operatorname{Re}\sum_{i=1}^{m-1}\sum_{k=i+1}^{m}\frac{q_{si}q_{sk}\exp(j\Delta\varphi_{sik})}{\sqrt{N_iN_k}}\frac{1}{2\pi}\int_{-\infty}^{\infty}\chi_i^*(\omega)\chi_k(\omega)|H(\omega)|^2$$

$$\times\exp[j\omega(\tau_{s1k}-\tau_{s1i})]\,d\omega\rightarrow\max(\Delta\varphi_{s12},\ldots,\Delta\varphi_{sm(m-1)},\tau_{s12},\ldots,\tau_{s1m}) \quad (11.38)$$

where it should be taken into account that $\tau_{s11}=0$. Assuming (as in Section 11.2) all phases, $\Delta\varphi_{sik}$, $i,k=\overline{1,m}$, $i<k$, to be mutually independent (such a signal model was discussed in Section 7.2) we can obtain their maximum likelihood estimates

$$\widehat{\Delta\varphi}_{sik}=-\arg\left\{\frac{1}{2\pi}\int_{-\infty}^{\infty}\chi_i^*(\omega)\chi_k(\omega)|H(\omega)|^2\exp[j\omega(\tau_{s1k}-\tau_{s1i})]\,d\omega\right\} \quad (11.39)$$

and substitute them for random phases $\Delta\varphi_{sik}$ in (11.38). Then the optimum signal TDOA estimation algorithm takes the form

$$L=\sum_{i=1}^{m-1}\sum_{k=i+1}^{m}\frac{q_{si}q_{sk}}{\sqrt{N_iN_k}}\left|\frac{1}{2\pi}\int_{-\infty}^{\infty}\chi_i^*(\omega)\chi_k(\omega)|H(\omega)|^2\right.$$

$$\left.\times\exp[j\omega(\tau_{s1k}-\tau_{s1i})]\,d\omega\right|\rightarrow\max(\tau_{s12},\ldots,\tau_{s1m}). \quad (11.40)$$

It can be seen that this algorithm is not much simpler than the one-stage algorithms (11.28) and (11.28′) though it does not yet localize a radiation source in space. This is why this approach is not widely used.

The optimal TDOA measurement algorithm according to *the second approach* mentioned above, can be derived in the same manner as (11.40). As a result we have instead of (11.40)

$$L=\sum_{i=1}^{m-1}\sum_{k=i+1}^{m}\frac{q_{si}q_{sk}}{\sqrt{N_iN_k}}\left|\frac{1}{2\pi}\int_{-\infty}^{\infty}\chi_i^*(\omega)\chi_k(\omega)|H(\omega)|^2\exp(j\omega\tau_{sik})\,d\omega\right|\to\max(\tau_{sik}).$$

At this step all TDOAs are assumed to be independent variables. Each term of the double sum contains only a single "its own" TDOA. Hence the sum reaches its maximum when all terms have their own maxima. The estimation algorithm takes the form

$$L=\left|\frac{1}{2\pi}\int_{-\infty}^{\infty}\chi_i^*(\omega)\chi_k(\omega)|H(\omega)|^2\exp(j\omega\tau_{sik})\,d\omega\right|\to\max(\tau_{sik});\quad i,k=\overline{1,m};\ i<k. \tag{11.41}$$

In accordance with (7.11), when input SNRs are sufficiently large ($q_{s\Sigma}^2 \gg 1$), the filters with the squared amplitude response $|H(\omega)|^2$ may be excluded. On the contrary, for weak input signals ($q_{s\Sigma}^2 \ll 1$), $|H(\omega)|^2=F_s(\omega-\omega_0)$, and this filter is necessary.

It follows from (11.41) that the required estimate of a signal TDOA at each pair of receiving stations is the delay value at one of the two correlator inputs which corresponds to a maximum of the estimate modulus of the cross-correlation of filtered signals and interferences received by that pair of stations. Apparently, this measurement procedure is much simpler than (11.40).

Having measured TDOAs, $\hat{\tau}_{s12},\ldots,\hat{\tau}_{s1m},\ \hat{\tau}_{s23},\ldots,\hat{\tau}_{s2m},\ldots,\hat{\tau}_{s(m-1)m}$, one can obtain optimum estimates of linearly independent TDOAs, $\hat{\tau}_{s12},\ldots,\hat{\tau}_{s1m}$. Furthermore, if the algorithm (11.41) is used with the filter $|H(\omega)|^2$ from (7.11) we may obtain *efficient* estimates $\hat{\tau}_{s12\,\mathrm{opt}},\ldots,\hat{\tau}_{s1m\,\mathrm{opt}}$, i.e. unbiased estimates with minimal attainable variances that are determined by the Cramér–Rao bound [240,241]. In fact, the vector $\boldsymbol{\tau}_s$ of all $m(m-1)/2$ TDOAs, $\boldsymbol{\tau}_s=[\tau_{s12},\ldots,\tau_{s1m},\ \tau_{s23},\ldots,\tau_{s2m},\ldots,\tau_{s(m-1)m}]^t$, is linearly related to the vector $\boldsymbol{\tau}_{s1}$ of linearly independent TDOAs, $\boldsymbol{\tau}_{s1}=(\tau_{s12},\ldots,\tau_{s1m})^t$, so that

$$\boldsymbol{\tau}_s=\mathbf{A}\boldsymbol{\tau}_{s1} \tag{11.42}$$

where the $m(m-1)/2\ \times(m-1)$ matrix $\mathbf{A}$ is given by

$$\mathbf{A}=\|A_{ik\,1n}\|=\begin{bmatrix} A_{12\,12} & A_{12\,13} & \cdots & A_{12\,1m}\\ \cdots & \cdots & \cdots & \cdots\\ A_{1m\,12} & A_{1m\,13} & \cdots & A_{1m\,1m}\\ A_{23\,12} & A_{23\,13} & \cdots & A_{23\,1m}\\ \cdots & \cdots & \cdots & \cdots\\ A_{2m\,12} & A_{2m\,13} & \cdots & A_{2m\,1m}\\ \cdots & \cdots & \cdots & \cdots\\ A_{(m-1)m\,12} & A_{(m-1)m\,13} & \cdots & A_{(m-1)m\,1m}\end{bmatrix};\quad A_{ik\,1n}=\delta_{kn}-\delta_{in}. \tag{11.43}$$

Let the covariance matrix of the unbiased estimates $\hat{\tau}_{sik}$, $i,k=\overline{1,m}$, $i<k$, obtained from (11.41) be denoted by $\tilde{\mathbf{B}}$. Then the optimum estimate of the vector $\boldsymbol{\tau}_{s1}$ is given by a known relationship derived from the least squares method (see Section 14.1 and [240,241])

$$\hat{\boldsymbol{\tau}}_{s1\,\text{opt}}=(\mathbf{A}^t\tilde{\mathbf{B}}^{-1}\mathbf{A})^{-1}\mathbf{A}^t\tilde{\mathbf{B}}^{-1}\hat{\boldsymbol{\tau}}_s \tag{11.44}$$

As it has been shown in [240,241], the vector $\hat{\boldsymbol{\tau}}_{s1\,\text{opt}}=(\hat{\tau}_{s12\,\text{opt}},\ldots,\hat{\tau}_{s1m\,\text{opt}})^t$ obtained according to (11.44) and (11.41), is *the vector of efficient estimates.* The covariance matrix $\mathbf{B}_{\text{eff}}$ of the estimates $\hat{\boldsymbol{\tau}}_{s1\,\text{opt}}$ is the inverse of the Fisher Information Matrix (FIM) and is given by

$$\mathbf{B}_{\text{eff}}=(\mathbf{A}^t\tilde{\mathbf{B}}^{-1}\mathbf{A})^{-1}. \tag{11.45}$$

The matrix $\tilde{\mathbf{B}}$ in (11.44) and (11.45) is a $m(m-1)/2\times m(m-1)/2$ symmetric matrix whose arbitrary entries, $b_{ik\,ln}$, are the variances and covariances of signal TDOA estimates, $\hat{\tau}_{sik}$ and $\hat{\tau}_{sln}$, i.e. $b_{ikln}=\overline{\hat{\tau}_{sik}\hat{\tau}_{sln}}$, $i,l=\overline{1,(m-1)}$, $k=\overline{(i+1),m}$, $n=\overline{(l+1),m}$. It is clear that $b_{ikln}=0$ when all the subscripts i,k,l,n are different, for such a pair of correlators has no common inputs. Besides, the sign of the covariance is changed if the correlator inputs at one of the correlators of each pair are switched, i.e. $b_{ikin}=\overline{\hat{\tau}_{sik}\hat{\tau}_{sin}}=-b_{ikni}=\overline{\hat{\tau}_{sik}\hat{\tau}_{sni}}$ and $b_{ikin}=\overline{\hat{\tau}_{sik}\hat{\tau}_{sin}}=-b_{kiin}=\overline{\hat{\tau}_{ski}\hat{\tau}_{sin}}$. Corresponding expressions for the variances, $b_{ikik}=\overline{\hat{\tau}_{sik}\hat{\tau}_{sik}}$, and the covariances, $b_{ikin}=\overline{\hat{\tau}_{sik}\hat{\tau}_{sin}}$, are presented in [241] for wideband stochastic processes[5]. For the assumed models of narrowband complex signals and white noises (see Section 4.1) and using our notations those expressions can be rewritten as follows

$$\sigma^2(\hat{\tau}_{sik})=\overline{\hat{\tau}_{sik}\hat{\tau}_{sik}}$$

$$=\frac{\pi}{T}\frac{\displaystyle\int_{-\infty}^{\infty}(\omega-\omega_0)^2|H(\omega)|^4[1+q_{si}^2F_s(\omega-\omega_0)+q_{sk}^2F_s(\omega-\omega_0)]\,d\omega}{q_{si}^2q_{sk}^2\left[\displaystyle\int_{-\infty}^{\infty}(\omega-\omega_0)^2|H(\omega)|^2F_s(\omega-\omega_0)\,d\omega\right]^2}; \tag{11.46}$$

$$\sigma(\hat{\tau}_{sik})\sigma(\hat{\tau}_{sin})\rho(\hat{\tau}_{sik},\hat{\tau}_{sin})=\overline{\hat{\tau}_{sik}\hat{\tau}_{sin}}$$

$$=\frac{\pi}{T}\frac{\displaystyle\int_{-\infty}^{\infty}(\omega-\omega_0)^2|H(\omega)|^4F_s(\omega-\omega_0)\,d\omega}{q_{si}^2\left[\displaystyle\int_{-\infty}^{\infty}(\omega-\omega_0)^2|H(\omega)|^2F_s(\omega-\omega_0)\,d\omega\right]^2}. \tag{11.47}$$

Example 11.2. Consider an application of the second approach to TDOAs optimum estimation. Let the signal PSD have a rectangular form, i.e. be determined by (7.27) for $\Omega_{s1}=0$. It follows from (7.11) that in this case $|H(\omega)|^2=1/(1+q_{s\Sigma}^2)$ and $|H(\omega)|^4=1/(1+q_{s\Sigma}^2)^2$ in the frequency band $(\omega_0-\Delta\omega_s/2,\omega_0+\Delta\omega_s/2)$. Substituting these quantities in (11.46) and (11.47) yields

$$\sigma^2(\hat{\tau}_{sik})=\frac{3(1+q_{si}^2+q_{sk}^2)}{2q_{si}^2q_{sk}^2\Delta f_sT\pi^2\Delta f_s^2};\quad \sigma(\hat{\tau}_{sik})\sigma(\hat{\tau}_{sin})\rho(\hat{\tau}_{sik},\hat{\tau}_{sin})=\frac{3}{2q_{si}^2\Delta f_sT\pi^2\Delta f_s^2}. \tag{11.48}$$

[5] It is worth reminding that the overbar at random quantities means averaging over the corresponding ensemble, whereas the overbar at integers means enumeration of possible values.

The correlation coefficient, $\rho(\hat{\tau}_{sik}, \hat{\tau}_{sin})$, is given from (11.48) by

$$\rho(\hat{\tau}_{sik}, \hat{\tau}_{sin}) = \frac{q_{sk} q_{sn}}{\sqrt{(1+q_{si}^2+q_{sk}^2)(1+q_{si}^2+q_{sn}^2)}}. \tag{11.49}$$

It is important to note that, as it follows from (11.49), when the input SNRs are small ($q_{si}^2 \ll 1, q_{sk}^2 \ll 1, q_{sn}^2 \ll 1$) signal TDOA estimates at different pairs of stations (with one common station in each pair) are weakly correlated. It should be also noted that small input SNRs not necessarily lead to low accuracy of TDOA estimates. It can be seen from (11.48) that a large accumulation (integration) factor, $\Delta f_s T$, may offset low input SNRs. For instance, even if $q_{si}^2 = q_{sk}^2 = q_{sn}^2 = -10\,\text{dB}$ and $\Delta f_s T = 10^3 - 10^4$ the attainable r.m.s error of signal TDOA estimates according to (11.48) do not exceed $0.135/\Delta f_s$–$0.043/\Delta f_s$, i.e. a small fraction of the correlation interval. This feature of correlation measurements is similar to the corresponding feature of correlation detection (see Chapter 7).

To simplify mathematical manipulations, we consider in the sequel a two-dimensional case and a passive MSRS containing three receiving stations with equal SNRs at the inputs of all stations, i.e. $q_{si}^2 = q_{sk}^2 = q_s^2$. Then the matrices $\mathbf{A}$ and $\tilde{\mathbf{B}}$ according to (11.43) and (11.48) are

$$\mathbf{A} = \begin{bmatrix} 1 & 0 \\ 0 & 1 \\ -1 & 1 \end{bmatrix}; \qquad \mathbf{A}' = \begin{bmatrix} 1 & 0 & -1 \\ 0 & 1 & 1 \end{bmatrix}; \tag{11.50}$$

$$\tilde{\mathbf{B}} = \frac{3}{2q_s^2 \Delta f_s T \pi^2 \Delta f_s^2} \begin{bmatrix} \dfrac{1+2q_s^2}{q_s^2} & 1 & -1 \\ 1 & \dfrac{1+2q_s^2}{q_s^2} & 1 \\ -1 & 1 & \dfrac{1+2q_s^2}{q_s^2} \end{bmatrix}. \tag{11.51}$$

It has been taken into account that $\rho(\hat{\tau}_{s12}, \hat{\tau}_{s13}) = \rho(\hat{\tau}_{s13}, \hat{\tau}_{s23}) = -\rho(\hat{\tau}_{s12}, \hat{\tau}_{s23})$. The inverse of $\tilde{\mathbf{B}}$ can be easily obtained:

$$\tilde{\mathbf{B}}^{-1} = \frac{2q_s^6 \Delta f_s T \pi^2 \Delta f_s^2}{3(1+3q_s^2)} \begin{bmatrix} \dfrac{1+4q_s^2+3q_s^4}{(1+3q_s^2)q_s^2} & -1 & 1 \\ -1 & \dfrac{1+4q_s^2+3q_s^4}{(1+3q_s^2)q_s^2} & -1 \\ 1 & -1 & \dfrac{1+4q_s^2+3q_s^4}{(1+3q_s^2)q_s^2} \end{bmatrix}. \tag{11.52}$$

Substituting (11.50) and (11.52) in (11.45) yields the covariance matrix, $\mathbf{B}_{\text{eff}}$, of optimum (efficient) estimates of the linearly independent TDOAs, $\hat{\tau}_{s12\,\text{opt}}$, $\hat{\tau}_{s13\text{opt}}$

$$\mathbf{B}_{\text{eff}} = (\mathbf{A}'\tilde{\mathbf{B}}^{-1}\mathbf{A})^{-1} = \frac{1+3q_s^2}{2q_s^4 \Delta f_s T \pi^2 \Delta f_s^2} \begin{pmatrix} 2 & 1 \\ 1 & 2 \end{pmatrix}. \tag{11.53}$$

It is interesting to compare the variances and covariances of the optimum estimates $\hat{\tau}_{s12\,\mathrm{opt}}, \hat{\tau}_{s13\,\mathrm{opt}}$ with (11.48) for individual the same TDOA estimates at each pair of stations. From (11.53) we have

$$\sigma^2_{\mathrm{eff}}(\hat{\tau}_{s12\,\mathrm{opt}}) = \sigma^2_{\mathrm{eff}}(\hat{\tau}_{s13\,\mathrm{opt}}) = \frac{1+3q_s^2}{q_s^4 \Delta f_s T \pi^2 \Delta f_s^2};$$

$$\sigma_{\mathrm{eff}}(\hat{\tau}_{s12\,\mathrm{opt}})\sigma_{\mathrm{eff}}(\hat{\tau}_{s13\,\mathrm{opt}})\rho_{\mathrm{eff}}(\hat{\tau}_{s12\,\mathrm{opt}}, \hat{\tau}_{s13\,\mathrm{opt}}) = \frac{1+3q_s^2}{2q_s^4 \Delta f_s T \pi^2 \Delta f_s^2}; \tag{11.54}$$

$$\rho_{\mathrm{eff}}(\hat{\tau}_{s12\,\mathrm{opt}}, \hat{\tau}_{s13\,\mathrm{opt}}) = 0.5.$$

From (11.48) for $i=1, k=2,3$ and $q_{s1}^2 = q_{s2}^2 = q_{s3}^2 = q_s^2$

$$\sigma^2(\hat{\tau}_{s12}) = \sigma^2(\hat{\tau}_{s13}) = \frac{3(1+2q_s^2)}{2q_s^4 \Delta f_s T \pi^2 \Delta f_s^2};$$

$$\sigma(\hat{\tau}_{s12})\sigma(\hat{\tau}_{s13})\rho(\hat{\tau}_{s12}, \hat{\tau}_{s13}) = \frac{3}{2q_s^2 \Delta f_s T \pi^2 \Delta f_s^2}; \quad \rho(\hat{\tau}_{s12}, \hat{\tau}_{s13}) = \frac{q_s^2}{1+2q_s^2}. \tag{11.55}$$

It can be seen that for $q_s^2 \ll 1$ (low input SNRs) $\sigma^2(\hat{\tau}_{s1k})/\sigma^2_{\mathrm{eff}}(\hat{\tau}_{s1k\,\mathrm{opt}}) \to 3/2$ and $\rho(\hat{\tau}_{s12}, \hat{\tau}_{s13}) \to 0$ while for $q_s^2 \gg 1$ (high input SNRs) $\sigma^2(\hat{\tau}_{s1k})/\sigma^2_{\mathrm{eff}}(\hat{\tau}_{s1k\,\mathrm{opt}}) \to 1$ and $\rho(\hat{\tau}_{s12}, \hat{\tau}_{s13}) \to 0.5$, i.e. $\hat{\tau}_{s1k} \to \hat{\tau}_{s1k\,\mathrm{opt}}$.

Using (11.53), (11.52) and (11.50) permits to obtain the optimum algorithm for $\hat{\tau}_{s12\mathrm{opt}}, \hat{\tau}_{s13\mathrm{opt}}$ estimation in accordance with (11.44)

$$\hat{\tau}_{s1\mathrm{opt}} = \begin{pmatrix} \hat{\tau}_{s12\,\mathrm{opt}} \\ \hat{\tau}_{s13\,\mathrm{opt}} \end{pmatrix} = \frac{1}{3}\begin{pmatrix} 2 & 1 & -1 \\ 1 & 2 & 1 \end{pmatrix}\begin{pmatrix} \hat{\tau}_{s12} \\ \hat{\tau}_{s13} \\ \hat{\tau}_{s23} \end{pmatrix} = \frac{1}{3}\begin{pmatrix} 2\hat{\tau}_{s12} + \hat{\tau}_{s13} - \hat{\tau}_{s23} \\ \hat{\tau}_{s12} + 2\hat{\tau}_{s13} + \hat{\tau}_{s23} \end{pmatrix}. \tag{11.56}$$

Clearly, the second approach to the optimum TDOAs estimation is simpler than the first one, since it allows us to replace simultaneous adjusting of all linearly independent delays at the inputs of correlators to achieve a total maximum of the sum (11.40) by individual TDOA measurements at the inputs of each pair of stations. Besides, this approach can yield efficient estimates of linearly independent TDOAs, whereas the first approach results in maximum likelihood and hence only asymptotically efficient estimates.

In the case of several measurements at different pairs of stations (which may be caused by both several radiation sources and false alarms), the correct association between radiation sources and corresponding measurements is necessary before the second step of the algorithm, i.e. before using (11.44) and (11.45). This problem is similar to that discussed in Section 11.3 in connection with the signal multiplication effect after external interference cancellation in MSRSs. The first approach described above, as well as the one-stage algorithms considered in Section 11.2, have their inherent "mechanism" to reject gross errors. When measurements originating from different sources or false alarms are used in (11.40) or (11.28), (11.28′) (in the case of wrong association) and there are redundant measurements, corresponding sums will be more likely less than for the correct association when all summed measurements originated from the same radiation source. It should be noted, however, that the degree of redundancy inherent in the first approach to the two-stage source localization algorithms is usually less than in one-stage ones. In fact, according to the first approach, $(m-1)$ linearly independent signal delays are to be estimated among $m(m-1)/2$ possible signal delays in a MSRS with m receiving stations. The ratio $[m(m-1)/2]/(m-1)=m/2$ determines the degree of redundancy. For the one-stage localization algorithms this ratio is $[m(m-1)/2]/3$ in the three-dimensional space and $[m(m-1)/2]/2$ in the two-dimensional space (on a plane). If $(m-1)$ is greater than the dimensionality of space considered, the one-stage algorithms will reject gross errors more effectively.

It is important to note that in general the second step of the algorithm just described is not to be performed necessarily. It may be skipped. Having obtained measurements, $\hat{\tau}_{s12},\ldots,\hat{\tau}_{s1m},\hat{\tau}_{s23},\ldots,\hat{\tau}_{s2m},\ldots,\hat{\tau}_{s(m-1)m}$ [see (11.41)], and using their covariance matrix, $\tilde{\mathbf{B}}$, one can calculate optimum estimates of the spatial coordinates of radiation sources directly (after correct measurement association) without computing optimum estimates of the linearly independent TDOAs, $\hat{\tau}_{s12\,\mathrm{opt}},\ldots,\tau_{s1m\,\mathrm{opt}}$, (see Chapter 14).

In practice, suboptimum TDOA estimation algorithms are commonly used at the first stage of two-stage radiation source localization algorithms. The simplest suboptimum algorithm implies measurements of only $(m-1)$ linearly independent TDOAs ignoring measurements of other TDOAs in the radiation source localization process. Measurements of linearly dependent TDOAs may only be utilized for interstation measurement association (see Section 15.4). A more complicated but more accurate algorithm requires measurements of all $m(m-1)/2$ TDOAs according to (11.41) but ignores the correlation between measurements. We have already drawn the reader's attention to the weak correlation between these measurements when input SNRs are low [see (11.49) and (11.55)].

Example 11.3. Let us evaluate the covariance matrix $\mathbf{B}'$ for optimal estimates $\hat{\tau}_{s12\,\mathrm{opt}},\hat{\tau}_{s13\,\mathrm{opt}}$ from Example 11.2 if the cross-correlation between measurements are ignored, i.e. if the measurement covariance matrix, $\tilde{\mathbf{B}}$, is set diagonal [see (11.48) and (11.51)]:

$$\tilde{\mathbf{B}}=\sigma^2(\hat{\tau}_{sik})\mathbf{I}=\frac{3(1+2q_s^2)}{2q_s^4\Delta f_sT\pi^2\Delta f_s^2}\mathbf{I},\quad \hat{\tau}_{sik}\in(\hat{\tau}_{12},\hat{\tau}_{s13},\hat{\tau}_{s23}) \tag{11.57}$$

where I is the identity matrix. Using (11.50) and inverting $\tilde{\mathbf{B}}$ yields

$$\mathbf{B}'=(\mathbf{A}'\tilde{\mathbf{B}}^{-1}\mathbf{A})^{-1}=\frac{1+2q_s^2}{2q_s^4\Delta f_sT\pi^2\Delta f_s^2}\begin{pmatrix}2 & 1\\ 1 & 2\end{pmatrix}. \tag{11.58}$$

It is seen that the matrix $\mathbf{B}'$ is similar to the matrix $\mathbf{B}_{\mathrm{eff}}$ (11.53) but the variances of the estimates $\hat{\tau}_{s12\,\mathrm{opt}},\hat{\tau}_{s13\,\mathrm{opt}}$ are here less than in (11.53). Their ratio is equal to $(1+2q_s^2)/(1+3q_s^2)$. It is clear, that ignoring cross-correlation between measurements leads to overvalued accuracy of the estimates $\hat{\tau}_{s12\,\mathrm{opt}},\hat{\tau}_{s13\,\mathrm{opt}}$. However, for low input SNRs ($q^2\ll 1$) this overvaluation may often be neglected.

Some other application examples for the algorithms considered above see in Section 14.3.

In this section, *we have presented two approaches to optimal TDOAs estimation at the first stage of two-stage radiation source position determination algorithms. For signal model with mutually independent unknown (random) phases of signal PSD* *(or covariance) matrix at the inputs of m spatially separated receiving stations* (*as in Section* 7.2), *we have synthesized the optimum* (*maximum likelihood*) *estimation algorithm* (11.40) *for* $m-1$ *linearly independent TDOAs. This is the first approach. Algorithm* (11.40) *requires to choose delays at the correlator inputs for all* $m(m-1)/2$ *pairs of stations so as to maximize the weighted sum of the moduli of all correlator outputs. Such a procedure is not much simpler that a one-stage radiation source localization algorithms considered in Section* 11.2. *The second approach implies much simpler individual TDOA measurement for all the* $m(m-1)/2$ *pairs of stations according to* (11.41) *and then optimum calculation of* $(m-1)$ *linearly independent TDOAs* [(11.42)–(11.44)]. *As it has been shown in* [240,241] *this technique permits to obtain efficient estimates of those TDOAs with the covariance matrix* (11.45) *which is the inverse of the Fisher information matrix. Example* 11.2 *shows an application of the second approach. Direct spatial coordinate estimation for a radiation source using signal TDOA measurements at all pairs of stations is also possible.*

It has been noted that in real situation correct measurement association is important in order to reject gross errors. For the second approach considered this can be done before efficient TDOA estimate calculation.

In practice, suboptimum TDOA estimation is often used at the first stage of two-stage algorithms for radiation source localization. Example 11.3 *shows that ignoring cross-correlation between* $m(m-1)/2$ *TDOA estimates leads to the optimistic accuracy evaluation of* $(m-1)$ *linearly independent TDOA estimates, but in practically important cases of low input SNRs this overvaluation is small and may often be neglected.*

12. MAXIMUM ATTAINABLE ACCURACY OF TEMPORAL PARAMETER MEASUREMENTS FOR SIGNALS CONTAINING STRAY PARAMETERS

12.1. MAXIMUM ATTAINABLE ACCURACY OF MAXIMUM LIKELIHOOD ESTIMATES OF INFORMATIVE PARAMETERS

The problem of target position and velocity estimation in active and passive MSRSs is a typical problem of multidimensional parameter estimation. It is accepted to characterize the accuracy of multidimensional estimators with the help of a vector of biases and an Error Covariance Matrix (ECM)[1]. Explicit analytical expressions for direct ECM calculation for specific estimation algorithms are usually difficult to derive. In those cases where ECM calculations are necessary (see, for instance, Section 13.2) computer simulation is to be used. At the same time it is known that maximum likelihood estimates are asymptotically unbiased and efficient under certain, usually satisfied, regularity conditions (e.g., [27]). Therefore, analysis of the ECM for efficient estimates is widely used for the accuracy analysis of maximum likelihood estimates of the same parameters, especially when requirements for estimation accuracy are sufficiently high so that the conditions of asymptotic efficiency are approximately satisfied. An ECM of efficient estimates presents the Cramér–Rao lower bound for estimate errors. It is a reliable approximation of attainable errors when the observation time is long enough (or, more correctly, when the product of the observation time by the signal bandwidth is large enough). The advantage of such an approach is that the ECM of efficient estimates is the inverse of the Fisher information matrix (FIM), so that it determines the maximum attainable accuracy inherent in Likelihood Functions or Functionals (LF) of estimated parameters regardless of specific estimation algorithms.

The "FIM technique" (or the "Cramér–Rao lower bound" technique) is usually employed in the cases where all unknown parameters of a LF are to be estimated. However, in practice LFs often contain not only informative (useful), but noninformative (stray) parameters. According to the "classical" procedure, a FIM for all parameters is to be calculated and inverted for obtaining the corresponding ECM. Only a part of this ECM will be the ECM of informative parameters. When the number of stray parameters is large (which is typical for MSRSs and many other multichannel systems), the dimension of the FIM turns out to be large too, and its inversion becomes difficult.

To overcome this difficulty, averaging of original LF over stray parameters is commonly employed. The averaged LF is used for both the synthesis of informative parameter estimators and the accuracy analysis with the help of FIM [51,52,72]. This technique is quite expedient when stray parameters are random with known probability distributions. However, it cannot be used when probability distributions are unknown or those parameters are not random though unknown. It should be

[1] This abbreviation should not be confused with Electronic Counter Measures.

noted that maximum likelihood estimates of informative parameters have been derived in Sections 11.2–11.4 by LF maximization with respect to all parameters (including stray ones) without averaging. It is desirable to simplify calculations of the ECM of efficient estimates with the help of "the FIM technique" without averaging over stray parameters.

A Technique for Deriving ECM of Informative Parameter Efficient Estimates by Inverting a Special FIM

This special FIM takes into account stray parameters but its dimension is determined by the number of informative parameters only [59].

Let a LF take the form $\Lambda[\mathbf{X}(t), \boldsymbol{\alpha}, \boldsymbol{\beta}]$ where $\mathbf{X}^t(t) = [X_1(t), \ldots, X_m(t)]$ is the vector of overall received signals, $\boldsymbol{\alpha}^t = (\alpha_1, \ldots, \alpha_N)$ is the vector of informative parameters, $\boldsymbol{\beta}^t = (\beta_1, \ldots, \beta_M)$ is the vector of stray parameters. In general estimates of $\boldsymbol{\alpha}$ are dependent on $\boldsymbol{\beta}$ and are obtained as result of joint solution of $N+M$ likelihood equations

$$\partial \ln \Lambda / \partial \alpha_i = 0, \quad i = \overline{1, N}; \qquad \partial \ln \Lambda / \partial \beta_k = 0, \quad k = \overline{1, M}. \tag{12.1}$$

The corresponding FIM for the estimates $\hat{\boldsymbol{\alpha}}$ and $\hat{\boldsymbol{\beta}}$ may be partitioned as follows:

$$\mathbf{J}^{(\alpha,\beta)} = \begin{pmatrix} \mathbf{J}^{(\alpha\alpha)} & \mathbf{J}^{(\alpha\beta)} \\ \mathbf{J}^{(\beta\alpha)} & \mathbf{J}^{(\beta\beta)} \end{pmatrix} \tag{12.2}$$

where $\mathbf{J}^{(\alpha\alpha)}, \mathbf{J}^{(\beta\beta)}$ are the $N \times N$ and $M \times M$ square matrices, respectively, with the entries

$$\begin{aligned} \mathbf{J}_{np}^{(\alpha\alpha)} &= -\mathrm{E}\{\partial^2 \ln \Lambda[\mathbf{X}(t), \boldsymbol{\alpha}_0, \boldsymbol{\beta}_0] / \partial \alpha_n \partial \alpha_p\}, \quad n, p = \overline{1, N}; \\ \mathbf{J}_{np}^{(\beta\beta)} &= -\mathrm{E}\{\partial^2 \ln \Lambda[\mathbf{X}(t), \boldsymbol{\alpha}_0, \boldsymbol{\beta}_0] / \partial \beta_n \partial \beta_p\}, \quad n, p = \overline{1, M}, \end{aligned} \tag{12.3}$$

and $\mathbf{J}^{(\alpha\beta)} = [\mathbf{J}^{(\beta\alpha)}]^t$ where $\mathbf{J}^{(\alpha\beta)}$ is the $N \times M$ rectangular matrix with the following elements

$$\mathbf{J}_{np}^{(\alpha\beta)} = -\mathrm{E}\{\partial^2 \ln \Lambda[\mathbf{X}(t), \boldsymbol{\alpha}_0, \boldsymbol{\beta}_0] / \partial \alpha_n \partial \beta_p\}, \quad n = \overline{1, N}, \; p = \overline{1, M}. \tag{12.4}$$

All derivatives are taken at the true values of parameters $\boldsymbol{\alpha}_0$ and $\boldsymbol{\beta}_0$. "E" means "expectation", i.e. the average over the ensemble of $\mathbf{X}(t)$. Such a notation is here more convenient than the overbar that was used in previous Sections and will be used in the sequel.

The inverse of the $(N+M) \times (N+M)$ matrix (12.2) is the ECM of the jointly efficient estimates $\hat{\boldsymbol{\alpha}}$ and $\hat{\boldsymbol{\beta}}$. We are interested in the upper right part only, i.e. in the $N \times N$ ECM matrix for the efficient estimates of informative parameters $\hat{\boldsymbol{\alpha}}$. Let us denote it by $\mathbf{D}^{(\alpha\alpha)}$. We may bypass the inversion of the matrix $\mathbf{J}^{(\alpha,\beta)}$ (12.2) using a known result on the inverse of a partitioned matrix. Then

$$\mathbf{D}^{(\alpha\alpha)} = \{\mathbf{J}^{(\alpha\alpha)} - \mathbf{J}^{(\alpha\beta)}[\mathbf{J}^{(\beta\beta)}]^{-1}\mathbf{J}^{(\beta\alpha)}\}^{-1}.$$

We denote the expression in braces by $\tilde{\mathbf{J}}^{(\alpha)}$:

$$\tilde{\mathbf{J}}^{(\alpha)}=\mathbf{J}^{(\alpha\alpha)}-\mathbf{J}^{(\alpha\beta)}[\mathbf{J}^{(\beta\beta)}]^{-1}\mathbf{J}^{(\beta\alpha)}. \tag{12.5}$$

so that

$$\mathbf{D}^{(\alpha\alpha)}=[\tilde{\mathbf{J}}^{(\alpha)}]^{-1}. \tag{12.6}$$

It is seen that the inversion of the $(N+M)\times(N+M)$ matrix $\mathbf{J}^{(\alpha,\beta)}$ (12.2) is replaced by the inversion first of the $M\times M$ matrix $\mathbf{J}^{(\beta\beta)}$ and then of the $N\times N$ matrix $\tilde{\mathbf{J}}^{(\alpha)}$.

The matrix $\tilde{\mathbf{J}}^{(\alpha)}$ may be considered as *a "FIM" of informative parameters taking into account the influence of stray parameters.* Calculation of $\tilde{\mathbf{J}}^{(\alpha)}$ may be in many cases significantly simplified in comparison with (12.5), first of all by excluding calculation and inversion of $\mathbf{J}^{(\beta\beta)}$. To do this, such a LF of informative parameters $\tilde{\Lambda}[\mathbf{X}(t),\boldsymbol{\alpha}]$ should be derived so as a FIM calculated according to the usual technique coincides with $\tilde{\mathbf{J}}^{(\alpha)}$, i.e.

$$-\mathrm{E}\{\partial^2\ln\tilde{\Lambda}[\mathbf{X}(t),\boldsymbol{\alpha}_0]/\partial\alpha_n\partial\alpha_p\}=\tilde{\mathbf{J}}^{(\alpha)}_{np},\quad n,p=\overline{1,N}. \tag{12.7}$$

To derive the required $\tilde{\Lambda}[\mathbf{X}(t),\boldsymbol{\alpha}]$ the likelihood equations (12.1) for $\boldsymbol{\beta}$ should be averaged over the ensemble of overall received signals $\mathbf{X}(t)$. Assume that having solved the system of equations

$$-\mathrm{E}\{\partial\ln\Lambda[\mathbf{X}(t),\boldsymbol{\alpha},\boldsymbol{\beta}]/\partial\beta_k\}=0,\quad k=\overline{1,M} \tag{12.8}$$

we can express $\boldsymbol{\beta}$ in terms of $\boldsymbol{\alpha}$. Denoting the solutions of (12.8) by $\boldsymbol{\beta}=\mathbf{u}(\boldsymbol{\alpha})$ we substitute them in the original LF for $\boldsymbol{\beta}$. Now it can be shown that at least near the true values of the estimated parameters, where the logarithm of the original LF can be well approximated by the first (linear and square) terms of Taylor-series expansion at the true point $(\boldsymbol{\alpha}_0,\boldsymbol{\beta}_0)$, the obtained LF $\Lambda[\mathbf{X}(t),\boldsymbol{\alpha},\mathbf{u}(\boldsymbol{\alpha})]$ is just the required LF $\tilde{\Lambda}[\mathbf{X}(t),\boldsymbol{\alpha}]$, i.e. the condition (12.7) is satisfied for it. Let the following equation be valid in the vicinity of the point $(\boldsymbol{\alpha}_0,\boldsymbol{\beta}_0)$:

$$\begin{aligned}\ln\Lambda[\mathbf{X}(t),\boldsymbol{\alpha},\boldsymbol{\beta}]&=\ln\Lambda(0)+[\partial\ln\Lambda(0)/\partial\boldsymbol{\alpha}]'(\boldsymbol{\alpha}-\boldsymbol{\alpha}_0)\\&\quad+[\partial\ln\Lambda(0)/\partial\boldsymbol{\beta}]'(\boldsymbol{\beta}-\boldsymbol{\beta}_0)+0.5(\boldsymbol{\alpha}-\boldsymbol{\alpha}_0)'[\partial^2\ln\Lambda(0)/\partial\boldsymbol{\alpha}^2](\boldsymbol{\alpha}-\boldsymbol{\alpha}_0)\\&\quad+0.5(\boldsymbol{\beta}-\boldsymbol{\beta}_0)'[\partial^2\ln\Lambda(0)/\partial\boldsymbol{\beta}^2](\boldsymbol{\beta}-\boldsymbol{\beta}_0)\\&\quad+(\boldsymbol{\alpha}-\boldsymbol{\alpha}_0)'[\partial^2\ln\Lambda(0)/\partial\boldsymbol{\alpha}\partial\boldsymbol{\beta}](\boldsymbol{\beta}-\boldsymbol{\beta}_0)\end{aligned} \tag{12.9}$$

where $\Lambda(0)=\Lambda[\mathbf{X}(t),\boldsymbol{\alpha}_0,\boldsymbol{\beta}_0]$ and in brackets are the vectors of the first derivatives and the matrices of the second derivatives of the LF with respect to parameters $\alpha_1,\ldots,\alpha_N$ and $\beta_1,\ldots,\beta_M$ at the point $(\boldsymbol{\alpha}_0,\boldsymbol{\beta}_0)$. We substitute (12.9) in (12.8) taking into account that

$$\mathrm{E}[\partial\ln\Lambda(0)/\partial\boldsymbol{\alpha}]=\mathrm{E}[\partial\ln\Lambda(0)/\partial\boldsymbol{\beta}]=0 \tag{12.10}$$

while the averaged matrices of the second derivatives coincide with the corresponding FIMs. As a result we have a system of linear equations from which the required matrix function $\boldsymbol{\beta}=\mathbf{u}(\boldsymbol{\alpha})$ can be found:

$$\boldsymbol{\beta}-\boldsymbol{\beta}_0=-[\mathbf{J}^{(\beta\beta)}]^{-1}\mathbf{J}^{(\alpha\beta)}(\boldsymbol{\alpha}-\boldsymbol{\alpha}_0). \tag{12.11}$$

Substituting (12.11) in (12.9) yields $\ln\Lambda[\mathbf{X}(t),\boldsymbol{\alpha},\mathbf{u}(\boldsymbol{\alpha})]$ which is equal to a logarithm of the desired LF $\ln\Lambda[\mathbf{X}(t),\boldsymbol{\alpha}]$. Indeed, if $\ln\Lambda[\mathbf{X}(t),\boldsymbol{\alpha},\mathbf{u}(\boldsymbol{\alpha})]$ is differentiated twice with respect to $\boldsymbol{\alpha}$ at the point $\boldsymbol{\alpha}_0$ and averaged over the ensemble of $\mathbf{X}(t)$, the matrix $\mathbf{J}^{(\alpha)}$ (12.5) will be obtained in the right-hand of (12.9). It means that the LF $\ln\Lambda[\mathbf{X}(t),\boldsymbol{\alpha},\mathbf{u}(\boldsymbol{\alpha})]$ satisfies the condition (12.7).

Thus *to derive a LF containing only informative parameters but taking into account the presence of stray parameters we have to solve averaged (over input received signals) likelihood equations for the stray parameters, then to express the stray parameters in terms of the informative ones and to substitute the obtained expression in the original LF for the stray parameters.*

Applications to Gaussian Likelihood Functionals

Now we apply the general approach developed above to Gaussian LFs which are used in signal parameter estimation for MSRSs. Let a sum of target echoes and interferences impinge upon the inputs of m receiving stations of an *active MSRS*. This sum is modelled as a realization of a complex Gaussian random process. *Unknown parameters are in the vector of mean values, i.e. of signals with a known waveform.* A system of averaged likelihood equations for $\boldsymbol{\beta}$ can be obtained from (4.44) replacing $\boldsymbol{\Theta}$ by $(\boldsymbol{\alpha},\boldsymbol{\beta})$

$$\mathrm{Re}\int_{-T/2}^{T/2}\int_{-T/2}^{T/2}[\partial\mathbf{S}^*(t_1,\boldsymbol{\alpha},\boldsymbol{\beta})/\partial\boldsymbol{\beta}]\mathbf{R}(t_1,t_2)[\mathbf{S}(t_2,\boldsymbol{\alpha}_0,\boldsymbol{\beta}_0)-\mathbf{S}(t_2,\boldsymbol{\alpha},\boldsymbol{\beta})]\,\mathrm{d}t_1\mathrm{d}t_2=0. \tag{12.12}$$

It has been taken into account in (12.12) that $\overline{\mathbf{X}(t)}=\mathrm{E}\{\mathbf{X}(t)\}=\mathbf{S}(t,\boldsymbol{\alpha}_0,\boldsymbol{\beta}_0)$ and $\mathbf{R}(t_1,t_2)=\mathbf{R}^*(t_2,t_1)$. Assume now that having solved the system of equations (12.12) for $\boldsymbol{\beta}$ we obtain the required matrix function $\boldsymbol{\beta}=\mathbf{u}(\boldsymbol{\alpha})$. Then an arbitrary element of the FIM for informative parameters according to (12.7) is given by

$$\mathcal{J}_{np}^{(\alpha)}=\mathrm{Re}\sum_{i=1}^{m}\sum_{k=1}^{m}\{\partial S_i^*[t_1,\boldsymbol{\alpha}_0,\mathbf{u}(\boldsymbol{\alpha}_0)]/\partial\alpha_n\}\{\partial S_k[t_2,\boldsymbol{\alpha}_0,\mathbf{u}(\boldsymbol{\alpha}_0)]/\partial\alpha_p\}$$

$$\times R_{ik}(t_1,t_2)\,\mathrm{d}t_1\,\mathrm{d}t_2,\quad n,p=\overline{1,N}. \tag{12.13}$$

If interferences may be considered as jointly stationary processes and the observation time, T, is much greater than their correlation intervals, it is convenient to go over from the time domain to the frequency domain. Then we obtain instead of (12.13)

$$\mathcal{J}_{np}^{(\alpha)}=\mathrm{Re}\sum_{i=1}^{m}\sum_{k=1}^{m}\frac{1}{2\pi}\int_{-\infty}^{\infty}\{\partial\Psi_i^*[\omega,\boldsymbol{\alpha}_0,\mathbf{u}_1(\boldsymbol{\alpha}_0)]/\partial\alpha_n\}$$

$$\times\{\partial\Psi_k[\omega,\boldsymbol{\alpha}_0,\mathbf{u}_1(\boldsymbol{\alpha}_0)]/\partial\alpha_p\}f_{ik}(\omega)\,\mathrm{d}\omega \tag{12.14}$$

where $\mathbf{u}_1(\boldsymbol{\alpha})$ are the solutions of the likelihood equations for stray parameters $\boldsymbol{\beta}$ averaged over received signals:

$$\mathrm{Re}\frac{1}{2\pi}\int_{-\infty}^{\infty}[\partial\boldsymbol{\Psi}^*(\omega,\boldsymbol{\alpha},\boldsymbol{\beta})/\partial\boldsymbol{\beta}]\mathbf{f}(\omega)[\boldsymbol{\Psi}(\omega,\boldsymbol{\alpha}_0,\boldsymbol{\beta}_0)-\boldsymbol{\Psi}(\omega,\boldsymbol{\alpha},\boldsymbol{\beta})]\,\mathrm{d}\omega=0. \qquad (12.15)$$

The remaining notations are the same as in (4.46), (4.48).

It can be seen from (12.12) and (12.15) that for a sum of target echoes (regular signals) and Gaussian interferences averaging of the likelihood equations over the overall received signals $\mathbf{X}(t)$ leads to expressions corresponding to the absence of interferences. Owing to the linearity of those equations in $\mathbf{X}(t)$ their averaging is equivalent to averaging of $\mathbf{X}(t)$. But the average of $\mathbf{X}(t)$ is the vector of target echoes with the true values of parameters: $\mathbf{S}(t,\boldsymbol{\alpha}_0,\boldsymbol{\beta}_0)$ with its Fourier transformation $\boldsymbol{\Psi}(\omega,\boldsymbol{\alpha}_0,\boldsymbol{\beta}_0)$.

Consider now a stochastic signal in *passive MSRS*, i.e. a m-dimensional complex Gaussian random process with zero mean. *Unknown parameters are in the covariance (correlation) matrix of a signal.* A system of likelihood equations averaged over received signal realizations takes the form

$$\sum_{i=1}^{m}\sum_{k=1}^{m}\int_{-T/2}^{T/2}\int_{-T/2}^{T/2}\{[\partial B^*_{\mathrm{sn}\,ik}(t_1,t_2,\boldsymbol{\alpha},\boldsymbol{\beta})/\partial\beta_l]R_{\mathrm{sn}\,ik}(t_1,t_2,\boldsymbol{\alpha},\boldsymbol{\beta})$$
$$+B^*_{\mathrm{sn}\,ik}(t_1,t_2,\boldsymbol{\alpha}_0,\boldsymbol{\beta}_0)[\partial R_{\mathrm{sn}\,ik}(t_1,t_2,\boldsymbol{\alpha},\boldsymbol{\beta})/\partial\beta_l]\}\,\mathrm{d}t_1\mathrm{d}t_2=0 \qquad (12.16)$$

and an arbitrary element of the FIM of informative parameters is given by

$$J^{(\alpha)}_{np}=-\sum_{i=1}^{m}\sum_{k=1}^{m}\int_{-T/2}^{T/2}\int_{-T/2}^{T/2}\{\partial B^*_{\mathrm{sn}\,ik}(t_1,t_2,\boldsymbol{\alpha}_0,\mathbf{u}_2(\boldsymbol{\alpha}_0))]/\partial\alpha_n\}$$
$$\times\{\partial R_{\mathrm{sn}\,ik}[t_1,t_2,\boldsymbol{\alpha}_0,\mathbf{u}_2(\boldsymbol{\alpha}_0))]/\partial\alpha_p\}\,\mathrm{d}t_1\mathrm{d}t_2,\quad n,p=\overline{1,N} \qquad (12.17)$$

where $\mathbf{u}_2(\boldsymbol{\alpha})$ is the vector of the solutions of the system (12.16). The remaining notations are the same as in (4.44) and (4.45). The subscript "sn" denotes a sum of "useful" signals and interferences. If this sum is a sum of jointly stationary processes and the observation time, T, is much greater than their correlation intervals, it is convenient to pass into the frequency domain (as for regular signals). Then we obtain instead of (12.17)

$$J^{(\alpha)}_{np}=-\sum_{i=1}^{m}\sum_{k=1}^{m}\frac{1}{2\pi}\int_{-}^{\infty}\{\partial\Phi^*_{\mathrm{sn}\,ik}[\omega,\boldsymbol{\alpha}_0,\mathbf{u}_3(\boldsymbol{\alpha}_0)]/\partial\alpha_n\}$$
$$\times\{\partial f_{\mathrm{sn}\,ik}[\omega,\boldsymbol{\alpha}_0,\mathbf{u}_3(\boldsymbol{\alpha}_0)]/\partial\alpha_p\}\,\mathrm{d}\omega,\quad n,p=\overline{1,N} \qquad (12.18)$$

where $\mathbf{u}_3(\boldsymbol{\alpha}_0)$ is the solution of the system of likelihood equations averaged over received signal realizations

$$\sum_{i=1}^{m}\sum_{k=1}^{m}\frac{1}{2\pi}\int_{-}\{[\partial\Phi^*_{\mathrm{sn}\,ik}(\omega,\boldsymbol{\alpha},\boldsymbol{\beta})/\partial\beta_l]f_{\mathrm{sn}\,ik}(\omega,\boldsymbol{\alpha},\boldsymbol{\beta})$$
$$+\Phi^*_{\mathrm{sn}\,ik}(\omega,\boldsymbol{\alpha}_0,\boldsymbol{\beta}_0)[\partial f_{\mathrm{sn}\,ik}(\omega,\boldsymbol{\alpha},\boldsymbol{\beta})/\partial\beta_l]\}\,\mathrm{d}\omega=0,\quad l=\overline{1,M}. \qquad (12.19)$$

Using the obtained general results we can analyze maximum attainable accuracies (Cramér–Rao bounds) for signal temporal (and, in the end, target spatial) parameter estimates in MSRSs.

In this section, we have considered an important problem of how to evaluate the error covariance matrix (ECM) of efficient estimates (the Cramér–Rao bound) of informative parameters when the corresponding likelihood function or functional (LF) contains many stray parameters so that inverting of the full Fisher information matrix (FIM) of a high order is an elaborate procedure. We have developed a general approach to the solution of this problem by constructing a special LF for informative parameters only which takes into account the influence of stray parameters. To obtain this special LF we have to solve averaged (over input received signals) likelihood equations (12.8) *for the stray parameters, then to express the stray parameters in terms of the informative ones, as in* (12.11), *and to substitute the obtained expression in the original LF* (12.9) *for the stray parameters. Using the special LF one can obtain (by the usual technique) a FIM whose dimension is determined only by the number of informative parameters in the original LF.*

This general approach has been applied to Gaussian functionals which are used for calculating such special FIMs. Their inverse determine the accuracy of efficient estimates of informative parameters in MSRSs. We have derived the averaged likelihood equations (12.12), (12.16) *in the time domain and* (12.15), (12.19) *in the frequency domain for active and passive MSRSs, respectively. Using solutions of those equations we have presented the expressions for an arbitrary elements of corresponding FIMs:* (12.13), (12.17) *in the time domain and* (12.14), (12.18) *in the frequency domain for active and passive MSRSs, respectively.*

12.2. MAXIMUM ATTAINABLE ESTIMATION ACCURACY OF SIGNAL TOAs IN ACTIVE MSRSs

Consider MSRSs with the short-term spatial coherence (where the external spatial correlated interference cancellation is possible, see Chapters 8 and 9) and spatially incoherent MSRSs. Let they consist of a single transmitting and m receiving stations (extension to m monostatic or bistatic radars is straightforward).

One-Stage Estimation Algorithms (See Section 11.2)

According to (11.10) a set of target echoes received by m stations from a single target contains three informative unknown parameters: signal TOAs, t_{s1}, t_{s2}, t_{s3}, and $2m$ stray unknown parameters: signal initial phases, $\varphi_{s1}, \ldots, \varphi_{sm}$, and r.m.s. values, $a_{s1}, \ldots, a_{sm}$. Each three TOAs, t_{s1}, t_{s2}, t_{s3}, are in a one-to-one correspondence with a point in three-dimensional physical space (singular cases are not considered here). Assume, as in Section 11.2, that the conditions under which the matrix $\mathbf{C}$ with the entries (8.35) is a diagonal one, are satisfied. Specifically, these conditions are satisfied in the absence of spatially correlated interferences (e.g., mainlobe jamming), see (5.32). Then the solutions of equations (12.15) for a_{si} do not depend on t_{si}. The derivatives $\partial\Psi_i(\omega)/\partial t_{sn}, n=1,2,3$, in (12.14) in accordance with (4.3) at $\Omega_{si}=0$ can be written as follows:

$$\frac{\partial \Psi(\omega)}{\partial t_{sn}} = -ja_{s1}^0 \Psi_0(\omega-\omega_0)\exp[-j(\omega t_{si}^0+\varphi_{si})]\left(\frac{\partial \varphi_{si}}{\partial t_{sn}}+\omega\frac{\partial t_{si}}{\partial t_{sn}}\right) \tag{12.20}$$

where $\partial\varphi_{si}/\partial t_{sn}$, $\partial t_{si}/\partial t_{sn}$ are taken at the point $\mathbf{t}_s^0$, i.e. for the true TOA values. Substituting (12.20) in (12.14) yields

$$J_{np}^{(t_s)} = \text{Re}\sum_{i=1}^{m}(a_{si}^0)^2\frac{1}{2\pi}\int_{-\infty}^{\infty}|\Psi_0(\omega-\omega_0)|^2 f_{ii}(\omega)$$

$$\times\left(\frac{\partial\varphi_{si}}{\partial t_{sn}}+\omega\frac{\partial t_{si}}{\partial t_{sn}}\right)\left(\frac{\partial\varphi_{si}}{\partial t_{sp}}+\omega\frac{\partial t_{si}}{\partial t_{sp}}\right)d\omega. \tag{12.21}$$

Here it is taken into account that after the substitution of (12.20) in (12.14) all terms in (12.14) are equal to zero for $i \neq k$ [under the same conditions as for the matrix $\mathbf{C}$ to be diagonal (see Section 8.3)]. The function $f_{ii}(\omega)$ is defined for the general case in (8.4)–(8.10). To express $\partial t_{si}/\partial t_{sn}$ we make use of (11.10) (δ_{in} is the Kronecker delta)

$$\frac{\partial t_{si}}{\partial t_{sn}} = \begin{cases}\delta_{in}, & i\leqslant 3;\\ \partial h_i(t_{s1},t_{s2},t_{s3})/\partial t_{sn}, & i>3,\end{cases} \quad n=1,2,3. \tag{12.22}$$

Now we have to solve the equations (12.15) for φ_{si} to find $\partial\varphi_{si}/\partial t_{sn}$. The corresponding non-averaged likelihood equations have been solved in (11.15). Using (8.35) for $G_{0i}=G_i$ in (11.15), ignoring the antenna influence [assuming $g_i(\alpha,\beta,\omega)=1$] and averaging over the ensemble of received signals by the substitution for $\chi_i(\omega)$ their mean values – target echo spectra with true parameter values, yields

$$\varphi_{si} = \varphi_{si}^0 - \arg\int_{-}^{\infty}|\Psi_0(\omega-\omega_0)|^2 f_{ii}(\omega)\exp[j\omega(t_{si}-t_{si}^0)]\,d\omega. \tag{12.23}$$

Let us denote the integral in (12.23) by $Z(t_{si})=Z(\alpha)$ The following identity is valid:

$$\frac{\partial \arg Z(\alpha)}{\partial\alpha} = \frac{\text{Im}[(\partial Z(\alpha)/\partial\alpha)Z^*(\alpha)]}{|Z(\alpha)|^2}. \tag{12.24}$$

Then from (12.23) taking into account (12.24) we can obtain for $t_{si}=t_{si}^0$

$$\frac{\partial\varphi_{si}}{\partial t_{sn}} = \begin{cases}\delta_{in}\bar{\omega}_n, & i\leqslant 3;\\ \bar{\omega}_i\dfrac{\partial h_i(t_{s1},t_{s2},t_{s3})}{\partial t_{sn}}, & i>3,\end{cases} \quad n=1,2,3. \tag{12.25}$$

where $\bar{\omega}_i$ is the mean frequency of the generalized signal spectrum:

$$\bar{\omega}_i = \frac{\dfrac{1}{2\pi}\displaystyle\int_{-\infty}^{\infty} \omega |\Psi_0(\omega-\omega_0)|^2 f_{ii}(\omega)\,d\omega}{\dfrac{1}{2\pi}\displaystyle\int_{-\infty}^{\infty} |\Psi_0(\omega-\omega_0)|^2 f_{ii}(\omega)\,d\omega}. \tag{12.26}$$

For narrowband signals and interferences $\bar{\omega}_i \approx \omega_0$. It is convenient to introduce also the output Generalized Signal-to-Interference-plus-Noise Ratio (GSINR), Q_i^2, and the mean squared bandwidth, $\overline{\Delta\omega_i^2}$, of the generalized spectrum at the ith station

$$Q_i^2 = \frac{(a_{si}^0)^2}{2\pi}\int_{-\infty}^{\infty} |\Psi_0(\omega-\omega_0)|^2 f_{ii}(\omega)\,d\omega; \tag{12.27}$$

$$\overline{\Delta\omega_i^2} = \frac{\dfrac{1}{2\pi}\displaystyle\int_{-\infty}^{\infty} (\omega-\bar{\omega}_i)^2 |\Psi_0(\omega-\omega_0)|^2 f_{ii}(\omega)\,d\omega}{\dfrac{1}{2\pi}\displaystyle\int_{-\infty}^{\infty} |\Psi_0(\omega-\omega_0)|^2 f_{ii}(\omega)\,d\omega}. \tag{12.28}$$

Now we can substitute (12.22) and (12.25) in (12.21) using (12.27) and (12.28). A final result takes the form

$$J_{np}^{(t_s)} = \delta_{np} Q_n^2 \overline{\Delta\omega_n^2} + \sum_{i=4}^{m} Q_i^2 \overline{\Delta\omega_i^2}\,\frac{\partial h_i(t_{s1},t_{s2},t_{s3})}{\partial t_{sn}}\,\frac{\partial h_i(t_{s1},t_{s2},t_{s3})}{\partial t_{sp}}. \tag{12.29}$$

In the absence of spatially correlated interferences $f_{ii}(\omega)=1/N_i$ [see (5.1)]. In this case $Q_i^2 = 2E_i/N_i$ where E_i is the target echo energy at the ith station, $\bar{\omega}_i$ and $\overline{\Delta\omega_i^2}$ are the mean frequency and the mean squared bandwidth of the signal spectrum $|\Psi_0(\omega-\omega_0)|^2$.

The expression (12.29) is significantly simplified if the GSINRs and the mean squared bandwidths are equal at all stations, i.e. $Q_i^2 = Q^2$ and $\overline{\Delta\omega_i^2} = \overline{\Delta\omega^2}$, $i=\overline{1,m}$. In this case

$$J_{np}^{(t_s)} = Q^2\overline{\Delta\omega^2}\left[\delta_{np} + \sum_{i=4}^{m} \frac{\partial h_i(t_{s1},t_{s2},t_{s3})}{\partial t_{sn}}\,\frac{\partial h_i(t_{s1},t_{s2},t_{s3})}{\partial t_{sp}}\right]. \tag{12.30}$$

Inverting the 3×3 matrix $\mathbf{J}^{(t_s)}$ with the elements (12.29), (12.30) yields the ECM of efficient estimates of t_{s1}, t_{s2}, t_{s3}, i.e. the Cramér–Rao lower bound on the errors of the informative TOA estimation in the presence of many stray parameters. As noted above, this ECM may usually be used as a reliable approximation of attainable errors for maximum likelihood estimates (when the observation time is long enough as compared with the reciprocal of the signal and interference bandwidths).

It can be seen that when $m \leqslant 3$ the FIM and hence the ECM are diagonal matrices. Then the lower bound on variances of TOA errors are

$$\sigma^2(t_{sn}) = \frac{1}{Q_n^2\overline{\Delta\omega_n^2}}, \quad n=1,2,3. \tag{12.31}$$

which coincides with a known expression for monostatic radars in the case where external interferences are absent. When $m>3$ information from redundant stations increases the accuracy of the estimates $\hat{t}_{s1}, \hat{t}_{s2}, \hat{t}_{s3}$. Using the obtained results for $\hat{t}_{s1}, \hat{t}_{s2}, \hat{t}_{s3}$ one can calculate the ECM for efficient estimates of spatial coordinates $\hat{x}, \hat{y}, \hat{z}$ of a target by making a change in variables with the help of relationships from Section 11.1. This is already not a statistical problem.

Owing to mutual single-valued relations between each triad of TOA estimates $\hat{t}_{s1}, \hat{t}_{s2}, \hat{t}_{s3}$ and the corresponding triad of spatial coordinates $\hat{x}, \hat{y}, \hat{z}$, the obtained results for TOA estimates may be easily transformed so as to describe target spatial coordinate estimates directly. To do this it is sufficient to replace the derivatives of t_{si} and φ_{si} with respect to t_{s1}, t_{s2}, t_{s3} by the derivatives of the same variables with respect to x, y and z. For a MSRS with a single transmitting and m receiving stations we have instead of (12.22) and (12.25)

$$\frac{\partial t_{si}}{\partial x}=\frac{1}{c}\frac{\partial R_{\Sigma i}}{\partial x};\quad \frac{\partial t_{si}}{\partial y}=\frac{1}{c}\frac{\partial R_{\Sigma i}}{\partial y};\quad \frac{\partial t_{si}}{\partial z}=\frac{1}{c}\frac{\partial R_{\Sigma i}}{\partial z},\qquad i=\overline{1,m}; \tag{12.22'}$$

$$\frac{\partial \varphi_{si}}{\partial x}=\bar{\omega}_i\frac{\partial t_{si}}{\partial x};\quad \frac{\partial \varphi_{si}}{\partial y}=\bar{\omega}_i\frac{\partial t_{si}}{\partial y};\quad \frac{\partial \varphi_{si}}{\partial z}=\bar{\omega}_i\frac{\partial t_{si}}{\partial z},\qquad i=\overline{1,m} \tag{12.25'}$$

where $R_{\Sigma i}$ is determined by (11.4). As a result an arbitrary element of the FIM $\mathbf{J}^{(x,y,z)}$, say $J_{xy}^{(x,y,z)}$, takes the form [(instead of (12.29)]

$$J_{xy}^{(x,y,z)}=\frac{1}{c^2}\sum_{i=1}^{m} Q_i^2\,\overline{\Delta\omega_i^2}\left(\frac{\partial R_{\Sigma i}}{\partial x}\frac{\partial R_{\Sigma i}}{\partial y}\right). \tag{12.29'}$$

Other elements of the 3×3 FIM $\mathbf{J}^{(x,y,z)}$ can be written in a similar manner. For $Q_i^2=Q^2$ and $\overline{\Delta\omega_i^2}=\overline{\Delta\omega^2}$, $i=\overline{1,m}$, (12.29′) is replaced by

$$J_{xy}^{(x,y,z)}=\frac{1}{c^2}Q^2\,\overline{\Delta\omega^2}\sum_{i=1}^{m}\left(\frac{\partial R_{\Sigma i}}{\partial x}\frac{\partial R_{\Sigma i}}{\partial y}\right). \tag{12.30'}$$

It is convenient to present (12.29′) and (12.30′) in a matrix form. Denoting

$$\Delta\mathbf{R}=\begin{bmatrix}\frac{\partial R_{\Sigma 1}}{\partial x} & \frac{\partial R_{\Sigma 1}}{\partial y} & \frac{\partial R_{\Sigma 1}}{\partial z}\\ \cdots & \cdots & \cdots\\ \cdots & \cdots & \cdots\\ \frac{\partial R_{\Sigma m}}{\partial x} & \frac{\partial R_{\Sigma m}}{\partial y} & \frac{\partial R_{\Sigma m}}{\partial z}\end{bmatrix};\quad [\boldsymbol{\sigma}^2(R)]^{-1}=\mathrm{diag}\left(\frac{Q_1^2\overline{\Delta\omega_1^2}}{c^2},\ldots,\frac{Q_m^2\overline{\Delta\omega_m^2}}{c^2}\right)$$

yields instead of (12.29′)

$$\mathbf{J}^{(x,y,z)}=\Delta\mathbf{R}'[\boldsymbol{\sigma}^2(R)]^{-1}\Delta\mathbf{R}. \tag{12.32}$$

Each row of the matrix $\Delta\mathbf{R}$ contains three projections of a gradient of the corresponding range sum $R_{\Sigma i}$. Inverting the 3×3 FIM (12.32) with the elements

(12.29′) yields the ECM of the efficient estimates (i.e. the Cramér–Rao lower bound on attainable errors) of spatial coordinates of a target through TOA measurements.

First Stage of Two-Stage Target Localization Algorithms (See Section 11.3)

Consider a MSRS of the same type as above. In this case we have m (instead of three) informative TOAs, $t_{s1},\ldots,t_{sm}$, and the same $2m$ stray parameters as before.

We can write the expression for an arbitrary element of the FIM $\mathbf{J}^{(t_s)}$ immediately from (12.21) taking into account that unlike (12.22), (12.25) $\partial t_{si}/\partial t_{sn}=\delta_{in}$, $\partial\varphi_{si}/\partial t_{sn}=\delta_{in}\bar{\omega}_n$, $i,n=\overline{1,m}$. Then

$$\mathbf{J}^{(t_s)}=\mathrm{diag}(Q_1^2\overline{\Delta\omega_1^2},\ldots,Q_m^2\overline{\Delta\omega_m^2}); \tag{12.33}$$

$$J_{np}^{(t_s)}=\delta_{np}Q_n^2\overline{\Delta\omega_n^2},\quad \sigma_{\mathrm{eff}}^2(t_{sn})=\left[Q_n^2\overline{\Delta\omega_n^2}\right]^{-1},\qquad n,p=\overline{1,m} \tag{12.33′}$$

where Q_n^2 and $\overline{\Delta\omega_n^2}$ are determined as earlier by (12.27), (12.28) and (12.26). It is seen that the ECM of these estimates is diagonal, and the variances of errors are determined by the expression (12.31) for $n=\overline{1,m}$.

The Influence of External Interferences on the Maximum Attainable Estimation Accuracy

Consider TOA measurements for the first stage of two-stage target localization algorithms. These algorithms are simpler than one-stage ones and more often used in practice. On the other hand, the general character of interference influence is similar on one-stage algorithms as well.

Optimum processing for MSRSs with the short-term spatial coherence includes coherent interstation processing with spatially correlated interference cancellation. The influence of remaining interferences after cancellation on the estimation maximum attainable accuracy is determined by the functions $f_{ii}(\omega)$, $i=\overline{1,m}$, in (12.26), (12.27) and (12.28). These functions are the diagonal elements of the matrix $\mathbf{f}(\omega)$ which is the inverse of the PSD matrix of the sum of external interferences and self-noises at the inputs of spatially separated receivers (see Section 4.1).

Let external spatially correlated interferences from a single source (e.g., mainlobe jammer) impinge on a MSRS with m receiving stations. Then $f_{ii}(\omega)$ takes the form (7.5) at $i=k$. For intensive input interferences ($q_\Sigma^2\gg1$) $f_{ii}(\omega)\approx(q_\Sigma^2-q_i^2)/N_iq_\Sigma^2$. Substituting it in (12.26)–(12.28) yields [compare with (12.33′)]

$$\sigma_{\mathrm{eff}}^2(\hat{t}_{si})\approx\left[\frac{2E_i}{N_i}\frac{q_\Sigma^2-q_i^2}{q_\Sigma^2}\overline{\Delta\omega^2}\right]^{-1},\quad i=\overline{1,m}. \tag{12.34}$$

Here $\overline{\Delta\omega^2}$ does not depend on the number of station "i".

The clearest result can be obtained for equal INRs at different stations ($q_i^2=q^2, i=\overline{1,m}$):

$$\sigma_{\mathrm{eff}}^2(\hat{t}_{si})=\left[\frac{2E_i}{N_i}\frac{m-1}{m}\overline{\Delta\omega^2}\right]^{-1}. \tag{12.34′}$$

It is interesting to compare (12.34) and (12.34') with the minimal attainable variance (the Cramér–Rao lower bound) when external interferences are absent so that $f_{li}(\omega)=1/N_i$ [see (7.5)]

$$\sigma^2_{\text{eff}}(\hat{t}_{si})=\left[\frac{2E_i}{N_i}\overline{\Delta\omega^2}\right]^{-1}. \tag{12.35}$$

The presence of external interferences from a single source, if the optimum processing with interference cancellation is performed, leads to the increase of the minimal attainable variance of TOA estimates, $\sigma^2_{\text{eff}}(\hat{t}_{si})$, only by $m/(m-1)$ times.

Let us now evaluate the variance of TOA efficient estimates in the presence of external interferences without interference cancellation. This may be in MSRSs with the short-term spatial coherence when interference cancellation is not performed or in spatially incoherent MSRSs where coherent interference cancellation is impossible. Using the expression (5.22) for PSD of the sum of external interference and self-noise at the ith station, assuming as before the interference to be strong (high INR, $q^2_{Ii}=q^2_i\gg 1$) and neglecting possible nonuniformities of the interference PSD $F_i(\omega-\omega_0)$ within the signal spectrum $|\Psi_0(\omega-\omega_0)|^2$ yields

$$\sigma^2_{\text{eff}}(\hat{t}_{si})\approx\left[\frac{2E_i}{N_i q_i^2}\overline{\Delta\omega_i^2}\right]^{-1}. \tag{12.36}$$

It is seen that for $q_i^2\gg 1$, the minimal attainable variance is much greater than both in the absence and in the presence of external interferences when optimum processing with interference cancellation is performed [(12.35) and (12.34), (12.34'), respectively].

It may be shown that when intensive spatially correlated interferences from M independent sources (e.g., jammers) enter mainlobes of m spatially separated station ADPs, $m>M$, and these sources are resolved in TDOAs, then the following approximate expression for the variance of efficient signal TOA estimates is valid:

$$\sigma^2_{\text{eff}}(\hat{t}_{si})=\left[\frac{2E_i}{N_i}\frac{m-M}{m}\overline{\Delta\omega^2}\right]. \tag{12.37}$$

The obtained results relate to nonfluctuating signals. For signals with fluctuating amplitudes the true r.m.s. value a^0_{si} in (12.20) should be replaced by σ^0_i [see (5.74) and explanation after (5.74)]. Then $(a^0_{si})^2$ and Q^2_i should be replaced by $\overline{(\sigma^0_i)^2}$ and $\overline{Q^2_i}$ in (12.27) and in the sequel. The SNRs $(2E_i/N_i)$ are to be replaced by $\bar{q}^2_{\text{out}\,i}=\bar{E}_i/N_i$ [see (5.75), (5.76)]. The overbars denote mean values.

In this section, *we have employed the general technique developed in Section* 12.1 *to the problem of maximum attainable accuracy for signal TOA estimation in active MSRSs when the corresponding likelihood functional contains many stray parameters. For one-stage algorithms considered in Section* 11.2, *we have derived the general expression* (12.29) *for arbitrary elements of the* 3×3 *Fisher information matrix* (*FIM*) *which can be easy inverted to obtain the error covariance matrix* (*ECM*) *for efficient TOA estimates* (*the Cramér–Rao lower bound on errors*). *Simplified formulas* (12.30), (12.31) *have been obtained from that general expression for some specific cases.*

We have shown that the derived relationships can be easily transformed so as to describe directly the FIM and ECM for efficient estimates of target spatial coordinates [*see* (12.29'), (12.30'), (12.31) *and* (12.32)]. *Such a transformation is possible owing to mutual single-valued relations between each triad of the TOA estimates* $(\hat{t}_{s1}, \hat{t}_{s2}, \hat{t}_{s3})$ *and the corresponding point in the three-dimensional physical space with the coordinates* $(\hat{x}, \hat{y}, \hat{z})$.

For the first stage of two-stage target localization algorithms considered in Section 11.3, *we have obtained expressions* (12.33) *and* (12.33′) *which determine the diagonal FIM and minimal attainable variances of TOA estimation at the inputs of m spatially separated stations.*

We have analyzed the influence of external interferences on the minimum attainable estimation errors. For a single source of spatially correlated interferences (e.g., a mainlobe jammer) we have presented expressions (12.34) *and* (12.34′) *indicating that when optimum interstation processing (with spatially correlated interference cancellation) is performed, the accuracy of efficient TOA estimates degrades not too much as compared with the case where external interferences are absent* (12.35). *However, when above mentioned optimum processing is rejected, the accuracy degradation may be prohibitively large* [*see* (12.36)].

12.3. MAXIMUM ATTAINABLE ESTIMATION ACCURACY OF SIGNAL TDOAs IN PASSIVE MSRSs

Now we apply the general technique developed in Section 12.1 to passive MSRSs. Consider a passive MSRS containing m receiving stations with the short-term spatial coherence. Assume that spatial correlation of external interferences are either absent or ignored (i.e. there is no intensive mainlobe jammers, see Example 9.6).

One-Stage Estimation Algorithms

A vector of unknown signal parameters contains three informative parameters, $\tau_{s12}, \tau_{s13}, \tau_{s14}$ (or spatial coordinates, x, y, z) and following stray parameters: $m(m-1)/2$ phases, $\Delta\varphi_{sik}$, of the signal covariance (correlation) matrix and m power SNRs, q_{si}^2, $i, k = \overline{1, m}$, $i<k$. In order to simplify the notation in subsequent derivations, we drop the subscript "s" denoting "signal". It can be shown that solutions of the averaged likelihood equations (12.19) for q_i^2 do not depend on τ_{ik}. Therefore, the general expression (12.18) for an arbitrary FIM element of the estimates $\hat{\tau}_{12}, \hat{\tau}_{13}, \hat{\tau}_{14}$ takes the form

$$J_{np}^{(\tau)} = -\sum_{i=1}^{m}\sum_{k=1}^{m}\frac{T}{2\pi}\int_{-\infty}^{\infty}\{\partial\Phi_{ik}^{*}[\omega, \boldsymbol{\tau}^0, \Delta\boldsymbol{\varphi}(\boldsymbol{\tau}^0)]/\partial\tau_{1(n+1)}\}$$

$$\times\{\partial f_{ik}[\omega, \boldsymbol{\tau}^0, \Delta\boldsymbol{\varphi}(\boldsymbol{\tau}^0)]/\partial\tau_{1(p+1)}\}\,d\omega, \quad n, p = 1, 2, 3. \tag{12.38}$$

Now we make use of expressions (7.2)–(7.5) for $\Phi_{snik}(\omega)$ and $f_{snik}(\omega)$. Substituting them in (12.38) yields

$$J_{np}^{(\tau)} = \sum_{i=1}^{m}\sum_{k=1}^{m}\frac{q_i^2 q_k^2 T}{2\pi}\int_{-\infty}^{\infty}\frac{F^2(\omega-\omega_0)}{1+F(\omega-\omega_0)q_\Sigma^2}$$

$$\times\left(\omega\frac{\partial\tau_{ik}}{\partial\tau_{1(n+1)}}+\frac{\partial\Delta\varphi_{ik}}{\partial\tau_{1(n+1)}}\right)\left(\omega\frac{\partial\tau_{ik}}{\partial\tau_{1(p+1)}}+\frac{\partial\Delta\varphi_{ik}}{\partial\tau_{1(p+1)}}\right)d\omega. \tag{12.39}$$

The dependence of $\Delta\varphi$ on τ can be determined from (12.19) setting $\beta_l=\Delta\varphi_{rs}$. Since all $\Delta\varphi_{rs}$ for different r,s are assumed to be independent parameters (such a model was discussed in Section 7.2), all terms in the double sum of (12.19) are equal to zero for each combination of r,s excluding two complex conjugate terms with subscripts "rs" and "sr". Besides, the first term under the integration sign in (12.19) is equal to zero for all i,k since $\Delta\varphi_{ik}$ and τ_{ik} are nonenergetic signal parameters. Then solving the equation (12.19) for $\Delta\varphi_{ik}$ yields

$$\Delta\varphi_{ik}=\Delta\varphi_{ik}^{0}-\arg\frac{T}{\pi}\int_{-\infty}^{\infty}\frac{F^{2}(\omega-\omega_{0})\exp[j\omega(\tau_{ik}-\tau_{ik}^{0})]}{1+F(\omega-\omega_{0})q_{\Sigma}^{2}}d\omega. \qquad (12.40)$$

To obtain the derivatives $\partial\Delta\varphi_{ik}/\partial\tau_{1(n+1)}$ from (12.40) we employ (12.24) and take into account that $\partial\Delta\varphi_{ik}/\partial\tau_{1(n+1)}=(\partial\Delta\varphi_{ik}/\partial\tau_{ik})[\partial\tau_{ik}/\partial\tau_{1(n+1)}]$. As a result we have

$$\frac{\partial\Delta\varphi_{ik}}{\partial\tau_{1(n+1)}}=-\tilde{\bar{\omega}}\frac{\partial\tau_{ik}}{\partial\tau_{1(n+1)}},\quad i,k=\overline{1,m};\ i<k;\ n=1,2,3. \qquad (12.41)$$

where $\tilde{\bar{\omega}}$ is the mean frequency of the generalized signal power spectrum

$$\tilde{\bar{\omega}}=\frac{1}{2\pi\Delta f_{e}}\int_{-\infty}^{\infty}\frac{\omega F^{2}(\omega-\omega_{0})}{1+F(\omega-\omega_{0})q_{\Sigma}^{2}}d\omega. \qquad (12.42)$$

and Δf_e may be considered as the "equivalent" signal bandwidth

$$\Delta f_{e}=\frac{1}{2\pi}\int_{-\infty}^{\infty}\frac{F^{2}(\omega-\omega_{0})}{1+F(\omega-\omega_{0})q_{\Sigma}^{2}}d\omega. \qquad (12.43)$$

It is convenient to introduce additionally the mean squared bandwidth of the generalized power spectrum

$$\overline{\overline{\Delta\omega^{2}}}=\frac{1}{2\pi\Delta f_{e}}\int_{-\infty}^{\infty}\frac{(\omega-\tilde{\bar{\omega}})^{2}F^{2}(\omega-\omega_{0})}{1+F(\omega-\omega_{0})q_{\Sigma}^{2}}d\omega. \qquad (12.44)$$

It is seen that Δf_e, $\tilde{\bar{\omega}}$ and $\overline{\overline{\Delta\omega^2}}$ depend on the total input SNR q_{Σ}^2 (7.7). This dependence vanishes for weak signals ($q_{\Sigma}^2\ll 1$) while $\tilde{\bar{\omega}}$ and $\overline{\overline{\Delta\omega^2}}$ do not also depend on q_{Σ}^2 for strong signals ($q_{\Sigma}^2\gg 1$) and for rectangular signal spectrum [(7.27) at $\Omega_{si}=0$] Let us substitute (12.41) in (12.39) taking into account (12.44) and the identity $\tau_{ik}=\tau_{1k}-\tau_{1i}$ [see (7.8)]. Then an arbitrary element of the FIM takes the form

$$J_{np}^{(\tau)}=2\Delta f_{e}T\,\overline{\overline{\Delta\omega^{2}}}\sum_{i=1}^{m-1}\sum_{k=i+1}^{m}q_{i}^{2}q_{k}^{2}\left(\frac{\partial\tau_{1k}}{\partial\tau_{1(n+1)}}-\frac{\partial\tau_{1i}}{\partial\tau_{1(n+1)}}\right)\left(\frac{\partial\tau_{1k}}{\partial\tau_{1(p+1)}}-\frac{\partial\tau_{1i}}{\partial\tau_{1(p+1)}}\right),$$

$$n,p=1,2,3. \qquad (12.45)$$

The derivatives are determined from (11.23):

$$\frac{\partial\tau_{1k}}{\partial\tau_{1(n+1)}}=\begin{cases}\delta_{k(n+1)}, & k\leqslant 4;\\ \dfrac{\partial\hbar_{1k}(\tau_{12},\tau_{13},\tau_{14})}{\partial\tau_{1(n+1)}}, & k>4.\end{cases} \tag{12.46}$$

The inverse of the 3×3 FIM with the elements (12.45) is the ECM of the efficient estimates $\hat{\tau}_{12},\hat{\tau}_{13},\hat{\tau}_{14}$, i.e. the Cramér–Rao lower bound on errors of the one-stage estimation of TDOAs $\hat{\tau}_{12},\hat{\tau}_{13},\hat{\tau}_{14}$.

When the number m of receiving stations is $2\leqslant m\leqslant 4$ so that the number of linearly independent TDOAs does not exceed the dimensionality of space, we obtain from (12.45) and (12.46)

$$J_{np}^{(\tau)}=2\Delta f_e T\,\overline{\overline{\Delta\omega^2}}\,q_{n+1}^2(\delta_{np}q_\Sigma^2-q_{p+1}^2),\quad n,p=\overline{1,m-1} \tag{12.47}$$

where δ_{np} is the Kronecker delta.

Owing to mutual single-valued relations between each triad $\hat{\tau}_{12},\hat{\tau}_{13},\hat{\tau}_{14}$ and the corresponding point in space with the coordinates $\hat{x},\hat{y},\hat{z}$, the ECM of efficient estimates $\hat{\tau}_{12},\hat{\tau}_{13},\hat{\tau}_{14}$ fully determines the ECM of efficient estimates $\hat{x},\hat{y},\hat{z}$. It may be presented in an explicit form as it has been done for TOA estimates in Section 12.2. In (12.45) the derivatives with respect to $\tau_{12},\tau_{13},\tau_{14}$ should be replaced by the derivatives with respect to x,y,z. Then we have instead of (12.46)

$$\frac{\partial\tau_{1k}}{\partial x}=\frac{1}{c}\frac{\partial\Delta R_{1k}}{\partial x};\quad \frac{\partial\tau_{1k}}{\partial y}=\frac{1}{c}\frac{\partial\Delta R_{1k}}{\partial y};\quad \frac{\partial\tau_{1k}}{\partial z}=\frac{1}{c}\frac{\partial\Delta R_{1k}}{\partial z},\qquad k=\overline{1,m} \tag{12.46'}$$

where c is the velocity of light and the relationships for range differences ΔR_{1k} are presented in Section 11.1. Substituting (12.46′) in (12.45) yields an arbitrary element of the FIM, $\mathbf{J}^{(x,y,z)}$, say $J_{xy}^{(x,y,z)}$

$$J_{xy}^{(x,y,z)}=\frac{2\Delta f_e T\overline{\overline{\Delta\omega^2}}}{c^2}\sum_{i=1}^{m-1}\sum_{k=i+1}^{m}q_i^2q_k^2\left(\frac{\partial\Delta R_{1k}}{\partial x}-\frac{\partial\Delta R_{1i}}{\partial x}\right)\left(\frac{\partial\Delta R_{1k}}{\partial y}-\frac{\partial\Delta R_{1i}}{\partial y}\right)$$

or in a slightly different form (since $\Delta R_{1k}-\Delta R_{1i}=\Delta R_{ik}$)

$$J_{xy}^{(x,y,z)}=\frac{2\Delta f_e T\,\overline{\overline{\Delta\omega^2}}}{c^2}\sum_{i=1}^{m-1}\sum_{k=i+1}^{m}q_i^2q_k^2\left(\frac{\partial\Delta R_{ik}}{\partial x}\,\frac{\partial\Delta R_{ik}}{\partial y}\right). \tag{12.45'}$$

It should be taken into account that $\Delta R_{11}=\Delta R_{ii}=0$. The remaining elements of the FIM $\mathbf{J}^{(x,y,z)}$ can be written by replacing the derivatives in parenthesis in according with (12.46′). Inverting this 3×3 FIM yields the ECM of efficient estimates of the radiation source spatial coordinates, i.e. the Cramér–Rao lower bound on errors of the one-stage estimation of x,y,z.

Example 12.1. Let us obtain the ECM of efficient TDOA estimates (the Cramér–Rao lower bound on errors) for a particular case where the number of linearly independent TDOAs does not exceed the dimensionality of space ($2\leqslant m\leqslant 4$) and signal PSD has a rectangular form within the bandwidth $(\omega_0-\Delta\omega/2,\omega_0+\Delta\omega/2)$ [see (7.27) at $\Omega_{si}=0$]. In this case from (12.43) and (12.44) we have $\Delta f_e=\Delta f/(1+q_\Sigma^2)$, $\overline{\overline{\Delta\omega^2}}=4\pi^2\Delta f^2/12$. Substituting in (12.47) and inverting the FIM yields the minimum attainable variances of efficient estimates, $\hat{\tau}_{12}, \hat{\tau}_{13}, \hat{\tau}_{14}$

$$\sigma^2_{\text{eff}}(\hat{\tau}_{1(n+1)}) = \frac{q_1^2 + q_{n+1}^2}{q_1^2 q_{n+1}^2} \frac{1+q_\Sigma^2}{q_\Sigma^2} \frac{1}{\Delta f T} \frac{3}{2\pi^2 \Delta f^2} \tag{12.48}$$

and the correlation coefficient of the efficient estimates

$$\rho_{\text{eff}}[\hat{\tau}_{1(n+1)}, \hat{\tau}_{1(p+1)}] = \frac{q_{n+1} q_{p+1}}{\sqrt{(q_1^2 + q_{n+1}^2)(q_1^2 + q_{p+1}^2)}}. \tag{12.49}$$

In (12.48) and (12.49) $n, p = \overline{1, m-1}$ and $2 \leqslant m \leqslant 4$. The increase of the number of stations leads to increasing of estimation accuracy owing to redundant measurements.

For equal SNRs at the inputs of stations, $q_1^2 = q_{n+1}^2 = q_{p+1}^2 = q^2$ we obtain from (12.48) and (12.49)

$$\sigma^2_{\text{eff}}(\hat{\tau}_{1(n+1)}) = \frac{(1+mq^2)}{mq^4} \frac{1}{\Delta f T} \frac{3}{\pi^2 \Delta f^2}; \qquad \rho_{\text{eff}}[\hat{\tau}_{1(n+1)}, \hat{\tau}_{1(p+1)}] = 0.5. \tag{12.50}$$

Note that for $m=3$ (12.50) coincides with the expression (11.54) from Example 11.2. It confirms that TDOA estimates derived in that Example are actually efficient estimates.

Consider now the maximum attainable accuracy of TDOA measurements *at the first stage of two-stage algorithms for radiation source localization by hyperbolic method*. As shown in Section 11.4, in a passive MSRS with m receiving stations $m-1$ linearly independent TDOAs are to be estimated at this stage. To obtain required expressions for an arbitrary element of the $(m-1) \times (m-1)$ FIM it is sufficient to transform (12.46) taking into account the linear independence of all TDOAs of the τ_{1k} type:

$$\frac{\partial \tau_{1k}}{\partial \tau_{1(n+1)}} = \delta_{k(n+1)}, \quad k = \overline{2, m}; \; n = \overline{1, m-1}. \tag{12.51}$$

Substituting (12.51) in (12.45) yields (12.47). However, the number of stations, m, may now be arbitrary, including the values of $m-1$ exceeding the dimensionality of space.

The $(m-1) \times (m-1)$ FIM with elements (12.47) can be easily inverted even when m is large. It may be presented as

$$\mathbf{J}^{(\tau)} = 2\Delta f_e T \overline{\overline{\Delta\omega^2}} [\mathbf{Q} - \mathbf{q}\mathbf{q}'] \tag{12.52}$$

where $\mathbf{Q}$ is a $(m-1) \times (m-1)$ diagonal matrix and $\mathbf{q}$ is the $(m-1) \times 1$ vector

$$\mathbf{Q} = q_\Sigma^2 \,\text{diag}(q_2^2, \ldots, q_m^2), \qquad \mathbf{q} = (q_2^2, \ldots, q_m^2)'. \tag{12.53}$$

We met matrices with the structure like (12.53) in Sections 7.1 and 8.1 [see (7.2) and (8.6)]. The inversion rule for such matrices was used there (e.g., [215]):

$$[\mathbf{J}^{(\tau)}]^{-1} = \mathbf{B}_{\text{eff}} = \frac{1}{2\Delta f_e T \overline{\overline{\Delta\omega^2}}} [\mathbf{Q}^{-1} + \gamma \mathbf{Q}^{-1} \mathbf{q}\mathbf{q}' \mathbf{Q}^{-1}] \tag{12.54}$$

where

$$\gamma=[1-\mathbf{q}'\mathbf{Q}^{-1}\mathbf{q}]^{-1}. \tag{12.55}$$

Substituting (12.53) in (12.54) yields the ECM of efficient estimates $\hat{\tau}_{12}, \ldots, \hat{\tau}_{1m}$ (the Cramér–Rao lower bound on estimate errors)

$$\mathbf{B}_{\text{eff}}=\frac{1}{2\Delta f_e T\overline{\overline{\Delta\omega^2}}q_\Sigma^2}\left\{\text{diag}\left(\frac{1}{q_2^2},\ldots,\frac{1}{q_m^2}\right)+\frac{1}{q_1^2}\begin{pmatrix}1 & \ldots & 1\\ \ldots & \ldots & \ldots\\ 1 & \ldots & 1\end{pmatrix}\right\}. \tag{12.56}$$

It can be seen that the FIM and ECM of efficient estimates, $\hat{\tau}_{12}, \hat{\tau}_{13}, \hat{\tau}_{14}$, for one-stage TDOA estimation considered above, are particular cases of (12.52), (12.53) and (12.56). When all input SNRs are equal to each other, i.e. $q_i^2=q^2, i=\overline{1,m}$, (12.56) takes the simplest form

$$\mathbf{B}_{\text{eff}}=\frac{1}{2\Delta f_e T\overline{\overline{\Delta\omega^2}}mq^4}\begin{bmatrix}2 & 1 & \ldots & 1\\ 1 & 2 & \ldots & 1\\ \ldots & \ldots & \ldots & \ldots\\ 1 & 1 & \ldots & 2\end{bmatrix}. \tag{12.57}$$

Example 12.2. Consider again signals with rectangular power spectrum [(7.27) at $\Omega_{si}=0$] as in Example 12.1, but for arbitrary number m of receiving stations. Substituting from (12.43) and (12.44) $\Delta f_e=\Delta f/(1+q_\Sigma^2)$, $\overline{\overline{\Delta\omega^2}}=4\pi^2\Delta f^2/12$ in (12.56) yields (12.48) and (12.49). Thus these expressions do not depend on the number of stations, m, and on whether $m-1$ exceeds the dimensionality of space. For equal input SNRs from (12.57) we have again (12.50).

It should be noted that for a MSRS with $m=3$ stations (12.57) coincides with (11.53), (11.54). It confirms that the second approach to the first stage of two-stage radiation source localization algorithm considered in Section 11.4 [see (11.44)], really yields efficient estimates of linearly independent TDOAs.

Maximum attainable accuracy of radiation source (e.g. jammer) localization in space using efficient estimates of linearly independent TDOAs, i.e. at the second stage of two-stage source localization, will be considered in Section 14.3.

In closing of this Section we consider the maximum attainable accuracy (the Cramér–Rao lower bound on errors) for *independent TDOA measurements by each pair of stations individually*. The minimum error variance can be obtained from (12.47) setting $m=2$. In this case $q_\Sigma^2=q_1^2+q_2^2$ and $[\mathbf{J}^{(\tau)}]^{-1}=\sigma_{\text{eff}}^2(\hat{\tau}_{12})$. However, such a relationship is valid for each pair of stations so that

$$\sigma_{\text{eff}}^2(\hat{\tau}_{ik})=(2\Delta f_e T\overline{\overline{\Delta\omega^2}}q_i^2q_k^2)^{-1} \tag{12.58}$$

where Δf_e and $\overline{\overline{\Delta\omega^2}}$ are defined in (12.43) and (12.44). We have taken into account in (12.58) that for each individual pair of the ith and kth stations $q_\Sigma^2=q_i^2+q_k^2$

Example 12.3. The accuracy of TDOA measurement by a single pair of receiving stations (or by a single correlator) is important for practice. Therefore, consider (12.58) for the rectangular signal PSD within the bandwidth Δf [(7.27) at $\Omega_{si}=0$]. Substituting again $\Delta f_e=\Delta f/(1+q_\Sigma^2)=\Delta f/(1+q_i^2+q_k^2)$, $\overline{\overline{\Delta\omega^2}}=4\pi^2\Delta f^2/12$ from

(12.43) and (12.44) in (12.58) yields

$$\sigma^2_{\text{eff}}(\hat{\tau}_{ik}) = \frac{3(1+q_i^2+q_k^2)}{2q_i^2 q_k^2 \Delta f T \pi^2 \Delta f^2}. \tag{12.59}$$

Note that for equal values of the input SNR ($q_i^2 = q_k^2 = q^2$) (12.59) coincides with (11.55). It means that a maximum likelihood TDOA estimate according to the algorithm (11.41) turns out to be an efficient estimate (with minimal variance) under the conditions of this Example.

In this section*, we have applied the general technique developed in Section* 12.1, *to the maximum attainable accuracy analysis of TDOA estimation, i.e. to deriving the Cramér–Rao lower bound on estimation errors, for one-stage and the first stage of two-stage radiation source localization by the hyperbolic method. We have developed the general expression* (12.45) *that together with* (12.46) *determines the* 3×3 *Fisher information matrix (FIM)* (12.47) *and hence its inverse, the error covariance matrix (ECM) of efficient TDOA estimates,* $\hat{\tau}_{12}, \hat{\tau}_{13}, \hat{\tau}_{14}$. *Owing to mutual single-valued correspondence between the set of triads,* $\hat{\tau}_{12}, \hat{\tau}_{13}, \hat{\tau}_{14}$, *and the set of points of the three-dimensional space, each FIM and ECM of efficient estimates* $\hat{\tau}_{12}, \hat{\tau}_{13}, \hat{\tau}_{14}$ *can be converted (by changing the variables) into corresponding FIM and ECM of efficient estimates of the radiation source spatial coordinates,* $\hat{x}, \hat{y}, \hat{z}$. *We have derived expression* (12.45′) *for an arbitrary element of the* 3×3 *FIM of efficient estimates* $\hat{x}, \hat{y}, \hat{z}$. *For practically important (though idealized) particular case of rectangular signal PSD we have obtained expressions* (12.48), (12.49), (12.50) *in Example* 12.1 *for variances and covariances of TDOA efficient estimates when the number of stations does not exceeds the dimensionality of space no more than by one.*

We have also considered the Cramér–Rao lower bound on estimation errors for the first stage of two-stage radiation source localization when all $m-1$ *linearly independent TDOAs are estimated. We have derived the expressions for the FIM* (12.52) *and the ECM* (12.56), (12.57) *of efficient estimates,* $\hat{\tau}_{12}, \ldots, \hat{\tau}_{1m}$. *We have shown in Example* 12.2 *that for the rectangular signal PSD expressions* (12.48), (12.49) *and* (12.50) *remain valid for arbitrary number of stations. Besides, this Example confirms (under the conditions considered) that the second approach discussed in Section* 11.4 *to the first stage of two-stage radiation source localization algorithms, really yields efficient estimates of linearly independent TDOAs.*

A particular case of obtained results relates to an important situation where a TDOA is to be measured by a pair of stations individually. For this case we have presented expressions (12.58) *and* (12.59) *for minimum attainable variance.*

13. DOPPLER FREQUENCY MEASUREMENT OF FLUCTUATING SIGNALS FOR TARGET VELOCITY VECTOR ESTIMATION

13.1. OPTIMUM DOPPLER FREQUENCY MEASUREMENT ALGORITHMS FOR POINT-LIKE TARGETS AND MAXIMUM ATTAINABLE ACCURACY

It was mentioned in Section 1.2 that target velocity vector measurements may be of great importance for different radar systems. To obtain the velocity vector of a moving target in a MSRS, Doppler frequencies (DFs) of target echoes can be measured at spatially separated stations (see Section 1.2). In connection with this, a problem of optimum DF measurement and of attainable measurement accuracy arises.

In this section we consider a model of point-like target. Such a model is conventionally used and is, as a rule, quite adequate for monostatic radars. However, as will be shown in Section 13.2, this model is often too idealized for MSRSs. Nevertheless, it allows us to obtain some reference relationships. A more complex and realistic model will be considered in Sections 13.2–13.5.

Optimum DF Measurement Algorithms

Let a point-like target move with a constant velocity $\mathbf{v}$ [within an observation interval $t \in (-T/2, T/2)$ which is of the order of signal duration] with respect to the conventional centre of a MSRS, consisting of a single transmitting and m receiving stations. The special features relevant to MSRSs containing m monostatic radars will be indicated below. We denote the unit vectors from the target to the transmitting station and to the ith receiving station by $\mathbf{r}_0$ and $\mathbf{r}_i$, respectively, and assume them to be constant within the interval $(-T/2, T/2)$.

The illuminating signal is a coherent train (a "coherent burst") consisting of M short pulses. Such a waveform is the most convenient for precise DF measurement by a pulse radar system. Since DF estimation with the help of pulse trains is based on phase shift measurements, phase ambiguity is inherent in this method. We assume here that phase ambiguity (leading to gross errors) is resolved by one of known techniques (e.g., [40,44] etc.). Assume that Doppler phase shifts during each pulse, τ_p, as well as differences of interpulse time interval (repetition period), T_R, caused by the Doppler effect in target echoes, may be neglected (the latter can be accommodated using rough velocity estimates). A target echo at the ith receiving station may be written in the form

$$S_i(t) = \exp\left[j\omega_0\left(t - \frac{R_0 + R_i}{c}\right)\right] \times \sum_{l=0}^{M-1} A_{il} s_0\left(t - lT_R - \frac{R_0 + R_i}{c}\right) \exp(j\Omega_i l T_R) \tag{13.1}$$

where ω_0 is the carrier frequency; R_0 and R_i are the target distances from the transmitting and receiving stations, respectively; c is the light velocity; A_{il} is the complex amplitude of the mth pulse: $A_{il}=\sqrt{2}a_{il}\exp(-j\varphi_{il})$; a_{il}, φ_{il} are the r.m.s. value and the initial phase of the lth pulse; $s_0(t)$ is the normalized complex envelope of an individual pulse so that

$$\int_{-T/2}^{T/2} s_0(t-lT_R)s_0^*(t-nT_R)\,dt=\begin{cases}\tau_p, & l=n;\\ 0, & l\neq n;\end{cases} \tag{13.2}$$

$$\Omega_i=\omega_0\frac{\mathbf{v}(\mathbf{r}_0+\mathbf{r}_i)}{c} \tag{13.3}$$

is the DF to be estimated. Note that the normalization in (13.1), (13.2) differs slightly from that adopted in Section 4.1 [$|A_{il}|=\sqrt{2}a_{il}$ and hence in the right side of (13.2) is τ_p instead of $2\tau_p$ as in (4.1)–(4.4)].

If a MSRS consists of m monostatic radars, R_0+R_i in (13.1) and $\mathbf{r}_0+\mathbf{r}_i$ in (13.3) should be replaced by $2R_i$, and $2\mathbf{r}_i$ respectively. In this case carrier frequencies ω_0 may be different, i.e. in (13.1) and (13.3) $\omega_0=\omega_i$, $i=\overline{1,m}$.

We consider here the most interesting for practice temporally coherent and spatially incoherent target echoes. It means that each station receives coherent pulse trains but these trains are mutually incoherent at the inputs of different stations. In other words, phase shifts between pulses of each train at each station are known (excluding the phase shifts caused by the Doppler effect which are to be estimated) while phase shifts between pulse trains at the inputs of different stations are unknown or random. An initial phase of each train is unknown or random too. We also assume that amplitudes of all pulses of each train may change in a known fashion (in the particular case they may be equal to each other) but may have a common unknown or random multiplier. These multipliers are mutually independent at different stations. More complex and realistic situations with not fully coherent fluctuating target echoes will be considered in Sections 13.2 and 13.3.

Interferences at the inputs of receiving stations are assumed to be spatially uncorrelated. If necessary, the spatial correlation of external interferences may be taken into account as in Chapters 11 and 12. Thus complex overall signals at the inputs of stations, $X_i(t)$, $i=\overline{1,m}$ are the sums of target echoes and white mutually independent noises with the PSDs N_i.

Consider the likelihood ratio functional (4.44) for the assumed model of noises omitting all terms independent of $\Omega_1,\ldots,\Omega_m$

$$\Lambda(\Omega_1,\ldots,\Omega_m)=\exp\left[\sum_{i=1}^{m}\frac{1}{N_i}\int_{-T/2}^{T/2}\operatorname{Re}X_i(t)S_i^*(t)\,dt\right]. \tag{13.4}$$

Assume that DFs, $\Omega_1,\ldots,\Omega_m$, do not depend on TOAs which we consider here to be known. Let us substitute (13.1) in (13.4) and denote the output of matched filtration of the lth pulse of each train (normalized to the noise PSD) by G_{il}

$$G_{il}=\tilde{G}_{il}\exp[j(\omega_0/c)(R_0+R_i)], \tag{13.5a}$$

$$\tilde{G}_{il}=\frac{1}{\sqrt{N_i}}\int_{-T/2}^{T/2}X_i(t)s_0^*\left(t-lT_R-\frac{R_0+R_i}{c}\right)\exp(-j\omega_0 t)\,dt. \tag{13.5b}$$

In (13.5a) and (13.5b) we use a slightly different expression for G_{il} and $\tilde{G}_{il}$ as compared with (5.31). It is convenient to introduce the normalized complex amplitudes

$$\tilde{A}_{il}=A_{il}/\sqrt{N_i}. \tag{13.6}$$

Taking into account (13.2),(13.5) and (13.6),(13.4) can be written in the following matrix form

$$\Lambda(\Omega_1,\ldots,\Omega_m)=\exp\{\mathrm{Re}[\mathbf{G}^*\mathbf{E}(\Omega_1,\ldots,\Omega_m)\tilde{\mathbf{A}}]\}. \tag{13.7}$$

where

$$\mathbf{G}^*=(\mathbf{G}_1^*,\ldots,\mathbf{G}_m^*);\quad \tilde{\mathbf{A}}^*=(\tilde{\mathbf{A}}_1^*,\ldots,\tilde{\mathbf{A}}_m^*);\quad \mathbf{E}=\mathrm{diag}(\mathbf{E}_1,\ldots,\mathbf{E}_m) \tag{13.8}$$

and

$$\mathbf{G}_i^*=[G_{i0}^*,\ldots,G_{i(M-1)}^*];\qquad \tilde{\mathbf{A}}_i^*=[\tilde{A}_{i0}^*,\ldots,\tilde{A}_{i(M-1)}^*];$$

$$\mathbf{E}_i=\mathrm{diag}\{1,\exp(\mathrm{j}\Omega_iT_\mathrm{R}),\ldots,\exp[\mathrm{j}\Omega_i(M-1)T_\mathrm{R}]\},\quad i=\overline{1,m}. \tag{13.9}$$

Now we should find such values of unknown DFs which maximize a logarithm of the likelihood ratio (13.7), i.e. optimum DF estimates according to the maximum likelihood optimality criterion.

As in Chapters 11 and 12, we can consider one-stage and two-stage algorithms for optimal DF estimation. One-stage algorithms take into account that only three independent target echo DFs (at the inputs of three arbitrary spatially separated stations) determine a target velocity vector in the three-dimensional space (singular cases are excluded). It should be noted, however, that (as it follows from Section 11.1) target echo DFs depend not only on the target velocity vector but on the target position as well. Therefore, target echo DFs measured at three spatially separated stations determine the velocity vector of the target only if its position is known. For a known target position (13.7) can be presented in the form

$$\ln\Lambda=L=\mathrm{Re}\{\mathbf{G}^*\mathbf{E}[\Omega_1,\Omega_2,\Omega_3,\Omega_4(\Omega_1,\Omega_2,\Omega_3),\ldots,\Omega_m(\Omega_1,\Omega_2,\Omega_3)]\tilde{\mathbf{A}}\} \tag{13.10}$$

or

$$\ln\Lambda=L=\mathrm{Re}\{\mathbf{G}^*\mathbf{E}[\Omega_1(v_x,v_y,v_z),\ldots,\Omega_m(v_x,v_y,v_z)]\tilde{\mathbf{A}}\} \tag{13.10'}$$

where functions $\Omega_i(\Omega_1,\Omega_2,\Omega_3)$, $i=\overline{4,m}$ and $\Omega_i(v_x,v_y,v_z)$, $i=\overline{1,m}$, express the DF at the ith station in terms of the three independent variables: the DFs at the three reference stations (numbering is arbitrary) or directly the Cartesian components of the target velocity vector $\mathbf{v}$. These functions depend on the target position in space [see (11.8)]. Optimum estimates, $\hat{\Omega}_1,\hat{\Omega}_2,\hat{\Omega}_3$ or $\hat{v}_x,\hat{v}_y,\hat{v}_z$, maximize (13.10) or (13.10′), respectively.

In practice simpler two-stage algorithms for target velocity vector estimation are more widely used. In this case all m target echo DFs, $\Omega_1,\ldots,\Omega_m$, at the inputs of spatially separated receiving stations (or monostatic radars) are assumed to be independent parameters. Transforming a logarithm of (13.7) into the scalar form yields

$$L=\mathrm{Re}\sum_{i=1}^{m}\sum_{l=0}^{M-1}\tilde{A}_{il}G_{il}^*\exp(\mathrm{j}\Omega_i lT_\mathrm{R}). \tag{13.11}$$

As mentioned above, we consider temporally coherent and spatially incoherent target echoes with common unknown or random amplitude multiplier of the whole pulse train at each station but with mutually independent amplitude multipliers at different stations. Thus the likelihood ratio logarithm (13.11) contains not only m informative parameters, $\Omega_1, \ldots, \Omega_m$ but $2m$ stray parameters: m initial phases and m amplitude multipliers. Denote these stray parameters by $\varphi_{10}, \ldots, \varphi_{m0}$ and $a_{10}, \ldots, a_{m0}$. These are the initial phases and r.m.s. values of the first pulse of the target echo train at all stations. Then (13.11) can be rewritten as follows

$$L = \operatorname{Re} \sum_{i=1}^{m} a_{i0} \exp(-j\varphi_{i0}) \sum_{l=0}^{M-1} Q_{il} G_{il}^{*} \exp(j\Omega_i l T_R)$$

$$\rightarrow \max(\Omega_1, \ldots, \Omega_m, \varphi_{10}, \ldots, \varphi_{m0}, a_{10}, \ldots, a_{m0}) \tag{13.12}$$

where

$$Q_{il} = \tilde{A}_{il} / a_{i0} \exp(-j\varphi_{i0}) \tag{13.13}$$

are known values.

Since all unknown parameters are mutually independent, the sum over i reaches its maximum when each summand has its maximum, so that (13.12) may be broken up into m separate expressions:

$$L_i = \operatorname{Re}\left\{ a_{i0} \exp(-j\varphi_{i0}) \sum_{l=0}^{M-1} Q_{il} G_{il}^{*} \exp(j\Omega_i l T_R) \right\} \rightarrow \max(\Omega_i, \varphi_{i0}, a_{i0}). \tag{13.14}$$

Obviously, a maximum likelihood estimate of φ_{i0} is

$$\hat{\varphi}_{i0} = \arg \sum_{l=0}^{M-1} Q_{il} G_{il}^{*} \exp(j\Omega_i l T_R). \tag{13.15}$$

Substituting (13.15) in (13.14) yields

$$L_i = a_{i0} \left| \sum_{l=0}^{M-1} Q_{il} G_{il}^{*} \exp(j\Omega_i l T_R) \right| \rightarrow \max(\Omega_i, a_{0i}). \tag{13.16}$$

It is seen that the value of Ω_i corresponding to the maximum of L_i does not depend on the value of a_{i0}. It means that the optimum (maximum likelihood) estimate of Ω_i is determined by

$$L_i = \left| \sum_{l=0}^{M-1} Q_{il} G_{il}^{*} \exp(j\Omega_i l T_R) \right| \rightarrow \max(\Omega_i), \quad i = \overline{1, m}. \tag{13.17}$$

Though Q_{il} according to (13.1) and (13.13) are complex values, their phases are known. These phases are equal (with the opposite sign) to the known initial phase shifts between each pulse of a target echo coherent train and the first pulse of this train $[-\varphi_{il} - (-\varphi_{i0})]$ so that multiplication of G_{il}^{*} by Q_{il} eliminates the phase shifts (excluding those caused by the Doppler effect). Without loss of generality we may assume the above mentioned initial phase shifts between pulses of the received coherent pulse train to be eliminated before. Then Q_{il} become real values and may be considered as weights that account for possible differences in amplitudes of pulses of each pulse train (a nonconstant pulse train envelope). If all pulses are expected to be of equal amplitude (a constant pulse train envelope or, which is the same, a rectangular pulse train), these weights turn out to be equal to one another and may be excluded. In this case we have

$$L_i=\left|\sum_{l=0}^{M-1} G_{il}\exp(-j\Omega_i lT_R)\right|\to\max(\Omega_i),\quad i=\overline{1,m}. \tag{13.18}$$

Thus, under assumed conditions, the optimum algorithm for DF $\Omega_1,\ldots,\Omega_m$ estimation at a MSRS is reduced to seeking a maximum of the modulus of the weighted or unweighted spectrum of a filtered received pulse train at each station individually. The quantities G_{il} are determined by (13.5).

When a pulse train consists of two pulses ($M=2$) an explicit expression for the maximum likelihood DF estimate can easily be derived from (13.18). It is clear that $|G_{i0}+G_{i1}\exp(-j\Omega_i T_R)|$ reaches its maximum over Ω_i when $\arg G_{i1}-\Omega_i T_R=\arg G_{i0}+2p\pi$. Therefore, the optimum estimate is

$$\hat{\Omega}_i=\frac{1}{T_R}[\arg G_{i1}-\arg G_{i0}+2p\pi],\quad i=\overline{1,m};\ p=0,\pm 1,\ldots \tag{13.19}$$

Thus, as can be expected, the optimum algorithm requires estimating the difference between phases of received pulses at each station. Phase ambiguity is reflected by the term $2p\pi$.

Maximum Attainable Accuracy of DF Measurements

As before, the maximum attainable accuracy will be characterized with the Cramér–Rao lower bound on estimation errors.

Since the optimum algorithms synthesized above require individual DF measurements at each receiving station (or at each monostatic radar included in a MSRS), it is sufficient to analyze the maximum attainable accuracy of DF measurements with the help of a coherent pulse train at a single station or at a single monostatic radar.

For simplicity we consider the case of a rectangular target echo pulse train so that the optimum algorithm is given by (13.18). Then (13.14) can be rewritten as follows

$$L_i=\frac{A_i}{\sqrt{N_i}}\operatorname{Re}\left\{\exp(-j\varphi_{i0})\sum_{l=0}^{M-1} G_{il}^*\exp(j\Omega_i lT_R)\right\}\to\max(\Omega_i,\varphi_{i0}). \tag{13.20}$$

In (13.20) $a_{i0}Q_i$ is replaced by $\tilde{A}_i=A_i/\sqrt{N_i}$ according to (13.13) and (13.6). Besides, we have taken into account first, that known phase shifts between pulses of a coherent train are eliminated so that A_i is a real value; second, that an optimum estimate of Ω_i does not depend on the common amplitude of all pulses of the train, a_{i0}.

The likelihood ratio logarithm (13.20) contains only a single informative parameter, Ω_i, and a single stray parameter, φ_{i0}. Therefore, it is easy to obtain the FIM and its inverse, the ECM of efficient estimates, directly. Differentiating L_i twice with respect to Ω_i and φ_{i0} at the point of true values of unknown parameters, using (13.5) for G_{il}^*, averaging over $X_i(t)$, i.e. replacing $X_i(t)$ by $S_i(t)$ from (13.1), and taking into account (13.2) yields the 2×2 FIM. The upper left term of its inverse presents the Cramér–Rao lower bound on the variance of DF estimation

$$\sigma_{\text{eff}}^2(\hat{\Omega}_i) = \frac{6}{(E_{ip}/N_i)T_R^2 M(M-1)(M+1)} \tag{13.21}$$

where E_{ip} is the energy of a single pulse at the ith station; N_i is the noise PSD at the ith station as before; T_R is the time interval between pulses (pulse repetition period within the train); M is the number of pulses in the train.

The same result can be derived with the help of the technique described in Chapter 12. Taking into consideration only a single, the ith, term of the double sum in (12.13) for $\alpha_n = \alpha_p = \Omega_i$, using $S_i[t, \boldsymbol{\alpha}_0, \mathbf{H}(\boldsymbol{\alpha}_0)] = S_i[t, \Omega_i, \varphi_{i0}(\Omega_i)]$ from (13.1) for $A_{il} = A_i$ and replacing $R_{ii}(t_1, t_2)$ by $(1/N_i)\delta(t_1 - t_2)$ yields

$$J^{(\Omega_i)} = \frac{A_i^2 \tau_p}{N_i} \sum_{l=0}^{M-1} \left[lT_R - \frac{\partial \varphi_{i0}(\Omega_i)}{\partial \Omega_i} \right]^2. \tag{13.22}$$

The derivative $\partial \varphi_{i0}(\Omega_i)/\partial \Omega_i|_{\Omega_i = \Omega_i^0}$ can be obtained from (13.15) for $Q_{il} = A_i/\sqrt{N_i}$ using (12.24), (13.5) and (13.1) as the average of $X_i(t)$. As a result we have

$$\left. \frac{\partial \varphi_{i0}(\Omega_i)}{\partial \Omega_i} \right|_{\Omega_i = \Omega_i^0} = \frac{M-1}{2} T_R. \tag{13.23}$$

Substituting (13.23) in (13.22), calculating the sum [10] $\sum_{l=0}^{M-1} [l^2 - (M-1)/2]^2 = M(M-1)(M+1)/12$, taking into account that $A_i^2 \tau_p = 2E_{ip}$ and inverting $J^{(\Omega_i)}$ yields (13.21).

The corresponding r.m.s. error for $\hat{F}_i = \hat{\Omega}_i/2\pi$ is

$$\sigma_{\text{eff}}(\hat{F}_i) = \frac{1}{2\pi T_R} \sqrt{\frac{6}{(E_{ip}/N_i)M(M-1)(M+1)}}. \tag{13.24}$$

Sometimes it is more convenient to express this minimal error through the total energy $E_{i\Sigma}$ and duration T_Σ of a pulse train. Then taking into account that $E_{i\Sigma} = ME_{ip}$ and $T_\Sigma = (M-1)T_R + \tau_p \approx (M-1)T_R$ we have from (13.24)

$$\sigma_{\text{eff}}(\hat{F}_i) = \frac{1}{2\pi T_\Sigma} \sqrt{\frac{6(M-1)}{(E_{i\Sigma}/N_i)(M+1)}}. \tag{13.25}$$

In the particular case, where $M = 2$, we obtain from (13.21), (13.24), (13.25) a well-known expression for the maximum attainable accuracy of DF estimation with the help of a pair of short pulses [47,51]. It is interesting to note that for fixed signal total energy $E_{i\Sigma}$ and duration T_Σ, the number of pulses M weakly influences the minimal attainable error. It can be seen from (13.25) that when M changes from $M = 2$ to $M \to \infty$ (continuous signal of the duration T_Σ) the minimal r.m.s. increases only by $\sqrt{3}$.

***In this section**, we have discussed the problem of optimum target echo Doppler frequency (DF) estimation in a MSRS in order to determine the vector velocity of a moving target that is modelled as a point-like object. We have employed a coherent pulse train as the most convenient waveform for precise DF measurement by a pulse* *radar system. We have considered the most interesting case for practice where a target echo coherent pulse train contains at each station an unknown or random initial phase and a common amplitude multiplier (r.m.s. value of the first pulse), while these phases and amplitude multipliers are mutually independent at different stations. It has been shown that, under those conditions, optimum (maximum likelihood) DF estimator in a MSRS with m stations is broken up into m estimators, for each station individually.*

We have synthesized the optimum algorithms (13.17) *and* (13.18) *for DF estimation in a MSRS. They are reduced to seeking a maximum of the modulus of the weighted or unweighted spectrum of a filtered received pulse train at each station individually. Especially simple form has the optimum estimation algorithm* (13.19) *for two-pulse "trains".*

We have analyzed the minimal attainable errors of DF estimation. The derived expressions (13.21), (13.24) *and* (13.25) *present variances and r.m.s. values of the efficient DF estimates (the Cramér–Rao lower bound on estimation errors).*

13.2. OPTIMUM DOPPLER FREQUENCY MEASUREMENT ALGORITHMS FOR KNOWN SPACE–TIME SIGNAL CORRELATION

The problem of target echo Doppler frequency (DF) measurement in MSRSs has certain special features. For target range (or range sum) measurements, waveforms with a moderate spectral bandwidth are, as a rule, used so that each resolution cell and measurement errors are greater than the target extension in the measured parameter. We mean here "usual" targets, aircrafts and so on, not spatially extended like dipole clouds etc. The attainable accuracy of down-range measurements with "low" resolution is sufficient for a monostatic radar whose measurement errors in cross-range directions are usually much greater. In MSRSs with not too small baselengths such down-range accuracy of each station is usually sufficient for increasing the accuracy of target localization in space significantly (as compared with monostatic radars). Therefore, for the synthesis and analysis of TOA measurement algorithms in Chapters 11 and 12 we ignored the target extension in range (though there may be cases where such an extension should be taken into account).

For target radial velocity measurement by the Doppler method in monostatic radars, targets may usually be considered as "point-like" objects, since allowable velocity errors are, as a rule, greater than the spread in radial velocities of different target elements ("flare spots" – FSs). However, for velocity vector estimation in such MSRSs where target ranges, R, are in most cases significantly greater than the effective baselengths between stations, L_{eff} (the so-called "MSRSs with short baselengths", see Section 1.1) this is no longer so. The point is that the same errors in DF measurements lead to much larger errors in tangential than in radial velocity estimates (approximately by $2R/L_{eff}$ times and more). To estimate the tangential target velocity with acceptable accuracy, precise DF measurements are required with the help of waveforms of a large duration. Under these conditions the spread of different FS velocities, caused by target motion about its centre of masses, results in the appearance of a spectrum of Doppler frequencies in target echoes at the receiver inputs, i.e. of so called "target Doppler noise". This effect is a fundamental limiting factor in achieving required high measurement accuracy for the velocity of translational target motion with respect to a MSRS. However, this accuracy may be expected to be increased if we take into account the fact that a target is really not a "point-like" object not only for the analysis of actual DF (and hence velocity) measurement accuracies but for the synthesis of estimation algorithms as well [69,70].

The target Doppler noise is closely connected with complex amplitude fluctuations of target echoes [3,46]. For the assumed model of Gaussian complex amplitude fluctuations at the inputs of spatially separated receiving stations, these fluctuations can be fully determined by their space–time correlation (covariance) matrix (see Section 4.2). Mutual dependence between the width of signal Doppler spectrum, on the one hand, and the character and the width of signal correlation matrix, on the other, was discussed in Section 10.3 for a similar clutter problem (see Figs. 10.8 and 10.9). In this Section we assume this matrix to be known. The situations where the space–time correlation matrix is unknown will be considered in Sections 13.3 and 13.5.

The General Case: There is Space–Time Correlation of Target Echo Complex Amplitudes

We consider the problem for a cell of a MSRS, specifically for arbitrary a single transmitting and two receiving stations. Let the centre of masses of a target move with a constant velocity $\mathbf{v}$ [within an observation interval $t \in (-T/2, T/2)$ which is of the order of signal duration] with respect to the conventional centre of the MSRS. The unit vectors from the target to the transmitting and receiving stations we denote by $\mathbf{r}_0, \mathbf{r}_1, \mathbf{r}_2$ and assume them to be constant within the interval $(-T/2, T/2)$. As in Section 13.1, we use a coherent pulse train as the illuminating waveform. Phase measurement ambiguity is resolved by the same techniques as for "point-like" targets. Assume, as in Section 13.1, that Doppler phase shifts during each pulse, τ_p, as well as differences of interpulse time interval (repetition period) T_R caused by the Doppler effect in target echoes, may be neglected (the latter can be accommodated using rough velocity estimates). In addition, we also assume that target "flare spots" (scattering centres) cannot be resolved in range (or in range sum).

We employ expressions (13.1)–(13.3) for the received target echo at the ith station, $S_i(t)$, where the DF, Ω_i, is to be estimated ($i=1,2$).

Unlike the case of point-like targets, the complex amplitudes A_{il} are random values here. Consider the conditional likelihood functional (for fixed A_{il}) relevant to Ω_1, Ω_2 which instead of (13.4) takes the form

$$\tilde{\Lambda}(\Omega_1, \Omega_2 | \tilde{\mathbf{A}}) = \exp\left\{\frac{1}{2}\sum_{i=1}^{2}\frac{1}{N_i}\int_{-T/2}^{T/2}\left[2\,\mathrm{Re}\,X_i(t)S_i^*(t) - |S_i(t)|^2\right]\mathrm{d}t\right\}. \tag{13.26}$$

As in Section 13.1, assume that DFs, Ω_1, Ω_2, do not depend on TOAs which we consider to be known. Let us substitute (13.1) in (13.26) using the notation G_{il} (13.5) for the lth pulse of each train (normalized to the noise PSD) at the output of the matched filter. Taking into account (13.2), (13.5) and (13.6), we can rewrite (13.26) in the following matrix form

$$\tilde{\Lambda}(\Omega_1, \Omega_2 | \tilde{\mathbf{A}}) = \exp[\mathrm{Re}(\mathbf{G}^* \mathbf{E} \tilde{\mathbf{A}}) - (\tau_p/2)\tilde{\mathbf{A}}^* \tilde{\mathbf{A}}]. \tag{13.27}$$

The notation in (13.27) is the same as in (13.8), (13.9) for $m=2$; τ_p is the pulse duration.

Assume the complex amplitudes of pulses, $\tilde{A}_{il}$, $i=1,2$, $l=\overline{0, M-1}$, to be Gaussian complex random variables with zero mean and the correlation (covariance) matrix

$$\mathbf{B} = \begin{pmatrix} \mathbf{B}_{11} & \mathbf{B}_{12} \\ \mathbf{B}_{12}^* & \mathbf{B}_{22} \end{pmatrix} \tag{13.28}$$

where $\mathbf{B}_{11}=\|B_{11ln}\|=0.5\|\overline{\tilde{A}_{1l}\tilde{A}^*_{1n}}\|$; $\mathbf{B}_{22}=\|B_{22ln}\|=0.5\|\overline{\tilde{A}_{2l}\tilde{A}^*_{2n}}\|$; $\mathbf{B}_{12}=\|B_{12ln}\|=0.5\|\overline{\tilde{A}_{1l}\tilde{A}^*_{2n}}\|$. The conditions under which such a representation is valid, were considered in Section 4.2. All matrices in (13.29) are Hermitian and Toeplitz ones.

The unconditional likelihood ratio can be obtained by averaging (13.27) over $\tilde{\mathbf{A}}$:

$$\Lambda(\Omega_1,\Omega_2)=[(2\pi)^{2M}\det\mathbf{B}]^{-1}\int_{-\infty}^{\infty}\exp[-0.5\tilde{\mathbf{A}}^*\mathbf{B}^{-1}\tilde{\mathbf{A}}]\tilde{\Lambda}(\Omega_1,\Omega_2|\tilde{\mathbf{A}})\,d\tilde{\mathbf{A}}. \quad (13.29)$$

Now we can substitute (13.27) in (13.29), transform the power of the exponential function so as to obtain a complete square, integrate and omit terms independent of Ω_1,Ω_2. Then a logarithm of the unconditional likelihood ratio takes the form

$$\ln\Lambda(\Omega_1,\Omega_2)=0.5\mathbf{G}^*\mathbf{E}(\mathbf{I}+\tau_p\mathbf{B})^{-1}\mathbf{B}\mathbf{E}^*\mathbf{G} \quad (13.30)$$

where $\mathbf{I}$ is the $2M\times 2M$ identity matrix. Assume the SNR values of all pulses at each station to be equal to one another while different at the first and second stations. It is convenient to introduce the reciprocal value of SNR at each pulse

$$h_i=\frac{N_i}{\sigma_i^2\tau_p},\quad i=1,2 \quad (13.31)$$

where $\sigma_i^2=0.5\overline{|A_{il}|^2}$, $l=\overline{0,M-1}$, is the power of a single pulse. Then (13.30) can be reduced to the form (arguments at $\ln\Lambda$ are dropped for simplicity)

$$\ln\Lambda=(2\tau_p)^{-1}\mathbf{G}^*\mathbf{E}(\boldsymbol{\rho}+\sqrt{h_1h_2}\,\mathbf{I})^{-1}\boldsymbol{\rho}\mathbf{E}^*\mathbf{G} \quad (13.32)$$

where

$$\boldsymbol{\rho}=\begin{pmatrix}\sqrt{h_2/h_1}\,\boldsymbol{\rho}_{11} & \boldsymbol{\rho}_{12}\\ \boldsymbol{\rho}^*_{12} & \sqrt{h_1/h_2}\,\boldsymbol{\rho}_{22}\end{pmatrix}. \quad (13.33)$$

In (13.33) $\boldsymbol{\rho}_{11}$, $\boldsymbol{\rho}_{12}$, and $\boldsymbol{\rho}_{22}$ are the $M\times M$ matrices of the correlation coefficients of the complex amplitudes A_{1l}, A_{2n}, $l,n=\overline{0,M-1}$. For strong input signals the complicated expression (13.32) can be simplified if the maximal correlation coefficient is not too close to one, i.e. if $\sqrt{h_1h_2}/(1-|\rho_{max}|^2)\ll 1$. Expanding the matrix $(\boldsymbol{\rho}+\sqrt{h_1h_2}\,\mathbf{I})^{-1}$ into a power series and retaining only linear terms yields

$$\ln\Lambda=-(\sqrt{h_1h_2}/2\tau_p)\mathbf{G}^*\mathbf{E}\boldsymbol{\rho}^{-1}\mathbf{E}^*\mathbf{G}=-(2\tau_p)^{-2}\mathbf{G}^*\mathbf{E}\mathbf{B}^{-1}\mathbf{E}^*\mathbf{G}. \quad (13.34)$$

Expressions (13.30), (13.32) and (13.34) can be replaced by a single common equation

$$\ln\Lambda=\mathbf{G}^*\mathbf{E}\mathbf{D}\mathbf{E}^*\mathbf{G} \quad (13.35)$$

where

$$\mathbf{D}=0.5(\mathbf{I}+\tau_p\mathbf{B})^{-1}\mathbf{B}=(2\tau_p)^{-1}(\boldsymbol{\rho}+\sqrt{h_1h_2}\,\mathbf{I})^{-1}\boldsymbol{\rho} \quad (13.36)$$

or, when $\sqrt{h_1h_2}/(1-|\rho_{max}|^2)\ll 1$,

$$\mathbf{D}=-(2\tau_p)^{-2}\mathbf{B}^{-1}=-(\sqrt{h_1h_2}/2\tau_p)\boldsymbol{\rho}^{-1}. \quad (13.37)$$

It is important to reveal the roles that are played by interstation (spatial) and interpulse (temporal) processings. To do this, let us substitute (13.8) for $m=2$ in (13.35). Then we obtain the algorithm for optimum (maximum likelihood) DF estimation

$$\ln \Lambda = L = \mathbf{G}_1^* \mathbf{E}_1(\Omega_1)\mathbf{D}_{11}\mathbf{E}_1^*(\Omega_1)\mathbf{G}_1 + 2\,\mathrm{Re}[\mathbf{G}_1^* \mathbf{E}_1(\Omega_1)\mathbf{D}_{12}\mathbf{E}_2^*(\Omega_2)\mathbf{G}_2]$$
$$+ \mathbf{G}_2^* \mathbf{E}_2(\Omega_2)\mathbf{D}_{22}\mathbf{E}_2^*(\Omega_2)\mathbf{G}_2 \to \max(\Omega_1, \Omega_2). \qquad (13.38)$$

Using the scalar notation yields instead of (13.38)

$$L = \mathrm{Re}\left\{ \sum_{l=0}^{M-2} \sum_{n=l+1}^{M-1} [G_{1l}^* G_{1n} D_{11ln} e^{j(l-n)\Omega_1 T_R} + G_{2l}^* G_{2n} D_{22ln} e^{j(l-n)\Omega_2 T_R}] \right.$$
$$\left. + \sum_{l=0}^{M-1} \sum_{n=0}^{M-1} G_{1l}^* G_{2n} D_{12ln} e^{j(l\Omega_1 - n\Omega_2)T_R} \right\} \to \max(\Omega_1, \Omega_2). \qquad (13.39)$$

Thus in the general case, *when there is space–time correlation of signal complex amplitudes, joint estimation of Ω_1 and Ω_2 is optimal*, i.e. joint processing of target echoes received by both stations is required.

Now we consider two extreme situations.

Interstation (Spatial) Correlation of Target Echo Complex Amplitudes is Absent

The problem of DF estimation in MSRSs with sufficiently large baselengths between stations when there is no spatial correlation between pulse trains received by different stations, is of great practical interest. In this case $\mathbf{B}_{12} = \boldsymbol{\rho}_{12} = \mathbf{D}_{12} = \mathbf{0}$. As can be seen from (13.38) and (13.39), *individual, independent DF estimation at each station turns out to be optimal*. Under these conditions we have from (13.39)

$$L = \mathrm{Re} \sum_{l=0}^{M-2} \sum_{n=l+1}^{M-1} G_{il}^* G_{in} D_{iiln} \exp[j(l-n)\Omega_i T_R] \to \max(\Omega_i), \quad i=1,2. \qquad (13.40)$$

The optimum estimate of Ω_i may be obtained by solving the following likelihood equations ($i=1,2$)

$$\arg\left\{ \sum_{l=0}^{M-2} \sum_{n=l+1}^{M-1} (n-l) G_{il}^* G_{in} D_{iiln} \exp[j(l-n)\Omega_i T_R] \right\} = 2p\pi, \quad p=0, \pm 1, \ldots \qquad (13.41)$$

It is seen that when spatial correlation of complex amplitude fluctuations is absent, temporal correlation is necessary for DF estimation. If temporal correlation is also absent, the matrices $\mathbf{D}_{ii}$ are diagonal so that $L=0$. In this case a total phase difference between each two pulses is uniformly distributed within the interval $(-\pi, \pi)$ so that the Doppler contributions, Ω_i / T_R, cannot be separated and measured. The periodicity of the solution of equation (13.41) reflects the known ambiguity in velocity measurements with the help of pulse trains.

Obviously, spatial coherence of a MSRS (see Section 1.1) is not required for individual measurement of Ω_1 and Ω_2 at each station, but at least short-term temporal coherence of a MSRS (within the time interval of the order of signal duration) is necessary. Any disturbances in this coherence are equivalent to decrease of received signal interpulse correlation.

In the simplest case of a two-pulse train ($M=2$) the explicit solution can be obtained from (13.41):

$$\hat{\Omega}_i=\frac{1}{T_R}[\arg G_{i1}-\arg G_{i0}+\arg D_{ii01}+2p\pi],\quad i=1,2;\; p=0,\pm 1,\ldots \qquad (13.42)$$

The optimal processing is reduced to measuring the phase difference between two pulses received at each station and taking into account the phase of the correlation coefficient of those pulses' complex amplitudes.

An explicit expression for the optimum DF estimates $\hat{\Omega}_i$ can easily be obtained for an arbitrary number of pulses in each train if their complex amplitudes may be considered to be elements of a Markov chain. In this case the matrix of correlation coefficients for the ith station takes the form

$$\boldsymbol{\rho}_{ii}=\begin{bmatrix} 1 & \rho_i & \ldots & \rho_i^{M-1} \\ \rho_i^* & 1 & \ldots & \rho_i^{M-2} \\ \ldots & \ldots & \ldots & \ldots \\ \rho_i^{*M-1} & \rho_i^{*M-2} & \ldots & 1 \end{bmatrix} \qquad (13.43)$$

where $\rho_i=(1/2\sigma_i^2)\overline{A_{il}A_{i(l+1)}^*}=\exp(-\alpha_i T_R+j\varphi_i)$ is the complex amplitude correlation coefficient of adjacent pulses.

The inverse matrix of (13.43) is a three-diagonal matrix

$$\boldsymbol{\rho}_{ii}^{-1}=\frac{1}{1-|\rho_i|^2}\begin{bmatrix} 1 & -\rho_i & 0 & \ldots & 0 \\ -\rho_i^* & 1+|\rho_i|^2 & -\rho_i & \ldots & 0 \\ 0 & -\rho_i^* & 1+|\rho_i|^2 & \ldots & 0 \\ \ldots & \ldots & \ldots & \ldots & \ldots \\ 0 & 0 & 0 & \ldots & 1 \end{bmatrix}. \qquad (13.44)$$

On the main diagonal the first and the last elements are equal to one whereas remaining elements are equal to $1+|\rho_i|^2$; on the adjacent right diagonal all elements are equal to $-\rho_i$, on the other adjacent (left) diagonal all elements are equal to $-\rho_i^*$. Using (13.44) we can obtain from (13.40) the explicit equation for the DF maximum likelihood estimate [for strong signals when (13.37) is valid]

$$\hat{\Omega}_i=\frac{1}{T_R}\left[\arg\rho_i+\arg\left(\sum_{l=0}^{M-2} G_{il}^* G_{i(l+1)}\right)+2p\pi\right],\quad i=1,2;\; p=0,\pm 1,\ldots \qquad (13.45)$$

It is interesting to apply the obtained results to the extreme case where *there is complete correlation between pulses of a target echo train at each station and no correlation between target echo trains at different stations.* This corresponds to a coherent pulse train at each station and independent pulse trains at different stations. Just such a problem was considered in Section 13.1. Under these conditions $\boldsymbol{\rho}_{ii}=\mathbf{1}$ where $\mathbf{1}$ is the $M\times M$ matrix with all elements equal to one, $\mathbf{B}_{ii}$ are singular matrices (and hence cannot be inverted) but from (13.36) we can obtain

$$\mathbf{D}_{11}=[2\tau_p(M+h_1)]^{-1}\mathbf{1};\quad \mathbf{D}_{22}=[2\tau_p(M+h_2)]^{-1}\mathbf{1};\quad \mathbf{D}_{12}=\mathbf{0}. \qquad (13.46)$$

Substituting (13.46) in (13.40), omitting the common multiplier $[2\tau_p(M+h_i)]^{-1}$ and replacing the square modulus by modulus (which does not influence the value of $\hat{\Omega}_i$ maximizing L), yields, as would be expected, the expression (13.18). When the waveform is a pair of pulses (a two-pulse "train", $M=2$), an explicit expression for the optimum DF estimate at each station can be derived from (13.42) and (13.46). Naturally, it coincides with (13.19).

It can be seen from the comparison of (13.18) with (13.40) that when *interpulse correlation decreases, the additional complex factors appear in the optimum DF estimation algorithm which compensate for phase shifts of interpulse correlation coefficients and weaken the influence of low correlated pulses on DF estimates.*

Now we turn to the other extreme situation.

Interpulse Correlation between Complex Amplitudes at Each Station is Absent

It means that an incoherent pulse train is received by each station. This situation corresponds to large time intervals between pulses of a coherent pulse train radiated by the transmitting station (see Section 4.2) or to the case where the transmitting station radiates an incoherent pulse train. In the latter case it is assumed that phases of radiated pulses are not measured so that coherent processing of received incoherent target echoes is impossible.

When a received pulse train at each station is incoherent, DFs cannot be measured. However, it may be expected from (13.38), (13.39) that the Differential Doppler Frequency (DDF) $\Delta\Omega=\Omega_1-\Omega_2$ can be measured. When the range of a target is significantly larger than baselengths between receiving stations, DDF is determined mostly by tangential velocity of the target. Therefore, its direct measurement is often of great practical importance.

The correlation (covariance) matrix $\mathbf{B}$ (13.28) for situation considered can be written as follows [taking into account (13.31) and (13.33)]

$$\mathbf{B}=(\tau_p\sqrt{h_1h_2})^{-1}\begin{pmatrix}\sqrt{h_2/h_1}\,\mathbf{I}_M & \rho_{12}\mathbf{I}_M\\ \rho_{12}^*\mathbf{I}_M & \sqrt{h_1/h_2}\,\mathbf{I}_M\end{pmatrix} \tag{13.47}$$

where $\mathbf{I}_M$ is the $M\times M$ identity matrix; ρ_{12} is the interstation (spatial) correlation coefficient of target echo complex amplitudes (the same for all pulses of trains). Using a known result for the inverse of a partitioned matrix

$$\begin{pmatrix}\mathbf{Q}_{11} & \mathbf{Q}_{12}\\ \mathbf{Q}_{21} & \mathbf{Q}_{22}\end{pmatrix}^{-1}=\begin{pmatrix}(\mathbf{Q}_{11}-\mathbf{Q}_{12}\mathbf{Q}_{22}^{-1}\mathbf{Q}_{21})^{-1} & (\mathbf{Q}_{21}-\mathbf{Q}_{22}\mathbf{Q}_{12}^{-1}\mathbf{Q}_{11})^{-1}\\ (\mathbf{Q}_{12}-\mathbf{Q}_{11}\mathbf{Q}_{21}^{-1}\mathbf{Q}_{22})^{-1} & (\mathbf{Q}_{22}-\mathbf{Q}_{21}\mathbf{Q}_{11}^{-1}\mathbf{Q}_{12})^{-1}\end{pmatrix}, \tag{13.48}$$

the entries of the matrix $\mathbf{D}$ from (13.36) can be presented in the form

$$\mathbf{D}_{11}=[(1+h_2-|\rho_{12}|^2)/\Delta_M]\mathbf{I}_M;\quad \mathbf{D}_{22}=[(1+h_1-|\rho_{12}|^2)/\Delta_M]\mathbf{I}_M;$$
$$\mathbf{D}_{12}=\mathbf{D}_{21}^*=(\rho_{12}\sqrt{h_1h_2}/\Delta_M)\mathbf{I}_M \tag{13.49}$$

where $\Delta_M=2\tau_p(1+h_1+h_2+h_1h_2-|\rho_{12}|^2)$.

Let us substitute (13.49) in (13.28) taking into account that $\mathbf{E}_1\mathbf{E}_1^*=\mathbf{E}_2\mathbf{E}_2^*=\mathbf{I}_M$ and denoting $\mathbf{E}_1\mathbf{E}_2^*=\Delta\mathbf{E}_{12}=\operatorname{diag}\{1,\exp(j\Delta\Omega T_R),\ldots,\exp[j(M-1)\Delta\Omega T_R]\}$. Then using the scalar notation, employing (13.5b), (13.6), (13.8) and dropping the common multiplier, yields

$$L = \mathrm{Re}\left\{\exp\left\{-\mathrm{j}\left[\arg\rho_{12} + \frac{\omega_0(R_1 - R_2)}{c}\right]\right\}\right.$$

$$\left.\times \sum_{l=0}^{M-1} \tilde{G}_{1l}\tilde{G}^*_{2l}\exp(-\mathrm{j}l\Delta\Omega T_{\mathrm{R}})\right\} \to \max(\Delta\Omega). \tag{13.50}$$

It is seen that the optimum DDF estimate, $\widehat{\Delta\Omega}$, is such a DDF value when the real part of the spectrum of the product $\tilde{G}_{1l}\tilde{G}^*_{2l}$ (with allowance for the phase of interstation correlation coefficient of complex amplitudes and for the phase difference caused by signal TDOA) reaches its maximum. It is reasonable to remind that $\tilde{G}_{1l}$ and $\tilde{G}_{2l}$ are the results of matched filtration of target echo pulses of the same number received by different stations. Note that the modulus of ρ_{12} does not influence optimum DDF estimates.

For a two-pulse "train", an explicit expression for the optimum DDF estimate can easily be derived from (13.50)

$$\widehat{\Delta\Omega} = -\frac{1}{T_{\mathrm{R}}}\left[\arg\tilde{G}_{21} - \arg\tilde{G}_{11} + \arg\rho_{12} + \frac{\omega_0(R_1 - R_1)}{c} + 2p\pi\right]. \tag{13.51}$$

Since $\arg\rho_{12}$ and $(\omega_0/c)(R_1 - R_1)$ are assumed to be known, it is sufficient to measure phase difference between the second pulses received by different stations (i.e. the interstation phase difference for the second pulses). Clearly, the optimum algorithms (13.50) and (13.51) can be implemented only in spatially coherent MSRSs (see Section 1.1). Temporal coherence of MSRSs is not required.

In this section, *we have considered the problem of target echo Doppler frequency (DF) measurement in MSRSs when a target cannot be assumed to be a point-like object. This problem arises when tangential components of the target velocity vector are required to be estimated with high accuracy, especially in a MSRS where target ranges are usually much greater than baselengths between stations (i.e. in a MSRS with so-called short baselengths, see Section* 1.1). *In such situations precise target echo DF measurements are necessary. Therefore, illuminating signals of long duration (usually in the form of pulse trains, or bursts) are employed. However, there is a fundamental limiting factor in achieving high DF measurement accuracy, namely "the target Doppler noise" caused by the small differences in velocities of target "flare spots" (scattering centres). As a result interpulse (temporal) and interstation (spatial) correlation of target echo fluctuations is reduced which leads to DF measurement accuracy degradation. To overcome these difficulties at least partly, DF measurement algorithms should take into account the incomplete interpulse and interstation correlation of target echoes.*

For simplicity we have analyzed the problem for a cell of a MSRS including a single transmitting and two receiving stations. Extension to arbitrary number of stations is straightforward. In this Section, we assumed interpulse and interstation target echo correlation to be known.

We have synthesized the optimum algorithm (13.38), (13.39) *for maximum likelihood DF estimation in the general case where both temporal and spatial correlation of target echo fluctuations is nonzero. The main attention has been paid to the practically important extreme cases where only temporal (interpulse) correlation at each station or spatial (interstation) correlation between received pulse trains is present. For the former case we have derived the optimum algorithms* (13.40), (13.41) *as well as* (13.42) *for the particular case where a pulse "train" consists of only two pulses. An explicit expression* (13.45) *for optimum DF estimates has been obtained for the case where target echo complex amplitudes may be approximated by a Markov chain.*

Having compared the synthesized algorithms with those obtained in Section 13.1 *for point-like targets, we have established that when interpulse correlation decreases, the additional complex factors appear in optimum DF estimation algorithms which compensate for phase shifts of interpulse correlation coefficients and weaken the influence of low correlated pulses on DF estimates.*

When only interstation target echo correlation is present, DF measurement at each station is impossible, but differential Doppler frequency (DDF) can be measured at each pair of stations. For such a situation we have derived the optimum estimation algorithm (13.50) *and explicit expression* (13.51) *for DDF estimates when a target is illuminated by a pulse "train" consisting of only two pulses.*

13.3. OPTIMUM DOPPLER FREQUENCY MEASUREMENT ALGORITHMS FOR UNKNOWN SPACE–TIME TARGET ECHO CORRELATION

Two extreme situations will be considered here: (1) *in the absence of temporal (interpulse) correlation there is nonzero but unknown spatial (interstation) correlation*; (2) *in the absence of spatial (interstation) correlation there is nonzero but unknown temporal (interpulse) correlation.*

Unknown Spatial Correlation in the Absence of Temporal Correlation

In this case in (13.50) and (13.51) $\arg \rho_{12}$ is unknown. A similar situation takes place when the phase shift caused by the TDOA, $(\omega_0/c)(R_1 - R_1)$, is unknown or/and when a MSRS is a system with short-term spatial coherence because of random uncontrolled phase shifts in equipment. Then we should find a maximum likelihood estimate of the unknown phase and substitute the obtained estimate for the unknown phase. Let us denote the unknown argument of the exponential function in (13.50) by $\Delta\psi_{12}$. Obviously, the maximum likelihood estimate of $\Delta\psi_{12}$ can be obtained from (13.50) in the form

$$\widehat{\Delta\psi}_{12} = \arg\left[\sum_{l=0}^{M-1} \tilde{G}_{1l}\tilde{G}_{2l}^{*} \exp(-jl\Delta\Omega T_R)\right] + 2p\pi, \quad p = 0, \pm 1, \ldots$$

Substituting $\widehat{\Delta\psi}_{12}$ in (13.50) instead of unknown $\Delta\psi_{12}$ yields the optimum estimation algorithm

$$L = \left|\sum_{l=0}^{M-1} \tilde{G}_{1l}\tilde{G}_{2l}^{*} \exp(-jl\Delta\Omega T_R)\right| \to \max(\Delta\Omega). \tag{13.52}$$

The desired maximum likelihood estimate of $\Delta\Omega$ is a point where the modulus of the spectrum of $\tilde{G}_{1l}\tilde{G}_{2l}^{*}$ reaches its maximum [compare (13.52) with (13.50)]. For two-pulse target echo waveform we have from (13.52) an explicit expression:

$$\acute{\Delta\Omega} = -\frac{1}{T_R}[\arg \tilde{G}_{21} - \arg \tilde{G}_{11} - (\arg \tilde{G}_{20} - \arg \tilde{G}_{10}) + 2p\pi], \quad p = 0, \pm 1, \ldots \tag{13.53}$$

The algorithm includes measuring interstation phase differences between the second and the first pulses of pulse trains received by both stations, calculating their difference (i.e. the second phase difference) and then dividing the result by the interpulse time interval T_R. We remind that $\tilde{G}_{il}$ is determined by (13.5b) and is the filtered lth pulse of received signal at the ith station. Thus we see that the maximum likelihood DDF estimate can be obtained in practically important situation where target echo interstation (spatial) correlation is unknown.

Unknown Temporal Correlation in the Absence of Spatial Correlation

Using a single received pulse train, a monostatic radar as well as a single receiving station of a MSRS cannot estimate interpulse correlation and employ this estimate to DF measurement. A Doppler phase shift and phases of interpulse correlation coefficients are inseparable. However, under certain conditions interpulse correlation coefficients may be assumed to be equal at different receiving stations of a MSRS, i.e. $\mathbf{\rho}_{11}=\mathbf{\rho}_{22}$ (see Section 4.2). In this case it is possible, in principle, to estimate interpulse correlation using target echo received by one station, and then to employ the obtained estimate for DF measurement at other stations.

A fundamental difficulty of this procedure is that an interpulse correlation matrix is to be estimated using only one received pulse train since this matrix may change significantly in time, i.e. during several subsequent repetition periods (because of the target motion). Under these conditions estimation accuracy of matrix elements turns out to be low even for the Toeplitz correlation matrices. To estimate the correlation between adjacent pulses of a M-pulse train, $M-1$ products of the type of $G_l G_{l+1}^*$, $l=\overline{0,M-2}$ can be used while for the first and the last pulses only one product $G_0 G_{M-1}^*$ can be utilized. Rough estimates of "outlying" elements corrupt the sample correlation matrix and its inverse. Not much better results may be obtained by using the techniques of direct estimation of the inverse of correlation matrices (e.g., [72]). Therefore, DF measurements based on interpulse correlation matrix estimation using a single pulse train at one of stations is in general unlikely expedient[1].

The situation is changed significantly when it is *a priori* known that the interpulse correlation matrix of complex amplitudes may be approximated by (13.43). In this case only one parameter is unknown. This is the phase of the correlation coefficient of adjacent pulses, $\arg \rho_i$.

In many cases when electrodynamic asymmetry of a target is expected to be not too large, the phases $\arg \rho_i$ are small and may be neglected [see (4.34) and Fig. 4.4]. Then we obtain from (13.45) the useful optimum algorithm

$$\hat{\Omega}_i=\frac{1}{T_R}\left[\arg\left(\sum_{l=0}^{M-2} G_{il}^* G_{i(l+1)}\right)+2p\pi\right],\quad i=1,2;\ p=0,\pm 1,\ldots \tag{13.54}$$

Note that (13.54) does not include unknown true values of the modulus $|\rho_i|$.

If an expected target is known to be essentially asymmetric in the electrodynamic sense, $\arg \rho_i$ may not be neglected. However, it can be estimated with satisfactory accuracy if the number of pulses M is not too small. Let, for instance, the first station be a "reference" one. Then a maximum likelihood estimate of the unknown phase

[1] Better results can be obtained if interpulse correlation coefficients are estimated using target echoes at several stations but only under the condition that these correlation coefficients may be considered to be equal at all stations.

$\arg\rho_1$ can be obtained from (13.44) in the form

$$\widehat{\arg\rho_1} = \Omega_1 T_R - \arg\left(\sum_{l=0}^{M-2} G_{1l}^* G_{1(l+1)}\right) + 2p\pi, \quad p=0, \pm 1, \ldots \tag{13.55}$$

Substituting (13.55) in (13.45) for the unknown $\arg\rho_2$ (assuming $\arg\rho_1 = \arg\rho_2$) yields the optimum adaptive algorithm for Doppler frequency estimation

$$\widehat{\Omega_1 - \Omega_2} = \widehat{\Delta\Omega} = \frac{1}{T_R}\left[\arg\left(\sum_{l=0}^{M-2} G_{1l}^* G_{1(l+1)}\right) - \arg\left(\sum_{l=0}^{M-2} G_{2l}^* G_{2(l+1)}\right) + 2p\pi\right], \quad p=0, \pm 1, \ldots \tag{13.56}$$

It is important to note that using estimates of unknown correlation between pulses obtained at one station, DFs at both stations cannot be measured. As it can be seen from (13.56), only differential Doppler frequency (DDF) with respect to the "reference" station can be measured. As mentioned in Section 13.2, in MSRSs where typical target ranges are much greater than the baselengths between stations, DDFs are mostly determined by tangential target velocities. Therefore, direct DDF measurement is important for such MSRSs. However, using only DDF measurements it is often difficult to estimate target radial velocity with required accuracy. Errors of radial velocity estimation using DDFs are proportional to the squared ratio of the target range to the effective baselength between receiving stations. In this case additional coherent pulse train of smaller duration may be employed for direct radial velocity measurement. As noted in Section 13.2, for radial velocity measurements, waveforms of smaller duration can provide required accuracy. Owing to small values of the interpulse time interval, T_R, of such a pulse train, high temporal coherence (between pulses) may remain in target echoes. Then the optimum algorithms (13.17) or (13.18) for a point-like target may be used.

The algorithms obtained above for two receiving stations can be easily extended to the case of arbitrary (known) number of stations, m.

In this section, *we have synthesized optimum Doppler frequency (DF) and differential Doppler frequency (DDF) estimation algorithms for two the most important for practice extreme cases:* (1) *in the absence of temporal (interpulse) correlation there is nonzero but unknown spatial (interstation) correlation;* (2) *in the absence of spatial (interstation) correlation there is nonzero but unknown temporal (interpulse) correlation.*

For the first situation the optimum DDF estimation algorithm (13.52) *is derived. In the particular case, where a pulse "train" consists of two pulses, the explicit expression* (13.53) *for optimum DDF estimate can be used.*

For the second situation it is possible, in principle, to estimate unknown interpulse correlation using target echo pulse train received at one, "reference", station and then to estimate the desired DDF at other stations with respect to the "reference" one. However, in the general case such estimates turn out to be of low accuracy. Much better results can be obtained when target echo pulse trains may be approximated by Markov chains. In this case only one correlation coefficient between adjacent pulses is unknown. For targets with small electrodynamic asymmetry the unknown phase of this interpulse correlation coefficient is usually small and may often be neglected. Then the optimum DF estimate can be obtained with the help of algorithm (13.54). *For targets having large electrodynamic asymmetry only DDF estimation is possible. We have presented the optimum algorithm* (13.56) *assuming the phases of interpulse correlation coefficients to be equal at both stations. It is important that the optimum algorithm* (13.54) *as well as* (13.56) *do not include unknown modulus of the interpulse correlation coefficient.*

13.4. MAXIMUM ATTAINABLE ACCURACY OF DOPPLER FREQUENCY ESTIMATION

In the general case where complex amplitudes of received target echoes are spatially and temporally correlated, the Fisher information matrix (FIM) $\mathbf{J} = \|J_{ik}\|$, $i, k = 1, 2$, can be obtained from (13.38) by a standard technique. Thus

$$J_{11} = 4T_R^2 \tau_p^2 \operatorname{Re} \sum_{l=0}^{M-1} \sum_{n=0}^{M-1} [l(l-n) B_{11ln}^* D_{11ln} + l^2 B_{12ln}^* D_{12ln}]$$

$$J_{22} = 4T_R^2 \tau_p^2 \operatorname{Re} \sum_{l=0}^{M-1} \sum_{n=0}^{M-1} [l(l-n) B_{22ln}^* D_{22ln} + l^2 B_{12ln} D_{12ln}^*] \tag{13.57}$$

$$J_{12} = J_{21} = -4T_R^2 \tau_p^2 \operatorname{Re} \sum_{l=0}^{M-1} \sum_{n=0}^{M-1} nl B_{12ln}^* D_{12ln}.$$

The notation in (13.57) is the same as in Sections 13.1, 13.2 and 13.3: T_R and τ_p are introduced in (13.1), (13.2); the matrices **B** and **D** see in (13.28) and (13.36), (13.37). Inverting **J** yields the error covariance matrix (ECM) of efficient estimates $\hat{\Omega}_1, \hat{\Omega}_2$, i.e. the Cramér–Rao lower bound on attainable errors.

As noted in Sections 13.2 and 13.3, of great practical interest are situations where *there is only temporal or only spatial correlation between target echo complex amplitudes of pulse trains at different stations.*

Only Temporal Correlation between Pulses at Each Station is Nonzero

In this case the matrices **B**, **D**, **J** are diagonal. The desired expression for r.m.s. values of errors of efficient DF estimates $\hat{F}_1 = \hat{\Omega}_1/2\pi$, $\hat{F}_2 = \hat{\Omega}_2/2\pi$ can be obtained from (13.57) ($i = 1, 2$)

$$\sigma_{\text{eff}}(\hat{F}_i) = \left[16\pi^2 T_R^2 \tau_p^2 \operatorname{Re} \sum_{l=0}^{M-2} \sum_{n=l+1}^{M-1} (l-n)^2 B_{iiln} D_{iiln}^* \right]^{-1/2}. \tag{13.58}$$

In the simplest case of two-pulse "trains" ($M = 2$) the minimal r.m.s. error is given by

$$\sigma_{\text{eff}}(\hat{F}_i) = \frac{1}{2\pi T_R} \sqrt{\frac{(1+h_i)^2 - |\rho_i|^2}{2|\rho_i|^2}} \tag{13.59}$$

where h_i is defined in (13.31) and ρ_i is the correlation coefficient between pulses of the target echo received by the ith station. For strong input signals when the approximate expression (13.37) is valid, the equation (13.58) can be simplified

$$\sigma_{\text{eff}}(\hat{F}_i) = \left[-8\pi^2 T_R^2 \operatorname{Re} \sum_{l=0}^{M-2} \sum_{n=l+1}^{M-1} (l-n)^2 B_{iiln} (\mathbf{B}_{ii}^{-1})_{ln}^* \right]^{-1/2} \tag{13.60}$$

where $(\mathbf{B}_{ii}^{-1})_{ln}$ is the element of the matrix $\mathbf{B}_{ii}^{-1}$ which is positioned at the intersection of the lth row and nth column ($l,n=\overline{0,M-1}$).

The approximate expression (13.60) is convenient to use for the accuracy analysis when complex amplitudes of a target echo train at each station may be approximated by a Markov chain (13.43). If in (13.60) $\mathbf{B}_{ii}$ is a Hermitian matrix and for $n \geqslant l$

$$B_{iiln}=(\sigma_i^2/N_i)\rho_i^{n-l}=(\sigma_i^2/N_i)\exp[(n-l)(-\alpha_i T_R+j\varphi_i)], \tag{13.61}$$

then the matrix $\mathbf{B}_{ii}^{-1}$ is a three-diagonal one, and for elements with $n>l$ [included in (13.60)] we have

$$(\mathbf{B}_{ii}^{-1})_{ln}^{*}=\begin{cases}0, & n\neq l+1,\ n>l;\\ -(N_i/\sigma_i^2)[\rho_i^{*}/(1-|\rho_i|^2)], & n=l+1\end{cases}. \tag{13.62}$$

Substituting (13.61) and (13.62) in (13.60) yields the minimum r.m.s. error of DF estimation for a train containing M pulses

$$\sigma_{\text{eff}}(\hat{F}_i)=\frac{1}{2\pi T_R}\sqrt{\frac{1-|\rho_i|^2}{2(M-1)|\rho_i|^2}}. \tag{13.63}$$

In the particular case where $M=2$, (13.63) coincides with (13.59) if we take into account in (13.59) that $h_i/(1-|\rho_i|^2)\ll 1$. Note that if (13.61) is valid, the minimum attainable error of DF estimation with the help of a train containing M pulses is the same as for $M-1$ independent measurements with two-pulse "trains".

It is seen from (13.58)–(13.60),(13.63) that r.m.s. errors of efficient DF estimates, $\sigma_{\text{eff}}(\hat{F}_i)$, are inversely proportional to the time interval between pulses (the pulse repetition period within trains), T_R. However, as T_R increases, correlation between pulses is reduced which in its turn leads to the increase of $\sigma_{\text{eff}}(\hat{F}_i)$. When the dependence of interpulse correlation on T_R is known, optimum value of T_R, i.e. $T_R=T_{R\,\text{opt}}$, may be found which minimizes the r.m.s. errors, $\sigma_{\text{eff}}(\hat{F}_i)$. Let the Markov model (13.61) and the inequality $h_i/(1-|\rho_i|^2)\ll 1$ be valid. Then substituting $|\rho_i|=\exp(-\alpha_i T_R)$ in (13.63) yields the optimum value $T_{R\,\text{opt}}\approx 0.80/\alpha_i$ and the optimum interpulse correlation coefficient $|\rho_i|_{\text{opt}}\approx 0.45$ when $\sigma_{\text{eff}}(\hat{F}_i)$ reaches its minimum: $\sigma_{\text{eff}\,\text{min}}(\hat{F}_i)\approx 0.28\alpha_i/\sqrt{M-1}$.

Example 13.1. Assume that the modulus of correlation coefficient is reduced down to 0.5 during the time interval 10 ms (because of target motion about its centre of masses). Then $\alpha_i=-(\ln 0.5)/0.01\approx 69.3\,\text{s}^{-1}$, $T_{R\,\text{opt}}\approx 11.5$ ms and $\sigma_{\text{eff}\,\text{min}}(\hat{F}_i)\approx 19.4/\sqrt{M-1}$ Hz. It should be noted that this optimum is relatively broad: variations of T_R within 20% lead to the increase of $\sigma_{\text{eff}}(\hat{F}_i)$ by 1–5% only.

For *complete interpulse correlation* of complex amplitudes we can obtain from (13.58) using (13.46)

$$\sigma_{\text{eff}}(\hat{F}_i)=\frac{1}{2\pi T_R}\sqrt{\frac{6h_i(M+h_i)}{M^2(M-1)(M+1)}}. \tag{13.64}$$

It is taken into account in (13.64) that [10] $\sum_{l=0}^{M-2}\sum_{n=l+1}^{M-1}(l-n)^2=M^2(M-1)(M+1)/12$. For high SNR when $h_i=(E_{ip}/N_i)^{-1}\ll 1$, (13.64) takes the form

$$\sigma_{\text{eff}}(\hat{F}_i)=\frac{1}{2\pi T_R}\sqrt{\frac{6}{(E_{ip}/N_i)M(M-1)(M+1)}} \tag{13.65}$$

where $E_{ip}=\sigma_i^2\tau_p$ is the energy of each target echo pulse at the ith station. As one would expect, (13.65) coincides with (13.24) for a point-like target. In fact, completely correlated pulses of a target echo pulse train (burst) correspond either to a target that does not move about its centre of masses and does not change its aspect angles relative to stations or to a point-like target. Therefore, (13.25) is also valid for the case of complete interpulse correlation.

Only Spatial Correlation between Pulse Trains at Different Stations is Nonzero

As shown in Sections 13.2 and 13.3, under this condition only differential Doppler frequency (DDF) can be measured, $\Delta F=F_1-F_2=\Delta\Omega/2\pi$. Minimum attainable r.m.s. error, $\sigma_{eff}(\Delta F)$, is impossible to obtain using elements (13.57) of the FIM, since this FIM turns out to be singular and cannot be inverted. We may use the likelihood ratio logarithm (13.50). Differentiating it twice with respect to $\Delta F=\Delta\Omega/2\pi$ and averaging over the set of received random signals yields the r.m.s. error of efficient DDF estimates in the form

$$\sigma_{eff}(\Delta F)=\frac{1}{2\pi T_R}\sqrt{\frac{3(1+h_1+h_2+h_1h_2-|\rho_{12}|^2)}{M(M-1)(2M-1)|\rho_{12}|^2}} \tag{13.66}$$

where ρ_{12} is the interstation (spatial) correlation coefficient that was introduced by (13.47). When $M=2$ and $h_1=h_2$, the minimum r.m.s. error (13.66) of DDF measurement by a pair of stations (in the case of no interpulse, temporal, correlation) is the same as the minimum r.m.s. error (13.59) of DF measurement at each station (in the case of no interstation, spatial, correlation), if, of course, $|\rho_i|=|\rho_{12}|$, $i=1,2$.

It should be noted that all expressions for r.m.s. errors of efficient estimates (the Cramér–Rao lower bound on estimation errors) derived in this Section, do not take into account the periodicity of phase differences. For instance, it follows from (13.59) and (13.63) that $\sigma_{eff}(\hat{F}_i)\to\infty$ when $|\rho_i|\to 0$. In reality, when $|\rho_i|\to 0$, the PDF of phase difference $\Delta\varphi$ tends to the uniform PDF within $(-\pi,\pi)$ with the r.m.s. value $\sigma(\Delta\varphi)=\pi/\sqrt{3}$, so that $\sigma_{eff\,max}(\hat{F}_i)=\sigma(\Delta\varphi)/2\pi T_R=1/2\sqrt{3}T_R\approx 0.289/T_R$. However, DF measurements are impossible under this condition since $\hat{F}_i$ takes equiprobably any value within the unambiguity interval of a pulse train signal: $(p-0.5)/T_R\leqslant\hat{F}_i\leqslant(p+0.5)/T_R$, $p=0,\pm1,\ldots$. It means that expressions for r.m.s. errors of efficient estimates may be used in practice only for such values of $|\rho_i|$ for which PDF of $\hat{F}_i$ is far from the uniform distribution and r.m.s. error $\sigma_{eff}(\hat{F}_i)$ is far from $0.289/T_R$. The smaller the number of pulses M of a pulse train, the greater values of $|\rho_i|$ required to satisfy this condition. Thus *the FIM technique cannot be applied to the maximum likelihood error evaluation for pulse trains with a small number of pulses, M*. Specifically, the optimum values of interpulse time interval, $T_{R\,opt}$, and of correlation coefficient, $|\rho_i|$, obtained above, are to be used in practice only for $M>8$.

This section *is devoted to the problem of maximum accuracy attainable in Doppler frequency (DF) and differential Doppler frequency (DDF) estimation with the help of a pulse train waveform in MSRSs. Maximum attainable accuracy is determined by means of the Cramér–Rao lower bound on estimation errors which does not depend on specific estimation algorithms.*

For the general case where there is space–time (interstation and interpulse) correlation between target echo complex amplitudes, we have obtained expression (13.57) *determining elements of the Fisher information matrix (FIM). Its inverse is the error covariance matrix (ECM) of efficient joint DF estimates at both stations. As in Section* 13.2, *the main attention has been paid to the practically important extreme cases where only temporal (interpulse) correlation at each station or spatial (interstation) correlation between received pulse trains is present.*

For the case where only temporal correlation is present, we have derived expression (13.58), *determining minimum attainable r.m.s. error in DF estimation, expression* (13.59) *for a pulse "train" consisting of two pulses and the simplified expression* (13.60) *applicable for strong input signals. A simple equation* (13.63) *has been obtained for a pulse train with M pulses when target echo complex amplitudes may be approximated by a Markov chain. We have shown that for such signals there is an optimum time interval between pulses minimizing the error of DF efficient estimates. Example* 13.1 *illustrates corresponding calculations. For complete interpulse correlation of complex amplitudes we have obtained expressions* (13.64) *and* (13.65) (*the latter for strong input signals*). *As one would expect,* (13.65) *coincides with* (13.24) *derived for a point-like targets.*

For the case where only spatial (interstation) correlation is present, we have derived expression (13.66), *determining minimum attainable r.m.s. error in DDF estimation.*

We have noted that the obtained r.m.s. errors of efficient DF and DDF estimates do not account for the periodicity of phase measurements inherent in pulse train waveforms. Therefore, the FIM technique (i.e. the Cramér–Rao lower bound on estimation errors) cannot be applied to the maximum likelihood error evaluation for pulse trains with a small number of pulses M.

13.5. ACCURACY OF DOPPLER FREQUENCY ESTIMATION ALGORITHMS

For DF and DDF measurement, pulse trains (bursts) containing a small number of pulses are often used. It may be expected that under this condition asymptotic properties of maximum likelihood estimates cannot yet manifest themselves. In such situations, accuracy of efficient estimates (the Cramér–Rao lower bound on errors) discussed in the previous Section, does not characterize adequately actual accuracy of the optimum estimation algorithms synthesized in Sections 13.2 and 13.3. Therefore performance analysis of specific estimation algorithms is necessary.

Accuracy of the Optimum DF and DDF Estimation Algorithms for Known Space–Time Correlation between Target Echo Complex Amplitudes (See Section 13.2)

Consider first *algorithms for independent DF estimation at each station which are optimum when there is no spatial (interstation) correlation* ($\rho_{12}=0$). The results of computer simulation of the algorithms (13.40) and (13.45) for the Markov fluctuation model (13.43), (13.44) are shown in Fig. 13.1. It is assumed for simplicity that the known phase of the interpulse correlation coefficient is equal to zero, i.e. $\arg \rho_i = 0$, $i = 1, 2$. Note that under this condition (13.45) coincides with (13.54). It is seen from Fig. 13.1 that lowering of the interpulse correlation coefficient, $|\rho_i|$, $i = 1, 2$, leads to significant degradation of the estimation accuracy. However, this degradation is less than when the usual algorithm (13.18) is used. The algorithm (13.18) is optimal for a point-like target, i.e., unlike the algorithms (13.40), (13.45) and (13.54), it does not take into account imperfect interpulse correlation of target echo pulse trains. The curves obtained by simulation according to the algorithm (13.18) are plotted in Figs. 13.2 and 13.3 for the pulse trains containing 16 and 32 pulses, respectively. In the same figures corresponding curves are repeated from Fig. 13.1, for comparison. It can be seen that the larger the number of pulses M, the greater the accuracy gain of

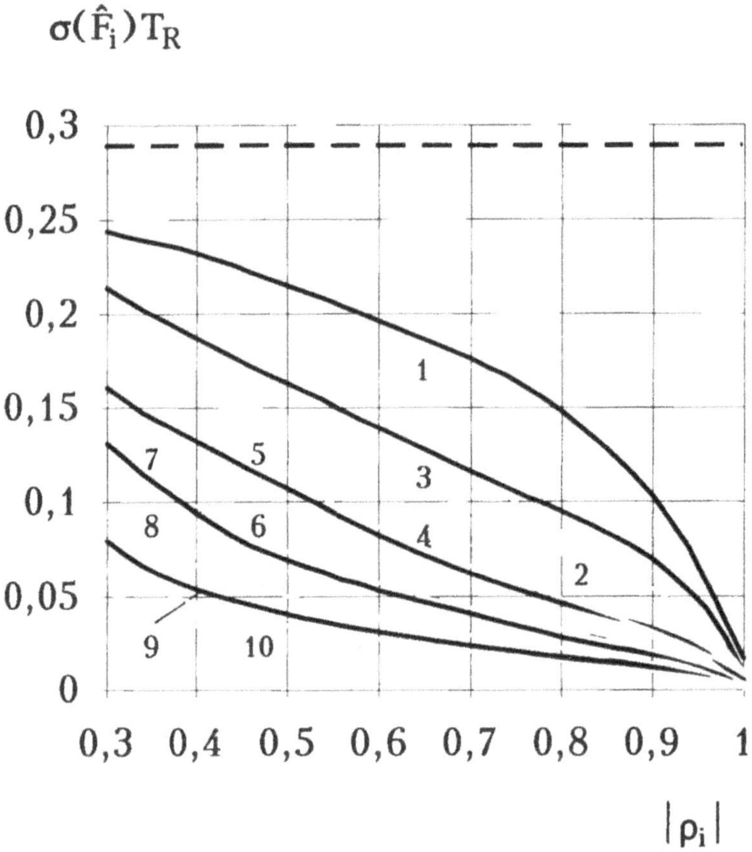

Figure 13.1 Dependences of the normalized r.m.s errors of DF estimates on the modulus of the interpulse (temporal) correlation coefficient for a Markovian fluctuation model and for different number of pulses, M; $q_s^2 = E_{ip}/N_i = 20\,\text{dB}$. Black lines: optimal estimates from (13.40), (13.45); grey lines: efficient estimates from (13.58); dashed straight line: maximum possible r.m.s. error corresponding to uniform error distribution; curves 1, 2: $M=2$, curves 3, 4: $M=4$, curves 5, 6: $M=8$, curves 7, 8: $M=16$, curves 9, 10: $M=32$

the algorithm (13.45) as compared with (13.18). When $M=16$ the accuracy of (13.45) is by 30–50% greater than that of (13.18) and when $M=32$ the accuracy of (13.45) is nearly twice as high as that of (13.18). For zero correlation, when DF estimates, $\hat{F}_i$, are uniformly distributed within the unambiguity interval $(-T_R/2, T_R/2)$, $\sigma(\hat{F}_i)T_R = 1/2\sqrt{3} \approx 0.289$ (see the dashed horizontal line in Fig. 13.1).

For "trains" containing only two pulses, $M=2$, (13.45) for $\arg \rho_i = 0$, i.e. (13.54), coincides with (13.18). In this case the optimum algorithms are reduced to phase difference estimation between two mutually correlated Gaussian variables with zero mean. Analytical expressions are known for the PDF and variance of this phase difference [33]. The obtained results by simulation and calculation (using formulas from [33]) turned out to be practically equal.

In Figs. 13.1–13.3 the curves for efficient estimates are shown too. They are calculated employing (13.58), (13.63). It should be noted that for chosen values of SNR at each pulse, $E_{ip}/N_i = 100$, i.e. $h_i = 0.01$, the approximate equation (13.63) leads

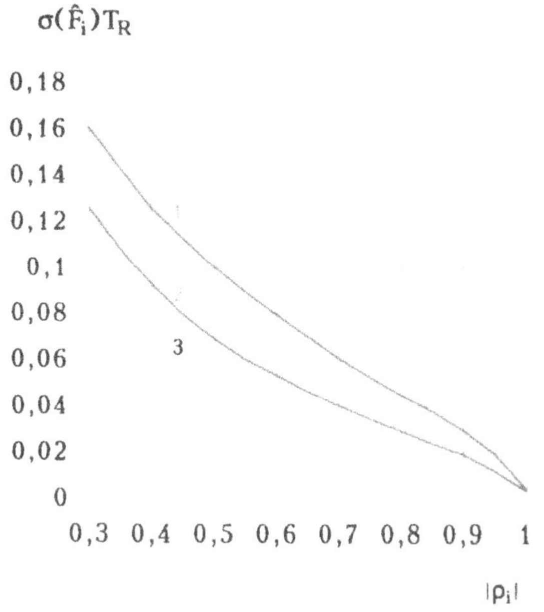

Figure 13.2 Comparison of DF estimation accuracy obtained by the optimum algorithm (13.40), (13.45) and by the conventional algorithm (13.18) for $M=16$, $q_s^2=E_{ip}/N_i=20\,\text{dB}$. Curve 1: conventional algorithm; curve 2: optimum algorithm: curve 3: efficient estimate

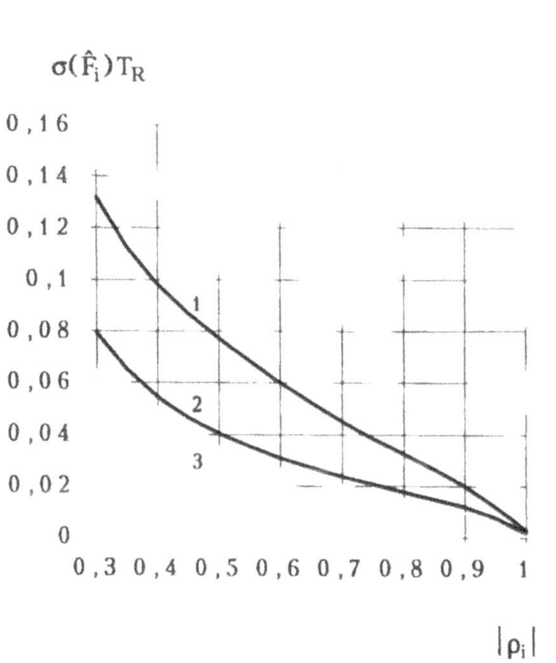

Figure 13.3 Comparison of DF estimation accuracy obtained by the optimum algorithm (13.40), (13.45) and by the conventional algorithm (13.18) for $M=32$, $q_s^2=E_{ip}/N_i=20\,\text{dB}$. Curve 1: conventional algorithm; curve 2: optimum algorithm: curve 3: efficient estimate

to errors not exceeding 5% for $|\rho_i| \leqslant 0.9$ and $M=2$, $|\rho_i| \leqslant 0.95$ and $M=4$, $|\rho_i| \leqslant 0.995$ and $M=16$.

As mentioned in Section 13.4, (13.58) does not take into account the periodicity of phase differences. Therefore, the curves of the errors of efficient estimates are plotted in Figs. 13.1–13.3 only for such values of $|\rho_i|$ where these errors increase with the decrease of $|\rho_i|$ not more rapidly than maximum likelihood errors. It is seen that for small M efficient estimates cannot serve as a good approximation to maximum likelihood ones. On the other hand, for $M=16$ and, especially, for $M=32$ the differences between accuracies of efficient and maximum likelihood estimates become small. For $M=32$ accuracy of the efficient estimates from (13.58),(13.63) is only slightly higher than that of the maximum likelihood estimates obtained by simulation according to (13.54).

Ambiguity of estimates $\hat{F}_i$ requires to know *a priori* true values of F_i with sufficient accuracy for resolving this ambiguity. It imposes constraints on the choice of the time interval between pulses, T_R, which has to be no greater than the reciprocal value of *a priori* DF unambiguity, i.e. if *a priori* true value of F_i is known with the maximal error $\pm\Delta_F$ Hz, then $T_R \leqslant 1/2\Delta_F$.

Example 13.2. Let the *a priori* true value of F_i be known with equiprobable error ± 100 Hz and the dependence $|\rho_i|$ on time interval between pulses, T_R, be the same as in Example 13.1: $|\rho_i| = \exp(-\alpha_i T_R)$ where $\alpha_i = 69.3\,\text{s}^{-1}$. It means that the modulus of the correlation coefficient is reduced down to 0.5 during the time interval 10 ms (as a result of target motion about the target's centre of mass). To cope with the DF ambiguity $|\Delta_F| = 100$ Hz the time interval between pulses has to be no greater than 5 ms, i.e. $T_R \leqslant 1/2|\Delta_F| = 5$ ms. Then $|\rho_i| \geqslant \exp(-69.3 \times 0.005) = 0.71$. It is seen from Fig. 13.1 that $\sigma(\hat{F}_i)T_R$ decreases more slowly with the increase of $|\rho_i|$ than $-\ln|\rho_i|$. Hence as T_R decreases, $\sigma(\hat{F}_i)T_R$ in Fig. 13.1 decreases not proportionally but more slowly. For this reason it is expedient to choose the maximum allowable value of T_R for achieving minimum DF r.m.s. error, $\sigma(\hat{F}_i)$. Thus let $T_R = 5$ ms and $|\rho_i| = 0.71$. Under this condition we can achieve different DF r.m.s. errors, $\sigma(\hat{F}_i)$ depending on the number of pulses in the employed pulse train. It is seen from Fig. 13.1 that for $|\rho_i| = 0.71$: when $M = 2, 4, 8, 16$ and 32, then $\sigma(\hat{F}_i)T_R = 0.174$, 0.114, 0.061, 0.039 and 0.023, respectively. It means that for $M = 2, 4, 8, 16$ error $\sigma(\hat{F}_i) = 34.8$ Hz, 22.8 Hz, 12.2 Hz, 7.8 Hz and 4.6 Hz, respectively. The *a priori* r.m.s. error $\sigma_a(\hat{F}_i) = 0.289 \times 200 = 57.8$ Hz. Therefore, using pulse trains of 2, 4, 8, 16 and 32 pulses yields reducing of initial *a priori* r.m.s. error approximately by 1.7, 2.5, 4.7, 7.4 and 12.6 times, respectively.

It is interesting to note that if we employ the usual algorithm (13.18) which is optimal for point-like targets, then for $M=16$ $\sigma(\hat{F}_i)T_R = 0.060$ and $\sigma(\hat{F}_i) = 12.0$ Hz instead of 7.8 Hz; for $M=32$ $\sigma(\hat{F}_i)T_R = 0.044$ and $\sigma(\hat{F}_i) = 8.8$ Hz instead of 4.6 Hz. Accuracy degradation as compared with the optimum algorithm is by 1.7 and 2.0 times, respectively.

Obviously, for a fixed T_R (i.e. for a fixed $|\rho_i|$) the increase of M leads to the increase of total pulse train duration, $T_\Sigma \approx (M-1)T_R$, and hence raises DF measurement accuracy. A maximum pulse train duration, T_Σ, is limited by the time interval where target echo complex amplitude fluctuations may be considered to be stationary and by transmitting station energy resources available .

It is important to note that though algorithm (13.45) is optimal for exponential interpulse correlation, its performance advantages [as compared with the usually employed algorithm (13.18)] are preserved for other correlation functions. For example, using (13.45) for Gaussian interpulse correlation of complex amplitudes [$0.5\overline{A_l A_{l+n}} = \sigma^2 \exp(-n^2 \ln|\rho_i|)$ and $\arg\rho_i = 0$] leads to a certain increase in r.m.s.

error (with respect to exponential correlation function) for high correlation coefficients and long pulse trains. However, the decrease of r.m.s. errors as compared with algorithm (13.18) is essential: by 30–55% for $M=8$ and by 50–75% for $M=16$. Obviously, lowering the influence of weakly correlated pulses on DF estimates leads to the increase of estimation accuracy regardless of the specific form of interpulse correlation functions.

Now we pass to the analysis of *the algorithm* (13.50) *which is optimal when there is interstation* (*spatial*) *correlation and no interpulse* (*temporal*) *correlation between target echo complex amplitudes.* The curves for r.m.s. errors obtained by computer simulation are shown in Fig. 13.4 (grey lines). Comparing the curves in Fig. 13.4 with those in Fig. 13.1, we can note a "thresholding effect" in Fig. 13.4 (excluding the case where $M=2$). For large values of $|\rho_{12}|$ and $M=8$, $M=16$ the r.m.s. DDF error, $\sigma(\widehat{\Delta F})T_R$, is much smaller than the r.m.s. DF error, $\sigma(\hat{F}_i)T_R$, at each station for the same values of $|\rho_i|$, $(i=1,2)$. Then as $|\rho_{12}|$ decreases, $\sigma(\Delta F)T_R$ sharply increases. This is caused by the phase shift, $2\pi\Delta F l T_R$, $l=\overline{0, M-1}$, in (13.50) linearly increasing from one pulse to another. The normalized (to $1/T_R$) width of the spectrum of $\tilde{G}_{1l}\tilde{G}^*_{2l}$, is

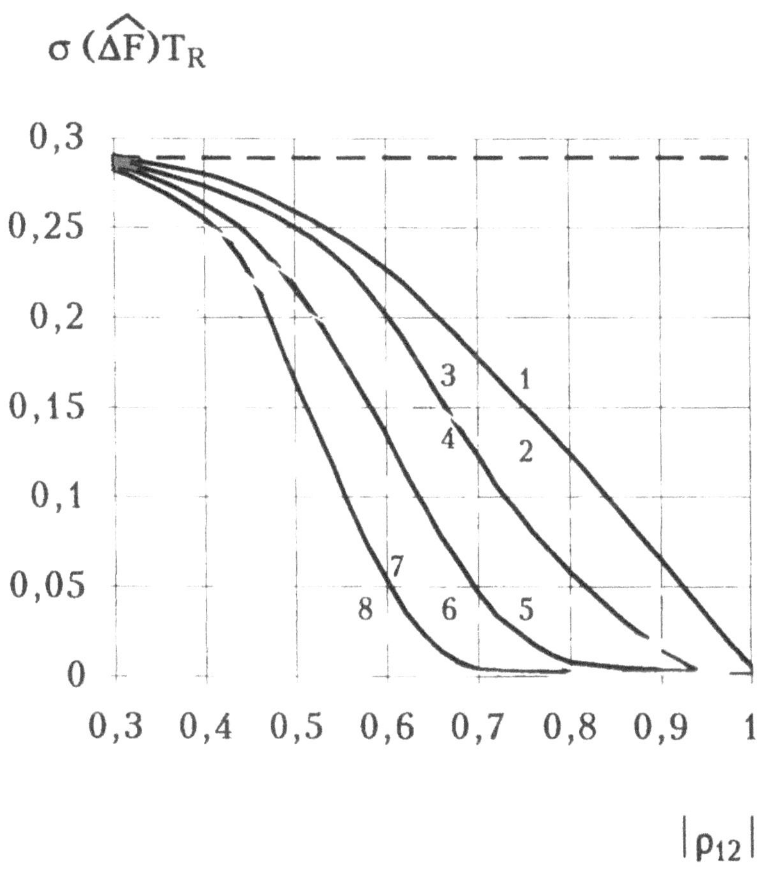

Figure 13.4 Dependences of the normalized r.m.s. errors of DDF estimates on the modulus of interstation (spatial) correlation coefficient for different number of pulses, M, $q_*^2=E_{i\rho}/N_i=20$ dB. Black lines: optimum adaptive estimates from (13.52) for unknown ρ_{12}; grey lines: optimum estimates from (13.50) for known ρ_{12}; curves 1, 2: $M=2$, curves 3, 4: $M=4$, curves 5, 6: $M=8$, curves 7, 8: $M=16$

inversely proportional to the number of pulses, $M-1$. Apparently, increasing M leads to narrowing of the spectrum of $\tilde{G}_{1l}\tilde{G}^*_{2l}$, $l=\overline{0,M-1}$, so that its maximum can be determined with higher accuracy. This is valid for high interstation correlation coefficients. However, when interstation correlation, $|\rho_{12}|$, becomes low (i.e. random and independent phase variations of products $\tilde{G}^*_{1l}\tilde{G}_{2l}$ become large) this spectrum "falls apart", especially for large M. In this case the phase shift $2\pi\Delta F l T_R$ may exceed the unambiguity interval $(-\pi,\pi)$ by many times and more often compensate for comparatively large random phase variations of products $\tilde{G}^*_{1l}\tilde{G}_{2l}$ which leads to greater DDF errors than for small M. This "thresholding effect" should be taken into account when possibilities of DDF measurement in a MSRS are evaluated.

Because of strong influence of phase periodicity, expression (13.66) for efficient estimates may be used only when $|\rho_{12}|$ is close to one. Therefore, corresponding curves for efficient estimates do not shown in Fig. 13.4.

Accuracy of Optimum Algorithms for Unknown Space–Time Target Echo Correlation (See Section 13.3)

Consider first a situation where *there is no interpulse (temporal) correlation; the interstation (spatial) correlation coefficient* ρ_{12} *and (or) the phase propagation delay* $(\omega_0/c)(R_1-R_1)$ *are unknown or a MSRS is a system with short-term spatial coherence* (see Section 1.1). In this situation the optimum (maximum likelihood) DDF estimate can be obtained with the help of (13.52). Since such a situation is usual in practice, the results of statistical simulation of that algorithm is shown in Fig. 13.4 by black lines. Comparing these curves with those for known ρ_{12} permits to reveal the price that has be to paid for *a priori* unknown interstation target echo phase difference.

The "thresholding effect" can be seen for algorithm (13.52) as for (13.50). The more rapid increase of the r.m.s. DDF error for $M=2$ with the decrease of $|\rho_{12}|$ is caused by the fact that a DDF estimate is determined here by the second but not the first phase difference [compare (13.53) with (13.51)]. Therefore the variance of estimates for the same values of $|\rho_{12}|$ is greater here.

Note that the interpulse time interval T_R does not influence the interstation correlation coefficient ρ_{12} (in a first approximation, see Section 4.2). Therefore, we often may employ waveforms with large T_R so as to decrease errors in DDF measurements. However, like the case of DF measurement at each station, increasing of T_R is limited by the ambiguity of DDF estimation with the help of pulse trains. If a true value of DDF is *a priori* known with the maximal error $\pm\Delta_{\Delta F}$, then allowable T_R has to be no greater than the reciprocal value of $2\Delta_{\Delta F}$, i.e. $T_R \leqslant 1/2\Delta_{\Delta F}$. Another constraints on T_R are imposed by the requirement that the assumed model of stationary Gaussian fluctuations of target echo complex amplitudes remains valid.

Now we consider a situation where *there is no interstation (spatial)* correlation and interpulse *(temporal) correlation is unknown.* In some cases unknown interpulse correlation is possible to be estimated using a target echo at one, "reference", station. Then DDFs can be measured at other stations with respect to the "reference" one (see Section 13.3). As mentioned in Section 13.3, interpulse correlation estimates obtained from a single target echo pulse train may be expected to be good when complex amplitudes of different pulses of each pulse train may be approximated by a Markov chain. For such signals the optimum algorithm (13.56) has been synthesized. The simulation results for that algorithm are presented in Fig. 13.5 (the correlation coefficients are assumed equal at both stations, i.e. $\rho_1=\rho_2$). In the same figure the curves for $\sigma(\hat{F}_i)T_R$ calculated using (13.45) for known $\arg\rho_i$, are repeated from Fig. 13.1 (grey lines). Since there is no interstation correlation, the first and the second terms in the right side of (13.56) are statistically independent. Hence

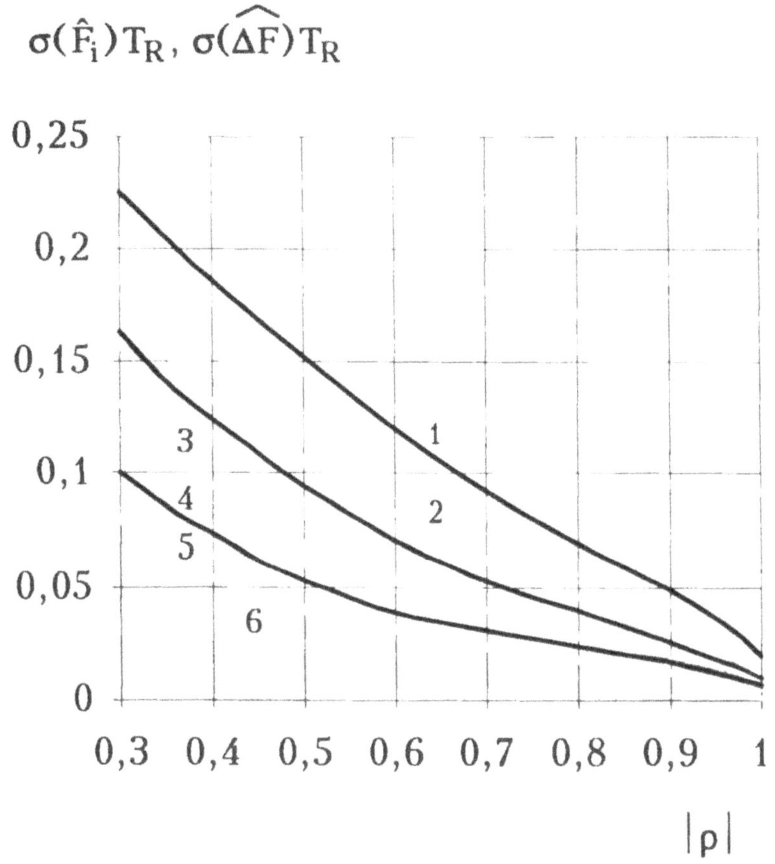

Figure 13.5 Comparison of r.m.s. errors of DF and DDF estimates, $q_s^2 = E_{ip}/N_i = 20\,\mathrm{dB}$. Black lines: optimum DDF adaptive estimates from (13.56); grey lines: optimum DF estimates from (13.45) for known interpulse correlation; curves 1, 2: $M=8$, curves 3, 4: $M=16$, curves 5, 6: $M=32$

it might be expected that r.m.s. errors of DDF estimates according to (13.56) would be by $\sqrt{2}$ greater than r.m.s. errors of DF estimates according to (13.45). This is approximately true but the phase difference periodicity can disturb this relationship. We may come to the same conclusion if we consider a DDF measurement as the estimation of the difference between statistically independent estimates $\hat{F}_1$ and $\hat{F}_2$ calculated using (13.45) for $\arg\rho_1 = \arg\rho_2$.

It is important to evaluate the sensitivity of the optimum adaptive algorithm (13.56) to deviations from the exponential interpulse correlation. Simulation has been performed for the Gaussian interpulse correlation of complex amplitudes: $0.5\overline{A_l A_{l+n}} = \sigma^2 \exp(-n^2 \ln|\rho_i|)$ and $\arg\rho_i = 0$. The increase of DDF r.m.s. error $\sigma(\widehat{\Delta F})T_R$ for $M \leqslant 32$ does not exceed 20–30% excluding the region of very high correlation for large M [when the usual algorithm (13.18) can be successfully employed so that there are no reasons to use algorithms optimal for imperfect interpulse correlation].

Probability distributions of the DF and DDF estimates, $\hat{F}_i, \Delta F$, are strongly dependent on the correlation coefficients $|\rho_i|$, $i=1,2$, and $|\rho_{12}|$. Two typical histograms for PDF of the normalized DF estimates $\hat{F}_i T_R$ are shown in Fig. 13.6(a) and (b) for $M=2$ and $M=8$, respectively. It is seen that as the correlation coefficient decreases, corresponding PDFs change from nearly the Gaussian to the uniform distribution. Therefore, evaluation of r.m.s. errors are often insufficient for the accuracy analysis. A more full and convenient characteristics are the error values that are not exceeded with a prescribed probability. In Figs. 13.7 and 13.8 confidence limits for the DF and DDF normalized error moduli and $M=2$, $M=8$ are shown. These curves are also obtained by computer simulation of the algorithms (13.45), (13.50) and (13.52) under the same conditions as before. The confidence levels (the prescribed probabilities) are chosen to be $P=0.7$ and $P=0.95$. Using the curves from Figs. 13.7 and 13.8 we can determine feasible combinations of the number of pulses M in a pulse train and the correlation coefficients $|\rho_i|$, $i=1,2$, or $|\rho_{12}|$ which provide the required refinement of *a priori* DF or DDF estimates with probability 0.7 or 0.95.

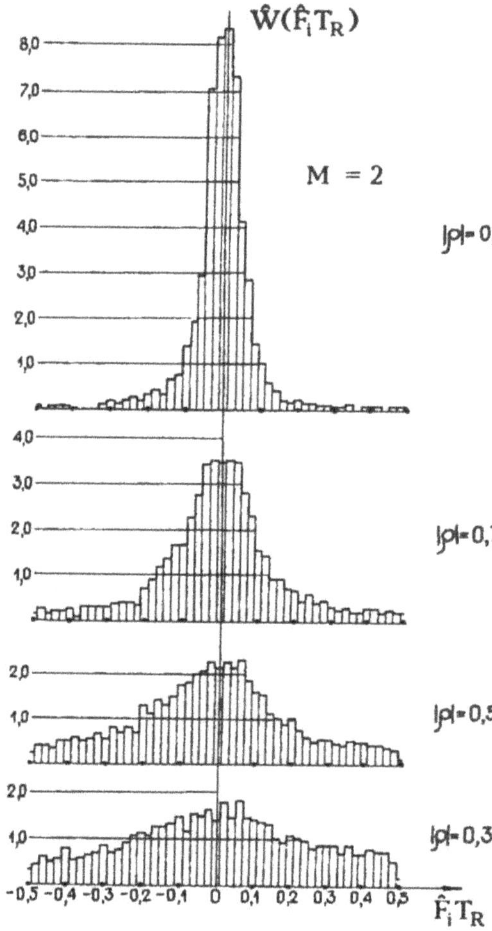

Figure 13.6(a)

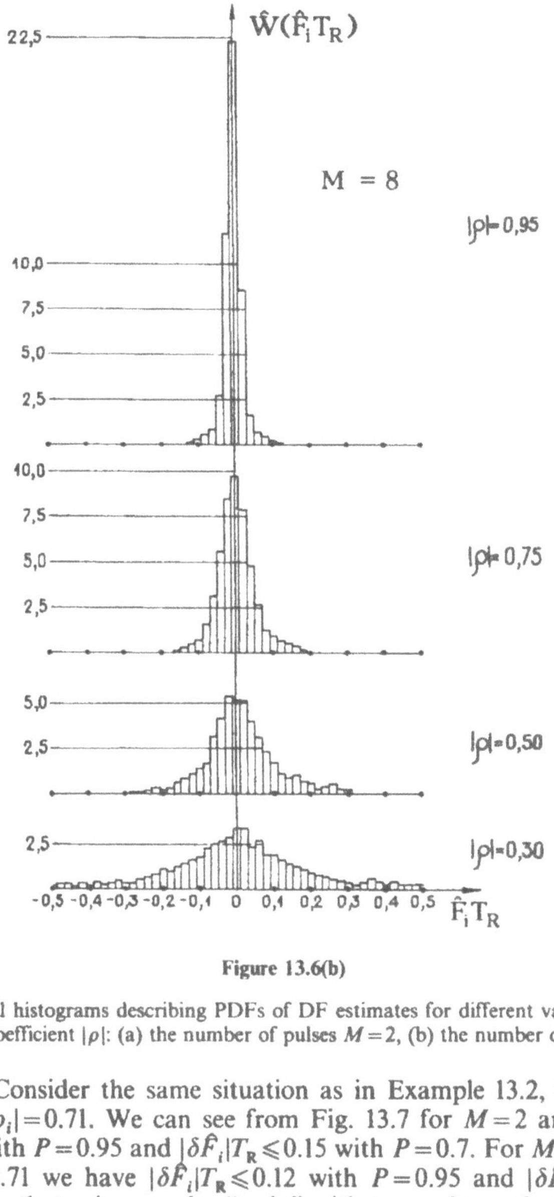

Figure 13.6(b)

Figure 13.6 Typical histograms describing PDFs of DF estimates for different values of the interpulse correlation coefficient $|\rho|$: (a) the number of pulses $M=2$, (b) the number of pulses $M=8$

Example 13.3. Consider the same situation as in Example 13.2, i.e. $|\Delta_F|=100\,\text{Hz}$, $T_R=5\,\text{ms}$ and $|\rho_i|=0.71$. We can see from Fig. 13.7 for $M=2$ and $|\rho_i|=0.71$ that $|\delta\hat{F}_i|T_R \leqslant 0.39$ with $P=0.95$ and $|\delta\hat{F}_i|T_R \leqslant 0.15$ with $P=0.7$. For $M=8$ and the same value of $|\rho_i|=0.71$ we have $|\delta\hat{F}_i|T_R \leqslant 0.12$ with $P=0.95$ and $|\delta\hat{F}_i|T_R \leqslant 0.053$ with $P=0.7$. It means that using a pulse "train" with two pulses only we obtain the DF measurement errors not exceeding $\pm 0.39\times 200=\pm 78\,\text{Hz}$ with the confidence level $P=0.95$ and $\pm 0.15\times 200=\pm 30\,\text{Hz}$ with the confidence level $P=0.7$. When we use a pulse train consisting of 8 pulses ($M=8$) we obtain the DF estimation errors within $\pm 0.12\times 200=\pm 24\,\text{Hz}$ with $P=0.95$ and $\pm 0.054\times 200=\pm 10.8\,\text{Hz}$ with $P=0.7$.

It is interesting to analyze the achievable accuracies of DF measurement when *a priori* DF accuracy is higher so that the time intervals between pulses, T_R, may be increased. Let a true DF value be known with the *a priori* maximal error $\pm 40\,\text{Hz}$. Then the maximum value of T_R is $T_R=1/2\times 40=12.5\,\text{ms}$. For the same target with $|\rho_i|=\exp(-\alpha_i T_R)$ where $\alpha_i=69.3\,\text{s}^{-1}$ we have $|\rho_i|=0.42$. In this case from Fig. 13.7 we have for $M=2$: $|\delta\hat{F}_i|T_R \leqslant 0.44$ with $P=0.95$ and $|\delta\hat{F}_i|T_R \leqslant 0.24$ with $P=0.7$.

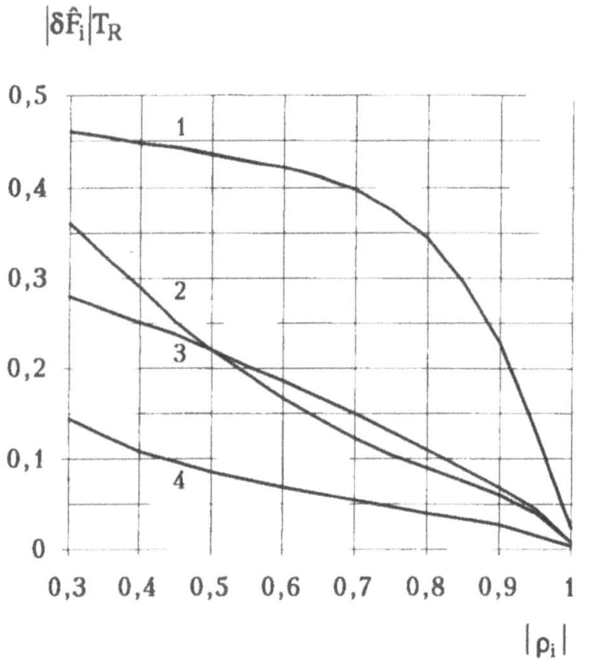

Figure 13.7 Confidence limits for the modulus of normalized DF estimates errors under the same conditions as in Fig. 13.1. Curves 1, 2: $P=0.95$, $M=2$ and 8, respectively; curves 3, 4: $P=0.7$, $M=2$ and 8, respectively

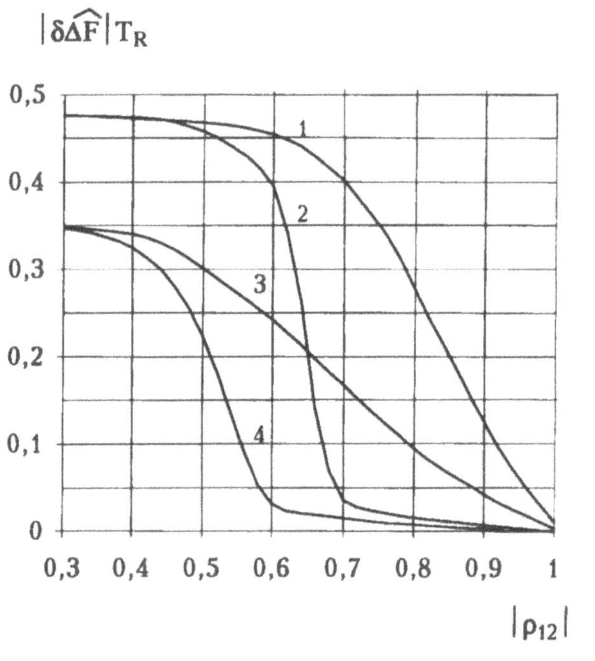

Figure 13.8 Confidence limits for the modulus of normalized adaptive DDF estimate errors under the same conditions as in Fig. 13.4 [algorithm (13.52)]. Curves 1, 2: $P=0.95$, $M=2$ and 8, respectively; curves 3, 4: $P=0.7$, $M=2$ and 8, respectively

For $M=8$ and the same value of $|\rho_i|=0.42$ we have $|\delta\hat{F}_i|T_R \leqslant 0.27$ with $P=0.95$ and $|\delta\hat{F}_i|T_R \leqslant 0.15$ with $P=0.7$. This yields the maximal DF estimation errors for $M=2$: ± 35.2 Hz with $P=0.95$ and ± 19.2 Hz with $P=0.7$; for $M=8$: ± 21.6 Hz with $P=0.95$ and ± 12.0 Hz with $P=0.7$. These results are essentially better than for $T_R=5$ ms when $M=2$ and close to those obtained above for $T_R=5$ ms when $M=8$. It can be explained by specific features of the error PDFs.

Example 13.4. For DDF estimation when, for example, there is no interpulse (temporal) correlation and interstation (spatial) correlation is unknown, we can evaluate the expected accuracy using Fig. 13.8. The interstation correlation coefficient, $|\rho_{12}|$, is determined by the angle between directions from a target to both receiving stations. Noticeable spatial target echo correlation is possible when this angle is significantly smaller than the averaged angular width of a lobe of the target scattering pattern (see Section 4.2). Under this condition DDFs are determined mainly by tangential target velocity with respect to a MSRS. Usual DDF values are, as a rule, much smaller than DF values at each station. Therefore, *a priori* DDF errors are small too. Taking it into account we consider two cases: $|\rho_{12}|=0.8$ and $|\rho_{12}|=0.5$. For both cases the *a priori* maximal DDF error is assumed to be ± 20 Hz. To cope with the DDF ambiguity we choose $T_R=1/40=25$ ms. It can be seen from Fig. 13.8 that for $|\rho_{12}|=0.8$ we have the following values of the normalized maximal DDF errors: $|\delta\widehat{\Delta F}|T_R=0.27$ and 0.094 for $M=2$ and the confidence levels $P=0.95$ and $P=0.7$, respectively; $|\delta\Delta\hat{F}|T_R=0.015$ and 0.008 for $M=8$ and the confidence levels $P=0.95$ and $P=0.7$, respectively. It means that the DDF errors for $M=2$ do not exceed: $\pm 0.27/25\times 10^{-3}=\pm 10.8$ Hz with $P=0.95$ and $\pm 0.094/25\times 10^{-3}=\pm 3.8$ Hz with $P=0.7$; for $M=8$ these errors do not exceed: $\pm 0.015/25\times 10^{-3}=\pm 0.6$ Hz with $P=0.95$ and $\pm 0.008/25\times 10^{-3}=\pm 0.32$ Hz with $P=0.7$. For the situation where $|\rho_{12}|=0.5$, we can see from Fig. 13.8 that $|\delta\Delta F|T_R=0.47$ and 0.30 for $M=2$ and the confidence levels $P=0.95$ and 0.7 respectively; $|\delta\Delta F|T_R=0.46$ and 0.22 for $M=8$ and the confidence levels $P=0.95$ and 0.7 respectively. Hence the maximal DDF errors for $M=2$ do not exceed: $\pm 0.47/25\times 10^{-3}=\pm 18.8$ Hz with $P=0.95$ and $\pm 0.30/25\times 10^{-3}=\pm 12.0$ Hz with $P=0.7$; for $M=8$ these errors do not exceed: $\pm 0.46/25\times 10^{-3}=\pm 18.4$ Hz with $P=0.95$ and $\pm 0.22/25\times 10^{-3}=\pm 8.8$ Hz with $P=0.7$. Almost equal maximal errors when $|\rho_{12}|=0.5$ as well as very large difference between errors when $|\rho_{12}|=0.8$ for $M=2$ and $M=8$ with $P=0.95$ may be attributed to the same "thresholding effect" in the behaviour of the curves in Fig. 13.8 which was discussed above with the connection of Fig. 13.4.

***In this section**, we have analyzed the optimum Doppler frequency (DF) and differential Doppler frequency (DDF) estimation algorithms synthesized in Sections* 13.2 *and* 13.3. *Their performance characteristics have been obtained by computer simulation (statistical tests).*

At first we have considered the optimum algorithms for known space–time target echo complex amplitude correlation (see Section 13.2).

For independent DF estimation at each station (when interstation correlation is absent) the simulation results according to the optimum algorithm (13.40) *for Markovian complex amplitudes, i.e.* (13.45), *are presented in Fig.* 13.1. *The same Fig.* 13.1 *corresponds to algorithm* (13.54) *from Section* 13.3 *for unknown modulus of interpulse correlation coefficient when its phase is equal or close to zero. The normalized r.m.s. error of DF estimates increases with the decrease of the interpulse correlation. However, this accuracy degradation of optimum algorithms is much less than that of the usually employed algorithm* (13.18), *especially when the number of pulses, M, is sufficiently large. This fact is illustrated by Fig.* 13.2 *for a pulse train with M* = 16 *and by Fig.* 13.3 *for* $M = 32$. *It is seen that errors of optimum estimates for* $M = 32$ *are only slightly greater than those of efficient estimates. Example* 13.2 *presents a numerical illustration of achievable accuracy of the optimum DF measurement algorithms. It is important that the performance advantages of algorithm* (13.45) (*which is optimal for exponential interpulse correlation function*) *with respect to the conventionally used algorithm* (13.18) *are valid for other interpulse correlation functions.*

For the case where only interstation target echo correlation is present, we have analyzed algorithm (13.50). *Simulation results for normalized r.m.s. DDF errors have been presented in Fig.* 13.4.

For practically important situation where interstation correlation is unknown and there is no interpulse correlation, the simulation results (*normalized r.m.s. DDF errors*) *corresponding to optimum algorithm* (13.52) *have been shown in the same Fig.* 13.4.

When only interpulse target echo correlation is present but unknown, the optimum algorithm (13.56) *can be used under certain conditions discussed in Section* 13.3. *The simulation results for this algorithm have been depicted in Fig.* 13.5.

As interpulse and interstation correlation coefficients decrease, probability distributions of DF and DDF estimates change significantly approaching to the uniform distribution. It is illustrated by two typical histograms in Fig. 13.6(*a*) *and* (*b*) for $M = 2$ *and* $M = 8$, *respectively. Under this condition maximal errors corresponding to certain confidence levels* (*with the prescribed probability*) *turn out to be more full accuracy characteristics than r.m.s. errors. Such curves for the DF and DDF optimum estimates have been obtained by simulation and are plotted in Figs.* 13.7 *and* 13.8, *respectively. Numerical Examples* 13.3 *and* 13.4 *illustrate practical calculations with the help of those curves.*

14. RESULTANT TARGET COORDINATE MEASUREMENT USING TWO-STAGE ALGORITHMS

14.1. RESULTANT COORDINATE MEASUREMENT FORMATION

The main function of the second stage of two-stage target coordinate estimation algorithms is to combine signal parameter [or local ("primary") coordinate] estimates obtained at the first stage with respect to individual stations of a MSRS, so as to form a resultant composite measurement – the global ("final") estimate of target coordinates and, possibly, their derivatives in a common for the MSRS reference coordinate system and referred to a certain time[1].

The general characteristic of a target at each time moment is *the vector of state*, $\boldsymbol{\alpha}$, which may include all the three target coordinates, their derivatives and other parameters [31,52,72,79,121]. In this Section we consider only such vectors $\boldsymbol{\alpha}$ whose components can be estimated within a short time interval, for example, using a single target illumination. These may be spatial coordinates and velocity components which are measured by the Doppler method. In contrast to the target tracking we do not take into account here measurements obtained at previous time instants.

When a set of local coordinate estimates determines all unknown components of $\boldsymbol{\alpha}$ unambiguously, i.e. a single point $\hat{\boldsymbol{\alpha}}$ in the state space of the target, this point is adopted as the estimate of $\boldsymbol{\alpha}$. In this case optimum estimates of local coordinates ensure an optimum estimate $\hat{\boldsymbol{\alpha}}$ according to the same optimality criterion. However, the total number of measured local coordinates of a target often exceeds the dimensionality of the vector $\boldsymbol{\alpha}$. Then resultant measurement formation becomes *a statistical problem* of $\boldsymbol{\alpha}$ estimation using redundant estimates of local target coordinates obtained by all stations of a MSRS.

This problem is usually solved by the maximum likelihood method (when local coordinate estimates are Gaussian random variables) or by the least squares method (when probability distributions of those estimates are non-Gaussian or unknown). According to both maximum likelihood and minimum r.m.s. error criteria, such an estimate $\hat{\boldsymbol{\alpha}}$ is optimum which minimizes the following quadratic form

$$L=[\hat{\boldsymbol{\xi}}-\mathbf{h}(\boldsymbol{\alpha})]'\mathbf{B}_{\xi}^{-1}[\hat{\boldsymbol{\xi}}-\mathbf{h}(\boldsymbol{\alpha})]\to\min(\boldsymbol{\alpha}). \tag{14.1}$$

Here $\hat{\boldsymbol{\xi}}$ is the vector of local coordinate estimates; $\mathbf{h}(\boldsymbol{\alpha})$ is the vector of known functions determining the dependence of the local coordinates $\boldsymbol{\xi}$ on the target state vector $\boldsymbol{\alpha}$, i.e. $\boldsymbol{\xi}=\mathbf{h}(\boldsymbol{\alpha})$; $\mathbf{B}_{\xi}^{-1}$ is the inverse of the covariance matrix (in the case of maximum likelihood criterion) or the weight matrix (in the case of least squares method) of the estimates $\hat{\boldsymbol{\xi}}$. Errors in $\hat{\boldsymbol{\xi}}$ are assumed to be additive. In general, $\hat{\boldsymbol{\xi}}$ includes estimates of different local coordinates: ranges, bearings, velocity

[1] The main results of Chapters 14 and 15 are valid not only for MSRSs (including radars of different types and frequency ranges) but for other multisensor systems with dissimilar sensors: radars, radio, optical and acoustical direction-finders, lidars and so on.

components and so on. The number of measured local coordinates may be unequal for different stations. The functions $\xi_i = h_i(\alpha)$ are, as a rule, nonlinear [see, for example, (11.1)–(11.8)]. To obtain $\hat{\alpha}$ from (14.1), two approaches are usually exploited: linearization of functions $\xi_i = h_i(\alpha)$ which yields estimates $\hat{\alpha}$ in an explicit form and iterative procedures (the successive approximation method). Their combination is also possible.

The Linearization Method with Parallel Processing of Measurements from All Stations of a MSRS

The linearization method is used when an approximate value of α is *a priori* known and may be assumed as a reference value, α_{ref}. Then *the purpose of measurements is to refine the estimate* α_{ref}. The difference between α_{ref} and the true value α_0 has to be not too large so as linear terms of a Taylor series of each function $\xi_i = h_i(\alpha)$ about $\alpha = \alpha_{ref}$ would approximate this function sufficiently good in the vicinity of α_{ref} including α_0. In this case

$$\mathbf{h}(\alpha) \approx \mathbf{h}(\alpha_{ref}) + \mathbf{H}(\alpha - \alpha_{ref}) \tag{14.2}$$

where $\mathbf{H} = \| \partial h_j(\alpha)/\partial \alpha_l \|_{\alpha = \alpha_{ref}}$ is the matrix of derivatives of functions $h_j(\alpha)$, $j = \overline{1, N}$ (N is the total number of measured local coordinates), with respect to all n components, α_l, $l = \overline{1, n}$, of the vector α at the point $\alpha = \alpha_{ref}$. Each row of this matrix represents projections of the gradient of one of the functions $h_j(\alpha)$. Substituting the linearized function (14.2) in (14.1) and solving the likelihood equations $\partial L/\partial \alpha = 0$ for α yield the *optimum* (*maximum likelihood*) *estimate*

$$\hat{\alpha} = \alpha_{ref} + (\mathbf{H}'\mathbf{B}_\xi^{-1}\mathbf{H})^{-1}\mathbf{H}'\mathbf{B}_\xi^{-1}[\hat{\xi} - \mathbf{h}(\alpha_{ref})]. \tag{14.3}$$

The expression (14.3) can be presented in the form

$$\hat{\alpha} = \alpha_{ref} + \mathbf{K}[\hat{\xi} - \mathbf{h}(\alpha_{ref})] \tag{14.4}$$

where

$$\mathbf{K} = (\mathbf{H}'\mathbf{B}_\xi^{-1}\mathbf{H})^{-1}\mathbf{H}'\mathbf{B}_\xi^{-1} \tag{14.5}$$

is the $n \times N$ matrix of optimal weights used for adding differences $\hat{\xi}_j - h_j(\alpha)$, $j = \overline{1, N}$, to each component α_{refl}, $l = \overline{1, n}$, of the vector α_{ref}. Each difference is the discrepancy between the really measured value of the local coordinate $\hat{\xi}_j$ and the value of ξ_j which would be measured if the target state vector was equal to α_{ref} and all measurement errors were absent. In other words, these are discrepancies between the measured values $\hat{\xi}_j$ and the calculated (true) values of ξ_j corresponding to the target state vector α_{ref}.

It is important to find out what is the influence of both measurement errors in $\hat{\xi}$ and linearization errors on resultant errors of $\hat{\alpha}$. Let us substitute $\hat{\xi} = \mathbf{h}(\alpha_0) + \mathbf{e}$ in (14.3) where $\mathbf{e}$ is the vector of random errors of local coordinate measurements. After some transformations (14.3) takes the form [117]

$$\hat{\alpha} = \alpha_0 + (\mathbf{H}'\mathbf{B}_\xi^{-1}\mathbf{H})^{-1}\mathbf{H}'\mathbf{B}_\xi^{-1}\{\mathbf{h}(\alpha_0) - [\mathbf{h}(\alpha_{ref}) + \mathbf{H}(\alpha_0 - \alpha_{ref})] + \mathbf{e}\}. \tag{14.6}$$

We can obtain a bias of the estimate $\hat{\boldsymbol{\alpha}}$ by averaging $\mathbf{e}$ in equation (14.6):

$$\mathbf{b}=\bar{\hat{\boldsymbol{\alpha}}}-\boldsymbol{\alpha}_0=(\mathbf{H}'\mathbf{B}_\xi^{-1}\mathbf{H})^{-1}\mathbf{H}'\mathbf{B}_\xi^{-1}\{\mathbf{h}(\boldsymbol{\alpha}_0)-[\mathbf{h}(\boldsymbol{\alpha}_{ref})+\mathbf{H}(\boldsymbol{\alpha}_0-\boldsymbol{\alpha}_{ref})]+\bar{\mathbf{e}}\}. \quad (14.7)$$

The bias arises if $\bar{\mathbf{e}}\neq\mathbf{0}$, and/or if there are linearization errors, i.e. $\mathbf{h}(\boldsymbol{\alpha}_0)-[\mathbf{h}(\boldsymbol{\alpha}_{ref})+\mathbf{H}(\boldsymbol{\alpha}_0-\boldsymbol{\alpha}_{ref})]\neq\mathbf{0}$.

When each local coordinate estimator at each station is calibrated but a measurement bias remains, and the dependence of this bias on unknown parameters (for instance, SNR values) is known, these unknown parameters may be additionally included in the vector $\boldsymbol{\alpha}$ to be estimated (if the dimensionality N of the measurement vector $\boldsymbol{\xi}$ is sufficiently large). When maximum likelihood estimators process signal samples of a large size so that asymptotic properties of the estimators manifest themselves, $\bar{\mathbf{e}}=\mathbf{0}$.

The *covariance matrix* of the estimate $\hat{\boldsymbol{\alpha}}$ we can obtain from (14.6)

$$\mathbf{B}_\alpha=\overline{(\hat{\boldsymbol{\alpha}}-\bar{\hat{\boldsymbol{\alpha}}})(\hat{\boldsymbol{\alpha}}-\bar{\hat{\boldsymbol{\alpha}}})'}=(\mathbf{H}'\mathbf{B}_\xi^{-1}\mathbf{H})^{-1}. \quad (14.8)$$

As can be seen from (14.3), calculating the estimate $\hat{\boldsymbol{\alpha}}$, requires calculating the matrix $\mathbf{B}_\alpha$ as a preliminary procedure, so that we obtain the covariance matrix, $\mathbf{B}_\alpha$, together with the estimate, $\hat{\boldsymbol{\alpha}}$. Since the nonlinear function $\mathbf{h}(\boldsymbol{\alpha})$ in (14.1) is replaced by the linear function (14.2), the maximum likelihood estimate (14.3), (14.4) is an *efficient* estimate $\hat{\boldsymbol{\alpha}}$ given $\boldsymbol{\xi}$. It means that inverting the Fisher information matrix (FIM)

$$\left\|\overline{\frac{\partial^2 L(\xi_1,\ldots,\xi_N,\alpha_1,\ldots,\alpha_n)}{\partial\alpha_p\partial\alpha_q}}\right\|, \quad p,q=\overline{1,n}$$

where L is determined by (14.1), yields (14.8) so that (14.8) is the Cramér–Rao lower bound on errors of vector $\boldsymbol{\alpha}$ estimates given $\boldsymbol{\xi}$.

If the vector of local coordinate estimates, $\boldsymbol{\xi}$, is a Gaussian random vector, then the vector $\hat{\boldsymbol{\alpha}}$ is also Gaussian (on account of a linear dependence of $\hat{\boldsymbol{\alpha}}$ on ξ). In this case the accuracy of $\hat{\boldsymbol{\alpha}}$ is convenient to be characterized by an error hyperellipsoid containing $\hat{\boldsymbol{\alpha}}$ with the prescribed probability P. The error hyperellipsoid equation is given by

$$(\boldsymbol{\alpha}-\bar{\hat{\boldsymbol{\alpha}}})'\mathbf{B}_\alpha^{-1}(\boldsymbol{\alpha}-\bar{\hat{\boldsymbol{\alpha}}})=k^2 \quad (14.9)$$

where the number k determines the size of the hyperellipsoid. This number is equal to the ratio of the principal semiaxes of the hyperellipsoid to the r.m.s. estimate errors of corresponding components of $\boldsymbol{\alpha}$ along these semiaxes. The principal semiaxes of the hyperellipsoid do not coincide in general with coordinate axes unless the matrix $\mathbf{B}_\alpha$ is a diagonal one. However, since $\mathbf{B}_\alpha$ is a real symmetric matrix, there exists an orthogonal matrix $\mathbf{A}$ that diagonalizes $\mathbf{B}_\alpha^{-1}$: $\mathbf{A}'\mathbf{B}_\alpha^{-1}\mathbf{A}=\tilde{\mathbf{B}}_\alpha^{-1}$ where $\tilde{\mathbf{B}}_\alpha^{-1}$ is a diagonal matrix. The number k, in its turn, is determined by the probability P. For a n-dimensional vector $\boldsymbol{\alpha}$, the probability that $\hat{\boldsymbol{\alpha}}$ falls into the hyperellipsoid (14.9) is determined by the following expression [72,117]:

$$P=P_n(k)=\frac{1}{\Gamma(n/2)}\int_0^{k^2/2}x^{n/2-1}\exp(-x)\,dx=\Gamma\left(\frac{k^2}{2},\frac{n}{2}\right) \quad (14.10)$$

where $\Gamma(n/2)$ is the gamma-function; $\Gamma(k^2/2,n/2)$ is the incomplete gamma-function. Graphics and tables for $\Gamma(k^2/2,n/2)$ are presented in [72] and [38], respectively.

In the particular cases of one-dimensional, two-dimensional and three-dimensional vector $\boldsymbol{\alpha}$, (14.10) is reduced to the form

$$P_1(k)=\operatorname{erf}(k/\sqrt{2}); \quad P_2(k)=1-\exp(-k^2/2);$$
$$P_3(k)=\operatorname{erf}(k/\sqrt{2})-k\sqrt{2/\pi}\exp(-k^2/2) \tag{14.11}$$

where erf(x) is determined by (5.15).

Setting a fixed value of the probability $P_n(k)$ we can calculate k from (14.10) or (14.11) and then using (14.9) construct an error ellipsoid (or hyperellipsoid) with principal semiaxes that are equal to $k\sigma_l$ where σ_l^2, $l=\overline{1,n}$, are diagonal elements of the diagonalized matrix $\tilde{\mathbf{B}}_\alpha$.

Accuracy of a multidimensional estimate $\hat{\boldsymbol{\alpha}}$ is often desirable to characterize by a scalar parameter. Such a scalar measure of accuracy may be the resultant r.m.s. error (or, more precisely, the square root of the resultant second initial moment)

$$\sigma_\Sigma(\hat{\boldsymbol{\alpha}})=\left[\overline{\sum_{l=1}^{n}(\hat{\alpha}_l-\alpha_{0l})^2}\right]^{1/2}=\left[\operatorname{Tr}\mathbf{B}_\alpha+\sum_{l=1}^{n}b_l^2\right]^{1/2} \tag{14.12}$$

where $\operatorname{Tr}\mathbf{B}_\alpha$ is the trace of the matrix $\mathbf{B}_\alpha$; b_l are the components of the bias vector (14.7) of estimates $\hat{\boldsymbol{\alpha}}$. When $\boldsymbol{\alpha}$ contains only three target coordinates, e.g., $\boldsymbol{\alpha}^t=(x,y,z)$ and $\hat{\boldsymbol{\alpha}}$ is unbiased, the quantity $\sigma_\Sigma(\hat{\boldsymbol{\alpha}})$ is often called *the radius of the spherical r.m.s. error*, r_{sph}. In a Cartesian coordinate system

$$r_{sph}=\sqrt{\overline{(\hat{x}-x_0)^2}+\overline{(\hat{y}-y_0)^2}+\overline{(\hat{z}-z_0)^2}}; \tag{14.13a}$$

in a spherical coordinate system

$$r_{sph}=\sqrt{\overline{(\hat{R}-R_0)^2}+R_0^2\cos^2\varepsilon_0\overline{(\hat{\beta}-\beta_0)^2}+R_0^2\overline{(\hat{\varepsilon}-\varepsilon_0)^2}}. \tag{14.13b}$$

Another scalar measure for target localization accuracy which often used in practice is the *Geometric Dilution Of Precision* (GDOP) which is defined either as the r.m.s. target position error {i.e. as (14.12) or (14.13) [247]} or more often as the ratio of this r.m.s. target position error to the r.m.s. ranging error [117,248,249]. The latter definition is

$$\mathrm{GDOP}=\sqrt{\operatorname{Tr}\mathbf{B}_\alpha}/\sigma(R). \tag{14.14}$$

Following the definition (14.14), the GDOP indicates how much the ranging error is magnified by the geometric relations among the target and station positions in a MSRS.

The main drawback of the linearization method is that an approximate value $\boldsymbol{\alpha}_{ref}$ of the estimated vector $\boldsymbol{\alpha}$ is necessary to be known a priori. As mentioned above, $\boldsymbol{\alpha}_{ref}$ must be sufficiently close to the true value $\boldsymbol{\alpha}_0$ so as the linear approximation (14.2) does not lead to a noticeable bias of estimates [see (14.7)]. It is important to note that *this difficulty may often be overcome in MSRSs*. When there is no external target designation with required accuracy, preliminary measurements can be performed by a part of the total number of stations (without redundancy) so as to obtain an unambiguity estimate of $\boldsymbol{\alpha}$. For instance, let the vector $\boldsymbol{\alpha}$ contain three spatial coordinates of a target and a MSRS consist of a single transmitting–receiving and

m receiving stations. If the transmitting–receiving station (a monostatic radar) can measure all the three target spatial coordinates, a measurement of these local coordinates may be used to form a reference point $\boldsymbol{\alpha}_{ref}$. A similar situation takes place if a MSRS consists of m monostatic radars. In other cases, it may be more expedient to calculate $\boldsymbol{\alpha}_{ref}$ using a range measurement by a monostatic radar and two range sum measurements by two receiving stations (or two other monostatic radars). In passive MSRSs $\boldsymbol{\alpha}_{ref}$ may be found by preliminary triangulation using measured bearings from two stations or by hyperbolic method using three linearly independent TDOA measurements. In all cases stations with maximum effective baselengths between them with respect to the target direction should be selected. This provides better accuracy of $\boldsymbol{\alpha}_{ref}$.

Using preliminary nonredundant measurements by a part of the total number of stations for calculating the reference value, $\boldsymbol{\alpha}_{ref}$, of the target state vector, $\boldsymbol{\alpha}$, permits in many cases to *refine the accuracy of resultant maximum likelihood estimates* (14.3)–(14.5) *taking into account* a priori *information about the accuracy of* $\boldsymbol{\alpha}_{ref}$. When $\boldsymbol{\alpha}_{ref}$ is a Gaussian vector with the known covariance matrix $\mathbf{B}_{ref}$ (including the case of external target designation) and errors of $\boldsymbol{\alpha}_{ref}$ are close by the order of magnitude to expected errors of the optimum estimates $\hat{\boldsymbol{\alpha}}$ [see (14.8)], then the matrix $\mathbf{K}$ (14.4) instead of (14.5) takes the form [72,121]

$$\mathbf{K}=(\mathbf{B}_{ref}^{-1}+\mathbf{H}'\mathbf{B}_{\xi}^{-1}\mathbf{H})^{-1}\mathbf{H}'\mathbf{B}_{\xi}^{-1}. \tag{14.15}$$

The inverse covariance matrix $\mathbf{B}_{ref}^{-1}$ (the "accuracy matrix" [72]) of $\boldsymbol{\alpha}_{ref}$ is additionally included in parentheses. Then the covariance matrix of the estimate $\hat{\boldsymbol{\alpha}}$ is given by [compare with (14.8)]

$$\mathbf{B}_{\alpha}=(\mathbf{B}_{ref}^{-1}+\mathbf{H}'\mathbf{B}_{\xi}^{-1}\mathbf{H})^{-1}. \tag{14.16}$$

The estimate $\hat{\boldsymbol{\alpha}}$ obtained from (14.4) where the matrix $\mathbf{K}$ is determined by (14.15) is an optimum estimate according to the maximum *a posteriori* probability criterion.

When signal complex amplitude fluctuations at different stations of a MSRS are mutually statistically independent, then "noise" *random errors of the target local coordinate measurements by different stations are statistically independent* too. This independence remains valid in the presence of external spatially correlated interferences (e.g., jamming) if target echoes and interferences are resolved in TDOA. Under the condition of statistically independent random errors the matrix $\mathbf{B}_{\xi}$ becomes diagonal, i.e. $\mathbf{B}_{\xi}=\mathrm{diag}(\mathbf{B}_{\xi_1},\ldots,\mathbf{B}_{\xi m})$, where $\mathbf{B}_{\xi i}$ is the covariance matrix of estimates $\hat{\boldsymbol{\xi}}_i$ obtained by the ith station. Instead of (14.4), (14.15) and (14.16) we have:

$$\hat{\boldsymbol{\alpha}}=\boldsymbol{\alpha}_{ref}+\sum_{i=1}^{m}\mathbf{K}_i[\hat{\boldsymbol{\xi}}_i-\mathbf{h}_i(\boldsymbol{\alpha}_{ref})]; \tag{14.17a}$$

$$\mathbf{K}_i=\left(\mathbf{B}_{ref}^{-1}+\sum_{i=1}^{m}\mathbf{H}_i'\mathbf{B}_{\xi i}^{-1}\mathbf{H}_i\right)^{-1}\mathbf{H}_i'\mathbf{B}_{\xi i}^{-1}; \tag{14.17b}$$

$$\mathbf{B}_{\alpha}=\left(\mathbf{B}_{ref}^{-1}+\sum_{i=1}^{m}\mathbf{H}_i'\mathbf{B}_{\xi i}^{-1}\mathbf{H}_i\right)^{-1}. \tag{14.18}$$

In (14.17b) and (14.18) $\mathbf{H}_i=\|\partial h_{ji}/\partial\alpha_l\|_{\alpha=\alpha_{ref}}$ is the matrix of derivatives of the functions $\xi_{ji}=h_{ji}(\boldsymbol{\alpha})$, $j=\overline{1,N_i}$, $i=\overline{1,m}$, at the point $\boldsymbol{\alpha}=\boldsymbol{\alpha}_{ref}$ reflecting the dependence of the

jth local coordinate measured by the ith station on $\boldsymbol{\alpha}$; N_i is the dimensionality of the measurement vector $\boldsymbol{\xi}_i = \mathbf{h}_i(\boldsymbol{\alpha})$ at the ith station.

By using (14.17a), (14.17b) and (14.18) one can simplify calculations owing to reduced dimensions of vectors and matrices as compared with (14.4), (14.15) and (14.16).

The Linearization Method with Sequential Processing of Measurements from Different Stations of a MSRS

For some MSRSs, it is convenient to include in signal processing one station after another sequentially. In this case accuracy of each estimate of a target state vector $\boldsymbol{\alpha}$ is refined gradually, step-by-step. At the first step, the initial approximation $\boldsymbol{\alpha}_{ref}$ is refined using the results of local coordinate measurement by the first station, $\hat{\boldsymbol{\xi}}_1$ (numbering of stations is arbitrary). At this step, repeating the same considerations as for parallel processing, yields

$$\hat{\boldsymbol{\alpha}}_1 = \boldsymbol{\alpha}_{ref} + \tilde{\mathbf{K}}_1[\hat{\boldsymbol{\xi}}_1 - \mathbf{h}_1(\boldsymbol{\alpha}_{ref})]; \tag{14.19}$$

where

$$\tilde{\mathbf{K}}_1 = (\mathbf{H}_1^{\prime}\mathbf{B}_{\xi_1}^{-1}\mathbf{H}_1)^{-1}\mathbf{H}_1^{\prime}\mathbf{B}_{\xi 1}^{-1}. \tag{14.20}$$

The remaining symbols in (14.19) and (14.20) are the same as in (14.17a), (14.17b) and (14.18). When the covariance matrix $\mathbf{B}_{ref}$ is known and elements of $\mathbf{B}_{ref}^{-1}$ may not be neglected in comparison with elements of $\mathbf{H}_1^{\prime}\mathbf{B}_{\xi 1}^{-1}\mathbf{H}_1$, then the matrix $\tilde{\mathbf{K}}_1$ in (14.19) should be determined not by (14.20) but by the equation similar to (14.15), i.e.

$$\tilde{\mathbf{K}}_1 = (\mathbf{B}_{ref}^{-1} + \mathbf{H}_1^{\prime}\mathbf{B}_{\xi 1}^{-1}\mathbf{H}_1)^{-1}\mathbf{H}_1^{\prime}\mathbf{B}_{\xi 1}^{-1}. \tag{14.21}$$

Accuracy of the estimate $\hat{\boldsymbol{\alpha}}_1$ is defined by the covariance matrix

$$\mathbf{B}_{\alpha 1} = (\mathbf{B}_{ref}^{-1} + \mathbf{H}_1^{\prime}\mathbf{B}_{\xi 1}^{-1}\mathbf{H}_1)^{-1}. \tag{14.22}$$

At the second step $\hat{\boldsymbol{\alpha}}_1$ is considered as an *a priori* estimate and is refined using the measurement $\hat{\boldsymbol{\xi}}_2$ obtained by the second station, and so on. After $i-1$ steps we have the estimate, $\hat{\boldsymbol{\alpha}}_{i-1}$, and its covariance matrix, $\mathbf{B}_{\alpha(i-1)}$, calculated using measurements from $i-1$ stations. This estimate is considered as an *a priori* one for the ith step where measurements from the ith station, $\hat{\boldsymbol{\xi}}_i$, and their covariance matrix, $\mathbf{B}_{\xi i}$, are used. Thus after i steps we have

$$\hat{\boldsymbol{\alpha}}_i = \hat{\boldsymbol{\alpha}}_{i-1} + \tilde{\mathbf{K}}_i[\hat{\boldsymbol{\xi}}_i - \mathbf{h}_i(\boldsymbol{\alpha}_{i-1})]; \tag{14.23}$$

$$\tilde{\mathbf{K}}_i = (\mathbf{B}_{\alpha(i-1)}^{-1} + \mathbf{H}_i^{\prime}\mathbf{B}_{\xi i}^{-1}\mathbf{H}_i)^{-1}\mathbf{H}_i^{\prime}\mathbf{B}_{\xi i}^{-1}. \tag{14.24}$$

The covariance matrix of the estimate $\hat{\boldsymbol{\alpha}}_i$ is given by

$$\mathbf{B}_{\alpha i} = (\mathbf{B}_{\alpha(i-1)}^{-1} + \mathbf{H}_i^{\prime}\mathbf{B}_{\xi i}^{-1}\mathbf{H}_i)^{-1}. \tag{14.25}$$

This recurrent procedure is carried on until measurements from all m stations are used. The resultant estimate, $\hat{\boldsymbol{\alpha}}_m$, of the target state vector, $\boldsymbol{\alpha}$, and its covariance matrix, $\mathbf{B}_{\alpha m}$, obtained by a MSRS can be calculated from (14.23) and (14.25) respectively, by substituting m for i.

Let us compare accuracy of the estimate $\hat{\alpha}_m$ with that of the estimate $\hat{\alpha}$ obtained by parallel measurement processing from all m stations. It follows from (14.25) that

$$\mathbf{B}_{\alpha m}^{-1} = \mathbf{B}_{\text{ref}}^{-1} + \sum_{i=1}^{m} \mathbf{H}_i' \mathbf{B}_{\xi i}^{-1} \mathbf{H}_i. \tag{14.26}$$

It is seen that for statistically independent errors at different stations [when the matrix $\mathbf{B}_\xi$ in (14.16) is a bloc-diagonal one], $\mathbf{B}_\alpha = \mathbf{B}_{\alpha m}$, i.e. the parallel and sequential processings of measurements from different stations yield the same accuracy. When measurement errors at different stations are mutually correlated, then the sequential processing algorithm, which does not include joint processing of measurements from all stations, yields in general a suboptimal estimate $\hat{\alpha}_m$ with lower accuracy. It is practically important that the algorithm of sequential processing reduces a problem of large dimensionality to several problems of the same type and of smaller dimensionality. The process of estimate refining may be stopped at any step when required estimate accuracy is achieved. At the first steps it is reasonable to employ measurements from those stations that can refine the estimate considered most significantly.

The Iteration Method

When we cannot "guess" the initial value, α_{ref}, sufficiently close to the true value, α_0, so as to avoid noticeable linearization errors [see (14.7)] iteration methods may be directly applied to equation (14.1). In principle, any extremum finding technique applicable to multidimensional nonlinear functions may be used. However, for large dimensionality of the target state vector α this problem becomes difficult. Several "semi-empirical" techniques have been proposed to solve the problem. In general, convergence of these iteration processes to the optimum estimate $\hat{\alpha}$ has not been proven. However, when such procedures are used for practical calculations, their convergence to $\hat{\alpha}$ is usually achieved [88,91]. Though an initial approximation, α_{ref}, is not necessary to be close to the true value, α_0, most iteration techniques really used the assumption that errors are "small" since linearized algorithms are employed at each step. Usually measurements of all stations are jointly taken into account (parallel measurement processing).

One group of iteration algorithms is based on equation (14.3) [88,118]:

$$\hat{\alpha}(k+1) = \hat{\alpha}(k) + [\mathbf{H}'(k)\mathbf{B}_\xi^{-1}\mathbf{H}(k)]^{-1}\mathbf{H}'(k)\mathbf{B}_\xi^{-1}\{\xi - \mathbf{h}[\hat{\alpha}(k)]\} \tag{14.27}$$

where k is the number of iteration, $k = 0, 1, 2, \ldots$; $\hat{\alpha}(0) = \alpha_{\text{ref}}$ is the initial approximation which can be chosen using *a priori* information or preliminary measurements. To ensure convergence of this procedure, $\hat{\alpha}(0)$ is desirable to choose not too far from α_0. The derivatives $\mathbf{H}(k)$ are taken at the point $\hat{\alpha}(k)$.

Another approach is based on the gradient method. In this case [91, 118]

$$\hat{\alpha}(k+1) = \hat{\alpha}(k) + \mu\mathbf{H}'(k)\mathbf{B}_\xi^{-1}\{\xi - \mathbf{h}[\hat{\alpha}(k)]\} \tag{14.28}$$

where μ is the scalar constant. Examples of successful applications of algorithms (14.27) and (14.28) are presented in [88,91].

An unusual approach to the target localization problem in passive triangulation systems is suggested in [268].

It is important to note that errors in stations' location lower the accuracy of resultant target state vector estimates. Therefore, stations' positions in a stationary MSRS have to be thoroughly referred to control points, for example by geodetic methods. Navigation methods using, for example, such global navigation systems as GLONASS or NAVSTAR are conventionally exploited for MSRSs with mobile stations. When station localization errors are random and of the same order of magnitude as other errors, they may be included in measurements of local coordinates. Then we may optimize the estimate $\hat{\alpha}$ including navigational errors [118]. It should be taken into account possible correlation between navigational errors for different stations. However, systematic errors (estimate biases) may have the most serious effect on the resultant target state estimation accuracy. Therefore, the errors due to station site uncertainties, antenna orientation and time calibration must be minimized. This process is called *registration*. For each MSRS it consists of two phases: independent individual station registration and relative station alignment. We do not consider here this problem in detail. There are many useful works analyzing the problem and suggesting different practical algorithms for its solution [110,118,153, 173,182,250–255].

In this section*, we have considered a problem of resultant composite measurement formation using target local coordinate measurements obtained by different stations of a MSRS. This resultant measurement is formed as an optimal estimate of the target state vector. Only those target state vectors have been studied in this Section which can be estimated in a short time interval (for example, using a single target illumination), i.e. without taking into account measurements at previous time instants (unlike for the tracking process). The general optimization algorithm* (14.1) *requires searching for an extremum of a multidimensional nonlinear function. To solve this difficult problem the linearization method is usually employed: nonlinear functions are replaced by linear ones according to* (14.2) *in the vicinity of an* a priori *known approximate value of the estimated target state vector. We have presented two approaches using the linearization method: parallel processing of measurements obtained by all stations* (14.3)–(14.8), (14.15), (14.16), *and sequential processing of measurements* (14.19)–(14.25) *which is much simpler to implement. In the latter case measurements from different stations are included sequentially, step-by-step. At each step measurements from only one station is added to refine the estimate obtained at the previous step. We have shown that in the case where measurement errors at different stations are statistically independent* [*see* (14.17), (14.18)] *both these approaches provide the same accuracy.*

Aside from linearization method, the iteration method may be used. We have presented two typical algorithms: (14.27) *and* (14.28).

We have also presented convenient measures for the estimate accuracy (14.10)–(14.14).

Actual accuracy of target state estimation by a MSRS is strongly influenced by local coordinate systematic errors (estimate biases). In connection with this we have emphasized the importance of registration process including independent individual station calibration and relative station alignment in a MSRS.

14.2. COMPARISON OF MAXIMUM ATTAINABLE ACCURACY OF ONE-STAGE AND TWO-STAGE ESTIMATION ALGORITHMS

The choice of one-stage or simpler two-stage target coordinate estimation algorithms depends on their relative accuracy. The algorithms considered in Section 14.1 optimize target state vector estimates, $\hat{\alpha}$, only at the second stage given estimates of local target coordinates $\hat{\xi}$. It is clear, that for arbitrary $\hat{\xi}$, output accuracy of $\hat{\alpha}$ is in general lower than that of optimum one-stage algorithms.

Let us compare the Cramér–Rao lower bound on errors of one-stage and two-stage algorithms in order to reveal maximum attainable accuracy of target state estimation[2] and to evaluate accuracy of specific algorithms. For the sake of simplicity, assume that each ith station, $i=\overline{1,m}$, can measure only one signal parameter, ξ_i (for example, TOA determining the target range or range sum). Extension to several measured parameters, when $\boldsymbol{\xi}_i$ is a vector, is not difficult. Let the dimensionality of the target state vector $\boldsymbol{\alpha}$ be equal to n. For *one-stage estimation algorithms* only n parameters ξ_i are mutually independent, $\boldsymbol{\xi}_{(n)}=(\xi_1,\ldots,\xi_n)^t$. The remaining parameters are their functions, i.e. $\xi_i=\xi_i(\boldsymbol{\xi}_{(n)})$ if $n<i\leqslant m$ (numbering of stations is arbitrary)[3]. Instead of independent parameters $\boldsymbol{\xi}_{(n)}=(\xi_1,\ldots,\xi_n)^t$ we may use a set of target spatial coordinates and their derivatives having the mutually single-valued relation to $\boldsymbol{\xi}_{(n)}$. However, for the general problem under consideration such a specification is not expedient. External interferences (in general, spatially correlated) and self-noises at the inputs of spatially separated receivers are assumed to be a Gaussian m-dimensional stationary random process, as before. Signals are regular (target echoes) and (for simplicity) without stray parameters.

Under these conditions an arbitrary element of the FIM relevant to estimates $\hat{\boldsymbol{\xi}}_{(n)}=(\hat{\xi}_1,\ldots,\hat{\xi}_n)^t$ can be obtained from (12.14) in the form $(r,s=\overline{1,n})$

$$\begin{aligned}\mathcal{J}_{rs}^{(\xi)}=\mathrm{Re}\Bigg\{\frac{1}{2\pi}\int_{-\infty}^{\infty}\Bigg\{&\frac{\partial\Psi_r^*(\omega,\xi_r)}{\partial\xi_r}\frac{\partial\Psi_s(\omega,\xi_s)}{\partial\xi_s}f_{rs}(\omega)\\&+\sum_{k=n+1}^{m}\Bigg[\frac{\partial\Psi_r^*(\omega,\xi_r)}{\partial\xi_r}\frac{\partial\Psi_k[\omega,\xi_k(\xi_1,\ldots,\xi_n)]}{\partial\xi_s}f_{rk}(\omega)\\&+\frac{\partial\Psi_s^*(\omega,\xi_s)}{\partial\xi_s}\frac{\partial\Psi_k[\omega,\xi_k(\xi_1,\ldots,\xi_n)]}{\partial\xi_r}f_{sk}(\omega)\Bigg]\\&+\sum_{i=n+1}^{m}\sum_{k=n+1}^{m}\frac{\partial\Psi_i^*[\omega,\xi_i(\xi_1,\ldots,\xi_n)]}{\partial\xi_r}\frac{\partial\Psi_k[\omega,\xi_k(\xi_1,\ldots,\xi_n)]}{\partial\xi_s}f_{ik}(\omega)\Bigg\}\mathrm{d}\omega\Bigg\}.\end{aligned}\tag{14.29}$$

Now we take into account that for $n<i\leqslant m$

$$\frac{\partial\Psi_i^*[\omega,\xi_i(\xi_1,\ldots,\xi_n)]}{\partial\xi_r}=\frac{\partial\Psi_i^*[\omega,\xi_i(\xi_1,\ldots,\xi_n)]}{\partial\xi_i}\frac{\partial\xi_i}{\partial\xi_r}.\tag{14.30}$$

Substituting (14.30) in (14.29) yields an element of the $n\times n$ FIM for the measurement vector $\boldsymbol{\xi}_{(n)}$ (used for one-stage target state vector estimation) in the final form

$$\mathcal{J}_{rs}^{(\xi)}=J_{rs}^{(\xi)}+\sum_{k=n+1}^{m}\left[\frac{\partial\xi_k}{\partial\xi_r}J_{sk}^{(\xi)}+\frac{\partial\xi_k}{\partial\xi_s}J_{rk}^{(\xi)}\right]+\sum_{i=n+1}^{m}\sum_{k=n+1}^{m}\frac{\partial\xi_i}{\partial\xi_r}\frac{\partial\xi_k}{\partial\xi_s}J_{ik}^{(\xi)}\tag{14.31}$$

[2] We remind that a target state vector $\boldsymbol{\alpha}$ in this section, as in Section 14.1, is determined only by current measurements, i.e. in contrast to the target tracking process does not take into account measurements at previous time instants.

[3] Of course, the correspondence between ξ_i and components of $\boldsymbol{\alpha}$ (to be estimated) is assumed, that is if, for instance, ξ_i are only TOAs or/and bearings, these components of $\boldsymbol{\alpha}$ may be only target spatial coordinates but not velocity.

where $J_{jl}^{(\xi)}$ is the element of the $m \times m$ FIM of estimates ξ_i when all ξ_i, $i=\overline{1,m}$, are assumed to be independent parameters

$$J_{jl}^{(\xi)} = \mathrm{Re}\,\frac{1}{2\pi}\int_{-\infty}^{\infty} \frac{\partial \Psi_j^*(\omega,\xi_j)}{\partial \xi_j}\,\frac{\partial \Psi_l(\omega,\xi_l)}{\partial \xi_l}\, f_{jl}(\omega)\,\mathrm{d}\omega. \tag{14.32}$$

The estimates $\hat{\boldsymbol{\xi}}_{(n)}=(\hat{\xi}_1,\ldots,\hat{\xi}_n)'$ are to be converted into the estimate $\hat{\boldsymbol{\alpha}}$ with the help of the known functions $\boldsymbol{\xi}_{(n)}=\mathbf{h}(\boldsymbol{\alpha})$. The vector $\hat{\boldsymbol{\xi}}_{(n)}$ may usually be considered to be Gaussian. (For example, maximum likelihood estimates obtained by using the algorithms from Section 11.2 are asymptotically Gaussian). Then a maximum likelihood function logarithm of $\boldsymbol{\alpha}$ is the quadratic form in (14.1) where the multiplier 0.5 is omitted. Let $\tilde{\mathbf{B}}_\xi$ be the covariance matrix of the unbiased estimate vector $\hat{\boldsymbol{\xi}}_{(n)}$ obtained by an arbitrary one-stage estimation algorithm. Then we can derive from (14.1) an element of the FIM for the target state vector estimate $\hat{\boldsymbol{\alpha}}=(\hat{\alpha}_1,\ldots,\hat{\alpha}_n)'$ given the estimate $\hat{\boldsymbol{\xi}}_{(n)}=(\hat{\xi}_1,\ldots,\hat{\xi}_n)'$

$$\begin{aligned} J_{pq}^{(\alpha)} &= 0.5\,\overline{\frac{\partial^2 L(\hat{\xi}_1,\ldots,\hat{\xi}_n|\alpha_1,\ldots,\alpha_n)}{\partial\alpha_p\partial\alpha_q}} \\ &= \sum_{r=1}^{n}\sum_{s=1}^{n} \frac{\partial h_r(\alpha_1,\ldots,\alpha_n)}{\partial \alpha_p}\,\frac{\partial h_s(\alpha_1,\ldots,\alpha_n)}{\partial \alpha_q}\,(\tilde{\mathbf{B}}_\xi^{-1})_{rs};\quad p,q=\overline{1,n} \end{aligned} \tag{14.33}$$

where $(\tilde{\mathbf{B}}_\xi^{-1})_{rs}$ is the (rs)th element (positioned at the intersection of the rth row and the sth column) of the inverse of the covariance matrix $\tilde{\mathbf{B}}_\xi$. If an estimation algorithm for $\boldsymbol{\xi}_{(n)}$ yields the efficient estimates $\hat{\boldsymbol{\xi}}_{(n)}$, then the matrix $\tilde{\mathbf{B}}_\xi^{-1}$ is the FIM of the estimates $\hat{\boldsymbol{\xi}}_{(n)}$ so that $(\tilde{\mathbf{B}}_\xi^{-1})_{rs}$ should be replaced by $J_{rs}^{(\xi)}$ from (14.31). As a result we have

$$J_{pq}^{(\alpha)} = \sum_{r=1}^{n}\sum_{s=1}^{n} \frac{\partial h_r(\alpha_1,\ldots,\alpha_n)}{\partial \alpha_p}\,\frac{\partial h_s(\alpha_1,\ldots,\alpha_n)}{\partial \alpha_q}\,J_{rs}^{(\xi)}. \tag{14.34}$$

When *two-stage target state vector estimation algorithms* are used, then local target coordinates, ξ_i, $i=\overline{1,m}$, are to be measured as functionally independent parameters at the first stage of these algorithms. Having obtained the estimates $\hat{\boldsymbol{\xi}}=(\hat{\xi}_1,\ldots,\hat{\xi}_m)'$ one can find the estimate $\hat{\boldsymbol{\alpha}}=(\hat{\alpha}_1,\ldots,\hat{\alpha}_n)'$. Usually we may assume $\hat{\boldsymbol{\xi}}$ to be unbiased and Gaussian. Then the maximum likelihood function logarithm from (14.1) may be exploited again taking into account that in the case considered the dimensionality of the vector $\boldsymbol{\xi}$ is larger than that of the vector $\hat{\boldsymbol{\alpha}}$, i.e. $m>n$. An arbitrary FIM element of the estimates $\hat{\boldsymbol{\alpha}}$ can be written in the form

$$\begin{aligned} J_{pq}^{(\alpha)} &= 0.5\,\overline{\frac{\partial^2 L(\hat{\xi}_1,\ldots,\hat{\xi}_m|\alpha_1,\ldots,\alpha_n)}{\partial\alpha_p\partial\alpha_q}} \\ &= \sum_{i=1}^{m}\sum_{k=1}^{m} \frac{\partial h_i(\alpha_1,\ldots,\alpha_n)}{\partial \alpha_p}\,\frac{\partial h_k(\alpha_1,\ldots,\alpha_n)}{\partial \alpha_q}\,(\mathbf{B}_\xi^{-1})_{ik};\quad p,q=\overline{1,n} \end{aligned} \tag{14.35}$$

where $(\mathbf{B}_\xi^{-1})_{ik}$ is the (ik)th element of the inverse $m \times m$ covariance matrix of the measurement vector $\hat{\boldsymbol{\xi}}$. If a measurement algorithm for $\boldsymbol{\xi}$ yields efficient estimates $\hat{\boldsymbol{\xi}}$, their FIM should be substituted in (14.35) for $\mathbf{B}_\xi^{-1}$ so that

$$J_{pq}^{(\alpha)} = \sum_{i=1}^{m} \sum_{k=1}^{m} \frac{\partial h_i(\alpha_1, \ldots, \alpha_n)}{\partial \alpha_p} \frac{\partial h_k(\alpha_1, \ldots, \alpha_n)}{\partial \alpha_q} J_{ik}^{(\xi)} \tag{14.36}$$

where $J_{ik}^{(\xi)}$ is determined by (14.32).

To evaluate the difference between maximum attainable accuracies (the Cramér–Rao lower bounds on errors) of one-stage and two-stage target state vector estimation algorithms, let us compare (14.36) with (14.34) where $\tilde{J}_{rs}^{(\xi)}$ is determined in (14.31). Note that the true values of ξ_i when $n<i\leqslant m$, are the functions of $\xi_{(n)}=(\xi_1, \ldots, \xi_n)'$, i.e. $\xi_i=\xi_i(\xi_1, \ldots, \xi_n)$. On the other hand, ξ_i are the functions of the state vector $\boldsymbol{\alpha}=(\alpha_1, \ldots, \alpha_n)'$, i.e. $\xi_i=h_i(\alpha_1, \ldots, \alpha_n)$. Then

$$\frac{\partial h_i(\alpha_1, \ldots, \alpha_n)}{\partial \alpha_p} = \sum_{r=1}^{n} \frac{\partial \xi_i}{\partial \xi_r} \frac{\partial \xi_r}{\partial \alpha_p} = \sum_{r=1}^{n} \frac{\partial \xi_i}{\partial \xi_r} \frac{\partial h_r(\alpha_1, \ldots, \alpha_n)}{\partial \alpha_p}, \quad i=\overline{1,m},\ p=\overline{1,n}. \tag{14.37}$$

Now let us separate the terms with $1\leqslant i\leqslant n$, $1\leqslant k\leqslant n$ in the double sum in (14.36) and substitute (14.37) into remaining terms. Then (14.36) will coincide with (14.34) if $\tilde{J}_{rs}^{(\xi)}$ is replaced by (14.31). It means that *the maximum attainable accuracies (the Cramér–Rao lower bounds on errors) of one-stage and two-stage target state vector estimation algorithms are the same*. The validity of this conclusion for passive MSRSs, when unknown signal parameters (for example, TDOAs) are included in the signal correlation matrix or PSD matrix, may be shown by a similar way.

Thus, as was to be expected, if efficient estimates, $\hat{\boldsymbol{\xi}}$, of local coordinates, $\boldsymbol{\xi}$, are obtained at the first stage of two-stage algorithms, and an efficient estimate, $\hat{\boldsymbol{\alpha}}$, of the target state vector $\boldsymbol{\alpha}$ is obtained at the second stage (from the estimates $\hat{\boldsymbol{\xi}}$), then the accuracy of this estimate $\hat{\boldsymbol{\alpha}}$ coincides with that of the efficient estimate $\hat{\boldsymbol{\alpha}}$ obtained by one-stage algorithms.

As mentioned in Section 14.1, maximum likelihood linearized algorithms of the second stage provide efficient estimates. In Section 11.4 we presented an example where efficient estimates of TDOAs were obtained at the first stage of a two-stage algorithm. Hence the resultant estimate of $\boldsymbol{\alpha}$ calculated by a linearized maximum likelihood algorithm and based on those efficient TDOA estimates, turns out to be efficient too. However, if we obtain maximum likelihood estimates $\hat{\boldsymbol{\xi}}$ at the first stage, they are in general only asymptotically efficient (see Sections 11.3, 11.4). Therefore, actual accuracy of two-stage algorithms with maximum likelihood estimates at the first stage tends to the maximum attainable accuracy of one-stage algorithms (the accuracy of efficient estimates $\hat{\boldsymbol{\alpha}}$) only asymptotically. On the other hand, when one-stage algorithms provide maximum likelihood estimates of $\boldsymbol{\alpha}$, they are only asymptotically efficient too. It means that though actual resultant accuracies of optimum (maximum likelihood) one-stage algorithms and of optimum two-stage algorithms may be in general different (because of different nonlinear processings), these differences are reduced with the increase of estimation accuracy (i.e. when the errors approach the Cramér–Rao lower bound). Thus in the cases where high estimation accuracy is required, more simpler two-stage algorithms are practically on a par with one-stage ones. When the dimensionality of $\boldsymbol{\xi}$ does not exceed the dimensionality of $\boldsymbol{\alpha}$, one-stage and two-stage algorithms coincide (see Sections 11.2–11.4).

***In this section**, we have compared maximum attainable accuracies of one-stage and two-stage target state vector estimation algorithms. We have derived equations* (14.33) *and* (14.35) *determining arbitrary elements of the Fisher information matrices* (*FIMs*) *for target state vector estimation by one-stage and two-stage algorithms, respectively. These equations express the above mentioned FIMs through the inverse covariance matrices of local coordinate measurements. When such measurements provide efficient estimates, their corresponding FIMs should be substituted for the inverse covariance matrices. We have obtained equations* (14.31) *and* (14.32) *for the FIMs of local coordinate estimates and equations* (14.34) *and* (14.36) *for the FIMs of the target state vector estimates for one-stage and two-stage algorithms, respectively.*

The obtained relationships show that, as was to be expected, if efficient estimates of local coordinates are obtained at the first stage of two-stage algorithms and efficient estimate of the target state vector is obtained at the second stage (given local coordinate estimates), then the resultant accuracy of the target state vector estimates coincides with that of the efficient estimates obtained by one-stage algorithms. However, maximum likelihood estimates of both one-stage algorithms and of the first stage of two-stage algorithms are in general only asymptotically efficient. Hence their accuracies may be different. Nevertheless this difference is reduced with the increase of estimation accuracy. Thus in the important cases for practice where high resultant estimation accuracy is required, more simpler two-stage algorithms are practically on a par in accuracy with one-stage ones.

14.3. EXAMPLES OF TWO-STAGE ALGORITHM APPLICATIONS TO TARGET COORDINATE ESTIMATION

In this section we present several numerical examples for better understanding the accuracy advantages of target position estimation in MSRSs and for revealing the contributions of different local coordinate measurements to the resultant target localization accuracy. These examples should be considered as illustrative ones.

We assume all estimates of local coordinates at the first stage of two-stage algorithms to be unbiased. Such an assumption permits to reveal the dependence of target position r.m.s. errors on the number of stations and system geometry as well as on the type of local coordinates. However, as was mentioned in Section 14.1, the problem of systematic errors caused by biases in local coordinate estimates is of great practical importance (see, e.g., [173]).

Example 14.1. Let a target be positioned at a point with the following spherical coordinates: range, $R=500$ km; azimuth, $\beta=0$; elevation angle, $\varepsilon=40°$. We consider a monostatic radar at the coordinate system origin as a "reference" variant. Assume that this monostatic radar can measure range, azimuth and elevation angle of the target. All measurements are unbiased and mutually uncorrelated. We also assume that antenna angular errors are independent of the antenna mainlobe direction and their r.m.s. values are $\sigma(\hat{\beta}_A)=\sigma(\hat{\varepsilon}_A)=5'$ in the coordinate system connected with the antenna ($\beta_A=0$, $\varepsilon_A=0$ correspond to the mainlobe direction). It means that in the common spherical coordinate system mentioned above r.m.s. angular errors are: $\sigma(\hat{\beta})=\sigma(\hat{\beta}_A)/\cos\varepsilon=5'/\cos\varepsilon$, $\sigma(\hat{\varepsilon})=\sigma(\hat{\varepsilon}_A)=5'$. The range r.m.s. error is $\sigma(\hat{R})=10$ m. Under these conditions the radius of the spherical r.m.s. error of the target localization according to (14.13) is $r_{sph}\approx 916$ m. The main contribution to this error make the angular errors. If the parameter k in the error ellipsoid equation (14.9) is assumed to be equal to 3, then the ellipsoid semiaxes are: $3R\cos\varepsilon\,\sigma(\hat{\beta})\approx 2180$ m, $3R\sigma(\hat{\varepsilon})\approx 2180$ m, $3\sigma(\hat{R})\approx 30$ m, i.e. the ellipsoid is essentially flattened in the range direction. The GDOP [according to (14.14)] is very large: $r_{sph}/\sigma(\hat{R})\approx 916/10=91.6$. Using (14.11) we can calculate the probability for an estimate to fall into the error ellipsoid whose semiaxes are equal to three r.m.s. values along all the three spatial coordinates. This probability is $P_3(k=3) \approx 0.97$. [Note that in the one-dimensional case $P_1(k=3) \approx 0.997$.]

Let now the same target be observed by a MSRS with one transmitting station at the coordinate system origin and six receiving stations arranged uniformly along the circle of radius $l = 50$ km in the horizontal plane with the centre at the coordinate origin (Fig. 14.1). *The receiving stations can measure only TOAs* with the same accuracy as the "reference" monostatic radar, i.e. the r.m.s. error of a range sum (the transmitting station–the target–any receiving station) measurement is $\sigma_0(\hat{R}_\Sigma) = 20$ m. Systematic errors (biases) are absent. Let us analyze accuracy of target localization if different number of receiving stations are included in the target coordinate estimation process. It is convenient to characterize the accuracy of interest by the radius of the spherical r.m.s. error (14.13), r_{sph}, and by the error ellipse in the cross-range plane where errors are much greater than in the range direction. The calculation results obtained from equation (14.8) for independent measurement errors at all stations are shown in Fig. 14.2 and Fig. 14.3. The corresponding characteristics of the "reference" monostatic radar are depicted too for comparison (variant 0).

It follows from the diagram of Fig. 14.2, that the r.m.s. spherical error radius, r_{sph}, is significantly reduced when we pass from the monostatic radar (with relatively high measurement accuracy) to the MSRS and the number of operating stations

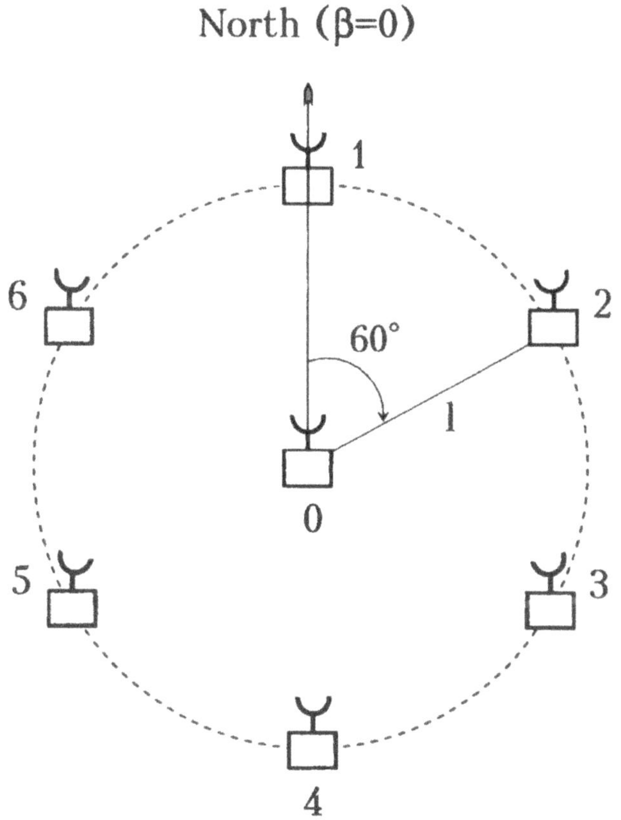

Figure 14.1 Configuration of the MSRS for Examples 14.1–14.8. 0: transmitting, transmitting–receiving or receiving station; 1–6: receiving stations; baselength l = 50 km

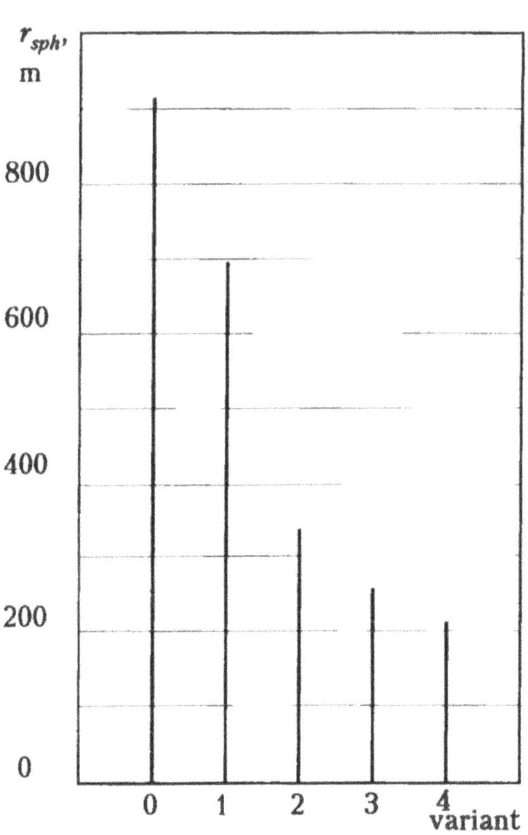

Figure 14.2 Radius of the spherical r.m.s. error of target position estimation for different variants of the MSRS (to Example 14.1). Variant 1: stations No 0, 1, 2, 6; variant 2: stations No 0, 1, 2, 6, 3; variant 3: stations No 0,1, 2, 6, 3, 5; variant 4: all stations; station No 0 for variants 1–4 is a transmitting one; variant 0: the single transmitting–receiving station No 0 (a monostatic radar)

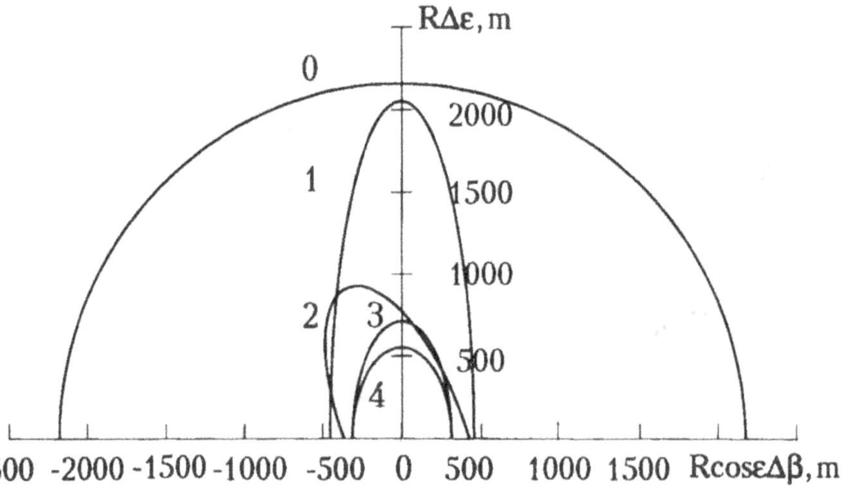

Figure 14.3 Halves of error ellipses in the cross-range plane; semiaxes are equal to three r.m.s. errors; numbers of curves correspond to the numbers of MSRS variants in Fig. 14.2

increases. The role played by different stations can be better understood from Fig. 14.3. In variant 1 the errors are only slightly reduced in the elevation direction but become essentially lower in the azimuthal direction as compared with variant 0. In fact, for the target with the angular coordinates $\beta=0$ and $\varepsilon=40°$ the stations with numbers 1, 2 and 6 form a system with relatively large effective baselengths in the azimuthal direction and small baselengths in the elevation direction. Including the station number 3 (variant 2) increases the effective baselengths in the elevation direction significantly which leads to noticeable error reduction in this direction. At the same time, including this station does not change essentially the effective baselengths in the azimuthal direction so that azimuthal errors decrease insignificantly. Because of the asymmetric station arrangement in variant 2, the azimuthal and the elevation errors are mutually correlated. Therefore, the error ellipse is turned approximately by 22.5°. In variant 3 all errors are smaller than in variant 2, and the mutual error correlation vanishes. In the 4th variant the highest accuracy is achieved but the azimuthal error is not changed since including the station number 4 increases the effective baselengths in the elevation direction only. It is reasonable to recall that the effective baselength is defined in Section 1.1 as *the length of the baseline's projection on the plane orthogonal to the bisector of the angle between directions from a target to stations of interest*.

Simple analytic expressions for elements of the error covariance matrix $\mathbf{B}_\alpha$ of target coordinate estimates $\hat{\boldsymbol{\alpha}}=(\hat{R},\hat{\beta},\hat{\varepsilon})^t$ can be derived from (14.8) for MSRSs with "short baselines" of the type shown in Fig. 14.1 when a single transmitting station is positioned at the spherical coordinate system origin and m receiving stations are arranged uniformly along the circle of the radius l, i.e. with the coordinates $L_i=l$, $\beta_i=2\pi(i-1)/m$, $\varepsilon_i=0$. The measurement vector in this case is $\hat{\boldsymbol{\xi}}=(\hat{R}_{\Sigma 1},\ldots,\hat{R}_{\Sigma m})^t$ with the error covariance matrix $\mathbf{B}_\xi=\boldsymbol{\sigma}_0(\hat{R}_\Sigma)\mathbf{I}$ where $\mathbf{I}$ is the identity matrix. We take into account that for $R\gg l$, as in the case considered, we have in accordance with (11.1), (11.2)

$$\xi_i=h_i(\boldsymbol{\alpha})=R_{\Sigma i}(\boldsymbol{\alpha})=R+\sqrt{R^2-2Rl\cos\varepsilon\cos(\beta-\beta_i)+l^2}$$

$$=2R-l\cos\varepsilon\cos(\beta-\beta_i)+O(l^2/2R). \qquad (14.38)$$

The elements of the matrix $\mathbf{H}=\|\partial R_{\Sigma i}(\boldsymbol{\alpha})/\partial\alpha_i\|$ take the form

$$\partial R_{\Sigma i}/\partial R=2;\quad \partial R_{\Sigma i}/\partial\beta=l\cos\varepsilon\sin(\beta-\beta_i);\quad \partial R_{\Sigma i}/\partial\varepsilon=l\sin\varepsilon\cos(\beta-\beta_i). \qquad (14.39)$$

Hence [see (14.8)]

$$\mathbf{B}_\alpha^{-1}=[1/\sigma_0^2(\hat{R}_\Sigma)]$$

$$\times\begin{pmatrix} 4m & 2l\cos\varepsilon\sum_{i=1}^{m}\sin(\beta-\beta_i) & 2l\sin\varepsilon\sum_{i=1}^{m}\cos(\beta-\beta_i) \\ 2l\cos\varepsilon\sum_{i=1}^{m}\sin(\beta-\beta_i) & l^2\cos^2\varepsilon\sum_{i=1}^{m}\sin^2(\beta-\beta_i) & 0.25l^2\sin 2\varepsilon\sum_{i=1}^{m}\sin 2(\beta-\beta_i) \\ 2l\sin\varepsilon\sum_{i=1}^{m}\cos(\beta-\beta_i) & 0.25l^2\sin 2\varepsilon\sum_{i=1}^{m}\sin 2(\beta-\beta_i) & l^2\sin^2\varepsilon\sum_{i=1}^{m}\cos^2(\beta-\beta_i) \end{pmatrix}.$$

$$(14.40)$$

Note that for $m \geqslant 3$ and $\beta_i = 2\pi(i-1)/m$ the matrix $\mathbf{B}_\alpha^{-1}$ and hence the matrix $\mathbf{B}_\alpha$ are diagonal. This is a consequence of the uniform (symmetric) station arrangement along the circle. Taking into account that for $m > 2$

$$\sum_{i=1}^{m} \sin^2[\beta - 2\pi(i-1)/m] = \sum_{i=1}^{m} \cos^2[\beta - 2\pi(i-1)/m] = m/2$$

yields

$$\mathbf{B}_\alpha = \sigma_0^2(\hat{R}_\Sigma)\,\mathrm{diag}\left(\frac{1}{4m}, \frac{2}{ml^2\cos^2\varepsilon}, \frac{2}{ml^2\sin^2\varepsilon}\right). \tag{14.41}$$

Thus within the frames of the approximation (14.38), (14.39), the range, azimuth and elevation angle estimates, $\hat{R}, \hat{\beta}, \hat{\varepsilon}$, of a target in such MSRSs are mutually uncorrelated, and their r.m.s. errors can be expressed in the form

$$\sigma(\hat{R}) = \frac{\sigma_0(\hat{R}_\Sigma)}{2\sqrt{m}}; \quad \sigma(\hat{\beta}) = \sqrt{\frac{2}{m}}\frac{\sigma_0(\hat{R}_\Sigma)}{l\cos\varepsilon}; \quad \sigma(\hat{\varepsilon}) = \sqrt{\frac{2}{m}}\frac{\sigma_0(\hat{R}_\Sigma)}{l\sin\varepsilon}. \tag{14.42}$$

Note that r.m.s. errors (14.42) are independent of the target azimuth, β, and range, R.

Using (14.42) one can derive expressions for the lengths of error ellipsoid semiaxes corresponding to a preset probability, P, for estimates to fall into this ellipsoid. For instance, when $P = 0.97$ these semiaxes are: $3\sigma(\hat{R})$, $3R\cos\varepsilon\sigma(\hat{\beta})$ and $3R\sigma(\hat{\varepsilon})$. We can also derive the following expression for the radius of the spherical r.m.s. error

$$r_{sph} = \sqrt{\sigma^2(\hat{R}) + R^2\cos^2\varepsilon\sigma^2(\hat{\beta}) + R^2\sigma^2(\hat{\varepsilon})} \approx \sqrt{\frac{2}{m}(1+\mathrm{cosec}^2\,\varepsilon)}\,\frac{R}{l}\,\sigma_0(\hat{R}_\Sigma). \tag{14.43}$$

For the assumed target coordinates $R = 500$ km, $\beta = 0$, $\varepsilon = 40°$, the baselength $l = 50$ km, and r.m.s. range sum error, $\sigma_0(\hat{R}_\Sigma) = 20$ m, the calculation results obtained from (14.42) and (14.43) correspond to those presented in Figs 14.2 and 14.3 for variant 4.

It should be noted that when the central station in Fig. 14.1 is not only a transmitting but a transmitting–receiving one, then within the frames of the approximation (14.38), (14.39) the angular coordinate r.m.s. errors and the spherical r.m.s. error radius are expressed by the same equations (14.42), (14.43) where m is the number of receiving stations along the circle as before. Only in the equation for the range r.m.s. error, $\sigma(\hat{R})$, m must be replaced by $m+1$ which reflects increased accuracy of target range estimation.

Example 14.2. Consider now a MSRS where *each station can measure not only target TOAs but target bearings as well*, and all these measurements are combined to form a resultant composite measurement for a target. Let for the same target position and station arrangement as in Example 14.1, a transmitting–receiving station (a monostatic radar) be located at the coordinate system origin. It can measure target range with the r.m.s. error $\sigma_0(\hat{R}) = 10$ m, and target angular local coordinates with the r.m.s. error $\sigma_0(\hat{\beta}_{A0}) = \sigma_0(\hat{\varepsilon}_{A0}) = 5'$. All receiving stations can measure target range sum with the r.m.s. error $\sigma_0(\hat{R}_{\Sigma i}) = 20$ m and both angular local coordinates with the same r.m.s. error $\sigma_0(\hat{\beta}_{Ai}) = \sigma_0(\hat{\varepsilon}_{Ai}) = 5'$. All errors are assumed to be unbiased as before. Characteristics presented in Figs 14.4 and 14.5 are similar to those shown in Figs 14.2 and 14.3. The calculations were performed using (14.8) and assuming errors of all measurements to be statistically mutually independent. It is seen that adding bearing measurements does not lead to a drastic accuracy increase. It is easily understood since the linear r.m.s. errors in cross-range directions at the range $R = 500$ km corresponding to the angular r.m.s. errors 5′ are about 730 m (one-third of the ellipse semiaxes, see Figs 14.3 and 14.5 for variant 0). At the same time range (range sum) measurements with the r.m.s. errors $\sigma(\hat{R}) = 10$ m $[\sigma_0(\hat{R}_{\Sigma i}) = 20$ m$]$ and with effective baselengths between stations 60–80 km provide linear r.m.s. errors in cross range directions of the order of 160–200 m (see Fig. 14.3). Passing from the monostatic radar to variant 1 of the MSRS yields the azimuthal error decrease (because of including an additional bearing measurement from the station number one) and the greater elevation error decrease due to range and range sum measurements from the stations number 0 and 1. The same effect can be seen when we pass from variant 1 to variant 2 but the elevation error decreases more sharply due to including the large effective baselength between stations number 1 and number 4. Variants 3 and 4 provide additional accuracy increase in the azimuthal direction.

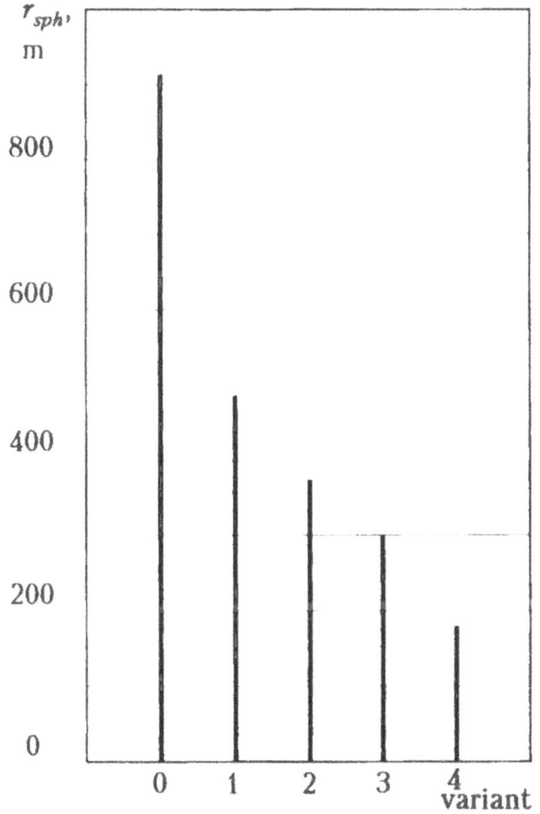

Figure 14.4 Radius of the spherical r.m.s. error of target position estimation for different variants of the MSRS (to Example 14.2). Variant 0: station No 0 only; variant 1: stations No 0 and 1; variant 2: stations No 0, 1, 4; variant 3: stations No 0, 1, 2, 6; variant 4: all stations

Example 14.3. Let us now employ the algorithm (14.23)–(14.25) of sequential local measurement processing for the successive target position estimate refinement by the MSRS considered in Example 14.1. An initial approximation, $\boldsymbol{\alpha}_{ref}$, can be found by using target range sum measurements at the stations number 1, number 2 and number 6. The accuracy of $\boldsymbol{\alpha}_{ref}$ is taken into account in the filter "gain" according to (14.21). Then measurements from the stations number 3, number 5 and number 4 are included sequentially. Five typical "trajectories" of estimates in the cross-range plane at the target are presented in Fig. 14.6. As the new measurements from additional stations are included, the resultant estimates fall into the error ellipse and gradually approach the true target position (of course, with some random deviations). Since all local coordinate errors are assumed to be mutually independent, the r.m.s. errors of target position estimates after each step of the algorithm coincide with the r.m.s. errors obtained in Example 14.1 for corresponding combination of stations and parallel processing.

Example 14.4. Let us now consider *a passive MSRS* determining target (radiation source) coordinates R, β and ε by *bearing measurements* (*the triangulation method or direction finding method*, see Section 11.1). The MSRS geometry is similar to that in Fig. 14.1. A central station is positioned at the spherical coordinate system origin and remaining m stations are arranged uniformly along the horizontal circle of the radius l so that their coordinates are: $L_i = l = 50\,\text{km}$, $\beta_i = 2\pi(i-1)/m$, $\varepsilon_i = 0$, $i = \overline{1,m}$. All $m+1$ stations are receiving ones and can measure only bearings of a target. The measurement vector has the form: $\hat{\boldsymbol{\xi}} = (\hat{\beta}_0, \hat{\varepsilon}_0, \hat{\beta}_1, \hat{\varepsilon}_1, \ldots, \hat{\beta}_m, \varepsilon_m)'$ with zero mean and the covariance matrix $\mathbf{B}_{\xi} = \sigma_A^2 \mathbf{I}$ where $\mathbf{I}$ is the identity matrix and $\sigma_A = \sigma_0(\hat{\beta}_{A0}) = \sigma_0(\hat{\varepsilon}_{A0}) = \sigma_0(\hat{\beta}_{Ai}) = \sigma_0(\hat{\varepsilon}_{Ai}) = 5'$ are the equal (for simplicity) r.m.s. errors of the target angular coordinate measurements by each station. The radiating target (e.g., jammer) is positioned at the same point as in the previous Examples: $R = 500\,\text{km}$, $\beta = 0$, $\varepsilon = 40°$.

By using expressions (14.8) and (14.13) all elements of the covariance matrix $\mathbf{B}_{\alpha}$ and values of the error ellipsoid semiaxes for the estimates $\hat{R}, \hat{\beta}, \hat{\varepsilon}$ as well as the spherical r.m.s. error radius, r_{sph}, were calculated. The numerical results relate to the different

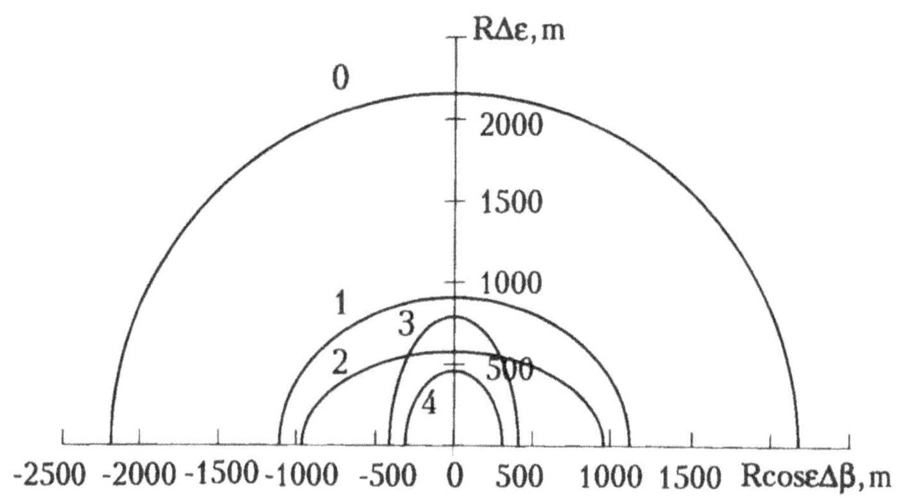

Figure 14.5 Halves of error ellipses similar to those in Fig. 14.3 but to Example 14.2. Variants of the MSRS are the same as in Fig. 14.4

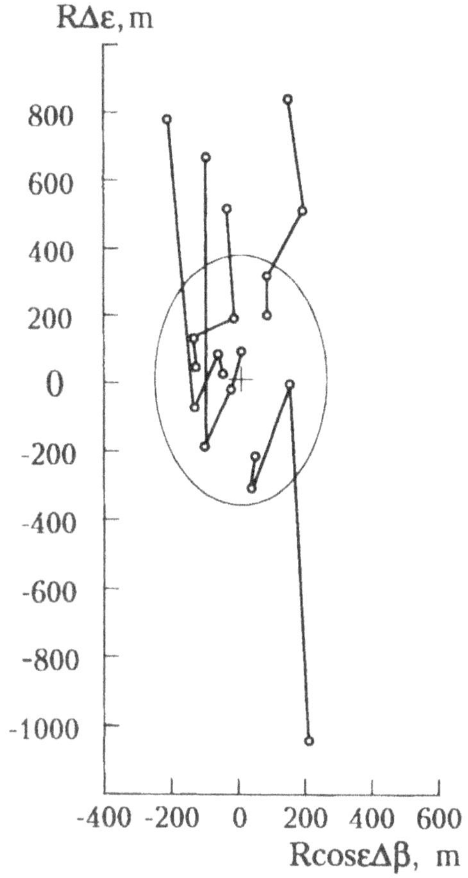

Figure 14.6 Five typical "trajectories" of target position estimates for sequential measurement processing (to Example 14.3). Notation: =obtained values of the estimates, + =the true target position; the semiaxes of the error ellipse are equal to two r.m.s. errors in the case where all stations from Example 14.1 are in operating condition

number of peripheral stations $m=3,4,5,6$ (variants 2–5) as in the previous examples. Besides, a new station arrangement is considered where there are $m=3$ peripheral stations positioned at the azimuths $\beta_1=0$, $\beta_2=60$, $\beta_3=-60$ (variant 1). The calculated values of r_{sph} are shown in Fig. 14.7. The quarters of error ellipses in the cross-range plane at the target are drawn in Fig. 14.8. The other parts of the ellipses (in other quadrants) are symmetrical. Since the resultant range error, $\sigma(\hat{R})$, is much greater than corresponding errors in cross-range directions, i.e. $\sigma(\hat{R}) \gg R\cos\varepsilon\sigma(\hat{\beta})$, $\sigma(\hat{R}) \gg R\sigma(\hat{\varepsilon})$, the values of the spherical r.m.s. error radius, r_{sph}, is determined mostly by the range errors so that $r_{sph} \approx \sigma(\hat{R})$ [see (14.13)]. Thus Fig. 14.7 shows the r.m.s. range errors as well. It follows from Figs 14.7 and 14.8 that the increase of the total number of stations from 4 to 7 improves the accuracy by (25–30)%. A small difference in errors between variants 1 and 2 for $m=3$ is caused by the angular difference between directions from the target to the stations numbers 2 and 3 with $\beta_2=60^\circ$, $\beta_3=-60$ (variant 1) and $\beta_2=120^\circ$, $\beta_3=240^\circ=-120$ (variant 2).

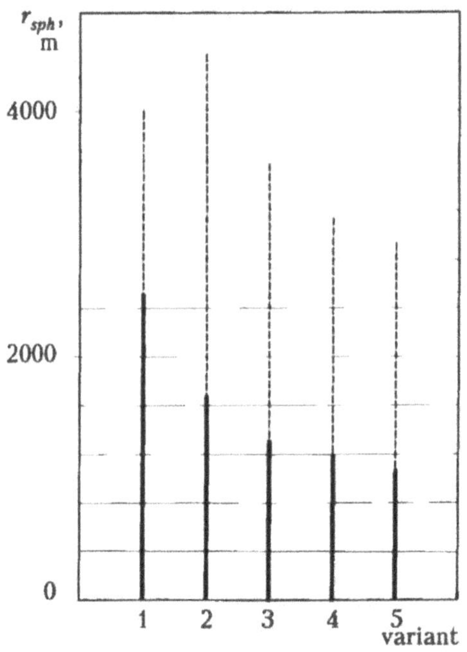

Figure 14.7 Radius of the spherical r.m.s. error of radiation source position estimation by the passive MSRS (to Examples 14.4 and 14.5). Dashed lines: triangulation system (Example 14.4); solid bold lines: hyperbolic system (Example 14.5); variants of the MSRS are described in the text

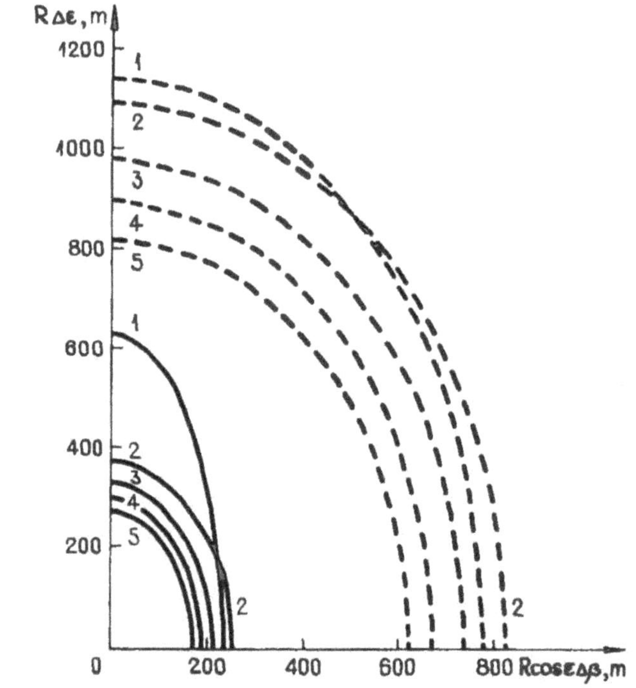

Figure 14.8 Quarters of the error ellipses in the cross-range plane to Examples 14.4 and 14.5. Dashed lines: triangulation system; solid bold lines: hyperbolic system; the numbers of curves correspond to the numbers of variants in Fig. 14.7

Example 14.5. Let *a passive MSRS* of the same geometry as in Example 14.4 estimate spatial coordinates of the same target (a radiation source) by the *hyperbolic method* (*range-difference-measurement method*, see Section 11.1). As noted in Section 11.2, $m(m+1)/2$ different TDOAs can be measured in a MSRS containing $m+1$ spatially separated receiving stations but only m TDOAs among them are linearly independent. In fact, the TDOAs taken around any closed circuit of stations add to zero [see (7.8)]. Following [240,241] we have shown in Section 11.4, that estimates of all $m(m+1)/2$ different TDOAs provide, under certain conditions, the efficient estimates of m linearly independent TDOAs. However, we assume here that only m linearly independent signal TDOAs between the central station and each of the peripheral one are directly measured and used for target localization. The remaining TDOAs are utilized, for instance, for solving the problem of association between measurements and targets (see Section 15.4). Thus the vector of measurements takes the form: $\hat{\xi}=(\widehat{\Delta R}_1,\ldots,\widehat{\Delta R}_m)^t$ where ΔR_i is the range difference from the target (the radiation source) to the central and the ith peripheral stations, i.e. $\Delta R_i = R - R_i$. Evidently, these range differences are proportional to corresponding TDOAs: $\Delta R_i = c\tau_{s0i}$ where c is the light velocity. It was shown in Section 11.4 that though maximum likelihood TDOA estimates $\hat{\tau}_{s0i}$ are mutually correlated for different $i=\overline{1,m}$, this correlation is weak for weak input signals (low input SNRs). Therefore, assuming input signals to be weak, we ignore here correlation between different estimates so that the covariance matrix of measurements is: $\mathbf{B}_\xi = \sigma^2(\widehat{\Delta R})\mathbf{I}$ where $\sigma^2(\widehat{\Delta R})$ is the range difference measurement error variance (for simplicity, the same for all stations) and $\mathbf{I}$ is the identity matrix.

The exact expressions for elements of the covariance matrix $\mathbf{B}_\alpha$ (determining accuracy of the estimates $\hat{R}$, $\hat{\beta}$ and $\hat{\varepsilon}$) can be obtained from (14.8) taking into account that

$$\xi_i = \Delta R_i = h_i(\alpha) = R\left[1-\sqrt{1-(2l/R)\cos\varepsilon\cos(\beta-\beta_i)+l^2/R^2}\right]. \tag{14.44}$$

For the considered MSRS (Fig. 14.1) with "short baselines" ($l \ll R$) we may expand the right side of (14.44) into a Taylor series retaining only first terms as in Example 14.1. Note that a linear approximation similar to (14.38) is usually sufficient for accuracy analysis of angular coordinate estimates. This approximation implies the signal wavefront to be planar for the MSRS as a whole. To evaluate range errors as well as correlation between angular and range errors, square approximation is necesary. As a result of calculations using equation (14.8) we have

for $m \geqslant 5$:

$$\sigma(\hat{R}) = \frac{2}{\sqrt{m}}\left(\frac{R}{L}\right)^2 \frac{1}{\sqrt{(1-0.5\cos^2\varepsilon)^2+0.125\cos^4\varepsilon}}\,\sigma(\Delta R)$$

$$\sigma(\hat{\beta}) = \sqrt{\frac{2}{m}}\,\frac{1}{l\cos\varepsilon}\,\sigma(\Delta R); \quad \sigma(\hat{\varepsilon}) = \sqrt{\frac{2}{m}}\,\frac{1}{l\sin\varepsilon}\,\sigma(\Delta R); \tag{14.45}$$

$$\rho(\hat{R},\hat{\beta}) = 0; \qquad \rho(\hat{\beta},\hat{\varepsilon}) = 0;$$

$$\rho(\hat{R},\hat{\varepsilon}) = \frac{3}{4\sqrt{2}}\,\frac{l}{R}\,\frac{\cos\varepsilon(4-3\cos^2\varepsilon)}{\sqrt{(1-0.5\cos^2\varepsilon)^2+0.125\cos^4\varepsilon}};$$

for $m=4$:

$$\sigma(\hat{R})=\left(\frac{R}{L}\right)^2\frac{1}{\sqrt{(1-0.5\cos^2\varepsilon)^2+0.125\cos^4\varepsilon(1+\cos 4\beta)}}\sigma(\Delta\hat{R});$$

$$\sigma(\hat{\beta})=\frac{1}{\sqrt{2}l\cos\varepsilon}\sigma(\Delta\hat{R});\qquad \sigma(\hat{\varepsilon})=\frac{1}{\sqrt{2}l\sin\varepsilon}\sigma(\Delta\hat{R});$$

$$\rho(\hat{R},\hat{\beta})=0;\qquad \rho(\hat{\beta},\hat{\varepsilon})=0; \tag{14.46}$$

$$\rho(\hat{R},\hat{\varepsilon})=\frac{3}{4\sqrt{2}}\frac{l}{R}\frac{\cos\varepsilon[4-\cos^2\varepsilon(3+\cos 4\beta)]}{\sqrt{(1-0.5\cos^2\varepsilon)^2+0.125\cos^4\varepsilon(1+\cos 4\beta)}}.$$

It is seen that, within the frames of assumed approximations, for $m\geqslant 5$ the r.m.s. errors do not depend on the target range R and azimuth β. The estimates $\hat{R}$ and $\hat{\beta}$ as well as $\hat{\beta}$ and $\hat{\varepsilon}$ are mutually uncorrelated while $\hat{R}$ and $\hat{\varepsilon}$ are weakly correlated. For $m\geqslant 4$ the r.m.s. errors of angular coordinates coincide with those for an active MSRS of the same structure if $\sigma(\hat{R}_\Sigma)$ is replaced by $\sigma(\widehat{\Delta R})$ [see (14.42)]. For $m=4$ the weak dependence of $\sigma(\hat{R})$ and $\rho(\hat{R},\hat{\beta})$ on the target azimuth, β, appears.

For the minimal total number of stations required to estimate all the three spatial coordinates of a radiation source, $m+1=4$, the MSRS in the Fig. 14.1 takes the form of a regular three-pointed star. For this case we have

$$\sigma(\hat{R})=\frac{2}{\sqrt{3}}\left(\frac{R}{l}\right)^2\frac{1}{1-0.5\cos^2\varepsilon}\sigma(\Delta\hat{R});$$

$$\sigma(\hat{\beta})=\frac{\sqrt{2}}{\sqrt{3}l\cos\varepsilon}\frac{\sqrt{(1-0.5\cos^2\varepsilon)^2+0.125\cos^4\varepsilon\sin^2 3\beta}}{1-0.5\cos^2\varepsilon}\sigma(\Delta\hat{R});$$

$$\sigma(\hat{\varepsilon})=\frac{\sqrt{2}}{\sqrt{3}l\sin\varepsilon}\frac{\sqrt{(1-0.5\cos^2\varepsilon)^2+0.125\cos^4\varepsilon\cos^2 3\beta}}{1-0.5\cos^2\varepsilon}\sigma(\Delta\hat{R});$$

$$\rho(\hat{R},\hat{\beta})=-\frac{\cos^2\varepsilon\sin 3\beta}{2\sqrt{2}\sqrt{(1-0.5\cos^2\varepsilon)^2+0.125\cos^4\varepsilon\sin^2 3\beta)}}; \tag{14.47}$$

$$\rho(\hat{\beta},\hat{\varepsilon})=-\{\cos\varepsilon\sin 3\beta[\cos^3\varepsilon\cos 3\beta-(2l/R)(4-\cos^2\varepsilon-0.75\cos^4\varepsilon)]\}\times$$

$$\left\{8\sqrt{[(1-0.5\cos^2\varepsilon)^2+0.125\cos^4\varepsilon\sin^2 3\beta][(1-0.5\cos^2\varepsilon)^2+0.125\cos^4\varepsilon\cos^2 3\beta]}\right\}^{-1};$$

$$\rho(\hat{R},\hat{\varepsilon})=-\{\cos\varepsilon[\cos\varepsilon\cos 3\beta-(l/2R)(12-7\cos^2\varepsilon)]\}$$

$$\times\left\{8\left\{(1-0.5\cos^2\varepsilon)^2+0.125\cos^4\varepsilon\cos^2 3\beta-(l/R)\cos\varepsilon\cos 3\beta\right.\right.$$

$$\left.\left.\times[1+\cos^2\varepsilon(1-0.875\cos^2\varepsilon)]\right\}\right\}^{-1/2}.$$

When $m=3$ correlation between the estimates appears, especially for low elevation angles, ε. It can also be seen that the r.m.s. angular errors are dependent on the source azimuth, β.

The values of spherical r.m.s. error radius, r_{sph}, and the quarters of the error ellipses in the cross-range plane calculated from (14.45)–(14.47), are shown in Figs. 14.7 and 14.8. These figures relate to the same variants of the number and arrangement of the receiving stations as in Example 14.4 including variant 1 where the azimuths of the peripheral stations are 0°, 60° and −60°. The coordinates of the radiation source (target) are as before: $R=500\,\mathrm{km}$, $\beta=0$, $\varepsilon=40°$. The r.m.s. error of the range difference estimates is assumed to be $\sigma(\Delta R)=10\,\mathrm{m}$ (which corresponds to the signal bandwidth of the order of several megahertz). Since $\rho(\hat{\beta},\hat{\varepsilon})=0$ (including the case $m=3$ for $\beta=0$), the principal axes of the ellipses are directed along the coordinate axes. It is seen that for the chosen parameters the hyperbolic method provide significantly higher accuracy than the triangulation method using target bearings only. However, even the hyperbolic method provides $\sigma(\hat{R})\gg R\cos\varepsilon\,\sigma(\hat{\beta})$ and $\sigma(\hat{R})\gg R\sigma(\hat{\varepsilon})$ so that the error ellipses are essentially extended in the range direction. It follows from (14.45)–(14.47) that $\sigma(\hat{R})$ *is proportional to the square of the ratio of the source range to the baselength between stations.* As noted in Section 1.2, the hyperbolic MSRS considered is approximately equivalent to a triangulation MSRS containing m bearing estimators that are positioned in the centres of the baselines between the central and each peripheral station and can measure angular coordinates with the r.m.s. errors $\sigma_0(\hat{\beta}_A)\approx\sigma(\widehat{\Delta R})/l_{\mathrm{eff}\,\beta}$ and $\sigma_0(\hat{\varepsilon}_A)\approx\sigma(\widehat{\Delta R})/l_{\mathrm{eff}\,\varepsilon}$ where $l_{\mathrm{eff}\,\beta}$ and $l_{\mathrm{eff}\,\varepsilon}$ are the effective baselengths with respect to corresponding coordinates.

Example 14.6. Let us evaluate maximum attainable accuracy of the radiation source localization positioned as in Examples 14.4 and 14.5 by the *passive MSRS* shown in Fig. 14.1 when at the first stage of a two-stage localization algorithm *all possible measurements of different TDOAs provide efficient estimates of linearly independent TDOAs* (see Section 11.4). We consider for brevity only one the simplest variant 2 where the MSRS is a regular three-pointed star (the azimuths of the peripheral stations are 0°, 120° and −120°). Let the signal have a rectangular PSD with the bandwidth Δf [see (7.27)], and the input SNRs, q^2, be equal at all the stations. It is easy to show according to (12.57), (12.50) that the ECM of range difference efficient estimates (the Cramér–Rao lower bound on errors) is given by

$$\mathbf{B}_{\Delta R\mathrm{eff}}=\frac{3(1+4q^2)c^2}{8q^4\,\Delta f T\pi^2\,\Delta f^2}\begin{pmatrix}2&1&1\\1&2&1\\1&1&2\end{pmatrix}. \tag{14.48}$$

where c is the velocity of light, T is the correlator integration time. To permit simple comparing with the results from Example 14.5 we assume r.m.s. error of each TDOA measurement to be $\sigma(\Delta R)=10\,\mathrm{m}$ as before, the input SNRs, $q^2=0.1$ at all the stations and other conditions to be equal to those of Example 14.5. Then using (11.55) we have

$$\mathbf{B}_{\Delta R\mathrm{eff}}=\frac{1+4q^2}{4(1+2q^2)}\sigma^2(\Delta R)\begin{pmatrix}2&1&1\\1&2&1\\1&1&2\end{pmatrix}=29.17\begin{pmatrix}2&1&1\\1&2&1\\1&1&2\end{pmatrix}[\mathrm{m}^2]. \tag{14.49}$$

Expanding (14.44) into a Taylor series and retaining terms up to second order yields the following equations for the range differences from the target to the central and to each of the three peripheral stations

$$\Delta R_{0i}=l\cos\varepsilon\cos(\beta-\beta_i)-(l^2/2R)[1-\cos^2\varepsilon\cos^2(\beta-\beta_i)] \tag{14.50}$$

where the notation is as in Example 14.5. The corresponding derivatives

$$\partial\Delta R_{0i}/\partial R=(l^2/2R^2)[1-\cos^2\varepsilon\cos^2(\beta-\beta_i)];$$
$$\partial\Delta R_{0i}/\partial\beta=-l\cos\varepsilon\sin(\beta-\beta_i)[1+(l/R)\cos\varepsilon\cos(\beta-\beta_i)]; \qquad (14.51)$$
$$\partial\Delta R_{0i}/\partial\varepsilon=-l\sin\varepsilon\cos(\beta-\beta_i)[1+(l/R)\cos\varepsilon\cos(\beta-\beta_i)].$$

Inverting the matrix (14.49), substituting (14.51) in (14.8) for $l=50\times10^3$ m, R = 500×10^3 m, $\beta=0$, $\varepsilon=40°$, $\beta_1=0$, $\beta_3=120°$ and $\beta_5=-120°$ yields

$$\mathbf{B}_\alpha=29.17\begin{pmatrix} 89948{,}139 & 0 & -4.4479\times10^{-3} \\ 0 & 4.9134\times10^{-10} & 0 \\ -4.4479\times10^{-3} & 0 & 1.8810\times10^{-9} \end{pmatrix}. \qquad (14.52)$$

It follows from (14.52) that

$$\sigma_{\text{eff}}(\hat{R})\approx1619\,\text{m};\quad \sigma_{\text{eff}}(\hat{\beta})\approx1.197\times10^{-4}\,\text{rad};\quad \sigma_{\text{eff}}(\hat{\varepsilon})\approx2.342\times10^{-4}\,\text{rad}.$$

Comparing the minimum attainable r.m.s. errors of efficient unbiased estimates of the radiation source coordinates from (14.52) with corresponding errors from (14.47) we can see the significant advantage in azimuth r.m.s. error (about 78%), the much smaller decrease in elevation angle r.m.s. error (about 13%) while the range r.m.s. error is nearly the same. Such a reduction of the azimuth r.m.s. error is caused by the contribution of the large effective baselength between stations number 3 and 5, i.e. by high sensitivity of the range difference ΔR_{35} to changes in the radiation source azimuth. This range difference was not included in the measurements in Example 14.5. At the same time the negative correlation between the range and elevation angular errors becomes higher: the correlation coefficient $\rho(R,\varepsilon)\approx-0.34$ instead of $\rho(R,\varepsilon)\approx-0.15$ from (14.47).

***In this section**, we have presented several illustrative numerical examples for better understanding the accuracy advantages of target position estimation in MSRSs and for revealing contributions of different local coordinate measurements to resultant target localization accuracy. We have considered a MSRS with one transmitting (or transmitting–receiving) station at the coordinate system origin and six receiving stations arranged uniformly along the circle of radius $l=50$ km in the horizontal plane with the centre at the coordinate origin (Fig. 14.1). For a given true position of the target (or the radiation source for the passive MSRS) we have calculated error covariance matrices for target (or radiation source) position estimates obtained for different number of operating stations and different measurement methods used. For each variant considered the radii of r.m.s. spherical errors* (14.13) *and the error ellipses* (14.11) *in the cross-range plane are shown. This allows to evaluate contributions of different stations to the resultant accuracy in a simple and easy-to-grasp form.*

Example 14.1 *shows the significant increase in target position estimation accuracy which provide measurements of signal TOAs (range sums) in the MSRS as compared with the monostatic radar (Figs.* 14.2 *and* 14.3*). Besides, analytical expressions* (14.40)–(14.42) *for the error covariance matrix and* (14.43) *for the r.m.s. spherical error radius have been derived which are applicable to the case where a target range is much* *greater than the baselengths between stations and the configuration of a MSRS is of the type depicted in Fig.* 14.1.

Example 14.2 *confirms that when all stations of a MSRS with "short baselengths" can measure signal TOAs (ranges or range sums), then adding bearing measurements usually does not lead to a significant increase of target position estimation accuracy (Figs.* 14.4 *and* 14.5).

Example 14.3 illustrates a step-by-step accuracy increase when measurements from different stations are included sequentially under the conditions of Example 14.1. *It is seen from Fig.* 14.6 *that as the new measurements from additional stations are included, the resultant estimates fall into the error ellipse and gradually approach the true target position.*

Examples 14.4, 14.5 *and* 14.6 *relate to passive MSRSs. From Examples* 14.4 *and* 14.5 *it is seen that the hyperbolic method based on signal TDOA measurements is usually superior in accuracy to the triangulation method (Figs.* 14.7, 14.8). *Analytical expressions for error covariance matrices have been presented for hyperbolic MSRSs having configuration of the type of Fig.* 14.1 *where radiation source range is much greater than baselengths between stations. Example* 14.6 *shows the advantage of using efficient estimates of TDOAs at the first stage of the two-stage algorithm for a passive hyperbolic MSRS.*

15. TARGET TRACKING BY MSRSs. PRINCIPLES OF INTERSTATION PLOT AND TRACK CORRELATION

15.1. PRINCIPLES OF TARGET TRACKING BY MSRSs. TARGET MOTION MODELS

The ultimate goal of moving target observation by a radar is usually track formation, i.e. target track (trajectory) parameter estimation. This is done by target state vector estimation at each time instant but, unlike the similar procedure considered in Chapter 14, these state vector estimates take into account measurements obtained at previous time instants, i.e. history of the target observation. Thus output target track parameters are results of target state vector filtration in time.

When a MSRS forms a resultant composite measurement or target state vector estimate using only current measurements (in a short time interval, for example, within each pulse repetition period, see Chapter 14), then such a MSRS is reduced to an "equivalent" monostatic radar from the target tracking point of view [13,31,52,72]. Higher accuracy of resultant measurements (as compared with local measurements by each station) provides higher accuracy of target tracking.

However, in many cases it is more expedient to build target tracks directly using local coordinate measurements coming from spatially separated stations (for example, when local coordinates of a target are measured by different stations in essentially different time instants). In these cases two approaches are possible to resultant target track formation: (1) fusion of local coordinate measurements (plots) relevant to the same target ("central-level tracking" [244]), and (2) fusion of local tracks (state vector estimates) built by different stations and relevant to the same target ("sensor-level tracking" [244]).

For both approaches, joint processing of information coming from spatially separated stations, may be broken down into three stages (Fig. 15.1).

At the first stage, *data coming from all stations* (local coordinate measurements or local track parameters) *are to be converted to a common* (*central*) *reference coordinate system*. This common reference system is, as a rule, a Cartesian one with the origin at a certain geographical point that is adopted as a conventional centre of the MSRS.

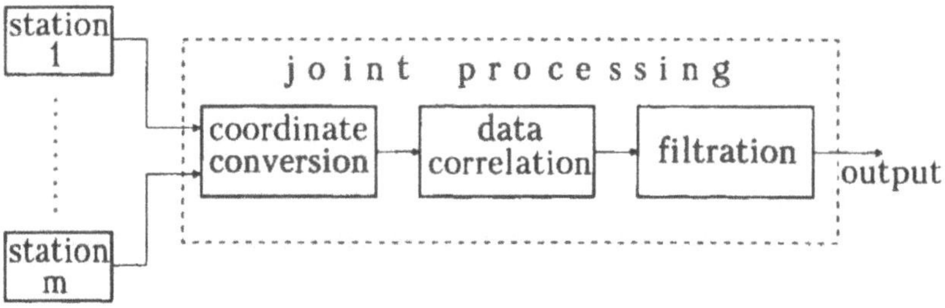

Figure 15.1 Three stages of joint information processing in MSRSs for target tracking

A main advantage of a Cartesian coordinate system is that the straight-line motion with a constant velocity can be described by differential or difference equations of the first order. Corresponding descriptions in other coordinate systems involve dummy "accelerations" that complicate data processing and may lead to additional errors [31,83,86]. However, mutually uncorrelated errors of different local coordinate measurements at each station (range, azimuth, elevation angle) become in general mutually correlated being converted to a common Cartesian coordinate system.

As in the case of resultant measurement formation considered in Chapter 14, errors in coordinate conversion may lead to significant degradation of tracking accuracy. Therefore, precise coordinate conversion is of key practical importance for achieving high target tracking accuracy by a MSRS [75,77,103]. Specific coordinate conversion algorithms depend on the actual geometry of a MSRS. Usually the curvature of Earth should be taken into account. Closed analytical expressions can be obtained only for simple "regular" arrangements of stations. Several examples have been presented in Section 14.3. However, spatially separated stations of an actual MSRS are unlikely arranged symmetrically, in a "regular" manner. Therefore, known general techniques for coordinate conversion well developed in navigation and geodesy are employed in practice.

It is clear that for precise coordinate conversion *registration process* must be performed in each MSRS. As mentioned in Section 14.1, it includes each station individual alignment and then relative alignment of all stations of the MSRS. The main task of registration is to minimize measurement errors (especially, systematic ones) caused by station misalignments and station position uncertainties. Corresponding registration techniques as well as the algorithms taking into account residual estimate biases and random errors are described in [110,118,153,173,182, 250–255].

At the second stage of joint information processing *data correlation* (or *data association*) is performed. It means that measurements (plots) or track parameters (state vector estimates) coming from each station are related to those from other stations and to previous data in order to determine if they originate from a common source. As a result of the association process, each measurement or state vector estimate must be related to one of three sets: the false alarm and clutter set, the new target set and the existing target set. False alarms and clutter may usually be ignored, new target data is used for track initiation and existing target data is used for updating of existing tracks. It is important to emphasize that the central problem at this stage of processing is the problem of determining from which target, if any, a particular measurement or state vector estimate originates.

At the third stage of joint processing, parameters of resultant ("central", "common") target tracks (trajectories) are estimated by *filtration*, i.e. tracking itself is performed. As a rule, recurrent algorithms are used at this stage. Specific algorithms depend on the purpose of a MSRS, the number and types of included transmitting and receiving facilities, computational resources.

The structure shown in Fig. 15.1 is of general character. In specific MSRSs different joint processing structures may be used which not necessary coincide with that of Fig. 15.1 (e.g. [103,257]).

Data processing in MSRSs is conventionally performed using general-purpose computers. For coordinate conversion and data correlation, specialized problem-oriented computers are often employed.

As with a monostatic radar, target tracking algorithms are based on *target motion mathematical models*. In many cases a model of target motion along a straight line with a constant velocity is exploited, e.g., in Air Traffic Control (ATC) systems and some air defense systems. Within the frames of this model target manoeuvres are considered to be random perturbations that can be described by Gaussian random processes of the white noise type. This is a Gaussian–Markovian target motion model [72]. Such a target motion is described by linear stochastic equations in a Cartesian coordinate system which permits utilizing well-developed linear filtration techniques.

If $\boldsymbol{\alpha}_k$ is the target state vector including, for instance, three spatial coordinates and their derivatives at the time instant t_k, i.e. $\boldsymbol{\alpha}_k=(x_k, y_k, z_k, \dot{x}_k, \dot{y}_k, \dot{z}_k)^t$, then

$$\boldsymbol{\alpha}_{k+1}=\mathbf{A}_k\boldsymbol{\alpha}_k+\boldsymbol{\mu}_k \tag{15.1}$$

where $\mathbf{A}_k$ is the "transition" or "prediction" matrix that can be expressed in a partitioned form

$$\mathbf{A}_k=\begin{pmatrix}\mathbf{I} & \tau_k\mathbf{I}\\ \mathbf{0} & \mathbf{I}\end{pmatrix}.$$

Here $\mathbf{0}$ and $\mathbf{I}$ are the 3×3 zero and identity matrices, respectively, $\tau_k=t_{k+1}-t_k$ is the time interval between two certain adjacent time instants (usually corresponding to the time instants of target illuminations), $\boldsymbol{\mu}_k$ is assumed to be a Gaussian 6×1 vector of random perturbations ("plant noise", "process noise") with zero mean and the covariance matrix

$$\overline{\boldsymbol{\mu}_k\boldsymbol{\mu}_n^t}=\delta_{kn}\mathbf{Q}_k. \tag{15.2}$$

In (15.2) δ_{kn} is the Kronecker symbol: $\delta_{kn}=1$ if $k=n$, $\delta_{kn}=0$ if $k\neq n$, i.e. only different components of $\boldsymbol{\mu}_k$ at the same time instant may be mutually correlated; at different time instants these vectors are uncorrelated (a process of the "white noise" type). The covariance matrix, $\mathbf{Q}_k$, is a symmetric positive definite or semidefinite one.

In the absence of target manoeuvres the matrix $\mathbf{Q}_k$ is assumed to be diagonal and independent of time (of k). In this case $\mathbf{Q}$ characterizes random stationary deviations from straight-line motion with a constant velocity (yawing etc.). Along the manoeuvre portions of target paths the matrix $\mathbf{Q}$ may be dependent on time. Then $\mathbf{Q}$ is replaced by $\mathbf{Q}_k$ that characterizes random accelerations. The vector $\boldsymbol{\mu}_k$ is often expressed in the form: $\boldsymbol{\mu}_k=\mathbf{U}_k\mathbf{v}$ where $\mathbf{v}^t=(v_x, v_y, v_z)$ is the acceleration vector, $\mathbf{U}_k$ is the 6×3 matrix describing the connection between accelerations and target state variables (elements of the state vector $\boldsymbol{\alpha}_k$)

$$\mathbf{U}_k=\begin{pmatrix}(\tau_k^2/2)\mathbf{I}\\ \tau_k\mathbf{I}\end{pmatrix}.$$

Then in accordance with (15.2) [258]

$$\mathbf{Q}_k=\mathbf{U}_k\overline{\mathbf{v}\mathbf{v}^t}\mathbf{U}_k^t=\sigma_a^2\begin{pmatrix}(\tau_k^4/4)\mathbf{I} & (\tau_k^3/2)\mathbf{I}\\ (\tau_k^3/2)\mathbf{I} & \tau_k^2\mathbf{I}\end{pmatrix} \tag{15.3}$$

where $\overline{\mathbf{v}\mathbf{v}^t}=\sigma_a^2\mathbf{I}$ is the 3×3 covariance acceleration matrix that usually assumed to be diagonal with equal variances in all the three spatial coordinates. Sometimes $\mathbf{Q}_k$ is more convenient to be expressed in terms of random velocity or path (instead of acceleration) variances [77,103].

When linear target motion models are inadequate (e.g., for ballistic targets, especially in the endoatmospheric portions of trajectories [79,121]), the initial linear equation (15.1) should be replaced by the nonlinear equation:

$$\boldsymbol{\alpha}_{k+1}=\mathbf{F}(\boldsymbol{\alpha}_k)+\boldsymbol{\mu}_k \tag{15.4}$$

where $\mathbf{F}(\cdot)$ is a known nonlinear vector function. In such cases the linearization method is usually exploited. Assuming errors of an optimum *a posteriori* estimate $\hat{\boldsymbol{\alpha}}_k$ after k observations to be small we may expand $\mathbf{F}(\boldsymbol{\alpha}_k)$ into a Taylor series at the point $\hat{\boldsymbol{\alpha}}_k$ retaining only linear terms. Then

$$\mathbf{F}(\boldsymbol{\alpha}_k)\approx\mathbf{F}(\hat{\boldsymbol{\alpha}}_k)+\mathbf{A}_k(\boldsymbol{\alpha}_k-\hat{\boldsymbol{\alpha}}_k) \tag{15.5}$$

where $\mathbf{F}(\hat{\boldsymbol{\alpha}}_k)=\hat{\boldsymbol{\alpha}}_{k+1|k}$ is the optimum predicted target state vector for the time instant t_{k+1}, based on the estimate $\hat{\boldsymbol{\alpha}}_k$; $\mathbf{A}_k=[\partial\mathbf{F}(\boldsymbol{\alpha}_k)/\partial\boldsymbol{\alpha}_k]|_{\boldsymbol{\alpha}_k=\hat{\boldsymbol{\alpha}}_k}$ is the matrix of the derivatives of the function $\mathbf{F}(\boldsymbol{\alpha}_k)$ at the point $\hat{\boldsymbol{\alpha}}_k$. The optimality holds according to following criteria: *a posteriori* probability maximum, average risk minimum for symmetric cost functions, and r.m.s. error minimum. Substituting (15.5) in (15.4) yields the linear equation for errors

$$\boldsymbol{\alpha}_{k+1}-\hat{\boldsymbol{\alpha}}_{k+1|k}=\mathbf{A}_k(\boldsymbol{\alpha}_k-\hat{\boldsymbol{\alpha}}_k)+\boldsymbol{\mu}_k. \tag{15.6}$$

When a target follows a deterministic trajectory (e.g., a ballistic trajectory without random perturbations) the random manoeuvre vector, $\boldsymbol{\mu}_k$, is assumed to be zero [31,52,79,121].

Obviously, target motion mathematical models do not depend on measuring systems. Therefore, these models are the same for both monostatic radars and MSRSs. However, because of differences in target measurement information obtained in monostatic radars and MSRSs, there are essential differences in multistatic (multiradar) tracking as compared with monoradar tracking. The characteristic features of target tracking by MSRSs will be considered in the subsequent Sections.

In this section, *we have presented a general structure of tracking algorithms for MSRSs which can be divided into three stages* (*Fig.* 15.1): *coordinate conversion, track/plot association* (*data correlation*) *and tracking itself* (*track parameter filtration*). *Since all target tracking algorithms are based on target motion mathematical models, we have considered most popular models. The linear model* (15.1) *is applicable to both straight-line motion with constant velocity and manoeuvre portions of target paths. A convenient expression* (15.3) *for the random manoeuvre covariance matrix has been presented. For the cases where nonlinear character of target motion has to be taken into account, the linear equation* (15.1) *is replaced by the nonlinear equation* (15.4). *However, in these cases the linearization method may usually be exploited so that the nonlinear target motion model* (15.4) *may be transformed into the linear model* (15.6) *for errors of target state vector estimates instead of the state vector itself.*

15.2. TRACKING BY LOCAL COORDINATE ESTIMATE FUSION

The simplest filters of the first order (in each coordinate) with constant smoothing parameters α and β (the so-called "α,β filters") are widely used in monostatic surveillance radars (e.g. [83,256]). In MSRSs local coordinate estimates (plots) may come from different stations to the FC asynchronously and with different accuracy.

Under such conditions α, β filters are often ineffective. Therefore, more complicated Kalman filters are usually employed in MSRSs though they require greater computational resources. Kalman filters based on linearized target motion models [see (15.6)] are usually called "extended Kalman filters" (e.g., [79,256,258]).

For each time instant, t_{k+1}, the Kalman filter forms the state vector smoothed estimate, $\hat{\boldsymbol{\alpha}}_{k+1}$, based on the estimate $\hat{\boldsymbol{\alpha}}_k$ (that has been obtained using k previous target observations at the time instants $t_1, t_2, \ldots, t_k$) and the new measurement of the state vector, $\hat{\boldsymbol{\alpha}}_{\text{mes}(k+1)}$, at the time instant t_{k+1} [13,31,72]:

$$\hat{\boldsymbol{\alpha}}_{k+1} = \hat{\boldsymbol{\alpha}}_{k+1|k} + \mathbf{K}_{k+1}[\hat{\boldsymbol{\alpha}}_{\text{mes}(k+1)} - \hat{\boldsymbol{\alpha}}_{k+1|k}]. \tag{15.7}$$

Here $\hat{\boldsymbol{\alpha}}_{k+1|k}$ is the prediction of the vector $\boldsymbol{\alpha}$ for the time instant t_{k+1} based on the estimate $\hat{\boldsymbol{\alpha}}_k$ and the assumed target motion model, i.e. $\hat{\boldsymbol{\alpha}}_{k+1|k}$ is the prediction of $\boldsymbol{\alpha}_{k+1}$ derived at time $t = t_k$. For linear state equations (15.1), $\hat{\boldsymbol{\alpha}}_{k+1|k} = \mathbf{A}_k\hat{\boldsymbol{\alpha}}_k$; for linearized state equations (15.4)–(15.6) $\hat{\boldsymbol{\alpha}}_{k+1|k} = \mathbf{F}(\hat{\boldsymbol{\alpha}}_k)$; $\mathbf{K}_{k+1}$ is the weight matrix (filter gain) with which the difference between the measured vector, $\hat{\boldsymbol{\alpha}}_{\text{mes}(k+1)}$, and the predicted vector, $\hat{\boldsymbol{\alpha}}_{k+1|k}$, is added to the predicted vector, $\hat{\boldsymbol{\alpha}}_{k+1|k}$, to obtain the estimate $\hat{\boldsymbol{\alpha}}_{k+1}$. These differences are usually called "innovations". The optimum weight matrix is given by

$$\mathbf{K}_{k+1} = \mathbf{B}_{\alpha(k+1)}\mathbf{B}^{-1}_{\alpha\,\text{mes}(k+1)} \tag{15.8}$$

where $\mathbf{B}_{\alpha(k+1)}$ is the covariance matrix of the resultant estimate, $\hat{\boldsymbol{\alpha}}_{k+1}$; $\mathbf{B}_{\alpha\,\text{mes}(k+1)}$ is the covariance matrix of the new measurement, $\hat{\boldsymbol{\alpha}}_{\text{mes}(k+1)}$. It is seen that owing to similarity of optimization criteria, (15.8) is similar to equations (14.5), (14.8). It can be shown that (e.g., [72])

$$\begin{aligned}\mathbf{B}_{\alpha(k+1)} &= [\mathbf{B}^{-1}_{\alpha(k+1|k)} + \mathbf{B}^{-1}_{\alpha\,\text{mes}(k+1)}]^{-1}\\ &= [(\mathbf{A}_k\mathbf{B}_{\alpha k}\mathbf{A}^t_k + \mathbf{Q}_k)^{-1} + \mathbf{B}^{-1}_{\alpha\,\text{mes}(k+1)}]^{-1}\end{aligned} \tag{15.9}$$

where $\mathbf{B}_{\alpha(k+1\,k)}$ is the covariance matrix of the predicted state vector, $\hat{\boldsymbol{\alpha}}_{k+1|k}$. It is seen that $\mathbf{B}_{\alpha(k+1|k)}$ is expressed in terms of the covariance matrix $\mathbf{B}_{\alpha k}$ of the previous estimate, $\hat{\boldsymbol{\alpha}}_k$, target transition matrix, $\mathbf{A}_k$, from (15.1) or (15.6) and the covariance matrix, $\mathbf{Q}_k$, of the random manoeuvre vector, $\boldsymbol{\mu}_k$.

Thus the optimal weight (gain) matrix, $\mathbf{K}_{k+1}$, depends on the accuracies of both the previous estimate, $\hat{\boldsymbol{\alpha}}_k$, and new measurement, $\hat{\boldsymbol{\alpha}}_{\text{mes}(k+1)}$, as well as on the target motion model.

Equations (15.7)–(15.9) together with (15.2) and (15.1) or (15.6) determine the Kalman algorithm for target track (trajectory) parameter filtration. For given initial conditions, $\boldsymbol{\alpha}_1$ and $\mathbf{B}_{\alpha 1}$, this algorithm provides at each step the optimal smoothed estimate (15.7) of the target state vector and the covariance matrix (15.9) of this estimate. The optimality takes place according to the criteria of *a posteriori* probability maximum and of average risk minimum for symmetric cost function. For non-Gaussian random vectors, $\boldsymbol{\mu}_k$, this filter provides minimum r.m.s. error among all linear filters.

The presented algorithms (15.7)–(15.9) are applicable in the cases where local coordinate measurements performed at each time instant can be transformed into a target state vector corresponding to the same time instant. However, in actual radar systems this is rather seldom possible. The point is, that the state vector of a moving target includes not only target spatial coordinates but their derivatives: velocities and, sometimes, accelerations. At the same time, local measurements include usually only target coordinate estimates. Clearly, from these estimates corresponding to one time instant, say t_k, an estimate of the state vector, $\hat{\boldsymbol{\alpha}}_{\mathrm{mes}k}$, containing derivatives cannot be derived. The technique described above may be employed in MSRSs where not only local coordinates are measured by spatially separated stations, but also range rates (or range sum rates), and accelerations are not included in target state vectors. In these cases local measurements are to be previously transformed into a state vector estimates, $\hat{\boldsymbol{\alpha}}_{\mathrm{mes}k}$, at each time instant, t_k, so that these state vector estimates undergo the filtration process. Different stations may contribute to the state vector estimate at different time instants.

Local coordinate measurements, $\boldsymbol{\xi}$, are more often used directly for target tracking. As mentioned in Section 15.1, when a MSRS forms a resultant composite measurement vector in a short time interval (for example, within each pulse repetition period), then such a MSRS is reduced to an "equivalent" monostatic radar from the target tracking point of view. We consider here another situation, where local coordinates of a target are measured by different stations at essentially different time instants and a resultant composite measurement vector is not formed. To obtain corresponding equations for these cases, it is sufficient to express the state vector measurement, $\hat{\boldsymbol{\alpha}}_{\mathrm{mes}(k+1)}$, in (15.7)–(15.9) in terms of local measurement, $\hat{\boldsymbol{\xi}}_{k+1}$. This vector is obtained for each target after interstation and measurement-to-track association (see Section 15.4). The latter procedure is usually performed in local coordinate systems connected with each station. To do this, predicted common state vector estimates are transformed from a common into local coordinate systems. For interstation correlation and filtration the associated local measurements are transformed into the common reference coordinate system. As mentioned in Section 14.1, local coordinates are, as a rule, nonlinear functions of a target state vector, i.e. for any time instant, t_{k+1}, we have $\boldsymbol{\xi}_{k+1}=\mathbf{h}_{k+1}(\boldsymbol{\alpha}_{k+1})$ where $\mathbf{h}_{k+1}(\cdot)$ is a nonlinear vector function. Let us use the linearization method [see (14.2)] and the predicted state vector, $\hat{\boldsymbol{\alpha}}_{k+1|k}$, at each step as a reference point, $\boldsymbol{\alpha}_{\mathrm{ref}}$, for linearization. Then we have instead of (14.2)

$$\mathbf{h}_{k+1}(\boldsymbol{\alpha}_{k+1}) \approx \mathbf{h}_{k+1}(\hat{\boldsymbol{\alpha}}_{k+1|k})+\mathbf{H}_{k+1}(\boldsymbol{\alpha}_{k+1}-\hat{\boldsymbol{\alpha}}_{k+1|k}). \tag{15.10}$$

The vector $\boldsymbol{\xi}_{k+1}=\mathbf{h}_{k+1}(\boldsymbol{\alpha}_{k+1})$ for different time instants, $k+1$, may include data from different stations and different local measurements.

Assuming measurement errors $\mathbf{w}_{k+1}$ of $\hat{\boldsymbol{\xi}}_{k+1}$ to be Gaussian with zero mean and the covariance matrix $\mathbf{B}_{\xi(k+1)}$ we have the following *measurement equation*

$$\hat{\boldsymbol{\xi}}_{k+1}=\mathbf{h}_{k+1}(\boldsymbol{\alpha}_{k+1})+\mathbf{w}_{k+1}=\mathbf{h}_{k+1}(\hat{\boldsymbol{\alpha}}_{k+1|k})+\mathbf{H}_{k+1}(\boldsymbol{\alpha}_{k+1}-\hat{\boldsymbol{\alpha}}_{k+1|k})+\mathbf{w}_{k+1};$$

$$\overline{\mathbf{w}_{k+1}}=0; \qquad \overline{\mathbf{w}_{k+1}\mathbf{w}^{t}_{k+1}}=\mathbf{B}_{\xi(k+1)}. \tag{15.10a}$$

Note that when $\boldsymbol{\xi}_{k+1}=\mathbf{h}_{k+1}(\boldsymbol{\alpha}_{k+1})$ is a linear function, i.e. when $\boldsymbol{\xi}_{k+1}=\mathbf{H}_{k+1}\boldsymbol{\alpha}_{k+1}$, we have a *linear measurement equation* with respect to the state vector itself

$$\hat{\boldsymbol{\xi}}_{k+1}=\mathbf{H}_{k+1}\boldsymbol{\alpha}_{k+1}+\mathbf{w}_{k+1}. \tag{15.10b}$$

Taking into account (15.10) we can obtain a maximum likelihood estimate, $\hat{\boldsymbol{\alpha}}_{\mathrm{mes}(k+1)}$, in a similar manner as in Section 14.1 [see (14.1)–(14.5)]. Denoting the matrix $\mathbf{K}$ from (14.4) and (14.5) by $\mathbf{C}$ [in order not to confuse with $\mathbf{K}$ in (15.7), (15.8)] we have

$$\hat{\boldsymbol{\alpha}}_{\mathrm{mes}(k+1)}=\hat{\boldsymbol{\alpha}}_{k+1|k}+\mathbf{C}_{k+1}[\hat{\boldsymbol{\xi}}_{k+1}-\mathbf{h}_{k+1}(\hat{\boldsymbol{\alpha}}_{k+1|k})] \tag{15.11}$$

where

$$\mathbf{C}_{k+1}=(\mathbf{H}_{k+1}^{t}\mathbf{B}_{\xi(k+1)}^{-1}\mathbf{H}_{k+1})^{-1}\mathbf{H}_{k+1}^{t}\mathbf{B}_{\xi(k+1)}^{-1}. \tag{15.12}$$

Besides, the covariance matrix of $\hat{\alpha}_{mes(k+1)}$ is given by [see (14.8)]

$$\mathbf{B}_{\alpha\,mes(k+1)}=(\mathbf{H}_{k+1}^{t}\mathbf{B}_{\xi(k+1)}^{-1}\mathbf{H}_{k+1})^{-1}. \tag{15.13}$$

Substituting (15.11)–(15.13) in (15.7)–(15.9) yields the Kalman algorithm for track (trajectory) parameter filtration when local coordinate measurements are used without preliminary transformation into state vector estimates

$$\hat{\alpha}_{k+1}=\hat{\alpha}_{k+1|k}+\mathbf{D}_{k+1}[\xi_{k+1}-\mathbf{h}_{k+1}(\hat{\alpha}_{k+1|k})] \tag{15.14}$$

where the new filter weight (gain) matrix $\mathbf{D}_{k+1}$ is given by

$$\mathbf{D}_{k+1}=\mathbf{B}_{\alpha(k+1)}\mathbf{H}_{k+1}^{t}\mathbf{B}_{\xi(k+1)}^{-1}. \tag{15.15}$$

The covariance matrix $\mathbf{B}_{\alpha(k+1)}$ of the output state vector estimate, $\hat{\alpha}_{k+1}$, is found from (15.9) and (15.13) in the form

$$\begin{aligned}\mathbf{B}_{\alpha(k+1)}&=[(\mathbf{A}_k\mathbf{B}_{\alpha k}\mathbf{A}_k^{t}+\mathbf{Q}_k)^{-1}+\mathbf{B}_{\alpha\,mes(k+1)}^{-1}]^{-1}\\ &=[(\mathbf{A}_k\mathbf{B}_{\alpha k}\mathbf{A}_k^{t}+\mathbf{Q}_k)^{-1}+\mathbf{H}_{k+1}\mathbf{B}_{\xi(k+1)}^{-1}\mathbf{H}_{k+1}]^{-1}.\end{aligned} \tag{15.16}$$

The algorithm (15.14)–(15.16), like the algorithm (15.7)–(15.9) derived above, provides an optimal (according to the same optimality criteria) smoothed estimate of the target state vector, i.e. of the target track (trajectory) parameters. It should be noted that equations (15.7)–(15.9) and (15.14)–(15.16) for a MSRS has the same form as for a monostatic radar. Differences between monostatic radars and MSRSs are taken into account in the specific forms of the functions $\xi_k=\mathbf{h}_k(\alpha_k)$, their derivatives, $\mathbf{H}_k$, and the covariance matrices $\mathbf{B}_{\xi k}$ at each step of the filtration algorithm. Naturally, these differences lead to corresponding differences in the filter outputs: the resultant estimate $\hat{\alpha}_{k+1}$ and its covariance matrix $\mathbf{B}_{\alpha(k+1)}$.

The coincidence of the matrix-vector representation of the Kalman filter algorithms for monostatic radars and MSRSs permits exploiting in MSRSs a lot of important theoretical and practical developments in the field of Kalman filtration accumulated in monostatic radar target tracking techniques. In particular, it relates to algorithms for manoeuvring target tracking, multitarget tracking in clutter, prediction of target motion along deterministic trajectories (for instance, ballistic targets) and so on (see, e.g., [13,31]). An excellent description and analysis of the Kalman filtration is presented in [256].

As mentioned in Section 14.1, local coordinate estimates obtained by different stations and relevant to the same target are often mutually independent. It means that the covariance matrix $\mathbf{B}_{\xi(k+1)}$ in (15.14)–(15.16) has a bloc diagonal form at each step of the filtration algorithm, i.e. $\mathbf{B}_{\xi(k+1)}=\mathrm{diag}\|\mathbf{B}_{\xi(k+1),i}\|$ where $\mathbf{B}_{\xi(k+1),i}$, $i=\overline{1,m_{k+1}}$, is the covariance matrix of the measurement obtained by the ith station at time $t=t_{k+1}$; m_{k+1} is the number of stations where local target coordinate estimates are obtained at $t=t_{k+1}$. In this case equations (15.14)–(15.16) can be transformed as follows [compare with (14.17), (14.18)]:

$$\hat{\alpha}_{k+1} = \hat{\alpha}_{k+1|k} + \sum_{i=1}^{m_{k+1}} \mathbf{D}_{k+1,i}[\xi_{k+1,i} - \mathbf{h}_{k+1,i}(\hat{\alpha}_{k+1|k})]; \tag{15.17}$$

$$\mathbf{D}_{k+1,i} = \mathbf{B}_{\alpha(k+1)}\mathbf{H}^{\prime}_{k+1,i}\mathbf{B}^{-1}_{\xi(k+1),i}; \tag{15.18}$$

$$\mathbf{B}_{\alpha(k+1)} = \left[(\mathbf{A}_k\mathbf{B}_{\alpha k}\mathbf{A}^{\prime}_k + \mathbf{Q}_k)^{-1} + \sum_{i=1}^{m_{k+1}} \mathbf{H}^{\prime}_{k+1,i}\mathbf{B}^{-1}_{\xi(k+1),i}\mathbf{H}_{k+1,i} \right]^{-1}. \tag{15.19}$$

Algorithm (15.17)–(15.19) may be called the filtration algorithm with *parallel processing of measurements coming from different stations*. If at time t_k the dimension of the measurement vector, $\xi_{k,i}$, from the ith station is equal to $n_{k,i}$, the total dimension of the complete measurement vector at the input of a Kalman filter at time t_k is equal to $M_k = \sum_{i=1}^{m_k} n_{k,i}$. It is assumed that data from different stations are collected synchronously.

As with resultant measurement formation (see Section 14.1) we can employ *the sequential processing of measurements coming from different stations* [121]. During each kth cycle of data updating, m_k steps of the algorithm are performed (m_k is the number of stations where measurements are available at time t_k). At each ith step the measurement vector $\xi_{k,i}$ of the dimension $n_{k,i}$ from the ith station only enters the input of a filter. During m_k steps, the smoothed estimate, $\hat{\alpha}_k$, is step-by-step refined as the result of sequential utilizing measurements from all m_k stations. Thus the algorithm of sequential data processing from different stations considered in Section 14.1 (for resultant coordinate measurement formation) is used within each kth cycle. Let us rewrite expressions (14.23)–(14.25) taking into account the dependence on the processing cycle's number

$$\hat{\alpha}_{k+1,i} = \hat{\alpha}_{k+1,i-1} + \tilde{\mathbf{K}}_{k+1,i}[\xi_{k+1,i} - \mathbf{h}_{k+1,i}(\hat{\alpha}_{k+1,i-1})]; \tag{15.20}$$

$$\tilde{\mathbf{K}}_{k+1,i} = \mathbf{B}_{\alpha(k+1),i}\mathbf{H}^{\prime}_{k+1,i}\mathbf{B}^{-1}_{\xi(k+1),i}; \tag{15.21}$$

$$\mathbf{B}_{\alpha(k+1),i} = (\mathbf{B}^{-1}_{\alpha(k+1),i-1} + \mathbf{H}^{\prime}_{k+1,i}\mathbf{B}^{-1}_{\xi(k+1),i}\mathbf{H}_{k+1,i})^{-1}. \tag{15.22}$$

We must also take into account that

$$\hat{\alpha}_{k+1,0} = \hat{\alpha}_{k+1|k}; \qquad \hat{\alpha}_{k+1,m_{k+1}} = \hat{\alpha}_{k+1}$$

$$\mathbf{B}_{\alpha(k+1),0} = \mathbf{B}_{\alpha(k+1)|k} = (\mathbf{A}_k\mathbf{B}_{\alpha k}\mathbf{A}^{\prime}_k + \mathbf{Q}_k)^{-1}; \ \mathbf{B}_{\alpha(k+1),m_{k+1}} = \mathbf{B}_{\alpha(k+1)}. \tag{15.23}$$

After m_{k+1} steps within the $(k+1)$th cycle of data updating, when all measurements available at $t = t_{k+1}$ have been utilized, we obtain

$$\hat{\alpha}_{k+1} = \hat{\alpha}_{k+1|k} + \sum_{i=1}^{m_{k+1}} \tilde{\mathbf{K}}_{k+1,i}[\xi_{k+1,i} - \mathbf{h}_{k+1,i}(\alpha_{k+1,i-1})]. \tag{15.24}$$

Apart from the parallel and sequential processing one can form a composite measurement vector by *the data compression* at the each cycle of filtration. At the time instant t_{k+1} the composite measurement vector, ξ_{k+1}, is formed. This procedure is similar to the resultant measurement formation considered in Section 14.1. The only difference is that this resultant measurement is not a target state vector estimate but only a combination of measurements coming from all stations. The dimension of the composite measurement vector is equal to that of the measurement vector from the station where this dimension is maximal. As a result, the dimension of $\hat{\xi}_{k+1}$ is significantly smaller than the sum of dimensions of measurement vectors from all stations. Such composite vector, ξ_{k+1}, of small dimension enters the input of the Kalman filter as in a monostatic radar [see (15.14)–(15.16)]. Data compression technique requires all the local coordinate measurements to be previously transformed into a common reference coordinate system. If dimensions of measurement vectors from some stations are smaller than from other stations, fictitious measurements should be included in those vectors in order to equalize dimensions of all local measurement vectors, $\hat{\xi}_{k+1,i}$, with the composite vector, ξ_{k+1}. Fictitious measurements have to be completely correlated with real measurements so as not to contain additional information. Assuming the transformed into a common coordinate system measurement vectors from different stations, $\xi_{k+1,i}$, to be mutually independent Gaussian variables with covariance matrices $\mathbf{B}_{\xi(k+1),i}$ we can write the likelihood function logarithm for the "compressed" composite measurement vector ξ_{k+1} in the form

$$\ln \Lambda(\xi_{k+1}) = -0.5 \sum_{i=1}^{m_{k+1}} (\hat{\xi}_{k+1,i} - \xi_{k+1})^t \mathbf{B}_{\xi(k+1),i}^{-1} (\hat{\xi}_{k+1,i} - \xi_{k+1}). \tag{15.25}$$

Differentiating $\ln \Lambda(\xi_{k+1})$ with respect to ξ_{k+1}, setting the result equal to zero and solving the likelihood equation for ξ_{k+1}, yield the optimum (maximum likelihood) estimate

$$\hat{\xi}_{k+1} = \left(\sum_{i=1}^{m_{k+1}} \mathbf{B}_{\xi(k+1),i}^{-1} \right)^{-1} \sum_{i=1}^{m_{k+1}} \mathbf{B}_{\xi(k+1),i}^{-1} \hat{\xi}_{k+1,i}. \tag{15.26}$$

This vector of small dimension with the covariance matrix

$$\mathbf{B}_{\xi(k+1)} = \left(\sum_{i=1}^{m_{k+1}} \mathbf{B}_{\xi(k+1),i}^{-1} \right)^{-1} \tag{15.27}$$

is used as the measurement vector at the $(k+1)$th step of the filtration algorithm (15.14)–(15.16). In (15.25)–(15.27) m_{k+1} is, as before, the number of stations where local target measurements are available at $t = t_{k+1}$.

It is seen that under the condition of independent local coordinate estimates coming from different stations, both the parallel and sequential processing techniques provide the same resulting state vector estimate with the same accuracy. This is valid for the data compression technique as well. Thus all the three described techniques provide optimum track parameter estimates according to the optimality criteria of the Kalman filter. At the same time required computational resources are essentially different. For synchronously collected measurements the data compression technique is computationally most efficient while the parallel processing requires the most computation. However, a significantly larger amount of computational resources is required when local measurement estimates are collected asynchronously (randomly). In this case filter updating is performed sequentially at random time instants when local coordinate estimate comes from either station. To avoid this additional computational burden the data–time alignment preprocessing is recommended (for instance, by means of a polynomial smoother) so as the data compression technique may be applied [121].

To further decrease of required computational resources and DTL capacity, different suboptimal filtration algorithms are of practical importance. In particular, *a measurement selection method* is often used in MSRSs for ATC systems. At each time instant a measurement (plot) associated with the same target from only one of the several stations is selected and used to create or update a common track for this target. Selection is made on the hierarchy of stations (usually, of monostatic radars) defined by a mosaic over the area controlled by an ATC centre [77,132]. According to a similar approach, a total set of redundant local measurements available in a MSRS at any time instant may be divided into several subsets of measurements without redundancy (i.e. determining two or three target coordinates). For track creating or updating only one subset is then chosen which provides the most accuracy of the filtered target state vector estimate. The measurement selection in this case can also be made off-line. The corresponding regions can be stored in the tracking computer to select the sensors to be aimed at certain targets [79,257].

Of course, measurement selection leads to information losses. However, in certain cases accuracy degradation may be not large. For example, when the range of a target is much greater than the baselengths of a MSRS, range measurements by spatially separated stations have, as a rule, a significantly greater effect on the increase of target position estimation accuracy than angular coordinate measurements by the same stations (see Fig. 1.2 in Section 1.2 and Example 14.2 in Section 14.3). Therefore, dropping redundant angle measurements does not lead to noticeable increase of target tracking errors. In other situations, on the contrary, the most important role may be played by angle measurements (when there are large baselengths as compared with target ranges, high range errors and low angular errors).

In some cases required computational resources may be reduced by using simplified filtration algorithms [261,262].

Useful examples for multitarget tracking by means of local measurement fusion are presented e.g., in [75,80,84–86,103,145,257,258,260,263].

***In this section**, we have considered the problem of common (central) track formation in MSRSs using local coordinate estimates obtained by spatially separated stations (including monostatic radars or other sensors). We have presented equations* (15.7)–(15.9) *that together with* (15.2) *and* (15.1) *or* (15.6) *determine the Kalman algorithm for target track (trajectory) parameter filtration in the cases where local coordinate measurements performed at each time instant can be transformed into a target state vector corresponding to the same time instant. For the cases (which more often occur in practice) where local coordinate measurements are directly included in the filtration process (after data correlation and coordinate conversion) we have derived equations* (15.14)–(15.16) *determining the corresponding Kalman filter algorithm. The presented Kalman filter equations for a MSRS has the same form as for a monostatic radar. This permits exploiting in MSRSs theoretical and practical developments in the field of the Kalman filtration accumulated in monostatic radar tracking techniques.*

For practically important cases where local coordinate estimates relevant to the same target are mutually independent and data from a group of stations are collected synchronously, we have considered three different tracking techniques: the parallel processing of measurements coming from different stations (15.17)–(15.19), *the sequential processing of these measurements* (15.20)–(15.24) *and the data compression technique* (15.25)–(15.27). *All the three techniques provide optimum track parameter estimates according to the optimality criteria of the Kalman filter but the data compression technique is most computationally efficient. When local measurements are collected asynchronously, the data-time alignment preprocessing is recommended.*

We also have briefly considered the measurement selection method that is often used in MSRSs for ATC systems to further decrease of required computational resources and DTL capacity. Though measurement selection leads to information losses, we have presented examples where accuracy degradation may be not large.

15.3. LOCAL TRACK FUSION

Resultant (common) track (trajectory) formation in MSRSs by fusion of local tracks (trajectories) obtained at each receiving station (or monostatic radar) using its "own" measurements ("sensor-level tracking" [244]) has certain practical advantages. First, requirements to DTL capacity are often reduced. For track fusion, only smoothed estimates of target local track parameters with their covariance matrices are to be transferred via DTLs to the FC, whereas "raw" measurements (including false alarms caused by clutter) are to be transferred to the FC when local coordinate measurement fusion is used for tracking. Therefore, when the expected number of targets is significantly smaller than that of false alarms, track fusion requires less DTL capacity [141]. Second, computational loading in a single processor can be reduced and certain advantage may be taken of parallel processing. Third, target tracking at each station allows for a tracking algorithm design specifically tailored to the individual stations. Finally, reliability and survivability of MSRSs are enhanced due to decentralization of data processing; such MSRSs have greater capabilities of graceful degradation in the case of cutoff of a part of equipment. On the other hand, track formation at each station and then track fusion lead in general to certain information losses as compared with local coordinate measurement fusion. In particular, mean errors (biases) of local tracks due to target manoeuvres cannot be averaged out by track fusion. Track fusion is most often used in systems containing autonomous and dissimilar sensors: 2-D and 3-D monostatic radars, direction finders, SSRs and optical facilities as well [77,131,175,176,244,251,262–265].

The simplest way to form a common target track in a MSRS is *to select one of the local tracks obtained at a certain station for each target and for each time instant*. Such a selection is performed on the basis of some prescribed track quality factor, for instance, of the track accuracy. As in the case of measurement selection (see Section 15.2), track selection may be made on the hierarchy of stations defined by a mosaic over the controlled space.

However, an essential part of target information may be lost because data from only a single station (one or another) is used for tracking all the time. Especially large errors may be caused by such a tracking method during target manoeuvres. Besides, "jumps" tend to occur in a common track when a target passes from a region where it is tracked by one station to a region where it is tracked by any other station. These "jumps" are very disturbing for automatic control systems using radar information (e.g., ATC systems, air defence systems) [132].

Higher quality of a common track in a MSRS can be obtained by a *weighted local track combination* though it requires greater computational resources. Let $\mathbf{z}_{k,i}$ be the target state vector obtained by the ith station in the local coordinate system at the time t_k ($i=\overline{1,m}, k=1,2,\ldots$). The optimum local track fusion is reduced to the optimum combination of the estimates, $\hat{\mathbf{z}}_{k,i}$, so as to obtain the target state estimate, $\hat{\boldsymbol{\alpha}}_k$, in the common coordinate system for a MSRS. Each local estimate, $\hat{\mathbf{z}}_{k,i}$, is a result of filtration. It means that all previous measurements (i.e. for $t \leqslant t_k$) obtained by the ith station relevant to the target considered have already been included in this estimate. Therefore, filtration in time (smoothing) is not required when combining $\hat{\mathbf{z}}_{k,i}$ to obtain $\hat{\boldsymbol{\alpha}}_k$. Track fusion deals only with current local track data.

Assume errors of the estimates $\hat{\mathbf{z}}_{k,i}$ to be Gaussian random variables with zero mean and the common covariance matrix $\mathbf{B}_{zk}$. Then the problem of obtaining an optimum estimate $\hat{\boldsymbol{\alpha}}_k$ is similar to that of resulting (composite) measurement formation in a MSRS (see Section 14.1). The difference is that the vector $\boldsymbol{\alpha}$ is to be estimated here by using the smoothed estimates $\hat{\mathbf{z}}_{k,i}$ instead of local coordinate measurements $\hat{\boldsymbol{\xi}}_i$, $i=\overline{1,m}$. Besides, nonlinear dependences $\boldsymbol{\xi}_i$ on $\boldsymbol{\alpha}$ [i.e. $\boldsymbol{\xi}_i=\mathbf{h}_i(\boldsymbol{\alpha})$] are here replaced by linear dependences $\mathbf{z}_{k,i}$ on $\boldsymbol{\alpha}_k$. Local tracks are usually built in Cartesian coordinate systems. Therefore, the vector $\boldsymbol{\alpha}_k$ in the common (reference) Cartesian coordinate system is connected with $\mathbf{z}_{k,i}$ by linear transformations including displacements and, possibly, rotations of coordinate axes. Let us introduce a rotation matrix $\mathbf{G}_i$ and a translation vector $\mathbf{T}_i$ for the ith station so that $\mathbf{z}_{k,i}=\mathbf{G}_i\boldsymbol{\alpha}_k+\mathbf{T}_i$. The maximum likelihood estimate $\hat{\boldsymbol{\alpha}}_k$ from the estimates $\hat{\mathbf{z}}_{k,i}$ can be obtained by minimizing the quadratic form:

$$L=\sum_{i=1}^{m}\sum_{j=1}^{m}[\hat{\mathbf{z}}_{k,i}-(\mathbf{G}_i\boldsymbol{\alpha}_k+\mathbf{T}_i)]^{\mathrm{t}}\mathbf{B}_{zk,ij}^{-1}[\hat{\mathbf{z}}_{k,j}-(\mathbf{G}_j\boldsymbol{\alpha}_k+\mathbf{T}_j)]\rightarrow\min(\boldsymbol{\alpha}_k). \quad (15.28)$$

Solving the likelihood equations $\partial L/\partial\boldsymbol{\alpha}_k=0$ for $\boldsymbol{\alpha}_k$ yields the algorithm

$$\hat{\boldsymbol{\alpha}}_k=\left(\sum_{i=1}^{m}\sum_{j=1}^{m}\mathbf{G}_i^{\mathrm{t}}\mathbf{B}_{zk,ij}^{-1}\mathbf{G}_j\right)^{-1}\sum_{i=1}^{m}\sum_{j=1}^{m}\mathbf{G}_i^{\mathrm{t}}\mathbf{B}_{zk,ij}^{-1}(\hat{\mathbf{z}}_{k,j}-\mathbf{T}_j). \quad (15.29)$$

In general, even if measurement errors are mutually independent, the target state estimates obtained by different stations are mutually correlated. This is a result of the common process noise, $\boldsymbol{\mu}_k$, entering into the target motion models (15.1) and (15.4). As discussed in Section 15.1, whenever the target manoeuvres or deviates from the process model, the deviation is modelled by the process noise, $\boldsymbol{\mu}_k$, which is the same for all stations. Therefore, target state estimates belonging to the same target and obtained by different stations turn out to be mutually correlated. This correlation is easily to derive for linear target motion model and linear measurement equation [243,180]. Taking into account equations [see (15.1), (15.10b)]

$$\mathbf{z}_{k,i}=\mathbf{G}_i\boldsymbol{\alpha}_k+\mathbf{T}_i;\quad \boldsymbol{\alpha}_k=\mathbf{A}_{k-1}\boldsymbol{\alpha}_{k-1}+\boldsymbol{\mu}_{k-1};\quad \hat{\boldsymbol{\xi}}_{k,i}=\mathbf{H}_{k,i}\boldsymbol{\alpha}_k+\mathbf{w}_{k,i} \quad (15.30)$$

we have

$$\begin{aligned}\hat{\boldsymbol{\alpha}}_{k,i}&=\mathbf{A}_{k-1}\hat{\boldsymbol{\alpha}}_{k-1,i}+\tilde{\mathbf{K}}_{k,i}(\hat{\boldsymbol{\xi}}_{k,i}-\mathbf{H}_{k,i}\mathbf{A}_{k-1}\hat{\boldsymbol{\alpha}}_{k-1,i})\\&=\mathbf{A}_{k-1}\hat{\boldsymbol{\alpha}}_{k-1,i}+\tilde{\mathbf{K}}_{k,i}[\mathbf{H}_{k,i}(\mathbf{A}_{k-1}\boldsymbol{\alpha}_{k-1}+\boldsymbol{\mu}_{k-1})+\mathbf{w}_{k,i}-\mathbf{H}_{k,i}\mathbf{A}_{k-1}\hat{\boldsymbol{\alpha}}_{k-1,i}] \quad (15.31)\end{aligned}$$

where $\tilde{\mathbf{K}}_{k,i}$ is the Kalman filter gain for the ith station and at the kth time instant. It can be expressed from (15.21) and (15.22), replacing the dependence on the station's number "i" by the dependence on the time instant "k", that

$$\tilde{\mathbf{K}}_{k,i}=\mathbf{B}_{\alpha k,i}\mathbf{H}_{k,i}^{\mathrm{t}}\mathbf{B}_{\xi k,i}^{-1};\quad \mathbf{B}_{\alpha k,i}=(\mathbf{B}_{\alpha(k-1),i}^{-1}+\mathbf{H}_{k,i}^{\mathrm{t}}\mathbf{B}_{\xi k,i}^{-1}\mathbf{H}_{k,i})^{-1}. \quad (15.32)$$

Using (15.30)–(15.32) we can obtain the covariance matrix for the target state vector estimates from two arbitrary stations, $\hat{\mathbf{z}}_{k,i}$, $\hat{\mathbf{z}}_{k,j}$, in the form

$$\mathbf{B}_{zk,ij}=\overline{(\mathbf{z}_{k,i}-\hat{\mathbf{z}}_{k,i})(\mathbf{z}_{k,j}-\hat{\mathbf{z}}_{k,j})^t}$$
$$=\mathbf{G}_i(\mathbf{I}-\tilde{\mathbf{K}}_{k,i}\mathbf{H}_{k,i})\mathbf{A}_{k-1}\mathbf{G}_i^{-1}\mathbf{B}_{zk-1,ij}(\mathbf{G}_j^{-1})^t\mathbf{A}_{k-1}^t(\mathbf{I}-\tilde{\mathbf{K}}_{k,j}\mathbf{H}_{k,j})^t\mathbf{G}_j^t$$
$$+\mathbf{G}_i(\mathbf{I}-\tilde{\mathbf{K}}_{k,i}\mathbf{H}_{k,i})\mathbf{Q}_{k-1}(\mathbf{I}-\tilde{\mathbf{K}}_{k,j}\mathbf{H}_{k,j})^t\mathbf{G}_j^t+\mathbf{G}_i\tilde{\mathbf{K}}_{k,j}\delta_{ij}\mathbf{B}_{\xi k,i}\tilde{\mathbf{K}}_{k,j}^t\mathbf{G}_j^t;\quad \mathbf{B}_{z0,ij}=0 \tag{15.33}$$

where $\mathbf{Q}_{k-1}$ is the covariance matrix of the process noise at time t_{k-1} [see (15.1)–(15.3)]; $\mathbf{B}_{\xi k,i}$ is the covariance matrix of the measurement noise $\mathbf{w}_{k,i}$ [see (15.10a); δ_{ij} is the Kronecker symbol; $\mathbf{I}$ is the identity matrix. The auto covariance matrices $\mathbf{B}_{zk,ii}$ and $\mathbf{B}_{zk,jj}$ can be immediately obtained from (15.33) setting $i=j$.

As mentioned in Section 15.1, the fused local tracks should be previously transformed into a common (central) coordinate system. It is also necessary for data correlation (association), see Section 15.4. If all target state vectors are considered after conversion in a common coordinate system, then $\hat{\mathbf{z}}_{k,i}$, $\hat{\mathbf{z}}_{k,j}$ should be replaced by $\hat{\boldsymbol{\alpha}}_{k,i}$, $\hat{\boldsymbol{\alpha}}_{k,j}$, respectively, and $\mathbf{z}_{k,i}$, $\mathbf{z}_{k,j}$ should be replaced by $\boldsymbol{\alpha}_k$ in (15.33). In this case $\hat{\boldsymbol{\alpha}}_{k,i}$, $\hat{\boldsymbol{\alpha}}_{k,j}$, are the state vectors estimates for the ith and jth local tracks (from the ith and jth stations) converted into the common reference coordinate system while $\boldsymbol{\alpha}_k$ is the true state vector of the same target in the same coordinate system. It follows from (15.30) that $\boldsymbol{\alpha}_k=\mathbf{G}_i^{-1}(\mathbf{z}_{k,i}-\mathbf{T}_i)=\mathbf{G}_j^{-1}(\mathbf{z}_{k,j}-\mathbf{T}_j)$. Then $\mathbf{B}_{\alpha k,ij}=\mathbf{G}_i^{-1}\mathbf{B}_{zk,ij}(\mathbf{G}_i^{-1})^t$ so that

$$\mathbf{B}_{\alpha k,ij}=\overline{(\boldsymbol{\alpha}_k-\hat{\boldsymbol{\alpha}}_{k,i})(\boldsymbol{\alpha}_k-\hat{\boldsymbol{\alpha}}_{k,i})^t}$$
$$=(\mathbf{I}-\tilde{\mathbf{K}}_{k,i}\mathbf{H}_{k,i})\mathbf{A}_{k-1}\mathbf{B}_{\alpha k-1,ij}\mathbf{A}_{k-1}^t(\mathbf{I}-\tilde{\mathbf{K}}_{k,j}\mathbf{H}_{k,j})^t$$
$$+(\mathbf{I}-\tilde{\mathbf{K}}_{k,i}\mathbf{H}_{k,i})\mathbf{Q}_{k-1}(\mathbf{I}-\tilde{\mathbf{K}}_{k,j}\mathbf{H}_{k,j})^t+\tilde{\mathbf{K}}_{k,j}\delta_{ij}\mathbf{B}_{\xi k,i}\tilde{\mathbf{K}}_{k,j}^t;\quad \mathbf{B}_{\alpha 0,ij}=0. \tag{15.34}$$

The covariance matrices (15.33), (15.34) are to be updated each time a new measurement is included in either station track estimate and also a new track estimate is combined with a common (resultant) track.

In practice, such a computational burden is seldom warranted. The practical alternative is to ignore the cross covariance between local target vector state estimates but limit the accuracy increase of common track estimate after track fusion [244]. In particular, an example presented in [244] shows that track fusion without using the cross covariance provides a slightly lower r.m.s. error and a noticeably larger mean error (bias) under the manoeuvre condition than in the case where the local track estimate cross covariance is used. On the other hand, even when the cross covariance is taken into account, mean error appeared in a local track cannot be reduced by track fusion. Thus mean errors turn out to be significantly larger after track fusion as compared with measurement (local coordinate estimate) fusion.

In certain cases, however, the estimates $\hat{\mathbf{z}}_{k,j}$ in (15.29) coming from different stations may really be considered as being statistically mutually independent. Such a situation takes place when a target moves along a deterministic (e.g., ballistic) trajectory so that the process noise $\boldsymbol{\mu}_k$ in the target motion model (15.1) or (15.6) is absent. The condition of mutually independent estimates $\hat{\mathbf{z}}_{k,j}$ for different j may be approximately satisfied when $\boldsymbol{\mu}_k \neq \mathbf{0}$ if a target moves along a straight-line portion of its path with a constant speed so that $\boldsymbol{\mu}_k$ is a stationary random process of small intensity in comparison with mutually independent measurement errors [$\mathbf{Q}_k \ll \mathbf{B}_{\xi k}$, see (15.2) and (15.10a)].

When $\hat{\mathbf{z}}_{k,i}$ for different i are assumed to be statistically independent, the matrix $\mathbf{B}_{zk}$ in (15.28), (15.29) is a bloc diagonal one: $\mathbf{B}_{zk,ij}=\mathbf{0}$ for $i \neq j$, i.e. $\mathbf{B}_{zk}=\mathrm{diag}\|\mathbf{B}_{zk,i}\|$ where $\mathbf{B}_{zk,i}=\mathbf{B}_{zk,ii}$ is the covariance matrix of $\hat{\mathbf{z}}_{k,i}$. Then algorithm (15.29) is simplified [compare with (15.26)]

$$\hat{\boldsymbol{\alpha}}_k=\left(\sum_{i=1}^{m}\mathbf{G}_i^{\prime}\mathbf{B}_{zk,i}^{-1}\mathbf{G}_i\right)^{-1}\sum_{i=1}^{m}\mathbf{G}_i^{\prime}\mathbf{B}_{zk,i}^{-1}(\hat{\mathbf{z}}_{k,i}-\mathbf{T}_i). \tag{15.35}$$

Both (15.29) and (15.35) are the optimal *parallel track fusion algorithms*. Often *sequential track fusion algorithms* are more preferable in practice. Such algorithms are similar to those considered in Section 14.1 for sequential measurement combining in MSRSs [see (14.23)–(14.25)]. Let $\hat{\boldsymbol{\alpha}}_{k,i-1\Sigma}$ be the estimate of $\boldsymbol{\alpha}_k$ obtained at the $(i-1)$th step when the data from stations number $1,\ldots,i-1$ have been fused. At the ith step the estimate $\hat{\mathbf{z}}_{k,i}$ from the ith station is added. Then

$$\hat{\boldsymbol{\alpha}}_{k,i\Sigma}=\hat{\boldsymbol{\alpha}}_{k,i-1\Sigma}+\check{\mathbf{C}}_{k,i}(\hat{\mathbf{z}}_{k,i}-\mathbf{G}_i\hat{\boldsymbol{\alpha}}_{k,i-1\Sigma}-\mathbf{T}_i) \tag{15.36}$$

where

$$\check{\mathbf{C}}_{k,i}=(\mathbf{B}_{\alpha k,i-1\Sigma}^{-1}+\mathbf{G}_i^{\prime}\mathbf{B}_{zk,i}^{-1}\mathbf{G}_i)^{-1}=\mathbf{B}_{\alpha k,i\Sigma}\mathbf{G}_i^{\prime}\mathbf{B}_{zk,i}^{-1} \tag{15.37}$$

and the covariance matrix for $\hat{\boldsymbol{\alpha}}_{k,i\Sigma}$ is given by

$$\mathbf{B}_{\alpha k,i\Sigma}=(\mathbf{B}_{\alpha k,i-1\Sigma}^{-1}+\mathbf{G}_i^{\prime}\mathbf{B}_{zk,i}^{-1}\mathbf{G}_i)^{-1}. \tag{15.38}$$

When local track parameters (target local state vectors) are preferably to be considered after coordinate conversion in the common coordinate system, $\hat{\mathbf{z}}_{k,i}$ should be replaced in (15.36)–(15.38) by $\mathbf{G}_i\hat{\boldsymbol{\alpha}}_{k,i}+\mathbf{T}_i$ [see (15.30)] and $\mathbf{B}_{zk,i}$ by $\mathbf{G}_i\mathbf{B}_{\alpha k,i}\mathbf{G}_i^{\prime}$.

It should be emphasized that systematic errors (estimate biases) may have a serious effect on the common (central) target track estimation accuracy. As mentioned above, such errors once developed in local tracks cannot be reduced by track fusion [244]. Therefore, the registration procedures must be performed before track fusion, i.e. errors due to station site uncertainties, antenna orientation and time calibration must be minimized. Residual registration errors may in some cases be incorporated as additional "target states" in estimation algorithms (e.g. [176]). In particular, coordinate conversion errors may be taken into account and minimized by the proper track (and measurement) fusion algorithm [173].

A comparative accuracy analysis of the local track fusion and local coordinate estimate fusion (see Section 15.2) shows that both methods are equivalent for deterministic trajectories [$\boldsymbol{\mu}_k=\mathbf{0}$ in the target motion models (15.1) and (15.6)] as well as for straight-line target motion with constant velocity. However, under manoeuvre condition track fusion is, as a rule, inferior to local coordinate estimate (measurement) fusion [132, 133]. The method of track fusion has been shown to be in general suboptimum [156]. A rigorous analysis for the case of two sensors shows that accuracy losses are of the order of several percents [154] and tend to reduce when the difference between the sensors' accuracy increases [175]. As noticed in [155] these losses are caused by the fact that the sufficient statistics for the global data set (in a MSRS as a whole) cannot be in general expressed in terms of the sufficient statistics of the local data sets. The track fusion is more sensitive to uncertainties (random errors) in communication networks [144]. In practice, accuracy losses become significant when a target manoeuvre at its initial stage is already detected by some stations while is not yet detected by other stations. At the stations that have detected the manoeuvre, the intensity of process noise in the target motion model is increased, i.e. the filter beamwidth is broadened, while in other stations the filter beamwidth remains narrow so that the data from the latter stations are fused with higher weights [77, 103, 257].

To overcome difficulties inherent in both fusion methods considered (local coordinate estimate fusion and local track fusion) several hybrid approaches have been proposed (e.g. [79, 244]). For example, when local coordinate estimate (measurement) fusion is used, local tracks are formed too. Only those plots are sent to the FC for common track updating which fall in the local track association gates. All unassociated plots are candidates for initiation of new common and local tracks [103]. Other track fusion simplifications are presented in [176].

In closing it should be emphasized that the choice between measurement fusion, track fusion or hybrid fusion methods in practice strongly depends on the specifics of the MSRS and its purposes.

In this section*, we have considered the problem of common (resultant, central) track formation in MSRSs by fusion of local tracks obtained in spatially separated stations (including monostatic radars or other sensors). We have discussed main advantages and drawbacks of "sensor-level tracking" as compared with "central-level tracking" considered in Section* 15.2. *The simplest way to form a common target track in a MSRS is to select one of the local tracks obtained at a certain station for each target. Such a selection is performed on the basis of some prescribed track quality factor. Using this technique may lead to significant losses of target information because data from only a single station is used for tracking all the time. Therefore, the main attention has been paid to optimal and suboptimal combining of local tracks.*

Since each local track estimate is a result of filtration in time, track fusion deals only with current local track data. Besides, since local tracks are usually built in Cartesian coordinate systems, target state vectors in a common Cartesian coordinate system is connected with local state vectors by linear transformations. We have derived the general equation (15.29) *for the maximum likelihood estimate of a common state vector.*

Because of the same process noise, local track estimates of a target obtained by different stations are in general mutually correlated even when measurements at those stations are mutually uncorrelated. We have presented equations (15.33), (15.34) *for the auto-covariance and cross-covariance of local track estimates from two arbitrary stations. However, large computational burden associated with taking into account this correlation is seldom warranted in practice. One of possible alternatives is to ignore the cross-covariance between local track estimates but limit the accuracy increase of common track estimate after track fusion. There are situations where those local track estimate may be really assumed to be mutually uncorrelated.*

For uncorrelated local track estimates, we have presented equations (15.35) *and* (15.36)–(15.38) *for the optimal parallel and sequential track fusion algorithms, respectively.*

A comparative accuracy analysis of local track fusion and local coordinate estimate fusion (see Section 15.2) *shows that both methods are equivalent for deterministic trajectories as well as for straight-line target motion with a constant velocity. However, under manoeuvre conditions, local track fusion is, as a rule, inferior to local measurement fusion. The difference may be especially essential when a part of stations of a MSRS have detected a target manoeuvre while other stations have not.*

15.4. PRINCIPLES OF INTERSTATION MEASUREMENT AND TRACK CORRELATION

When there are several targets within the region under responsibility of a MSRS or even there is a single target but under the condition of high false alarm rate (for instance, in clutter), the problem of *interstation data* (*measurement or track*) *correlation* arises. The essence of this problem is the identification of measurements or tracks from different stations (including monostatic radars or, may be, other sensors) associated with the same target.

Since all data obtained by spatially separated stations in MSRSs are statistical estimates of target parameters, i.e. are random variables, the data correlation problem is statistical in character.

A similar problem of *data association* exists in monostatic radars. This is the problem of determining from which target, if any, each new measurement originates. The association process must relate each measurement to one of three sets of data: the false alarm set, the new target set and the existing set of previous measurements related to any of previously detected target [191]. This is valid for MSRSs too but interstation measurement or track correlation is an essential additional procedure characteristic of MSRSs. Errors in interstation data correlation lead to gross errors in target state estimation.

Interstation data correlation is not required in MSRSs with signal fusion (information integration at the signal level). For each measurement, such a MSRS is "tuned" for signal reception from a specific spatial resolution cell (or several specific cells). Signals coming from targets that are outside this cell (or these cells) are weakened significantly in the fusion process[1]. Such conditions are typical for spatially coherent MSRSs as well as for MSRSs with short spatial coherence and spatially incoherent MSRSs (see Section 1.1) when one-stage estimators are used and the total number of signal informative parameters is greater than the number of target state vector parameters to be estimated. However, one-stage target state estimation algorithms are used in practice, as a rule, in spatially coherent MSRSs [42].

Techniques for interstation data correlation in MSRSs are similar to data association techniques in monostatic radars especially under multitarget tracking conditions.

Data correlation (association) in MSRSs may be performed in two stages. At the first stage "spatial" (interstation) correlation is carried out for data from different stations corresponding to approximately the same time instant (or short time interval). At the second stage "temporal" correlation is realized: data associated with certain targets are now to be associated with existing tracks of those targets or used for new track initiation. One-stage correlation-association is also possible: measurements or target state vector (track) estimates from different stations may be associated with common (central) tracks directly taking into account target motion models (e.g., [163,169,176,177,191,267,273,275]). This may be called "spatial–temporal" data correlation.

[1] In this case another problem arises: elimination of gross errors caused by signal reception through sidelobes of the resultant signal reception pattern (RSRP) of a MSRS. However, this problem may be solved by other techniques and is not considered here.

Let in a MSRS with m receiving stations (including monostatic or bistatic radars) N targets be detected and resolved by each station. In this case there are $(N!)^{m-1}$ different ways to distribute all measurements between targets and stations, i.e. to form N groups of measurements so as each of these groups would contain m measurements and at least two groups would not coincide. In other words, there are $(N!)^{m-1}$ interstation correlation hypotheses which differ one from another in the content of at least two groups. Only one among these hypotheses is true. It is clear that the total number of hypotheses increases rapidly with the increase of the number of measurements, N, at each station. Therefore, before using "fine" statistical decision techniques, "coarse" and simple techniques should be employed to reject apparently false hypotheses, as many as possible. This may be achieved by "gating" in target coordinates and other parameters (e.g., target attributes). The preliminary interstation data correlation by gating is, in essence, a method of exploiting redundant measurements and *a priori* dependences between true values of target local coordinates. It can be seen an apparent similarity to the one-stage target coordinate estimation with the redundant number of measured informative signal parameters (as compared with the dimensionality of space). For example, if three spatial coordinates of a target are measured by one station of a MSRS, gates (validation regions) in corresponding coordinates can be formed for other stations so that measurements are expected to fall within these gates with high probability if these measurements (with their expected errors) originate, indeed, from the same target. When spatial–temporal data correlation is used, such gates are placed about each predicted estimate of target state vector or measurement vector taking into account the estimate errors. Those hypotheses for which measurements do not fall within gates, are rejected as obviously false ones. In active and active–passive MSRSs operating in active mode where target ranges (or range sums) can be directly measured using target echoes, interstation data correlation gates may be of comparatively small spatial size. In this case only few measurements usually fall within each corresponding gate at different stations. As a result the number of data correlation hypotheses decreases drastically. If all measurements are distributed between several nonoverlapping gates, then the total number of remaining hypotheses turns out to be equal only to the sum of hypotheses corresponding to measurements fell within each individual gate.

Efficiency of gating depends on target spatial density, accuracy of target local coordinate measurement by each station of a MSRS and baselengths between stations.

Example 15.1. Consider a simple MSRS consisting of two monostatic radars with the baselength L between stations (Fig. 15.2). Let the radar TR_1 positioned at the coordinate system origin can measure the target range, R_1, and the angular coordinate, Θ_1, in the plane "TR_1–the target–TR_2". Each measurement by TR_1 determines in this plane an error ellipse so that the target is located within this ellipse with high probability. As a rule, the ellipse is flattened in the range direction (see Section 14.3). For any target within the ellipse, the range with respect to the second monostatic radar, TR_2, can lie between the limits R_2 and $R_2+\Delta R$ with the same high probability. Therefore, the measurement (R_1, Θ_1) from the first radar is to be correlated (associated) only with those measurements from TR_2 which fall within the range gate $(R_2, R_2+\Delta R)$. The size (width) of this gate is given by

$$\Delta R=\sqrt{R_1^2+L^2-2R_1L\cos(\Theta_1+\Delta\Theta/2)}-\sqrt{R_1^2+L^2-2R_1L\cos(\Theta_1-\Delta\Theta/2)} \quad (15.39)$$

where $\Delta\Theta$ is the angular width of the error ellipse (the "thickness" of the ellipse in range direction may be neglected). For the sake of clarity we assume that $R_1 \gg L$

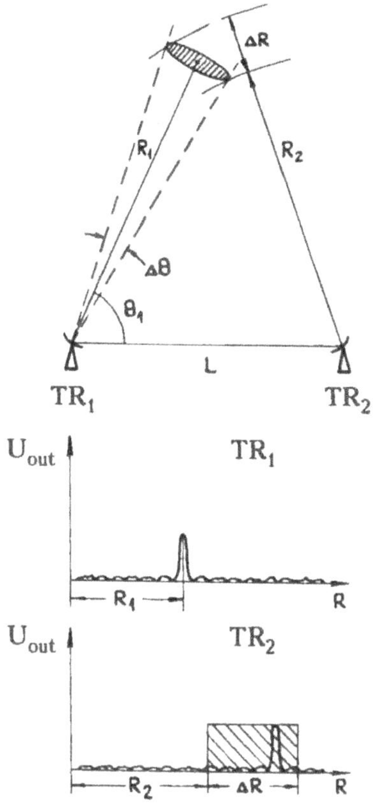

Figure 15.2 To the evaluation of a gate size for preliminary interstation measurement correlation

and $\Delta\Theta/2 \ll 1$ so that $\cos \Delta\Theta/2 \approx 1$, $\sin \Delta\Theta/2 \approx \Delta\Theta/2$. Then

$$\Delta R \approx L \sin \Theta_1 \Delta\Theta \approx L_{eff} \Delta\Theta. \tag{15.40}$$

Thus the size ΔR of the gate is approximately proportional to the product of the effective baselength between stations by the doubled maximal error of target angular coordinate measurements in the plane passing through the target and baseline (or, what is the same, to the product of the effective baselength and the angular width of the error ellipse in the plane considered). In three-dimensional space if the radar TR_1 can measure the target range, R, azimuth, β, and elevation angle, ε, then under the same conditions as before ($R \gg L$, $\Delta\beta \ll 1$, $\Delta\varepsilon \ll 1$)

$$\Delta R \approx L_{eff\beta} \Delta\beta + L_{eff\varepsilon} \Delta\varepsilon \tag{15.41}$$

where $L_{eff\beta} = L \cos\varepsilon \sin\beta$, $L_{eff\varepsilon} = L \sin\varepsilon \cos\beta$ are the approximate values of the effective baselengths in the azimuthal and elevation angle directions, respectively; $\Delta\beta, \Delta\varepsilon$ are the angular width of the error ellipsoid in corresponding directions[2].

[2] When range estimation errors of TR_2, (with r.m.s. value $\sigma(\hat{R})$) should be taken into account, then $6\sigma(\hat{R})$ is to be added in the right parts of (15.40) and (15.41).

When baselengths in a MSRS are not large and angular accuracy of each station is not too low, a gate for preliminary interstation data correlation covers only few range cells so that many target measurements cannot fall into each gate. Let, for instance, $L_{eff\beta} = L_{eff\varepsilon} = 20\,\text{km}$ and $\Delta\beta = \Delta\varepsilon = 30'$ [for $\sigma(\beta) = \sigma(\varepsilon) = 5'$ the probability for a measurement to fall within the error ellipsoid $P_3(k=3) \approx 0.97$, see (14.11)]. Then $\Delta R \approx 350\,\text{m}$. For radars with the resolution capability in range $\delta R = 50-100\,\text{m}$ such a gate covers only from seven to four adjacent range resolution cells. It is scarcely probable to fall more than 2–3 measurements within such a gate.

Interstation data correlation is much more difficult for passive MSRSs and passive modes of active–passive MSRSs though preliminary data correlation by gating is applicable to such MSRSs too. For example, when the hyperbolic method is used for radiation source localization (see Section 11.1), additional angular coordinate measurements by spatially separated stations permits to form certain gates in TDOAs (signal delays) within which TDOA measurements associated with one and the same source can fall. However, since direct source range measurements cannot be performed, the spatial size of corresponding gates turn out to be large. When the target (radiation source) spatial density is high, many measurements may fall within those gates.

In any case available redundant information should be used. For example, if in a triangulation system consisting of two stations each station can measure both angular coordinates, one measurement is redundant which permits to associate bearings with corresponding radiation sources (e.g., jammers). Specifically, one may exploit the fact that true bearings of the same source from both stations and the baseline between stations lie in one plane [6,22]. To determine three spatial coordinates of a radiation source by the hyperbolic method at least four stations are necessary. However, such a MSRS with four stations can measure six TDOAs (see Sections 11.1, 11.4). Three redundant measurements help in data correlation process. The relationship (7.8) can be used for this purpose. According to this relationship the sum of TDOAs for any three stations is equal to zero. This is valid for differential Doppler frequencies (DDFs) too.

A more expensive way to facilitate data correlation is the use of redundant receiving or/and transmitting stations in MSRSs. Although the total number of data correlation hypotheses increases with the increase of the number of stations, efficiency of false hypothesis rejection grows. For the redundant number of stations and high target detection probability, a collection of measurements belonging to the same target, determines a small region in space (taking into account measurement errors). For false hypotheses this condition is not satisfied: corresponding measurements do not fall within that small region. Using redundant stations may be considered as a way to reduce the spatial size of data correlation gates. This is especially important when each station can measure only one or two spatial coordinates of a target since data correlation gates based on such "incomplete" measurements turn out to be large. The greater the number of redundant stations, the higher probability of true data correlation.

For false hypothesis rejection in data correlation process, it is reasonable to utilize maximum *a priori* information concerning possible values of target velocity and acceleration [46,47] as well as other available information: target RCS, signatures etc. [163–165]. For example, when triangulation is used for jammer localization, mutual correlation processing of received signals can be performed to assign bearings to corresponding jammers since interferences from different sources are mutually uncorrelated (e.g. [46]).

The general approach to the problem of data correlation is based on statistical properties of target coordinate and other parameter estimates (e.g. [24,72,125,147, 157,169,177,180,243,244,264,266,267]). In the recent paper [269] the fuzzy logic is successfully used for multisensor–multitarget correlation.

We consider first *spatial* data correlation that may be used either for stationary targets or for moving targets as the first stage of a two-stage correlation (association) process (for instance, when a resultant measurement is formed for each detected target, see Chapter 14). Let an arbitrary pair of stations (number i and number j, $i, j=\overline{1,m}$) of a MSRS obtain measurements from several targets. To determine from which target a particular measurement originates, the following quadratic form can be calculated:

$$\Gamma_{ij}^{(ln)}=(\xi_{il}-\xi_{jn})^{t}(\mathbf{B}_{\xi il}+\mathbf{B}_{\xi jn})^{-1}(\xi_{il}-\xi_{jn}) \qquad (15.42)$$

where ξ_{il}, ξ_{jn} are the vectors of the lth measurement from the ith station and the nth measurement from the jth station, respectively; $\mathbf{B}_{\xi il}, \mathbf{B}_{\xi jn}$ are the covariance matrices associated with the corresponding measurements. All measurement vectors are assumed to be of equal dimension, converted to a common coordinate system and aligned in time; measurement errors are mutually independent at different stations and Gaussian distributed with zero mean. If ξ_{il} and ξ_{jn} relate to the same target, the probability distribution of the decision variable, $\Gamma_{ij}^{(ln)}$, is a central chi-square distribution with the number of degrees of freedom equal to the dimension of the measurement vectors. Using this well-known distribution and assigning a low probability for a true hypothesis rejection we can determine the threshold with which $\Gamma_{ij}^{(ln)}$ is to be compared for different hypotheses. If $\Gamma_{ij}^{(ln)}$ exceeds the threshold, the corresponding hypothesis is considered as a false one and has to be rejected. When ξ_{il} and ξ_{jn} relate to different targets, the chi-square distribution becomes non-central so that the probability of threshold exceeding increases.

Now we turn to the problem of *spatial–temporal* data correlation. As mentioned above, there are two possible approaches: *measurement-to-track correlation* and *track-to-track correlation*. In the first case measurements from different stations are to be associated with common (central) tracks stored at the FC of the MSRS. In the second case local tracks from different stations (target state estimates) are to be associated to common (central) tracks at the FC. When measurement-to-track correlation is used, then differences between predicted and measured vectors in measurement equations are analyzed. This procedure is usually performed in local coordinate systems. For nonlinear measurement equations a difference vector for the ith station between the predicted vector for the jth target and the vector of the actual lth measurement at the kth time instant (an innovation vector) takes the form

$$\mathbf{r}_{k,ijl}=\xi_{k,il}-\mathbf{h}_{k,i}(\hat{\alpha}_{k|k-1,j}), \quad i=\overline{1,m}. \qquad (15.43)$$

The decision variables are given by the following quadratic form

$$\Gamma_{ijl}^{(k)}=\mathbf{r}_{k,ijl}^{t}\mathbf{W}_{k,ijl}^{-1}\mathbf{r}_{k,ijl} \qquad (15.44)$$

where $\mathbf{W}_{k,ijl}$ is the covariance matrix of the differences, $\mathbf{r}_{k,ijl}$. The quantities $\Gamma_{ijl}^{(k)}$ behave like $\Gamma_{ij}^{(ln)}$ in (15.42). Therefore, comparing each $\Gamma_{ijl}^{(k)}$ with a prescribed threshold calculated by using the chi-square distribution, allows us to reject false hypotheses, i.e. to declare that the lth measurement from the ith station does not associated with the jth track in the common coordinate system of the MSRS.

Since measurements coming from different stations are, as a rule, mutually statistically independent (see Section 14.1), $\mathbf{W}_{k,ijl}$ is given by

$$\mathbf{W}_{k,ijl}=\mathbf{H}_{k,ij}\mathbf{B}_{\alpha k|k-1,j}\mathbf{H}'_{k,ij}+\mathbf{B}_{\xi k,il} \qquad (15.45)$$

where $\mathbf{B}_{\alpha k|k-1,j}$ is the covariance matrix of a predicted estimate of the jth target state vector in the common coordinate system, $\hat{\boldsymbol{\alpha}}_{k|k-1,j}$; $\mathbf{B}_{\xi k,il}$ is the covariance matrix of the lth local coordinate measurement obtained at the ith station at the time instant t_k; $\mathbf{H}_{k,ij}=\partial\mathbf{h}_{k,i}/\partial\boldsymbol{\alpha}$ is the matrix of derivatives of the functions $\mathbf{h}_{k,i}(\cdot)$ at the point $\boldsymbol{\alpha}=\hat{\boldsymbol{\alpha}}_{k|k-1,j}$.

In the case of track-to-track correlation the analyzed differences are similar to those for spatial data correlation (15.42) but target state vectors estimates of local tracks at each time instant are used instead of local coordinate measurements. To test whether local state vector estimates obtained at the kth time instant from the ith and the jth stations and converted to a common coordinate system, $\hat{\boldsymbol{\alpha}}_{k,i}$ and $\hat{\boldsymbol{\alpha}}_{k,j}$, originate from the same target, the following quadratic form should be calculated

$$\tilde{\Gamma}_{ij}^{(k)}=(\hat{\boldsymbol{\alpha}}_{k,i}-\hat{\boldsymbol{\alpha}}_{k,j})'\,\tilde{\mathbf{W}}_{k,ij}^{-1}(\hat{\boldsymbol{\alpha}}_{k,i}-\hat{\boldsymbol{\alpha}}_{k,j}). \qquad (15.46)$$

As mentioned in Section 15.3, local track (target state vector) estimates for the same target obtained by different stations are in general mutually correlated because of the common process noise in the target motion model. Therefore, the covariance matrix $\tilde{\mathbf{W}}_{k,ij}$ in (15.46) can be written as [243]

$$\begin{aligned}\tilde{\mathbf{W}}_{k,ij}&=\overline{(\hat{\boldsymbol{\alpha}}_{k,i}-\hat{\boldsymbol{\alpha}}_{k,j})(\hat{\boldsymbol{\alpha}}_{k,i}-\hat{\boldsymbol{\alpha}}_{k,j})'}\\&=\overline{[\hat{\boldsymbol{\alpha}}_{k,i}-\boldsymbol{\alpha}_{k}-(\hat{\boldsymbol{\alpha}}_{k,j}-\boldsymbol{\alpha}_{k})][\hat{\boldsymbol{\alpha}}_{k,i}-\boldsymbol{\alpha}_{k}-(\hat{\boldsymbol{\alpha}}_{k,j}-\boldsymbol{\alpha}_{k})]'}\\&=\mathbf{B}_{\alpha k,ii}+\mathbf{B}_{\alpha k,jj}-\mathbf{B}_{\alpha k,ij}-\mathbf{B}_{\alpha k,ji}\end{aligned} \qquad (15.47)$$

where $\boldsymbol{\alpha}_k$ is the true value of the target state vector; $\mathbf{B}_{\alpha k,ii}, \mathbf{B}_{\alpha k,jj}, \mathbf{B}_{\alpha k,ij}, \mathbf{B}_{\alpha k,ji}$ are the auto-covariance and cross-covariance matrices for local target state vectors estimates converted to a central coordinate system. These matrices are determined in (15.34). As in the previous cases, if $\hat{\boldsymbol{\alpha}}_{k,i}$ and $\hat{\boldsymbol{\alpha}}_{k,j}$ relate to the same target, the probability distribution of the decision variable $\tilde{\Gamma}_{ij}^{(k)}$ is a central chi-square distribution where the number of degrees of freedom is equal to the dimension of the target state vectors. Using this distribution and assigning a low probability for a true hypothesis rejection we can determine the threshold with which $\tilde{\Gamma}_{ij}^{(k)}$ is to be compared. If $\tilde{\Gamma}_{ij}^{(k)}$ exceeds the threshold, the tested hypothesis is considered as a false one and has to be rejected. When $\hat{\boldsymbol{\alpha}}_{k,i}$ and $\hat{\boldsymbol{\alpha}}_{k,j}$ relate to different targets, the chi-square distribution becomes noncentral so that the probability of threshold exceeding increases.

The quantities $\Gamma_{ij}^{(ln)}$ in (15.42), $\Gamma_{ij}^{(k)}$ in (15.44) and $\tilde{\Gamma}_{ij}^{(k)}$ in (15.46) may be treated as the weighted distances between corresponding estimates. Rejection of hypotheses associating "distant" estimates is equivalent to gating in multidimensional space. More complicated tests than (15.42), (15.44) and (15.46) can be used. For example, such tests may take into account differences like (15.43) but simultaneously for several stations and several time instants.

For a dense target and clutter environment, one-to-one associations of measurements to tracks described above have large computational requirements. The method of Probabilistic Data Association (PDA) and Joint Probabilistic Data Association (JPDA) were developed to overcome these difficulties. The essence of the JDPA method for a monostatic radar is the computation of association probabilities for every track with every measurement in the present scan, and the subsequent use of those probabilities as weighting coefficients in the formation of weighted-average measurement for updating each track. The JPDA method has been extended to multiradar (multisensor) systems (see, e.g., [147,272,273]).

Example 15.2. Let two equal monostatic radars ($m=2$), TR_1 and TR_2, be positioned at the points 2 and 6 of Fig. 14.1. Assume the distance between each radar and the coordinate system origin (at the point 0) to be $l=20$ km. Unlike the examples in Section 14.3 we take here into account the curvature of Earth. Let six targets ($N=6$) are located at the range $R \approx 200$ km and at the height $H \approx 15$ km. The actual arrangement of the targets are shown in Fig. 15.3 (a view in plan). The azimuth of the target No 1 is equal to zero and the range $R_0=200$ km is that of the target No 4. The target's heights, H_i, differ by 100 m: $H_1=15.0$ km, $H_2=15.1$ km, $H_3=15.2$ km, $H_4=15.3$ km, $H_5=15.4$ km, $H_6=15.5$ km. Both radars measure ranges, R_{1i} and R_{2i}, azimuths, β_{1i} and β_{2i}, as well as elevation angles, ε_{1i} and ε_{2i}, of

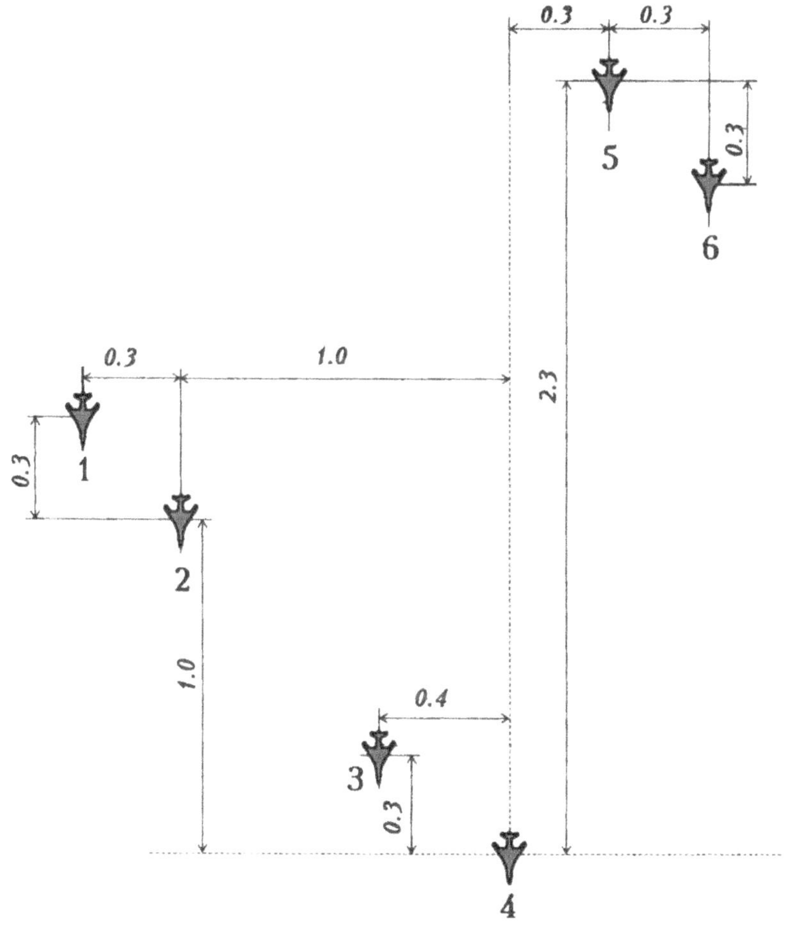

Figure 15.3 Relative arrangement of the targets in plan to Example 15.3. All distances are indicated in kilometers

each target ($i=\overline{1,6}$). All measurement errors are assumed to be Gaussian with zero mean, mutually statistically independent with r.m.s. values $\sigma_1(\hat{R})=\sigma_2(\hat{R})=\sigma(\hat{R})$; $\sigma_1(\hat{\beta})=\sigma_2(\hat{\beta})=\sigma(\hat{\beta})$; $\sigma_1(\hat{\varepsilon})=\sigma_2(\hat{\varepsilon})=\sigma(\hat{\varepsilon})$. To form composite measurement for each target, interstation (spatial) measurement correlation is to be performed before.

For six measurements at each station (if we do not take into account false alarms and signal misses), the total number of data correlation hypotheses is equal to $(N!)^{m-1}=6!=720$. Only one hypothesis from 720 is true. To decrease the total number of hypotheses to be analyzed we use preliminary gating. Consider the target No 1. Let the station TR_1 obtain an estimate of the target position $\hat{\alpha}_{11}=(\hat{R}_{11},\hat{\beta}_{11},\hat{\varepsilon}_{11})$. We can build an error ellipsoid with the centre at the point $\hat{\alpha}_{11}$ and the semiaxes equal, for instance, $3\sigma(\hat{R}), 3\hat{R}_{11}\cos\hat{\varepsilon}\sigma(\hat{\beta}), 3\hat{R}_{11}\sigma(\hat{\varepsilon})$. The location and size of this ellipsoid determine the region of the target's true position with the probability $P_3(3)=0.97$ [see (14.11)]. To obtain a spatial gate for the station TR_2 similar error ellipsoids for estimates of TR_2 should be built around all points of the first ellipsoid. Then measurements by the station TR_2, $\hat{R}_{21}$, $\hat{\beta}_{21}$ and $\hat{\varepsilon}_{21}$, must fall within the sum of all ellipsoids with the probability $P=0.97\times0.97\approx0.94$. For the considered MSRS with "short baselines", it is easier to add to each corresponding measurement of the station $TR_1 \pm\ [6\sigma(\hat{R})+3l\sqrt{3}\sigma(\hat{\beta})]$, $\pm 6\hat{R}_{11}\cos\hat{\varepsilon}\sigma(\hat{\beta})$ and $\pm 6\hat{R}_{11}\sigma(\hat{\varepsilon})$ so that we obtain a figure bounded by corresponding arcs and radii. Coordinates of the vertices of this figure in the reference system associated with TR_2 determine the gate of interest in corresponding coordinates R_{21}, β_{21} and ε_{21}. The probability that a measurement obtained by TR_2 from the target No 1 falls within the figure $P>0.94$ since the built figure covers the error ellipsoids.

Simulation results of such a gating in the station TR_2 for all the six targets are shown in Table 15.1 for different values of r.m.s. errors $\sigma(\hat{\beta})$, $\sigma(\hat{\varepsilon})$ and the fixed value of r.m.s. error $\sigma(\hat{R})=10\,\text{m}$. In this table, the estimates of the probability that measurements from one target and two targets [$\hat{P}(1)$ and $\hat{P}(2)$, respectively] fall within the gates of TR_2 are presented. Each gate was built as described above for each target which coordinates were measured by TR_1. Each probability estimate was calculated using 1000 statistical trials. Measurements from three or more targets did not fall within the gates in all the 1000 trials.

When only one measurement (i.e. from only a single target) falls within each gate, the problem of interstation measurement correlation turns out to be solved. When measurements from two targets fall within the gate formed for a certain target, then the measurement interstation correlation process should be continued. However, the number of correlation hypotheses is drastically reduced since for each measurement by TR_1 no more than two variants of association with measurements by TR_2 remain.

It is seen from Table 15.1 that for $\sigma(\hat{\beta})=\sigma(\hat{\varepsilon})=2.5'$ the gating described permits to solve the interstation correlation problem nearly always. For $\sigma(\hat{\beta})=\sigma(\hat{\varepsilon})=3.5'$

Table 15.1 Probability estimates $\hat{P}(1)$ and $\hat{P}(2)$ for measurements by radar TR_1 to fall within the gate of radar TR_2

r.m.s angular error (min)	Number of the target observed by radar TR_1											
	1		2		3		4		5		6	
	$\hat{P}(1)$	$\hat{P}(2)$	$\hat{P}(1)$	$\hat{P}(2)$	$\hat{P}(1)$	$\hat{P}(2)$	$\hat{P}(1)$	$\hat{P}(2)$	$\hat{P}(1)$	$\hat{P}(2)$	$\hat{P}(1)$	$\hat{P}(2)$
2.5	1.000	0.000	0.996	0.004	0.995	0.005	1.000	0.000	0.999	0.001	0.998	0.002
3.5	0.889	0.111	0.889	0.111	0.900	0.100	0.903	0.097	0.891	0.109	0.884	0.116
5	0.475	0.525	0.520	0.479	0.502	0.498	0.463	0.535	0.484	0.515	0.498	0.502

additional efforts required in approximately (10–12)% of trials and for $\sigma(\hat{\beta})=\sigma(\hat{\varepsilon})=5'$ – in approximately 50% of trials. [For $\sigma(\hat{\beta})=\sigma(\hat{\varepsilon})=5'$ and for targets No 2, 4 and 5 we can see that $\hat{P}(1)+\hat{P}(2)<1$ since the measurements obtained by TR_2 from these targets turned out to be outside the gates in one–two trials from 1000. If it is desirable to reduce the probability of such events, the gates should be broadened at the cost of some decrease of the probability of true correlation by gating.]

In the cases where measurements from two targets fall within the gate for a certain target, algorithm (15.42) may be used. The corresponding simulation results are presented in Table 15.2. Estimates for the probabilities of true and false correlation ($\hat{P}_{tr}$ and $\hat{P}_{fs}$, respectively) are shown for $\sigma(\hat{\beta})=\sigma(\hat{\varepsilon})=5'$. Besides, when the quadratic form (15.42) does not exceed the threshold for measurements from both the same target and different targets, a certain decision cannot be made. The estimate of the probability of such an event, P_{uncert}, is shown in Table 15.2 too. The threshold level was chosen so as the probability to reject a true hypothesis be equal to 0.01. For convenience the values of $\hat{P}(2)$ are also repeated in Table 15.2 from the last row of Table 15.1.

It follows from Tables 15.2 and 15.1 that, under assumed conditions, when TR_1 measured, for instance, coordinates of the target No 2 and a corresponding gate was formed in TR_2, the interstation correlation problem was correctly solved: by gating in 520 trials from 1000; additionally in $0.724\times480\approx348$ trials by using algorithm (15.42) in remaining 480 trials, i.e. in 868 trials from 1000 altogether. In $0.004\times480=2$ trials a false decision was made and in $0.271\times480\approx130$ trials no decision was made.

Algorithm (15.42) may be used without the comparison with a prescribed threshold. As in the case where thresholding is used, the quadratic form (15.42) is calculated for each measurement from TR_1 (converted to the coordinate system of TR_2) with each of the two measurements that have fallen within a corresponding gate in TR_2. That measurement obtained by TR_2 for which the quadratic form value turns out to be smaller than for the other measurement, is decided to originate from the same target as the considered measurement from TR_1. We may treat such an algorithm as the correlation algorithm using minimum "weighted distance" between the points corresponding to measurements from TR_1 and TR_2. The results of exploiting the algorithm described for remaining measurements after gating are presented in Table 15.3. According to this algorithm, situations where no decision is

Table 15.2

Number of the target	1	2	3	4	5	6
$\hat{P}(2)$	0.525	0.479	0.498	0.535	0.515	0.502
$\hat{P}_{tr}$	0.731	0.724	0.713	0.738	0.740	0.737
$\hat{P}_{fs}$	0.000	0.004	0.000	0.002	0.000	0.002
$\hat{P}_{uncert}$	0.269	0.271	0.287	0.260	0.260	0.261

Table 15.3

Number of the target	1	2	3	4	5	6
$\hat{P}(2)$	0.487	0.502	0.500	0.499	0.520	0.510
$\hat{P}_{tr}$	0.953	0.988	0.978	0.958	0.977	0.949
$\hat{P}_{fs}$	0.047	0.012	0.022	0.042	0.023	0.051

made are impossible. Therefore, in Table 15.3 estimates of the probabilities of only true and false correlation, $\hat{P}_{tr}$ and $\hat{P}_{fs}$, are shown. As in Table 15.2, the estimates of the probability $\hat{P}(2)$ in these new 1000 trials are also presented.

It follows from Table 15.3 that in a new 1000 trials preliminary gating solved the interstation correlation problem correctly, for instance, for the same target No 2 in 502 trials from 1000. Applying the algorithm of minimum weighted distance [(15.42) without thresholding] to remaining 498 trials yielded correct results additionally in $0.988 \times 498 \approx 492$ trials. Thus in 994 trials from 1000 a correct interstation correlation was achieved. In 6 trials from 1000 erroneous decisions were made. It is seen that though this algorithm provides higher probability of correct correlation, the probability of erroneous decisions also increases. The situations where no decisions are made (when thresholding is used) are often preferable than those where erroneous decisions are made, since in the former case the possibilities of using some other algorithms to achieve higher probability of correct correlation yet remain.

Example 15.3 shows that simple interstation correlation algorithms based on the comparison of local coordinate measurements can yield satisfactory results. Further improving the efficiency of interstation correlation is possible, as mentioned above, by utilizing additional *a priori* target information. For target tracking, when resultant (composite) measurements are not formed in MSRSs, direct spatial–temporal data correlation (association) algorithms should be used.

In this section*, we have discussed some features of data correlation (association) in MSRSs. First of all, a problem of interstation data correlation arises. We have noted that two different approaches to the solution of this problem are possible: interstation (spatial) measurement correlation and direct measurement-to-track or track-to-track- (spatial–temporal) correlation. In the cases where a resultant (composite) measurement is formed (see Sections* 14.1, 14.3)*, preceding interstation measurement correlation is necessary. Taking into account the enormous growth of the total number of correlation hypotheses with the increase of the number of stations and especially the number of local measurements at each station, we have emphasized the importance of preliminary "gating" as a simple and efficient way to reduce drastically the number of hypotheses to be analyzed with the help of "fine" statistical techniques. A simple Example* 15.1 *(with Fig.* 15.2) *illustrates the fact that in active MSRSs with short baselines between stations the spatial size of gates (validation regions) may be not large so that only few measurements can fall within corresponding gates in different stations. Then more complicated statistical decision techniques are to be applied to these few measurements only. The interstation data correlation problem is more difficult to solve in passive MSRSs where each station cannot measure target range directly. We have noted that available redundant measurements and* a priori *target information should be used in any case.*

We have considered statistical decision techniques based on calculating corresponding quadratic forms. We have presented algorithm (15.42) *for interstation (spatial) measurement correlation and algorithm* (15.43)–(15.45) *for measurement-to-track spatial–temporal correlation. We have also derived algorithm* (15.46), (15.47) *for track-to-track-spatial–temporal correlation taking into account mutual correlation between local track (local target state vector) estimates.*

In Example 15.2 *(with Fig.* 15.3) *we have considered a situation close to practical one where there is a compact group of targets observed by a MSRS consisting of two monostatic radars, and interstation measurement correlation is to be performed before resultant (composite) measurements formation. We have presented simulation results which allow to evaluate the efficiency of gating for different accuracy of local coordinate measurements. These results also demonstrate the efficiency of applying algorithm* (15.42) *(in two modifications) to measurements which remain nonassociated by gating.*

REFERENCES

1. Averyanov, V.Ya., *Spatially Separated Radars and Radar Systems*, Nauka i Technika, Minsk, 1978 (in Russian).
2. Bakulev, P.A. and Stepin, V.M., *Methods and Devices for Moving Target Indication*, Radio i Svyaz, Moscow, 1986 (in Russian).
3. Barton, D.K. and Ward, H.R., *Handbook of Radar Measurement*, Prentice-Hall, Englewood Cliffs, NJ, 1969.
4. Belavin, O.V., *Fundamentals of Radionavigation*, 2nd edition, Sov. Radio, Moscow, 1977 (in Russian).
5. Born, M. and Wolf, E., *Principles of Optics*, 4th edition, Pergamon Press, 1968.
6. Bulychev, Yu.G. and Taran, V.N., Invariant-group method for bearing correlation in passive triangulation multistatic radar systems, *Radiotekhnika i Elektronika*[1], Vol. 32, No. 4, 1987, pp. 755–765 (in Russian).
7. Burlakov, Yu.G., Ivanov, V.A. and Chernyak, V.S., Quasioptimal spatial–temporal filters–detectors with preliminary signal filtration, *Radiotekhnika i Elektronika*, Vol. 19, No. 2, 1974, pp. 432–435 (in Russian).
8. Van Trees, H.L., *Detection, Estimation and Modulation Theory*, Part 1, John Wiley & Sons, 1969.
9. Voevodin, V.V. and Tyrtyshnikov, E.E., Calculations with Toeplitz's matrices, in *Computational Processes and Systems*, No. 1, ed. by G.I. Marchuk, Nauka, Moscow, 1983 (in Russian).
10. Gradshtein, I.S. and Ryzhik, I.M., *Tables of Integrals, Sums, Series and Products*, Nauka, Moscow, 1971 (in Russian).
11. Gribanov, Yu. I., Veselova, G.P. and Andreev, V.N., *Automatic Digital Correlators*, Energiya, Moscow, 1971 (in Russian).
12. Gusev, V.G. and Cherenkova, E.V., Performance analysis and choice of parameter values for a space–time filter of the optimum detection system for a multidimensional stochastic process, *Radiotekhnika i Elektronika*, Vol. 32, No. 2, 1987, pp. 300–308 (in Russian).
13. Zhdanyuk, B.F., *Fundamentals of Trajectory Measurement Statistical Processing*, Sov. Radio, Moscow, 1978 (in Russian).
14. Zaslavsky, L.P. and Slavkin, V.L., Some issues on characteristic calculations for system with square summation, *Voprosy radioelektroniki, seriya obshchetekhnicheskaya*[2], No. 1, 1963, pp. 3–10 (in Russian).
15. Maksimov, M.V. *et al.*, *Protection from Radiointerferences*, Sov. Radio, Moscow, 1976 (in Russian). See also translation into English: *Radar Anti-Jamming Techniques*, Artech House.
16. Kazakov, E.L. and Shishkin, Yu.M., About the polarization eigenbasis of a radar target observed by a system with spatially separated receivers, *Radiotekhnika*[3], 1984, No. 7, pp. 77–81 (in Russian).
17. Kanareykin, D.B., Pavlov, N.F. and Potekhin, V.A., *Polarization of Radar Signals*, Sov. Radio, Moscow, 1986 (in Russian).
18. Karavaev, V.V. and Sazonov, V.V., *Statistical Theory of Passive Location*, Radio i Svyaz, Moscow, 1987 (in Russian).
19. Karavaev, V.V. and Sazonov, V.V., Signal detection in a background of chaotic scatterers by a bistatic system, *Radiotekhnika i Elektronika*, Vol. 28, No. 1, 1983, pp. 67–73 (in Russian).
20. Kobak, V.O., *Radar Reflectors*, Sov. Radio, Moscow, 1975 (in Russian).
21. Kell, R.E., On the derivation of bistatic RCS from monostatic measurements, *Proc. IEEE*, Vol. 53, No. 8, 1965, pp. 963–970.
22. Kolessa, A.E., Coordinate estimation of a collection of objects by a multistatic system of direction finders, *Radiotekhnika i Elektronika*, Vol. 32, No. 12, 1987, pp. 2534–2541 (in Russian).
23. Kondratyev, V.C., Kotov, A.F. and Markov, L.N., *Multistatic Radio Systems*, ed. by V.V. Tsvetnov, Radio i Sviaz, Moscow, 1986 (in Russian).

---

[1] This journal is translated into English and is known as *Radioengineering and Electronic Physics* and recently as *Journal of Communication Technology and Electronics*.

[2] *Issues of Radioelectronics, General Engineering Series*.

[3] This journal is translated into English and is known as *Telecommunication and Radioengineering* or (recently) *Radio and Communication Technology*.

24. Kontorov, D.S. and Golubev-Novozhilov, Yu.S., *Introduction to the Radar System Engineering*, Sov. Radio, Moscow, 1971 (in Russian).
25. Korn, G.A. and Korn, T.M., *Mathematical Handbook for Scientists and Engineers*, 2nd edition, McGraw-Hill, 1968.
26. Katsenbogen, M.S., *Detection Characteristics*, Sov. Radio, Moscow, 1965 (in Russian).
27. Cramér, H., *Mathematical Methods of Statistics*, Princeton University Press, 1946.
28. Kremer, I.Ya., Petrov, V.M. and Shapiro, S.M., Efficiency comparison of optimum and matched processing for spatial discrimination of signal and noise sources in Freshnel zone, *Radiotekhnika i Elektronika*, Vol. 30, No. 7, 1985, pp. 1341–1347 (in Russian).
29. Kremer, I.Ya. and Tabatsky, V.A., Analysis of the quasioptimum algorithm for spatial signal processing in clutter from precipitations, in *Sbornik nauchnykh trudov Moskovskogo energeticheskogo instituta*[4], No. 156, Moscow, 1988, pp. 35–40 (in Russian).
30. Crispin, J.W., Jr. and Maffet, A.L., Radar cross section estimation for complex shapes, *Proc. IEEE*, Vol. 53, No. 8, 1965, pp. 972–982.
31. Kuzmin, S.Z., *Design Fundamentals of Radar Information Digital Processing*, Radio i Svyaz, Moscow, 1986 (in Russian).
32. Kouyoumjian, R.G., Asymptotic high-frequency methods, *Proc. IEEE*, Vol. 53, No. 8, 1965, pp. 864–876.
33. Levin, B.R., *Theoretical Foundations of Statistical Radioengineering*, Part 1, Sov. Radio, Moscow, 1966 (in Russian).
34. Mishina, A.P. and Proskuryakov, I.V., *Higher Algebra*, ed. by P.K. Rashevsky, Fizmatgiz, Moscow, 1962 (in Russian).
35. Monzingo, R.A. and Miller, T.W., *Introduction to Adaptive Arrays*, J. Wiley & Sons, 1980.
36. Nakhmanson, G.S., Estimation of target coordinates and their derivatives in multistatic measurement systems with wideband signals in a background of internal and external noises, *Izvestiya vuzov SSSR, Radioelektronika*[5], Vol. 30, No. 4, 1987, pp. 37–43 (in Russian).
37. Nikol'sky, V.V., *Electrodynamics and Propagation of Radiowaves*, Nauka, Moscow, 1973 (in Russian).
38. Pagurova, V.I., *Tables of Incomplete Gamma-Function*, Computer Centre of the Academy of Science USSR, 1963 (in Russian).
39. Papoulis, A., *Systems and Transforms with Applications in Optic*, McGraw-Hill, 1969.
40. Penzin, K.V., Synthesis of multiscaled measurement systems, *Radiotekhnika i Elektronika*, Vol. 32, No. 2, 1987, pp. 347–355 (in Russian).
41. Petrov, V.M., Shapiro, S.M. and Tabatsky, V.A., Optimum reception of stochastic signals in multistatic systems, *Izvestiya vuzov SSSR, Radioelektronika*, Vol. 30, No. 4, 1987, pp. 32–37 (in Russian).
42. Kremer, I.Ya. *et al.*, *Space–Time Signal Processing*, ed. by I.Ya. Kremer, Radio i Svyaz, Moscow, 1984 (in Russian).
43. Repin, V.G. and Tartakovsky, G.P., *Statistical Synthesis Under Condition of a priori Uncertainty and Adaptation of Information Systems*, Sov. Radio, Moscow, 1977 (in Russian).
44. Sobtsov, N.V., Maximum likelihood estimation for a multiscaled phase measurement system, *Radiotekhnika i Elektronika*, Vol. 18, No. 6, 1973, pp. 1180–1186 (in Russian).
45. *Modern Radar*, ed. by R.S. Berkowitz, J. Wiley & Sons, 1965.
46. Caspers, J.W., Bistatic and multistatic radar, in *Radar Handbook*, Editor-in-chief M.I. Skolnik, McGraw-Hill, 1970.
47. Shirman, Ya.D. *et al.*, *Theoretical Foundations of Radar*, ed. by Ya.D. Shirman, Sov. Radio, Moscow, 1970 (in Russian).
48. Korostelev, A.A. *et al.*, *Theoretical Foundations of Radar*, ed. by V.E. Dulevich, Sov. Radio, Moscow, 1970 (in Russian).
49. Tikhonov, V.I., *Statistical Radioengineering*, 2nd edition, Radio i Svyaz, Moscow, 1982 (in Russian).
50. Ufimtsev, P.Ya., *Boundary Wave Method in the Physical Diffraction Theory*, Sov. Radio, Moscow, 1962 (in Russian). This book was translated into English.
51. Fal'kovich, S.E., *Signal Parameter Estimation*, Sov. Radio, Moscow, 1970 (in Russian).
52. Fal'kovich, S.E. and Khomyakov, E.N., *Statistical Theory of Measurement Radiosystems*, Radio i Svyaz, 1981 (in Russian).
53. Feldman, Yu.I. and Mandurovsky, I.A., *Fluctuation Theory of Radar Signals Reflected by Extended Targets*, ed. by Yu.I. Feldman, Radio i Svyaz, Moscow, 1981 (in Russian).
54. Helstrom, C.W., *Statistical Theory of Signal Detection*, Pergamon Press, 1960.
55. Cheremisov, A.K., Statistical characteristics of bistatic radar cross section of a body, *Radiotekhnika i Elektronika*, Vol. 32, No. 12, 1985, pp. 2516–2524 (in Russian).

[4] *Proceedings of Moscow Power Institute.*
[5] *Proc. Higher Educational Institutions of USSR, Radioelectronics.*

56. Chernyak, V.S., Optimal signal detection in systems with spatially separated receivers, *Voprosy radioelektroniki, seriya obshchetekhnicheskaya*, No. 11, 1970, pp. 17–28 (in Russian).
57. Chernyak, V.S., Optimal detection of Gaussian stochastic signals by a system with spatially separated receivers, *Radiotekhnika i Elektronika*, Vol. 13, No. 10, 1968, pp. 1874–1879 (in Russian).
58. Chernyak, V.S., Synthesis of optimum signal temporal parameter estimators for a system with spatially separated receivers, in *Proc. 4th Conf. on Information Transmission and Coding Theory, Section 2*, Moscow-Tashkent, 1969, pp. 221–225 (in Russian).
59. Chernyak, V.S., About the use of the Fisher information matrix for maximum attainable accuracy analysis of maximum likelihood estimates in the presence of stray parameters, *Radiotekhnika i Elektronika*, Vol. 16, No. 6, 1971, pp. 956–966 (in Russian).
60. Chernyak, V.S., Space-frequency signal filtration in multichannel receiving systems in a background of stochastic interferences, *Radiotekhnika i Elektronika*, Vol. 18, No. 5, 1973, pp. 959–969 (in Russian).
61. Chernyak, V.S., Adaptive linear filtration of random processes in multichannel systems, in *Proc. 6th Conf. on Information Transmission and Coding Theory*, Part 6, Moscow-Tomsk, 1975, pp. 162–167 (in Russian).
62. Chernyak, V.S., Radioimaging of objects, *Radiotekhnika i Elektronika*, Vol. 24, No. 12, 1979, pp. 2454–2463 (in Russian).
63. Chernyak, V.S., Radioimaging of objects by a receiving system with partial channels, *Radiotekhnika i Elektronika*, Vol. 28, No. 3, 1983, pp. 479–490 (in Russian).
64. Chernyak, V.S., Maximum attainable efficiency of spatial processing for signal detection in clutter, *Radiotekhnika i Elektronika*, Vol. 29, No. 6, 1984, pp. 1110–1119 (in Russian).
65. Chernyak, V.S., Multistatic detectors for fluctuating signals in a background of spatially correlated interferences, *Radiotekhnika i Elektronika*, Vol. 32, No. 2, 1987, pp. 334–346 (in Russian).
66. Chernyak, V.S., Efficiency of multistatic detectors for fluctuating signals in a background of spatially correlated interferences, *Radiotekhnika i Elektronika*, Vol. 32, No. 3, 1987, pp. 559–573 (in Russian).
67. Chernyak, V.S., Zaslavsky, L.P. and Osipov, L.V., Multistatic radars and systems (review), *Zarubezhnaya radioelektronika*[6], No. 1, 1987, pp. 9–69 (in Russian).
68. Chernyak, V.S., Preface of the scientific editor, *Zarubezhnaya radioelektronika*[6], No. 1, 1987, pp. 3–8 (in Russian).
69. Chernyak, V.S., Doppler frequency shift measurement of fluctuating signals in a multistatic system, *Radiotekhnika i Elektronika*, Vol. 34, No. 8, 1989, pp. 1655–1664 (in Russian).
70. Chernyak, V.S., Accuracy of Doppler frequency shift measurement of fluctuating signals in a multistatic system, *Radiotekhnika i Elektronika*, Vol. 34, No. 9, 1989, pp. 1861–1871 (in Russian).
71. Chernyak, Yu.B., Correlators with ideal limiters, *Radiotekhnika*, No. 3, 1965, pp. 70–77 (in Russian).
72. Shirman, Ya.D. and Manzhos, V.N., *Theory and techniques of radar information processing in a background of interferences*, Radio i Svyaz, Moscow, 1981 (in Russian).
73. Shirman, Ya.D., *Resolution and compression of signals*, Sov. Radio, Moscow, 1974 (in Russian).
74. Ancker, C.J., Airborne detection finding – the theory of navigation errors, *IRE Trans. on Aeronautical and Navigational Electronics*, Vol. ANE-5, No. 4, 1958, pp. 199–210.
75. Barale,G., Frashetty, G. and Pardini, S., The multiradar tracking in the ATC system of the Rome FIR, in *Proc. Int. Radar Conf. "Radar '82"*, London, 1982, pp. 296–299.
76. Bath, W.G., Association of multistatic radar data in the presence of large navigation and sensor alignment errors, in *Proc. Int. Radar Conf. "Radar '82"*, London, 1982, pp. 169–173.
77. Bonnefoy, I. *et al.*, The development of true multi-radar tracking system, in *Proc. Int. Conf. on Radar*, Paris, 1978, pp. 109–117.
78. Borison, S.L., Bistatic scattering cross section of a randomly oriented dipole, *IEEE Trans. on Antennas and Propagation*, Vol. AP-15, No. 2, 1967, pp. 320–321.
79. Buchner, M.R., A multistatic track filter with optimal measurement selection, in *Proc. Int. Radar Conf. "Radar '77"*, London, 1977, pp. 72–75.
80. Cantrell, B. and Grindlay, A., Multiple site radar tracking system, in *Proc. IEEE 1980 Int. Radar Conf.*, Arlington, Va, 1980, pp. 348–354.
81. Ewing, E.F., The applicability of bistatic radar to short range surveillance, in *Proc. Int. Radar Conf. "Radar '77"*, London, 1977, pp. 53–58.
82. Ewing, E.F. and Dicken, L.W., Some application of bistatic and multistatic radars, in *Proc. Int. Conf. on Radar*, Paris, 1978, pp. 222–231.
83. Farina, A. and Pardini, S., Survey of radar data-processing techniques in air-traffic-control and surveillance systems, *IEE Proceedings*, Pt. F, Vol. 127, No. 3, 1980, pp. 190–204.
84. Farina, A. and Hanle, E., Position accuracy in netted monostatic and bistatic radar, *IEEE Trans. on Aerospace and Electronic Systems*, Vol. AES-19, No. 4, 1983, pp. 513–520.

[6] *Foreign Radioelectronics.*

85. Farina, A. and Pardini, S., Multiradar tracking system using radial velocity measurements, *IEEE Trans. on Aerospace and Electronic Systems*, Vol. AES-15, No. 4, 1979, pp. 513–520.
86. Farina, A., Tracking function in bistatic and multistatic radar systems, *IEE Proceedings*, Pt. F, Vol. 133, No. 7, 1986, pp. 630–637.
87. Fawcette, J., Bistatic radar may a "sanctuary" in space, *Electronic Warfare/Defence Electronics*, Vol. 10, No. 1, 1978, pp. 84–86, 88.
88. Foy, W.H., Position–location solutions by Taylor series estimation, *IEEE Trans. on Aerospace and Electronic Systems*, Vol. AES-12, No. 2, 1976, pp. 187–193.
89. Glaser, J.I., Bistatic radar hold promise for future systems, *MSN*, Vol. 16, No. 11, 1984, pp. 119–136.
90. Glaser, J.I., Bistatic RCS of complex objects near forward scatter, *IEEE Trans. on Aerospace and Electronic Systems*, Vol. AES-21, No. 1, 1985, pp. 70–78.
91. Groginsky, H.L., Position estimation using only multiple simultaneous range measurements, *IRE Trans. on Aeronautical and Navigational Electronics*, Vol. ANE-6, No. 5, 1959, pp. 178–187.
92. Hanle, E., Distance considerations for multistatic radar, in *Proc. IEEE 1980 Int. Radar Conf.*, Arlington, Va, 1980, pp. 100–105.
93. Heath, G.E., Bistatic scattering reflection asymmetry, polarization reversal asymmetry and polarization reversal reflection symmetry, *IEEE Trans. on Antennas and Propagation*, Vol. AP-29, No. 3, 1981, pp. 429–434.
94. Heath, G.E., Properties of the linear polarization bistatic scattering matrix, *IEEE Trans. on Antennas and Propagation*, Vol. AP-29, No. 3, 1981, pp. 523–525.
95. Heimiller, R.C., Belyea, J.E. and Tomlinson, P.G., Distributed array radar, *IEEE Trans. on Aerospace and Electronic Systems*, Vol. AES-19, No. 6, 1983, pp. 831–839.
96. Knittel, G.H., Spoerry, S. and Morse, G.B., The netted radar demonstration at Fort Sill, Oklahoma, in *Proc. EASCON' 81*, Washington, D.C., 1981, pp. 79–88.
97. Knoppik, N. *et al.*, Simultaneous automatic tracking in multiple radar networks, in *Proc. Int. Conf. on Radar*, Paris, 1978, pp. 100–108.
98. Lockheed wins Air Force approval to build Precision Location and Strike System, *Microwaves*, Vol. 16, No. 11, 1977, p. 14.
99. Lorti, D.C. and Bowman, J.J., Will tactical aircraft use bistatic radar? *MSN*, Vol. 8, No. 9, 1978, pp. 49–54.
100. Low-cost ABM radar given emphasis, *Aviation Week & Space Technology*, Vol. 116, No. 9, 1982, pp. 74–75.
101. Matthiesen, D.J., Performance characteristics of optimal and quantized, phase only time-of-arrival systems, in *Proc. IEEE 1980 Int. Radar Conf.*, Arlington, Va, 1980, pp. 438–444.
102. Milne, K., Principles and concepts of multistatic surveillance radar, in *Proc. Int. Radar Conf. "Radar '77"*, London, 1977, pp. 46–52.
103. Morley, A.R. and Wilsdon, A.S., Multiradar tracking in a multisite environment, in *Proc. Int. Radar Conf. "Radar '77"*, London, 1977, pp. 66–71.
104. Napier, P.J., Thompson, A.R. and Ekers, R.D., The Very Large Array: design and performance of a modern synthesis radio telescope, *Proc. IEEE*, Vol. 71, No. 11, 1983, pp. 1295–1320.
105. Peebles, P.Z., Jr., Bistatic radar cross section of chaff, *IEEE Trans. on Aerospace and Electronic Systems*, Vol. AES-20, No. 2, 1984, pp. 128–140.
106. Retzer, G., A concept for signal processing in bistatic radar, in *Proc. IEEE 1980 Int. Radar Conf.*, Arlington, Va, 1980, pp. 288–293.
107. Salah, J.E. and Moriello, J.E., Development of a multistatic measurement system, in *Proc. IEEE* 1980 *Int. Radar Conf.*, Arlington, Va, 1980, pp. 88–93.
108. Schultheiss, P.M. and Weinstein, E., Passive localization of a moving source, in *Proc. EASCON '78*, Arlington, Va, 1978, pp. 258–266.
109. Siegel, K.M. *et al.*, Bistatic radar cross sections of surfaces of revolution, *J. Applied Phys.*, Vol. 26, No. 3, 1955, pp. 297–305.
110. Simcox, L.N., A method of automatic alignment of radars with overlapping cover, in *Proc. Int. Radar Conf. "Radar '77"*, London, 1977, pp. 66–71.
111. Skolnik, M.J., An analysis of bistatic radar, *IRE Trans. on Aeronautical and Navigational Electronics*, Vol. ANE-8, No.1, 1961, pp. 19–27.
112. Srinivasan, R., Distributed radar detection theory, *IEE Proceedings*, Pt F, Vol. 133, No. 1, 1986, pp. 55–60.
113. Srinivasan, R., Theory of distributed detection, *Signal Processing*, No. 4, 1986, pp. 319–327.
114. Steinberg, B.D., *Principles of Aperture and Array Systems*, J. Wiley & Sons, 1976.
115. Steinberg, B.D. *et al.*, First experimental results from the Valley Forge radio camera program, *Proc. IEEE*, Vol. 67, No. 9, 1979, pp. 1370–1371.
116. Steinberg, B.D., Radar imaging from a distorted array: the radio camera algorithm and experiments, *IEEE Trans. on Antennas and Propagation*, Vol. AP-29, No. 5, 1981, pp. 740–748.

117. Torrieri, D.J., Statistical theory of passive location systems, *IEEE Trans. on Aerospace and Electronic Systems*, Vol. AES-20, No. 2, 1984, pp. 183–198.
118. Wax, M., Position location from sensors with position uncertainty, *IEEE Trans. on Aerospace and Electronic Systems*, Vol. AES-19, No. 5, 1983, pp. 658–662.
119. Weinstein, E., Optimal source localization and tracking from passive array measurements, *IEEE Trans. on Acoustics, Speech and Signal Processing*, Vol. ASSP-30, No. 1, 1982, pp. 69–76.
120. Wernersson, E., Coherent multi-static radar: stochastic signal theory and performance evaluation, in *Proc. Int. Radar Conf. "Radar '82"*, London, 1982, pp. 179–182.
121. Willner, D., Chang, C.B. and Dunn, K.P., Kalman filter algorithms for a multisensor system, *Proc. 15th IEEE Conf. on Decision and Control*, Clearwater, Fl, 1976, pp. 570–574.
122. Zasada, N., Multistatic radar system for aircraft defense, *Signal*, Vol. 34, No. 8, 1980, pp. 65–75.
123. Vishin, G.M., *Multifrequency radar*, Voenizdat, Moscow, 1973 (in Russian).
124. Dzhun, V.I. and Shchesnyak, S.S., Adaptive antenna systems with mainlobe interference suppression, *Zarubezhnaya Radioelektronika*, No. 4, 1988, pp. 3–15 (in Russian).
125. Usachev, V.V. and Fedorov, I.B., Object flow correlation in a system of estimators, *Izvestiya vuzov SSSR, Radioelektronika*, Vol. 23, No. 11, 1980, pp. 32–37 (in Russian).
126. Conte, E. *et al.*, Multistatic radar detection: synthesis and comparison of optimum and suboptimum receivers, *IEE Proceedings*, Pt. F, Vol. 130, No. 6, 1983, pp. 484–494.
127. Gurov, G.B., Ryndin, Yu.G. and Sukovatkin, N.N., Detection of a point-like object by a spatially separated system with signal reradiation, *Radiotekhnika i Elektronika*, Vol. 32, No. 5, 1987, pp. 945–953 (in Russian).
128. Bol'shakov, I.A., *Signal flow extraction from noise*, Sov. Radio, Moscow, 1969 (in Russian).
129. Trukhachev, A.A., Threshold level determination for given exceeding probability by a sum of random variables, *Radiotekhnika i Elektronika*, Vol. 11, No. 8, 1966, pp. 1486–1488 (in Russian).
130. Filer, E.H., Active Swept Frequency Interferometer Radar (ASFIR), in *IEEE Int. Conv. Rec.*, Pt. 7, 1964, pp. 252–258.
131. Mirkin, M.I., Schwartz, C.E. and Spoerry, S., Automated tracking with netted ground surveillance radars, in *Proc. IEEE Int. Radar Conf.*, Arlington, Va, 1980, pp. 371–379.
132. Thomas, H.W., Maignan, G. and Storey, J.T., Tracking in a multisensor environment, *IEE Proceedings*, Pt. F, Vol. 123, No. 3, 1976, pp. 191–194.
133. Miller, J.T., Jr. and Berry, J.R., Multisensor utilization investigation, in *Proc. Int. Radar Conf. "Radar '77"*, London, 1977, pp. 248–252.
134. Berry, J.T., Multisensor airspace surveillance: the results and conclusions of two experiments, in *Proc. Int. Conf. on Radar*, Paris, 1978, pp. 93–99.
135. Anderson, S.J., Stereoscopic and bistatic radars: assessment of capabilities and limitations, in *Proc. RADARCON 90*, Adelaide, Australia, 1990, pp. 305–313.
136. Cameron, A., The Jindalee operational radar network: its architecture and surveillance capability, in *Proc. Int. Radar Conf. "Radar 95"*, Washington DC, 1995, pp. 692–697.
137. Kewley, D.J. and Dall, I.W., Performance assessment criteria for OTH radar, in *Proc. Int. Radar Conf. "Radar 92"*, London, 1992, session 1A, pp. 1–4.
138. Furcolo, B. *et al.*, Multiradar tracking in the new ATC system of the Mazatlan (Mexico) area control center, in *Proc. IEEE Int. Radar Conf.*, Arlington, Va, 1985, pp. 403–408.
139. Farina, A. and Studer, F.A., Radar and sensor netting: present and future, *Microwave Journal*, Vol. 29, No. 1, 1986, pp. 97, 98, 100, 104, 106.
140. French antistealth radar, *Signal*, Vol. 44, No. 9, 1990, p. 12.
141. Farina, A. and Pardini, S., Introduction to multiradar tracking systems, *Rivista Tecnica Selenia*, Vol. 8, No. 1, 1982, pp. 14–26.
142. Farina, A. and Studer, F.A., A review of CFAR detection techniques in radar systems, *Microwave Journal*, Vol. 29, No. 9, 1986, pp. 115–128.
143. Nitzberg, R., Clutter map CFAR analysis, *IEEE Trans. on Aerospace and Electronic Systems*, Vol. AES-22, No. 4, 1986, pp. 419–421.
144. Lang Hong, Centralized and distributed multisensor integration with uncertainties in communication networks, *IEEE Trans. on Aerospace and Electronic Systems*, Vol. AES-27, No. 2, 1991, pp. 370–379.
145. Chen, Y. and Sun, Zh., Location and tracking of bistatic system in jamming, in *Proc. CIE Int. Conf. of Radar, ICR '96*, Beijing, China, 1996, pp. 43–47.
146. Ferri, M. *et al.*, Advanced airport surveillance and imaging using the surface miniradar network, in *Proc. CIE Int. Conf. of Radar, ICR '96*, Beijing, China, 1996, pp. 246–249.
147. Hu, W. and Mao, Sh., A probabilistic data association algorithm for multisensor multitarget tracking, in *Proc. CIE Int. Conf. of Radar, ICR '96*, Beijing, China, 1996, pp. 475–479.
148. Tol, J., van Genderen, P. and Ligthart, L.P., Improvements of the SSR mode S data link using a distributed groundsystem, in *Proc. CIE Int. Conf. of Radar, ICR '96*, Beijing, China, 1996, pp. 519–522.
149. Que, W. *et al.*, An approach to radar netting, in *Proc. CIE Int. Conf. of Radar, ICR '96*, Beijing, China, 1996, pp. 573–577.

150. Cui, N. *et al.*, Improvement of multisensor data fusion on track loss in clutter, in *Proc. CIE Int. Conf. of Radar, ICR '96*, Beijing, China, 1996, pp. 719–722.
151. Sidorov, *et al.*, Equipment for meteor synchronization and communications, *Proc. Symp. "Metrology of time and space" (MVP '94)*, Mendeleevo, Moscow Region, 1994, pp. 405–410 (in Russian).
152. Paradowski, L., Computational procedure for two-dimensional direction finding, its characteristics and application perspective, in *Proc. Int. Conf. on Radar*, Paris, 1994, pp. 662–667.
153. Paradowski, L., Position estimation from netted radar systems with sensor position uncertainty, in *Proc. CIE Int. Conf. of Radar, ICR '96*, Beijing, China, 1996, pp. 609–613.
154. Roecker, J.A. and McGillem, C.D., Comparison of two-sensor tracking methods based on state vector fusion and measurement fusion, *IEEE Trans. on Aerospace and Electronic Systems*, Vol. AES-24, No. 4, 1988, pp. 447–449.
155. Bar-Shalom, Y., Comments on "Comparison of two-sensor tracking methods based on state vector fusion and measurement fusion", *IEEE Trans. on Aerospace and Electronic Systems*, Vol. AES-24, No. 4, 1988, pp. 456–457.
156. Willsky, *et al.*, Combining and updating of local estimates and regional maps along sets of one-dimensional tracks, *IEEE Trans. on Automatic Control*, Vol. AC-27, No. 8, 1982, pp. 799–813.
157. Kurien, T. and Liggins, M.E., Report-to-target assignment in multisensor multitarget tracking, in *Proc. 27th IEEE Conf. Decision and Control*, Austin,TX, 1988, pp. 2484–2488.
158. Oshchepkov, P.K., *Life and Dream*, Moskowsky rabochy, Moscow, 1967 (in Russian).
159. Hanle, E., Survey of bistatic and multistatic radar, *IEE Proceedings*, Pt. F, Vol. 133, No. 7, 1986, pp. 587–595.
160. Golubev, O.V. *et al.*, *The Russian Ballistic Missile Defence System*, Technoconsult, Moscow, 1994 (in Russian).
161. Chernyak, V.S., Doctoral dissertation, Moscow, 1972 (in Russian).
162. Glaser, J.I., Fifty years of bistatic and multistatic radar, *IEE Proceedings*, Pt F, Vol. 133, No. 7, 1986, pp. 596–603.
163. Bowman, C.L., Morefield, C.L. and Murphy, M.S., Multisensor multitarget recognition and tracking, in *Rec. 13th Asilomar Conf. on Circuits, Systems and Computers*, Pacific Groove, CA, 1979, pp. 329–333.
164. Bowman, C.L., Multisensor track file correlation using attributes and kinematics, in *IEEE Nat. Aerospace and Electron. Conf., NAECON '80, Record*, Dayton O., 1980, pp. 1181–1186.
165. Bowman, C.L. and Morefield, C.L., Multisensor fusion of target attributes and kinematics, in *Proc. 19th IEEE Conf. on Decision and Control*, Albuquerque N. Mex., 1980, pp. 837–839.
166. Dall, I.W. and Kewley, D.J., Track association in the presence of multimode, in *Proc. Int. Radar Conf. "Radar 92"*, London, 1992, pp. 70–73.
167. Dall, I.W. and Shellshear, A.J., Performance of an improved model based data fusion algorithm for OTHR data, in *Proc. Int. Conf. on Radar*, Paris, 1994, pp. 625–629.
168. Abramovich Yu.I. *et al.*, Over-the-horizon radiolocation in Russia and Ukraine, in *Proc. Int. Conf. on Radar*, Paris, 1994, pp. 232–236.
169. Chang, C.B. and Youens, L.C., Measurement correlation for multiple sensor tracking in a dense target environment, *IEEE Trans. on Automatic Control*, Vol. AC-27, No. 6, 1982, pp. 1250–1252. (See also *Proc. of the 20th IEEE Conf. on Decision and Control*, San Diego, CA, 1981, pp. 830–831).
170. Koch, V. and Westphal, R., A new approach to a multistatic passive radar sensor for air defense, in *Proc. Int. Radar Conf. "Radar 95"*, Washington DC, 1995, pp. 22–28.
171. Dunsmore, M.R.B., Bistatic radars, *Alta Frequenza*, Vol. 58, No. 2, 1989, pp. 53–79.
172. Lobanov, M.M., *From the Past of Radar*, Voenizdat, Moscow, 1969 (in Russian).
173. Dana, M.P., Registration: a prerequisite for multiple sensor tracking, in *Multitarget–Multisensor Tracking: Advanced Applications*, ed. by Y. Bar-Shalom, Vol. 1, Artech House, Norwood MA, 1990, pp. 155–185.
174. Kurien, T., Issues in the design of practical multitarget tracking algorithms, in *Multitarget-Multisensor Tracking: Advanced Applications*, ed. by Y. Bar-Shalom, Vol. 1, Artech House, Norwood MA, 1990, pp. 43–82.
175. Haimovich, A.M. *et al.*, Fusion of sensors with dissimilar measurement tracking accuracies. *IEEE Trans. on Aerospace and Electronic Systems*, Vol. AES-29, No. 1, 1993, pp. 245–249.
176. Bowman, C.L., Multisensor integration for defensive fire control surveillance, in *Nat. Aerospace and Electron. Conf., NAECON*, Dayton, O, 1979, pp. 176–180.
177. Bowman, C.L., Maximum likelihood track correlation for multisensor integration, in *Proc. 18th IEEE Conf. on Decision and Control*, Fort Lauderdale, Fla, 1979, pp. 374–376.
178. Griffiths, H.D. and Long, N.R.W., Television-based bistatic radar, *IEE Proceedings*, Pt. F, Vol. 133, No. 7, 1986, pp. 649–657.
179. Hartman, R., Budget cuts miss most electronic programs, *Defense Electronics*, Vol. 14, No. 10, 1982, pp. 125–127.

180. Bar-Shalom, Y. and Campo, L., The effect of the common process noise on the two-sensor fused track covariance, *IEEE Trans. on Aerospace and Electronic Systems*, Vol. AES-22, No. 6, 1986, pp. 803–805.
181. Lockheed prepared for TP-1 production, *Aviation Week & Space Technology*, Vol. 110, No. 26, 1979, p. 60.
182. Jónsdóttir, I. and Hauksdóttir, A.S., Integrity monitoring and estimation of systematic errors in radar data systems, in *Proc. Int. Radar Conf. "Radar 95"*, Washington DC, 1995, pp. 310–316.
183. Marlow, H.C. *et al.*, The RAT SCAT cross-section facility, *Proc. IEEE*, Vol. 53, No. 8, 1965, pp. 946–954.
184. Gumble, B., Air Force upgrading defenses at NORAD, *Defense Electronics*, Vol. 17, 1985, p. 86–100.
185. Chung-Chi Cha, Michels, J. and Starszewski, E., An RCS analysis of generic airborne vehicles depending on frequency and bistatic angle, *Proc. IEEE Nat. Radar Conf.*, Ann Arbor, Mi, 1988, pp. 214–219.
186. Kuznetsov, Yu.A., Starikovsky, I.M. and Tanygin, A.A., From Ostekhburo to All-Russian Scientific Research Institute of Radioengineering, *Radioengineering Industry*, Issue 1–2, 1995, pp. 59–71 (in Russian).
187. Kuriksha, A.A. *et al.*, Location of space targets by radars, *Radioengineering Industry*, No. 1–2, 1995, pp. 44–51 (in Russian).
188. Guo, Y. and Überall, H., Bistatic radar scattering by a chaff cloud, *IEEE Trans. on Antennas and Propagation*, Vol. AP-40, No. 7, 1992, pp. 837–841.
189. Palermo, C.J. and Bauer, L.H., Bistatic scattering cross section of chaff dipoles with applications to communications, *Proc. IEEE*, Vol. 53, No. 8, 1965, p. 1119–1121.
190. Jackson, M.C., The geometry of bistatic radar system, *IEE Proceedings*, Pt F, Vol. 133, No. 7, 1986, pp. 604–612.
191. Waltz, E. and Llinas, J., *Multisensor data fusion*, Artech House, Norwood, MA, 1990.
192. Willis, N.J., *Bistatic Radar*, Artech House, Norwood, MA, 1991.
193. Ormsby, J.F.-A. *et al.*, Analitic coherent radar techniques for target mapping, *IEEE Trans. on Aerospace and Electronic Systems*, Vol. AES-6, No. 3, 1970, pp. 295–305.
194. Rockmore, A.J., Denton, R.V. and Friedlander, B., Direct three-dimensional image reconstruction, *IEEE Trans. on Antennas and Propagation*, Vol. AP-27, No. 2, 1979, pp. 239–241.
195. Rozenbaum-Raz, S., On scatterer reconstruction from far-field data, *IEEE Trans. on Antennas and Propagation*, Vol. AP-24, No. 1, 1976, pp. 66–70.
196. Detlefsen, J., The resolution limits of the imaging of conducting bodies using multistatic scattering, *IEEE Trans. on Antennas and Propagation*, Vol. AP-28, No. 3, 1980, pp. 377–380.
197. Fiore, M., Olivetti, R. and Pardini, S., Experimental results in the ATC multiradar system of the Rome FIR, *Proc. Int. Conf. on Radar*, Paris, 1984, pp. 79–84.
198. Bethke, K.-H., Roede, B. and Schroth, A., A novel near-range radar network for the guidance and control of vehicles on airport manoeuvring areas, in *Int. Radar Conf. "Radar 92"*, London, 1992, pp. 17–22.
199. Schroth, A. *et al.*, The DRL near-range experimental radar system for airport surface movement guidance and control, in *Proc. Int. Radar Conf. "Radar 95"*, Washington DC, 1995, pp. 505–510.
200. Galati, G., Ferri, M., Mariano, P. and Marti, F., Advanced integrated architecture for airport ground movements surveillance, in *Proc. Int. Radar Conf. "Radar 95"*, Washington DC, 1995, pp. 282–287.
201. Bogomolov, A.F. and Pobedonostsev, K.A., Contribution of the Special Designers' Office of the Moscow Power Institute in the development of our country's missile and space radioelectronics, *Radioengineering Notebooks*, No. 7 (special issue), 1995, pp. 8–16 (in Russian).
202. Marcum, J.I., A statistical theory of target detection by pulsed radar. Mathematical appendix, *IRE Trans. on Information Theory*, Vol. IT-6, No. 2, 1960, pp. 145–207.
203. Swerling, P., Probability of detection for fluctuating targets, *IRE Trans. on Information Theory*, Vol. IT-6, No. 2, 1960, pp. 269–308.
204. Reibman, A.R. and Nolte, L.W., Optimal detection and performance evaluation of distributed sensor systems, *IEEE Trans. on Aerospace and Electronic Systems*, Vol. AES-23, No. 1, 1987, pp. 24–30.
205. Reibman, A.R. and Nolte, L.W., Design and performance comparison of distributed detection networks, *IEEE Trans. on Aerospace and Electronic Systems*, Vol. AES-23, No. 6, 1987, pp. 789–797.
206. Chair, Z. and Varshney, P.K., Optimum data fusion in multiple sensor detection systems, *IEEE Trans. on Aerospace and Electronic Systems*, Vol. AES-22, No. 1, 1986, pp. 98–101.
207. Hashemi, H.R. and Rhodes, I.B., Distributed sequential detection, *IEEE Trans. on Information Theory*, Vol. IT-35, No. 4, 1989, pp. 509–520.
208. Chong, C.-Y., Mori, S. and Chang, K.-C., Distributed multitarget multisensor tracking, in *Multitarget–Multisensor Tracking: Advanced Applications*, ed. by Y. Bar-Shalom, Vol. 1, Artech House, Norwood MA, 1990, pp. 247–295.

209. Viswanathan, R., Thomopoulos, S.C.A. and Tumuluri, R., Optimal series distributed decision fusion, *IEEE Trans. on Aerospace and Electronic Systems*, Vol. AES-24, No. 4, 1988, pp. 366–375.
210. Swaszek, P.F., On the performance of serial networks in distributed detection, *IEEE Trans. on Aerospace and Electronic Systems*, Vol. AES-29, No. 1, 1993, pp. 254–260.
211. Srinivasan, R., Distributed detection with decision feedback, *IEE Proceedings*, Pt F, Vol. 137, No. 6, 1990, pp. 427–432.
212. Shalaby, H.M.H. and Paramarcon, A., A Note on the asymptotics of distributed detection with feedback, *IEEE Trans. on Information Theory*, Vol. IT-38, No. 3, 1992, pp. 633–640.
213. Swaszek, P.F. and Willett, P., Parley as an approach to distributed detection, *IEEE Trans. on Aerospace and Electronic Systems*, Vol. AES-31, No. 1, 1995, pp. 447–457.
214. Alhakeem, S. and Varshney, P.K., A unified approach to the design of decentralized detection systems, *IEEE Trans. on Aerospace and Electronic Systems*, Vol. AES-31, No. 1, 1995, pp. 9–20.
215. Faddeev, D.K. and Faddeeva, V.N., *Calculation Methods of the Linear Algebra*, Fizmatgiz, Moscow, 1960 (in Russian).
216. Borovikov, V.A., *Uniform Stationary Phase Method*, IEE Electromagnetic Wave Series, No. 40, London, 1994.
217. Chernyak, V.S., Adaptive mainlobe jamming cancellation and target detection in multistatic radar systems, in *Proc. CIE Int. Conf. of Radar, ICR '96*, Beijing, China, 1996, pp. 297–300.
218. Finn, H.M. and Johnson, R.S., Adaptive detection mode with threshold control as a function of spatially sampled clutter-level estimates, *RCA Review*, Vol. 29, No. 3, 1968, pp. 414–464.
219. Al-Hussaini, E.K. and Ibrahim, B.M., Comparison of adaptive cell-averaging detectors for multiple-target situations, *IEE Proceedings*, Pt F, Vol. 133, No. 3, 1986, pp. 217–223.
220. Barkat, M., Himonas, S.D. and Varshney, P.K., CFAR detection for multiple target situations, *IEE Proceedings*, Pt F, Vol. 136, No. 5, 1989, pp. 193–209.
221. Gandhi, P.P. and Kassam, S.A., Analysis of CFAR processors in nonhomoheneous background, *IEEE Trans. on Aerospace and Electronic Systems*, Vol. AES-24, No. 4, 1988, pp. 427–445.
222. Meng, X. *et al.*, Performance analysis of a new greatest of selection CFAR detector, in *Proc. CIE Int. Conf. of Radar, ICR '96*, Beijing, China, 1996, pp. 401–406.
223. Rickard, J.T. and Dillard, G.M., Adaptive detection algorithm for multiple target situations, *IEEE Trans. on Aerospace and Electronic Systems*, Vol. AES-13, No. 4, 1977, pp. 338–343.
224. Moore, J.D. and Lawrence, N.B., Comparison of two CFAR methods used with square law detection of Swerling 1 targets, in *Proc. IEEE Int. Radar Conf.*, Arlington, Va, 1980, pp. 403–409.
225. Weiss, M., Analysis of some modified cell-averaging CFAR processors in multiple-target situations, *IEEE Trans. on Aerospace and Electronic Systems*, Vol. AES-18, No. 1, 1982, pp. 102–113.
226. Hansen, V.G., Constant false alarm rate processing in search radars, in *Proc. Int. Radar Conf. "Radar '73"*, London, 1973, pp. 325–332.
227. Trunk, G.V., Range resolution of targets using automatic detection, *IEEE Trans. on Aerospace and Electronic Systems*, Vol. AES-14, No. 5, 1978, pp. 750–755.
228. Rohling, H., Radar CFAR thresholding in clutter and multiple target situations, *IEEE Trans. on Aerospace and Electronic Systems*, Vol. AES-19, No. 4, 1983, pp. 608–621.
229. Rohling, H., New CFAR processor based on ordered statistic, in *Proc. Int. Conf. on Radar*, Paris, 1984, pp. 38–42.
230. Ritcey, J.A., Censored mean-level detector analysis, *IEEE Trans. on Aerospace and Electronic Systems*, Vol. AES-22, No. 4, 1986, pp. 443–454.
231. He, Y. *et al.*, A new CFAR detector based on ordered statistics and cell averaging, in *Proc. CIE Int. Conf. of Radar, ICR '96*, Beijing, China, 1996, pp. 106–108.
232. Elias-Fuste, A.R., De Mercado, M.G. and Davo, E.R., Analysis of some modified order statistic CFAR: OSGO and OSSO CFAR, *IEEE Trans. on Aerospace and Electronic Systems*, Vol. AES-26, No. 1, 1990, pp. 197–202.
233. He, Y., Performance of some generalised modified order statistics CFAR detectors with automatic censoring technique in multiple target situation, *IEE Proceedings*, Pt F, Vol. 141, No. 4, 1994, pp. 205–212.
234. He, Y. and Guan, J., A new CFAR detector with greatest of selection, in *Proc. Int. Radar Conf. "Radar 95"*, Washington DC, 1995, pp. 589–591.
235. Rohling, H. and Mende, R., OS CFAR performance in a 77 GHz radar sensor for car application, in *Proc. CIE Int. Conf. of Radar, ICR '96*, Beijing, China, 1996, pp. 109–114.
236. Barkat, M. and Varshney, P.K., Decentralized CFAR signal detection, *IEEE Trans. on Aerospace and Electronic Systems*, Vol. AES-25, No. 2, 1989, pp. 141–149.
237. Barkat, M. and Varshney, P.K., Adaptive cell-averaging CFAR detection in distributed sensor networks, *IEEE Trans. on Aerospace and Electronic Systems*, Vol. AES-27, No. 3, 1991, pp. 424–429.
238. Elias-Fuste, A.R. *et al.*, CFAR data fusion center with inhomogeneous receivers, *IEEE Trans. on Aerospace and Electronic Systems*, Vol. AES-28, No. 1, 1992, pp. 276–284.

239. Egau, P.C., Correlation systems in radio astronomy and related fields, *IEE Proceedings*, Pt F, Vol. 131, No. 1, 1984, pp. 32–39.
240. Hahn, W.R. and Tretter, S.A., Optimum processing for delay-vector estimation in passive signal arrays, *IEEE Trans. on Information Theory*, Vol. IT-19, No. 5, 1973, pp. 608–614.
241. Hahn, W.R., Optimum signal processing for passive sonar range and bearing estimation, *J. Acoust. Soc. Am*, Vol. 58, No. 1, 1975, pp. 201–207.
242. Chan, Y.T. and Ho, K.C., A simple and efficient estimator for hyperbolic location, *IEEE Trans. on Signal Processing*, Vol. SP-42, No. 8, 1994, pp. 1905–1915.
243. Bar-Shalom, Y., On track-to-track correlation problem, *IEEE Trans. on Automatic Control*, Vol. AC-26, No. 4, 1981, pp. 571–572.
244. Blackman, S.S., Association and fusion of multiple sensor data, in *Multitarget-MultisensorTracking: Advanced Applications*, ed. by Y. Bar-Shalom, Vol. 1, Artech House, Norwood MA, 1990, pp. 187–218.
245. Üner, M.K. and Varshney, P.K., Distributed CFAR detection in homogeneous and non-homogeneous backgrounds, *IEEE Trans. on Aerospace and Electronic Systems*, Vol. AES-32, No. 1, 1996, pp. 84–96.
246. Blum, R.S. and Qiao, J., Threshold optimization for distributed order-statistics CFAR signal detection, *IEEE Trans. on Aerospace and Electronic Systems*, Vol. AES-32, No. 1, 1996, pp. 368–377.
247. Sivazlian, B.D. and Green, R.E., Effect of instrument siting and coordinate selection on GDOP in target tracking, in *IEEE Nat. Aerospace and Electron. Conf., NAECON '76, Record*, Dayton O, pp. 142–147.
248. Lee, H.B., Accuracy limitations of hyperbolic multilateration systems, *IEEE Trans. on Aerospace and Electronic Systems*, Vol. AES-11, No. 1, 1975, pp. 16– 29.
249. Yin, Ch., Xu, Sh. and Wang, D., Location accuracy of multistatic radars (*TRn*) based on ranging information, in *Proc. CIE Int. Conf. of Radar, ICR '96*, Beijing, China, 1996, pp. 34–38.
250. Giordano, R., Farina, A. and Pardini, S., Algorithms for the compensation of errors of disalignment in multiradar system, *Rivista Tecnica Selenia*, Vol. 7, No. 2, 1981, pp. 10–15.
251. Bath, W.G., Association of multistatic radar data in the presence of large navigation and sensor alignment errors, in *Proc. Int. Radar Conf. "Radar '82"*, London, 1982, pp. 169–173.
252. Leung, H., Blanchette, M. and Harrison, C., Least squares fusion of multiple radar data, in *Proc. Int. Conf. on Radar*, Paris, 1994, p. 364–369.
253. Paradowski, L., Position determination from modern radar systems – new theoretical results and their practical aspects, in *Proc. Int. Radar Conf. "Radar 95"*, Washington DC, 1995, pp. 131 135.
254. Poor, W., Statistical estimation of navigational errors, *IEEE Trans. on Aerospace and Electronic Systems*, Vol. AES-28, No. 2, 1992, pp. 428–438.
255. Carter, C.R. and Mangel, M., Three bearing method for passive triangulation in systems with unknown deterministic biases, *IEEE Trans. on Aerospace and Electronic Systems*, Vol. AES-17, No. 6, 1981, pp. 814–819.
256. Farina, A. and Studer, F.A., *Radar data processing*, Vol. 1 – *Introduction and Tracking*. Research Studies Press, Lechtworth, Herts, England; John Wiley & Sons, N.Y., 1985.
257. Farina, A. and Studer, F.A., *Radar Data Processing*, Vol. 2 – *Advanced topics and applications*, Chapter 5, Research Studies Press, Lechtworth, Herts, England; John Wiley & Sons, N.Y., 1986.
258. Kenefic, R.J. and Goulette, P.L., Sensor netting via the discrete time extended Kalman filter, *IEEE Trans. on Aerospace and Electronic Systems*, Vol. AES-17, No. 4, 1981, pp. 482–488.
259. Manolakis, D.E., Efficient solution and performance analysis of 3-D position estimation by trilateration, *IEEE Trans. on Aerospace and Electronic Systems*, Vol. AES-32, No. 4, 1996, pp. 1239–1248.
260. Farina, A., Multistatic tracking and comparison with netted monostatic systems, in *Proc. Int. Radar Conf. "Radar '82"*, London, 1982, pp. 183–186.
261. Casner, P.G. and Prengaman, R.J., Integration and automation of multiple co-located radars, in *Proc. Int. Radar Conf. "Radar '77"*, London, 1977, pp. 145–149.
262. Bath, W.G., Castella, F.R. and Haase, S.F., Techniques for filtering range and angle measurements from colocated surveillance radars, in *Proc. of the IEEE Int. Radar Conf. "Radar 80"*, Arlington, Va, 1980, pp. 355–360.
263. Dufour, F. and Mariton, M., Tracking a 3D maneuvering target with passive sensors, *IEEE Trans. on Aerospace and Electronic Systems*, Vol. AES-27, No. 4, 1991, pp. 725–738.
264. Morefield, C.L., Decision directed multitarget tracking, in *Proc. 17th IEEE Conf. Decision and Control*, San Diego, CA, 1979, pp. 1195–1201.
265. Knoppik, N. *et al.*, Simultaneous automatic tracking in multiple radar networks, in *Proc. Int. Conf. on Radar*, Paris, 1978, pp. 100–108.
266. Martinerie, F., Data fusion and tracking using HMMs in a decentralized sensor network, *IEEE Trans. on Aerospace and Electronic Systems*, Vol. AES-33, No. 1, 1997, pp. 11–28.
267. Deb, S. *et al.*, A generalized S-D assignment algorithm for multisensor–multitarget state estimation, *IEEE Trans. on Aerospace and Electronic Systems*, Vol. AES-33, No. 2, 1997, pp. 523–532.

268. Taff, L.G., Target localization from bearings-only observations, *IEEE Trans. on Aerospace and Electronic Systems*, Vol. AES-33, No. 1, 1997, pp. 1–9.
269. Singh, R.-N.P. and Bayley, W.H., Fuzzy logic applications to multisensor-multitarget correlation, *IEEE Trans. on Aerospace and Electronic Systems*, Vol. AES-33, No. 3, 1997, pp. 752–769.
270. Amirmehrabi, H. and Viswanathan, R., A new distributed constant false alarm detector, *IEEE Trans. on Aerospace and Electronic Systems*, Vol. AES-33, No. 1, 1997, pp. 85–97.
271. Gini, F., Lombardini, F. and Verrazzani, L., Decentralized CFAR detection with binary integration in Weibull clutter, *IEEE Trans. on Aerospace and Electronic Systems*, Vol. AES-33, No. 2, 1997, pp. 396–407.
272. Fitzgerald, R.J., Development of practical PDA logic for multitarget tracking by multiprocessor, in *Multitarget-Multisensor Tracking: Advanced Applications*, ed. by Y. Bar-Shalom, Vol. 1, Artech House, Norwood MA, 1990, pp. 1–23.
273. Bar-Shalom, Y. and Fortmann, T.E., *Tracking and Data Association*, Academic Press, N.Y. 1988.
274. Longo, M. and Lops, M., OS-CFAR thresholding in decentralized radar systems, *IEEE Trans. on Aerospace and Electronic Systems*, Vol. AES-32, No. 4, 1996, pp. 1257–1267.
275. Baltes, R. and van Keuk, G., Tracking multiple manoeuvring targets in a network of passive radars, in *Proc. Int. Radar Conf. "Radar 95"*, Washington DC, 1995, pp. 304–309.

For Product Safety Concerns and Information please contact our EU
representative GPSR@taylorandfrancis.com
Taylor & Francis Verlag GmbH, Kaufingerstraße 24, 80331 München, Germany

www.ingramcontent.com/pod-product-compliance
Ingram Content Group UK Ltd.
Pitfield, Milton Keynes, MK11 3LW, UK
UKHW050339191125
9046UKWH00043B/25
* 9 7 8 9 0 5 6 9 9 1 6 5 4 *